Drought Stress Tolerance in Plants, Vol 1

Mohammad Anwar Hossain
Shabir Hussain Wani
Soumen Bhattacharjee
David J. Burritt
Lam-Son Phan Tran
Editors

Drought Stress Tolerance in Plants, Vol 1

Physiology and Biochemistry

Springer

Editors
Mohammad Anwar Hossain
Department of Genetics and Plant Breeding
Bangladesh Agricultural University
Mymensingh, Bangladesh

Soumen Bhattacharjee
Department of Botany
University of Burdwan
Burdwan, West Bengal, India

Lam-Son Phan Tran
RIKEN Center for Sustainable
Resource Science
Yokohama, Japan

Shabir Hussain Wani
Department of Genetics and Plant Breeding
Sher-e-Kashmir University of Agricultural
Sciences and Technology of Kashmir
Srinagar, Jammu and Kashmir
India

David J. Burritt
Department of Botany
University of Otago
Dunedin, Otago, New Zealand

ISBN 978-3-319-28897-0 ISBN 978-3-319-28899-4 (eBook)
DOI 10.1007/978-3-319-28899-4

Library of Congress Control Number: 2016932885

Printed on acid-free paper

This Springer imprint is published by Springer Nature
The registered company is Springer International Publishing AG Switzerland

Preface

Plants are subjected to a wide range of abiotic stresses, such as drought, salinity, extreme temperatures, pollution, UV radiation, etc. Abiotic stress adversely affects crop production worldwide, causing yield reductions for most major crops. Among the various abiotic stresses, drought is considered to be the most serious. Due to an increasing global population, drought may lead to a serious food shortage by 2050, when the world's population is expected to reach ten billion. This situation may be worsened due to global climate change that may multiply the frequency, duration, and severity of water deficit. Hence, there is an urgent need to improve our understanding of the complex mechanisms associated with drought tolerance and to develop elite crop varieties that are more resilient to drought without affecting other agronomic and quality parameters. Identification of novel genes responsible for drought tolerance in crop plants will contribute to our understanding of the molecular mechanisms behind drought tolerance. The discovery of novel genes, the analysis of their expression patterns in response to drought, and the determination of their potential functions in drought adaptation will provide the basis for effective breeding strategies to enhance crop drought tolerance. The general effects of drought on plant growth are well known, but the effects of water deficit at the biochemical and molecular levels are not well understood. Although we do not have a complete understanding of the biological mechanisms associated with tolerance to drought, tolerance can to some extent be explained on the basis of ion homeostasis mediated by stress adaptation effectors, toxic radical scavenging, osmolyte biosynthesis, water transport, and the coordination of long-distance signaling mechanisms. Complete elucidation of the physiological, biochemical, and molecular mechanisms by which plants respond to drought, including signal perception and transduction, as well as adaptation, is still a challenge for plant biologists.

In this book we present a collection of 21 chapters written by recognized experts in the field of plant drought responses, tolerance, and crop improvement. This volume deals with an array of topics in the broad area of drought responses and tolerance in plants and focuses on plant "physiology and biochemistry." The information presented in this book demonstrates how plants respond to drought and will ultimately lead to both conventional and biotechnological approaches for improvement of crop productivity under drought stress and for sustainable agricultural production. We trust that the information covered in this volume will be useful in building strategies to counter the negative impacts of drought. Hopefully this volume will serve as a major source of information and knowledge to graduate and postgraduate students and researchers investigating abiotic stresses. We also believe that it will be of interest to a wide range of plant scientists, including agronomists, physiologists, biotechnologists, molecular biologists and plant breeders who have concerns about the drought responses of plants and improving the drought tolerance of crop plants.

As editors of this volume, we are grateful to the authors of various chapters of this book for writing their chapters meticulously and enabling us to produce this volume in time. We would also like to extend our thanks to Dr. Kenneth Teng and the editorial staff of Springer, New York, who enabled us to initiate this book project. Finally, our special thanks to Springer, Switzerland, for publishing this volume. We fervently believe that the information covered in this book will make a sound contribution to this fascinating area of research.

Mymensingh, Bangladesh — Mohammad Anwar Hossain
Srinagar, Kashmir, India — Shabir Hussain Wani
West Bengal, India — Soumen Bhattacharjee
Dunedin, New Zealand — David J. Burritt
Yokohama, Japan — Lam-Son Phan Tran

Contents

Contributors

Walid Abuelsoud Institute of Botany, Leibniz University Hannover, Hannover, Germany

Botany Department, Faculty of Science, Cairo University, Giza, Egypt

Manashi Aditya Plant Physiology and Biochemistry Research Laboratory, Department of Botany, UGC Centre for Advanced Study, The University of Burdwan, Burdwan, West Bengal, India

Imrul Mosaddek Ahmed Department of Agronomy, College of Agriculture and Biotechnology, Zijingang Campus, Zhejiang University, Hangzhou, China

Rubén Alcázar Department of Natural Products, Plant Biology and Soil Science, Faculty of Pharmacy, University of Barcelona, Barcelona, Spain

Durna R. Aliyeva Department of Fundamental Problems of Biological Productivity, Institute of Molecular Biology and Biotechnology, Azerbaijan National Academy of Sciences, Baku, Azerbaijan

Jalal A. Aliyev Department of Fundamental Problems of Biological Productivity, Institute of Molecular Biology and Biotechnology, Azerbaijan National Academy of Sciences, Baku, Azerbaijan

Department of Plant Physiology and Biotechnology, Research Institute of Crop Husbandry, Ministry of Agriculture of Azerbaijan Republic, Baku, Azerbaijan

Sruthy Maria Augustine Sugarcane Breeding Institute (ICAR), Coimbatore, India

Hasan G. Babayev Department of Fundamental Problems of Biological Productivity, Institute of Molecular Biology and Biotechnology, Azerbaijan National Academy of Sciences, Baku, Azerbaijan

Hamideh Bakhshayeshan-Agdam Department of Plant Sciences, Faculty of Natural Sciences, University of Tabriz, Tabriz, Iran

Nabanita Banik Department of Botany, Plant Physiology and Biochemistry Research Laboratory, UGC Centre for Advanced Study, The University of Burdwan, Burdwan, West Bengal, India

Soumen Bhattacharjee Department of Botany, University of Burdwan, West Bengal, India

Eliana Bianucci Departamento de Ciencias Naturales, Facultad de Ciencias Exactas, Físico-Químicas y Naturales, Universidad Nacional de Río Cuarto, Córdoba, Argentina

Marian Brestic Department of Plant Physiology, Slovak University of Agriculture, Nitra, Slovak Republic

Patrick Burgess Department of Plant Biology and Pathology, Rutgers University, New Brunswick, NJ, USA

Cohen Ana Carmen Laboratorio de Bioquímica Vegetal, Instituto de Biología Agrícola de Mendoza, Consejo Nacional de Investigaciones Científicas y Técnicas-Universidad Nacional de Cuyo, Chacras de Coria, Argentina

Stella Castro Departamento de Ciencias Naturales, Facultad de Ciencias Exactas, Físico-Químicas y Naturales, Universidad Nacional de Río Cuarto, Córdoba, Argentina

Ananya Chakrabarty Department of Botany, Plant Physiology and Biochemistry Research Laboratory, UGC Centre for Advanced Study, The University of Burdwan, Burdwan, West Bengal, India

Sharmila Chattopadhyay Plant Biology Lab, CSIR-Indian Institute of Chemical Biology, Kolkata, India

Aanchal Choudhary National Institute of Plant Genome Research, New Delhi, India

Jong Ahn Chun Climate Change Research Team, Department of Climate Research, APEC Climate Center, Busan, Republic of Korea

Nivedita Dey Department of Botany, Plant Physiology and Biochemistry Research Laboratory, UGC Centre for Advanced Study, The University of Burdwan, Burdwan, West Bengal, India

Ana Laura Furlan Departamento de Ciencias Naturales, Facultad de Ciencias Exactas, Físico-Químicas y Naturales, Universidad Nacional de Río Cuarto, Córdoba, Argentina

Ajit Ghosh Plant Molecular Biology Group, International Centre for Genetic Engineering and Biotechnology (ICGEB), New Delhi, India

Department of Biochemistry and Molecular Biology, Shahjalal University of Science and Technology, Sylhet, Bangladesh

Mohammad Rokebul Hasan Plant Molecular Biology Group, International Centre for Genetic Engineering and Biotechnology (ICGEB), New Delhi, India

Dagmar Hillová Department of Planting Design and Maintenance, Faculty of Horticulture and Landscape Engineering, Slovak University of Agriculture, Nitra, Slovak Republic

Felix Hirschmann Institute of Botany, Leibniz University Hannover, Hannover, Germany

Thi My Linh Hoang Centre for Tropical Crops and Biocommodities, Queensland University of Technology (QUT), Brisbane, QLD, Australia

Bingru Huang Department of Plant Biology and Pathology, Rutgers University, New Brunswick, NJ, USA

Irada M. Huseynova Department of Fundamental Problems of Biological Productivity, Institute of Molecular Biology and Biotechnology, Azerbaijan National Academy of Sciences, Baku, Azerbaijan

Charanpreet Kaur Plant Molecular Biology Group, International Centre for Genetic Engineering and Biotechnology (ICGEB), New Delhi, India

Stress Physiology and Molecular Biology Laboratory, School of Life Sciences, Jawaharlal Nehru University, New Delhi, India

Sanai Li Climate Change Research Team, Department of Climate Research, APEC Climate Center, Busan, Republic of Korea

Xiangnan Li Department of Plant and Environmental Sciences, Faculty of Science, University of Copenhagen, Taastrup, Denmark

Fulai Liu Department of Plant and Environmental Sciences, Faculty of Science, University of Copenhagen, Taastrup, Denmark

Emily Merewitz Department of Plant, Soil, and Microbial Sciences, Michigan State University, East Lansing, MI, USA

Sagadevan G. Mundree Centre for Tropical Crops and Biocommodities, Queensland University of Technology (QUT), Brisbane, QLD, Australia

Umme Aktari Nadira Department of Agronomy, College of Agriculture and Biotechnology, Zijingang Campus, Zhejiang University, Hangzhou, China

Mariela Odjakova Department of Biochemistry, Faculty of Biology, Sofia University, Sofia, Bulgaria

Veena Pandey Department of Plant Physiology, College of Basic Sciences and Humanities, G.B. Pant University of Agriculture and Technology, Pantnagar, Uttarakhand, India

Prachi Pandey National Institute of Plant Genome Research, JNU Campus, New Delhi, India

Jutta Papenbrock Institute of Botany, Leibniz University Hannover, Hannover, Germany

Ashwani Pareek Stress Physiology and Molecular Biology Laboratory, School of Life Sciences, Jawaharlal Nehru University, New Delhi, India

Piccoli Patricia Laboratorio de Bioquímica Vegetal, Instituto de Biología Agrícola de Mendoza, Consejo Nacional de Investigaciones Científicas y Técnicas-Universidad Nacional de Cuyo, Chacras de Coria, Argentina

Bottini Rubén Laboratorio de Bioquímica Vegetal, Instituto de Biología Agrícola de Mendoza, Consejo Nacional de Investigaciones Científicas y Técnicas-Universidad Nacional de Cuyo, Chacras de Coria, Argentina

Samira M. Rustamova Department of Fundamental Problems of Biological Productivity, Institute of Molecular Biology and Biotechnology, Azerbaijan National Academy of Sciences, Baku, Azerbaijan

Seyed Yahya Salehi-Lisar Department of Plant Sciences, Faculty of Natural Sciences, University of Tabriz, Tabriz, Iran

Muthappa Senthil-Kumar National Institute of Plant Genome Research, JNU Campus, New Delhi, India

Miren Sequera-Mutiozabal Department of Natural Products, Plant Biology and Soil Science, Faculty of Pharmacy, University of Barcelona, Barcelona, Spain

Alok Shukla Department of Plant Physiology, College of Basic Sciences and Humanities, G.B. Pant University of Agriculture and Technology, Pantnagar, Uttarakhand, India

Sneh Lata Singla-Pareek Plant Molecular Biology Group, International Centre for Genetic Engineering and Biotechnology (ICGEB), New Delhi, India

Oksana Sytar Department of Plant Physiology and Ecology, Taras Shevchenko National University of Kyiv, Kyiv, Ukraine

Denitsa Teofanova Department of Biochemistry, Faculty of Biology, Sofia University, Sofia, Bulgaria

Antonio F. Tiburcio Department of Natural Products, Plant Biology and Soil Science, Faculty of Pharmacy, University of Barcelona, Barcelona, Spain

Salomon María Victoria Laboratorio de Bioquímica Vegetal, Instituto de Biología Agrícola de Mendoza, Consejo Nacional de Investigaciones Científicas y Técnicas-Universidad Nacional de Cuyo, Chacras de Coria, Argentina

Qingguo Wang Climate Change Research Team, Department of Climate Research, APEC Climate Center, Busan, Republic of Korea

Brett Williams Centre for Tropical Crops and Biocommodities, Queensland University of Technology (QUT), Brisbane, QLD, Australia

Feibo B. Wu Department of Agronomy, College of Agriculture and Biotechnology, Zijingang Campus, Zhejiang University, Hangzhou, China

Lyuben Zagorchev Department of Biochemistry, Faculty of Biology, Sofia University, Sofia, Bulgaria

Guoping P. Zhang Department of Agronomy, College of Agriculture and Biotechnology, Zijingang Campus, Zhejiang University, Hangzhou, China

Marek Zivcak Department of Plant Physiology, Slovak University of Agriculture, Nitra, Slovak Republic

About the Editors

Mohammad Anwar Hossain is a professor in the Department of Genetics and Plant Breeding, Bangladesh Agricultural University, Mymensingh-2202, Bangladesh. He received his B.Sc. in agriculture and M.S. in genetics and plant breeding from Bangladesh Agricultural University, Bangladesh. He also received an M.Sc. in agriculture from Kagawa University, Japan, in 2008 and a Ph.D. in abiotic stress physiology and molecular biology from Ehime University, Japan, in 2011. In November 2015 he moved to Tokyo University, Japan, as a postdoctoral scientist to work on isolating low phosphorus stress tolerant genes/QTLs from rice. He has published 25 research articles, 15 book chapters, and 5 review articles on important aspects of plant physiology and breeding, plant stress responses and tolerance mechanisms, and exogenous chemical priming-induced abiotic stress tolerance. Recently, he edited a book entitled *Managing Salt Tolerance in Plants: Molecular and Genomic Perspectives* published by CRC press, Taylor and Francis Group, USA. He has attended several international and national conferences for presenting his research findings. He is a professional member of the Bangladesh Society of Genetics and Plant Breeding, Bangladesh Association for Plant Tissue Culture and Biotechnology, and the Seed Science Society of Bangladesh.

Shabir Hussain Wani is an assistant professor in the Department of Genetics and Plant Breeding, Sher-e-Kashmir University of Agricultural Sciences and Technology of Kashmir, Srinagar, Jammu and Kashmir, India. He received his B.Sc. in agriculture from BhimRao Agricultural University Agra, India, and M.Sc. in genetics and plant breeding from Central Agricultural University, Manipur, India, and Ph.D. in plant breeding and genetics on "transgenic rice for abiotic stress tolerance" from the Punjab Agricultural University Ludhiana, India. After obtaining his Ph.D. he worked as research associate in the Biotechnology Laboratory, Central Institute of Temperate Horticulture (ICAR), Rangreth, Srinagar, India, for 2 years, up to October 2011. In November 2011 he joined the Krishi Vigyan Kendra (Farm Science Centre) as program coordinator (i/c) at Senapati, Manipur, India. He teaches courses related to plant breeding, seed science and technology, and stress breeding and has published more than 80 papers/chapters in journals and books of international and national repute. He has also edited several books on current topics in crop improvement including *Managing Salt Tolerance in Plants: Molecular and Genomic Perspectives* published by CRC press, Taylor and Francis Group, USA, in 2015. His Ph.D. research won first prize in the North Zone Competition, at national level, in India. He was awarded a Young Scientist Award from the Society for Promotion of Plant Sciences, Jaipur, India, in 2009. He is a fellow of the Society for Plant Research, India. Recently he also received Young Scientist Award (Agriculture) 2015 from Society for Plant Research, Meerut, India. He has been selected for one year University Grants Commission funded RAMAN fellowship for Post Doc Research in Michigan State University USA for the year 2016-17. He has attended several international and national conferences, presenting his research.

Soumen Bhattacharjee is presently head of UGC Centre for Advanced Studies, Department of Botany, the University of Burdwan, West Bengal. He completed his master's in botany and Ph.D. on the abiotic stress physiology of plants at the University of Burdwan, West Bengal, India. Later, he started his teaching career as a faculty member in the Department of Botany in Delhi University Constituent College. After serving almost two and a half years, he joined the West Bengal Education Service and worked mainly in the Post Graduate Department of Botany, Hooghly Mohsin College, West Bengal, as lecturer, reader, and associate professor. In 2007, Dr. Bhattacharjee was selected by the Indian Council of Agricultural Research (ICAR) as a senior scientist and joined the Vivekananda Institute of Hill Agriculture, Almora, India. In 2013 he joined the University of Burdwan as associate professor. His research interests center around plant redox biology, particularly understanding the relationship between oxidative stress and plant growth, and the role of ROS signaling in stress acclimation, characterization of redox-regulatory mechanisms during germination of rice under abiotic stress, and understanding the physiological basis of antioxidant accumulation in underutilized medicinal plants. He has published 34 research papers in international peer-reviewed journals, 12 review articles in various national and international journals, and 14 book chapters and has 6 edited books and journal volumes. He is also a member of several professional research bodies and is a guest editor and reviewer for several international peer-reviewed journals.

David J. Burritt is an associate professor in the Department of Botany, the University of Otago, Dunedin, New Zealand. He received his B.Sc. and M.Sc. (Hons.) in botany and his Ph.D. in plant biotechnology from the University of Canterbury, Christchurch, New Zealand. His research interests include oxidative stress and redox biology, plant-based foods and bioactive molecules, plant breeding and biotechnology, cryopreservation of germplasm, and the stress biology of plants, animals, and algae. He has over 90 peer-reviewed publications.

Lam-Son Phan Tran is head of the Signaling Pathway Research Unit at RIKEN Center for Sustainable Resource Science, Japan. He obtained his M.Sc. in biotechnology in 1994 and Ph.D. in biological sciences in 1997, from Szent Istvan University, Hungary. After doing his postdoctoral research at the National Food Research Institute (1999–2000) and the Nara Institute of Science and Technology of Japan (2001), in October 2001, he joined the Japan International Research Center for Agricultural Sciences to work on the functional analysis of transcription factors and osmosensors in *Arabidopsis* plants under stress. In August 2007, he moved to the University of Missouri–Columbia, USA, as a senior research scientist to coordinate a research team working to discover soybean genes to be used for genetic engineering of drought-tolerant soybean plants. His current research interests are elucidation of the roles of phytohormones and their interactions in abiotic stress responses, as well as translational genomics of legume crops with the aim to enhance crop productivity under adverse environmental conditions. He has published over 90 peer-reviewed papers with more than 70 research and 20 review articles, contributed 7 book chapters to various book editions published by Springer, Wiley-Blackwell, American Society of Agronomy, Crop Science Society of America, and Soil Science Society of America. He has also edited 5 book volumes for Springer, including this one.

Chapter 1
Drought Stress in Plants: Causes, Consequences, and Tolerance

Seyed Yahya Salehi-Lisar and Hamideh Bakhshayeshan-Agdam

1.1 Introduction

Under both natural and agricultural conditions, plants are often exposed to various environmental stresses. Water accounts for between 80–95 % of the fresh biomass of nonwoody plants and plays an important role in many aspects of plant growth, development, metabolism, and so on [16, 33]. Drought is one of the most important and prevalent stress factors for plants in many parts of the world, especially in arid and semiarid areas [23]. There are several reasons for a water deficit in plants; these include low rainfall, salinity, high and low temperatures, high intensity of light, among others. On the other hand, in many conditions there is enough water in the soil, but plants cannot uptake it. This type of water stress is called a pseudo-drought or physiological drought [3, 4, 33]. Drought stress is a multidimensional stress and generally leads to changes in the physiological, morphological, ecological, biochemical, and molecular traits of plants [8, 14, 35]. In addition, it can negatively affect the quantity and quality of plant growth and yield [17, 27, 39]. Plant responses to a water deficit depend on the length and severity of the water deficiency as well as the plant species, age, and developmental stage [23]. Many plants have developed resistance mechanisms to tolerate drought stress, but these mechanisms are varied and depend on the plant species. There are several options in drought tolerance in plants, including developmental, physiological, morphological, ecological, biochemical, and molecular mechanisms. Typically, the mechanisms involved in plant tolerance to drought follow a general plan: maintaining cell water homeostasis under drought conditions. This is possible mainly by prohibiting water loss and increasing the water inlet to the cells, which eventually leads to normal cell

S.Y. Salehi-Lisar (✉) • H. Bakhshayeshan-Agdam (✉)
Department of Plant Sciences, Faculty of Natural Sciences, University of Tabriz, Tabriz, Iran
e-mail: y_salehi@tabrizu.ac.ir; h_bakhshayeshan@tabrizu.ac.ir

M.A. Hossain et al. (eds.), *Drought Stress Tolerance in Plants, Vol 1*,
DOI 10.1007/978-3-319-28899-4_1

functions. In addition to drought tolerance, drought avoidance is another common drought resistance mechanism in annual plants [4, 23, 33]. Scientists have been tested different methods for improving plants' capacity for drought resistance. However, each method has some problems and limitations because of the complexity of drought effects on plants and the plants' responses to the drought. In addition, several strategies for drought management in agricultural fields could be useful in order to minimize the effects of drought on plants, especially on crops. "Drought" is a general term usually used to describe a period without rainfall and derives from an agricultural context [33]. Although the terms "drought," "water deficit," "dehydration," and "water stress"can address different issues, in this text we will use these terms to mean an inadequate water supply for plants.

1.2 Definition of Drought

Plants are sessile organisms often exposed to various environmental stresses [18, 29, 30, 39] including biotic and abiotic stresses [18, 29, 30, 39]. Drought is one of the most important abiotic stresses that negatively influences plant growth and development [29, 30, 39]. Drought is a normal recurrent feature of the climate [12, 22, 27, 30] that occurs in almost all areas, especially arid and semiarid regions, and its characteristics may be very different from one region to another [9, 30]. "Drought" is a general term for the description of atmospheric or weather phenomena and is commonly explained as a period without rainfall [9, 12, 17, 35]. Drought is difficult to define; it can be described from several viewpoints, such as through meteorological, agricultural, hydrological, and socioeconomic lenses [5, 12, 17, 29]. Generally, from agricultural and physiological viewpoints, drought stress occurs when the available water for plants in the soil is decreased due to low soil moisture at a certain time [12, 18]. On the other hand, water stress (deficiency) in plant occurs when the transpiration rate from leaf surfaces is higher than the water uptake by roots [33]. This imbalance in water uptake and water losses from plants mainly occurs when the water potential of the soil is lower than the water potential of plant roots. Many plants, such as the water spender, water collector, and water saver xerophytes, can grow under drought conditions of deserts without encountering water stress. Therefore, scientists must consider that drought certainly is not equal to water deficiency in plants. Mostly, the atmospheric conditions cause a continuous water deficit by transpiration or evaporation [12, 24, 35, 36]. Therefore, an agricultural drought comes after a meteorological drought [5, 12, 17]. Usually, under normal conditions, drought isn't a disaster in many regions, but it could be an important problem when human beings are wasteful with water [21]. In addition, in some regions rainfall is adequate but nonuniform precipitation leads to water stress in plants. Drought occurs in both developing and developed countries, and all societies are vulnerable to this natural phenomenon [12].

1.3 Causes of Drought Stress in Plants

Today alterations in rainfall patterns in many regions occur due to global climate changes that are leading to increases in temperature and atmospheric CO_2 levels [3, 12, 25, 27]. Global climate alterations are the main factor triggering drought stress worldwide [25, 30]. However, there are many other reasons for droughts, such as high temperature, high intensity of light, and dry wind, all of which increase evaporation of water from soil. In addition, these factors increase water losses from plants and subsequently facilitate plant exposure to water stress [12, 24, 33, 35, 36]. Sometimes drought doesn't occur truly because of a water deficit in the environment. In some cases there is enough water in the soil but several soil factors, such as salinity, low soil temperatures, and flooding, prevent or decrease water uptake by roots and subsequently lead to water stress in plants. This type of drought is called pseudo-drought or physiological drought and the atmospheric conditions are not determining factors [3, 33].

1.4 Drought Symptoms in Plants

The symptoms of drought in plants vary depending on the plant species, developmental stage, growth conditions, and other environmental factors [3, 8, 27]. Drought severity, drought length, soil physicochemical conditions, and plant vigor are other factors influencing drought symptoms in plants. Generally, drought symptoms include loss of leaf turgor, drooping, wilting, etiolation, yellowing, and premature leaf downfall [2, 7, 8, 14, 17, 34, 38]. Also, some unusual symptoms include bark and twig crack, branch dieback, thinning tree and shrub canopy, necrosis, and poor and stunted growth. Finally, under extreme conditions, plant death occurs [3, 14, 34].

1.5 Drought Effects on Plants

1.5.1 Plant Growth and Development

Drought can severely reduce plant growth and development [8, 29, 35]. Drought is a multidimensional stress for plants; therefore, it can influence different aspects of plant growth and development [8, 14, 35]. In addition, drought can negatively affect the quantity and quality of growth and yield of plants, especially crops [17, 27, 39]. Plant growth and development are dependent on cell division, elongation, and differentiation. All of these phases are affected under drought conditions by loss of turgor, disordered enzyme activities, and decreased energy supply from photosynthesis [8, 13, 17, 18, 28, 35]. Plant water potential and turgor are reduced in dehydration conditions; therefore, plant cells can't perform their normal functions [18, 29].

Turgor reduction leads to suppressed cell expansion and growth. Cell expansion and growth are necessary phenomena for the initial phase of plant growth and establishment [8, 38]. The following factors are extremely important under water-deficit conditions: the stress severity; the duration and timing of the stress; the responses of plants after the stress removal [14, 39].

1.5.2 Morphological and Anatomical Characteristics

Drought can influence many aspects of plants' morphological and anatomical characteristics. The anatomy of a leaf and its ultrastructure are altered by water stress [16, 23, 33]. A decrease in leaves' size, a lower aperture and decrease in the number of stomata, cell wall thickening, cutinization of the leaf surface and developed conductive system (increase in the number of large vessels), submersion of stomata in succulent and xerophyte plants, and the formation of tube leaves in cereals are some alterations that occur in plants exposed to drought [11, 17, 18, 24, 27, 33, 35]. Additionally, premature leaf senescence increases in water-deficit situations [14, 35]. Optimal leaf area development and stomatal opening are essential factors for optimal photosynthesis in plants [17]. Therefore, net photosynthesis in water-deficit plants is reduced due to a low leaf area, a higher resistance for gas exchange in stomata, and an increase in leaf senescence [13, 25, 35, 38]. The main effect of drought stress on plant morphology is size reduction. A low photosynthesis rate is one of the most important factors in the reduction of plant size and biomass production [14, 15, 35, 38]. Decreasing chlorophyll content is a typical symptom under drought stress that could change the morphology of plants [3, 8, 21, 25, 29]. In order to increase water uptake under dehydration conditions, plants expand their roots and produce a ramified root system [2, 8, 14, 15, 17, 29]. An increased biomass allocation to roots under drought situations and an expansion of the plant's root system generally lead to a higher capacity for water uptake [8, 14, 15, 35]. Accordingly, despite reducing the shoot growth, the root growth isn't significantly reduced under a mild water deficit. Therefore, under dehydration conditions, the root-to-shoot ratios of plants usually increase; however, the total biomasses of plants are reduced considerably [2, 17, 33, 35, 38].

1.5.3 Plant–Water Relationships

The relative water content (RWC), leaf water potential, stomatal resistance, transpiration rate, leaf temperature, and canopy temperature are important factors in plant–water relationships [8, 14, 18, 35, 39]. An RWC reduction is the earliest effect of drought on plants [14]. A low RWC decreases the leaf water potential and leads to stomatal closing. A higher stomatal resistance decreases the transpiration rate and finally leads to increases in the leaf temperature because transpiration is the main factor controlling the leaf temperature. An increase in the stomatal resistance is an important reason for a high leaf temperature, especially when the light intensity is high. Therefore, there is a positive feedback effect between the leaf temperature and the stomatal resistance. However, stomatal closure increased leaf temperature overly at first [3, 14, 24, 34]. Higher temperatures of leaves can lead to denaturation of proteins, especially enzymes. In addition, changes in membrane flexibility are another effect of higher temperatures, which can influence different aspects of metabolism. These alterations are the most important reasons for a disturbance in cell metabolic functions such as photosynthesis, respiration, ion uptake, and mineral nutrition, the synthesis of important macromolecules such as amino acids and proteins, and others [8, 30, 33, 34, 39].

1.5.4 Photosynthesis

A reduction and/or inhibition of photosynthesis is one of the main effects of drought in higher plants [8, 9, 18, 27]. There are many reasons for this effect, including a decrease in the leaf expansion rate and a low leaf surface, an increased leaf temperature, impaired photosynthetic machinery, and premature leaf senescence [8, 14, 38]. Stomatal and nonstomatal factors can be effective in inhibiting photosynthesis under water-deficit situations [9, 34, 39]. Carbon dioxide limitations due to prolonged stomatal closure, especially under light saturation conditions, lead to the accumulation of reduced photosynthetic electron transport components. The accumulation of these compounds can reduce molecular oxygen and give rise to the production of reactive oxygen species (ROS) such as superoxide and hydroxyl radicals as well as H_2O_2, thus causing oxidative damage in chloroplasts [3, 8, 28, 33, 35, 37, 39]. In addition, low CO_2 uptake due to stomatal closure is the primary stomatal-dependent factor that decreases the photosynthesis rate due to reduced activity of enzymes involved in CO_2 reduction (Calvin cycle, dark reactions). The lower activity of dark reactions could lead to imbalances between the light and dark reactions of photosynthesis and ROS accumulation in chloroplasts [8, 14, 27, 33]. The ROS can damage the photosynthetic apparatus, including thylakoid membranes, photosynthetic pigments, and enzymes [8, 14, 33]. A decrease in the chlorophyll content of leaves under water stress is another factor involved in reduction of the photosynthesis rate [18, 29, 34]. A decrease in chlorophyll content during drought stress depends on the duration and severity of the drought and implies a lowered capacity for light harvesting [18, 33, 34]. According to reports in the literature, carotenoids are less sensitive to water stress than chlorophylls. However, unlike chlorophylls, an increase in xanthophyll pigments such as zeaxanthin and antheraxanthin in plants under water stress has been reported. Xanthophyll pigments play a protective role in plants under stress, and some of these pigments are involved in the xanthophyll cycle, which is involved in ROS detoxification [11, 14, 17, 27, 33]. The key enzyme for carbon metabolism in the Calvin cycle is ribulose-bisphosphate carboxylase/oxygenase (RuBisCO) [14, 33]. The level of RuBisCO in leaves is controlled by the rate of its biosynthesis and degradation. The amount and activity of RuBisCO decrease rapidly under water-deficit conditions. This effect is evident in all plants studied, but the severity of the decrease is species-dependent [11, 14, 33]. A decline in RuBisCO activity is caused by the acidification of chloroplast stroma, a lack of the substrate for carboxylation (CO_2 and ribulose-bisphosphate), a reduction in the amount and/or activity of the coupling factor (ATPase, ATP synthase), structural alterations of chloroplasts and RuBisCO, and release of RuBisCO from damaged plastids [2, 8, 11, 14, 33, 39]. In addition to RuBisCO, activities of some other enzymes involved in carbon metabolism, such as phosphoenolpyruvate carboxylase, NADP-malic enzyme, fructose-1,6-bisphosphatase, NADP-glyceraldehyde phosphate dehydrogenase, phosphoribulokinase, sucrose phosphate synthase, and pyruvate orthophosphate dikinase, decrease linearly with lowered leaf water potential under drought conditions [11, 14, 33]. Drought stress also disrupts the cyclic and noncyclic types of electron transport in the light reactions of photosynthesis [8, 33].

A lower electron transport rate negatively affects the photophosphorylation process (ATP biosynthesis) [2, 8, 11, 33] as well as the NADPH/H^+ reduction [11, 14, 33]. These alterations cumulatively disrupt the photosynthetic apparatus under water stress conditions [8, 11, 33]. Both of the photosystems PSI and PSII in chloroplasts are affected by water-deficit conditions mainly due to a lower electron transport rate and the accumulation of ROS [8, 11]. The responses of adaptive plants to resist drought-induced damage to the photosynthetic apparatus include thermal dissipation of light energy, photo destruction of the D1 protein of PSII, triggering of and increased xanthophyll cycle activity, water–water cycle, and dissociation of the light-harvesting complexes from photosynthetic reaction centers [8, 11, 33].

1.5.5 Respiration

Drought tolerance is a costly phenomenon for plants, and the quantity of energy used to cope with it is enormous [8, 14]. The major consumer of fixed carbon in photosynthesis is the root for growth and maintenance [14]. In addition to plant growth and development, environmental conditions also influence the respiration rate. Under water stress conditions, a change can occur in carbon metabolism as a result of diminished photosynthesis and active respiration. A plant's growth rate is determined precisely by photosynthetic CO_2 assimilation and the respiration ratio [8, 13, 14, 18]. Drought-sensitive plants use a relatively greater amount of energy resources to absorb water from soil, especially under severe drought stress. Under drought stress, the tricarboxylic acid (TCA) cycle and ATP biosynthesis are negatively affected and lead to a decreased respiration rate [3, 8, 14]. However, limited root respiration rate and root biomass production under a severe soil water-deficit can improve the growth and physiological activity of plants [3, 8, 14, 15]. There are two mitochondrial electron transport pathways from ubiquinone to oxygen in plants. The alternative pathway branches from the cytochrome pathway and transfers electrons to oxygen directly by alternative oxidase [8, 14]. When plants are exposed to drought stress, they produce ROS in the mitochondria. These free radicals could damage cellular components [3, 22]. Alternative oxidase activity could be useful in maintaining normal levels of metabolites and reducing ROS production by transferring electrons to O_2 and reducing H_2O_2 [2, 8, 14].

1.5.6 Mineral Nutrition

Water stress affects plant mineral nutrition and disrupts ion homeostasis in plant cells [2, 9, 21]. Generally, decreasing water availability under water stress conditions limits the total nutrient availability in soil, decreases the nutrient uptake by roots, and finally reduces their tissue concentrations in plants [14, 21, 33]. Changing nutrient uptake by the root and their transport to the shoots is an important effect of water deficit on plants. Generally, drought stress leads to an increase in N, causes a reduction in the P concentration, and has no definitive effects on the K concentration in plants [2, 14, 35]. A decrease in the Ca content of plants has been reported by many researchers as well [2, 8, 33]. The cell membrane is one of the earliest targets of many stresses such as drought; membrane stability in the roots plays an essential role in the appropriate mineral nutrition of plants. Therefore, preservation of the membrane stability is a very important factor in plant resistance to drought. Damage of cell membranes under water-deficit conditions is an important factor leading to disruption of ion homeostasis in plants [14, 21, 29, 33].

1.5.7 Hormonal Balance

Hormones play key roles in the regulation of plant processes. Some hormones are involved in plant interactions with environmental stresses such as drought [7, 21]. Abscisic acid (ABA) is one of the most effective hormones in plant response to drought stress [2, 7, 8]. After plants are exposed to drought, ABA is synthesized in roots and translocates to shoots, especially leaves. Furthermore, water stress induces ABA synthesis in chloroplasts. In addition, the plasma membrane ATPase (PM-ATPase) activity decreases under water-deficit conditions due to a lower ATP supply by photosynthesis and respiration. Low PM-ATPase increases the apoplastic (cell wall) pH and leads to the conversion of ABA to its anionic form (ABA^-). ABA^- cannot cross the plasma membrane of the leaf cells and translocates toward the gourd cells of stomata by a transpiration stream in the leaf apoplast. ABA translocation to stomata induces stomatal closure and decreases the stomatal conductance capacity. A higher stomatal resistance leads to lower water losses from the leaf surface, which is one of the earliest plant responses for resistance to water stress. However, low CO_2 uptake by stomata leads to a reduction in the photosynthesis rate in leaves [2, 7, 9, 28, 31, 33]. ABA plays a key role in the regulation of aquaporin's activity as well [8, 14]. It is well known that ABA accumulation under drought conditions reduces ethylene production [8, 9]. In contrast, auxins act as negative regulators of drought tolerance in plants because indole-3-acetic acid (IAA) downregulation facilitates the accumulation of late embryogenesis abundant (LEA) mRNA. ABA induces the accumulation of LEA proteins, which are involved in plant adaptation to drought stress, especially in seeds [7, 8, 13, 27]. Endogenous cytokinin (zeatin) and gibberellin (GA3) levels of plants decline rapidly under water stress situations. Cytokinins have been shown to delay senescence; hence, those could lead to better adaptation of plants by delaying drought-induced senescence [8, 11]. Generally, drought leads to an increase in brassinosteroid (BR) accumulation in plants. Brassinosteroids increase water uptake and cell membranes stability and can also reduce ion leakage from membrane under drought stress conditions [8, 29].

1.5.8 Protein, Amino Acids, and Mineral Content

Plants synthesize compounds such as proteins and amino acids and accumulate some minerals in response to drought stress [27, 29]. Drought conditions change the quantity and quality of plant proteins [11, 14, 29]. Generally, the protein content decreases under a water deficit due to suppression of their synthesis. Gene expression changes during drought stress; hence, the synthesis of drought-related proteins and mRNAs changes consequently [7, 11, 27, 33]. However, the synthesis of some proteins and enzymes—such as LEA proteins, proteases, enzymes required for the biosynthesis of various osmotic-compatible compounds (osmoprotectants), enzymes involved in the detoxification of ROS [e.g., superoxide dismutase (SOD), catalase (CAT), peroxidase (POD), ascorbate peroxidase (APX), and glutathione reductase (GR)] and protein factors involved in the regulation of signal transduction and gene expression—increase under drought stress [13, 14, 22, 27, 29, 33, 37, 39].

The accumulation of compatible solutes (osmoprotectants in some texts) in order to provide osmotic adaptation (osmotic regulation and osmotic adjustment) is a well-known mechanism for plant resistance to drought and some other stress such as salinity [4, 16, 23, 33]. Compatible solutes have a low molecular weight and can accumulate at high concentrations without having damaging effects on the cell components and metabolism [29, 37]. The accumulation of compatible solutes increases the cellular osmotic pressure and triggers water uptake from soil. In addition, compatible solutes regulate the osmotic balance between the vacuole and the cytosol, maintain the turgor pressure and water content of cells, and protect against water loss from plants because of their high lipophilicity. Also, they might replace water molecules around nucleic acids, proteins (like enzymes), and membranes during water shortages. Compatible solutes might prevent interactions between ions (at a high concentration) with cellular components by replacing the water molecules around these components and protecting against the destabilization of important macromolecules [4, 16, 23, 33]. Proline is one of the standard amino acids known as osmoprotectants [3, 14]. Drought increases cell proline levels in two ways: by increasing proline synthesis and by decreasing the activity of enzymes involved in its degradation. Low turgor pressure is the first reason for proline accumulation under drought stress. There are close relationships between proline accumulation and plant resistance to drought stress [18, 29]. Many researchers have reported that proline has an important role in osmotic regulation. Proline accumulation as well as that of other osmoprotectants lead to a lower water potential of cells and hence help water uptake from soil under drought conditions [9, 14]. In addition, proline protects cell components from oxidative stress, and its biosynthesis and degradation process play important roles in balancing the energy between chloroplasts and mitochondria [33]. During proline generation and destruction pathways, NADPH/H^+ oxidizes to $NADP^+$ in chloroplasts and NAD^+ reduces to NADH/H in mitochondria, respectively. The NADPH/H^+ oxidation in chloroplasts reduces the ROS generation because of the consumption of excess electrons. In addition, NADH/H^+ reduction in mitochondria is necessary for energy supply for cells as well as for recovery processes after stress [24, 29, 33]. Proline isn't the only compatible solute or osmoprotectant whose production and accumulation are induced under water-deficit conditions. Compatible solutes are divided into four major groups: (1) sugars, including monosaccharides (e.g., fructose and glucose) and di- and oligo-saccharides (e.g., sucrose, trehalose, and raffinose); (2) amino acids (e.g., proline and citrulline); (3) onium compounds, including tertiary and quaternary ammonium as well as sulfonium compounds (e.g., glycine-betaine and 3-dimethylsulfoniopropionate); and (4) polyols and sugar alcohols (e.g., mannitol, pinitol, glycerol, and sorbitol) [3, 9, 13, 14, 31, 33, 37].

In addition to compatible compounds, in some cases plants accumulate specific minerals such as NaCl in order to maintain the intracellular water potential. Although mineral accumulation isn't always compatible with metabolism, some plants, such as halophytes, accumulate some minerals and are resistant to their damages due to specific mechanisms. Generally, plants accumulate minerals in the vacuole and compatible compounds in the cytosol in order to balance the water potential of the two compartments [32, 33].

1.5.9 Lipids

Lipids are the most abundant component of cell membranes and play an important role in the resistance of plant cells to environmental stresses [8, 29, 33]. Generally, drought stress leads to a disturbance in the association between membrane lipids and proteins as well as decreases the membrane-bound enzyme activity and transport capacity of the bilayer [14, 21]. Monogalactosyldiacylglycerol (MGDG) is a major leaf glycolipid that decreases after plant exposure to drought. MGDG is the most important component of the chloroplast membrane; accordingly, its lower content leads to destruction of the chloroplast membrane and negatively affects photosynthesis. Lipid peroxidation due to oxidative damage is the well-known effect of drought and many other environmental stresses in plants [14, 24, 29, 33].

1.5.10 Oxidative Stress as a Secondary Stress

Exposure of plants to many environmental stresses such as drought leads to the generation of ROS, including superoxide radical (O_2^-), hydroxyl radical (OH), hydrogen peroxide (H_2O_2), alkoxy radicals (RO), and singlet oxygen. Oxidative stress is known as a secondary stress and causes oxidative damage in cells [8, 14, 22, 27, 28, 29, 33, 39]. ROS may react with proteins, lipids, and other important macromolecules and can denaturize the structure and function of the macromolecules [3, 8, 21, 24, 33]. Many cell compartments produce ROS under drought stress, such as chloroplasts, mitochondria, peroxisomes, and others [14, 22, 28]. The generation of ROS in biological systems is represented by both nonenzymatic and enzymatic mechanisms, which are dependent on some factor such as oxygen concentration in the cells [14]. Generally, ROS accumulation leads to DNA nicking, oxidation of amino acids, protein and photosynthetic pigments, lipid peroxidation, and so on [14, 27, 33]. Plants have developed some mechanisms to avoid ROS damage. All these mechanisms form an antioxidant defense system, which includes both enzymatic and nonenzymatic components. SOD, CAT, POD, APX, and GR are some enzymes involved in the antioxidant responses of plants [3, 9, 13, 14, 17, 22, 27, 33, 39]. Glutathione, ascorbic acid, carotenoids, and α-tocopherol are some compounds involved in the antioxidant defense system of plants [8, 14, 17, 33].

1.5.11 Molecular Effects

A complex set of genes participates in plant responses to drought stress [7, 8, 14]. Many gene expression patterns change when plants are exposed to drought [7, 8]. First, the expression of genes involved in early responses—such as signal transduction, transcription, and translation factors—has been changed. Next, changes in the expression of genes involved in late responses—such as water transport, osmotic balance, oxidative stress, and the damage-repair process—have occurred [8, 28, 37]. Drought sensing and signal transduction are still not clearly known. Generally, drought signaling is closely joined with ABA signal transduction. ABA plays a key role in plant drought responses and gives rise to drought-induced genes [3, 7, 9, 19, 33, 37]. Plant gene expression is controlled at different levels, including the transcriptional, posttranscriptional, translational, and posttranslational phases [8, 9, 28]. Apparently, the regulation of plant response mechanisms to abiotic stresses including drought stress is controlled at two levels: the transcriptional and translational levels [8, 14, 37]. Bioinformatics analyses have identified several transcription factors (TF) induced under drought stress. TFs are classified in several families, including MYB/MYC, zinc-finger protein, and NAC [9, 13, 26, 27, 28, 37]. Translational control is another mechanism involved in plant responses to drought and controls the protein production [37]. Molecular biology research has shown that plants respond to stress not only at the cells' mRNA or protein level, but also at the posttranscriptional phase [8, 27]. MicroRNAs (miRNAs) are a class of small RNAs that are recognized as important modulators of gene expression at the post-transcriptional level [6, 8]. Previously, many RNA molecules were counted, such as miR474, miR528, miR167, miR160, miR390, miR166, miR397, miR398, miR393, miR159, miR169, miR172, miR395, NAT-siRNAs, and tasiRNAs, which are involved in plant response and resistance to drought [6, 13]. Studies have shown that these miRNA molecules are involved in responses mediating with ABA, auxin signaling, cell growth, antioxidant defense, osmotic adjustment, photosynthesis, and respiration under drought [6, 13, 14].

1.6 Plant Responses to Drought Stress

Plants are sessile organisms and must tolerate environmental stresses; hence, they have developed various mechanisms for resistance to the stresses. Moreover, as plants are multicellular organisms, their responses to environmental stresses such as drought are complex [8, 19, 27, 30, 33]. Generally, plant resistance to environmental stress is divided into two main strategies: stress avoidance and stress tolerance. Plant adaptation to a water deficit is made possible by physiological, morphological, phenological, biochemical, and molecular responses. The responses can range from being at a molecular level to being at a whole plant level. Plant strategies to cope with drought are summarized in the next three subsections, escape, avoidance, and tolerance. Although escape is generally a part of plants' avoidance strategy, plants that escape from drought actually are not exposed to a water deficit. Therefore, in this chapter we explain it in a separate section.

1.6.1 Escape

Escape from drought is possible because of a shortened life cycle or growing seasonally and allowing plants to reproduce before the environment becomes dry [2, 9, 14]. A short life cycle can lead to drought escape due to early flowering, which is considered a form of adaptation to drought by stress avoidance [2, 14, 19]. The plant life cycle is dependent on the plant genotype and the environmental conditions. Drought escape occurs when the phenological development matches periods during which soil moisture is available. Therefore, early maturity and consequently early flowering help plants avoid drought stress although the yield is generally decreased [2, 9, 14].

1.6.2 Avoidance

The main aim of this strategy is the preservation of a high water potential in plants. The chief characteristic of this strategy is reducing water loss from plants by stomatal control of transpiration and maintaining water uptake from the soil by an extensive and prolific root system [9, 14]. A deep and thick root system is helpful for exploring water from a considerable soil depth and at a large distance from the plant [2, 14, 15]. The cuticle and hairy leaves help to maintain a high tissue water potential within plant and are considered a xeromorphic trait for drought tolerance. The production of these structures leads to a decreased plant yield due to the energy consumed to produce them. Therefore, plants that use the avoidance strategy to maintain a relatively high water potential are generally small in size [14, 19, 33].

1.6.3 Tolerance

Plants that use a tolerance strategy for drought resistance limit the number and area of leaves in response to water deficit; however, this strategy leads to yield loss [2, 9]. In addition, these plants show some xeromorphic traits such as hairy leaves and the production of trichomes on both sides of leaves [14, 19, 33]. Hairiness reduces the leaf temperature, while transpiration increases light reflectance and minimizes water loss by increasing the boundary layer resistance to water vapor movement away from the leaf surface. Inter- and intracellular changes in leaves are visible [9, 14, 33]. The root is the main tissue to uptake water from the soil. Hence, the root growth rate, density, proliferation, and size are key factors influencing plant responses to drought stress. Studies have shown that an alteration in the root system architecture is the main factor in plant tolerance, especially when tolerance is defined as the ability of a plant to maintain its leaf area and growth rate during a prolonged vegetative stage [14, 33]. The accumulation of compatible solute and osmotic adaptation, the induction of an antioxidant system, an alteration in metabolic pathways, an increase in the root/shoot ratio, and closure of the stomata are other mechanisms involved in plant tolerance to drought.

1.7 Improved Drought Tolerance in Crops and Drought Management

Scientists have tested many techniques to improve drought tolerance in crop plants [8, 9, 26, 33, 37]. The production of transgenic plants is one of the well-known methods for this purpose [19, 26, 33, 37]. The wide range of drought-related genes in the plant genome has opened amazing opportunities for crop improvement [19, 20, 26, 37]. With all these interpretations, in practice the generation of transgenic plants cannot be completely effective for the production of drought-tolerant plants, because it requires a very complex and expensive laboratory method and generally its success rate is low [19, 26, 27, 33, 37]. Traditionally, there have been several efforts to generate drought-tolerant crop plants through usual breeding methods [27, 30, 33]. In this method, two groups of plants with desirable traits are selected and crossed to exchange their genes; therefore, the offspring have new genetic arrangements [19, 37]. Important traits to use in plant breeding might include water-extraction efficiency, water-use efficiency, hydraulic conductance, osmotic and elastic adjustments, and modulation of leaf area [8, 13, 14, 26, 27, 30, 33, 37]. Genetic data can improve the efficiency of the breeding method. Genetic improvement can assist by using recognizable tags to target genes; these are known as polymorphisms based on molecular markers that occur naturally in the DNA sequence [37]. Different methods are employed to recognize linked markers, including restriction fragment length polymorphisms (RFLPs), sequence characteristic amplified regions (SCARs), random amplified polymorphic DNA (RAPDs), simple sequence repeats (SSRs), amplified fragment length polymorphism (AFLPs), and others [19, 20, 37]. The genetic factors involved in quantitative characteristics of phenotypes are called quantitative trait loci (QTLs) [3, 8, 27, 30, 37].

The use of plant breeding methods has an enormous potential to accelerate drought-tolerant plant production and help drought management assist these plants [14, 37]. In addition, there are several strategies for drought management in agricultural fields on a number of levels. Useful strategies include irrigating during periods of low soil moisture, especially for young plants, using modern and effective methods, selecting the appropriate place and imitating good planting practices, selecting native plants or matching plant species to site conditions, using mulch to maintain soil moisture, eliminating any dead or weak tissues to resist secondary problems such as insects and herbivore invasions [14, 19, 27, 30], and inoculating plants with symbiotic microorganisms such as arbuscular mycorrhizal fungi [1, 10].

1.8 Conclusion

Drought is a prevalent stress factor especially in arid and semiarid areas and can affect different aspects of plant growth, development, and metabolism. Drought is a multidimensional stress factor and hence its effects on plants are complex. It effects on plants can occur on a molecular level up to a whole plant level. There are several reasons for drought in nature, including low rainfall, salinity, high temperature, and high intensity of light, among others. Plants have developed some mechanisms for resistance to drought; they are generally classified as avoidance and tolerance strategies. Plants have several options they can use for drought tolerance, including developmental, physiological, morphological, ecological, biochemical, and molecular mechanisms. The production of tolerant plants by traditional breeding methods as well as the generation of transgenic plants by gene manipulation are useful

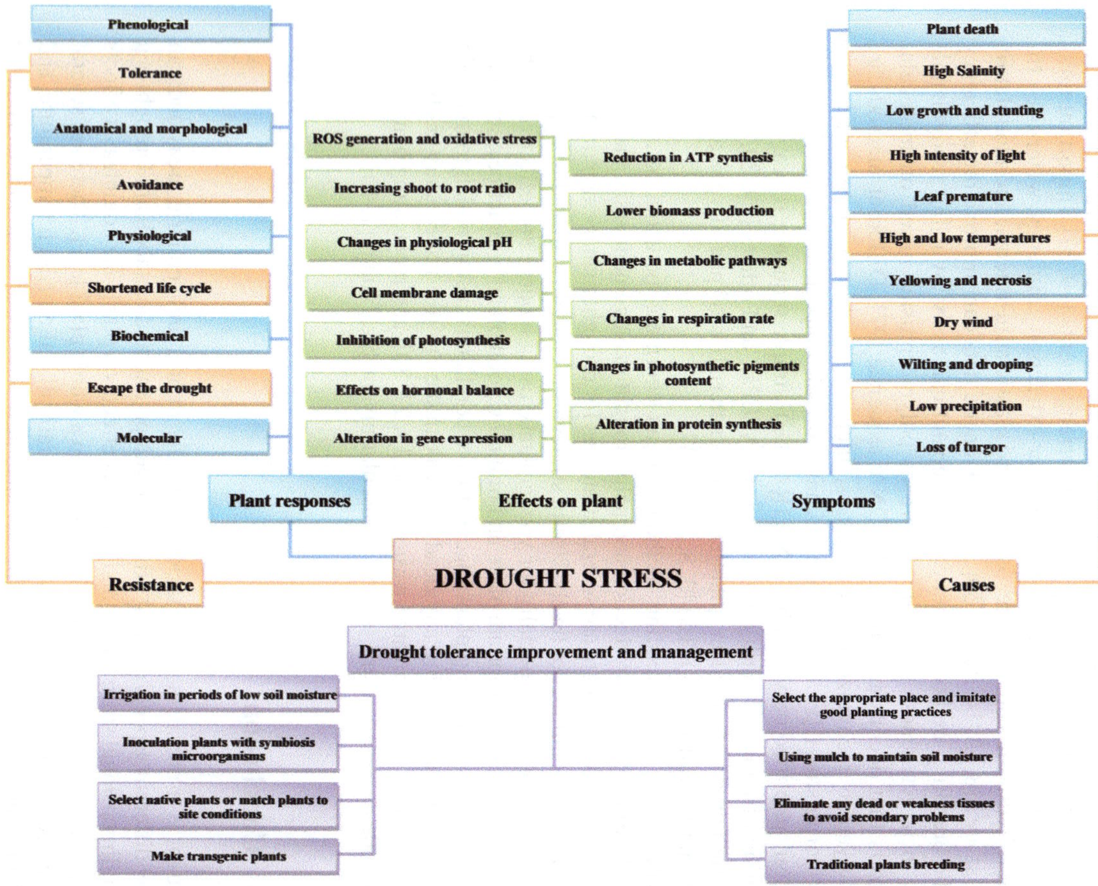

Fig. 1.1 Causes of drought and its effects on plants, symptoms of water stress, plant responses to drought and mechanisms involved in resistance, and some useful strategies for drought management

procedures in order to minimize the negative effects of drought on plants. In addition, several strategies for drought management in agricultural fields on multiple levels can be effective. The causes of drought, drought effects and its symptoms in plants, plant responses in order to resist drought, and some strategies that can be useful for drought management are summarized in Fig. 1.1.

References

1. Abdelmoneim TS, Tarek AAM, Almaghrabi OA, Alzahrani HS, Abdelbagi I. Increasing plant tolerance to drought stress by inoculation with arbuscular mycorrhizal fungi. Life Sci J. 2014;11:10–7.
2. Akhtar I, Nazir N. Effect of waterlogging and drought stress in plants. Int J Water Res Environ Sci. 2013;2:34–40.
3. Arbona V, Manzi M, de Ollas C, Gómez-Cadenas A. Metabolomics as a tool to investigate abiotic stress tolerance in plants. Int J Mol Sci. 2013;14:4885–911.
4. Ashraf M, Ozturk M, Athar HR. Salinity and water stress, improving crop efficiency. the Netherlands: Springer; 2009.
5. Barriopedro D, Gouveia C, Trigo RM, Wang L. The 2009/10 drought in China: possible causes and impacts on vegetation. Amer Meteor Soc. 2012;13:1251–67.
6. Bej S, Basak J. MicroRNAs: the potential biomarkers in plant stress response. Amer J Plant Sci. 2014;5:748–59.
7. Bernacchia G, Furini A. Biochemical and molecular responses to water stress in resurrection plants. Physiol Plant. 2004;121:175–81.
8. Bhargava S, Sawant K. Drought stress adaptation: metabolic adjustment and regulation of gene expression. Plant Breeding. 2013;132:21–32.
9. Bray EA. Plant response to water-deficit stress. Encycl Life Sci. 2001. 10.1002/9780470015902.a0001298.pub2.
10. Chepsergon J, Mwamburi L, KipkemboiKassim M. Mechanism of drought tolerance in plants using *Trichoderma* spp. Int J Sci Res. 2012;3:1592–5.
11. Chernyad'ev II. Effect of water stress on the photosynthetic apparatus of plants and the protective role of cytokinins: a review. Appl Biochem Microbiol. 2005;41:115–28.
12. Dai A. Drought under global warming: a review. Wires Clim Chg. 2012;2:45–65.
13. Ding Y, Tao Y, Zhu C. Emerging roles of MicroRNAs in the mediation of drought stress response in plants. J Exp Bot. 2013;64:3077–86.
14. Farooq M, Wahid A, Kobayashi N, Fujita D, Basra SMA. Plant drought stress: effects, mechanisms and management. Agron Sustain Dev. 2009;29:185–212.
15. Franco JA. Root development under drought stress. Technol Knowl Transf e-Bull. 2011;2:1–3.
16. Hirt H, Shinozaki K, editors. Plant responses to abiotic stress. Germany: Springer; 2004.
17. Jaleel CA, Manivannan P, Wahid A, Farooq M, Somasundaram R, Panneerselvam R. Drought stress in plants: a review on morphological characteristics and pigments composition. Int J Agric Biol. 2009;11:100–5.
18. Keyvan S. The effects of drought stress on yield, relative water content, proline, soluble carbohydrates and chlorophyll of bread wheat cultivars. J Anim Plant Sci. 2010;8:1051–60.
19. Khan MA, Iqbal M, Jameel M, Nazeer W, Shakir S, Aslam MT, Iqbal B. Potentials of molecular based breeding to enhance drought tolerance in wheat (*Triticum aestivum* L.). Afr J Biotechnol. 2011;10:11340–4.
20. Khan MA, Iqbal M, Akram M, Ahmad M, Hassan MW, Jameel M. Recent advances in molecular tool development for drought tolerance breeding in cereal crops: a review. Zemdirbyste. 2013;100:325–34.

21. Kheradmand MA, Shahmoradzadeh Fahraji S, Fatahi E, Raoofi MM. Effect of water stress on oil yield and some characteristics of *Brassica napus*. Intl Res J Appl Basic Sci. 2014;8:1447–53.
22. Labudda M, SafiulAzam FM. Glutathione-dependent responses of plants to drought: a review. Acta Soc Bot Pol. 2014;83:3–12.
23. Madhava Rao KV, Raghavendra AS, Janardhan Reddy K, editors. Physiology and molecular biology of stress tolerance in plants. the Netherlands: Springer; 2006.
24. Mafakheri A, Siosemardeh A, Bahramnejad B, Struik PC, Sohrabi E. Effect of drought stress on yield, proline and chlorophyll contents in three chickpea cultivars. Amer J Chin Stud. 2010;4:580–5.
25. Mishra AK, Singh VP. Drought modeling—a review. J Hydrol. 2011;403:157–75.
26. Nakashima K, Yamaguchi-Shinozaki K, Shinozaki K. The transcriptional regulatory network in the drought response and its crosstalk in abiotic stress responses including drought, cold, and heat. Front Plant Sci. 2014;5:1–8.
27. Nezhadahmadi A, Hossain Prodhan Z, Faruq G. Drought tolerance in wheat. Scientific World J. 2013;2013:1–12.
28. Osakabe Y, Osakabe K, Shinozaki K, Tran LSP. Response of plants to water stress. Front Plant Sci. 2014;5:1–7.
29. Rahdari P, Hoseini SM. Drought stress: a review. Intl J Agron Plant Prod. 2012;3:443–6.
30. Rana RM, Rehman SU, Ahmed J, Bilal M. A comprehensive overview of recent advances in drought stress tolerance research in wheat (*Triticum aestivum* L.). Asian J Agric Biol. 2013;1:29–37.
31. Rangan P, Subramani R, Kumar R, Singh AK, Singh R. Recent advances in polyamine metabolism and abiotic stress tolerance. BioMed Res Int. 2014;2014:1–9.
32. Rauf M, Shahzad K, Ali R, Ahmad M, Habib I, Mansoor S, Berkowitz GA, Saeed NA. Cloning and characterization of Na^+/H^+antiporter (LfNHX1) gene from a halophyte grass *Leptochloa fusca* for drought and salt tolerance. Mol Biol Rep. 2014;41:1669–82.
33. Salehi-lisar SY, Motafakkerazad R, Hossain MM, Rahman IMM. Water stress in plants: causes, effects and responses, water stress. In: Ismail Md. Mofizur Rahman, editor. InTech, 2012.
34. Sapeta H, Costa M, Lourenc T, Marocod J, Van der Linde P, Oliveiraa MM. Drought stress response in *Jatropha curcas*: growth and physiology. Environ Exp Bot. 2013;85:76–84.
35. Shao HB, Chu LY, Jaleel CA, Zhao CX. Water-deficit stress-induced anatomical changes in higher plants. C R Biol. 2008;331:215–25.
36. Trenberth KE, Dai A, van der Schrier G, Jones PD, Barichivich J, Briffa KR, Sheffield J. Global warming and changes in drought. Nat Clim Chg. 2014;4:17–22.
37. Xoconostle-Cazares B, Ramirez-Ortega FA, Flores-Elenes L, Ruiz-Medrano R. Drought tolerance in crop plants. Am J Plant Physiol. 2010;5:241–56.
38. Zare M, Azizi MH, Bazrafshan F. Effect of drought stress on some agronomic traits in ten barley (*Hordeum vulgare*) cultivars. Tech J Eng Appl Sci. 2011;1:57–62.
39. Zlatev Z, Lidon FC. An overview on drought induced changes in plant growth, water relations and photosynthesis. Emir J Food Agric. 2012;24:57–72.

Chapter 2
Drought Stress Memory and Drought Stress Tolerance in Plants: Biochemical and Molecular Basis

Xiangnan Li and Fulai Liu

2.1 Introduction

Global warming will not only affect air temperature but also influence the amount and distribution of precipitation possibly leading to more frequent drought spells in the future (Wang et al. 2014a). Drought is one of the major threats to plants, as water deficit affects the plant–water relations at all levels from molecular, cellular, and organ to the whole plant (Li et al. 2014a; Muscolo et al. 2015). Drought depresses plant growth and development, which results in the production of smaller organs, and hampered flower production and grain filling. Following drought, stomata close progressively with a parallel decline in net photosynthesis and water-use efficiency (Farooq et al. 2009a, b). Stomatal conductance is controlled not only by soil water condition, but by a complex interaction of intrinsic and extrinsic factors (Liu et al. 2006). Depending on the availability of soil moisture, activities of the enzymes of carbon assimilation and the enzymes involved in adenosine triphosphate synthesis are decreased (Farooq et al. 2009a, b). One of the major factors responsible for impaired plant growth and productivity under drought stress is the production of reactive oxygen species in organelles including chloroplasts, mitochondria, and peroxisomes (Farooq et al. 2009a, b; Wei et al. 2015). The overproduction of reactive oxygen species (ROS) results in the peroxidation of cellular membrane lipids and degradation of enzyme proteins and nucleic acids (Li et al. 2013).

A number of physiological and biochemical processes at molecular, tissue, organ, and whole-plant levels are involved in drought tolerance mechanism. For instance, the plant water loss is reduced by increasing stomatal resistance, and the water uptake is increased by developing large and deep root systems (Liu et al. 2006). Among plant growth substances, salicylic acid, melatonin and abscisic acid

X. Li • F. Liu (✉)
Department of Plant and Environmental Sciences, Faculty of Science, University of Copenhagen, Højbakkegaard Allé 13, Taastrup 2630, Denmark
e-mail: fl@plen.ku.dk

M.A. Hossain et al. (eds.), *Drought Stress Tolerance in Plants, Vol 1*,
DOI 10.1007/978-3-319-28899-4_2

were reported to play an important role in drought tolerance. Scavenging of reactive oxygen species by enzymatic and nonenzymatic systems, cell membrane stability, and expression of stress proteins are also vital mechanisms of drought tolerance (Farooq et al. 2009a, b). Drought stress effects can be managed by production of most appropriate plant genotypes, seed priming, plant growth regulators, use of osmoprotectants, and some other strategies.

2.2 Priming, Stress Memory, and Drought Tolerance

The increased climatic variability and more frequent episodes of extreme conditions also result in plants being exposed to not only one single drought event but also multiple abiotic stresses at different periods. Although the abiotic stresses occurring at different stages result in a higher risk of injury, earlier stress events may prime the plant to protect it against later stresses. A large body of evidence has shown that a previous exposure to different types of stress can affect the subsequent responses and eventually prepare the plants to more quickly or actively respond to future stresses (Ramírez et al. 2015; Walter et al. 2011; Li et al. 2014a). The trigger for stress tolerance (the early moderate stress event) is referred to "priming." Priming has been known as a potential way to enhance the stress tolerance of plant (Bruce et al. 2007), which is related to stress memory. Stress memory involves multiple modifications at physiological, proteomic, transcriptional levels and epigenetic mechanisms in plants (Kinoshita and Seki 2014), which can occur in any periods of the life cycle, including seed germination, vegetative growth, and reproductive growth (Ramírez et al. 2015; Munné-Bosch and Alegre 2013). Recently, many studies have focused on exploring the mechanisms of the priming effects and stress memory in the formation of drought tolerance in different plant species (Ramírez et al. 2015; Walter et al. 2011; Wang et al. 2014c, 2015; Shukla et al. 2015; Li et al. 2015b). In this chapter, we summarized recent advancements in physiological, biochemical and molecular and cellular research related to drought tolerance formation in plants. The mechanisms of drought stress memory and the possible priming-induced cross-tolerance to other abiotic stresses are discussed.

2.2.1 Seed Priming

Seed priming is different from plant priming, although both could result in increased stress tolerance. As stated above, priming is a process where a first exposure to a moderate stress enables plants to be more tolerant to subsequent stress events (Conrath 2011). Seed priming is pre-sowing partial hydration of seeds without allowing radicle emergence to improve germination rate and stress tolerance of germinating seeds, and even to improve the seedling establishment. A hypothetical model has been proposed to illustrating the cellular physiology of priming-induced

Table 2.1 Various seed priming methods adopted for developing drought tolerance in plants

Plant species	Priming methods	References
Cicer arietinum	Hydropriming	Kaur et al. (2002)
Cicer arietinum	Osmopriming with mannitol	Elkoca et al. (2007)
Saccharum officinarum	Halopriming with NaCl	Pandita et al. (2010)
Oryza sativa	Hydropriming, Osmopriming with KNO_3	Basra et al. (2005)
Oryza sativa	Osmopriming with KCl and $CaCl_2$	Farooq et al. (2010)
Oryza sativa	Osmopriming with PEG	Yuan-Yuan et al. (2010)
Spinacea oleracea	Osmopriming with PEG	Chen and Arora (2011)
Spinacea oleracea	Osmopriming	Chen et al. (2013)
Lesquerella fendleri	Osmopriming with PEG	Windauer et al. (2007)
Zea mays and *Spinacea oleracea*	Osmopriming with PEG	Chen et al. (2012)
Zea mays	Ascorbic Acid, Salicylic Acid, and Hydrogen Peroxide	Ahmad et al. (2015)
Triticum aestivum	Hydropriming	Meena et al. (2014)
Triticum aestivum	Chemical priming (KH_2PO_4, H_2O_2, NO)	Giri and Schillinger (2003)
Bromus	Osmopriming with PEG	Tavili et al. (2011)

stress tolerance, which involves two strategies (Chen and Arora 2012). First, seed priming sets in motion activities involved in seed germination, such as respiration, endosperm weakening and seed reserve (starch) degradation, which facilitate the transition of quiescent dry seeds into germinating state and increase the germination potential (Chen and Arora 2012; Li et al. 2013). Second, imposing abiotic stress to germinating seeds to stimulate stress responses (e.g., activation of ROS scavenging systems and accumulation of stress response proteins), hence inducing cross-tolerance (Chen and Arora 2012).

Various priming methods, including hydropriming, osmopriming, chemical priming, hormonal priming, biological priming, redox priming, and solid matrix priming, have been reported to improve seed germination under osmotic stress and promote the drought tolerance in seedlings (Jisha et al. 2012) (Table 2.1). Hydropriming significantly increases the root and shoot length compared with seedlings obtained from non-primed seeds in drought condition (Kaur et al. 2002). In addition, hydropriming has been used as an easy seed invigoration treatment for maize inbred lines under salinity and drought stress (Janmohammadi et al. 2008). Also, osmopriming with PEG can improve the germination of *Bromus* seeds under drought (Tavili et al. 2011). Comparing to hydropriming, priming with PEG has a better effect on seed germination and seedling growth under drought (Yuan-Yuan et al. 2010). In *Agropyron elongatum*, osmopriming with gibberellin (GA) and abscisic acid (ABA) increased CAT and SOD activities, and enhanced the drought tolerance, in relation to unprimed seeds (Eisvand et al. 2010). In addition, seed priming with triazoles affects turf grass growth and response to drought (Shahrokhi et al. 2011).

2.2.2 *Plant Acclimation to Drought*

Plant priming, which differs from seed priming, is a process that an earlier exposure to biotic stress enhances plants' tolerance to later abiotic or biotic insult (Chen and Arora 2012; Bruce et al. 2007). The similar process, with abiotic stress as the first exposure instead of biotic stress, is considered as acclimation or hardening (Chen et al. 2012; Bruce et al. 2007). However, in some studies, the earlier exposure to abiotic stress that favors the tolerance to later stress is also called plant priming (Li et al. 2014a, 2015b; Wang et al. 2014b, c). Plant priming induces stress memory, which is mediated by protein, transcription factors and the modifications in epigenetics, and this process is always accompanied by compromised plant growth (Chinnusamy and Zhu 2009; Chen and Arora 2012). The plants can acclimate to the drought event by modifications at morphological, metabolic, subcellular, proteomic and transcriptional levels, and even changes in microRNA expression.

2.2.2.1 Stomatal Regulation and Drought Acclimation

Morphological changes are consequence of a wide spectrum of physiological and molecular programs evolved to acclimate to drought conditions (Valdes et al. 2013). The morphological acclimation strategy usually includes smaller leaf area to decrease the transpiration and larger root system to enhance the water uptake capacity, and both are related to water-use efficiency (WUE). Some studies reported that WUE increases at moderate drought when water consumption dropped while photosynthetic rate remained high, that is, moderate drought often results in higher WUE (Varga et al. 2015; Peuke et al. 2006). It has also been documented that water supplies well below the optimum level led to a reduction in WUE as a consequence of stress effects, while above-optimum water supplies would result in the opposite tendency (Varga et al. 2013). Most recently, Varga et al. (2015) suggested that WUE decreased in some cultivars even in response to water deficit stress during early growth stages, while it dropped significantly when water was withheld at heading or grain filling except in the case of early maturing cultivars in wheat.

Stomatal regulation is one of the key mechanisms allowing plants to optimize CO_2 assimilation versus evaporative water loss (Tombesi et al. 2015). The stomatal density (SD) has been closely related to WUE and drought tolerance (Hepworth et al. 2015). Modification of SD in response to drought is contingent on the severity of drought, which varies among plant species (Hamanishi et al. 2012). For instance, drought decreased the stomatal numbers in wheat (Quarrie and Jones 1977), *Squash cotyledons* (Sakurai et al. 1986), and *Phytolacca dioica* (Silva et al. 2009). However, increased stomatal density was observed in grass with moderate drought stress (Xu and Zhou 2008). Plants with lower SD have significantly reduced levels of transpiration, and were able to grow continuously under drought condition (Doheny-Adams et al. 2012). The plants with reduced SD were also found to have significantly higher WUE (Franks et al. 2015). With *Arabidopsis thaliana* plant lines which have stomatal densities ranging from c. 50 to 250 % of normal levels, Hepworth et al. (2015) found that plants with less than half of their normal complement of stomata, and correspondingly reduced levels of transpiration, conserve soil moisture and are able to avoid drought stress but show little or no reduction in shoot nitrogen concentrations especially when water availability is restricted. By contrast, plants with over twice the normal density of stomata have a greater capacity for nitrogen uptake, except when water supply is limited (Hepworth et al. 2015). However, the stomatal development in response to drought is complicated and influenced by the expressions of many genes, such as *STOMAGEN, ERECTA (ER), STOMATA DENSITY AND DISTRIBUTION 1 (SDD1), and FAMA* (Hamanishi et al. 2012). More factors are probably involved in this process (Hamanishi et al. 2012); further exploration of these players in the stomatal development pathway would provide an increased insight into the long-term modulation of stomatal morphology in response to drought stress.

2.2.2.2 Metabolic Responses and Drought Tolerance

Plants reprogramming their metabolic pathways to acclimate to drought stress could result in changes in the upstream production and downstream utilization of metabolites (Baerenfaller et al. 2012; Suseela et al. 2015). Recent extensive and elegant metabolomics approaches have revealed that stressed plants invest in the production of important metabolites such as amino acids, organic acids, phenolic acids, polyamines, and lipids that partially mitigate stress by acting as osmoregulators, antioxidants, and defense compounds (Rivas-Ubach et al. 2012). Drought stress can also alter the content and composition of leaf proteins leading to changes in the proportion of structural and soluble proteins (Suseela et al. 2015). In oat (*Avena sativa* L.), the key processes involved in drought tolerance have been defined by metabolomic approach (Sanchez-Martin et al. 2015). During a time course of increasing water deficit, metabolites from leaf samples were profiled using direct infusion-electrospray mass spectroscopy (DI-ESI-MS) and high-performance liquid chromatography (HPLC) ESI-MS/MS and the data were analyzed using principal component analysis (PCA) and discriminant function analysis (DFA). The involvement of metabolite pathways was confirmed through targeted assays of key metabolites and physiological experiments in oat (Sanchez-Martin et al. 2015). This metabolomics experiment highlights a drought tolerance mechanism based on salicylate signaling pathways and the changes in carbon, antioxidant, and photooxidative metabolism. To identify the metabolic traits related to drought tolerance, a metabolomics and phenotypic study with four contrasting lentil accessions was carried out during germination and early growth stages (Muscolo et al. 2015). It was found that metabolic differences in the stress tolerance of the different genotypes were related to a reduction in the levels of tricarboxylic acid (TCA) cycle intermediates. In addition, ornithine and asparagine were identified as drought stress-specific metabolite indicators. In wheat, some metabolic parameters were also identified as good indicators of drought stress tolerance, such as the total protein content, glutamine synthetase (GS) enzyme activities, and the presence of GS isoforms.

Recently, the ^1H Nuclear magnetic resonance (NMR) spectroscopy is applied to monitor and quantify the degree of metabolic impact induced by drought, since NMR can bring high-throughput spectroscopic/structural information on a wide range of metabolites simultaneously with high analytical precision. The main advantage is that it can avoid biases against various classes of compounds (Silvente et al. 2012). The metabolic profiles in two soybean genotypes under short-term drought stress demonstrate critical differences in physiological responses between the genotypes. Metabolic changes in response to drought stress highlight the pools of metabolites that play key roles in the adjustment of metabolism and physiology of the soybean genotypes to response to drought stress (Silvente et al. 2012).

2.2.2.3 Photosynthetic Adaptation to Drought

The increased rate of photorespiration in plant that is observed during the onset of drought stress can be seen as an acclimation process to avoid an over-excitation of PSII under more severe drought conditions (Massacci et al. 2008). In cotton, photosynthetic electron transport is promoted during the onset of drought stress due to a higher efficiency of the open PSII reaction centers (Massacci et al. 2008). The additional energy is used to increase the rate of photorespiration while photosynthesis is kept constant or slightly decreases (Massacci et al. 2008). Chlorophyll fluorescence measurement has been proven as an efficient and reproducible tool for evaluating plant susceptibility index to drought (Su et al. 2015; Mishra et al. 2012). It can be used in selection of drought-tolerant cultivar and comparison of photosynthetic electron transport among cultivars with contrasting drought tolerance. The experiment with different wheat cultivars released in different years documented that the modern and intermediate cultivars had more sensitive stomata to water shortage, but the decreased activity of the PSII reaction center helped avoid damage from photoinhibition in these cultivars (Guan et al. 2015).

Chlororespiration is a respiratory electron transport chain in the thylakoid membrane of chloroplasts, which interacts with photosynthetic electron transport, involving both the non-photochemical reduction and plastoquinones oxidation with the corresponding consumption of oxygen (Ibáñez et al. 2010). It was found that the chlororespiration and the cyclic electron pathways play important roles in the tolerance to drought, and the different adaptive mechanisms to drought stress were indicated in sun and shade plants. In addition, the nitrate nutrition-induced chloroplast downsizing also significantly affects the mesophyll conductance and photosynthesis of rice in response to drought stress (Li et al. 2012). Recently, the proteomic and enzymatic studies documented that the main regulatory mechanisms for high drought tolerance of apple plants include the maintaining of Calvin cycle function by increasing key enzymes and stabilization of photosynthetic electron transfer, thus enhance net photosynthesis rate (Zhou et al. 2015). In addition, the response of signal regulatory proteins and abiotic stress-responsive proteins to drought also helps plants to cope with drought stress.

2.2.2.4 Mitochondrial Acclimation to Drought

One of the important factors determining the effect of water stress on plant productivity is its impact on mitochondrial respiration in different organs. Although specific rates of respiration are typically lower than the rate of net photosynthesis, the respirations by roots and shoots play key roles in determining the carbon balance and productivity of plants (Atkin and Macherel 2009). Of the CO_2 fixed each day by net photosynthesis in well-watered plants, 30–70 % is released back into atmosphere by the respiration of plants (Atkin et al. 2006; Loveys et al. 2002), and 50–70 % of whole-plant respiration occurs in the leaves (Atkin et al. 2007). The maintenance of mitochondrial respiration can also play several positive roles in helping plants grow and survive, both in normal and water-stress conditions (Atkin and Macherel 2009). Atkin and Macherel (2009) summarized that mitochondria and chloroplasts are closely connected by metabolic and signaling networks, and that, in intact leaves, photosynthesis depends to a large extent on mitochondrial functions. This reliance on mitochondria has the potential to be further enhanced under water-stress conditions. The mitochondria show a high flexibility in electron transfer and energy dissipation, which is very important for optimizing the energy balance in plants under drought stress. Alternative oxidase (AOX) constitutes a non-energy conserving branch of the mitochondrial electron transport chain. AOX activity may be important to avoid reactive oxygen species (ROS) generation by the chain under water stress (Wang and Vanlerberghe 2013). A study compared leaf *AOX1a* transcript and AOX protein amounts in wild-type (WT) *Nicotiana tabacum* plants experiencing mild to severe drought, and found mild to moderate drought resulted in a progressive and modest increase in AOX amount, accompanied by a progressive increased expression of different ROS-scavenging components (Wang and Vanlerberghe 2013). Under these conditions, transgenic plants with suppressed AOX amount, due to an RNA interference construct, were not compromised in their ability to manage ROS load and prevent cellular damage. Under severe drought condition, plants lacking AOX suffered more cellular damage than did WT and, at the most severe stage, were found to downregulate rather than upregulate the transcript level of several important ROS-scavenging components (Wang and Vanlerberghe 2013). In addition, WT plants could still recover rapidly after rewatering, but the recoverability of AOX knockdown plants was strongly compromised. However, a priority for future studies should be to clarify the ability of AOX affecting the functions of mitochondrion under drought stress.

2.2.2.5 Proteomic Acclimation to Drought

High-throughput methods have facilitated the identifying key regulatory processes, genes, and proteins that provide a theoretical basis for breeding drought-tolerant plant varieties. Proteomics has proven to be a good tool to explore biochemical pathways and the complex response mechanism of plants to drought stress (Zhou et al. 2015). The proteomic study with a drought-tolerant apple (*Malus domestica* Borkh) cultivar suggested that the main regulatory mechanisms for high WUE under moderate drought stress included the maintaining of Calvin cycle function by increasing the activity of key enzymes, stabilizing photosynthetic electron transfer, and keeping reactive oxygen species at normal level by regulating the photosynthetic electron transfer chain, photorespiration and ROS scavenging capability, thus preventing photoinhibition, reducing ROS production, and enhancing net photosynthesis rate. In addition, studies have indicated that some of the signal regulatory proteins and abiotic stress-responsive proteins also help plants to cope with drought stress (Zhou et al. 2015). Our recent finding showed that prior mild drought priming contributed to the homeostasis of oxidative metabolism and relatively better photosynthesis, and modification of oxidative stress defense, C metabolism and photosynthesis related proteins, hence to less grain yield loss caused by later spring low temperature stress (Fig. 2.1) (Li et al. 2014c).

Drought-inducible proteins are divided into two main groups: abiotic stress tolerance proteins including chaperones, detoxification enzymes, and mRNA-binding proteins; and regulatory proteins such as protein kinases, protein phosphatases, or other signal-related proteins (Ashoub et al. 2013). Different plant organs (e.g., root, stem, and leaf) contain different drought-inducible proteins and show distinct responses to drought (Hao et al. 2015). Differentially expressed proteins (DEPs) in roots, intermediate sections between roots and leaves (ISRL), and leaves in wheat showed significant changes in expression in response to drought stress and recovery. Numerous DEPs associated with cell defense and detoxifications were significantly regulated in roots and ISRLs, while in leaves, DEPs related to photosynthesis showed significant changes in expression. Expression of six Heat shock proteins (HSPs) potentially related to drought tolerance was significantly upregulated under drought conditions, and these proteins were involved in a complex protein–protein interaction network (Hao et al. 2015). Also, in wheat, proteomic analysis showed an increased abundance of proteins related to defense and oxidative stress responses such as GLPs, GST, and SOD, and those related to protein processing such as small HSPs in roots of both genotypes in response to drought stress (Faghani et al. 2014). In common bean, the majority of identified proteins in response to drought stress are classified into functional categories that include energy metabolism, photosynthesis, ATP interconversion, protein synthesis and proteolysis, stress and defense related proteins (Zadraznik

Fig. 2.1 Global presentation on response of winter wheat to the combination of freeze and water stress (drought and waterlogging). The numbers in parentheses indicate the protein spots in Table 2.2. The up- and downregulation of pathways in different stress treatments as compared with the CC were indicated by *red* and *blue triangles*, respectively. *APX* ascorbate peroxidase, *AsA* Ascorbic acid, *ATPase* ATP synthase, *CAT* catalase, *CYS* cysteine, *CYSase* cysteine synthase, *Cyt* b_6f cytochrome, *DHA* dehydroascorbate, *DHAR* monodehydroascorbate reductase, *Fd* ferredoxin, *FNR* ferredoxin-$NADP^+$ reductase, *Glc-6-P* Glucose-6-phosphate, *GPX* glutathione peroxidase, *GR* glutathione reductase, *GSH* glutathione, *GSSG* oxidized glutathione, *G-3-PD* glyceraldehyde-3-phosphate dehydrogenase, *MDA* monodehydroascorbate, *MDH* malate dehydrogenase, *PC* plasltocyanin, *PEP* phosphoenolpyruvate, *PEPC* phosphoenolpyruvate carboxylase, *P-GY* 3-phosphoglycerate, *PQ* plastoquinone, P_{680} and P_{700} PSII and PSI reaction center pigments, Q_A and Q_B, PSII primary and secondary plastoquinone electron acceptors, *RB* Ribulose bisphosphate, *RET* respiratory electron transport, *R-1, 5 BCA* ribulose-1,5-bisphosphate carboxylase activase, *SAM* S-adenosyl-L-methionine, *SAMase* S-adenosyl- L-methionine synthesis, *SOD* superoxide dismutase, *Suc* sucrose, *Triose-P* Triosephosphate, *UDP-Glc* Uridine diphosphate glucose, $^1O_2$ singlet oxygen

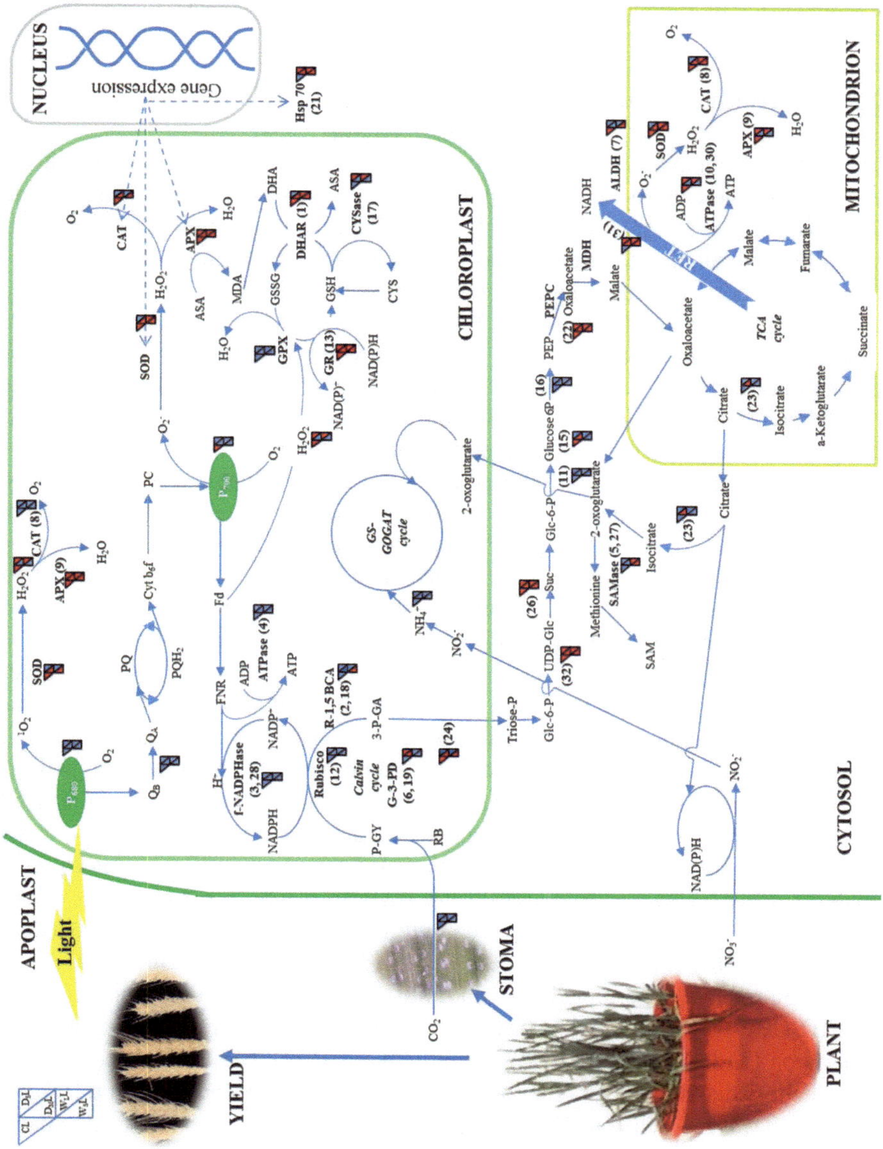
NUCLEUS
Gene expression
Hsp 70 (21)
CHLOROPLAST
MITOCHONDRION
CYTOSOL
APOPLAST
Light
STOMA
PLANT
YIELD
GS-GOGAT cycle
Calvin cycle
TCA cycle
RET
SOD
CAT (8)
APX (9)
DHAR (1)
GR (13)
GPX
CYSase (17)
ATPase (4)
f-NADPHase (3,28)
Rubisco (12)
R-1,5 BCA (2,18)
G-3-PD (6,19)
ALDH (7)
ATPase (10,30)
MDH
PEPC
SAMase (5,27)
Oxaloacetate
Malate
Fumarate
Succinate
Citrate
Isocitrate
a-Ketoglutarate
2-oxoglutarate
Methionine
SAM
Triose-P
Glc-6-P
UDP-Glc
Suc
Glucose 6P
PEP
NADH
NAD(P)H
NADPH
Fd
FNR
PC
PQ
Cyt b6f
ASA
MDA
DHA
GSSG
GSH
CYS
P-GY
RB
3-P-GA

et al. 2013). In addition to providing new information on the response to water deprivation, the proteomic study offers opportunities to pursue the breeding of wheat with enhanced drought tolerance using identified candidate genetic markers (Hajheidari et al. 2007). The proteomic studies also provide the basic insight needed to further investigate the molecular regulatory mechanism of drought tolerance.

2.2.2.6 Transcriptional Acclimation to Drought

Technological innovations over the past decades have made it possible to measure changes in gene expression (transcript levels) on genome-wide scales (Zhang et al. 2014; Udvardi et al. 2007; Urano et al. 2010). This enables an unprecedented overview of the global molecular changes occurring under drought stress. There are many published reports on transcriptomic variation induced by drought treatments in a variety of plant species (Zhang et al. 2014). In *Medicago*, the plants were subjected to a progressive drought stress over 14 days by withholding of water followed by rewatering to expose the plant to mild, moderate, and severe drought stress before rehydration. Transcriptome analysis of roots and shoots from control, mildly, moderately and severely stressed, and rewatered plants, identified many thousands of genes that were altered in expression in response to drought. Many genes with expression tightly coupled to the plant water potential (i.e., drought intensity), including eight NACs, eight MYBs, six AP2/EREBPs, six bZIPs, five HDs, four bHLHs, and other TFs, were identified suggesting their involvement in *Medicago* drought adaptation responses (Zhang et al. 2014). In chickpea, an oligonucleotide microarray was used for analyzing the transcriptomic profiles of unigenes in leaf and root under drought stress (Wang et al. 2012a), revealing that 4815 differentially expressed unigenes were either $\geq$2-fold up- or $\leq$0.5-fold downregulated in at least one of the five time points during drought stress. 2623 and 3969 unigenes were time-dependent differentially expressed in root and leaf, respectively. In this study, 110 pathways in two tissues were found to respond to drought stress. Compared to the control, 88 and 52 unigenes were expressed only in drought-stressed root and leaf, respectively, while nine unigenes were expressed in both the tissues (Wang et al. 2012a). Transcriptome analyses using transgenic Arabidopsis and soybean plants showed that the downstream genes of GmDREB1B;1 included numerous soybean-specific stress-inducible genes that encode an ABA receptor family protein, GmPYL21, and translation-related genes, such as ribosomal proteins, indicating that soybean DREB1/CBF-type transcription factors function in drought stress-responsive gene expression (Kidokoro et al. 2015). Recently, many genes and transcription factors were identified as key players for conferring ABA sensitivity and drought tolerance. In wheat, it was reported that an R2R3 MYB transcription factor, TaPIMP1, mediates drought stresses through regulation of defense- and stress-related genes (Zhang et al. 2012). In *Arabidopsis thaliana*, overexpression of *FTL1/DDF1*, an AP2 transcription factor, enhances tolerance to drought stresses (Kang et al. 2011). In rice, OsbZIP23 is documented as a major player of the bZIP family in rice for conferring ABA-dependent drought and salinity tolerance and has high potential usefulness in genetic improvement of stress tolerance (Xiang et al. 2008).

Table 2.2 A list of studies on the roles of miRNAs in plant tolerance to environmental stresses

miRNAs	Plant species	Environmental stress	References
169	*Solanum lycopersicum*	Drought	Zhang et al. (2011)
169	*Arabidopsis*	Drought	Li et al. (2008)
319	*Agrostis capillaris*	Salinity and drought	Zhou et al. (2013)
394	*Arabidopsis*	Drought	Ni et al. 2012)
395	*Arabidopsis*	Salinity and drought	Kim et al. (2010b)
396	*Arabidopsis*	Salinity and drought	Gao et al. (2010)
398	*Arabidopsis*	Salinity and drought	Zhou et al. (2007)
402	*Arabidopsis*	Salinity, drought, and cold	Kim et al. (2010a)
168 and 528	*Zea mays*	Drought	Wei et al. (2009)

2.2.2.7 microRNAs Involved in Drought Tolerance

MicroRNAs (miRNAs) are an extensive class of endogenous, small RNA molecules that sit at the heart of regulating gene expression in multiple developmental and signaling pathways (Zhang 2015). Recent studies have shown that drought induces aberrant expression of many miRNAs, thus suggesting that miRNAs may be a new target for genetically improving plant tolerance to drought stresses (Table 2.2). miRNA expression response to drought stress is genotype-dependent. For instance, the expression of miR168 and miR396 was induced in *Arabidopsis* (Liu et al. 2008) and tobacco (Frazier et al. 2011), but was inhibited in rice by drought treatment (Zhou et al. 2010). However, some species-specific miRNAs are found in response to drought. In switchgrass, 17 drought-specific miRNAs were identified, of which four were conserved and 13 were switchgrass-specific miRNAs (Xie et al. 2014). In addition, novel species-specific miRNAs (hvu-miRX33, hvu-miRX34, and hvu-miRNA35) were found in barley induced by drought stress (Hackenberg et al. 2015). To identify genome-wide drought-responsive miRNAs in root and leaf, four small RNA libraries from both control and drought treated leaf and root samples were constructed and the small RNA populations were thoroughly sequenced (Eldem et al. 2012). The authors reported that drought significantly induced the aberrant expression of 263 and 368 miRNAs in leaf and root tissues, respectively. In cotton, it was also found that the expression changes of miRNAs are dose- and tissue-dependent under drought stress (Wang et al. 2013a).

Among these drought-responsive miRNAs, miR169 is one of the largest miRNA families that is conserved in all plant species (Zhang 2015). In *Arabidopsis*, miR169 can be inhibited by drought stress, and the drought-repressed miR169 expression was through an ABA-dependent pathway (Li et al. 2008). The target of miR169, nuclear factor Y (NF-Y) subunit A 5 (NFYA5), was strongly induced by drought stress when miR169 was inhibited (Li et al. 2008). Constitutive overexpression of miR169 in transgenic tomato significantly enhanced plant tolerance to drought stress after 7 days of drought treatment (Zhang et al. 2011). Under drought stress, non-transgenic wild-type tomato plants showed clear dehydration symptoms, while the transgenic plants that overexpressed miR169 grew very well. Thus, miR169 should be a promising target for improving plant tolerance to drought stress.

2.2.2.8 Hormonal Regulation

Abscisic Acid

It has long been recognized that the production of abscisic acid (ABA) in drying roots and its transport to the leaves play a key role in regulating plant water status during drought (Zhang and Davies 1990). In addition, ABA was found to be a dominant player in mediating the adaptation of the plant to other abiotic stresses, including drought and low temperature, by improving oxygen scavenging efficiency, increasing sugar accumulation and upregulating the expression of key enzymes (Jiang and Zhang 2001; Liu et al. 2013). In wheat, the involvement of ABA in drought (Ali et al. 1998) and low temperature stress (Lalk and Dörfling 1985) adaptation has been reported. Exogenous application of abscisic acid (ABA) significantly increased the tolerance of wheat seedlings suffering from 5 days of 15 % polyethylene glycol (PEG)-stimulated drought stress, as exemplified by increased shoot lengths and shoot and root dry weights, and decreased contents of hydrogen peroxide (H_2O_2) and malondialdehyde (MDA) (Wei et al. 2015). Under drought stress, ABA significantly increased contents of Glutathione (GSH) and ascorbate (ASA) in both leaves and roots. ABA temporally regulated the transcript levels of genes encoding ASA–GSH cycle enzymes. Moreover, these genes exhibited differential expression patterns between the root and leaf organs of ABA-treated wheat seedlings during drought stress. In wheat, the exogenous ABA application during grain filling stage decreased seed weight and slightly reduced seed set and seed number per spike; however, the seedlings from seeds of ABA-treated plants performed better under temperature stress, which is related to higher endogenous ABA level, and increased activities of the antioxidant enzymes (Li et al. 2014b), which can be expected to enhance drought tolerance as well (Fig. 2.2).

Partial root-zone drying (PRD) is a water-saving irrigation strategy, which involves irrigating only part of the root zone, leaving the other part to dry to a predetermined level before the next irrigation (Wang et al. 2012b; Plauborg et al. 2010; Liu et al. 2005a, 2006). PRD allows the induction of the ABA-based root-to-shoot chemical signaling system to regulate growth and water use and thereby increase WUE (Jacobsen et al. 2009; Liu et al. 2005b; Plauborg et al. 2010). Our results indicated that at mild soil water deficits, stomatal conductance of potato was controlled by root-originated ABA (Topbjerg et al. 2015; Sun et al. 2014, 2015; Liu et al. 2015; Kaminski et al. 2015). As a consequence of photosynthesis rate being less sensitive than stomatal conductance to soil water deficit, photosynthetic water-use efficiency was improved under mild soil water deficits (Liu et al. 2005b). ABA-based drought stress chemical signaling plays a key role in regulating crop vegetative and reproductive development and crop drought adaptation (reviewed by Liu et al. 2005b). Increased concentrations of ABA in the root induced by soil drying may maintain root growth and increase root hydraulic conductivity; both lead to an increase in water uptake and thereby postpone the development of water deficit in the shoot (Liu et al. 2005a). Root ABA is also transported in the xylem to the shoot and is perceived at the acting sites, where it causes stomatal closure and reduced

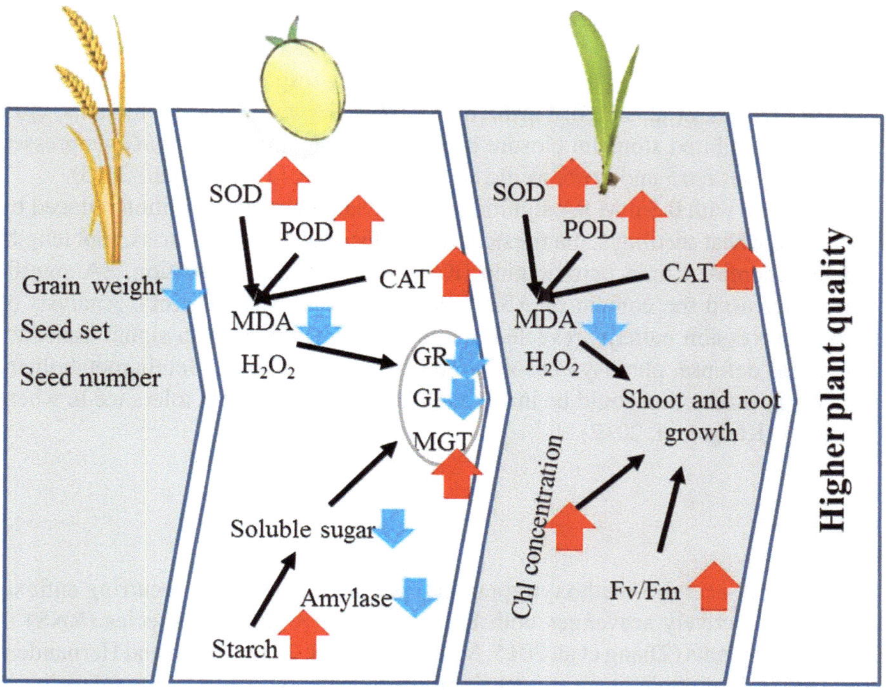

Fig. 2.2 Schematic representation of temperature and water stress tolerance in offspring induced by ABA during grain filling stage in wheat

leaf expansion, thereby preventing dehydration of leaf tissues and enhancing the chance for survival under prolonged drought (Liu et al. 2005a). ABA-based chemical signaling can be amplified by several factors, particularly increased pH in the xylem/apoplast, which retains anionic ABA (Liu et al. 2004, 2005a). Such an increase in xylem pH detected in field-grown maize might have been brought about by reduced nitrate uptake by plants during soil drying (Wang et al. 2012c, d). However, more attention should be paid to the network of ABA signaling of plants in response to drought stress.

Salicylic Acid

An early accumulation of salicylic acid (SA) affects stomatal opening, photorespiration, and antioxidant defenses before any detectable change in the relative water content (Sanchez-Martin et al. 2015). These changes are likely to maintain plant water status, with any photo-inhibitory effect being counteracted by an efficient antioxidant capacity, thereby representing an integrated mechanism of drought tolerance in oats (Sanchez-Martin et al. 2015). It was also documented that *SIZ1*-mediated endogenous SA accumulation plays an important role in stomatal closure and drought tolerance (Miura et al. 2013). In this study, the *siz1* mutation showed drought tolerance, while *nahG siz1* decreased the tolerance to drought stress. Drought stress also induced expression of SA-responsive genes, such as *PR1* and *PR2* (Miura et al. 2013). Furthermore, other SA-accumulating mutants, *cpr5* and *acd6*, exhibited stomatal closure and drought tolerance, and *nahG* suppressed the phenotypes *cpr5* and *acd6*, as did *siz1* and *nahG siz1* (Miura et al. 2013).

Treatment with 0.5 mM SA significantly alleviated growth inhibition induced by drought in wheat seedlings, manifested by less decreased plant biomass, root length, and less increased lipid peroxidation (Kang et al. 2013). In addition, SA significantly increased the content of ASA and GSH under drought stress. Analysis of protein expression patterns revealed that proteins associated with signal transduction, stress defense, photosynthesis, carbohydrate metabolism, protein metabolism, and energy production could be involved in SA-induced drought tolerance in wheat seedlings (Kang et al. 2012).

Melatonin

Melatonin (*N*-acetyl-5-methoxytryptamine) is a potent, naturally occurring antioxidant that effectively scavenges both ROS and reactive nitrogen species (RNS) in animals and plants (Zhang et al. 2015; Manchester et al. 2015; Arnao and Hernandez-Ruiz 2015). Melatonin pretreatment significantly increases the drought tolerance of both drought-tolerant *Malus prunifolia* and drought-sensitive *M. hupehensis* plants. Melatonin application results in better water conservation in leaves, less electrolyte leakage, steady chlorophyll contents, and greater photosynthetic performance under stress conditions (Li et al. 2015a). In addition, melatonin selectively downregulates *MdNCED3*, an ABA synthesis gene, and upregulates its catabolic genes, *MdCYP707A1* and *MdCYP707A2*, thereby reducing ABA contents in drought-stressed plants. Melatonin also directly scavenges H_2O_2 and enhances the activities of antioxidant enzymes to detoxify H_2O_2 indirectly (Li et al. 2015a).

N-acetylserotonin-*O*-methyltransferase (ASMT) is a specific enzyme required for melatonin synthesis (Lee et al. 2015). An ASMT gene was cloned from apple rootstock (*Malus zumi* Mats) and designated as *MzASMT1* (KJ123721). The *MzASMT1* expression in apple leaves can be induced by drought stress. Melatonin levels in *MzASMT1* transgenic *Arabidopsis* plants were 2–4 times higher than those in the wild type. The transgenic *Arabidopsis* plants had significantly lower intrinsic ROS than the wild type and therefore these plants exhibited greater tolerance to drought stress than that of wild type (Zuo et al. 2014). When melatonin was added to soils under drought conditions, the resultant oxidative stress was eased and leaf senescence was delayed (Wang et al. 2013b). Transgenic Micro-Tom tomato plants overexpressing the homologous ovine arylalkylamine N-acetyltransferase (AANAT) and hydroxyindole-O-methyltransferase (HIOMT) genes display loss of apical dominance and enhanced drought tolerance (Wang et al. 2014a). The melatonin application significantly reduces chlorophyll degradation and suppresses the upregulation of senescence-associated gene 12 (*SAG12*) and pheophorbide a oxygenase (*PAO*).

It also alleviates the inhibition of photosynthesis brought on by drought stress (Wang et al. 2013b). In order to better understand the roles of this molecule in induction of drought tolerance, further investigations are needed.

Nitric Oxide

Nitric oxide (NO), a key signaling molecule, is involved in mediation of drought stress-induced physiological responses in plants (Fig. 2.3). In marigold (*Tagetes erecta* L.), it was found that the promoting effect of NO on rooting under drought stress was dose-dependent. NO treatment attenuated the destruction of mesophyll cell ultrastructure by drought stress, and increased leaf chlorophyll content, maximal PSII efficiency and quantum efficiency of PSII electron transport, and hypocotyls soluble carbohydrate and protein content (Liao et al. 2012). It is suggested that the protection of mesophyll cell ultrastructure by NO under drought conditions improves the photosynthetic performance of leaves and alleviates the negative effects of drought. In rice, drought tolerance of plants was strongly related to the maintenance of tissue water potential and enhanced capacity of antioxidants, improved stability of cellular membranes and enhanced photosynthetic capacity, plausibly by signaling action of NO (Farooq et al. 2009a, b). In addition, the exogenous sodium nitroprusside (SNP, nitric oxide donor) treatment could significantly alleviate the stress injury and accelerate the progress of recovery (Wang et al. 2011). Using two NO donors, SNP and S-nitroso-N-acetylpenicillamine, it was documented that NO induces stomatal closure and

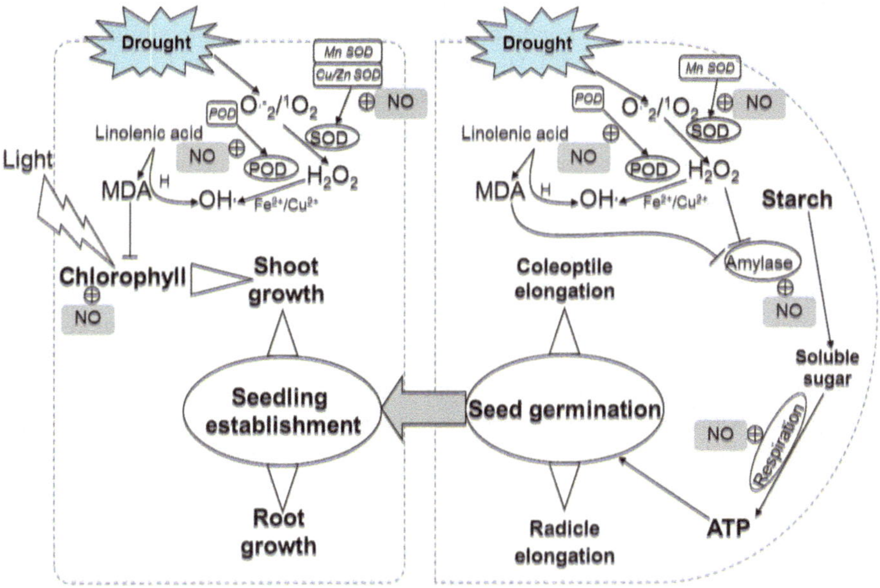

Fig. 2.3 Schematic representation of drought tolerance induced by nitric oxide during germination and the seedling growth stages in wheat

enhances the adaptive plant responses against drought stress (Mata and Lamattina 2001). In SNP-treated leaves of *Tradescantia sp.*, the stomatal closure was associated with a 10 % increase of RWC. Ion leakage, a cell injury index, was 25 % lower in SNP-treated wheat leaves compared with the controls after the recovery period (Mata and Lamattina 2001). A study on the interaction between polyamine (PA) and nitric oxide signaling in adaptive responses to drought in cucumber showed that seedlings pretreated with PAs and subjected to water deficit possessed early and transient NO production (Arasimowicz-Jelonek et al. 2009). However, NO donor administration preceding drought had no effect on endogenous PA levels but was positively correlated with an alleviation of water deficit-induced membrane permeability and lipid peroxidation (Arasimowicz-Jelonek et al. 2009).

2.3 Drought Stress Memory

Drought and other abiotic stresses are recurring environmental stresses experienced by plants throughout their life (Avramova 2015). To survive the repeated stresses, plants can response to later stress events in a way that may be different from their response during the first encounter with the stress. A different response to a similar stress represents the concept of "stress memory" (Avramova 2015). During this process, a coordinated reaction at the organismal, cellular, and genome levels is considered contribute to the improved tolerance in plants. In order to test the long-term stress memory on tuber yield and drought tolerance related traits in potato under drought, seed tubers produced by plants grown under well-water (non-primed tubers) and drought (primed tubers) conditions were sown and exposed to similar watering treatments (Ramírez et al. 2015). Higher tuber yield was produced by primed plants under both nonrestricted and restricted water regimes. The decrease in tuber yield and tuber carbon isotope discrimination with water restriction was lower in primed plants. In addition, it was also found that long-term stress memory consequently appears to be highly genotype-dependent in potato (Ramírez et al. 2015). In grasses, the responses of *Arrhenatherum elatius* plants under a second, later drought (pre-exposed to an earlier drought), to plants exposed to a single (only later) drought were compared (Walter et al. 2011). The results showed that the percentage of living biomass after a late drought was higher for plants that were exposed to drought earlier in the growing season compared to single-stressed plants, even after harvest and resprouting after the first drought. Recently, the effects of drought priming were exemplified with sustaining ROS homeostasis, increasing photosynthetic rate, and higher grain yield when plants exposed to later abiotic stress, such as low temperature, drought, and heat (Li et al. 2015b; Wang et al. 2014c, 2015). Interestingly, *Arabidopsis* plants subjected to a daily dehydration cycle display physiological and transcriptional stress memory: previously stressed plants showed partially closed stomata during a watered recovery period, facilitating water conservation during a subsequent dehydration stress (Virlouvet and Fromm 2015).

Many experiments have proven that the short- and long-term drought stress memory exist in plants (Hu et al. 2015a, b; Berry and Dean 2015; Avramova 2015;

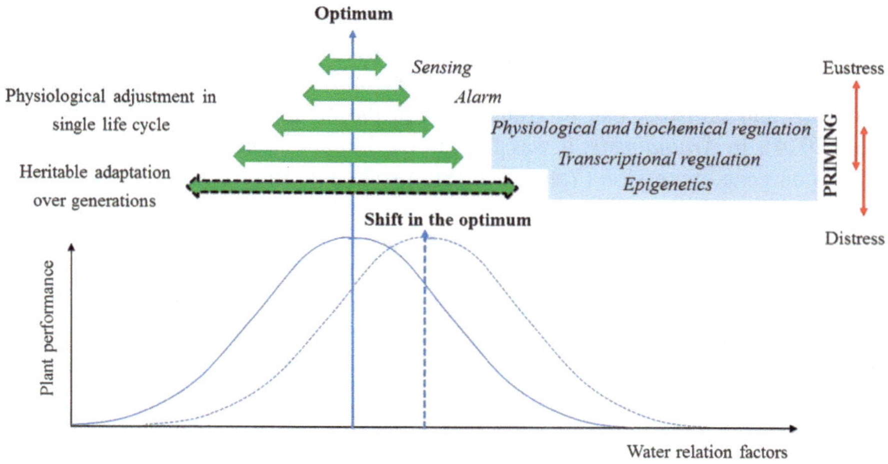

Fig. 2.4 Sensing and adjustment of plants to drought stress by regulations in physiological, transcriptional, and epigenetic levels

Thellier and Luttge 2012; Guan et al. 2012; Pecinka et al. 2009; Molinier et al. 2006). It should be noted that the mechanisms establishing short- or long-term acquisition of stress-induced states may be different (Fig. 2.4). For short-term drought memory, the mechanisms related to morphological adaptation, physiological and biochemical changes, and transcriptional modifications have been reviewed above. In order to understand the mechanism of long-term drought memory in *Arabidopsis*, chromatin marks, such as histone modifications, have been tested in primed and non-primed plants. Primed plants are identical to non-primed plants in growth and development, yet they display enhanced drought tolerance after a second stress exposure (Sani et al. 2013). ChIP-seq analysis of four histone modifications revealed that the priming treatment altered the epigenomic landscape; the changes were small but they were specific for the treated tissue, varied in number and direction depending on the modification, and preferentially targeted transcription factors (Sani et al. 2013). Although some of the mechanisms underlying stress memory in plants have been illustrated, such as morphological changes, the accumulation of specific transcription factors and protective metabolites, and epigenetic modifications, it is still important to understand cross-stress tolerance and stress memory from gene to ecosystem (Munné-Bosch and Alegre 2013).

2.4 Molecular Mechanisms of Drought Memory and Epigenetics

Transcriptional evidences for drought stress memory indicate that the molecular mechanisms regulating production of different transcript amounts in response to single stress stimulation and repeated stress stimulation are different (Avramova 2015;

Berry and Dean 2015). Epigenetic mechanisms are now known to play a critical role in regulating gene expression through small RNAs, histone modifications, and DNA methylation (Kinoshita and Seki 2014). These are inherited through mitotic cell divisions and probably can be transmitted to the next generation (Kinoshita and Seki 2014). In *Arabidopsis*, it was found that drought signals are transduced into effects on gene expression (Yamaguchi-Shinozaki and Shinozaki 2005). The study on changes of histone tails in response to dehydration showed that *RD29A*, *RD20*, and *AtGOLS2* transcripts accumulate under drought stress, while the amounts of these transcripts fall to the basal level when rehydration (Kim et al. 2008, 2012). In addition, alterations to trimethylated histone H3 lysine 4 (H3K4me3), which is correlated with active transcription, suggest that this chromatin mark may play a role in transcription memory for these genes since it is enriched by drought stress and maintained at same levels during the rehydration process (Kim et al. 2012). It was also reported that multiple exposures to drought stress enable plants to respond to a new stress by more rapid adaptive changes to gene expression patterns compared with plants not previously exposed to a drought stress (Ding et al. 2012, 2014; Liu et al. 2014; Alvarez-Venegas et al. 2014). Genome-wide DNA methylation profiles were investigated with regard to a possible role in memory of drought stress; however, no correlation has been identified between gene expression patterns and DNA methylation levels in *Arabidopsis* (Colaneri and Jones 2013).

2.5 Cross-Stress Memory

It is well known that temperate plants including wheat have the ability to obtain cold tolerance by cold acclimation (Theocharis et al. 2012). Cold acclimation in plant is a complex process involving many morphological, physiological, and biochemical changes (Theocharis et al. 2012), including a significant reduction in tissue hydration during cold hardening (Rajashekar and Panda 2014). Besides, evidence shows that drought stress alone, in the absence of low temperatures, can also induce cold tolerance in different plant species such as Arabidopsis, wheat, oats, rye, and strawberry (Rajashekar and Panda 2014). As similar effects and plant responses were noticed at cellular and transcriptional levels, the hardening and stress memory mechanisms may be connected (Mahajan and Tuteja 2005). Thus, hardening and acquisition of stress memory will also prevent attack of other stresses (Walter et al. 2013). Recent studies found that preceding exposure to a drought stress could enhance later frost tolerance in several plant species (Kreyling et al. 2012). For example, the freezing tolerance in Norway spruce (*Picea abies*) progenies was physiologically correlated with drought tolerance (Blödner et al. 2005). In our study, drought priming at vegetative stages alleviated photodamage due to drought and heat stresses during reproductive stage (Wang et al. 2015). Compared to the non-hardened plants, the hardened plant obtained higher grain yield, which was mainly attributed to a higher kernel number under drought stress and to a higher kernel weight under heat stress (Wang et al. 2014c). In addition, our recent study found that drought priming at

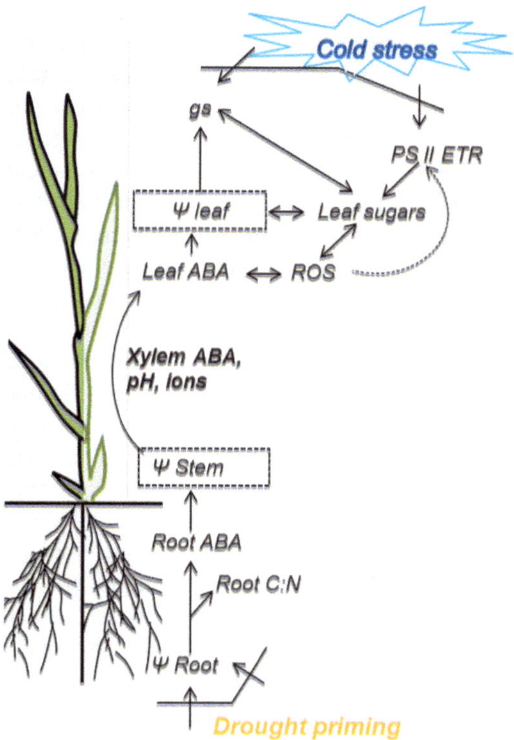

Fig. 2.5 Drought priming improves the cold tolerance by modification of antioxidant capacity and photosynthesis performance in wheat

vegetative stage improves the antioxidant capacity and photosynthesis performance of wheat exposed to a short-term low temperature stress at jointing stage (Fig. 2.5) (Li et al. 2015b). Another kind of cross-stress tolerance towards herbivore induced by drought was caused by modifications of secondary compounds (Herms and Mattson 1992). However, the mechanisms underlying the cross-stress memory remain largely unknown.

2.6 Transgenerational Stress Memory

Sufficient evidence indicates that the modifications induced by former stress events could stimulate a faster immune mechanism to improve stress tolerance in face of repeated events, while some modifications could be inherited to the next generation (Molinier et al. 2006). Transgenerational transmission of information about stress exposure is manifested as an increase in the somatic homologous recombination frequency in plants. Upregulated activity of antioxidative enzymes in maternal plants under drought was reported to be inherited to the next generation of *Arabidopsis thaliana* (Ćuk et al. 2010). The activity of catalase was significantly decreased in the irradiated plants in comparison to the non-irradiated control plants, while the activity of guaiacol peroxidase was increased. In irradiated plants, there was an induction of a new HSP70 protein isoform. In the non-irradiated progeny of irradiated plants, a significant decrease in catalase and ascorbate peroxidase activity was noticed in comparison to plants whose parents were not irradiated (Ćuk et al. 2010). There was no significant change in guaiacol peroxidase activity or induction of HSP70 isoforms in the progeny. This indicates that results indicate that, besides the already known increase in frequency of somatic homologous recombination, transmission of information about stress exposure can also include changes in activities of antioxidative enzymes catalase and ascorbate peroxidase. The transgenerational stress memory was also found in mild heat stress, where F3 generation plant showed a heat-specific fitness enhancement after parental plant and F1 generation had been treated with mild heat (Whittle et al. 2009).

It has been suggested that the epigenetic mechanisms, such as histone modifications and DNA methylation and acetylation, can be inherited through mitotic or meiotic cell divisions (Chinnusamy and Zhu 2009), which support the transgenerational stress memory in cellular biochemistry. Recent evidence suggests that exposure of *Arabidopsis* plants to abiotic stresses, including salt, UV, flooding and extreme temperatures (heat and cold), led to an increased homologous recombination frequency (HRF) and global genome methylation, and showed higher tolerance to the abiotic stress in their untreated progeny (Boyko et al. 2010; Pecinka et al. 2009). It was also proved that the stress-induced transgenerational responses in *Arabidopsis* depend not only on altered DNA methylation but also on smRNA silencing pathways, revealed by using *dcl2* and *dcl3* deficiency mutants (Boyko et al. 2010; Pecinka et al. 2009). However, the heritability of epigenetic change induced by stress exposure was not confirmed in Arabidopsis, indicating strict requirement to specific conditions of transgenerational epigenetic memory (Pecinka et al. 2009).

2.7 An Integrated View and Future Prospects

As one of main focuses in plant-abiotic stress research, studies on plant drought priming and stress memory are still rare. To date, most of results on plant drought priming were obtained in controlled lab experiments, which might be different from the natural conditions. Thus, in future studies, a combination of experiments from controlled lab evaluations with observations and simulation under field conditions should be performed. In addition, certain environmental responses in many seed crops can persist in the next sexual generation. These transgenerational effects have potential significance in agronomy, thus it is essential to elucidate the mechanisms of transgenerational stress memory and to understand the possible regulation pathways. Collectively, to further understanding the processes and mechanisms of priming effects, ecophysiologists and molecular biologists should work together in order to reveal the complete regulation network at different levels and scales, such that management strategies could be developed to sustain crop productivity under future climate changes scenarios.

Acknowledgement This work was supported by Villum Foundation (107903).

References

Ahmad I, Basra SMA, Hussain S, Hussain SA, Rehman HU, Rehman A, Ali A. Priming with ascorbic acid, salicylic acid and hydrogen peroxide improves seedling growth of spring maize at suboptimal temperature. J Environ Agr Sci. 2015;3:14–22.

Ali M, Jensen CR, Mogensen VO. Early signals in field grown wheat in response to shallow soil drying. Funct Plant Biol. 1998;25:871–82.

Alvarez-Venegas R, Abdallat AA, Guo M, Alfano JR, Avramova Z. Epigenetic control of a transcription factor at the cross section of two antagonistic pathways. Epigenetics. 2014;2:106–13.

Arasimowicz-Jelonek M, Floryszak-Wieczorek J, Kubiś J. Interaction between polyamine and nitric oxide signaling in adaptive responses to drought in cucumber. J Plant Growth Regul. 2009;28:177–86.

Arnao MB, Hernandez-Ruiz J. Functions of melatonin in plants: a review. J Pineal Res. 2015;59:133–50.

Ashoub A, Beckhaus T, Berberich T, Karas M, Brüggemann W. Comparative analysis of barley leaf proteome as affected by drought stress. Planta. 2013;237:771–81.

Atkin OK, Macherel D. The crucial role of plant mitochondria in orchestrating drought tolerance. Ann Bot. 2009;103:581–97.

Atkin O, Scheurwater I, Pons T. High thermal acclimation potential of both photosynthesis and respiration in two lowland *Plantago* species in contrast to an alpine congeneric. Glob Chang Biol. 2006;12:500–15.

Atkin O, Scheurwater I, Pons T. Respiration as a percentage of daily photosynthesis in whole plants is homeostatic at moderate, but not high, growth temperatures. New Phytol. 2007;174:367–80.

Avramova Z. Transcriptional 'memory' of a stress; transient chromatin and memory (epigenetic) marks at stress response genes. Plant J. 2015;83:149–59.

Baerenfaller K, Massonnet C, Walsh S, Baginsky S, Bühlmann P, Hennig L, Hirsch-Hoffmann M, Howell KA, Kahlau S, Radziejwoski A. Systems-based analysis of *Arabidopsis* leaf growth reveals adaptation to water deficit. Mol Syst Biol. 2012;8:606.

Basra S, Farooq M, Tabassam R, Ahmad N. Physiological and biochemical aspects of pre-sowing seed treatments in fine rice (*Oryza sativa* L.). Seed Sci Technol. 2005;33:623–8.

Berry S, Dean C. Environmental perception and epigenetic memory: mechanistic insight through FLC. Plant J. 2015;83:133–48.

Blödner C, Skroppa T, Johnsen Ø, Polle A. Freezing tolerance in two Norway spruce (*Picea abies* L. Karst.) progenies is physiologically correlated with drought tolerance. J Plant Physiol. 2005;162:549–58.

Boyko A, Blevins T, Yao Y, Golubov A, Bilichak A, Ilnytskyy Y, Hollander J, Frederick Meins J, Kovalchuk I. Transgenerational adaptation of *Arabidopsis* to stress requires DNA methylation and the function of Dicer-Like proteins. PLoS One. 2010;5, e9514.

Bruce TJA, Matthes MC, Napier JA, Pickett JA. Stressful "memories" of plants: evidence and possible mechanisms. Plant Sci. 2007;173:603–8.

Chen K, Arora R. Dynamics of the antioxidant system during seed osmopriming, post-priming germination, and seedling establishment in Spinach (*Spinacia oleracea*). Plant Sci. 2011;180:212–20.

Chen K, Arora R. Priming memory invokes seed stress-tolerance. Environ Exp Bot. 2012;94: 33–45.

Chen K, Fessehaie A, Arora R. Selection of reference genes for normalizing gene expression during seed priming and germination using qPCR in *Zea mays* and *Spinacia oleracea*. Plant Mol Biol Rep. 2012;30:478–87.

Chen K, Fessehaie A, Arora R. Aquaporin expression during seed osmopriming and post-priming germination in spinach. Biol Plantarum. 2013;57:193–8.
Chinnusamy V, Zhu J-K. Epigenetic regulation of stress responses in plants. Curr Opin Plant Biol. 2009;12:133–9.
Colaneri AC, Jones AM. Genome-wide quantitative identification of DNA differentially methylated sites in Arabidopsis seedlings growing at different water potential. PLoS One. 2013;8, e59878.
Conrath U. Molecular aspects of defence priming. Trends Plant Sci. 2011;16:524–31.
Ćuk K, Gogalo M, Tkalec M, Vidaković-Cifrek Ž. Transgenerational stress memory in *Arabidopsis thaliana* (L.) Heynh.: antioxidative enzymes and HSP70. Acta Bot Croat. 2010;69:183–97.
Ding Y, Fromm M, Avramova Z. Multiple exposures to drought 'train' transcriptional responses in Arabidopsis. Nat Commun. 2012;3:740.
Ding Y, Virlouvet L, Liu N, Riethoven JJ, Fromm M, Avramova Z. Dehydration stress memory genes of *Zea mays*; comparison with *Arabidopsis thaliana*. BMC Plant Biol. 2014;14:141.
Doheny-Adams T, Hunt L, Franks PJ, Beerling DJ, Gray JE. Genetic manipulation of stomatal density influences stomatal size, plant growth and tolerance to restricted water supply across a growth carbon dioxide gradient. Philos Trans R Soc Lond B Biol Sci. 2012;367:547–55.
Eisvand H, Tavakkol-Afshari R, Sharifzadeh F, Maddah Arefi H, Hesamzadeh Hejazi S. Effects of hormonal priming and drought stress on activity and isozyme profiles of antioxidant enzymes in deteriorated seed of tall wheatgrass (*Agropyron elongatum* Host). Seed Sci Technol. 2010;38:280–97.
Eldem V, Akcay UC, Ozhuner E, Bakır Y, Uranbey S, Unver T. Genome-wide identification of miRNAs responsive to drought in peach (*Prunus persica*) by high-throughput deep sequencing. PLoS One. 2012;7(12):502–98.
Elkoca E, Haliloglu K, Esitken A, Ercisli S. Hydro-and osmopriming improve chickpea germination. Acta Agric Scand Sect B Soil Plant Sci. 2007;57:193–200.
Faghani E, Gharechahi J, Komatsu S, Mirzaei M, Khavarinejad RA, Najafi F, Farsad LK, Salekdeh GH. Comparative physiology and proteomic analysis of two wheat genotypes contrasting in drought tolerance. J Proteomics. 2014;114:1–15.
Farooq M, Basra SMA, Wahid A, Rehman H. Exogenously applied nitric oxide enhances the drought tolerance in fine grain aromatic rice (*Oryza sativa* L.). J Agron Crop Sci. 2009a;195:254–61.
Farooq M, Wahid A, Kobayashi N, Fujita D, Basra SMA. Plant drought stress: effects, mechanisms and management. Agron Sustain Dev. 2009b;29:185–212.
Farooq M, Basra SM, Wahid A, Ahmad N. Changes in nutrient-homeostasis and reserves metabolism during rice seed priming: consequences for seedling emergence and growth. Agr Sci China. 2010;9:191–8.
Franks PJ, Doheny-Adams WT, Britton-Harper ZJ, Gray JE. Increasing water-use efficiency directly through genetic manipulation of stomatal density. New Phytol. 2015;207:188–95.
Frazier TP, Sun G, Burklew CE, Zhang B. Salt and drought stresses induce the aberrant expression of microRNA genes in tobacco. Mol Biotechnol. 2011;49:159–65.
Gao P, Bai X, Yang L, Lv D, Li Y, Cai H, Ji W, Guo D, Zhu Y. Over-expression of *osa-MIR396c* decreases salt and alkali stress tolerance. Planta. 2010;231:991–1001.
Giri GS, Schillinger WF. Seed priming winter wheat for germination, emergence, and yield. Crop Sci. 2003;43:2135–41.
Guan Q, Haroon S, Bravo DG, Will JL, Gasch AP. Cellular memory of acquired stress resistance in *Saccharomyces cerevisiae*. Genetics. 2012;192:495–505.
Guan XK, Song L, Wang TC, Turner NC, Li FM. Effect of drought on the gas exchange, chlorophyll fluorescence and yield of six different-era spring wheat cultivars. J Agron Crop Sci. 2015;201:253–66.
Hackenberg M, Gustafson P, Langridge P, Shi BJ. Differential expression of microRNAs and other small RNAs in barley between water and drought conditions. Plant Biotechnol J. 2015;13:2–13.
Hajheidari M, Eivazi A, Buchanan BB, Wong JH, Majidi I, Salekdeh GH. Proteomics uncovers a role for redox in drought tolerance in wheat. J Proteome Res. 2007;6:1451–60.

Hamanishi ET, Thomas BR, Campbell MM. Drought induces alterations in the stomatal development program in *Populus*. J Exp Bot. 2012;63:4969–71.
Hao P, Zhu J, Gu A, Lv D, Ge P, Chen G, Li X, Yan Y. An integrative proteome analysis of different seedling organs in tolerant and sensitive wheat cultivars under drought stress and recovery. Proteomics. 2015;15:1544–63.
Hepworth C, Doheny-Adams T, Hunt L, Cameron DD, Gray JE. Manipulating stomatal density enhances drought tolerance without deleterious effect on nutrient uptake. New Phytol. 2015. doi:10.1111/nph.13598.
Herms DA, Mattson WJ. The dilemma of plants: to grow or defend. Q Rev Biol. 1992;67:283–335.
Hu T, Jin Y, Li H, Amombo E, Fu J. Stress memory induced transcriptional and metabolic changes of perennial ryegrass (*Lolium perenne*) in response to salt stress. Physiol Plant. 2015a. doi:10.1111/ppl.12342.
Hu T, Liu SQ, Amombo E, Fu JM. Stress memory induced rearrangements of HSP transcription, photosystem II photochemistry and metabolism of tall fescue (*Festuca arundinacea* Schreb.) in response to high-temperature stress. Front Plant Sci. 2015b;6:403–25.
Ibáñez H, Ballester A, Muñoz R, José Quiles M. Chlororespiration and tolerance to drought, heat and high illumination. J Plant Physiol. 2010;167:732–8.
Jacobsen S-E, Liu F, Jensen CR. Does root-sourced ABA play a role for regulation of stomata under drought in quinoa (*Chenopodium quinoa* Willd.). Sci Hortic. 2009;122:281–7.
Janmohammadi M, Dezfuli PM, Sharifzadeh F. Seed invigoration techniques to improve germination and early growth of inbred line of maize under salinity and drought stress. Genet Appl Plant Physiol. 2008;34:215–26.
Jiang M, Zhang J. Effect of abscisic acid on active oxygen species, antioxidative defence system and oxidative damage in leaves of maize seedlings. Plant Cell Physiol. 2001;42:1265–73.
Jisha KC, Vijayakumari K, Puthur JT. Seed priming for abiotic stress tolerance: an overview. Acta Physiol Plant. 2012;35:1381–96.
Kaminski KP, Kørup K, Kristensen K, Nielsen KL, Liu F, Topbjerg HB, Kirk HG, Andersen MN. Contrasting water-use efficiency (WUE) responses of a potato mapping population and capability of modified ball-berry model to predict stomatal conductance and WUE measured at different environmental conditions. J Agron Crop Sci. 2015;201:81–94.
Kang H-G, Kim J, Kim B, Jeong H, Choi SH, Kim EK, Lee H-Y, Lim PO. Overexpression of FTL1/DDF1, an AP2 transcription factor, enhances tolerance to cold, drought, and heat stresses in *Arabidopsis thaliana*. Plant Sci. 2011;180(4):634–41.
Kang G, Li G, Xu W, Peng X, Han Q, Zhu Y, Guo T. Proteomics reveals the effects of salicylic acid on growth and tolerance to subsequent drought stress in wheat. J Proteome Res. 2012;11:6066–79.
Kang G, Li G, Liu G, Xu W, Peng X, Wang C, Zhu Y, Guo T. Exogenous salicylic acid enhances wheat drought tolerance by influence on the expression of genes related to ascorbate-glutathione cycle. Biol Plantarum. 2013;57:718–24.
Kaur S, Gupta A, Kaur N. Effect of osmo- and hydropriming of chickpea seeds on seedling growth and carbohydrate metabolism under water deficit stress. Plant Growth Regul. 2002;37:17–22.
Kidokoro S, Watanabe K, Ohori T, Moriwaki T, Maruyama K, Mizoi J, Myint Phyu Sin Htwe N, Fujita Y, Sekita S, Shinozaki K, Yamaguchi-Shinozaki K. Soybean DREB1/CBF-type transcription factors function in heat and drought as well as cold stress-responsive gene expression. Plant J. 2015;81:505–18.
Kim J-M, To TK, Ishida J, Morosawa T, Kawashima M, Matsui A, Toyoda T, Kimura H, Shinozaki K, Seki M. Alterations of lysine modifications on the histone H3 N-tail under drought stress conditions in *Arabidopsis thaliana*. Plant Cell Physiol. 2008;49:1580–8.
Kim JY, Kwak KJ, Jung HJ, Lee HJ, Kang H. MicroRNA402 affects seed germination of Arabidopsis thaliana under stress conditions via targeting DEMETER-LIKE Protein3 mRNA. Plant Cell Physiol. 2010a;51:1079–83.
Kim JY, Lee HJ, Jung HJ, Maruyama K, Suzuki N, Kang H. Overexpression of microRNA395c or 395e affects differently the seed germination of *Arabidopsis thaliana* under stress conditions. Planta. 2010b;232:1447–54.

Kim J-M, To TK, Ishida J, Matsui A, Kimura H, Seki M. Transition of chromatin status during the process of recovery from drought stress in *Arabidopsis thaliana*. Plant Cell Physiol. 2012;53:847–56.

Kinoshita T, Seki M. Epigenetic memory for stress response and adaptation in plants. Plant Cell Physiol. 2014;55:1859–63.

Kreyling J, Wiesenberg GLB, Thiel D, Wohlfart C, Huber G, Walter J, Jentsch A, Konnert M, Beierkuhnlein C. Cold hardiness of *Pinus nigra* Arnold as influenced by geographic origin, warming, and extreme summer drought. Environ Exp Bot. 2012;78:99–108.

Lalk I, Dörfling K. Hardening, abscisic acid, praline and freezing resistance in two winter wheat varieties. Physiol Plant. 1985;63:287–92.

Lee HY, Byeon Y, Tan D-X, Reiter RJ, Back K. Arabidopsis serotonin N-acetyltransferase knockout mutant plants exhibit decreased melatonin and salicylic acid levels resulting in susceptibility to an avirulent pathogen. J Pineal Res. 2015;58:291–9.

Li W-X, Oono Y, Zhu J, He X-J, Wu J-M, Iida K, Lu X-Y, Cui X, Jin H, Zhu J-K. The *Arabidopsis NFYA5* transcription factor is regulated transcriptionally and posttranscriptionally to promote drought resistance. Plant Cell. 2008;20:2238–51.

Li Y, Ren B, Yang X, Xu G, Shen Q, Guo S. Chloroplast downsizing under nitrate nutrition restrained mesophyll conductance and photosynthesis in rice (*Oryza sativa* L.) under drought conditions. Plant Cell Physiol. 2012;53:892–900.

Li X, Jiang H, Liu F, Cai J, Dai T, Cao W, Jiang D. Induction of chilling tolerance in wheat during germination by pre-soaking seed with nitric oxide and gibberellin. Plant Growth Regul. 2013;71:31–40.

Li X, Cai J, Liu F, Dai T, Cao W, Jiang D. Cold priming drives the sub-cellular antioxidant systems to protect photosynthetic electron transport against subsequent low temperature stress in winter wheat. Plant Physiol Biochem. 2014a;82:34–43.

Li X, Cai J, Liu F, Dai T, Cao W, Jiang D. Exogenous abscisic acid application during grain filling in winter wheat improves cold tolerance of offspring's seedlings. J Agron Crop Sci. 2014b;200:467–78.

Li X, Cai J, Liu F, Dai T, Cao W, Jiang D. Physiological, proteomic and transcriptional responses of wheat to combination of drought or waterlogging with late spring low temperature. Funct Plant Biol. 2014c;41:690–703.

Li C, Tan DX, Liang D, Chang C, Jia D, Ma F. Melatonin mediates the regulation of ABA metabolism, free-radical scavenging, and stomatal behaviour in two *Malus* species under drought stress. J Exp Bot. 2015a;66:669–80.

Li X, Topbjerg HB, Jiang D, Liu F. Drought priming at vegetative stage improves the antioxidant capacity and photosynthesis performance of wheat exposed to a short-term low temperature stress at jointing stage. Plant and Soil. 2015b;393:307–18.

Liao WB, Huang GB, Yu JH, Zhang ML. Nitric oxide and hydrogen peroxide alleviate drought stress in marigold explants and promote its adventitious root development. Plant Physiol. 2012;58:6–15.

Liu F, Andersen MN, Jensen CR. Root signal controls pod growth in drought-stressed soybean during the critical, abortion-sensitive phase of pod development. Field Crops Res. 2004;85:159–66.

Liu F, Jensen CR, Andersen MN. A review of drought adaptation in crop plants: changes in vegetative and reproductive physiology induced by ABA-based chemical signals. Aust J Agric Res. 2005a;56:1245–52.

Liu F, Jensen CR, Shahanzari A, Andersen MN, Jacobsen S-E. ABA regulated stomatal control and photosynthetic water use efficiency of potato (*Solanum tuberosum* L.) during progressive soil drying. Plant Sci. 2005b;168:831–6.

Liu F, Shahnazari A, Andersen MN, Jacobsen SE, Jensen CR. Physiological responses of potato (*Solanum tuberosum* L.) to partial root-zone drying: ABA signalling, leaf gas exchange, and water use efficiency. J Exp Bot. 2006;57(14):3727–35.

Liu H-H, Tian X, Li Y-J, Wu C-A, Zheng C-C. Microarray-based analysis of stress-regulated microRNAs in *Arabidopsis thaliana*. RNA. 2008;14(5):836–43.

Liu L, Cang J, Wang X, Huang R, Wang J, Yu J, Lu B. Effects of exogenous abscisic acid on carbohydrate metabolism and the expression levels of correlative key enzymes in winter wheat under low temperature. Biosci Biotech Biochem. 2013;77:516–25.

Liu N, Ding Y, Fromm M, Avramova Z. Different gene-specific mechanisms determine the 'revised-response' memory transcription patterns of a subset of *A. thaliana* dehydration stress responding genes. Nucleic Acids Res. 2014;42:5556–66.

Liu C, Rubæk GH, Liu F, Andersen MN. Effect of partial root zone drying and deficit irrigation on nitrogen and phosphorus uptake in potato. Agric Water Manage. 2015;159:66–76.

Loveys B, Scheurwater I, Pons T, Fitter A, Atkin O. Growth temperature influences the underlying components of relative growth rate: an investigation using inherently fast-and slow-growing plant species. Plant Cell Environ. 2002;25:975–88.

Mahajan S, Tuteja N. Cold, salinity and drought stresses: an overview. Arch Biochem Biophys. 2005;444:139–58.

Manchester LC, Coto-Montes A, Boga JA, Andersen LP, Zhou Z, Galano A, Vriend J, Tan DX, Reiter RJ. Melatonin: an ancient molecule that makes oxygen metabolically tolerable. J Pineal Res. 2015. doi:10.1111/jpi.12267.

Massacci A, Nabiev SM, Pietrosanti L, Nematov SK, Chernikova TN, Thor K, Leipner J. Response of the photosynthetic apparatus of cotton (*Gossypium hirsutum*) to the onset of drought stress under field conditions studied by gas-exchange analysis and chlorophyll fluorescence imaging. Plant Physiol Biochem. 2008;46:189–95.

Mata CG, Lamattina L. Nitric oxide induces stomatal closure and enhances the adaptive plant responses against drought stress. Plant Physiol. 2001;126:1196–204.

Meena RP, Sendhil R, Tripathi S, Chander S, Chhokar R, Sharma R. Hydro-priming of seed improves the water use efficiency, grain yield and Net economic return of wheat under different moisture regimes. SAARC J Agr. 2014;11:149–59.

Mishra KB, Iannacone R, Petrozza A, Mishra A, Armentano N, La Vecchia G, Trtílek M, Cellini F, Nedbal L. Engineered drought tolerance in tomato plants is reflected in chlorophyll fluorescence emission. Plant Sci. 2012;182:79–86.

Miura K, Okamoto H, Okuma E, Shiba H, Kamada H, Hasegawa PM, Murata Y. *SIZ1* deficiency causes reduced stomatal aperture and enhanced drought tolerance via controlling salicylic acid-induced accumulation of reactive oxygen species in Arabidopsis. Plant J. 2013;73:91–104.

Molinier J, Ries G, Zipfel C, Hohn B. Transgeneration memory of stress in plants. Nature. 2006;442.

Munné-Bosch S, Alegre L. Cross-stress tolerance and stress "memory" in plants: an integrated view. Environ Exp Bot. 2013;94:1–2.

Muscolo A, Junker A, Klukas C, Weigelt-Fischer K, Riewe D, Altmann T. Phenotypic and metabolic responses to drought and salinity of four contrasting lentil accessions. J Exp Bot. 2015;66:5467–80.

Ni Z, Hu Z, Jiang Q, Zhang H. Overexpression of *gma-MIR394a* confers tolerance to drought in transgenic *Arabidopsis thaliana*. Biochem Biophys Res Commun. 2012;427:330–5.

Pandita V, Anand A, Nagarajan S, Seth R, Sinha S. Solid matrix priming improves seed emergence and crop performance in okra. Seed Sci Technol. 2010;38:665–74.

Pecinka A, Rosa M, Schikora A, Berlinger M, Hirt H, Luschnig C, Scheid OM. Transgenerational stress memory is not a general response in Arabidopsis. PLoS One. 2009;4, e5202.

Peuke A, Gessler A, Rennenberg H. The effect of drought on C and N stable isotopes in different fractions of leaves, stems and roots of sensitive and tolerant beech ecotypes. Plant Cell Environ. 2006;29:823–35.

Plauborg F, Abrahamsen P, Gjettermann B, Mollerup M, Iversen BV, Liu F, Andersen MN, Hansen S. Modelling of root ABA synthesis, stomatal conductance, transpiration and potato production under water saving irrigation regimes. Agr Water Manage. 2010;98:425–39.

Quarrie S, Jones H. Effects of abscisic acid and water stress on development and morphology of wheat. J Exp Bot. 1977;28:192–203.

Rajashekar CB, Panda M. Water stress is a component of cold acclimation process essential for inducing full freezing tolerance in strawberry. Sci Hortic. 2014;174:54–9.

Ramírez DA, Rolando JL, Yactayo W, Monneveux P, Mares V, Quiroz R. Improving potato drought tolerance through the induction of long-term water stress memory. Plant Sci. 2015;238:26–32.

Rivas-Ubach A, Sardans J, Perez-Trujillo M, Estiarte M, Penuelas J. Strong relationship between elemental stoichiometry and metabolome in plants. Proc Natl Acad Sci U S A. 2012;109:4181–6.

Sakurai N, Akiyama M, Kuraishi S. Irreversible effects of water stress on growth and stomatal development in cotyledons of etiolated squash seedlings. Plant Cell Physiol. 1986;27:1177–85.

Sanchez-Martin J, Heald J, Kingston-Smith A, Winters A, Rubiales D, Sanz M, Mur LA, Prats E. A metabolomic study in oats (*Avena sativa*) highlights a drought tolerance mechanism based upon salicylate signalling pathways and the modulation of carbon, antioxidant and photo-oxidative metabolism. Plant Cell Environ. 2015;38:1434–52.

Sani E, Herzyk P, Perrella G, Colot V, Amtmann A. Hyperosmotic priming of Arabidopsis seedlings establishes a long-term somatic memory accompanied by specific changes of the epigenome. Genome Biol. 2013;14:R59.

Shahrokhi M, Tehranifar A, Hadizadeh H, Selahvarzi Y. Effect of drought stress and paclobutrazol-treated seeds on physiological response of *Festuca arundinacea* L. master and *Lolium perenne* L. barrage. J Biol Environ Sci. 2011;5:77–85.

Shukla N, Awasthi RP, Rawat L, Kumar J. Seed biopriming with drought tolerant isolates of Trichoderma harzianum promote growth and drought tolerance in *Triticum aestivum*. Ann Appl Biol. 2015;166:171–82.

Silva EC, Nogueira RJ, Vale FH, Araújo FP, Pimenta MA. Stomatal changes induced by intermittent drought in four umbu tree genotypes. Brazil J Plant Physiol. 2009;21:33–42.

Silvente S, Sobolev AP, Lara M. Metabolite adjustments in drought tolerant and sensitive soybean genotypes in response to water stress. PLoS One. 2012;7(6), e38554.

Su L, Dai Z, Li S, Xin H. A novel system for evaluating drought-cold tolerance of grapevines using chlorophyll fluorescence. BMC Plant Biol. 2015;15(1):82.

Sun Y, Yan F, Cui X, Liu F. Plasticity in stomatal size and density of potato leaves under different irrigation and phosphorus regimes. J Plant Physiol. 2014;171:1248–55.

Sun Y, Cui X, Liu F. Effect of irrigation regimes and phosphorus rates on water and phosphorus use efficiencies in potato. Sci Hortic. 2015;190:64–9.

Suseela V, Tharayil N, Xing B, Dukes JS. Warming and drought differentially influence the production and resorption of elemental and metabolic nitrogen pools in *Quercus rubra*. Glob Chang Biol. 2015. doi:10.1111/gcb.13033.

Tavili A, Zare S, Moosavi SA, Enayati A. Effects of seed priming on germination characteristics of Bromus species under salt and drought conditions. Amer-Eurasian J Agric Environ Sci. 2011;10:163–8.

Thellier M, Luttge U. Plant memory: a tentative model. Plant Biol. 2012;15:1–12.

Theocharis A, Clement C, Barka EA. Physiological and molecular changes in plants grown at low temperatures. Planta. 2012;235:1091–105.

Tombesi S, Nardini A, Frioni T, Soccolini M, Zadra C, Farinelli D, Poni S, Palliotti A. Stomatal closure is induced by hydraulic signals and maintained by ABA in drought-stressed grapevine. Sci Rep. 2015;5:12449.

Topbjerg HB, Kaminski KP, Kørup K, Nielsen KL, Kirk HG, Andersen MN, Liu F. Screening for intrinsic water use efficiency in a potato dihaploid mapping population under progressive drought conditions. Acta Agric Scand B Plant Sci. 2015. doi:10.1080/09064710.2015.1015439.

Udvardi MK, Kakar K, Wandrey M, Montanari O, Murray J, Andriankaja A, Zhang J-Y, Benedito V, Hofer JM, Chueng F. Legume transcription factors: global regulators of plant development and response to the environment. Plant Physiol. 2007;144:538–49.

Urano K, Kurihara Y, Seki M, Shinozaki K. 'Omics' analyses of regulatory networks in plant abiotic stress responses. Curr Opin Plant Biol. 2010;13:132–8.

Valdes AE, Irar S, Majada JP, Rodriguez A, Fernandez B, Pages M. Drought tolerance acquisition in *Eucalyptus globulus* (Labill.): a research on plant morphology, physiology and proteomics. J Proteomics. 2013;79:263–76.

Varga B, Varga-László E, Bencze S, Balla K, Veisz O. Water use of winter cereals under well-watered and drought-stressed conditions. Plant Soil Environ. 2013;59:150–5.

Varga B, Vida G, Varga-László E, Bencze S, Veisz O. Effect of simulating drought in various phenophases on the water use efficiency of winter wheat. J Agron Crop Sci. 2015;201:1–9.

Virlouvet L, Fromm M. Physiological and transcriptional memory in guard cells during repetitive dehydration stress. New Phytol. 2015;205:596–607.

Walter J, Nagy L, Hein R, Rascher U, Beierkuhnlein C, Willner E, Jentsch A. Do plants remember drought? Hints towards a drought-memory in grasses. Environ Exp Bot. 2011;71:34–40.

Walter J, Jentsch A, Beierkuhnlein C, Kreyling J. Ecological stress memory and cross stress tolerance in plants in the face of climate extremes. Environ Exp Bot. 2013;94:3–8.

Wang J, Vanlerberghe GC. A lack of mitochondrial alternative oxidase compromises capacity to recover from severe drought stress. Physiol Plant. 2013;149:461–73.

Wang Y, Suo B, Zhao T, Qu X, Yuan L, Zhao X, Zhao H. Effect of nitric oxide treatment on antioxidant responses and psbA gene expression in two wheat cultivars during grain filling stage under drought stress and rewatering. Acta Physiol Plant. 2011;33:1923–32.

Wang X, Liu Y, Jia Y, Gu H, Ma H, Yu T, Zhang H, Chen Q, Ma L, Gu A, Zhang J, Shi S. Transcriptional responses to drought stress in root and leaf of chickpea seedling. Mol Biol Rep. 2012a;39:8147–58.

Wang Y, Liu F, Jensen CR. Comparative effects of deficit irrigation and alternate partial root-zone irrigation on xylem pH, ABA and ionic concentrations in tomatoes. J Exp Bot. 2012b;63:1907–17.

Wang Z, Kang S, Jensen CR, Liu F. Alternate partial root-zone irrigation reduces bundle-sheath cell leakage to CO2 and enhances photosynthetic capacity in maize leaves. J Exp Bot. 2012c;63:1145–53.

Wang Z, Liu F, Kang S, Jensen CR. Alternate partial root-zone drying irrigation improves nitrogen nutrition in maize (*Zea mays* L.) leaves. Environ Exp Bot. 2012d;75:36–40.

Wang M, Wang Q, Zhang B. Response of miRNAs and their targets to salt and drought stresses in cotton (*Gossypium hirsutum* L.). Gene. 2013a;530:26–32.

Wang P, Sun X, Li C, Wei Z, Liang D, Ma F. Long-term exogenous application of melatonin delays drought-induced leaf senescence in apple. J Pineal Res. 2013b;54:292–302.

Wang L, Zhao Y, Reiter RJ, He CJ, Liu GS, Lei Q, Zuo BX, Zheng XD, Li QT, Kong J. Changes in melatonin levels in transgenic 'Micro-Tom' tomato overexpressing ovine AANAT and ovine HIOMT genes. J Pineal Res. 2014a;56:134–42.

Wang X, Cai J, Liu F, Dai T, Cao W, Wollenweber B, Jiang D. Multiple heat priming enhances thermo-tolerance to a later high temperature stress via improving subcellular antioxidant activities in wheat seedlings. Plant Physiol Biochem. 2014b;74:185–92.

Wang X, Vignjevic M, Jiang D, Jacobsen S, Wollenweber B. Improved tolerance to drought stress after anthesis due to priming before anthesis in wheat (*Triticum aestivum* L.) var. Vinjett. J Exp Bot. 2014c;65:6441–56.

Wang X, Vigjevic M, Liu F, Jacobsen S, Jiang D, Wollenweber B. Drought priming at vegetative growth stages improves tolerance to drought and heat stresses during grain filling in spring wheat (*Triticum aestivum* L. cv. Vinjett). Plant Growth Regul. 2015;75:677–87.

Wei L, Zhang D, Xiang F, Zhang Z. Differentially expressed miRNAs potentially involved in the regulation of defense mechanism to drought stress in maize seedlings. Int J Plant Sci. 2009;170:979–89.

Wei L, Wang L, Yang Y, Wang P, Guo T, Kang G. Abscisic acid enhances tolerance of wheat seedlings to drought and regulates transcript levels of genes encoding ascorbate-glutathione biosynthesis. Front Plant Sci. 2015;6:458.

Whittle C, Otto S, Johnston M, Krochko J. Adaptive epigenetic memory of ancestral temperature regime in *Arabidopsis thaliana*. Botany. 2009;87(6):650–7.

Windauer L, Altuna A, Benech-Arnold R. Hydrotime analysis of Lesquerella fendleri seed germination responses to priming treatments. Ind Crops Products. 2007;25:70–4.

Xiang Y, Tang N, Du H, Ye H, Xiong L. Characterization of *OsbZIP23* as a key player of the basic leucine zipper transcription factor family for conferring abscisic acid sensitivity and salinity and drought tolerance in rice. Plant Physiol. 2008;148:1938–52.

Xie F, Stewart CN, Taki FA, He Q, Liu H, Zhang B. High-throughput deep sequencing shows that microRNAs play important roles in switchgrass responses to drought and salinity stress. Plant Biotechnol J. 2014;12:354–66.

Xu Z, Zhou G. Responses of leaf stomatal density to water status and its relationship with photosynthesis in a grass. J Exp Bot. 2008;59:3317–25.

Yamaguchi-Shinozaki K, Shinozaki K. Organization of cis-acting regulatory elements in osmotic- and cold-stress-responsive promoters. Trends Plant Sci. 2005;10:88–94.

Yuan-Yuan S, Yong-Jian S, Ming-Tian W, Xu-Yi L, Xiang G, Rong H, Jun M. Effects of seed priming on germination and seedling growth under water stress in rice. Acta Agron Sin. 2010;36:1931–40.

Zadraznik T, Hollung K, Egge-Jacobsen W, Meglic V, Sustar-Vozlic J. Differential proteomic analysis of drought stress response in leaves of common bean (*Phaseolus vulgaris* L.). J Proteomics. 2013;78:254–72.

Zhang B. MicroRNA: a new target for improving plant tolerance to abiotic stress. J Exp Bot. 2015;66:1749–61.

Zhang J, Davies W. Changes in the concentration of ABA in xylem sap as a function of changing soil water status can account for changes in leaf conductance and growth. Plant Cell Environ. 1990;13:277–85.

Zhang X, Zou Z, Gong P, Zhang J, Ziaf K, Li H, Xiao F, Ye Z. Over-expression of microRNA169 confers enhanced drought tolerance to tomato. Biotechnol Lett. 2011;33:403–9.

Zhang Z, Liu X, Wang X, Zhou M, Zhou X, Ye X, Wei X. An R2R3 MYB transcription factor in wheat, *TaPIMP1*, mediates host resistance to *Bipolaris sorokiniana* and drought stresses through regulation of defense- and stress-related genes. New Phytol. 2012;196:1155–70.

Zhang JY, Cruz DECMH, Torres-Jerez I, Kang Y, Allen SN, Huhman DV, Tang Y, Murray J, Sumner LW, Udvardi MK. Global reprogramming of transcription and metabolism in Medicago truncatula during progressive drought and after rewatering. Plant Cell Environ. 2014;37:2553–76.

Zhang N, Sun Q, Zhang H, Cao Y, Weeda S, Ren S, Guo YD. Roles of melatonin in abiotic stress resistance in plants. J Exp Bot. 2015;66:647–56.

Zhou X, Wang G, Zhang W. UV-B responsive microRNA genes in *Arabidopsis thaliana*. Mol Syst Biol. 2007;3:103.

Zhou L, Liu Y, Liu Z, Kong D, Duan M, Luo L. Genome-wide identification and analysis of drought-responsive microRNAs in *Oryza sativa*. J Exp Bot. 2010;61:4157–68.

Zhou M, Li D, Li Z, Hu Q, Yang C, Zhu L, Luo H. Constitutive expression of a miR319 gene alters plant development and enhances salt and drought tolerance in transgenic creeping bentgrass. Plant Physiol. 2013;161:1375–91.

Zhou S, Li M, Guan Q, Liu F, Zhang S, Chen W, Yin L, Qin Y, Ma F. Physiological and proteome analysis suggest critical roles for the photosynthetic system for high water-use efficiency under drought stress in *Malus*. Plant Sci. 2015;236:44–60.

Zuo B, Zheng X, He P, Wang L, Lei Q, Feng C, Zhou J, Li Q, Han Z, Kong J. Overexpression of *MzASMT* improves melatonin production and enhances drought tolerance in transgenic *Arabidopsis thaliana* plants. J Pineal Res. 2014;57:408–17.

Chapter 3
Mechanisms of Hormone Regulation for Drought Tolerance in Plants

Patrick Burgess and Bingru Huang

Abbreviations

ABA	Abscisic acid
APX	Ascorbate peroxidase
ARF	Auxin response factor
ASA	Ascorbic acid
CAT	Catalase
CK	Cytokinin
ERF	Ethylene response factor
GA	Gibberellic acid
GR	Glutathione reductase
GSH	Glutathione
IAA	Indole-3-acetic acid
JA	Jasmonic acid
MDA	Malondialdehyde
MeJA	Methyl jasmonate
POD	Peroxidase
ROS	Reactive oxygen species
SA	Salicylic acid
SOD	Superoxide dismutase

P. Burgess • B. Huang (✉)
Department of Plant Biology and Pathology, Rutgers University,
New Brunswick, NJ 08901, USA
e-mail: huang@aesop.rutgers.edu

M.A. Hossain et al. (eds.), *Drought Stress Tolerance in Plants, Vol 1*,
DOI 10.1007/978-3-319-28899-4_3

3.1 Introduction

Drought stress caused by lack of rainfall or declining fresh water supplies for irrigation imposes significant limitations to growth and productivity of many plant species across different climatic areas. Predictive models of global climate change have shown that frequency of precipitation events and net volumes have changed drastically during the past one hundred years, suggesting that certain regions may experience drought episodes more frequently and of longer durations in the future (Solomon et al. 2007). The far-reaching effects of climate change in conjunction with an increasing global population will likely contribute to greater instability in food security and underscores the need for greater knowledge of the specific mechanisms underlying drought tolerance across major and novel plant species. The physiological effects of long- or short-term drought stress have been well characterized for the major grain crops such as maize (*Zea mays*), wheat (*Triticum* spp.), rice (*Oryza sativa*), and barley (*Hordeum vulgare*) and have also been investigated to a lesser extent in certain novel or underutilized crop species (Graham and Vance 2003; Hanjra and Qureshi 2010; Kang et al. 2009; Ruiz et al. 2014; Zwart and Bastiaanssen 2004). Drought stress can also impose functional limitations upon non-crop species such as trees, shrubs, and perennial grasses (Abrams 1994; Bréda et al. 2006; Condit et al. 1995; Hacke et al. 2000; Tester and Bacic 2005).

The extent of damage sustained during drought stress depends on factors including plant species or variety, developmental stage, rate of soil water decline, frequency of drought events, and duration of water deficit (Mahajan and Tuteja 2005; Reddy et al. 2004). These interacting factors cause significant changes at the physiological, cellular, biochemical, and molecular levels preempting the observed decline in plant performance or net yield (Huang 2003; Huang et al. 2014). Substantial progress has been made to better understand plant drought tolerance mechanisms through research on physiological processes (water relations, carbon metabolism), protein metabolism, and genomic factors (Atkinson and Urwin 2012; Cattivelli et al. 2008). More recently, there have been significant advancements in the ability to accurately detect and quantify low-concentration plant hormones to elucidate on how hormone metabolism may regulate whole-plant stress responses (Peleg and Blumwald 2011; Robert-Seilaniantz et al. 2011; Vanstraelen and Benková 2012). Investigating the earliest mechanisms through which the drought-response cascade is initiated in plants may aide in breeding and selecting for drought-tolerant lines and in developing new management techniques to minimize plant damages when water for irrigation is limited.

This chapter focuses on the recent advancements in plant hormone metabolism in relation to drought tolerance with the following aims: (1) to provide an overview of several major physiological drought responses that are highly regulated by plant hormones, including leaf senescence and antioxidant metabolism, carbon metabolism, and stomatal movement; (2) to discuss the roles of different hormones including abscisic acid, auxins, cytokinins, gibberellins, jasmonates, salicylates, and ethylene regulating these physiological responses during plant drought responses; (3) to describe the current knowledge of interactions or cross talk between various hormones or between hormones and other plant metabolites regulating physiological responses to drought stress; and (4) to summarize and propose future research perspectives for enhancing our understanding of hormone regulatory mechanisms conferring plant drought tolerance.

3.2 Major Physiological Responses to Drought Stress

3.2.1 Leaf Senescence and Antioxidant Metabolism

Leaf senescence is a key developmental process which occurs naturally during plant maturation and is also a common result of prolonged abiotic stress. The coordinated breakdown and translocation of leaf cellular constituents increases the likelihood for plant survival during short-term stress periods and leaf senescence may be reversed if stress conditions are relieved within a certain time period (Buchanan-Wollaston 1997). Chlorophyll degradation is preempted by protein and RNA degradation mobilizing amino acids and nutrients towards other actively growing tissues or storage organs, thereby enhancing likelihood for drought survival (Buchanan-Wollaston et al. 2003; Hörtensteiner and Feller 2002).

In addition, there are extensive reviews detailing the relationship between oxidative stress agents and the antioxidative mechanisms which mitigate cellular damage to chlorophyll (Apel and Hirt 2004; Blokhina et al. 2003; Mittler 2002). The balance between reactive oxygen species (superoxide radical, hydrogen peroxide, hydroxyl radical) and enzymatic (CAT, POD, SOD, APX, and GR) or nonenzymatic (GSH and ASA) antioxidants, as well as carotenoids and tocopherol, determine the extensiveness of lipid peroxidation leading to chlorophyll membrane failure and eventual leaf senescence (Apel and Hirt 2004; Prochazkova et al. 2001). As opposed to other abiotic stresses such as salinity or UV-B radiation, oxidative stress during drought periods may increase tocopherol, carotenoid, and glutathione, while ascorbate pools tend to decrease (Munné-Bosch and Alegre 2000; Smirnoff 1993). Additionally, plants under drought stress may suppress production of reactive oxygen species and mitigate leaf senescence by decreasing cytochrome respiration and utilizing alternative respiratory pathways, though the influence of plant hormones on distinct respiratory pathways is not well known (Bartoli et al. 2005; Vanlerberghe 2013). Recent studies which use the systems biology approach have begun to shed light on how specific hormones influence the balance between oxidative stressors and antioxidant agents which mitigate their damaging effects on leaf senescence and are discussed below (Jibran et al. 2013).

3.2.2 Carbon Metabolism

Carbon metabolism or carbohydrate production during photosynthesis supplies the substrates needed to drive an array of growth, energy, and signaling processes in plants. The extent to which drought stress impairs carbon metabolic processes depends on the intensity, duration, and onset rate of the stress and varies based on plant species, maturation, and specific tissue type (Jaleel et al. 2009). It is well known that prolonged drought stress impairs photosynthesis either by decreasing stomatal aperture size limiting CO_2 diffusion into the mesophyll cells or by indirectly inhibiting associated biochemical and photochemical processes (i.e., RuBisCO deactivation or slowed RuBP regeneration) (Bota et al. 2004; Chaves et al. 2002, 2009; Flexas and Medrano 2002; Lawlor 2002). However, despite these limitations, carbon-rich molecules such as soluble carbohydrates (hexose, sucrose, trehalose, mannitol), amino acids (proline), organic acids (malate, fumarate, citrate), and structural compounds (cellulose and lignin) typically increase within plant tissues during drought stress (Muller et al. 2011).

Additionally, many of these compounds act as compatible solutes within cells and protect subcellular structures against damaging effects of water deficit, a topic which has been reviewed in detail (Chaves et al. 2002; Farooq et al. 2009; Yordanov et al. 2000). As soil water deficit is prolonged and level of drought stress becomes more severe, plant growth rates will decrease which lessens net carbon demand while net photosynthetic rates temporarily remain less affected which maintains net carbon gain within the plant system (Poorter and Nagel 2000). Coinciding with the significant reductions in stomatal aperture size and cellular water content, many plants seek to sustain photosynthesis by altering metabolic aspects such as Rubisco activity or activity of sugar-cleaving enzymes, among other enzymes (Muller et al. 2011). The underlying mechanisms by which carbon-containing molecules interact with hormonal stress-signaling pathways to initiate specific growth processes during drought stress have been of particular interest to researchers over the past several decades and current knowledge is discussed below.

3.2.3 Stomatal Movement

Stomatal closure is the primary line of defense by which plants decrease transpirational water loss to maintain cellular water content during drought stress and is induced by either hydropassive or hydroactive mechanisms (Murata and Mori 2014). Hydropassive stomatal closure occurs most often in environments of low humidity and/or high air currents and is characterized by guard cells quickly losing turgor due to rapid evaporative water loss without timely replenishment of water from adjacent epidermal cells (Wang et al. 2001). Alternatively, hydroactive stomatal closure is a more complex process, occurring as a result of whole-plant (root and shoot) water deficit and involves solute expulsion from guard cells increasing their osmotic potential and causing them to become flaccid and close (Kaiser and Legner 2007). The factors which contribute to hydroactive stomatal movement (opening or closing) under a variety of abiotic stresses have been of particular interest to researchers within the context of global climate change and limited water supplies for irrigation.

Hormone profiling techniques coupled to genetic approaches such as transcript profiling and plant mutants have offered valuable insight into the signal transduction pathways preempting initiation of stomatal closure during drought stress (Daszkowska-Golec and Szarejko 2013; Dodd 2003). Specifically, ABA-mediated stomatal closure through signal transduction pathways and downstream effects on cellular ion content during drought stress has been investigated over the past several decades, though important factors including mechanisms of drought stress perception activating abscisic acid (ABA) as well as upstream genes in the ABA-signaling pathway remain to be discovered (see ABA section below). Furthermore, recent advances in transcriptomics and next-generation sequencing techniques have suggested critical functions of other hormones and metabolites interacting with ABA and further contributing to stomatal closure during drought stress.

3.3 Roles of Hormones Regulating Physiological Responses to Drought Stress

3.3.1 Abscisic Acid

Among all classes of drought-responsive endogenous hormones currently known to exist within the plant system, ABA has been implicated as the primary chemical signal initiating stomatal responses to drought stress (Wilkinson and Davies 2002). Provided the abundant current knowledge, emphasis of ABA (and subsequent hormones) research within this chapter will be limited to research conducted during the past fifteen years. It is a well-known fact that ABA concentrations increase in response to drought imposition and the series of chemical reactions involving carotenoids for ABA biosynthesis and catabolism have been previously elucidated through a serious of experiments (Ikegami et al. 2009; Schwartz et al. 2003; Schroeder and Nambara 2006; Sridha and Wu 2006).

The mechanisms or intermediate signaling components by which ABA induces stomatal closure have also been characterized, though little was previously known regarding the earliest events in ABA perception initiating specific signal transduction pathways (Hirayama and Shinozaki 2007; Zhang et al. 2006). A recently discovered family of proteins known as PYRABACTIN (4-bromo-N-[pyridin-2-yl methyl] naphthalene-1-sulfonamide) RESISTANCE (PYR)/REGULATORY COMPONENT OF ABA RECEPTOR (RCAR) has opened additional avenues for ABA-signaling research and continues to prompt new questions about how the ABA-signaling network operates (Cutler et al. 2010). PYR/RCARs are ABA-binding proteins which interact with two other protein classes, Protein Phosphatase 2Cs (PP2Cs) and SNF1-related protein kinase 2s (SnRK2s), to initiate ABA recognition and signaling cascades. Specifically, PYR/RCARs are ABA receptors while PP2Cs and SnRK2s are negative and positive regulators of the signaling pathway, respectively (Hubbard et al. 2010). When ABA is present and associates with PYR/

RCARs, PP2Cs are inhibited which subsequently allow SnRK2s to become active and phosphorylate downstream transcriptional factors such as ABSCISIC ACID RESPONSIVE ELEMENTS-BINDING FACTOR 2 (ABF2) and ABI5-regulating downstream effects on target proteins gene expression, secondary messenger productions, and ion transport (Fujii et al. 2009; Ren et al. 2010). Alternatively, when ABA is not present, PP2Cs are active and inhibit SnRK2s activity thereby preventing downstream ABA responses. For in-depth descriptions of the PYR/RCAR–PP2C–SnRK2-signaling module and associated questions about the hormonal response pathways, see reviews by Hubbard et al. (2010), Joshi-Saha et al. (2011), Melcher et al. (2010), and Raghavendra et al. (2010).

Technological advancements in next-generation sequencing during recent years have produced extensive transcriptome data sets from plants under a variety of abiotic stresses and with or without interacting hormone factors (exogenous or endogenous) (Cramer et al 2011; Huang et al. 2008; van der Graaff et al. 2006; Zawaski and Busov 2014). Comparatively to other classes of plant hormones, the ABA-regulated genomic changes are two to six times greater and can be in excess of 10 % of the genome in *Arabidopsis* seedlings (Cutler et al. 2010). ABA-induced genes influence a variety of stress-promoting factors including enzymes to detoxify reactive oxygen species, enzymes for compatible solute metabolism maintaining cell turgidity, protein transporters, transcription factors, and enzymes contributing to phospholipid signaling (Cutler et al. 2010; Nakashima et al. 2009; Nemhauser et al. 2006; Zhu 2002). From the transcriptomic data, there has been an increased focus on ABA-induced changes to WRKY transcription factors, one of the largest families of transcriptional regulators spanning across many different plant processes (Chen et al 2010). For example, rice lines overexpressing OsWRKY11 were more drought tolerant due to enhanced accumulation of compatible solutes such as raffinose (Wu et al. 2009). It was suggested that ABO3/AtWRKY63 functions in ABA-mediated drought stress response pathways since the *abo3* mutation impairs ABA-induced stomatal closure (Ren et al. 2010). Overexpression of the ABA-inducible OsWRKY45 gene in *Arabidopsis* conferred enhanced tolerance to salt and drought stress possibly due to the plants having a higher proportion of closed stomates (Jiang et al. 2012; Qiu and Yu 2009). While the direct link between WRKY transcription factors and stomatal movement remains to be proven, it is well known that stomatal closure upon drought onset is initiated by ABA-mediated calcium increases in the cytosol which stimulate potassium efflux and increased water potential causing guard cell flaccidity (Himmelbach et al. 2003; Sridha and Wu 2006). Identifying additional ion channels or transporters which contribute to polarization state of plasma membranes for potassium movement has been a recent research focus and new evidence is beginning to suggest that inhibitors of protein kinases and protein phosphatases may further affect transporter capabilities (see review by Sirichandra et al. 2009).

The drought-induced increase in endogenous ABA has also been correlated to leaf senescence and carbon remobilization in wheat (Yang et al. 2001b, 2003). The increase in leaf senescence due to ABA has been associated with induction of lipid peroxidation caused mainly by increased hydrogen peroxide content within leaf cells of rice (Hung and Kao 2003). However, low to moderate ABA concentrations may mitigate downstream oxidative damages preempting leaf senescence by increasing SOD, CAT, APX, and GR activity as well as carotenoid and tocopherol contents in wheat, though these beneficial effects were no longer evident at excessively high exogenous ABA concentrations (Jiang and Zhang 2001). Jiang and Zhang (2002) also suggested an interrelationship between drought stress-induced ABA production and ROS production stimulating an upregulation of antioxidant defense system. While it may be inferred that ABA-induced oxidative stress and antioxidant metabolism governing leaf senescence as well as stomatal conductance governing carbon influx for photosynthesis will likely have downstream effects on plant growth, ABA has recently been implicated for direct effects on various aspects of plant growth during drought stress. The historical view of ABA affecting plant growth implied that higher concentrations within the plant system would inhibit shoot growth due to stomatal regulation of water status during prolonged drought stress (Trewavas and Jones 1991). However, systems biology research has utilized ABA-deficient mutants of maize, tomato, and Arabidopsis to suggest that ABA sustains growth of plant organs, namely roots, through antagonism with drought stress-induced ethylene production (Sharp 2002). The potential implications of this hormone to hormone cross talk for enhancing plant drought tolerance are discussed below.

3.3.2 Auxins

The naturally occurring auxin indole-3-acetic acid (IAA) is synthesized within the rapidly dividing tissues of root and shoot apical meristems and young leaves across virtually all plant species (Ljung et al. 2001, 2005). Historically, the biological connection between auxin and stress-induced leaf senescence has been variable depending on plant species or tissue maturity, and the specific leaf responses may be dependent not only upon auxin concentration but also cellular responsiveness or sensitivity (Schippers et al. 2007). Some of the earliest studies showed that leaf senescence progresses as IAA concentrations decline towards similar levels between stems and leaves in beans, whereas applying IAA exogenously to the distal or proximal end of abscission zone will delay or promote abscission, respectively (Addicott and Lynch 1951; Shoji et al. 1951). The gradient-dependent manner by which auxin delays leaf senescence has been partly explained by studying AUXIN RESPONSE FACTOR1 (ARF1) and ARF2 genes using *Arabidopsis arf1* and *arf2* mutants (Ellis et al. 2005; Lim et al. 2010). *ARF1* and *ARF2* mutants, as well as *NPH4/ARF7* and *ARF19* mutants, all displayed some degree of delayed leaf senescence, suggesting that the respective ARF transcription factors repress auxin signaling and are positive regulators of leaf senescence, or, auxin is involved in the negative regulation of leaf senescence. Lim et al. (2010) further demonstrated that *arf2* was more tolerant to oxidative stress since the mutants maintained chlorophyll content and PSII activity compared to wild-type plants. The presence of oxidizing ROS agents can induce auxin degradation, alter auxin transport and distribution, relocate PIN proteins for auxin transport, and induce auxin conjugation (Tognetti et al. 2012). Concurrently, gene expression associated with auxin response factors, transporters, and biosynthetic enzymes has been shown to be stimulated by ethylene, which itself displays antagonism with auxin and is discussed further below (Peleg and Blumwald 2011).

A link between auxin content and antioxidant capacity was hypothesized when various grass species displayed increased abiotic stress tolerance following exogenous treatment with humic acids possessing auxin-like activity (Zhang and Schmidt 1999, 2000; Zhang et al. 2003, 2007). More recently, a mutation of *Arabidopsis CATALASE2* resulted in cross talk between hydrogen peroxide and auxin signaling as mediated by changes in glutathione redox status resulting in a hyponastic phenotype (Gao et al. 2014). Csiszár et al. (2004) showed that auxin autotrophic tobacco callus may resist oxidative damages (less cellular hydrogen peroxide and malondialdehyde) by increasing GPX, GST, and GSH-PX activities, while heterotrophic tobacco callus did so similarly via increases to SOD and CAT activity. Finally, transgenic *Arabidopsis* with higher endogenous IAA content or wild-type plants treated with exogenous IAA were more drought tolerant due to enhanced SOD, CAT, POD, and GR activity for accelerated ROS mitigation (Shi et al. 2014). These transgenic lines also displayed improved root growth or lateral rooting for water uptake, maintained metabolic homeostasis, and positively modulated specific stress-related genes (RAB18, RD22, RD29A, RD29B, DREB2A, and DREB2B) during drought stress.

Auxin has also been implicated in altering hydrogen peroxide dynamics with downstream signaling effects on stomatal movement and root morphology, both of which are important contributors to whole-plant drought tolerance. It was suggested that hydrogen peroxide contributes to the auxin-dependent responses of plasma membrane H^+-ATPase and cytoplasmic pH controlling inward and outward potassium movement to guard cells (Song et al. 2006). Similarly, there is increasing evidence that auxins influence nitric oxide levels within guard cells possibly stimulating ion movement via these cross-membrane channels (Xiao-Ping and Xi-Gui 2006). More specifically, low auxin concentrations induced potassium influx and guard cell opening whereas increasing auxin concentrations induced potassium efflux and closing of guard cells (Acharya and Assmann 2009; Daszkowska-Golec and Szarejko 2013). However, this can be interpreted as contradictory to the observed effects of exogenous auxins which counteract ABA-induced stomatal closure, possibly through interactions with ethylene in *Arabidopsis* (Tanaka et al. 2006). Throughout the literature, there exist contradictory reports regarding stomatal responses to auxins stemming from organic versus synthetic forms applied, concentration dependencies, species or organ-specific responses, and potential cross talk or interactions with other plant hormones (Daszkowska-Golec and Szarejko 2013; Pospíšilová 2003).

Joo et al. 2001 suggested a novel role for auxin-induced ROS in root gravitropism by which unilateral application of auxin caused transient increases in ROS to mediate directional root growth and may also be interdependent with calcium signaling. Alternatively, ROS-mitigating GSH may be closely linked to auxin transport since *Arabidopsis* treated with the GSH inhibitor, buthionine sulphoximine (BSO), displayed a loss of PIN1, PIN2, and PIN7 auxin carriers (Koprivova et al. 2010). Auxin has been implicated across a wide array of plant developmental processes including organogenesis and upregulating the AVP1 H+-pyrophosphatase accelerates auxin fluxes and pyrophosphate-driven cation transport into root vacuoles which increases root biomass for enhanced drought tolerance (Li et al. 2005; Park et al. 2005). Similarly, moderate water stress may stimulate auxin transport into root tips and increase plasma membrane H^+-ATPase activity for enhanced proton secretion driving root elongation and root hair development (Xu et al. 2013). Along with drought perception, a myriad of environmental response pathways converge on auxin signal transduction with downstream effects for each stage of lateral root development including cell initiation, emergence of the lateral root primordial, and lateral root growth and orientation (Casimiro et al. 2003; Malamy 2005). Maintaining auxin homeostasis is an essential component of lateral root growth during drought stress and there is increasing evidence that cross talk interactions between auxin and other hormones, such as ABA, may inhibit auxin-stimulated root growth, as discussed further below.

3.3.3 Cytokinins

Cytokinins are well known to influence many biological functions throughout the plant system, which one of the most well-known positive regulators of senescence, as demonstrated by exogenous applications or endogenous manipulations suppressing the rate of natural or stress-induced leaf senescence (Lim et al. 2003; Taiz and Zeiger 2010). As CK levels decrease during stress-induced leaf senescence, genes for CK synthase and adenosine phosphate isopentenyl-transferase (*IPT*) are downregulated, genes for CK oxidase are upregulated, and until recently, little was known about which gene(s) directly influence leaf senescence by means of CK signaling (Lim et al. 2007). Unlike other hormone-signaling pathways (except ethylene), CK signaling comprises a histidyl-aspartyl (His-Asp) phosphorelay system by which histidine kinases (HKs) serve as cytokinin receptors which transfer a phosphoryl group to nuclear type-B response regulators (RRs) activating the type-A RR primary response genes (Imamura et al. 1998; To and Kieber 2008). While six distinct HKs have been identified in *Arabidopsis*, AHK2-4 (A for *Arabidopsis*) are localized on the endoplasmic reticulum, serve CK receptor functions, and have distinct roles in various aspects of plant growth and development (see comprehensive review by Ha et al. 2012).

The process of stress-induced leaf senescence can be divided into three distinct phases; the initiation phase which involves stress perception and signal transductions, the reorganization phase during which changes in gene expression inducing a wide range of biochemical and metabolic changes including hormonal changes, and the terminal phase during which permanent or nonreversible cell death occurs (Munné-Bosch and Alegre 2004). Drought-induced leaf senescence typically coincides with decreasing endogenous CK concentrations, though the hypothesis that low CK content directly triggers leaf senescence may not be accurate since CK-deficient mutants typically display delayed chlorophyll degradation compared to wild-types suggesting that other factors such as altered source–sink responses or antioxidant profiles may be responsible for the observed senescence during drought stress (Werner et al. 2003). For example, transgenic tobacco plants expressing the *IPT* gene driven by a maturation- and stress-induced senescence-associated receptor protein kinase (*SARK*) promoter maintained higher photosynthetic rates and delayed leaf senescence during drought stress and the improved drought tolerance was associated with larger pools of ascorbate and glutathione accounting for the lower levels of hydrogen peroxide (Rivero et al. 2007). Similarly, transgenic creeping bentgrass containing the *IPT* gene driven by an auto-regulated senescence-activated (*SAG12*) promoter displayed less lipid peroxidation maintaining cellular integrity and photochemical efficiency during drought stress, possibly associated with maintenance of SOD, POD, and CAT activities (Merewitz et al. 2011). Despite these examples and numerous other reports correlating CK (exogenous or endogenous) to enhanced antioxidant capacity during drought stress, the stress-induced changes to signaling pathways linking cytokinins (CKs) to antioxidant metabolism, either directly or indirectly, are largely undefined and deserve further investigation.

Historically, CKs were generally regarded as antagonists to ABA throughout the plant system, though the literature provides contradictory reports by which CKs enhance, mitigate, or have neutral effects on stomatal apertures depending upon plant species, CK type, and concentration dependencies (see reviews by Acharya and Assmann (2009) and Pospíšilová (2003)). In certain cases, increasing concentrations of CKs in xylem sap may reduce stomatal sensitivity to ABA and promote stomatal opening or delay the drought-induced decrease of stomatal aperture (Havlova et al. 2008; Wilkinson and Davies 2002). Alternatively, a reduction in endogenous CK content may not imply enhanced sensitivity to ABA as demonstrated by CK-deficient *Arabidopsis* lines which regulated the endogenous ABA:CK ratio and maintained stomatal aperture for photosynthetic carbon dioxide uptake (Nishiyama et al. 2011). From this, it was suggested that there may exist mutual regulatory mechanisms between CKs and other plant hormones which collectively mediate stomatal responses upon onset of adverse environmental conditions (discussed further below). Environmental cues such as drought stress may induce synergism between CKs and ABA to collectively exert antagonism upon auxin resulting in suppression of lateral root formation and promotion of primary root growth into deeper soil profiles in search of water supplies (Ha et al. 2012). Such effects may have been evident for *SAG12-ipt* and *HSP18.2-ipt* transgenic creeping bentgrass plants with greater root:shoot and CK:ABA ratios upon drought stress, though direct correlations between the two parameters were not clear (Merewitz et al. 2010). There are apparently discrepancies on CK effects on root growth between dicots, such as *Arabidopsis* and tobacco, and inhibitory effects on monocots, such as creeping bentgrass; however, the mechanisms underlying the differential responses of roots to CKs between different types of plant species are unknown, which is likely influenced by a variety of plant growth factors and associated signaling pathways influencing hormonal, nutritional, and/or source–sink relationships throughout different plant organs, and different sensitivity of plants to endogenous levels of CKs (Werner et al. 2010).

3.3.4 *Gibberellins*

Gibberellins (GAs) are a large class of diterpenoid plant hormones with over one hundred different chemical structures currently known to exist, though only a select few are biologically active and influence a variety of plant growth and developmental processes including seed germination, stem and root elongation, leaf expansion, transition from juvenile to adult phases, sex determination, and floral initiation (Sun and Gubler 2004; Taiz and Zeiger 2010; Yamaguchi 2008). In comparison to other plant hormones, there is far less information regarding contribution of GAs to drought tolerance and how GA content changes upon increasing level of drought stress (Pospíšilová 2003). Despite GA typically classified as antagonistic to ABA, the few studies investigating GA contribution to stomatal function suggested that GA may not contribute to stomatal movement since exogenous GA had little or no effect on stomatal closure in *Arabidopsis* and GA-deficient mutants had similar transpiration rates compared to wild-type plants during drought stress (Acharya and Assmann 2009). The direct effect of GA on stomatal movement by means of exogenous applications or endogenous manipulation is a particular research area which deserves further attention. Alternatively, Saibo et al. (2003) demonstrated that GA is the main hormonal signal inducing stomata formation on *Arabidopsis* hypocotyls and the GA-induced developmental signal is further regulated through interactions with auxin and ethylene. It would seem inherent that leaf stomatal density strictly regulated by GA would be a major determinant of transpiration rates and leaf water status during drought stress, though this remains to be empirically proven. There is also considerable evidence suggesting that stomatal aperture responding to drought onset is regulated through multiple signaling pathways or cross talk among various plant hormones, including GA, as discussed in subsequent sections.

Along with the previously mentioned contribution of CKs to leaf senescence, there is increasing evidence suggesting that GAs also serve important regulatory functions during natural or stress-induced leaf senescence. For example, Rosenvasser et al. (2006) summarized that GA_1, GA_4, and GA_9 content all decreased in *Alstromeria* and lettuce leaves upon dark-induced leaf senescence, while exogenous GA delayed leaf senescence in *Pelargonium*, *Taraxacum*, *Rumex*, *Nasturtium*, and *Alstromeria*. Similarly, leaf senescence was mitigated by exogenous applications of GA_4 and GA_7 in *Lilium* plants following low-temperature storage in darkness and the beneficial effects were associated with increased CAT activity and decreased lipid peroxidation and proteolysis (Ranwala and Miller 2000). GA_3 applied as a foliar spray or soil drench enhanced the antioxidant potential of *Catharanthus roseus* by stimulating production of the indole alkaloid ajmalicine and also alleviated the toxic effects of cadmium in a separate study with the same plant species (Pandey et al. 2007; Jaleel et al. 2007). GA_3 applications mitigated oxidative stress and slowed the rate of *Paris polyphylla* leaf senescence by increasing endogenous GA_4 and GA_7 with potential downstream effects on lipid peroxidation, hydrogen peroxide content, activities of SOD, POD, and APX, and while a potential antagonistic interaction with ABA was suggested, the actual mechanisms underlying these changes remain unclear (Yu et al. 2009). The inhibitory effects of drought stress on various morphological aspects of maize growth were reduced for plants treated with GA_3 during the vegetative growth stage, though underlying mechanisms were not explored during the study (Akter et al. 2014). Finally, the beneficial senescence-mitigating effects of GA_3 were associated with a significant enhancement of SOD activity in marigold (*Calendula officinalis*) during drought stress (Sedghi et al. 2012). While we can only speculate on the direct link between antioxidant metabolism and GA for delaying leaf senescence during drought periods, it is possible that WRKY transcription factors associated with the gibberellin-signaling pathway are involved, which may exert downstream effects on ROS signaling and hydrogen peroxide accumulation for antioxidant response, though much more work is needed to verify the actual signaling process (Jo and Hyun 2011).

The growth-promoting effects of GA by means of downstream effects on cellular elongation rates have been recognized for many years, though only until recently have advancements in molecular biology techniques been able to shed light on signaling pathways preempting the GA-induced cellular expansion process (Fleet and Sun 2005; Olszewski et al. 2002; Richards et al. 2001; Schwechheimer and Willige 2009). Moreover, despite the well-known adverse effects of drought stress on root and shoot growth rates, the direct influence of GA on the mechanisms underlying cellular expansion under drought stress are not well known. One particular mechanism by which GA may influence cellular expansion under drought stress involves GA_3 upregulating expansin genes *EXPA4* and *EXPB4* and xyloglucan endotransglycosylase (XET) genes *XET1* and *XET1* to maintain leaf elongation rates as demonstrated in tall fescue (*Festuca arundinacea*) under chemically induced drought stress (Xu et al. 2016). A GA-responsive transcription factor, OsPIL1, was repressed by drought stress and was associated with downregulation of expansin in the internode of rice plants (Todaka et al. 2012). A comprehensive study by Ribeiro et al. (2012) investigated changes in *Arabidopsis* translatome and metabolome as triggered by GA and suggested that there exists a close interaction between energy metabolic processes and GA-mediated growth with downstream effects on cell wall extension, secondary metabolism, and lipid metabolism. Specifically, GA and paclobutrazol (PAC; GA-inhibitor) had opposing effects on the expression of genes encoding expansins and xyloglucan endotransglucosylase/endohydrolases (XTHs), products of primary metabolism including nitrates, total amino acids, and protein, as well as carbon allocation governing growth rates. Based on these studies which demonstrate that potential interactions between GA and carbon metabolism and growth may likely exist, it would be useful for future research to begin identifying which genes are responsible and identify new markers for growth potential (Ribeiro et al. 2012). Additionally, similar studies should be conducted investigating translatomic and metabolomic changes conferred by GA induction or suppression and which changes may contribute to enhanced drought tolerance in Arabidopsis or other model plant species.

3.3.5 Ethylene

The gaseous plant hormone ethylene induces a triple response on plant development encompassing radial swelling, inhibited elongation of the epicotyl, and horizontal growth of the epicotyl, and is also implicated in downstream effects on various aspects of plant stress responses (Chaves et al. 2003; Sharp and LeNoble 2002; Taiz and Zeiger 2010). The current model of ethylene signaling suggests that ethylene molecules are sensed by a family of receptors acting as negative regulators of the ethylene-responsive pathway (Guo and Ecker 2004). More specifically, ethylene binds to the receptors and inactivates the receptor-CTR1 complex which allows EIN3 and EIN3-like transcription factors to accumulate in the nucleus and express transcription factor genes *ERF1-4* which initiate activation or repression of hundreds of downstream genes, though until recently, little was known about how these signaling networks may contribute to ethylene responses during drought stress (Stepanova and Alonso 2009). Additionally, signaling pathways mediated by ethylene may also involve cross talk between calcium-dependent protein kinases (CDPKs) and mitogen-activated protein kinases (MAPKs) preempting the downstream activation of stress-response genes (Fujita et al. 2006; Ludwig et al. 2005; Nakagami et al. 2005).

Recent studies have suggested that expression of specific ethylene response factors (ERFs) exert downstream effects in various plant species responding to osmotic stresses, including drought or salinity, and the potential mechanisms by which ERFs promote stress tolerance have been suggested in several cases. Transgenic sugarcane (*Saccharum officinarum*) plants overexpressing *SodERF3* displayed improved drought tolerance as noted by enhanced plant height, leaf weight, and flower production following 20-day water withholding as compared to wild-type plants, though the underlying mechanisms facilitating the enhanced growth were not investigated (Trujillo et al. 2008). Alternatively, *OsDERF1* is induced by drought stress and while transgenic rice plants overexpressing *OsDERF1* were more drought-susceptible at seedling stage, *OsDERF1* knockdown lines had enhanced drought tolerance at seedling and tillering stages associated with lower MDA accumulation and higher accumulation of sugars and proline suggesting specificity of ERF regulation in drought response (Wan et al. 2011). Several studies which regulated ERF expression levels to induce downstream changes in stress tolerance have shown that enhanced drought tolerance may be attributed to ERFs activating transcription of specific stress-responsive genes regulating the observed physiological changes during water withholding. Rice plants overexpressing *JERF3* displayed improved drought tolerance similarly associated with proline and sugar accumulation and the increased proline content was most likely due to upregulation of *OsP5CSs* encoding two key enzymes in proline synthesis (Zhang et al. 2010a, b). In the same study, *JERF3* also upregulated three stress-inducible genes, *WCOR413-like*, *OsEnol*, and *OsSPDS2*, which were attributed to maintaining membrane stability under drought stress. In addition to upregulating osmotic stress genes, upregulation of the ethylene-responsive *JERF3* has also been shown to reduce ROS by enhancing expression of antioxidative SOD genes and further contributed to drought tolerance by activating photosynthetic carbon assimilation/metabolism genes (Wu et al. 2008). ROS-responsive genes may contain ethylene-responsive *cis* elements, as was shown for *Zat7*, *Zat12*, *WRKY25*, and *Apx1* in *Arabidopsis* (Miller et al. 2010). Overexpressing *OsWR1*, a homolog of the wax/cutin synthesis regulatory gene *WIN1/SHN1*, regulated the expression of wax-related genes *OsLACS2* and *OsFAE1'-L*, as well as genes related to oxidative stress response and membrane integrity, all of which likely contributed to improved drought tolerance in transgenic rice plants (Wang et al. 2012).

Ethylene contributes to long-distance stress signaling upon perception of soil drying by means of the ethylene precursor 1-aminocyclopropane-1-carboxylic acid (ACC) transported from roots to shoots, though the literature provides inconsistent results as to when and under which specific conditions ACC transport and ethylene production occurs (see review by Wilkinson and Davies 2010). Wilkinson and Davies (2010) also summarized that plants respond to ethylene biphasically with low or high ethylene concentrations increasing or decreasing plant growth, respectively, though there is still no clear role of ethylene in maintaining shoot growth under drought stress conditions. The well-defined effects of ethylene on leaf senescence and growth inhibition under drought stress has retracted from research interests of potential contributions of ethylene to drought tolerance, even though in the absence of detrimental growth effects under induced by high ethylene concentrations, lower concentrations of ethylene may maintain stomatal apertures for leaf cooling and carbon uptake during mild to moderate drought stress (Wilkinson et al. 2012). ERFs such as ETR1 (ethylene response 1) may facilitate signaling functions for stomatal movement, glucose-sensing, and hydrogen peroxide biosynthesis suggesting a potential link between ROS, sugars, and hormone pathways (Pinheiro and Chaves 2011). Furthermore, the ethylene-mediated reductions in shoot growth as well as stomatal responses under drought stress are highly dependent upon ABA accumulation in shoots, since ABA and ethylene exert antagonism upon each other as discussed below (Chaves et al. 2003).

3.3.6 Salicylates

Salicylic acid (SA) is an endogenous phenolic plant hormone which serves diverse regulatory roles in plant metabolism and has been implicated in modulating specific plant responses to oxidative stress, such as the signaling cascades and regulation of chloroplast biogenesis with subsequent promotive effects on photosynthesis (Hayat et al. 2010). SA contributes to the initial development of stress responses and higher concentrations of SA within the plant system tend to induce the beneficial responses promoting tolerance to osmotic stresses such as salinity or drought (Horváth et al. 2007). Exogenous SA applications enhanced the drought tolerance of tomato (*Lycopersicon esculentum*) by increasing photosynthetic parameters, membrane integrity, leaf water potential, chlorophyll content, and activity of nitrate reductase carbonic anhydrase (Hayat et al. 2008). Similarly, higher water content, dry mass accumulation, and chlorophyll content associated with maintenance of carboxylase activity and SOD activity were observed in drought-stressed wheat seedlings following SA application (Singh and Usha 2003). Higher antioxidant enzyme activities limited hydrogen peroxide accumulation and lipid peroxidation in droughted wheat leaves previously sprayed with SA (Agarwal et al. 2005). The stimulative effect of exogenous SA on plant antioxidant components is likely concentration-dependent as SA-deficient mutants lack the ability to mitigate ROS, low concentrations (0.01–0.05 mM) of SA induce slight stimulation of AOX and HSPs, while optimum concentrations (0.1–0.5 mM) of SA initially increase ROS which themselves act as secondary messengers to dramatically enhance CAT, APX, SOD, GR, AOX, and HSP activities conferring the observed drought tolerance (Yuan and Lin 2008). A similar dose-dependent effect of SA was observed in tomato and bean (*Phaseolus vulgaris*) plants which displayed enhanced drought tolerance for plants grown from seed imbibed in low SA concentrations but not for plants grown from seed imbibed in high SA concentrations (Senaratna et al. 2000).

Interactions between SA and hydrogen peroxide affect the rate of ROS accumulation within plant tissues which stimulates oxidative stress-induced PR gene expression and downstream systemic acquired response (SAR) responses (Horváth et al. 2007; Lee et al. 2006; Mateo et al. 2006). SA-mediated ROS accumulation may also influence stomatal aperture during drought responses as demonstrated by *siz1* Arabidopsis mutants lacking *SIZ1*-mediated endogenous SA accumulation, though this is one particular area which deserves further investigation (Miura et al. 2013). Similarly, increasing SA concentrations by means of endogenous manipulation or exogenous applications stimulates nitric oxide (NO) synthesis, another key component in stress-responsive signaling cascades (Zottini et al. 2007). MAPKs have also been shown to be stimulated by SA and initiate various downstream defense responses including expression of key enzymes for defense signaling and initiation of abiotic and biotic stress responses (Bowler and Fluhr 2000; Yang et al. 2001a; Zhang and Liu 2001). Despite considerable work investigating SA contribution to other oxidative stresses including salinity, ozone, and UV-B radiation, there has been far less investigation of which SA-responsive genes contribute to drought tolerance in plants. Exogenous SA application promoted drought tolerance in wheat seedlings by enhancing the transcription of *GST1*, *GST2*, *GR*, and *MDHAR* which facilitate the detoxification of ROS, though the focus of this study was narrowed towards genes involved in the ASA-GSH cycle (Kang et al. 2013). The interaction between GSH and SA regulating ROS production under stress conditions may also regulate a variety of other plant processes due to the effects on cell redox states (Horváth et al. 2007). *Tobacco stress-induced gene1* (*Tsi1*) was shown to be induced by exogenous SA application and subsequently increased expression of drought stress-responsive target genes *PR1*, *PR2*, *PR3*, *osmotin*, and *SAR8.2* (Park et al. 2001). Transcriptional profiling of the *WRKY* gene family showed that genes encoding certain WRKY transcription factors are upregulated by SA application while others are upregulated by drought stress and that a specific *WRKY* gene, *12g02400*, was upregulated by both SA and drought stress (Ramamoorthy et al. 2008). However, it remains to be determined as to which downstream plant responses are regulated by SA- or drought-induced *WRKY* gene regulation. It would be interesting to know whether there exists a link between SA-induced WRKY gene expression and downstream proteomic changes, as SA-induced growth and drought tolerance of wheat was associated with altered expression patterns of proteins facilitating signal transduction, stress defense, photosynthesis, carbohydrate metabolism, protein metabolism, and energy production during drought stress (Kang et al. 2012).

3.3.7 Jasmonates

Jasmonic acid (JA) and methyl jasmonate (MeJA) are biologically active lipid derivatives formed by fatty acid oxidation and contribute to the regulation of various stress responses in plants including leaf senescence, ROS and NO signaling, antioxidant metabolism, and stomatal movement (Balbi and Devoto 2008; Murata and Mori 2014; Taiz and Zeiger 2010; Wasternack 2007). Similar to the current state of SA research, jasmonates have been implicated in promoting tolerance to abiotic stresses including salinity, ozone, or UV-B irradiance through downstream effects on antioxidant metabolism, whereas less research has been conducted regarding the contribution of jasmonates to antioxidant-facilitated drought tolerance (Kumari et al. 2006). Nevertheless, several studies which have been conducted suggest jasmonates mitigate the drought-induced oxidative burst in a similar manner as for other oxidative stresses, by means of increased antioxidant enzyme activities. For example, soybean plants treated with 50 μM MeJA had decreased lipid peroxidation associated with increased activities of SOD, POD, and CAT while the observed increase in proline concentration may have further facilitated the higher leaf water content for MeJA-treated plants during irrigation withholding (Anjum et al. 2011). Cellular water retention is also enhanced following JA application by increasing betaine aldehyde dehydrogenase (BADH) for enhanced betaine accumulation and subsequent osmotic adjustment in pear (*Pyrus bretschneideri*) leaf cells (Gao et al. 2004). The increased activities of SOD, POD, CAT, APX, and GR collectively detoxified hydrogen peroxide and, in conjunction with higher proline and soluble sugar content, enhanced the drought persistence of cauliflower (*Brassica oleracea*) seedlings following MeJA or coronatine (COR; a phytotoxin that mimics some biological activities of MeJA) application (Wu et al. 2012). A microarray analysis of over two thousand selected *Arabidopsis* genes showed that the abundance of 221 mRNAs was highly upregulated following MeJA application and the upregulated mRNAs served putative functions spanning oxidative stress responses, cellular maintenance, as well as low and high molecular weight defense signaling (Schenk et al. 2000). More specifically, the reduction in transcript levels of L-galactono-1,4-lactone dehydrogenase (GalLDH), APX, GR, dehydroascorbate reductase (DHAR), and monodehydroascorbate reductase (MDHAR) during drought stress was mitigated in crested wheatgrass (*Agropyron cristatum*) leaves following JA application, reinforcing the notion that JA also serves critical roles in regulating ascorbate and glutathione metabolism during drought periods (Brossa et al. 2011; Shan and Liang 2010). Alternatively, the increase in jasmonate content upon drought stress may induce specific NAC transcription factor gene (i.e., *ANAC019* and *ANAC055*) expression with potential downstream effects on abiotic-stress cellular networks, though the ways in which JA-induced NACs may be utilized to promote abiotic stress tolerance remain largely unknown (Bu et al. 2008; Puranik et al. 2012).

Given that jasmonate concentrations increase in a similar manner as ABA during drought onset, it was hypothesized that the biologically active jasmonate derivatives may positively regulate stomatal closure as drought stress severity increases (Acharya and Assmann 2009). Furthermore, whether or not jasmonates regulate similar mechanisms as ABA to influence stomatal movement was of particular interest. Studies which utilized *jar1* (MeJA-insensitive) mutants have shown that MeJA-mediated stomatal closure involves guard cell alkalization, ROS and NO production, potassium efflux, and slowed anion channels, all of which are similarly associated with ABA-induced stomatal closure (Evans 2003; Munemasa et al. 2007; Suhita et al. 2004). Further research investigating how MeJA and fluridon (ABA-inhibitor) influence stomatal movement in ABA-deficient mutants suggested that endogenous ABA is required to activate calcium signaling during MeJA-induced stomatal closure (Hossain et al. 2011). Additionally, stomatal closure will not occur for Arabidopsis mutants lacking the coronatine-insensitive1 (*COI1*) gene likely due to little change in ROS and NO production or anion efflux following MeJA application whereas *coi1* mutants will close stomates following ABA application, indicating that *COI1* is upstream of ROS and NO in MeJA signaling (Munemasa et al. 2007). Synergism exists between JA and NO in stimulating stomatal closure in broad bean (*Vicia faba*) leaves such that JA enhances NO synthesis in guard cells and both JA and NO induce stomatal closure in a dose-responsive manner (Liu et al. 2005). A review by Hadiarto and Tran (2011) suggested that JA may serve important regulatory roles during the ABA-dependent drought responses in plants since JAZ (Jasmonate ZIM-domain), ABA-dependent, and drought-inducible AtMYC2 transcription factors all regulate gene expression in jasmonate pathway.

In comparison to the extensive research regarding jasmonate-stimulated antioxidant metabolism or stomatal movement, there is less known regarding how jasmonates directly influence growth and photosynthetic processes during drought stress. One research area which may be of particular interest is in how jasmonates influence homeostasis of various energy-consuming processes during drought stress, as hormone balance likely controls metabolic and physiological stabilization during periods of abiotic stress (Harb et al. 2010). For example, MeJA pretreatment has been shown to have reversible effects on nitrogen uptake inhibition and remobilization of RuBisCO subunits in field-grown oilseed rape (*Brassica napus*), though whether JA-induced changes in these parameters confers drought acclimation remains unknown (Rossato et al. 2001). Furthermore, despite abundant knowledge detailing jasmonate contributions to many different plant physiological processes (i.e., floral development, senescence induction, growth inhibition, root morphogenesis) under non-stress conditions, little is known as to how these parameters may be individually affected by JA under short- or long-term drought stress treatment (Santino et al. 2013).

3.4 Interactions Between Hormones and Plant Metabolites During Drought Stress

3.4.1 Hormone to Hormone Interactions

As discussed above, multiple hormones may be involved in regulating a particular growth trait or physiological responses to drought stress through synergistic or antagonistic interactions, although each hormone play unique roles. The analysis of *Arabidopsis* mutant phenotypes in conjunction with transcriptomic profiling studies has provided convincing evidence supporting the theory that cross talk between plant hormones results in antagonistic or synergistic effects on various phenotypic responses to abiotic stress (Depuydt and Hardtke 2011). Cross talk signals derived from hormone to hormone or hormone to secondary messenger (i.e., calcium or ROS) interactions may converge upon or be transduced by MAPK modules to regulate gene expression by means of transcription factor modulation (Fujita et al. 2006; Smékalová et al. 2014). For example, two specific MAPKs, *OsMPK5* and *OsEIN2*, have been shown to facilitate antagonism between ABA and ethylene in that RNAi suppression of *OsMPK5* reduces rice sensitivity to ABA, increases endogenous ethylene, and reduces drought tolerance, while suppression of *OsEIN2* reduces sensitivity to ethylene, increases hypersensitivity to ABA, and enhances drought tolerance (Sharma et al. 2013). Ethylene has also been shown to regulate many auxin-related genes including ARFs, transporters, and genes encoding biosynthetic enzymes, while genes encoding rate-limiting enzymes in ethylene biosynthesis are conversely regulated by auxin (Peleg and Blumwald 2011). Given the well-known contribution of ABA to stomatal closure during plant drought response, many studies have investigated how other hormones influence ABA-mediated stomatal closure upon water deficit. Thus far, it is generally accepted that auxins, CKs, and ethylene are antagonistic with ABA and counteract stomatal closure while SA and jasmonates are in synergism with ABA and positively regulate stomatal closure during drought, though all of these hormones may differentially modulate the downstream expression of stress-related genes (Acharya and Assmann 2009; Nilson and Assmann 2007). For example, one of the ABA-regulated bZIP transcription factors (ABI5-Like1) is typically induced by drought or salinity but may also be regulated by auxin to then activate a variety of stress response genes including ABRE-containing genes related to auxin metabolism (Yang et al. 2011). Wang et al. (2011) also suggested that cross talk between ABA and MeJA occurs at the transcript level and is consistent with the downstream effects of cross talk at the physiological level both in guard cells and other tissues. Synergism may exist between CKs and auxins, specifically IAA, in that CKs are positive regulators of auxin biosynthesis and the two hormones may establish a homeostatic feedback regulatory loop to maintain proper proportions in developing root tissues (Jones et al. 2010). Alternatively, the well-known antagonism between ABA and CKs contributing to drought-induced stress responses may be in part facilitated by CK-receptor histidine kinases (*AHK2*, *AHK3*, and *CRE1*) acting as negative regulators of ABA and osmotic stress signaling, whereas another non-ethylene histidine kinase (*AHK1*) is a positive regulator of these same processes (Tran et al. 2007). It was also proposed that nitrate transporters facilitating nitrate uptake may serve in hormone cross talk since NRT2.6 is regulated by auxin, CKs, and ABA, whether be individually or interactively (Krouk et al. 2011). GA may be synergistic with SA as exogenous GA applications increased expression levels for two genes encoding SA-synthesis genes, plus transgenic *Arabidopsis* plants overexpressing a GA-responsiveness gene had higher endogenous SA content and were more tolerant to oxidative stress (Alonso-Ramírez et al. 2009a). Cross talk between GA and SA by which GA induces both SA production and action may also contribute to changes in source–sink relationships during drought stress, most notably through the effects on photosynthesis, mobilizing resources, and sink strength (Alonso-Ramírez et al. 2009b).

The recent discovery of the JAZ (JASMONATE-ZIM DOMAIN) family proteins, acting as JA co-receptors and transcriptional repressors in JA signaling, has suggested that JAZ proteins facilitate JA-mediated cross talk with auxins, ethylene, SA, and interestingly may be antagonistic or synergistic with GA depending on which plant function is of focus (Kazan and Manners 2012). For example, GA and JA are antagonistic with respect to plant growth and defense but synergistic in that both are required for jasmonate- and GA-mediated stamen development and male fertility (Cheng et al. 2009; Pauwels et al. 2009; Navarro et al. 2008). JAZ proteins facilitate synergism between JA and ethylene supporting plant defense functions and antagonism between jasmonates and SA or auxins by which SA- or auxin-mediated signaling is regulated by jasmonates (Broekaert et al. 2006; Leon-Reyes et al. 2010; Sun et al. 2009). Observations of similar developmental changes responding to distinct abiotic stress signals suggests that redundant signaling intermediates, such as DELLA proteins (negative regulators of GA signaling), facilitate cross talk between different phytohormones (Kohli et al. 2013). For example, JA was shown to interfere with DELLA–PIF3/4 interactions and inhibit GA-mediated hypocotyl elongation (Lyons et al. 2013). DELLAs have also been implicated in orchestrating GA and ABA signaling cross talk controlling *Arabidopsis* seed germination and seedling development under oxidative stress conditions (Yuan et al. 2011). During drought stress, increased ABA and ethylene concentrations exert antagonism on GA signaling and GA-mediated growth and this cross talk occurs by means of DELLA proteins, though GA interacting with other plant hormones such as SA also contributes to changes in growth under drought stress (Kohli et al. 2013; Wolters and Jürgens 2009). As described previously, ethylene exerts strict control upon drought-induced leaf senescence by controlling gene expression of EIN transcription factors and, more specifically, the EIN2 transcription factor has been shown to be similarly regulated by ABA and MeJA, suggesting a means for cross talk between the three hormones controlling downstream expression of stress-responsive genes (Kim et al. 2011). Elucidating on how specific points in ethylene pathway interact with other plant hormones and whether similar mechanisms are involved across different interactions to confer the plant drought responses at the physiological level, such as stomatal movement and growth processes, continues to be a primary focus of researchers (Vandenbussche and Van Der Straeten 2007).

3.4.2 Hormone to Sugar Interactions

Hormone regulation of plant growth and responses to drought stress not only involve interactions among hormones, but also interaction with other metabolites, such as sugars, as found in recent research. The ongoing and extensive research into cross talk between multiple hormone classes also suggests that sugars may exert influence upon biosynthesis or response pathways of other plant hormones, such as those associated with auxin or CK signaling. Sugars have been recognized to serve integral signaling functions modulating a range of growth processes throughout the plant life cycle and, in an attempt to understand why various genes respond to specific sugars or sugar phosphorylations, it was noted that *Arabidopsis* mutants with altered sugar responses displayed phenotypes similar to plant-hormone biosynthesis or signaling mutants suggesting the existence of links between sugar- and hormone-signaling pathways (Gibson 2005; Hanson and Smeekens 2009; León and Sheen 2003; Pinheiro and Chaves 2011). The initial comparable mutant screens coinciding with subsequent genetic and functional analyses suggested that there is extensive overlap between sugar, ABA, and ethylene signaling preempting various downstream plant processes such as root development (Eveland and Jackson 2012). For example, mutants lacking genes encoding for ABA biosynthesis or sensitivity (*aba* or *abi*, respectively) are similarly insensitive to high concentrations of glucose and the potential link facilitating sugars and ABA-perception cross talk might be *ABI4*, which encodes an AP2 transcription factor required for normal sugar response (Arenas-Huertero et al. 2000). Co-expression of a sucrose synthase gene and *ABI3* occurred under stress conditions, as did *ABI1* with one neutral invertase, two sucrose synthases, and one β-amylase, all of which may serve to amplify the signaling capacity and phenotypic responses (i.e., stomatal closure) under drought stress (Pinheiro and Chaves 2011). Alternatively, mutants lacking genes encoding for ethylene perception (*etr1*, *ein2*, *ein3*) are hypersensitive to glucose while a mutant with negative regulation of ethylene signaling (*ctr1*) is insensitive to glucose and the antagonistic relationship between ethylene and glucose may similarly be mediated through repression of ABA biosynthetic genes (Ghassemian et al. 2000; Yanagisawa et al. 2003). Studies conducted on ABA and ethylene mutants, *Arabidopsis* mutants (*hxk*) unable to catalyze glucose phosphorylation were resistant to exogenous auxin, insensitive to high glucose concentrations, and the *hxk*-based signaling negatively interacted with CKs (Moore et al. 2003). Tobacco transgenic lines with reduced levels of ASR (ABA-stress-ripening) protein displayed limited glucose metabolism and altered ABA and GA levels with downstream effects on leaf senescence, suggesting that *Asr* may be a central signaling component between glucose, ABA, and GA (Dominguez et al. 2013). Despite the identification of several novel genes possibly serving as links between sugars and auxin, the overall complexity of cross talk between hormones, sugars, and interacting secondary metabolites establishes the need for more in-depth genomic studies to show how gene expression levels change across thousands of genes, of which distinct changes can be associated with sugar or hormone signaling under different abiotic stress conditions (Eveland and Jackson

2012; Kissoudis et al. 2014). Mishra et al. (2009) performed genome-wide expression profiling of *Arabidopsis* seedlings which showed that over two-thirds of genes affected by auxin were regulated by glucose and that glucose and auxin establish either antagonistic or synergistic mechanisms to regulate transcription. Furthermore, the auxin-deficient mutants receiving exogenous glucose displayed phenotypes indicative of various defects in root development, suggesting that glucose contributes to proper root development by means of auxin-based signaling functions.

3.5 Concluding Remarks

There has been increasing evidence supporting the critical roles of various plant hormones involved in regulating plant growth and physiological responses to drought stress in the last decade, although this has been a research area that have been studied for many decades. Among various drought responses, leaf senescence, antioxidant metabolism, carbon metabolism, and stomatal movements are directly impacted or indirectly mediated by a particular or multiple hormones. Physiological and metabolic regulation of drought responses by hormones are well known as demonstrated by extensive research in related areas discussed throughout the chapter. Various transcription factors and downstream genes controlling hormone synthesis, degradation, and responses or sensitivities have been identified through transcriptomic analysis and confirmation through genetic transformation of overexpressing and silencing or mutating specific genes. Through the analysis of transcription factors, recent research is beginning to unravel signaling pathways of a single or multiple hormones and interactions among hormones and between hormones and other metabolites such as sugars, which coordinately mediate drought responses. However, the events or molecules in the perception of a specific hormone initiating specific signal transduction pathways are not completely understood. Furthermore, cross talk signals derived from hormone to hormone or hormone to sugars or secondary messenger (i.e., calcium or ROS) are not yet clear. Further research addressing those critical questions regarding hormone-signaling perception and cross talk among hormones and other metabolites will provide further insights into molecular factors controlling hormone regulation of plant tolerance to drought stress.

References

Abrams MD. Genotypic and phenotypic variation as stress adaptations in temperate tree species: a review of several case studies. Tree Physiol. 1994;14(7):833–42.

Acharya BR, Assmann SM. Hormone interactions in stomatal function. Plant Mol Biol. 2009;69(4):451–62.

Addicott FT, Lynch RS. Acceleration and retardation of abscission by indoleacetic acid. Science. 1951;114(2974):688–9.

Agarwal S, Sairam RK, Srivastava GC, Meena RC. Changes in antioxidant enzymes activity and oxidative stress by abscisic acid and salicylic acid in wheat genotypes. Biol. Plant. 2005;49(4):541–50.

Akter N, Islam MR, Karim MA, Hossain T. Alleviation of drought stress in maize by exogenous application of gibberellic acid and cytokinin. J Crop Sci Biotech. 2014;17(1):41–8.

Alonso-Ramírez A, Rodríguez D, Reyes D, Jiménez JA, Nicolás G, López-Climent M, Gómez-Cadenas A, Nicolás C. Evidence for a role of gibberellins in salicylic acid-modulated early plant responses to abiotic stress in Arabidopsis seeds. Plant Physiol. 2009a;150(3):1335–44.

Alonso-Ramírez A, Rodríguez D, Reyes D, Jiménez JA, Nicolás G, López-Climent M, Gómez-Cadenas A, Nicolás C. Cross-talk between gibberellins and salicylic acid in early stress responses in Arabidopsis thaliana seeds. Plant Signal Behav. 2009b;4(8):750–1.

Anjum SA, Wang L, Farooq M, Khan I, Xue L. Methyl jasmonate-induced alteration in lipid peroxidation, antioxidative defence system and yield in soybean under drought. J Agron Crop Sci. 2011;197(4):296–301.

Apel K, Hirt H. Reactive oxygen species: metabolism, oxidative stress, and signal transduction. Annu Rev Plant Biol. 2004;55:373–99.

Arenas-Huertero F, Arroyo A, Zhou L, Sheen J, Leon P. Analysis of Arabidopsis glucose insensitive mutants, gin5 and gin6, reveals a central role of the plant hormone ABA in the regulation of plant vegetative development by sugar. Genes Dev. 2000;14(16):2085–96.

Atkinson NJ, Urwin PE. The interaction of plant biotic and abiotic stresses: from genes to the field. J Exp Bot. 2012;63(10):3523–43.

Balbi V, Devoto A. Jasmonate signalling network in *Arabidopsis thaliana*: crucial regulatory nodes and new physiological scenarios. New Phytol. 2008;177(2):301–18.

Bartoli CG, Gomez F, Gergoff G, Guiamét JJ, Puntarulo S. Up-regulation of the mitochondrial alternative oxidase pathway enhances photosynthetic electron transport under drought conditions. J Exp Bot. 2005;56(415):1269–76.

Blokhina O, Virolainen E, Fagerstedt KV. Antioxidants, oxidative damage and oxygen deprivation stress: a review. Ann Bot. 2003;91(2):179–94.

Bota J, Medrano H, Flexas J. Is photosynthesis limited by decreased Rubisco activity and RuBP content under progressive water stress? New Phytol. 2004;162(3):671–81.

Bowler C, Fluhr R. The role of calcium and activated oxygens as signals for controlling cross-tolerance. Trends Plant Sci. 2000;5(6):241–6.

Bréda N, Huc R, Granier A, Dreyer E. Temperate forest trees and stands under severe drought: a review of ecophysiological responses, adaptation processes and long-term consequences. Ann Forest Sci. 2006;63(6):625–44.

Broekaert WF, Delauré SL, De-Bolle MF, Cammue BP. The role of ethylene in host-pathogen interactions. Annu Rev Phytopathol. 2006;44:393–416.

Brossa R, López-Carbonell M, Jubany-Marí T, Alegre L. Interplay between abscisic acid and jasmonic acid and its role in water-oxidative stress in wild-type, ABA-deficient, JA-deficient, and ascorbate-deficient Arabidopsis plants. J Plant Growth Regul. 2011;30(3):322–33.

Bu Q, Jiang H, Li C-B, Zhai Q, Zhang J, Wu X, Sun J, Xie Q, Li C. Role of the *Arabidopsis thaliana* NAC transcription factors ANAC019 and ANAC055 in regulating jasmonic acid-signaled defense responses. Cell Res. 2008;18(7):756–67.

Buchanan-Wollaston V. The molecular biology of leaf senescence. J Exp Bot. 1997;48(2):181–99.

Buchanan-Wollaston V, Earl S, Harrison E, Mathas E, Navabpour S, Page T, Pink D. The molecular analysis of leaf senescence: a genomics approach. Plant Biotech J. 2003;1(1):3–22.

Casimiro I, Beeckman T, Graham N, Bhalerao R, Zhang H, Casero P, Sandberg G, Bennett MJ. Dissecting Arabidopsis lateral root development. Trends Plant Sci. 2003;8(4):165–71.

Cattivelli L, Rizza F, Badeck FW, Mazzucotelli E, Mastrangelo AM, Francia E, Maré C, Tondelli A, Stanca AM. Drought tolerance improvement in crop plants: an integrated view from breeding to genomics. Field Crops Res. 2008;105(1):1–14.

Chaves MM, Pereira JS, Maroco J, Rodrigues ML, Ricardo CPP, Osório ML, Carvalho I, Faria T, Pinheiro C. How plants cope with water stress in the field? Photosynthesis and growth. Ann Bot. 2002;89(7):907–16.

Chaves MM, Maroco JP, Pereira JS. Understanding plant responses to drought-from genes to the whole plant. Func Plant Biol. 2003;30(3):239–64.
Chaves MM, Flexas J, Pinheiro C. Photosynthesis under drought and salt stress: regulation mechanisms from whole plant to cell. Ann Bot. 2009;103(4):551–60.
Chen H, Lai Z, Shi J, Xiao Y, Chen Z, Xu X. Roles of Arabidopsis WRKY18, WRKY40 and WRKY60 transcription factors in plant responses to abscisic acid and abiotic stress. BMC Plant Biol. 2010;10(1):281–96.
Cheng H, Song S, Xiao L, Soo HM, Cheng Z, Xie D, Peng J. Gibberellin acts through jasmonate to control the expression of MYB21, MYB24, and MYB57 to promote stamen filament growth in Arabidopsis. PLoS Genet. 2009;5(3):e1000440.
Condit R, Hubbell SP, Foster RB. Mortality rates of 205 neotropical tree and shrub species and the impact of a severe drought. Ecol Mono. 1995;65(4):419–39.
Cramer GR, Urano K, Delrot S, Pezzotti M, Shinozaki K. Effects of abiotic stress on plants: a systems biology perspective. BMC Plant Biol. 2011;11(1):163–77.
Csiszár J, Szabó M, Erdei L, Márton L, Horváth F, Tari I. Auxin autotrophic tobacco callus tissues resist oxidative stress: the importance of glutathione S-transferase and glutathione peroxidase activities in auxin heterotrophic and autotrophic calli. J Plant Physiol. 2004;161(6):691–9.
Cutler SR, Rodriguez PL, Finkelstein RR, Abrams SR. Abscisic acid: emergence of a core signaling network. Annu Rev Plant Biol. 2010;61:651–79.
Daszkowska-Golec A, Szarejko I. Open or close the gate-stomata action under the control of phytohormones in drought stress conditions. Front Plant Sci. 2013;4(138):1–16.
Depuydt S, Hardtke CS. Hormone signalling cross talk in plant growth regulation. Curr Biol. 2011;21(9):R365–73.
Dodd IC. Hormonal interactions and stomatal responses. J Plant Growth Regul. 2003;22(1):32–46.
Dominguez PG, Frankel N, Mazuch J, Balbo I, Iusem N, Fernie AR, Carrari F. ASR1 mediates glucose-hormone cross talk by affecting sugar trafficking in tobacco plants. Plant Physiol. 2013;161(3):1486–500.
Ellis CM, Nagpal P, Young JC, Hagen G, Guilfoyle TJ, Reed JW. AUXIN RESPONSE FACTOR1 and AUXIN RESPONSE FACTOR2 regulate senescence and floral organ abscission in Arabidopsis thaliana. Development. 2005;132(20):4563–74.
Evans NH. Modulation of guard cell plasma membrane potassium currents by methyl jasmonate. Plant Physiol. 2003;131(1):8–11.
Eveland AL, Jackson DP. Sugars, signalling, and plant development. J Exp Bot. 2012;63(9):3367–77.
Farooq M, Wahid A, Kobayashi N, Fujita D, Basra SMA (2009) Plant drought stress: effects, mechanisms and management. In Sustainable Agriculture, Springer, The Netherlands. p. 153–188.
Fleet CM, Sun TP. A DELLAcate balance: the role of gibberellin in plant morphogenesis. Curr Opin Plant Biol. 2005;8(1):77–85.
Flexas J, Medrano H. Drought-inhibition of photosynthesis in C3 plants: stomatal and non-stomatal limitations revisited. Ann Bot. 2002;89(2):183–9.
Fujii H, Chinnusamy V, Rodrigues A, Rubio S, Antoni R, Park S-Y, Cutler SR, Sheen J, Rodriguez PL, Zhu J-K. In vitro reconstitution of an abscisic acid signalling pathway. Nature. 2009;462(7273):660–4.
Fujita M, Fujita Y, Noutoshi Y, Takahashi F, Narusaka Y, Yamaguchi-Shinozaki K, Shinozaki K. Crosstalk between abiotic and biotic stress responses: a current view from the points of convergence in the stress signaling networks. Curr Opin Plant Biol. 2006;9(4):436–42.
Gao XP, Wang XF, Lu YF, Zhang LY, Shen YY, Liang Z, Zhang DP. Jasmonic acid is involved in the water-stress-induced betaine accumulation in pear leaves. Plant Cell Environ. 2004;27(4):497–507.
Gao X, Yuan HM, Hu YQ, Li J, Lu YT. Mutation of Arabidopsis CATALASE2 results in hyponastic leaves by changes of auxin levels. Plant Cell Environ. 2014;37(1):175–88.
Ghassemian M, Nambara E, Cutler S, Kawaide H, Kamiya Y, McCourt P. Regulation of abscisic acid signaling by the ethylene response pathway in Arabidopsis. Plant Cell. 2000;12(7):1117–26.
Gibson SI. Control of plant development and gene expression by sugar signaling. Curr Opin Plant Biol. 2005;8(1):93–102.

Graham PH, Vance CP. Legumes: importance and constraints to greater use. Plant Physiol. 2003;131(3):872–7.

Guo H, Ecker JR. The ethylene signaling pathway: new insights. Curr Opin Plant Biol. 2004;7(1):40–9.

Ha S, Vankova R, Yamaguchi-Shinozaki K, Shinozaki K, Tran LSP. Cytokinins: metabolism and function in plant adaptation to environmental stresses. Trends Plant Sci. 2012;17(3):172–9.

Hacke UG, Sperry JS, Pittermann J. Drought experience and cavitation resistance in six shrubs from the Great Basin, Utah. Basic Appl Ecol. 2000;1(1):31–41.

Hadiarto T, Tran LSP. Progress studies of drought-responsive genes in rice. Plant Cell Rep. 2011;30(3):297–310.

Hanjra MA, Qureshi ME. Global water crisis and future food security in an era of climate change. Food Policy. 2010;35(5):365–77.

Hanson J, Smeekens S. Sugar perception and signaling-an update. Curr Opin Plant Biol. 2009;12(5):562–7.

Harb A, Krishnan A, Ambavaram MM, Pereira A. Molecular and physiological analysis of drought stress in Arabidopsis reveals early responses leading to acclimation in plant growth. Plant Physiol. 2010;154(3):1254–71.

Havlova M, Dobrev PI, Motyka VAC, Štorchová H, Libus J, Dobra J, Malbeck J, Gaudinova A, Vankova R. The role of cytokinins in responses to water deficit in tobacco plants over-expressing trans-zeatin O-glucosyltransferase gene under 35S or SAG12 promoters. Plant Cell Environ. 2008;31(3):341–53.

Hayat S, Hasan SA, Fariduddin Q, Ahmad A. Growth of tomato (*Lycopersicon esculentum*) in response to salicylic acid under water stress. J Plant Interact. 2008;3(4):297–304.

Hayat Q, Hayat S, Irfan M, Ahmad A. Effect of exogenous salicylic acid under changing environment: a review. Environ Exp Bot. 2010;68(1):14–25.

Himmelbach A, Yang Y, Grill E. Relay and control of abscisic acid signaling. Curr Opin Plant Biol. 2003;6(5):470–9.

Hirayama T, Shinozaki K. Perception and transduction of abscisic acid signals: keys to the function of the versatile plant hormone ABA. Trends Plant Sci. 2007;12(8):343–51.

Hörtensteiner S, Feller U. Nitrogen metabolism and remobilization during senescence. J Exp Bot. 2002;53(370):927–37.

Horváth E, Szalai G, Janda T. Induction of abiotic stress tolerance by salicylic acid signaling. J Plant Growth Regul. 2007;26(3):290–300.

Hossain MA, Munemasa S, Uraji M, Nakamura Y, Mori IC, Murata Y. Involvement of endogenous abscisic acid in methyl jasmonate-induced stomatal closure in Arabidopsis. Plant Physiol. 2011;156(1):430–8.

Huang B. Recent advances in drought and heat stress physiology of turfgrass: a review. Int Conf Turfgrass Manage Sci Sports Fields. 2003;661:185–92.

Huang D, Wu W, Abrams SR, Cutler AJ. The relationship of drought-related gene expression in Arabidopsis thaliana to hormonal and environmental factors. J Exp Bot. 2008;59(11):2991–3007.

Huang B, DaCosta M, Jiang Y. Research advances in mechanisms of turfgrass tolerance to abiotic stresses: from physiology to molecular biology. Crit Rev Plant Sci. 2014;33(2-3):141–89.

Hubbard KE, Nishimura N, Hitomi K, Getzoff ED, Schroeder JI. Early abscisic acid signal transduction mechanisms: newly discovered components and newly emerging questions. Genes Dev. 2010;24(16):1695–708.

Hung KT, Kao CH. Nitric oxide counteracts the senescence of rice leaves induced by abscisic acid. J Plant Physiol. 2003;160(8):871–9.

Ikegami K, Okamoto M, Seo M, Koshiba T. Activation of abscisic acid biosynthesis in the leaves of Arabidopsis thaliana in response to water deficit. J Plant Res. 2009;122(2):235–43.

Imamura A, Hanaki N, Umeda H, Nakamura A, Suzuki T, Ueguchi C, Mizuno T. Response regulators implicated in His-to-Asp phosphotransfer signaling in Arabidopsis. Proc Natl Acad Sci U S A. 1998;95(5):2691–6.

Jaleel CA, Gopi R, Manivannan P, Sankar B, Kishorekumar A, Panneerselvam R. Antioxidant potentials and ajmalicine accumulation in Catharanthus roseus after treatment with gibberellic acid. Colloids Surf B Biointerfaces. 2007;60(2):195–200.

Jaleel CA, Manivannan P, Wahid A, Farooq M, Al-Juburi HJ, Somasundaram R, Panneerselvam R. Drought stress plants: a review on morphological characteristics and pigments composition. Int J Agric Biol. 2009;11:100–5.

Jiang M, Zhang J. Effect of abscisic acid on active oxygen species, antioxidative defence system and oxidative damage in leaves of maize seedlings. Plant Cell Physiol. 2001;42(11):1265–73.

Jiang M, Zhang J. Water stress-induced abscisic acid accumulation triggers the increased generation of reactive oxygen species and up-regulates the activities of antioxidant enzymes in maize leaves. J Exp Bot. 2002;53(379):2401–10.

Jiang Y, Liang G, Yu D. Activated expression of WRKY57 confers drought tolerance in Arabidopsis. Mol Plant. 2012;5(6):1375–88.

Jibran R, Hunter DA, Dijkwel PP. Hormonal regulation of leaf senescence through integration of developmental and stress signals. Plant Mol Biol. 2013;82(6):547–61.

Jo Y, Hyun TK. Genome-wide identification of antioxidant component biosynthetic enzymes: comprehensive analysis of ascorbic acid and tocochromanols biosynthetic genes in rice. Comput Biol Chem. 2011;35(5):261–8.

Jones B, Gunnerås SA, Petersson SV, Tarkowski P, Graham N, May S, Dolezal K, Sandberg G, Ljung K. Cytokinin regulation of auxin synthesis in Arabidopsis involves a homeostatic feedback loop regulated via auxin and cytokinin signal transduction. Plant Cell. 2010;22(9):2956–69.

Joo JH, Bae YS, Lee JS. Role of auxin-induced reactive oxygen species in root gravitropism. Plant Physiol. 2001;126(3):1055–60.

Joshi-Saha A, Valon C, Leung J. A brand new START: abscisic acid perception and transduction in the guard cell. Sci Signal. 2011;4:1–13.

Kaiser H, Legner N. Localization of mechanisms involved in hydropassive and hydroactive stomatal responses of Sambucus nigra to dry air. Plant Physiol. 2007;143(2):1068–77.

Kang Y, Shahbaz K, Xiaoyi M. Climate change impacts on crop yield, crop water productivity and food security: a review. Proc Natl Acad Sci U S A. 2009;19(12):1665–74.

Kang G, Li G, Xu W, Peng X, Han Q, Zhu Y, Guo T. Proteomics reveals the effects of salicylic acid on growth and tolerance to subsequent drought stress in wheat. J Proteome Res. 2012;11(12):6066–79.

Kang GZ, Li GZ, Liu GQ, Xu W, Peng XQ, Wang CY, Zhu YJ, Guo TC. Exogenous salicylic acid enhances wheat drought tolerance by influence on the expression of genes related to ascorbate-glutathione cycle. Biologia Planta. 2013;57(4):718–24.

Kazan K, Manners JM. JAZ repressors and the orchestration of phytohormone cross talk. Trends Plant Sci. 2012;17(1):22–31.

Kim JH, Chung KM, Woo HR. Three positive regulators of leaf senescence in Arabidopsis, ORE1, ORE3 and ORE9, play roles in cross talk among multiple hormone-mediated senescence pathways. Genes Genom. 2011;33(4):373–81.

Kissoudis C, van de Wiel C, Visser RG, van der Linden G. Enhancing crop resilience to combined abiotic and biotic stress through the dissection of physiological and molecular cross talk. Front Plant Sci. 2014;5(207):125. doi:10.3389/fpls.2014.00207.

Kohli A, Sreenivasulu N, Lakshmanan P, Kumar PP. The phytohormone cross talk paradigm takes center stage in understanding how plants respond to abiotic stresses. Plant Cell Rep. 2013;32(7):945–57.

Koprivova A, Mugford ST, Kopriva S. Arabidopsis root growth dependence on glutathione is linked to auxin transport. Plant Cell Rep. 2010;29(10):1157–67.

Krouk G, Ruffel S, Gutierrez RA, Gojon A, Crawford NM, Coruzzi GM, Lacombe B. A framework integrating plant growth with hormones and nutrients. Trends Plant Sci. 2011;16(4):178–82.

Kumari GJ, Reddy AM, Naik ST, Kumar SG, Prasanthi J, Sriranganayakulu G, Reddy PC, Sudhakar C. Jasmonic acid induced changes in protein pattern, antioxidative enzyme activities and peroxidase isozymes in peanut seedlings. Biologia Planta. 2006;50(2):219–26.

Lawlor DW. Limitation to photosynthesis in water-stressed leaves: stomata vs. metabolism and the role of ATP. Ann Bot. 2002;89(7):871–85.

Lee SC, Kim SH, An SH, Yi SY, Hwang BK. Identification and functional expression of the pepper pathogen-induced gene, CAPIP2, involved in disease resistance and drought and salt stress tolerance. Plant Mol Biol. 2006;62(1-2):151–64.

León P, Sheen J. Sugar and hormone connections. Trends Plant Sci. 2003;8(3):110–6.
Leon-Reyes A, van der Does D, De-Lange ES, Delker C, Wasternack C, van Wees SCM, Ritsema T, Pieterse CMJ. Salicylate-mediated suppression of jasmonate-responsive gene expression in Arabidopsis is targeted downstream of the jasmonate biosynthesis pathway. Planta. 2010; 232(6):1423–32.
Li J, Yang H, Peer WA, Richter G, Blakeslee J, Bandyopadhyay A, Titapiwantakun B, Undurraga S, Khodakovskaya M, Richards EL, Krizek B, Murphy AS, Gilroy S, Gaxiola R. Arabidopsis H^+-PPase AVP1 regulates auxin-mediated organ development. Science. 2005;310(5745):121–5.
Lim PO, Woo HR, Nam HG. Molecular genetics of leaf senescence in Arabidopsis. Trends Plant Sci. 2003;8(6):272–8.
Lim PO, Kim HJ, Nam HG. Leaf senescence. Annu Rev Plant Biol. 2007;58:115–36.
Lim PO, Lee IC, Kim J, Kim HJ, Ryu JS, Woo HR, Nam HG. Auxin response factor 2 (ARF2) plays a major role in regulating auxin-mediated leaf longevity. J Exp Bot. 2010. doi:10.1093/jxb/erq010.
Liu X, Shi W, Zhang S, Lou C. Nitric oxide involved in signal transduction of Jasmonic acid-induced stomatal closure of Vicia faba L. Chinese Sci Bull. 2005;50(6):520–5.
Ljung K, Bhalerao RP, Sandberg G. Sites and homeostatic control of auxin biosynthesis in Arabidopsis during vegetative growth. Plant J. 2001;28(4):465–74.
Ljung K, Hull AK, Celenza J, Yamada M, Estelle M, Normanly J, Sandberg G. Sites and regulation of auxin biosynthesis in Arabidopsis roots. Plant Cell. 2005;17(4):1090–104.
Ludwig AA, Saitoh H, Felix G, Freymark G, Miersch O, Wasternack C, Boller T, Jones JDG, Romeis T. Ethylene-mediated cross-talk between calcium-dependent protein kinase and MAPK signaling controls stress responses in plants. Proc Nat Acad Sci U S A. 2005;102(30):10736–41.
Lyons R, Manners JM, Kazan K. Jasmonate biosynthesis and signaling in monocots: a comparative overview. Plant Cell Rep. 2013;32(6):815–27.
Mahajan S, Tuteja N. Cold, salinity and drought stresses: an overview. Arch Biochem Biophys. 2005;444(2):139–58.
Malamy JE. Intrinsic and environmental response pathways that regulate root system architecture. Plant Cell Environ. 2005;28(1):67–77.
Mateo A, Funck D, Mühlenbock P, Kular B, Mullineaux PM, Karpinski S. Controlled levels of salicylic acid are required for optimal photosynthesis and redox homeostasis. J Exp Bot. 2006;57(8):1795–807.
Melcher K, Zhou XE, Xu HE. Thirsty plants and beyond: structural mechanisms of abscisic acid perception and signaling. Curr Opin Struct Biol. 2010;20(6):722–9.
Merewitz EB, Gianfagna T, Huang B. Effects of SAG12-ipt and HSP18. 2-ipt expression on cytokinin production, root growth, and leaf senescence in creeping bentgrass exposed to drought stress. J Amer Soc Hort Sci. 2010;135(3):230–9.
Merewitz EB, Gianfagna T, Huang B. Protein accumulation in leaves and roots associated with improved drought tolerance in creeping bentgrass expressing an ipt gene for cytokinin synthesis. J Exp Bot. 2011. doi:10.1093/jxb/err166.
Miller G, Suzuki N, Ciftci-Yilmaz S, Mittler R. Reactive oxygen species homeostasis and signalling during drought and salinity stresses. Plant Cell Environ. 2010;33(4):453–67.
Mishra BS, Singh M, Aggrawal P, Laxmi A. Glucose and auxin signaling interaction in controlling Arabidopsis thaliana seedlings root growth and development. PLoS One. 2009;4(2), e4502. doi:10.1371/journal.pone.0004502.
Mittler R. Oxidative stress, antioxidants and stress tolerance. Trends Plant Sci. 2002;7(9):405–10.
Miura K, Okamoto H, Okuma E, Shiba H, Kamada H, Hasegawa PM, Murata Y. SIZ1 deficiency causes reduced stomatal aperture and enhanced drought tolerance via controlling salicylic acid-induced accumulation of reactive oxygen species in Arabidopsis. Plant J. 2013;73(1):91–104.
Moore B, Zhou L, Rolland F, Hall Q, Cheng W-H, Liu Y-X, Hwang I, Jones T, Sheen J. Role of the Arabidopsis glucose sensor HXK1 in nutrient, light, and hormonal signaling. Science. 2003;300(5617):332–6.
Muller B, Pantin F, Génard M, Turc O, Freixes S, Piques M, Gibon Y. Water deficits uncouple growth from photosynthesis, increase C content, and modify the relationships between C and growth in sink organs. J Exp Bot. 2011;62(6):1715–29.

Munemasa S, Oda K, Watanabe-Sugimoto M, Nakamura Y, Shimoishi Y, Murata Y. The coronatine-insensitive 1 mutation reveals the hormonal signaling interaction between abscisic acid and methyl jasmonate in Arabidopsis guard cells. Specific impairment of ion channel activation and second messenger production. Plant Physiol. 2007;143(3):1398–407.

Munné-Bosch S, Alegre L. Changes in carotenoids, tocopherols and diterpenes during drought and recovery, and the biological significance of chlorophyll loss in Rosmarinus officinalis plants. Planta. 2000;210(6):925–31.

Munné-Bosch S, Alegre L. Die and let live: leaf senescence contributes to plant survival under drought stress. Funct Plant Biol. 2004;31(3):203–16.

Murata Y, Mori IC. Stomatal regulation of plant water status. In: Jenks MA, Hasegawa PM, editors. Plant abiotic stress. 2nd ed. Ames, IO: John Wiley and Sons; 2014. p. 47–67.

Nakagami H, Pitzschke A, Hirt H. Emerging MAP kinase pathways in plant stress signalling. Trends Plant Sci. 2005;10(7):339–46.

Nakashima K, Ito Y, Yamaguchi-Shinozaki K. Transcriptional regulatory networks in response to abiotic stresses in Arabidopsis and grasses. Plant Physiol. 2009;149(1):88–95.

Navarro L, Bari R, Achard P, Lisón P, Nemri A, Harberd NP, Jones JD. DELLAs control plant immune responses by modulating the balance of jasmonic acid and salicylic acid signaling. Curr Biol. 2008;18(9):650–5.

Nemhauser JL, Hong F, Chory J. Different plant hormones regulate similar processes through largely nonoverlapping transcriptional responses. Cell. 2006;126(3):467–75.

Nilson SE, Assmann SM. The control of transpiration. Insights from Arabidopsis. Plant Physiol. 2007;143(1):19–27.

Nishiyama R, Watanabe Y, Fujita Y, Le DT, Kojima M, Werner T, Vankova R, Yamaguchi-Shinozaki K, Shinozaki K, Kakimoto T, Sakakibara H, Schmülling T, Tran L-SP. Analysis of cytokinin mutants and regulation of cytokinin metabolic genes reveals important regulatory roles of cytokinins in drought, salt and abscisic acid responses, and abscisic acid biosynthesis. Plant Cell. 2011;23(6):2169–83.

Olszewski N, Sun TP, Gubler F. Gibberellin signaling biosynthesis, catabolism, and response pathways. Plant Cell. 2002;14(1):61–80.

Pandey S, Gupta K, Mukherjee AK. Impact of cadmium and lead on Catharanthus roseus-A phytoremediation study. J Environ Biol. 2007;28(3):655–62.

Park JM, Park C-J, Lee S-B, Ham B-K, Shin R, Paek K-H. Overexpression of the tobacco Tsi1 gene encoding an EREBP/AP2–type transcription factor enhances resistance against pathogen attack and osmotic stress in tobacco. Plant Cell. 2001;13(5):1035–46.

Park S, Li J, Pittman JK, Berkowitz GA, Yang H, Undurraga S, Morris J, Hirschi KD, Gaxiola RA. Up-regulation of a H^+-pyrophosphatase (H^+-PPase) as a strategy to engineer drought-resistant crop plants. Proc Nat Acad Sci U S A. 2005;102(52):18830–5.

Pauwels L, Inzé D, Goossens A. Jasmonate-inducible gene: what does it mean? Trends Plant Sci. 2009;14(2):87–91.

Peleg Z, Blumwald E. Hormone balance and abiotic stress tolerance in crop plants. Curr Opin Plant Biol. 2011;14(3):290–5.

Pinheiro C, Chaves MM. Photosynthesis and drought: can we make metabolic connections from available data? J Exp Bot. 2011;62(3):869–82.

Poorter H, Nagel O. The role of biomass allocation in the growth response of plants to different levels of light, CO2, nutrients and water: a quantitative review. Aust J Plant Physiol. 2000;27:595–607.

Pospíšilová J. Participation of phytohormones in the stomatal regulation of gas exchange during water stress. Biologia Planta. 2003;46(4):491–506.

Prochazkova D, Sairam RK, Srivastava GC, Singh DV. Oxidative stress and antioxidant activity as the basis of senescence in maize leaves. Plant Sci. 2001;161(4):765–71.

Puranik S, Sahu PP, Srivastava PS, Prasad M. NAC proteins: regulation and role in stress tolerance. Trends Plant Sci. 2012;17(6):369–81.

Qiu Y, Yu D. Over-expression of the stress-induced OsWRKY45 enhances disease resistance and drought tolerance in Arabidopsis. Environ Exp Bot. 2009;65(1):35–47.

Raghavendra AS, Gonugunta VK, Christmann A, Grill E. ABA perception and signalling. Trends Plant Sci. 2010;15(7):395–401.

Ramamoorthy R, Jiang SY, Kumar N, Venkatesh PN, Ramachandran S. A comprehensive transcriptional profiling of the WRKY gene family in rice under various abiotic and phytohormone treatments. Plant Cell Physiol. 2008;49(6):865–79.

Ranwala AP, Miller WB. Preventive mechanisms of gibberellin 4+7 and light on low-temperature-induced leaf senescence in Lilium cv. Stargazer. Postharvest Biol Tech. 2000;19(1):85–92.

Reddy AR, Kolluru VC, Munusamy V. Drought-induced responses of photosynthesis and antioxidant metabolism in higher plants. J Plant Physiol. 2004;161(11):1189–202.

Ren X, Zhizhong C, Yue L, Hairong Z, Min Z, Qian L, Xuhui H, Zhu J-K, Zhizhong G. ABO3, a WRKY transcription factor, mediates plant responses to abscisic acid and drought tolerance in Arabidopsis. Plant J. 2010;63(3):417–29.

Ribeiro DM, Araújo WL, Fernie AR, Schippers JH, Mueller-Roeber B. Translatome and metabolome effects triggered by gibberellins during rosette growth in Arabidopsis. J Exp Bot. 2012. doi:10.1093/jxb/err463.

Richards DE, King KE, Ait-ali T, Harberd NP. How gibberellin regulates plant growth and development: a molecular genetic analysis of gibberellin signaling. Annu Rev Plant Physiol Plant Mol Biol. 2001;52(1):67–88.

Rivero RM, Kojima M, Gepstein A, Sakakibara H, Mittler R, Gepstein S, Blumwald E. Delayed leaf senescence induces extreme drought tolerance in a flowering plant. Proc Natl Acad Sci U S A. 2007;104(49):19631–6.

Robert-Seilaniantz A, Murray G, Jones JDG. Hormone cross talk in plant disease and defense: more than just jasmonate-salicylate antagonism. Annu Rev Plant Physiol Plant Mol Biol. 2011;49:317–43.

Rosenvasser S, Mayak S, Friedman H. Increase in reactive oxygen species (ROS) and in senescence-associated gene transcript (SAG) levels during dark-induced senescence of Pelargonium cuttings, and the effect of gibberellic acid. Plant Sci. 2006;170(4):873–9.

Rossato L, Laine P, Ourry A. Nitrogen storage and remobilization in *Brassica napus* L. during the growth cycle: nitrogen fluxes within the plant and changes in soluble protein patterns. J Exp Bot. 2001;52(361):1655–63.

Ruiz KB, Biondi S, Oses R, Acuña-Rodríguez IS, Antognoni F, Martinez-Mosqueira EA, Coulibaly A, Canahua-Murillo A, Pinto M, Zurita-Silva A, Bazile D, Jacobsen S-E, Molina-Montenegro MA. Quinoa biodiversity and sustainability for food security under climate change: a review. Agron Sustain Dev. 2014;34(2):349–59.

Saibo NJ, Vriezen WH, Beemster GT, van Der Straeten D. Growth and stomata development of Arabidopsis hypocotyls are controlled by gibberellins and modulated by ethylene and auxins. Plant J. 2003;33(6):989–1000.

Santino A, Taurino M, De-Domenico S, Bonsegna S, Poltronieri P, Pastor V, Flors V. Jasmonate signaling in plant development and defense response to multiple (a)biotic stresses. Plant Cell Rep. 2013;32(7):1085–98.

Schenk PM, Kazan K, Wilson I, Anderson JP, Richmond T, Somerville SC, Manners JM. Coordinated plant defense responses in Arabidopsis revealed by microarray analysis. Proc Natl Acad Sci U S A. 2000;97(21):11655–60.

Schippers JH, Jing HC, Hille J, Dijkwel PP. Developmental and hormonal control of leaf senescence. In: Gan S, editor. Senescence processes in plants. Oxford: Blackwell Publishing; 2007. p. 145–70.

Schroeder JI, Nambara E. A quick release mechanism for abscisic acid. Cell. 2006;126(6):1023–5.

Schwartz SH, Qin X, Zeevaart JA. Elucidation of the indirect pathway of abscisic acid biosynthesis by mutants, genes, and enzymes. Plant Physiol. 2003;131(4):1591–601.

Schwechheimer C, Willige BC. Shedding light on gibberellic acid signalling. Curr Opin Plant Biol. 2009;12(1):57–62.

Sedghi M, Sharifi RS, Pirzad AR, Amanpour-Balaneji B. Phytohormonal regulation of antioxidant systems in petals of drought stressed pot marigold (*Calendula officinalis* L.). J Agric Sci Tech. 2012;14(4):869–78.

Senaratna T, Touchell D, Bunn E, Dixon K. Acetyl salicylic acid (Aspirin) and salicylic acid induce multiple stress tolerance in bean and tomato plants. Plant Growth Regul. 2000;30(2):157–61.

Shan C, Liang Z. Jasmonic acid regulates ascorbate and glutathione metabolism in Agropyron cristatum leaves under water stress. Plant Sci. 2010;178(2):130–9.

Sharma R, De-Vleesschauwer D, Sharma MK, Ronald PC. Recent advances in dissecting stress-regulatory cross talk in rice. Mol Plant. 2013;6(2):250–60.

Sharp RE. Interaction with ethylene: changing views on the role of abscisic acid in root and shoot growth responses to water stress. Plant Cell Environ. 2002;25(2):211–22.

Sharp RE, LeNoble ME. ABA, ethylene and the control of shoot and root growth under water stress. J Exp Bot. 2002;53(366):33–7.

Shi H, Chen L, Ye T, Liu X, Ding K, Chan Z. Modulation of auxin content in Arabidopsis confers improved drought stress resistance. Plant Physiol Biochem. 2014;82:209–17.

Shoji K, Addicott FT, Swets WA. Auxin in relation to leaf blade abscission. Plant Physiol. 1951;26(1):189–91.

Singh B, Usha K. Salicylic acid induced physiological and biochemical changes in wheat seedlings under water stress. Plant Growth Regul. 2003;39(2):137–41.

Sirichandra C, Wasilewska A, Vlad F, Valon C, Leung J. The guard cell as a single-cell model towards understanding drought tolerance and abscisic acid action. J Exp Bot. 2009;60(5):1439–63.

Smékalová V, Doskočilová A, Komis G, Šamaj J. Cross talk between secondary messengers, hormones and MAPK modules during abiotic stress signalling in plants. Biotech Adv. 2014;32(1):2–11.

Smirnoff N. Tansley Review No. 52. The role of active oxygen in the response of plants to water deficit and desiccation. New Phytol. 1993;12:27–58.

Solomon S, Qin D, Manning M, Marquis M, Averyt K, Tignor M. IPCC: climate change 2007: the physical science basis. In: Miller HL, Chen Z, editors. Contribution of working groups I, II, and III to the fourth assessment report of the intergovernmental panel on climate change. Cambridge, UK: Cambridge University Press; 2007.

Song XG, She XP, He JM, Huang C, Song TS. Cytokinin-and auxin-induced stomatal opening involves a decrease in levels of hydrogen peroxide in guard cells of Vicia faba. Funct Plant Biol. 2006;33(6):573–83.

Sridha S, Wu K. Identification of AtHD2C as a novel regulator of abscisic acid responses in Arabidopsis. Plant J. 2006;46(1):124–33.

Stepanova AN, Alonso JM. Ethylene signaling and response: where different regulatory modules meet. Curr Opin Plant Biol. 2009;12(5):548–55.

Suhita D, Raghavendra AS, Kwak JM, Vavasseur A. Cytoplasmic alkalization precedes reactive oxygen species production during methyl jasmonate-and abscisic acid-induced stomatal closure. Plant Physiol. 2004;134(4):1536–45.

Sun TP, Gubler F. Molecular mechanism of gibberellin signaling in plants. Annu Rev Plant Biol. 2004;55:197–223.

Sun J, Xu Y, Ye S, Jiang H, Chen Q, Liu F, Zhou W, Chen R, Li X, Tietz O, Wu X, Cohen JD, Palme K, Li C. Arabidopsis ASA1 is important for jasmonate-mediated regulation of auxin biosynthesis and transport during lateral root formation. Plant Cell. 2009;21(5):1495–511.

Taiz L, Zeiger E. Plant physiology. 5th ed. Sunderland, MA: Sinauer Associates; 2010.

Tanaka Y, Sano T, Tamaoki M, Nakajima N, Kondo N, Hasezawa S. Cytokinin and auxin inhibit abscisic acid-induced stomatal closure by enhancing ethylene production in Arabidopsis. J Exp Bot. 2006;57(10):2259–66.

Tester M, Bacic A. Abiotic stress tolerance in grasses. From model plants to crop plants. Plant Physiol. 2005;137(3):791–3.

To JP, Kieber JJ. Cytokinin signaling: two-components and more. Trends Plant Sci. 2008;13(2):85–92.

Todaka D, Nakashima K, Maruyama K, Kidokoro S, Osakabe Y, Ito Y, Matsukura S, Fujita Y, Yoshiwara K, Ohme-Takagi M, Kojima M, Sakakibara H, Shinozaki K, Yamaguchi-Shinozaki K. Rice phytochrome-interacting factor-like protein OsPIL1 functions as a key regulator of internode elongation and induces a morphological response to drought stress. Proc Natl Acad Sci U S A. 2012;109(39):15947–52.

Tognetti VB, Mühlenbock PER, van Breusegem F. Stress homeostasis-the redox and auxin perspective. Plant Cell Environ. 2012;35(2):321–33.

Tran LSP, Urao T, Qin F, Maruyama K, Kakimoto T, Shinozaki K, Yamaguchi-Shinozaki K. Functional analysis of AHK1/ATHK1 and cytokinin receptor histidine kinases in response to abscisic acid, drought, and salt stress in Arabidopsis. Proc Natl Acad Sci U S A. 2007;104(51):20623–8.

Trewavas AJ, Jones HG. An assessment of the role of ABA in plant development. In: Davies WJ, Jones HG, editors. Abscisic acid: physiology and biochemistry. Oxford: Bio Scientific Publishers pp; 1991. p. 169–88.

Trujillo LE, Sotolongo M, Menendez C, Ochogavia ME, Coll Y, Hernandez I, Borras-Hidalgo O, Thomma BPHJ, Vera P, Hernandez L. SodERF3, a novel sugarcane ethylene responsive factor (ERF), enhances salt and drought tolerance when overexpressed in tobacco plants. Plant Cell Physiol. 2008;49(4):512–25.

van der Graaff E, Schwacke R, Schneider A, Desimone M, Flügge UI, Kunze R. Transcription analysis of Arabidopsis membrane transporters and hormone pathways during developmental and induced leaf senescence. Plant Physiol. 2006;141(2):776–92.

Vandenbussche F, Van Der Straeten D. One for all and all for one: cross-talk of multiple signals controlling the plant phenotype. J Plant Growth Regul. 2007;26(2):178–87.

Vanlerberghe GC. Alternative oxidase: a mitochondrial respiratory pathway to maintain metabolic and signaling homeostasis during abiotic and biotic stress in plants. Int J Mol Sci. 2013; 14(4):6805–47.

Vanstraelen M, Benková E. Hormonal interactions in the regulation of plant development. Ann Rev Cell Dev Biol. 2012;28:463–87.

Wan L, Zhang J, Zhang H, Zhang Z, Quan R, Zhou S, Huang R. Transcriptional activation of OsDERF1 in OsERF3 and OsAP2-39 negatively modulates ethylene synthesis and drought tolerance in rice. PLoS One. 2011;6(9), e25216. doi:10.1371/journal.pone.0025216.

Wang GX, Zhang J, Liao JX, Wang JL. Hydropassive evidence and effective factors in stomatal oscillations of Glycyrrhiza inflata under desert conditions. Plant Sci. 2001;160(5):1007–13.

Wang RS, Pandey S, Li S, Gookin TE, Zhao Z, Albert R, Assmann SM. Common and unique elements of the ABA-regulated transcriptome of Arabidopsis guard cells. BMC Genomics. 2011;12(1):216–40.

Wang Y, Wan L, Zhang L, Zhang Z, Zhang H, Quan R, Zhou S, Huang R. An ethylene response factor OsWR1 responsive to drought stress transcriptionally activates wax synthesis related genes and increases wax production in rice. Plant Mol Biol. 2012;78(3):275–88.

Wasternack C. Jasmonates: an update on biosynthesis, signal transduction and action in plant stress response, growth and development. Ann Bot. 2007;100(4):681–97.

Werner T, Motyka V, Laucou V, Smets R, Van Onckelen H, Schmülling T. Cytokinin-deficient transgenic Arabidopsis plants show multiple developmental alterations indicating opposite functions of cytokinins in the regulation of shoot and root meristem activity. Plant Cell. 2003;15(11):2532–50.

Werner T, Nehnevajova E, Köllmer I, Novák O, Strnad M, Krämer U, Schmülling T. Root-specific reduction of cytokinin causes enhanced root growth, drought tolerance, and leaf mineral enrichment in Arabidopsis and tobacco. Plant Cell. 2010;22(12):3905–20.

Wilkinson S, Davies WJ. ABA-based chemical signalling: the co-ordination of responses to stress in plants. Plant Cell Environ. 2002;25(2):195–210.

Wilkinson S, Davies WJ. Drought, ozone, ABA and ethylene: new insights from cell to plant to community. Plant Cell Environ. 2010;33(4):510–25.

Wilkinson S, Kudoyarova GR, Veselov DS, Arkhipova TN, Davies WJ. Plant hormone interactions: innovative targets for crop breeding and management. J Exp Bot. 2012;63(9):3499–509.

Wolters H, Jürgens G. Survival of the flexible: hormonal growth control and adaptation in plant development. Nat Rev Genet. 2009;10(5):305–17.

Wu L, Zhang Z, Zhang H, Wang XC, Huang R. Transcriptional modulation of ethylene response factor protein JERF3 in the oxidative stress response enhances tolerance of tobacco seedlings to salt, drought, and freezing. Plant Physiol. 2008;148(4):1953–63.

Wu X, Kishitani S, Ito Y, Toriyama K. Accumulation of raffinose in rice seedlings overexpressing OsWRKY11 in relation to desiccation tolerance. Plant Biotechnol. 2009;26(4):431–4.

Wu H, Wu X, Li Z, Duan L, Zhang M. Physiological evaluation of drought stress tolerance and recovery in cauliflower (*Brassica oleracea* L.) seedlings treated with methyl jasmonate and coronatine. J Plant Growth Regul. 2012;31(1):113–23.

Xiao-Ping S, Xi-Gui S. Cytokinin-and auxin-induced stomatal opening is related to the change of nitric oxide levels in guard cells in broad bean. Physiologia Planta. 2006;128(3):569–79.

Xu W, Jia L, Shi W, Liang J, Zhou F, Li Q, Zhang J. Abscisic acid accumulation modulates auxin transport in the root tip to enhance proton secretion for maintaining root growth under moderate water stress. New Phytol. 2013;197(1):139–50.

Xu Q, Burgess P, Huang B (2016) Stress-inhibition and gibberellin-mitigation of leaf elongation associated with up-regulation of genes controlling cell expansion. Physiologia Planta.

Yamaguchi S. Gibberellin metabolism and its regulation. Annu Rev Plant Biol. 2008;59:225–51.

Yanagisawa S, Yoo SD, Sheen J. Differential regulation of EIN3 stability by glucose and ethylene signalling in plants. Nature. 2003;425(6957):521–5.

Yang KY, Liu Y, Zhang S. Activation of a mitogen-activated protein kinase pathway is involved in disease resistance in tobacco. Proc Natl Acad Sci U S A. 2001a;98(2):741–6.

Yang J, Zhang J, Wang Z, Zhu Q, Liu L. Water deficit–induced senescence and its relationship to the remobilization of pre-stored carbon in wheat during grain filling. Agron J. 2001b;93(1):196–206.

Yang JC, Zhang JH, Wang ZQ, Zhu QS, Liu LJ. Involvement of abscisic acid and cytokinins in the senescence and remobilization of carbon reserves in wheat subjected to water stress during grain filling. Plant Cell Environ. 2003;26(10):1621–31.

Yang X, Yang YN, Xue LJ, Zou MJ, Liu JY, Chen F, Xue HW. Rice ABI5-Like1 regulates abscisic acid and auxin responses by affecting the expression of ABRE-containing genes. Plant Physiol. 2011;156(3):1397–409.

Yordanov I, Velikova V, Tsonev T. Plant responses to drought, acclimation, and stress tolerance. Photosynthetica. 2000;38(2):171–86.

Yu K, Wei J, Ma Q, Yu D, Li J. Senescence of aerial parts is impeded by exogenous gibberellic acid in herbaceous perennial Paris polyphylla. J Plant Physiol. 2009;166(8):819–30.

Yuan S, Lin HH. Minireview: Role of Salicylic Acid in Plant Abiotic Stress. Zeitschrift für Naturforschung C. 2008;63(5-6):313–20.

Yuan K, Rashotte AM, Wysocka-Diller JW. ABA and GA signaling pathways interact and regulate seed germination and seedling development under salt stress. Acta Physiologiae Planta. 2011;33(2):261–71.

Zawaski C, Busov VB. Roles of gibberellin catabolism and signaling in growth and physiological response to drought and short-day photoperiods in Populus trees. PLoS One. 2014;9(1):1–12.

Zhang S, Liu Y. Activation of salicylic acid–induced protein kinase, a mitogen-activated protein kinase, induces multiple defense responses in tobacco. Plant Cell. 2001;13(8):1877–89.

Zhang X, Schmidt RE. Antioxidant response to hormone-containing product in Kentucky bluegrass subjected to drought. Crop Sci. 1999;39(2):545–51.

Zhang X, Schmidt RE. Hormone-containing products' impact on antioxidant status of tall fescue and creeping bentgrass subjected to drought. Crop Sci. 2000;40(5):1344–9.

Zhang X, Ervin EH, Schmidt RE. Physiological effects of liquid applications of a seaweed extract and a humic acid on creeping bentgrass. J Amer Soc Hort Sci. 2003;128(4):492–6.

Zhang J, Jia W, Yang J, Ismail AM. Role of ABA in integrating plant responses to drought and salt stresses. Field Crops Res. 2006;97(1):111–9.

Zhang X, Ervin EH, Evanylo GK, Haering K. Drought assessment of auxin-boosted biosolids. Proc Water Environ Federation. 2007;2007(3):150–65.

Zhang H, Wu L, Wan L, Li F, Dai L, Li D, Zhang Z, Huang R. Functional analyses of ethylene response factor JERF3 with the aim of improving tolerance to drought and osmotic stress in transgenic rice. Transgenic Res. 2010a;19(5):809–18.

Zhang Z, Li F, Li D, Zhang H, Huang R. Expression of ethylene response factor JERF1 in rice improves tolerance to drought. Planta. 2010b;232(3):765–74.

Zhu JK. Salt and drought stress signal transduction in plants. Ann. Rev. Plant Biol. 2002;53:247–73.

Zottini M, Costa A, De-Michele R, Ruzzene M, Carimi F, Schiavo FL. Salicylic acid activates nitric oxide synthesis in Arabidopsis. J Exp Bot. 2007;58(6):1397–405.

Zwart SJ, Bastiaanssen WGM. Review of measured crop water productivity values for irrigated wheat, rice, cotton and maize. Agric Water Manage. 2004;69(2):115–33.

Chapter 4
Chemical Priming-Induced Drought Stress Tolerance in Plants

Emily Merewitz

4.1 Introduction

Major agricultural industries have few technological defenses against crop loss due to abiotic stresses, particularly drought stress. Drought stress is a complex stress since it often coincides with high light stress and high temperature stress to plant canopies. Nutrient limitation is also commonly encountered by plants under water limited conditions. These secondary stresses occur because major physiological processes such as photosynthesis, transpiration, and respiration may become limited during drought conditions. In addition to drought having an effect on whole-plant responses, cellular level processes are also severely limited due to lack of adequate turgor pressure and other cellular damage. Cell division, elongation, and differentiation are the major processes that determine plant growth and productivity. These cellular processes are highly drought sensitive. Due to the complex effects of drought stress on multiple plant processes and the quantitative nature of genes involved in plant tolerance, plant breeding aimed to improve plant germplasm for drought tolerance is difficult.

Breeding practices, whether classical or via biotechnological methods, are known to be relatively slow in producing new germplasm on the market. Additionally, new germplasm available on the market is not the only answer to the problem of abiotic stresses in agriculture. Many areas worldwide do not often have access to new varieties, may not have the funds for such varieties, may have restrictions on genetically modified materials, and certain new germplasm may not function as predicted in all climates. Additionally, regardless of location, farmers or crop managers need quicker strategies or solutions for existing perennial crop species.

E. Merewitz (✉)
Department of Plant, Soil, and Microbial Sciences, Michigan State University,
East Lansing, MI 48824, USA
e-mail: merewitz@msu.edu

M.A. Hossain et al. (eds.), *Drought Stress Tolerance in Plants, Vol 1*,
DOI 10.1007/978-3-319-28899-4_4

For all of these reasons, plant priming technologies aimed to improve drought tolerance with various exogenous compounds is of extreme importance to aid in reducing crop losses and allow for agriculture to support growing demands from the human population.

Plant priming in this chapter is defined as the application of any exogenous compound, microorganism, or abiotic stress in order to activate or enhance a plant's own natural defenses in order for the plant to be better prepared for a future stress. In scientific literature, compounds that have a direct effect on acquired systemic immunity are most commonly referred to as priming agents. A primed state of plants is a unique physiological state that has specific molecular patterns associated with the state in response to pathogens, some beneficial microbes, and chemical compounds. This has been termed defense priming (Conrath 2011). But for purposes of this chapter the term priming will be used more broadly in order to expand the relevance of the term priming to many practices that aim to promote drought tolerance. Using a broad definition of priming in this chapter allows for a discussion of many types of priming practices that may not necessarily cause the same physiological as does defense priming. Thus, it allows for discussion of many methods used for seed, foliar, and root priming of plants for drought tolerance.

Current priming strategies are relatively underutilized compared to the large availability of information on various chemistries available in scientific literature. For plant priming for drought tolerance to be an effective solution to promote agricultural yields, more widespread information is needed on how these compounds are effective, what plants they are effective on, under what conditions are the compounds effective, what is the duration of the effectiveness, and if applicable, what trade or brand names they are available under. What specific physiological mechanism each priming compound or method is targeting and how that affects whole plant functioning needs to be elucidated and disseminated. In other words, many priming technologies still require further research in order to move from the realm of basic science to an effective, widespread applied agricultural technology. This chapter provides a broad overview of priming technologies utilized and specific compounds used exogenously to improve plant drought tolerance. Specifically, the chapter discusses various types of inorganic or organic chemistries applied exogenously that induce drought responses or tolerance mechanisms in plants.

4.2 Priming Methods

Priming of plants is a general concept that can include multiple strategies to increase tolerance to a variety of abiotic and biotic stresses. Predisposing plants to a limited degree of an abiotic or biotic stress has been shown to increase future tolerance to a given stress or a different stress. Plant priming is an effective way to promote drought tolerance because priming methods often harness the plant's natural defenses and stress memory. For instance, Ding et al. (2012) have demonstrated that exposure of plants to several drought episodes increased transcriptional activity related to drought tolerance in a subsequent stress period. A plant's ability to be better prepared for a subsequent stress following an initial one is likely due to epigenetic changes or other metabolic responses (Kinoshita and Seki 2014). More details about a plant's drought memory for drought survival are provided in Chap. 2.

Chemical priming may also effectively alter a plant's memory, supplement the plant's natural response, or provide a new resource for the plant that it does not naturally utilize as a survival mechanism. In this chapter, chemical priming will refer to the exogenous application of various types of compounds, whether it is a simple compound or complex formulation, which may harness different signaling pathways or mechanisms to promote stress tolerance. Plant priming with microorganisms will not be discussed here as plant–microbial interactions are the focus of Chap. 12. The most common types of chemical application methods for plant priming for abiotic or biotic stresses are seed, root, and leaf based methods. The best priming strategy to be used is largely dependent on the type of compound that is in question and the industry in which that compound will be utilized.

4.2.1 Seed Priming

Seed priming is a very common practice in the agricultural seed industry. The seed industry has various types of priming technologies available including chemopriming, hydropriming, osmopriming, solid matrix priming, biopriming, thermopriming, and halo-priming (Paparella et al. 2015; Mondal and Bose 2014). The goals of the seed priming and treatment type used can vary based on the end use of the seed stock. Commercial seed lots are most commonly primed for enhanced germination potential. Other establishments such as seed banks may prime seeds to supplement preservation practices or prolong dormancy under less than optimal conditions.

The goals of seed priming vary but all are targeting physiological processes that occur prior to seed germination. These processes include seed defense mechanisms that preserve seed health for germination such as DNA repair and reduction of oxidative stress (Paparella et al. 2015). Seed priming methods include chemical, physical, and biotic treatments. Generally, priming seeds to promote crop performance by enhanced tolerance of abiotic and biotic stresses is less common than other seed priming practices. However, since adverse environmental growing conditions and water resource limitations may become more and more problematic in the future, priming of seeds to promote crop tolerance of abiotic stresses is of growing interest in agriculture (Jisha et al. 2013). Chemical priming of seeds for abiotic stress tolerance has been suggested to cause mild to moderate stress on seeds to allow for better preparedness for response to future stresses (Gallardo et al. 2002). Examples of specific seed priming treatments that promote abiotic stress tolerance that are also may be used as foliar priming agents are discussed within Sect. 4.4 for each respective compound.

4.2.2 Foliar Priming

Exogenous application of various compounds to above-ground plants parts is a common practice in plant research and in agricultural practice to enhance the performance of a given crop. Foliar spray priming methods may be adopted if seed or root incorporation methods are not feasible, based on the intended result of the priming practice, or if the compound requires leaf or stem exposure for the compound to be taken up by the plant. Perennial crop plants may not benefit as much long-term from seed priming practices compared to annual crop plants that may be re-seeded every year. Root zone based priming practices may be difficult to incorporate into already established perennial crops or no-till annual cropping systems. Additionally, both annual and perennial crops often demonstrate drought resistance strategies that are in contrary to good yield production, such as escape and avoidance mechanisms. For all of these reasons, foliar priming practices to improve both annual and perennial crop drought tolerance are of great importance to agriculture. Foliar priming practices can be done at any stage of the plant life cycle. Some priming compounds may have a degree of cell toxicity associated with them, so whether priming with a given compound should be performed on a mature plant compared to a seedling may vary among technologies.

4.2.3 Root Priming

Priming of the root zone can be effective for priming compounds that are not readily taken up by above-ground plant parts or are more naturally occurring or absorbed by plant roots. Some chemical compounds are also best to apply to the root zone as they stimulate microorganism associations with roots that promote abiotic stress tolerance. Various plant developmental stages are also targets for root priming practices, from seedlings to mature plants. Priming of the roots is also a common research practice as the plants can be under uniform exposure to the compound for optimal evaluation of the effects of the compound, such as in hydroponics, tissue culture, or in various potting media types. Many of the compounds discussed below have been evaluated by root priming techniques.

4.3 Physiological Targets of Chemical Priming

When a plant is in a primed state, it has a higher level of fitness or readiness to take on a given stress. The primed plants are typically able to respond more rapidly and more effectively for stress protection. The rapid response is largely triggered by efficient plant hormone and other stress signaling systems such as reactive oxygen species accumulation. All of the priming agents mentioned in the chapter have many physiological targets in common or that are unique to each compound. Several of the priming agents largely have an effect on antioxidant responses. Some of the compounds are able to illicit systemic acquired resistance (SAR) and induced systemic resistance (ISR) pathways, which are highly targeted by several plant priming methods. Abscisic acid, ethylene, auxins, and other stress associated hormones may also be regulated by plant priming strategies.

SAR responses are primarily associated with pathogen infection, particularly biotrophic pathogens. When a biotrophic pathogen infects a plant, salicylic acid (SA) accumulation serves as a systemic signal to activate defense response genes in the whole plant. SA signaling stimulates the production of pathogenesis-related (PR) proteins, which inhibit pathogen development and spread (Van Loon and Van Strien 1999). This increases plant tolerance of the primary infection and may reduce secondary pathogen attack (Beckers and Conrath 2007). Interestingly, some necrotrophic pathogens are thought to have the ability to use the SA pathway against the plant to promote disease development (Rahman et al. 2012). The increase in SA and subsequent changes in SAR-inducible genes has been termed a primed state of plants.

Relative to biotic pathogens, less is known about how drought stress affects SAR responses. Abscisic acid is a key regulator of drought tolerance and is the hormone that stimulates stomatal closure to restrict water loss. In Arabidopsis, abscisic acid (ABA) and SAR were shown to act antagonistically since ABA accumulation suppressed SAR responses and SAR suppressed ABA biosynthesis (Yasuda et al. 2008). This may explain in part why plants could be more susceptible to pathogen attach while under drought stress. A plant's ability to maintain or accumulate SA or SAR responses while under drought stress are often associated with drought stress tolerance. SA, either applied exogenously or maintained endogenously, have been associated with improved drought tolerance (Larkindale and Huang 2004; Senaratna et al. 2000a, b; Horváth et al. 2007; Krishnan and Merewitz 2015). Conversely, SA application was also shown to decrease drought tolerance in corn (*Zea mays*; Nemeth et al. 2002). The differences could be related to species or be concentration dependent.

Gibberellic acid (GA) may also be a factor in regulating SA pathways. GA biosynthesis promotes the degradation of proteins known as DELLA proteins in plants. Higher concentrations of DELLA proteins are known to increase the resistance of plants to various stressors by improving salicylic acid (SA) defense pathways (Alonso-Ramírez et al. 2009). If drought reduces GA concentration in plant tissues, this could regulate SAR responses. More work in this area is needed to better understand SA accumulation, SA exposure, and SAR activity with abiotic stress tolerance in relationship to other plant hormones. A better understanding of these responses could lead to better use recommendations for priming agents and new priming agent chemistries.

Several chemical priming agents are known to be SAR activators. For instance, 2,6-dichloroisonicotinic acid (INA) and acibenzolar-S-methyl (BHT) are well-known SAR activators (Conrath 2009). Exogenous application of SA has also been shown to regulate SAR responses (Ryals et al. 1996). More detailed information about different SAR activators and drought tolerance is described below.

Plants can also be primed via the ISR pathway. Whereas SAR is largely regulated by SA, the ISR pathway is thought to be controlled by jasmonic acid (JA) and ethylene. JA is a relatively newly classified plant hormone that has been associated with the regulation of growth and promoting stress defenses in plants (Delker et al. 2006). JA may also promote defense against pathogens, particularly those that are necrotrophic (Rahman et al. 2012). Increases in JA accumulation promotes defense against pathogens and is involved in drought stress signaling (Hase et al. 2008; Yang et al. 2012). ISR priming genes are mostly associated with plant–microbe interactions or applications of beneficial organisms that may promote tolerance to drought stress (Cho et al. 2011). More details about priming with microorganisms are given in Chap. 12. Specific effects of JA on drought tolerance are discussed below.

After the priming agent has signaled a plant response either by SAR, ISR, or other pathways, whole plant physiological responses occur that may reduce water loss or reduce cellular damage during drought. The increase in effectiveness of resistance mechanisms in the primed state varies by which priming technology is utilized. Several different stress resistance mechanisms may play a role in conferring the priming-induced resistance. The mechanisms used by plants include those related to stomatal closure, maintaining turgor pressure, antioxidant defense, membrane stability, and other mechanisms (Beckers and Conrath 2007).

4.4 Major Priming Compounds and Their Effects on Drought Tolerance

4.4.1 Inorganic Compounds

Adequate nutrient status of plants has long been known to be a major factor in plant survival of abiotic and biotic stress. Deficiency in any essential nutrient may reduce plant tolerance of drought significantly. Ensuring that adequate fertilization practices are in place and optimal soil conditions exist are essential in promoting plant tolerance of drought stress. Priming with inorganic compounds or plant nutrients is considered a practice separate from typical fertilization regimes, as the goals of the priming are specifically to promote tolerance to a given stress.

Chemical applications of macronutrients, micronutrients, or supplemental nutrients is done in order to provide plants with nutrients that may become limited during drought stress, are in high demand during drought periods, or have stress protective properties. Each type of salt generally has specific functions in plants and has been studied for protective effects specifically during drought stress either as a pre- or posttreatment during recovery. As there are too many nutrients required by plants to discuss all in depth here, a few select salts (K, Si, and Se) relevant to drought tolerance are discussed. The section will also discuss several other compounds used as priming agents.

Cellular requirements for K are significantly elevated under drought stress due to the role of K in regulating stomatal aperture, osmotic relations, photosynthesis, and reactive oxygen species signaling (Cakmak 2005). Rice plants containing transgenes for increased K absorption via K transporters have been shown to be more drought tolerant compared to non-transgenic plants (Song et al. 2014). Addition of K by foliar or root based applications to plants prior to drought stress has been shown in many studies to improve drought tolerance via both drought avoidance and tolerance mechanisms. For instance, in two wheat varieties contrasting in drought tolerance, K supplementation to the roots played a major role in promoting antioxidant responses, photosynthesis, and overall biomass during drought (Wei et al. 2013). Improvements in root biomass and depth by K supplementation can have significant impact on drought survival. Foliar application post-drought stress of K improved stomatal opening and recovery of photosynthesis in kentucky bluegrass (*Poa pratensis*) during recovery from drought stress (Hu et al. 2013). In addition to studies done on drought stress, sources of K such as potassium phosphite have been shown to induce defense responses against various pathogens (Araujo et al. 2015). Thus, in addition to K playing a role in well-characterized drought responses, K-induced drought tolerance could also be associated with induced responses of secondary metabolites and plant protective compounds. Such assumptions need further research.

Silicon (Si) and selenium (Se) are plant nutrients that are often required in trace amounts that are becoming more recognized for promoting health of both plants and animals. The amount required can differ significantly among plant species. Many studies on major crop species have shown that Si plays a role in drought resistance under lab and field conditions (Nolla et al. 2012; Shen et al. 2010; Chen et al. 2011). Unlike some of the other compounds discussed in this chapter, Si can regulate both plant metabolism and have major structural effects on plants. Si can incorporate into leaf epidermal tissues and in plant cell walls, which have been shown to have an effect on stomatal aperture, cuticle thickness, and xylem rigidity, for example (Gao et al. 2006). Other important factors related to Si-induced drought resistance include changes in photosynthesis rate, maintained plant water status, increased root growth, decreased transpiration rates, and enhanced nutrient uptake (Gao et al. 2006; Gong and Chen 2012; Chen et al. 2011; Habibi 2014). Recent evidence suggests that Si may regulate endogenous plant compounds that are considered to be priming agents themselves, such as polyamines (Yin et al. 2014). Relative to the extensive amount of research done on Si related to drought survival, more work needs to be done on Si regulation of many biochemical pathways. Regardless, Si is a promising drought priming agent that increases plant readiness for drought survival.

Much like Si, researchers have investigated the use of Se as a seed, root, or foliar treatment. Se has been shown to enhance the growth of plants under both optimal and stress conditions (Hasanuzzaman et al. 2010). For agronomic and forage crops, high levels of Se accumulation in plant tissues is a concern for human and animal health. Exposure to too much Se can also be toxic to plants. Thus, the intended crop, concentrations of applications, and purpose of the crop is a major consideration when investigation of Se as a priming agent is proposed. But many crops not destined for consumption, those crops that do not uptake high levels of Se, or crops with specific Se recommendations for priming are available may be the best targets for Se treatment or priming. Applications of different forms of Se have several benefits to plants including enhancing seed germination, plant growth, and plant survival of stresses (Ahmad et al. 2015).

Stress protective properties of Se on a metabolic or tolerance level appear to be most related to its function as a component of antioxidant enzymes (Hasanuzzaman and Fujita 2011). Morphologically for drought avoidance, Se stimulates root growth resulting in enhanced water uptake in the roots during drought incidence (Kuznetsov et al. 2003). Plants have tolerance mechanisms to deal with high levels of Se, which have been found to be similar to mechanisms used in salt stress conditions. Se tolerance has also been shown to involve ethylene and jasmonic acid pathways (Van Hoewyk et al. 2008). Thus, priming with Se, particularly to Se sensitive plant species, could elicit a mild stress response that primes plants for a future stress. However, more research on molecular pathways regulated by Se that may specifically relate to drought survival is needed.

In addition to foliar or root zone applications of inorganic salts to mature plants or seedlings, seed priming technologies commonly use inorganic compounds as primary priming agents. Seed priming is thought to impose a stress on seeds or give them a head start with germination without the development of the radical. Priming of seeds has been deemed a type of false germination since genes and proteins involved in the two processes are largely similar (Chen and Arora 2011). Typically, inorganic salts of sodium, potassium, and magnesium or other compounds used to adjust the osmotic potential of solutions, such as PEG or sugar alcohols, are used for osmopriming of seeds.

Commonly, polyethylene glycol (PEG) is used as a seed osmopriming agent alone or in combination with inorganic salts such as KCl. Seed priming or osmopriming with KCl improved drought tolerance of wheat (*Triticum aestivum*; Eivazi 2011), Chinese cabbage (*Brassica rapa;* Yan 2015), cucumber (*Cucumis sativus*; İşeri et al. 2015), and barley (*Hordeum vulgare;* Ajouri et al. 2004). PEG, although not an inorganic salt, is commonly used to simulate the action of a salt that may impart osmotic or drought stress to plants. For instance, in research laboratories PEG is commonly added to hydroponic systems or tissue culture to simulate drought stress by causing osmotic stress to roots. PEG-primed plants may benefit from several different physiological mechanisms. PEG priming has been shown to have an effect on gene and protein expression of proteins involved in stress response or protection. For instance, catalase, a major antioxidant enzyme, has been reported to be upregulated due to PEG priming of Arabidopsis (Gallardo et al. 2001) and sunflower (*Helianthus annuus*; Kibinza et al. 2011) seeds. In spinach seeds (*Spinacia oleracea*), catalase and superoxide dismutase were downregulated during germination whereas ascorbate peroxidase activity was stimulated (Chen and Arora 2011). It seems clear that osmopriming has an effect on antioxidant responses during seedling germination and growth but the specific antioxidant pathways that may be important may be species specific or based on the priming technique utilized.

In addition to antioxidant regulation, osmopriming influence on drought tolerance seems to be associated with specific stress associated proteins such as dehydrins or late embryogenesis associated proteins (LEAs), aquaporins, heat shock proteins, and enzymes related to nutrient uptake such as nitrate reductase. In spinach, expression of specific dehydrin genes and the associated protein contents were found to be highly regulated by priming of seeds in response to chilling and dehydration (Chen et al. 2012b). LEA proteins, dehydrins, and heat shock proteins have all been shown to be differentially regulated by osmopriming practices with either inorganic salts or PEG (Cortez-Baheza et al. 2008; Chen et al. 2012a, b). Aquaporins, protein channels that regulate water movement across plant membranes, are also upregulated by osmopriming practices (Chen et al. 2013). Nitrogen metabolism through enhanced nitrate reductase activity in primed seedlings may also play a role in performance and stress tolerance, as was demonstrated in tomato seeds (*Lycopersicon esculentum*) in response to priming with K salts and PEG (Lara et al. 2014). Other processes such as ROS signaling and hormone regulation (such as regulation of the ABA-to-GA ratio) also are key regulators of priming induced changes to seeds. Thus, priming practices seem to have a wide range of effects on plant mechanisms involved in tolerance of dehydration.

4.4.2 Amino Acids

Several amino acids such as proline naturally accumulate to high concentration in plant tissues during abiotic stress periods. Shifts in metabolism towards proline may play a role in osmotic adjustment, ROS scavenging, maintain cellular pH and redox balance, and play a role in stress signaling (Hare and Cress 1991; Ashraf and Foolad 2007). Wheat plants expressing the key enzyme in proline biosynthesis, Δ-pyrroline-5-carboxylate synthetase (*P5CS*), exhibited enhanced drought tolerance when the gene was under the control of a drought stress-induced promoter. The improvement in drought tolerance was attributed to prolines role in protecting against oxidative stress (Vendruscolo et al. 2007). Since many protective effects of proline were observed in plant tissues, researchers attempted to simulate such responses with exogenous application. Priming of plants with proline has been shown to protect plants from drought and other stresses.

Priming with proline has similar effects on plants as those that naturally occur due to proline accumulation in plant tissues. For instance, improved photosystem functioning and enhanced soluble sugars were shown to accumulate under drought stress following proline application in Arabidopsis (Moustakas et al. 2011). Similar results were found due to proline application to soybean (*Glycine max*) under salt stress conditions (Yan et al. 2000). Compared to other stresses such as salt, the effects of exogenous application of proline specifically on promoting drought stress tolerance is lacking for many plant species in the literature. Additional work in this area would be beneficial so that rates and frequency of use are determined for various crop species. This is particularly true because cellular damage has been associated with too much proline accumulated in plant tissues (Vendruscolo et al. 2007) or applied exogenously (Hare et al. 2002). Therefore, proline should not be considered an inert osmoprotectant when used as a priming agent.

Proline is also important in other priming practices and technologies. For instance, osmopriming of *Brassica* seeds was shown to significantly enhance proline accumulation of germinating seeds under salt stress (Kubala et al. 2015). A similar response of proline was noted after seed treatments with irradiation from a laser, known as laser priming of durum wheat (*Triticum turgidum*) seeds (Zare et al. 2014). Thus, not only is proline an effective priming agent but it is also integrally related to other priming practices.

4.4.3 Nonprotein Amino Acids

Compounds known as nonprotein amino acids are those amino acids that are not directly used in translational processes resulting in protein synthesis. They are also known as nonproteinogenic or non-coded amino acids. Nonprotein amino acids can be incorporated into proteins posttranslationally but largely they are metabolic intermediates within biochemical pathways or are products of proteolysis (Bell 2003). Some commonly investigated nonprotein amino acids related to priming for drought tolerance in plants include glycine betaine (GB), γ- aminobutyric acid (GABA), β-aminobutyric acid (BABA), and 5-aminolevulinic acid (ALA).

GB is an osmoprotectant that is readily taken up by plant leaves, transported, and remains stable in plant tissues (Ashraf and Foolad 2007). The positive effects associated with GB in plant tissues under stress periods have made GB an interesting candidate for use as a priming agent. In response to salinity stress, GB has been associated with protection of plants by osmotic adjustment, protection or stabilization of important photosynthetic enzymes such as RuBisCo, and may be a free radical scavenger.

Many successes have been demonstrated by priming with GB for drought stress, particularly in plants that are not naturally high GB producers such as rice (*Oryza sativa*), potato (*Solanum tuberosum*), soybean (*Glycine max*), and tomato (*Lycopersicum esculentum*) (Agboma et al. 1997; Wani et al. 2013). For instance, supplementation with a pretreatment of exogenous GB was shown to improve physiological responses including relative water content and biomass production during drought stress (Rezaei et al. 2012). In soybean, positive effects of glycine betaine were exhibited during drought, which lead to increased yield, photosynthetic activity, nitrogen fixation, and leaf biomass (Agboma et al. 1997). The positive effects of GB have also been demonstrated by transgenic means of increasing endogenous GB either constitutively or in response to a given stress in several species such as wheat (*Triticum aestivum*) (He et al. 2011). It is important to note that the concentration of GB used is a significant factor in the success of using GB as a priming agent. Some species can be more sensitive to the effects of priming with GB than others (Rezaei et al. 2012). For instance, detrimental effects on plant growth and development of GB application were found for tomato (Heuer 2003).

The nonprotein amino acid GABA is most well known as a neurotransmitter in animal systems (Bown and Shelp 1997; Shelp et al. 2012; Bouché and Fromm 2004) and as the key component of the GABA shunt pathway. The function of GABA is less well understood in plants. GABA is synthesized from glutamate and then converted to succinic semi-aldehyde and succinate before entering the tricarboxylic acid cycle (Shelp et al. 2012). In addition to GABA playing a major role in regulating carbon and nitrogen metabolic processes in plants, GABA is involved in numerous other cellular processes in plants ranging from specialized functions such as pollen tube development to broader functions such as being an osmolyte for abiotic stress protection (Fait et al. 2008).

Under drought stress conditions GABA rapidly accumulates in many plant species. For instance, GABA has been shown to accumulate in soybean (Serraj et al. 1998), barley (*Hordeum vulgare*; Guo et al. 2009), and creeping bentgrass (*Agrostis stolonifera*, Merewitz et al. 2012). Other abiotic and biotic stresses that GABA may be involved in providing stress protection for include salt, wounding, hypoxia, heat shock, and pathogen infection (Fait et al. 2005; Kinnersley and Turano 2000; Shelp et al. 1999). There is mounting evidence that GABA may act as a signal molecule acting in concert with phytohormones (Lancien and Roberts 2006; Renault et al. 2011). GABA may also limit cellular elongation under stress conditions (Renault et al. 2011). How GABA serves as a signaling molecule and how it may regulate both primary and secondary metabolism pathways in plants is still not fully understood but are becoming clearer. It is clear that GABA plays a major role in the regulation of C and N metabolism, in stress signaling, and that these two phenomena are likely closely linked (Michaeli and Fromm 2015).

Priming with GABA has proven to be an effective method of enhancing plant tolerance to several abiotic stresses. Under osmotic, stress, GABA has been shown to improve plant performance of black pepper plants (*Piper nigrum*; Vijayakumari and Puthur 2015). Priming peach (*Prunus persica*) fruit in GABA enhanced the antioxidant activity in response to chilling stress (Yang et al. 2011). Less information is available regarding priming using GABA specifically for drought stress tolerance. In perennial ryegrass (*Lolium perenne*) plants, GABA application to leaves prior to drought incidence was effective in improving leaf water content, reducing lipid peroxidation, and enhancing some antioxidant enzyme activities during drought (Krishnan et al. 2013). GABA was not an effective priming method in Arabidopsis plants; however, an isomer of GABA, known as BABA, was shown to be the most effective priming compound during drought (Jakab et al. 2005).

A soil drench method of priming plants with BABA seems to be the most effective and commonly reported method. In potato plants, a soil drench BABA treatment caused a reduction of water loss during drought compared to plants that were not primed (Sós-Hegedűs et al. 2014). The large degree of interest in BABA has lead to significant insight into the mechanism of why priming with BABA is effective. BABA has an effect on expression of genes related to hormonal regulation of biotic and abiotic stress responses such as ethylene receptors and ethylene induced genes (Sós-Hegedűs et al. 2014), salicylic acid dependent defense genes and abscisic acid signaling (Ton et al. 2005). However, some researchers have noted that the effects of priming with these compounds can be short lived. Despite this limitation, the mounting evidence that priming with BABA provides broad spectrum protection to abiotic and biotic stresses via systemic mechanisms makes BABA an attractive candidate for many applied uses in agriculture.

Aminolevulinic acid (ALA) is a tetrapyrrole precursor to various photosynthetic pigments and may have distinct regulatory effects on photosynthesis and overall plant growth (Rosenthal 1982). ALA has been found to have plant growth regulator and beneficial effects such as increasing yield under both optimal and stressed conditions (Akram and Ashraf 2013). ALA has been shown to have several different physiological effects on various plant species to reduce stress damage. Positive effects that have been demonstrated from priming with ALA prior to drought stress include improvements in gas exchange, maintenance of chlorophyll content, photochemical health, and enhanced enzymatic or nonenzymatic antioxidant activities (Li et al. 2011; Liu et al. 2011). In addition to drought, priming with ALA has been shown to cause similar responses or enhancements in plant performance under salt stress (Yang et al. 2014), heavy metal stress (Ali et al. 2013), and cold stress (Korkmaz et al. 2010; Balestrasse et al. 2010).

Many other nonprotein amino acids are also being investigated as potential stress preventative priming treatments such as *p*-aminophenylalanine, L-azetidine-2-carboxylic acid, δ-,4-aminobenzoic acid, ornithine, citruline, homoserine, L-3,4-dihydroxyphenylalanine (L-DOPA), and 5-hydroxy-L-tryptophan (5-HTP) (Jakab et al. 2005).

4.4.4 Polyamines

Free polyamines putrescine, spermidine, and spermine are long chain amine compounds primarily synthesized from S-adenosyl methionine (SAM). Polyamines have hormone like properties since they are involved in cell division, differentiation, and DNA replication processes for plant cell growth (Kusano et al. 2008). They are also closely associated with other hormones such as ethylene. Polyamines have been shown to act antagonistically to ethylene since they are competing for SAM substrates and tend to promote opposite cellular processes such as those related to growth and senescence (Bitrián et al. 2012; Torrigiani et al. 2012).

There are major differences among species in how specific stresses regulate endogenous PA content, which is likely due to different resistance strategies (i.e., tolerance, avoidance, or escape) (Alcazar et al. 2011). Differences in environmental stresses implementation and severity among studies may also make it difficult to compare among studies. Polyamines also exist in different states in plant cells, with some as free polyamines, convalently conjugated (also called bound polyamines), or non-covalently conjugated to other macromolecules or cell structures such as proteins, membranes, and DNA. Due to these reasons, the regulation of polyamines under drought stress remains relatively uncertain. For instance, increased spermidine and spermine and a reduction in putrescine in sugarcane plants under water stress were associated with drought sensitivity (Zhang et al. 1996). In leaves of a drought tolerant wheat compared to a sensitive one under dehydration stress, higher conjugated spermidine and spermine levels were detected in the drought tolerant type (Liu et al. 2006). Due to the dynamic flux through polyamine pathways, detailed research on how to specifically exploit the benefits of each polyamine compound as priming technologies has been needed.

Interestingly, polyamines appear to be naturally occurring priming agents. There is recent evidence that there are PA exuding rhizobacteria (*Bacillus subtilis*) that appear to affect plant growth and other responses (Xie et al. 2014). Manmade foliar or root based exogenous application of PAs has been shown to be an effective strategy to increase plant tolerance of various stresses and regulate plant growth. In relation to abiotic stresses, the mechanisms of PA improvements in tolerance are not fully elucidated but many studies have shown the beneficial effects of exogenous application of PA or endogenous upregulation of PA on drought survival.

Like many of the compounds discussed so far, beneficial effects of exogenous application of PA are readily observed for not only drought stress but also other abiotic stresses such as salt stress (Shi et al. 2013; Shi and Chan 2014). Seed priming with polyamines has been shown to promote establishment and growth in several plant species (Farooq et al. 2008; Khan et al. 2012), particularly during non-optimal germination conditions such as low or high temperatures or under salt stress (Korkmaz et al. 2005). For drought stress, wheat plants exposed to seed priming and foliar spray with polyamines were shown to be more drought tolerant than those not primed (Farooq et al. 2009). Additional research on priming for drought tolerance is needed to fully elucidate the mechanism and determine utility in more plant species. Relative to their major effects on plants either exogenously or endogenously, relatively few species beyond model plant species have been evaluated for PA priming. Thus, it would be desirable for more research on PA regulation, function, and effects on important crop species. See Chap. 10 for a more detailed discussion on polyamine metabolism related to drought stress tolerance in plants.

4.4.5 Reactive Oxygen and Nitrogen Compounds

Priming directly with reactive oxygen species (ROS) or compounds that have an effect on reactive oxygen species signaling or scavenging can also regulate drought induced oxidative stress tolerance. Hydrogen peroxide (H_2O_2) is the one most commonly researched ROS as a priming agent due to its greater longevity and diffusion in plants compared to other ROS. H_2O_2 is naturally produced during many cellular processes and in times of stress. It acts as a signaling molecule for various cellular processes (Mittler et al. 2011). H_2O_2 priming has been shown to improve plant tolerance of heat, salt, chilling, and drought stress.

The improvements in plant performance under abiotic stresses are thought to be due to enhanced expression of heat shock proteins, photosynthesis, proline biosynthesis, and antioxidant pathways (Hossain et al. 2015). In mustard (*Brassica juncea*) seedlings, roots were primed with H_2O_2 and showed enhanced antioxidant and methylglyoxal detoxification enzymes during simulated drought stress using polyethylene glycol treatment for 48 h (Hossain and Fujita 2013). Relative to the other aforementioned stresses, less is known about the specific mechanisms of H_2O_2 mediated improvements in drought tolerance. However, it does seem that H_2O_2 priming causes tolerance to drought stress as opposed to drought avoidance due to stomatal closure. Specific responses of plants to H_2O_2 have been revealed such as improved osmolyte accumulation (Ishibashi et al. 2011); however, more detailed effects on plant responses under drought stress and whether those effects are lasting beyond a few days or few hours is needed.

Nitric oxide (NO) is also a signaling molecule that regulates oxidative stress responses in plants. This regulation of reactive oxygen species is thought to be intimately involved in controlling cellular responses to ABA to regulate stomatal closure during drought stress. Plants produce NO through several known enzymes in several different cellular compartments or organelles and in both leaves and roots. The known enzymes include nitrate reductase, nitric oxide synthase, nitrite-NO reductase, and xanthine oxidoreductase (Sidana et al. 2015). Research performed with NO donors and NO scavengers on various plant species has revealed that exogenous supply of NO induces stomatal closure whereas scavengers reverse this process (Neill et al. 2008). Genetic studies have also shown that NO is required for ABA-induced stomatal closure. For instance, transgenic rice with a gene for NO synthase enhanced plant NO activity and accumulation. The plants exhibited greater drought and salt tolerance as demonstrated by higher water content, less membrane damage, and greater levels of proline (Cai et al. 2015). The severity and rapidity of cellular dehydration does seem to play a role in determining whether the plant utilizes NO as a signaling component in conjunction with ABA. For instance, drying of detached leaves did not seem to stimulate NO-regulated stomatal closure (Neill et al. 2008).

NO plays a major role in protection from various other stress conditions in addition to drought including heavy metal stresses. Thus, it would seem that NO would serve to have additional protective effects in addition to stomatal regulation. The other protective effect is thought to be related to regulation of oxidative stress or antioxidant enzyme regulation. The effects of NO on plant antioxidant systems is complex since NO has been shown to regulate both antioxidants and pro-oxidants (Groß et al. 2013). Exposure of plants to high levels of NO can cause oxidative damage such as to plant membranes. Lower levels of NO provide beneficial signaling effects to prime plants for stress tolerance (Fukuto et al. 2000). Thus, effective use of NO as a stress priming agent requires knowledge of appropriate concentrations for various plant species.

In addition to stomatal regulation and antioxidant regulation, NO plays a role in plant developmental processes. NO regulation of adventitious roots has been shown to play a role in drought survival in several plant species. In root tips, NO seems to interact with auxin regulated genes that induce cellular elongation and root formation (Sidana et al. 2015). The wide range of biochemical processes affected by NO that play a beneficial role in promoting drought tolerance makes it a potentially valuable plant priming agent. Despite the ample literature on NO available, more information on how NO is a regulator of plant processes and whether NO may have utility in long term drought or applied agricultural processes is still needed.

4.4.6 *Antioxidant Compounds*

Since many other priming compounds have a major effect on the antioxidant systems in plants and many abiotic stresses cause the accumulation of reactive oxygen species, one would expect that priming with various antioxidant compounds could be an effective method of promoting stress tolerance. Ascorbate and glutathione are two examples of antioxidant compounds that have been used in plant priming research. Ascorbate, or vitamin C, priming has also been called vitamin priming. Seed priming of rice plants for improved salt tolerance was more effective with ascorbate compared to other hormonal priming compounds tested including kinetin and SA (Afzal et al. 2013). Wheat plants also benefited from ascorbate priming of seeds, seedlings exposed to drought stress after ascorbate priming exhibited greater chlorophyll content, leaf area, and overall dry weight. These increases in growth were associated with increased proline content under drought conditions (Farooq et al. 2013). Proteomic analysis of ascorbate priming of wheat seeds for salt stress tolerance has shown that ascorbate priming has a significant effect on various antioxidant enzymes in embryos and other defense related proteins (Fercha et al. 2014). Exogenous application of ascorbate through plant roots has been shown to reduce the effects of salt stress on wheat plants, but the results were specific to only a more tolerant cultivar used in the study (Athar et al. 2008). Similarly, seed priming with ascorbate likely may not be effective for all crop plants.

Glutathione is a major and powerful antioxidant component in plant cells. It can directly scavenge free radicals and reactive oxygen compounds. It also serves to regulate other endogenous compounds in the cell that are also used as priming agents, such as NO. Some hormonal priming practices have been shown to have an effect on endogenous glutathione transferases content or in the redox state of glutathione (Csiszár et al. 2014; Horváth et al. 2015). Glutathione plays a major role in drought tolerance of plants and may play a role in regulating ABA signaling. Exogenous application of glutathione to plants increased ABA accumulation in Arabidopsis (Chen et al. 2012a, b). In bean plants, treatment of seedlings with glutathione enhanced antioxidant responses and drought performance, but the effects were seemingly very short term (Nahar et al. 2015). Relative to many of the other priming compounds and its recent attention in the literature for promoting stress tolerance, little information is available regarding the utility of glutathione as a priming practice specifically for drought tolerance. Several other antioxidant compounds exist in plants and also could be exploited further.

4.4.7 *Plant Hormones*

Many of the priming practices or agents mentioned in this chapter have a significant effect on plant hormone interactions and signaling in plants. It cannot go without mention that plant hormones themselves can effectively be used to prime plants for drought stress tolerance. This has been called hormonal priming. Hormonal priming has been used within all priming methods including seeds, foliar, and root based methods. An in depth evaluation of plant hormones and drought stress responses is covered in Chap. 3. Here, a brief discussion of the latest knowledge of several hormones applied exogenously and how they have been shown to improve drought resistance is provided.

ABA is well known to be a primary drought signal from roots to shoots to cause stomatal closure. Exogenous application of ABA has been thoroughly shown to act as an antitranspirant for plants and can activate many stress responsive genes. ABA interacts with genes containing ABA response elements, MYC-like and MYB-like gene elements (Bray 2002). Recently, the function of ABA in regulating guard cells has become clearer. ABA acts as a cellular priming agent in that it primes guard cells to respond more readily to increased cytosolic calcium levels (Kim and Maik 2010). ABA is also involved in signaling cascades of many other priming compounds. For instance, BABA primed Arabidopsis plants were shown to have enhanced salt and drought tolerance, which was associated with higher expression of SA and ABA dependent genes (Jakab et al. 2005). Thus, ABA is a highly important regulatory agent of drought stress in relation to priming practices. It can act alone to better prepare plants for drought survival via stomatal closure or can be an important endogenous signal that is regulated by other priming agents.

ABA is highly linked to JA during drought stress. JA and the methyl ester form, methyl jasmonate, have been shown to play a role in drought stress tolerance and be required for drought responsive cellular processes. Through molecular studies using chemical inhibitors of JA and ABA, Ollas et al. (2012) showed that a rapid and transient increase in JA was required for ABA accumulation in roots under severe drought stress in citrus (*Citrus paradisi* × *Poncirus trifoliate*). Exogenous JA has been shown to improve drought tolerance in multiple plant species. Treatment of seedlings of various *Brassica* species with 0.5 nM JA improved antioxidant activities and growth under drought stress (Alam et al. 2014). JA is also known to accumulate or have a transient accumulation in many plant species due to drought stress (Ollas et al. 2012; Krishnan and Merewitz 2015). However, in tomato and a *Brassica* species JA did not specifically accumulate due to drought stress however the JA precursor 12-oxo-phytodienoic acid did and was shown to be involved in stomatal regulation in conjunction with ABA (Savchenko et al. 2014). More research on JA regulation and function specifically for drought stress is needed.

As described above, JA and SA are intricately connected in plant defense signaling. SA applied exogenously has been shown to induce stomatal closure (Waseem et al. 2006) and improve antioxidant responses (Mori et al. 2001). SA stimulation of defense proteins known as pathogenesis related proteins (PR) could offer better performance of plants under drought stress, since often pathogens may attack plants while weakened by drought stress. Thus, an indirect promotion of tolerance to abiotic stress is possible. It is clear that SA supplementation provides protection against many biotic and abiotic stresses. However, with respect to drought stress, whether SA plays a direct protective role is less clear (Horváth et al. 2007). A decrease in drought tolerance was observed in maize seedlings (Nemeth et al. 2002). However, many other studies have demonstrated that SA improves drought tolerance via different methods including soaking of seeds, foliar treatment of seedlings, and root based applications (Bandurska 2005; Senaratna et al. 2000a, b). Many priming agents are criticized for their effects being short lived. SA seems to have more prolonged effects as a priming agent (Kadioglu et al. 2011). With more research on rates and timing of use of SA as a priming agent in more important plant species, long term drought stress benefits could be exploited with SA priming.

Another group of hormones is also highly related to JA. JA synthesis may be stimulated by another large group of plant hormones known as brassinosteroids (BR) (Müssig et al. 2000). There are a relatively new group of hormones with over 60 BR, with the most bioactive ones being brassinolide, 28-homobrassinolide, and 24-epibrassinolide (Vardhini and Anjum 2015). Priming or exogenous application of brassinolides has been shown to promote drought tolerance in several plant species including maize (Anjum et al. 2011), soybean (Zhang et al. 2008), and apple (*Malus domestica;* Kairong et al. 2006). As many of the priming agents, the protective effects of BR are associated with the promotion of the antioxidant system and accumulation of osmolytes such as proline and glycine betaine in plants (Vardhini 2014).

Many other examples of plant hormones promoting drought tolerance via priming technologies have been reported. For instance, promotion of rooting by auxins, cytokinin-induced improvements in photosynthetic attributes, and GA effects on seed germination have all been shown to enhance plant performance under drought (Eisvand et al. 2010; Merewitz et al. 2011; Akter et al. 2014). GA effects on drought tolerance are less clear and may be species dependent. Application of various levels of GA to maize plants prior to drought stress improved leaf water content, chlorophyll content, and other parameters compared to plants not treated with GA (Kaya et al. 2006). Negative regulators of GA, one such gene is known as *SPINDLY*, were shown to exhibit a reduction in drought tolerance in Arabidopsis. This was demonstrated by the vast improvement in drought tolerance of Arabidopsis that were mutant for the *SPINDLY* genes (Qin et al. 2011). However, many synthetic compounds that suppress or enhance endogenous GA, known as plant growth regulators, seem to promote drought tolerance in plants (Bian et al. 2009; Krishnan and Merewitz 2015). Suppression of GA biosynthesis in mature plants could alter morphology and create a microenvironment for some plant species that contributes to better water use efficiency. More details about GA and drought stress are reviewed in depth in Colebrook et al. (2014). Thus, it seems there are multiple ways to exploit the functions of GA and other hormones used as priming agents or as targets for priming practices in order to improve plant performance under drought stress.

4.4.8 Organic Alcohols

Organic alcohols including polyols or sugar alcohols such as mannitol, glycol, sorbitol, myo-inositol, and glycerol are often found to accumulate in many plant species during times of stress. Some alcohols such as ethanol are toxic or damaging to plant cells and largely accumulate due to cellular stress. However, there is evidence that several of the alcohols present in plant tissues that accumulate to higher levels under stress conditions may play stress protective roles. For example, mannitol accumulation, whether naturally or by transgenic modification, may be an osmolyte involved in osmotic adjustment (Karakas et al. 1997). Much information is available regarding the production of mannitol as an osmolyte, biosynthesis within photosynthetic processes, and species specific evidence of which plants actively accumulate mannitol. For mannitol and many of the polyols, less information is available regarding the effects of these compounds used exogenously or as a stress priming agent.

There is evidence that several alcohols may be useful as stress priming agents. A study done on wheat roots under salt stress demonstrated that exogenous mannitol was effective in reducing oxidative damage due to salt stress via enhanced activation of antioxidant enzymes. Wheat plants do not naturally accumulate high levels of mannitol under stress conditions (Seckin et al. 2009). Sorbitol had positive effects on membrane stability of salt sensitive types of rice seedlings when applied to the growth media (Theerakulpisut and Gunnula 2012). This study demonstrated that the effects were cultivar specific as little effects were seen with sorbitol treatment on salt tolerant types. Like many priming compounds, the effects of organic alcohols can be species, cultivar, stress, and environmentally specific. Therefore, additional research and testing for efficacy of use of various alcohols is needed.

4.4.9 Volatile Organic Compounds

Plant volatile organic compounds (VOCs) are generally classified as small or heavier VOCs. Small VOCs include ethylene (discussed in hormone section), methanol, isoprene acrolein, and others. Heavier VOCs are compounds such as terpenes, methyl jasmonate, methyl salicylate, and other green-leaf volatiles. The size or potential for diffusion is thought to play a major role in where within plant canopies the compounds may effectively signal (Baldwin et al. 2006). The amount of plant VOCs emitted has been shown to increase under drought stress conditions. This has been attributed to shifts in carbon allocation and metabolism. Under drought stress, carbon supplies may become limited to the plant due to extended stomatal closure. VOCs are emitted due to drought stress because of this carbon starvation as well as other factors such as increased tissue temperatures, changes in respiration rates, and as a mechanism to eliminate toxic compounds to preserve cellular health. However, this phenomenon is not yet fully understood. VOCs also play a major role in plant response to herbivores and other pathogens. VOCs released by plants during drought may cause insects to become more attracted to the stressed plants (McDowell et al. 2008). There is also evidence that VOC signals are involved in plant–plant interactions. Baldwin et al. (2006) demonstrated that trees activated stress defenses after perceiving signals from stressed neighboring trees. Whether VOCs plays a role in plant–plant interactions during drought stress and if priming with VOCs may be an effective way to alter abiotic stress tolerance has not been thoroughly investigated.

Priming with VOCs has been shown to alter the attractiveness or have an effect on disease progression in plant-insect or plant–microbe interactions. Butenolide, a lactone compound that is a component of smoke, is one that has been tested for seed priming purposes (Ton et al. 2007). Butenolide has exhibited effects to improve plant defense against pathogens, low temperatures, seed vigor, and growth (Jisha et al. 2013). Little to no information is available on whether volatile organic compounds such as butenolide may have utility as priming agents for improving plant responses to drought stress. Z-3-Hexenyl acetate (Z-3-HAC) is another VOC that has been used for priming. In wheat plants, Z-3-HAC enhanced the defense against *Fusarium graminearum*. The response was largely regulated by SA and JA pathways (Ameye et al. 2015). As SA and JA are also known to be involved in abiotic stress tolerance, experimenting with priming with VOCs for drought tolerance may be warranted.

Microbial produced VOCs also may have an effect on plant tolerance to drought and other abiotic stresses. For instance, 2R,3R-butanediol is produced by the rhizobacteria *Pseudomonas chlororaphis*. Applied exogenously, 2R,3R-butanediol caused an effect on SAR and ISR hormones and improved drought tolerance, which may be associated with improved stomatal closure (Cho et al. 2008). Further research into this has shown that 2R,3R-butanediol can induce drought tolerance by inducing hydrogen peroxide and NO biosynthesis in plants (Cho et al. 2011). Both hydrogen peroxide and NO are known to be regulators of stomatal aperture (Xie et al. 2014). The implications of 2R,3R-butanediol induction of drought tolerance have not been fully exploited. *P. chlororaphis* has been studied as a biocontrol agent for several diseases but the priming for drought tolerance using bacteria that produce 2R,3R-butanediol needs to be moved beyond model species such as Arabidopsis. Specific use of purified VOCs as chemical priming agents seems to be a promising possibility for the future.

4.5 Conclusion

As with any technology, priming technologies have potential pros and cons of use. Not all priming agents work effectively on all crops and not all priming agents have long term effects. Different plants have different natural tolerance mechanisms and thus will differ greatly in which priming compounds may be beneficial to supplement to plants. The utility of some potential chemical priming agents is likely limited by the ease of application, cost effectiveness, potential harmful effects to humans, or the environment. For example, priming with NO is done using nitroprusside, a NO donor, which is a classified drug with known harmful effects on humans. Granted, many commonly used chemicals in agriculture are thought to have potentially harmful effects to humans or the environment. Thus, if not done already, the risks and benefits will have to be weighed for each priming agent.

With the large amount of research going in to plant priming practices, more specific information will become available about the benefits of priming with different chemical agents for drought tolerance in more crop species. The future of priming for drought tolerance in plants as a method to improve crop production worldwide relies on such research. Fortunately, with the ever increasing speed of gene discovery, advances in gene expression technologies, and advances in other biochemical analysis such as proteomics, new priming compounds with new gene targets will likely become available. Since many agents can be combined to increase the impact of priming, for instance using both hormonal and salt priming compounds simultaneously, there is a great potential for research to reveal the most effective combinations for various crop species.

In addition to the compounds and methods mentioned in this chapter, many other chemicals exist that are being researched for use as priming agents. For instance, various lactone compounds, compounds isolated from plant pathogens such as chitosan, additional salts such as copper and zinc sulfate, and several other metabolites are being investigated for potential in mitigating drought stress damage in plants. Regardless of the chemical agent under investigation, the ultimate goal of priming for drought tolerance should be to find priming methods that harness the power of several drought resistance mechanisms including avoidance and tolerance mechanisms for a truly broad spectrum type of improvement in drought tolerance.

References

Afzal I, Basra SMA, Cheema MA, et al. Seed priming: a shotgun approach for alleviation of salt stress in wheat. Int J Agric Biol. 2013;15:1199–203.

Agboma PC, Sinclair TR, Jokinen K, et al. An evaluation of the effect of exogenous glycinebetaine on the growth and yield of soybean: timing of application, watering regimes and cultivars. Field Crops Res. 1997;54:51–64.

Ahmad R, Waraich EA, Nawaz F, et al. Selenium (Se) improves drought tolerance in crop plants—a myth or fact? J Sci Food Agric. 2015. doi:10.1002/jsfa.7231.

Ajouri A, Haben A, Becker M. Seed priming enhances germination and seedling growth of barley under conditions of P and Zn deficiency. J Plant Nutr Soil Sci. 2004;167:630–6.

Akram NA, Ashraf M. Regulation in plant stress tolerance by a potential plant growth regulator, 5-aminolevulinic acid. J Plant Growth Regul. 2013;32:663–79.

Akter N, Islam MR, Karim MA, et al. Alleviation of drought stress in maize by exogenous application of gibberellic acid and cytokinin. J Crop Sci Biotechnol. 2014;17:41–8.

Alam MM, Nahar K, Hasanuzzaman M, Fujita M. Exogenous jasmonic acid modulates the physiology, antioxidant defense and glyoxalase systems in imparting drought stress tolerance in different Brassica species. Plant Biotechnol Rep. 2014;8:279–93.

Alcazar R, Bitrian M, Bartels D, Koncz C, Altabella T, Tiburcio AF. Polyamine metabolic canalization in response to drought stress in Arabidopsis and the resurrection plant *Craterostigma plantagineum*. Plant Sgl Behavior. 2011;6:243–50.

Ali B, Wang B, Ali S, et al. 5-Aminolevulinic acid ameliorates the growth, photosynthetic gas exchange capacity, and ultrastructural changes under cadmium stress in *Brassica napus* L. J Plant Growth Regul. 2013;32:604–14. doi:10.1007/s00344-013-9328-6.

Alonso-Ramírez A, Rodríguez D, Reyes D, Jiménez JA, Nicolás G, López-Climent M, Gómez-Cadenas A, Nicolás C. Evidence for a role of gibberellins in salicylic acid-modulated early plant responses to abiotic stress in Arabidopsis seeds. Plant Physiol. 2009;150:1335–44.

Ameye M, Audenaert K, De Zutter N, et al. Priming of wheat with the green leaf volatile z-3-hexenyl acetate enhances defense against fusarium graminearum but boosts deoxynivalenol production. Plant Physiol. 2015;167:1671–84. doi:10.1104/pp.15.00107.

Anjum SA, Wang LC, Farooq M, et al. Brassinolide application improves the drought tolerance in maize through modulation of enzymatic antioxidants and leaf gas exchange. J Agron Crop Sci. 2011;197:177–85.

Araujo L, Bispo WS, Rios VS, et al. Induction of the phenylpropanoid pathway by acibenzolar-s-methyl and potassium phosphite increases mango resistance to ceratocystis fimbriata infection. Plant Dis. 2015;99:447–59.

Ashraf M, Foolad MR. Roles of glycine betaine and proline in improving plant abiotic stress resistance. Environ Expt Bot. 2007;59:206–16.

Athar H, Khan A, Ashraf M. Exogenously applied ascorbic acid alleviates salt-induced oxidative stress in wheat. Environ Expt Bot. 2008;63:224–31.

Baldwin IT, Halitschke R, Paschold A, et al. Volatile signaling in plant-plant interactions: "talking trees" in the genomics era. Science. 2006;311:812–5.

Balestrasse KB, Tomaro ML, Batlle A, et al. The role of 5-aminolevulinic acid in the response to cold stress in soybean plants. Phys Chem Chem Phys. 2010;71:2038–45.

Bandurska H. The effect of salicylic acid on barley response to water deficit. Acta Physiologiae Plantarum. 2005;27:379–86.

Beckers GJM, Conrath U. Priming for stress resistance: from the lab to the field. Curr Oper Plant Biol. 2007;10:425–31.

Bell AE. Nonprotein amino acids of plants: significance in medicine, nutrition, and agriculture. J Agric Food Chem. 2003;51:2854–65.

Bian X, Merewitz E, Huang B. Effects of trinexapac-ethyl on drought responses in creeping bentgrass associated with water use and osmotic adjustment. J Amer Soc Hort Sci. 2009;134:505–10.

Bitrián M, Zarza X, Altabella T, et al. Polyamines under abiotic stress: metabolic crossroads and hormonal crosstalks in plants. Metabolites. 2012;2:516–28. doi: 10.3390/metabo2030516.

Bouché N, Fromm H. GABA in plants: Just a metabolite? Trends Plant Sci. 2004;9:110–5.

Bown AW, Shelp BJ. The metabolism and functions of γ–aminobutyric acid. Plant Physiol. 1997;115:1–5.

Bray EA. Abscisic acid regulation of gene expression during water-deficit stress in the era of the Arabidopsis genome. Plant Cell Environ. 2002;25:153–61.

Cai W, Liu W, Wang WS, et al. Overexpression of rat neurons nitric oxide synthase in rice enhances drought and salt tolerance. PLoS One. 2015;10(6), e0131599. doi:10.1371/journal.pone.0131599.

Cakmak I. The role of potassium in alleviating detrimental effects of abiotic stresses in plants. J Plant Nutr Soil Sci. 2005;168:521–30. doi:10.1002/jpln.200420485521.

Chen K, Arora R. Dynamics of the antioxidant system during seed osmopriming, post-priming germination, and seedling establishment in spinach (*Spinacia oleracea*). Plant Sci. 2011;180:212–20.

Chen W, Yao X, Cai K, et al. Silicon alleviates drought stress of rice plants by improving plant water status, photosynthesis and mineral nutrient absorption. Biol Trace Element Res. 2011;142:67–76.

Chen JH, Jiang HW, Hsieh EJ, et al. Drought and salt stress tolerance of an Arabidopsis glutathione s-transferase u17 knockout mutant are attributed to the combined effect of glutathione and abscisic acid. Plant Physiol. 2012a;158:340–51.

Chen K, Fessehaie A, Arora R. Dehydrin metabolism is altered during seed osmopriming and subsequent germination under chilling and desiccation in Spinacia oleracea L. cv. Bloomsdale: possible role in stress tolerance. Plant Sci. 2012b;183:27–36. doi:10.1016/j.plantsci.2011.11.002.

Chen K, Fessehaie A, Arora R. Aquaporin expression during seed osmopriming and post-priming germination in spinach. Biologia Plantarum. 2013;57:193–8.

Cho SM, Kang BR, Han SH, et al. 2R,3R-butanediol, a bacterial volatile produced by pseudomonas chlororaphis o6, is involved in induction of systemic tolerance to drought in Arabidopsis thaliana. Mol Plant Microbe Interact. 2008;21:1067–75. doi:10.1094/MPMI-21-8-1067.

Cho SM, Park JY, Han SH, et al. Identification and transcriptional analysis of priming genes in *Arabidopsis thaliana* induced by root colonization with pseudomonas chlororaphis O6. Plant Pathol J. 2011;27:272–9.

Colebrook EH, Thomas SG, Phillips AL, et al. The role of gibberellin signalling in plant responses to abiotic stress. J Exp Biol. 2014;217:67–75. doi:10.1242/jeb.089938.

Conrath U. Priming of induced plant defense responses. In: Van Loon LC, editor. Plant innate immunity, advances in botanical research, vol. 51. Oxford: Elsevier; 2009. p. 361–95.

Conrath U. Molecular aspects of defence priming. Trends Plant Sci. 2011;16:524–31.

Cortez-Baheza E, Cruz-Fernandez F, Peraza-Luna F, et al. A new Lea gene is induced during osmopriming of *Capsicum annuum* L. seeds. Int J Bot. 2008;4:77–84.

Csiszár J, Horváth E, Váry Z, et al. Glutathione transferase supergene family in tomato: salt stress-regulated expression of representative genes from distinct GST classes in plants primed with salicylic acid. Plant Physiol Biochem. 2014;78:15–26.

Delker C, Stenzel I, Hause B, Miersch O, Feussner I, Wasternack C. Jasmonate biosynthesis in Arabidopsis thaliana—enzymes, products, regulation. Plant Biol. 2006;8:297–306. doi:10.1007/s00344-011-9216-x.

Ding Y, Fromm M, Avramova Z. Multiple exposures to drought "train" transcriptional responses in Arabidopsis. Nat Commun. 2012;3:740. doi:10.1038/ncomms1732.

Eisvand HR, Tavakkol-Afshari R, Sharifzadeh F, et al. Effects of hormonal priming and drought stress on activity and isozyme profiles of antioxidant enzymes in deteriorated seed of tall wheatgrass (*Agropyron elongatum* host). Seed Sci Technol. 2010;38:280–97.

Eivazi. Induction of drought tolerance with seed priming in wheat cultivars (Triticum aestivum L.). Int Res J Appl Basic Sci. 2011;2:95–105. doi:10.2478/v10014-012-0003-6.

Fait A, Yellin A, Fromm H. GABA shunt deficiencies and accumulation of reactive oxygen intermediates: insight from Arabidopsis mutants. FEBS Lett. 2005;579:415–20.

Fait A, Fromm H, Walter D, et al. Highway or byway: the metabolic role of the GABA shunt in plants. Trends Plant Sci. 2008;13:14–9.

Farooq M, Basra SMA, Rehman H, et al. Seed priming with polyamines improves the germination and early seedling growth in fine rice. J New Seeds. 2008;9:145–55.

Farooq M, Wahid A, Lee DJ. Exogenously applied polyamines increase drought tolerance of rice by improving leaf water status, photosynthesis and membrane properties. Acta Physiol Plant. 2009;31:937–45.

Farooq M, Irfan M, Aziz T, et al. Seed priming with ascorbic acid improves drought resistance of wheat. J Agron Crop Sci. 2013;199:12–22.

Fercha A, Capriotti AL, Caruso G, Cavaliere C, Samperi R, Stampachiacchiere S, Laganà A. Comparative analysis of metabolic proteome variation in ascorbate-primed and unprimed wheat seeds during germination under salt stress. J Proteomics. 2014;108:238–57.

Fukuto JM, Cho JY, Sitzer CH. The chemical properties of nitric oxide and related nitrogen oxides. In: Ignarro LJ, editor. Nitric oxide: biology and pathobiology. 2mdth ed. San Diego: Academic; 2000.

Gallardo K, Job C, Groot S, et al. Proteomic analysis of Arabidopsis seed germination and priming. Plant Physiol. 2001;126:835–48.

Gallardo K, Job C, Groot SPC, et al. Proteomics of Arabidopsis seed germination and priming. In: Nicholas G et al., editors. The biology of seeds: recent advances. Cambridge: CABI; 2002. p. 199–209.

Gao X, Zou C, Wang L, et al. Silicon decreases transpiration rate and conductance from stomata of maize plants. J Plant Nutr. 2006;29:1637–47.

Gong H, Chen K. The regulatory role of silicon on water relations, photosynthetic gas exchange, and carboxylation activities of wheat leaves in field drought conditions. Acta Physiologiae Plantarum. 2012;34(4):1589–94.

Groß F, Durner J, Gaupels F. Nitric oxide, antioxidants and prooxidants in plant defence responses. Front Plant Sci. 2013;4:419.

Guo P, Baum M, Grando S, et al. Differentially expressed genes between drought-tolerant and drought-sensitive barley genotypes in response to drought stress during the reproductive stage. J Expt Bot. 2009;60:3531–44.

Habibi G. Silicon supplementation improves drought tolerance in canola plants. Russian J Plant Physiol. 2014;61:784–91.

Hare PD, Cress WA. Metabolic implications of stress-induced proline accumulation in plants. Plant Growth Regul. 1991;21:79–102.

Hare PD, Cress WA, Staden JV. Disruptive effects of exogenous proline on chloroplast and mitochondrial ultrastructure in Arabidopsis leaves. South African J Bot. 2002;68:393–6.

Hasanuzzaman M, Fujita M. Selenium pretreatment upregulates the antioxidant defense and methylglyoxal detoxification system and confers enhanced tolerance to drought stress in rapeseed seedlings. Biol Trace Elem Res. 2011;143:1758–76.

Hasanuzzaman M, Hossain MA, Fujita M. Selenium-induced up-regulation of the antioxidant defense and methylglyoxal detoxification system reduces salinity-induced damage in rapeseed seedlings. Biol Trace Elem Res. 2010. doi:10.1007/s12011-011-8958-4.

Hase S, Takahashi S, Takenaka S, et al. Involvement of jasmonic acid signaling in bacterial wilt disease resistance induced by biocontrol agent Pythium oligandrum in tomato. Plant Path. 2008;57:870–6.

He C, Zhang W, Gao Q, et al. Enhancement of drought resistance and biomass by increasing the amount of glycine betaine in wheat seedlings. Euphytica. 2011;177:151–67.

Heuer B. Influence of exogenous application of proline and glycinebetaine on growth of salt-stressed tomato plants. Plant Sci. 2003;165:693–9.

Horváth E, Szalai G, Janda T. Induction of abiotic stress tolerance by salicylic acid signaling. J Plant Growth Regul. 2007;26:290–300.

Horváth E, Brunner S, Bela K, et al. Exogenous salicylic acid-triggered changes in the glutathione transferases and peroxidases are key factors in the successful salt stress acclimation of *Arabidopsis thaliana*. Func Plant Biol. 2015;42(12):1129–40.

Hossain MA, Fujita M. Hydrogen peroxide priming stimulates drought tolerance in mustard (*Brassica juncea* L.). Plant Gene Trait. 2013;4:109–23.

Hossain MA, Bhattacharjee S, Armin SM, et al. Hydrogen peroxide priming modulates abiotic oxidative stress tolerance: insights from ROS detoxification and scavenging. Front Plant Sci. 2015;6:420. doi:10.3389/fpls.2015.00420.

Hu L, Wang Z, Huang B. Effects of cytokinin and potassium on stomatal and photosynthetic recovery of Kentucky Bluegrass from drought stress. Crop Sci. 2013;53:221–31.

İşeri OD, Körpe DA, Sahin FI, et al. Seed priming to increase germination, drought tolerance and yield of cucumber. Adv Appl Agric Sci. 2015;03:42–53.

Ishibashi Y, Yamaguchi H, Yuasa T, Iwaya-Inoue M, Arima S, Zheng SH. Hydrogen peroxide spraying alleviates drought stress in soybean plants. J Plant Physiol. 2011;168:1562–7.

Jakab G, Ton J, Flors V, et al. Enhancing Arabidopsis salt and drought stress tolerance by chemical priming for its abscisic acid responses. Plant Physiol. 2005;139:267–74. doi:10.1104/pp.105.065698.

Jisha KC, Vijayakumari K, Puthur JT. Seed priming for abiotic stress tolerance: an overview. Acta Physiol Plant. 2013;35:1381–96. doi:10.1007/s11738-012-1186-5.

Kadioglu A, Saruhan N, Sağlam A, et al. Exogenous salicylic acid alleviates effects of long term drought stress and delays leaf rolling by inducing antioxidant system. Plant Growth Regul. 2011;64:27–37.

Kairong L, Huike L, Jian W. Effect of natural brassinolide on drought resistance and yield of red fuji apple. Acta Hort Sinica. 2006;33:1059.

Karakas B, Ozias-akins P, Stushnoff C, et al. Salinity and drought tolerance of mannitol-accumulating transgenic tobacco. Plant Cell Environ. 1997;20:609–26.

Kaya C, Tuna AL, Alves AAC. Gibberellic acid improves water deficit tolerance in maize plants. Acta Physiol Plant. 2006;28:331–7. doi:10.1007/s11738-006-0029-7.

Khan HA, Ziaf K, Amjad M, et al. Exogenous application of polyamines improves germination and early seedling growth of hot pepper. Chilean J Agric Res. 2012;72:429–33.

Kibinza S, Bazin J, Bailly C, et al. Catalase is a key enzyme in seed recovery from ageing during priming. Plant Sci. 2011;181:309–15. doi:10.1016/j.plantsci.2011.06.003.

Kim TH, Maik BĆ. Guard cell signal transduction network: advances in understanding abscisic acid, CO2, and Ca2+ signaling. Annu Rev Plant Physiol Plant Mol Biol. 2010;61:561–91.

Kinnersley AM, Turano FJ. Gamma aminobutyric acid (GABA) and plant response to stress. Crit Rev Plant Sci. 2000;19:479–509.

Kinoshita T, Seki M. Epigenetic memory for stress response and adaptation in plants. Plant Cell Physiol. 2014;55:1859–63. doi:10.1093/pcp/pcu125.

Korkmaz A, Ozbay N, Tiryaki I, et al. Combining priming and plant growth regulators improves muskmelon germination and emergence at low temperatures. Euro J Hort Sci. 2005;70:29–34.

Korkmaz A, Korkmaz Y, Demirkiran AR. Enhancing chilling stress tolerance of pepper seedlings by exogenous application of 5-aminolevulinic acid. Environ Expt Bot. 2010;67:495–501.

Krishnan S, Merewitz E. Drought stress and trinexapac-ethyl modify phytohormone content within kentucky bluegrass leaves. J Plant Growth Regul. 2015;34:1–12.

Krishnan S, Laskowski K, Shukla V, et al. Mitigation of drought stress damage by exogenous application of a non-protein amino acid γ– aminobutyric acid on perennial ryegrass. J Amer Soc Hort Sci. 2013;138:358–66.

Kubala S, Wojtyla Ł, Met Q, et al. Enhanced expression of the proline synthesis gene P5CSA in relation to seed osmopriming improvement of Brassica napus germination under salinity stress. J Plant Physiol. 2015;183:1–12.

Kusano T, Berberich T, Tateda C, et al. Polyamines: essential factors for growth and survival. Planta. 2008;228:367–81.

Kuznetsov VV, Kholodova VP, Kuznetsov VV, et al. Selenium regulates the water status of plants exposed to drought. Doklady Biol Sci. 2003;390:266–8.

Lancien M, Roberts MR. Regulation of *Arabidopsis thaliana* 14–3–3 gene expression by γ–aminobutyric acid. Plant Cell Environ. 2006;29:1430–6.

Lara TS, Lira JMS, Rodrigues AC, et al. Potassium nitrate priming affects the activity of nitrate reductase and antioxidant enzymes in tomato germination. J Agric Sci. 2014;6:72–9.

Larkindale J, Huang B. Thermotolerance and antioxidant systems in *Agrostis stolonifera*: involvement of salicylic acid, abscisic acid, calcium, hydrogen peroxide, and ethylene. J Plant Physiol. 2004;161:405–13.

Li DM, Zhang J, Sun WJ, et al. 5-Aminolevulinic acid pretreatment mitigates drought stress of cucumber leaves through altering antioxidant enzyme activity. Sci Hortic. 2011;130:820–8.

Liu H, Ji X, Yu B, et al. Relationship between osmotic stress and polyamines conjugated to the deoxyribonucleic acid-protein in wheat seedling roots. Sci China. 2006;49:12–7.

Liu D, Pei ZF, Naeem MS, et al. 5-aminolevulinic acid activates antioxidative defence system and seedling growth in brassica napus l. under water-deficit stress. J Agron Crop Sci. 2011;197:284–95.

McDowell N, Pockman WT, Allen CD, et al. Mechanisms of plant survival and mortality during drought: why do some plants survive while others succumb to drought? New Phytol. 2008;178:719–39.

Merewitz E, Gianfagna T, Huang B. Photosynthesis, water use, and root viability under water stress as affected by expression of *SAG12-ipt* controlling cytokinin synthesis in *Agrostis stolonifera*. J Expt Bot. 2011;62:383–95. doi:10.1093/jxb/erq285.

Merewitz EB, Du H, Yu Q, et al. Elevated cytokinin content in ipt transgenic creeping bentgrass promotes drought tolerance through regulating metabolite accumulation. J Expt Bot. 2012;63:1315–28.

Michaeli S, Fromm H. Closing the loop on the GABA shunt in plants: are GABA metabolism and signaling entwined? Front Plant Sci. 2015;6:419. doi:10.3389/fpls.2015.00419.

Mittler R, Vanderauwera S, Suzuki N, et al. ROS signaling: the new wave? Trends Plant Sci. 2011;16:300–9. doi:10.1016/j.tplants.2011.03.007.

Mondal S, Bose B. An impact of seed priming on disease resistance: a review. In: Kharwar, editor. Microbial diversity biotechnology food seconds. New York: Springer; 2014. p. 193–203.

Mori IC, Pinontoan R, Kawano T, et al. Involvement of superoxide generation in salicylic acid-induced stomatal closure in *Vicia faba*. Plant Cell Physiol. 2001;42(12):1383–8.

Moustakas M, Sperdouli I, Kouna T, et al. Exogenous proline induces soluble sugar accumulation and alleviates drought stress effects on photosystem II functioning of *Arabidopsis thaliana* leaves. Plant Growth Regul. 2011;65:315–25.

Müssig C, Biesgen C, Lisso J, et al. A novel stress-inducible 12-oxophytodienoate reductase from Arabidopsis thaliana provides a potential link between brassinosteroid-action and jasmonic-acid synthesis. J Plant Physiol. 2000;157:143–52.

Nahar K, Hasanuzzaman M, Alam M, et al. Glutathione-induced drought stress tolerance in mung bean: coordinated roles of the antioxidant defence and methylglyoxal detoxification systems. AoB Plants. 2015;7:69–75.

Neill S, Barros R, Bright J, et al. Nitric oxide, stomatal closure, and abiotic stress. J Expt Bot. 2008;59:165–76. doi:10.1093/jxb/erm293.

Nemeth M, Janda T, Horvath E, et al. Exogenous salicylic acid increases polyamine content but may decrease drought tolerance in maize. Plant Sci. 2002;162:569–74.

Nolla A, de Faria RJ, Korndörfer GH, et al. Effect of silicon on drought tolerance of upland rice. J Food Agric Environ. 2012;10:269–72.

Ollas C, Hernando B, Arbona V, et al. Jasmonic acid transient accumulation is needed for abscisic acid increase in citrus roots under drought stress conditions. Physiol Plantarum. 2012;147:296–306.

Paparella S, Araujo SS, Rossi G, et al. Seed priming: state of the art and new perspectives. Plant Cell Rep. 2015;34(8):1281–93.

Qin F, Kodaira KS, Maruyama K, et al. SPINDLY, a negative regulator of gibberellic acid signaling, is involved in the plant abiotic stress response. Plant Physiol. 2011;157:1900–13. doi:10.1104/pp.111.187302.

Rahman TA, Oirdi ME, Gonzalez-Lamothe R, et al. Necrotrophic pathogens use the salicylic acid signaling pathway to promote disease development in tomato. Mol Plant Microbe Interact. 2012;25:1584–93. doi:10.1094/MPMI-07-12-0187-R.

Renault H, Amrani AE, Palanivelu R, et al. GABA accumulation causes cell elongation defects and a decrease in expression of genes encoding secreted and cell wall-related proteins in *Arabidopsis thaliana*. Plant Cell Physiol. 2011;52:894–908.

Rezaei MA, Jokar I, Ghorbanli M, et al. Morpho-physiological improving effects of exogenous glycine betaine on tomato (*Lycopersicum esculentum* Mill.) cv. PS under drought stress conditions. Plant Omics J. 2012;5:79–86.

Rosenthal GA. Plant nonprotein amino and imino acids. New York: Academic; 1982.

Ryals JA, Neuenschwander UH, Willits MG, et al. Systemic acquired resistance. Plant Cell. 1996;8:1809–19.

Savchenko T, Kolla VA, Wang CQ, et al. Functional convergence of oxylipin and abscisic acid pathways controls stomatal closure in response to drought. Plant Physiol. 2014;164:1151–60.

Seckin B, Sekman AH, Turkan I. An enhancing effect of exogenous mannitol on the antioxidant enzyme activities in roots of wheat under salt stress. J Plant Growth Regul. 2009;28:12–20.

Senaratna T, Touchell D, Bunn E, et al. Acetyl salicylic acid (Aspirin) and salicylic acid induce multiple stress tolerance in bean and tomato plants. Plant Growth Regul. 2000a;30:157–61.

Senaratna T, Touchell D, Bunn E, et al. Acetyl salicylic acid (aspirin) and salicylic acid induce multiple stress tolerance in bean and tomato plants. Plant Growth Regul. 2000b;30:157–61.

Serraj R, Shelp BJ, Sinclair TR. Accumulation of γ-aminobutyric acid in nodulated soybean in response to drought stress. Physiol Plant. 1998;102:79–86.

Shelp BJ, Bown AW, McLean MD. Metabolism and functions of gamma—aminobutyric acid. Trends Plant Sci. 1999;4:446–52.

Shelp BJ, Bozzo GG, Trobacher CP, et al. Strategies and tools for studying the metabolism and function of γ–aminobutyrate in plants. I Pathway structure. Botany. 2012;90:651–68.

Shen X, Zhou Y, Duan L, et al. Silicon effects on photosynthesis and antioxidant parameters of soybean seedlings under drought and ultraviolet-B radiation. J Plant Physiol. 2010;167:1248–52.

Shi H, Chan Z. Improvement of plant abiotic stress tolerance through modulation of the polyamine pathway. J Integr Plant Biol. 2014;56:114–21.

Shi H, Ye T, Zhulong C. Comparative proteomic and physiological analyses reveal the protective effect of exogenous polyamines in the bermudagrass (*Cynodon dactylon*) response to salt and drought stresses. J Proteome Res. 2013. doi:10.1021/pr400479k.

Sidana S, Bose J, Shabala L, et al. Nitric oxide in drought stress signaling and tolerance in plants. In: Khan M, Mobin M, Mohammad F, Corpas FJ, editors. Nitric oxide action in abiotic stress responses in plants. Switzerland: Springer International; 2015.
Song ZZ, Yang SY, Zuo J, et al. Over-expression of ApKUP3 enhances potassium nutrition and drought tolerance in transgenic rice. Biologia Plantarum. 2014;58:649–58.
Sós-Hegedűs A, Juhás Z, Poór P, et al. Soil drench treatment with ß-aminobutyric acid increases drought tolerance of potato. PLoS One. 2014;9(12), e114297. doi:10.1371/journal.pone.0114297.
Theerakulpisut P, Gunnula W. Exogenous sorbitol and trehalose mitigated salt stress damage in salt-sensitive but not salt-tolerant rice seedlings. Asian J Crop Sci. 2012;4:165–70.
Ton J, Jakab G, Toquin V, et al. Dissecting the β-aminobutyric acid–induced priming phenomenon in Arabidopsis. Plant Cell. 2005;17:987–99. doi:10.1105/tpc.104.029728.
Ton J, D'Alessandro M, Jourdie V, et al. Priming by airborne signals boosts direct and indirect resistance in maize. Plant J. 2007;49:16–26.
Torrigiani P, Bressanin D, Ruiz KB. Spermidine application to young developing peach fruits leads to a slowing down of ripening by impairing ripening-related ethylene and auxin metabolism and signaling. Physiol Plant. 2012;146:86–98.
Van Hoewyk D, Takahashi H, Inoue E, et al. Transcriptome analyses give insights into selenium-stress responses and selenium tolerance mechanisms in Arabidopsis. Physiol Plant. 2008;132:236–53.
Van Loon LC, Van Strien EA. The families of pathogenesis-related proteins, their activities, and comparative analysis of PR-1 type proteins. Phys Mol Plant Pathol. 1999;55:85–97.
Vardhini BV. Brassinosteroids' role for amino acids, peptides and amines modulation in stressed plants—a review. In: Anjum NA et al., editors. Plant adaptation to environmental change: significance of amino acids and their derivatives. Wallingford, CT: CAB International; 2014.
Vardhini BV, Anjum NA. Brassinosteroids make plant life easier under abiotic stresses mainly by modulating major components of antioxidant defense system. Front Environ Sci. 2015;2:67.
Vendruscolo ECG, Schuster I, Pileggi M, et al. Stress-induced synthesis of proline confers tolerance to water deficit in transgenic wheat. J Plant Physiol. 2007;164:1367–76.
Vijayakumari K, Puthur JT. γ-Aminobutyric acid (GABA) priming enhances the osmotic stress tolerance in *Piper nigrum* Linn. plants subjected to PEG-induced stress. Plant Growth Regul. 2015;12:2–11. doi:10.1007/s10725-015-0074-6.
Wani SH, Singh NB, Haribhushan A, et al. Compatible solute engineering in plants for abiotic stress tolerance - role of glycine betaine. Curr Genom. 2013;14:157–65.
Waseem M, Athar HUR, Ashraf M. Effect of salicylic acid applied through rooting medium on drought tolerance of wheat. Pak J Bot. 2006;38:1127–36.
Wei J, Li C, Li Y, et al. Effects of external potassium (k) supply on drought tolerances of two contrasting winter wheat cultivars. PLoS One. 2013;8(7), e69737.
Xie SS, Wu HJ, Zang HY, et al. Plant growth promotion by spermidine-producing Bacillus subtilis OKB105. Mol Plant Microbe Interact. 2014;27:655–63.
Yan M. Seed priming stimulate germination and early seedling growth of Chinese cabbage under drought stress. South African J Bot. 2015;99:88–92.
Yan H, ZhiGang L, YunZhao C, YuGuo W. Effects of exogenous proline on the physiology of soyabean plantlets regenerated from embryos in vitro and on the ultrastructure of their mitochondria under NaCl stress. Soybean Sc. 2000;19:314–9.
Yang A, Cao S, Yang Z, et al. γ– Aminobutyric acid treatment reduces chilling injury and activates the defense response of peach fruit. Food Chem. 2011;129:1619–22.
Yang DL, Yao J, Mei CS, et al. Plant hormone jasmonate prioritizes defense over growth by interfering with gibberellin signaling cascade. Proc Natl Acad Sci U S A. 2012;109(9):152–99. doi:10.1073/pnas.1201616109.
Yang Z, Chang Z, Sun L, et al. Physiological and metabolic effects of 5-aminolevulinic acid for mitigating salinity stress in creeping bentgrass. PLoS One. 2014. doi:10.1371/journal.pone.0116283.

Yasuda M, Ishikawa A, Jikumaru Y, et al. Antagonistic interaction between systemic acquired resistance and the abscisic acid-mediated abiotic stress response in Arabidopsis. Plant Cell. 2008;20:1678–92. doi:10.1105/tpc.107.054296.

Yin L, Wang S, Liu P, et al. Silicon-mediated changes in polyamine and 1-aminocyclopropane-1-carboxylic acid are involved in silicon-induced drought resistance in *Sorghum bicolor* L. Plant Physiol Biochem. 2014;80:268–77.

Zare N, Noori S, Ahmad S, et al. Effect of laser priming on accumulation of free proline in spring durum wheat (*Triticum turgidum* L.) under salinity stress. Intl Transaction J Eng Mgt App Sci Tech. 2014;5:119–75.

Zhang MQ, Chen RK, Yu SL. Changes of polyamine metabolism in drought-stressed sugarcane leaves and their relation to drought resistance. Acta Phytophysiologia Sinica. 1996;22:327–32.

Zhang M, Zhai Z, Tian X, et al. Brassinolide alleviated the adverse effect of water deficits on photosynthesis and the antioxidant of soybean (*Glycine max* L.). Plant Growth Regul. 2008;56:257–64.

Chapter 5
Osmotic Adjustment and Plant Adaptation to Drought Stress

Marek Zivcak, Marian Brestic, and Oksana Sytar

5.1 Introduction

Drought negatively affects the productivity of crops and terrestrial plant ecosystems (Passioura 2007; Ciais et al. 2005) as well as plant biodiversity (Engelbrecht et al. 2007). It can have a global influence on carbon gain (Buermann et al. 2007), as it is a global phenomenon; a varying extent of drought can be found in almost all climatic zones. It is mostly connected with insufficient precipitation, but the plant water deficit can have many other causes. Drought highly depends on abiotic, biotic, and human activity–related factors, such as temperature, air humidity, winds, vegetation, and soil management, among others. A water deficit often leads to losses of crop yields, which involves the drought among the main global problems. Plants are exposed to drought stresses in two ways: by limiting root water uptake and by exceeding the transpiration rate over the threshold limit (Anjum et al. 2011a, b).

A water deficit usually affects plants, mainly through its effects on assimilation. However, photosynthetic drought responses are very complex and strongly depend on the plant development (Chaves et al. 2009). Drought stress effects can be direct or indirect. A decrease in carbon dioxide (CO_2) availability due to a decrease in stomatal and mesophyll conductance represents a direct effect (Lal et al. 1996; Chaves et al. 2002; Flexas et al. 2012; Zivcak et al. 2013). Indirect effects are connected with changes in photosynthetic metabolism (Tezara et al. 1999; Maroco et al. 2002; Lawlor and Cornic 2002; Parry et al. 2002, among others). In conditions of limited CO_2 diffusion, photorespiration lowers the energetic efficiency of photosynthesis in

M. Zivcak (✉) • M. Brestic
Department of Plant Physiology, Slovak University of Agriculture,
Tr. A. Hlinku 2, Nitra 94976, Slovak Republic
e-mail: marek.zivcak@uniag.sk

O. Sytar
Department of Plant Physiology and Ecology, Taras Shevchenko National University of Kyiv,
Volodymyrs'ka St. 64, Kyiv 01601, Ukraine

M.A. Hossain et al. (eds.), *Drought Stress Tolerance in Plants, Vol 1*,
DOI 10.1007/978-3-319-28899-4_5

C3 plants (Ogren 1984). Other indirect effects go through oxidative stress, contributing to the nonstomatic limitation of photosynthesis (Ort 2001; Chaves and Oliveira 2004; Foyer and Noctor 2009). Drought can lead to long-term alterations in the content, structure, and activity of the individual components of the photosynthetic apparatus (Tezara et al. 1999; Balaguer et al. 2002; Zivcak et al. 2008, 2014; Kohzuma et al. 2009; Brestic and Zivcak 2013).

In addition to photosynthesis, all the key processes, such as the biosynthesis of cellular elements, proteosynthesis, energy and lipid metabolism, membrane permeability, and activities of enzymes, are ceased by drought (Liu et al. 2011; Filippou et al. 2014). Moreover, plants adapt various changes in response to drought stress, such as alterations in their growth rate and plant morphology and an improvement in defense mechanisms (Duan et al. 2007). One of the most important adaptive mechanisms is also an accumulation of osmotically active compounds (osmolytes), called *osmotic adjustment* (OA) (Hsiao et al. 1976; Munns 1988; Zivcak et al. 2009).

OA represents the lowering of the osmotic potential caused by the accumulation of solutes in response to a water deficit (Condon 1982; Chandra Babu et al. 1999; Zhang et al. 1999). The term "osmotic adjustment" can be used when new solutes are accumulated but not when the decrease in the osmotic potential is caused by a concentration of existing solutes due to water loss (Chandra Babu et al. 1999). OA is considered an important feature of plant drought tolerance (Levitt 1980; Jones et al. 1981; Blum 1988; Chandra Babu et al. 1999), but it is also a general cellular response to water shortage (Zhang et al. 1999). In ecosystems limited by water availability, plants capable of accumulating osmolytes will maintain a higher vitality and better survival rates (Pintó-Marijuan and Munné-Bosch 2013). The topic of OA has received much attention in previous decades, and despite some controversies, an agreement on several protective functions of accumulated osmolytes against detrimental effects of drought was achieved.

5.2 Physiological Functions of OA

Although the accumulation of osmotically active compounds represents a general stress response, from a functional viewpoint, two distinct roles of OA can be recognized: osmoregulation and osmoprotection. These two terms are often improperly used as synonyms mainly because the majority of osmotically active compounds play both osmoprotective and osmoregulatory functions in plant tissues. The term "osmotic adjustment" in early studies fully corresponded to osmoregulatory function (Condon 1982; Ludlow and Muchow 1990). However, for practical reasons it is advantageous to dispose of a general term covering all physiological effects associated with the accumulation of osmolytes. Therefore, in the next two sections we present osmoregulation and osmoprotection separately, as the two distinct functions of OA in plants.

5.2.1 Osmoregulation

Osmoregulation, as a part of OA, represents the active lowering of the osmotic potential in plant cells via the accumulation of osmotically active compounds. It is considered an effective and beneficial component of drought resistance in plants (Ludlow and Muchow 1990; Chandra Babu et al. 1999) mostly because the turgor pressure, which is associated with a host of metabolic and physiological processes for which the presence of cell turgor is crucial, is maintained (Jones et al. 1981; Sharp et al. 1990). The decrease in osmotic potential is caused by the accumulation of different inorganic and organic osmolytes. The contribution of individual osmolytes to the total osmoregulation depends strongly on the plant species and environmental conditions. For example, inorganic ions such as K^+ contribute most to the decrease in osmotic potential under water deficit in cereal crops (Munns et al. 1979; Condon 1982; Morgan 1992). However, the contribution of organic solutes can also be important in some species; in particular, the soluble sugars can contribute considerably to the osmoregulatory function of OA. For example, sugars were reported to contribute more than 50 % to the decrease in osmotic potential in *Sorghum* (Jones et al. 1981). Unlike osmoprotective effects, osmoregulatory effects depend more on the total capacity to decrease the osmotic potential than on the composition of osmotically active compounds. For example, shifts in the osmotic potential as a result of the accumulation of osmolytes reached up to 2 MPa in wheat (Morgan 1977, 1983; Blum et al. 1999), 1.5 MPa in rice (Lilley and Ludlow 1996), and 1.7 MPa in sorghum (Basnayake et al. 1993). These values demonstrate the enormous physical effects of osmoregulation in plant cells, resulting in several important physiological effects of osmoregulation, described next.

5.2.1.1 Turgor Maintenance and Delayed Stomata Closure

Advantages of accumulating osmotically active solutes include the fact that tissues that accumulate more solutes via the process of OA can maintain positive turgor over a wider water potential (Ψ_w) range and incur less shrinkage below the turgor loss point, thus preventing damage to the integrity of cell membranes and other cell constituents and avoiding cell death associated with stress-induced leaf necrosis (Melkonian et al. 1982; Setter 2012). Maintaining cell turgor pressure might contribute to sustaining physiological processes such as stomatal opening, photosynthesis, and expansion growth (Condon 1982; Blum et al. 1983; Morgan 1984; Ludlow and Muchow 1990; Blum 1996; Kamoshita et al. 2001). Delayed stomatal closure due to a higher level of osmolyte accumulation was clearly associated with a higher net assimilation rate and yields in wheat exposed to postanthesis drought stress (Zivcak et al. 2009), which is well documented in Table 5.1.

The results of drought stress experiments with wheat genotypes of different origin grown in big pots indicated that the genotypes with a higher capacity for OA showed a delayed stomata closure, had a higher net assimilation rate over a longer

Table 5.1 Results of correlation analyses between capacity for osmotic adjustment and selected parameters in winter wheat genotypes (Zivcak et al. 2009)

Parameter[a]	*r*	*p*
Stomatal conductance	0.590	0.019
Net assimilation rate	0.789	0.035
Grain yield	0.510	0.053

[a] *r* correlation index, *p* probability value

time, and, finally, had a lower yield decrease compared to genotypes with a lower accumulation of osmotically active compounds. These results suggest that if the drought stress occurs typically in the late growth stages (e.g., postanthesis drought stress in cereals), the extended period in which the carbon assimilation is not limited by stomatal closure (by virtue of OA) may provide some benefits to the crop yield (Zivcak et al. 2009).

On the other hand, the delayed stomatal closure can lead to faster consumption of soil water, and thus, the real positive effect on crop yield in some environments can be negligible (Serraj and Sinclair 2002). Plants with an accumulation of osmolytes show a delayed carbon assimilation decrease at a lower water potential compared to plants without OA. The level of osmolyte accumulation and water stress may be important factors that determine the balance between the benefits of turgor maintenance and continued carbon assimilation and stomatal control (Gebre and Tschaplinski 2000). In addition, turgor maintenance can be crucial for the development and filling of grains in the spikes and panicles of cereal crops, which may positively influence the grain yield (Ludlow et al. 1990; Serraj and Sinclair 2002).

5.2.1.2 Effects of Osmoregulation at the Root Level

Another physical effect of decreasing the root water potential (which occurs because of osmolyte accumulation) is continuing water movement from the soil into plants as a result of maintaining a water potential gradient. Although the physical principle of this effect is clear, in many cases very little additional water can be extracted from the soil compared to plant needs (Serraj and Sinclair 2002).

On the other hand, OA can positively influence root growth, as turgor maintenance is necessary for root elongation (Voetberg and Sharp 1991). Root depth is crucial for yields in dry environments (Sinclair and Muchow 2001). Results obtained on wheat clearly showed a greater depth of water extraction in lines with a high capacity for osmolyte accumulation compared to low osmotic-adjusting lines (Morgan and Condon 1986; Morgan 1995). OA could be greater in roots than in leaves (Sharp and Davies 1979; Westgate and Boyer 1985; Hsiao and Xu 2000). In an extreme opinion, Matyssek et al. (1991) suggested diverting water flow from other plant organs into the roots, resulting in sustained root growth in dry soils, as a consequence of high osmolyte accumulation in the root tips. OA in roots seems to be an unappreciated topic that deserves more attention (Serraj and Sinclair 2002).

5.2.1.3 Survival of Extreme Drought

Osmoregulation was shown to enhance dehydration tolerance (Turner and Jones 1980; Hsiao et al. 1984; Morgan 1984), extending plant survival under severe drought (Ludlow and Muchow 1990; Sinclair 2000). Although survival is not an important feature for annual crops during grain filling, where the crop yield is the measure of the plant's success (Serraj and Sinclair 2002), it is extremely important in perennials. Moreover, even for annual crops, plant survival is important during a short period of severe stress or during stress in the vegetative stage. It was shown to be an important trait for trees, too (Braatne et al. 1992; Gebre and Kuhns 1991. Moreover, a high capacity for OA was shown to be beneficial for recovery after drought stress. Plants with a high accumulation of osmotically active compounds also exhibited a faster recovery from a drought period compared to nonadjusting plants (Barlow et al. 1977; Tan et al. 1992; Prasertsak and Fukai 1997).

The effect of OA on plant survival (as well as on other functions mentioned above) is probably associated not only with osmoregulation but also with the second function of osmolytes, that is, osmoprotection, which stabilizes cell functions and has many other specific effects, as described in detail next.

5.2.2 Osmoprotection and Specific Roles of Osmolytes

One of the complex mechanisms enabling the survival of the osmotic effects of abiotic stresses is the osmoprotection provided by the accumulation of compatible solutes. The mechanism of osmoprotection is based on associations of nontoxic compounds with different cellular components (Rhodes and Samaras 1994; Rathinasabapathi 2000; Koyro et al. 2012; Ranganayakulu et al. 2013). Osmoprotectants accumulate mostly in the cytosol and chloroplast; they are synthesized in response to stress, are degraded after stress relief, and play a protective role in plant cells. They contribute to the stabilization of enzymes and proteins (Wani et al. 2013), mitigate damaging risks caused by free radicals, and thus protect the cell membrane integrity against the detrimental effects of stress factors at the cellular level. Osmoprotectants should not interfere with the normal metabolic functions in plant cells (Tabaeizadeh 1998; Valliyodan and Nguyen 2006; Le and McQueen-Mason 2006; Galvani 2007; Bohnert and Jensen 1996; Ashraf and Foolad 2007; Conde et al. 2011; Pintó-Marijuan and Munné-Bosch 2013).

According to Singh et al. (2015), osmoprotectants eliminate the negative effects of drought in plants via two different mechanisms: improving the antioxidant defense system and sustaining ion homeostasis. To minimize injury from oxidative stress triggered by drought or other abiotic stresses, plants induce a response by activating antioxidant enzymes, including peroxidase, superoxide dismutase, ascorbate peroxidase, and catalase, which, together with antioxidants such as glutathione, ascorbate, and carotenoids, may quench the toxic effects of reactive oxygen species in plant cells (Gill and Tuteja 2010; Kadioglu et al. 2011; Vardharajula et al. 2011;

Kubis et al. 2014; Brestic et al. 2014). In stress conditions, osmoprotectants such as polyamines, glycine-betaine, soluble sugars, or proline have been shown to stimulate the activity of antioxidant enzymes or protect the enzymatic system, which contributes to eliminating the negative effects of free radicals in plant cells (Ashraf and Foolad 2007; Wei et al. 2009; Hossain and Fujita 2010; Koyro et al. 2012; Theerakulpisut and Gunnula 2012; Kaya et al. 2013; Filippou et al. 2014).

Ion homeostasis is a major strategy to mitigate the toxic effects indirectly caused by drought. Plants accumulate different osmoprotectants, providing ion homeostasis via the ion exchange activity (Ranganayakulu et al. 2013). Moreover, ion homeostasis is ensured by the transportation potential of membrane proteins (Osakabe et al. 2014), regulating ion influx and efflux (Niu et al. 1995). The accumulation of K^+ into the vacuole against an electrochemical gradient is necessary to induce sufficient turgor for stomatal opening (Zhu 2001). Although these protective effects have particular importance in salinity stress (Zhu 2001; Munns and Tester 2008; Parihar et al. 2014), drought stress also leads to an impairment of the ionic balance in cells. In water-deficit conditions, anions and cations, such as K^+ and Cl^-, water transport systems in the plasma membrane and tonoplast induce turgor pressure changes in guard cells, which result in stomatal closure (Kim et al. 2010; Wani et al. 2013). In addition to effects of ions, the exogenous application of some organic osmolytes (glycine betaine, proline) was shown to be beneficial to maintaining ion homeostasis under conditions of drought stress (Ashraf and Foolad 2007; Zhang et al. 2014).

Osmoprotectants represent low molecular weight, highly soluble, electrically neutral compounds (Ahn et al. 2011) such as sugars, amino acids (proline), quaternary ammonium compounds (e.g., glycine betaine), and polyols. However, some osmoprotective effects can also be identified in inorganic ions such as potassium. The specific effects of individual groups of osmolytes, including references to research studies, are described in subsequent sections of this chapter.

5.2.2.1 Soluble Sugars

Sugars are accumulated in the leaves of different crops in conditions of stress (Turner et al. 1978). Sugars may represent a major part of all osmolytes in plant cells; for example, sugars contributed 40–50 % to the osmotic potential of plant tissues of sorghum (Jones et al. 1981) and 30 % in wheat (Condon 1982). Sucrose is the most important soluble sugar; it can account for 70–90 % of the total sugars in leaves (Munns and Weir 1981). Sugars especially accumulate in conditions when the utilization of carbohydrates by metabolism is reduced relative to that of photosynthesis (Munns and Weir 1981). The presence of sugars (trehalose, sucrose, etc.) may protect the cell membrane integrity in conditions of drought (Bohnert et al. 1995). Sugars provide the carbon and energy required for normal functions of plant metabolism and for the regulation of plant growth and development. Sugars act as osmoprotectants, regulating the cell osmotic status, protecting the membranes, and contributing to the scavenging of free radicals in plant cells (Kerepesi and Galiba

2000; Murakeozy et al. 2003; Livingston et al. 2009; Koyro et al. 2012). In particular, fructan may serve as a reserve source of carbohydrates and can protect plants against severe drought stress (Pilon-Smits et al. 1995). Similarly, trehalose is typical sugar important for drought resistance in desiccation-tolerant plants (Vinocur and Altman 2005). It acts as an osmoprotectant or osmolyte, protecting proteins and membranes; more specifically, in denatured proteins it leads to a decrease in aggregation (Ashraf and Harris 2004; Koyro et al. 2012). Some of the most recent studies dealing with soluble sugars in crop plants are listed in Table 5.2.

5.2.2.2 Proline

The accumulation of proline is a typical response not only in plants, but also in eubacteria, protozoa, and marine invertebrates exposed to various stresses. In plants, proline accumulation was found in conditions of salt, drought, low temperature, high temperature, heavy metal stresses, anaerobiosis, UV irradiation, atmospheric pollution, and nutrient deficiency as well as after pathogen infection (Hare and Cress 1997; Saradhi et al. 1995; Siripornadulsil et al. 2002). This makes proline accumulation one of the most universal stress responses. The level of proline in stress conditions can increase to be 100 times greater than that in the control, but the accumulation capacity varies from species to species (Verbruggen and Hermans 2008). The biosynthesis of proline in plants runs in the cytosol and chloroplast, but the degradation runs in mitochondria (Ashraf and Foolad 2007). The upregulation of proline biosynthesis depends on the activity of enzymes such as pyrroline-5-carboxylate reductase (P5CR) and pyrroline-5-carboxylate synthetase (Nounjana et al. 2012). However, the level of proline in plants is controlled by degradation (proline dehydrogenase activity), which is inhibited in stress conditions and the content of proline thus increases (Delauney and Verma 1993; Peng et al. 1996).

Proline is expected to have adaptive roles in plants, contributing significantly to stress tolerance. This amino acid acts as a compatible osmolyte, but it also represents the means to store nitrogen and carbon (Hare and Cress 1997). Proline was shown to be an efficient scavenger of free radicals (Smirnoff and Cumbes 1989). Moreover, proline has been proposed as a molecular chaperone functioning in stabilization of the protein structure and as a component contributing to buffering of cytosolic pH, thus being important for a balance of redox status in plant cells (Sharma and Dietz 2006;Hoque et al. 2008; Verbruggen and Hermans 2008; Filippou et al. 2014). Proline accumulation is part of signaling in a plant cell, which is important for adaptive responses of plants (Maggio et al. 2002; Brestic et al. 2014). Proline accumulates in the cytosol and chloroplast and contributes to the protection of proteins, membranes, and enzymes against stress. It may also contribute to alleviating the acidosis of the cytoplasm necessary to maintain the $NADP^+$/NADPH ratio (Hoque et al. 2008). Proline may also serve as a source of organic carbon, nitrogen, and energy during stress recovery (Tyagi and Sairam 2004).

Table 5.2 The most recent studies revealing the osmotic effects of soluble sugars in crop species

Crop species	References
Wheat (*Triticum aestivum*)	Asthir et al. (2014), Iqbal et al. (2012), He et al. (2011), Charkazi et al. (2010), Javadian et al. (2010)
Maize (*Zea mays*)	Nikolaeva et al. (2015), He et al. (2013), Javadmanesh et al. (2012)
Soybean (*Glycine max*)	Grümberg et al. (2015)
Rice (*Oryza sativa*)	Khan et al. (2015a, b), Joseph et al. (2015), Pandey and Shukla (2015), Abdelgawad et al. (2014), Deyanira et al. (2012), Todaka et al. (2012), Shehab et al. (2010)
Common bean (*Phaseolus vulgaris*)	Talaat et al. (2015), Abass and Mohamed (2011)
Chickpea (*Cicer arietinum* L.)	Arefian et al. (2014), Boukraâ et al. (2013)
Vetch (*Vicia faba*)	Dawood and El-Awadi (2015)
Canola (*Brassica napus*)	Tookalloo (2011)
Arabidopsis (*Arabidopsis thaliana*)	Dai et al. (2011)
Tobacco (*Nicotiana tabacum*)	Zhong et al. (2014)
Flax (*Linum usitatissimum*)	Gaikwad et al. (2014)
Field bean (*Dolichos lablab*)	D'Souza Myrene and Devaraj (2013)
Tomato (*Solanum lycopersicum*)	Khavari-Nejad et al. (2013), Loukehaich et al. (2012)
Pea (*Pisum sativum*)	Lahuta and Dzik (2011)
Sugar beet (*Beta vulgaris*)	Wu et al. (2014)
Potato (*Solanum tuberosum*)	Pino et al. (2013), Farhad et al. (2011)
Pepper (*Capsicum annum*)	Sziderics et al. (2010)

To improve plant tolerance in agriculture, the genetic manipulation of osmolytes such as proline was done (Hong et al. 2000; Seki et al. 2007; Székely et al. 2008; Rivero et al. 2007; Verbruggen and Hermans 2008). The results have shown that high levels of proline were not always correlated with osmotolerance (Szoke et al. 1992; Lui and Zhu 1997). On the other hand, soybeans overexpressing the *P5CR* gene, producing a very high level of proline, exhibited significantly improved drought and heat tolerance compared to wild-type cultivars (De Ronde et al. 2004). There are numerous studies evaluating proline accumulation in response to drought; the most recent works are listed in Table 5.3.

Table 5.3 Recent studies dealing with proline accumulation during drought in crops

Crop species	References
Wheat (*Triticum aestivum*)	Sarafraz-Ardakani et al. (2014), Maevskaya and Nikolaeva (2013), Bowne et al. (2012), He et al. (2011)
Maize (*Zea mays*)	Nikolaeva et al. (2015), Zadehbagheri et al. (2014), Ali et al. (2013a, b), Ali and Ashraf (2011), Anjum et al. (2011a, b)
Soybean (*Glycine max*)	Bîrsan et al. (2015), Silvente et al. (2012)
Rice (*Oryza sativa*)	Joseph et al. (2015), Khan et al. (2015a, b), Gurumoorthy and Singh (2014), Khomdram et al. (2013), Liu et al. (2012), Jha et al. (2011), Shehab et al. (2010), Summart et al. (2010), Somboonwatthanaku et al. (2010)
Barley (*Hordeum vulgare*)	Chéour et al. (2014)
Sunflower (*Helianthus annuus*)	Manivannan et al. (2015), Rabert et al. (2014)
Arabidopsis (*Arabidopsis thaliana*)	Zhao et al. (2015), Kumar et al. (2015), Aleksza et al. (2014), Kesari et al. (2012)
Tobacco (*Nicotiana tabacum*)	Borgo et al. (2015), Zhong et al. (2014), Vanková et al. (2012), Dobra et al. (2010)
Flax (*Linum usitatissimum*)	Gaikwad et al. (2014)
Tomato (*Solanum lycopersicum*)	Montesinos-Pereira et al. (2014), Ali et al. (2012a, b), Sánchez-Rodríguez et al. (2010)
Sorghum	Gosavi et al. (2014)
Amaranthus	Slabbert and Krüger (2014))
Sugar beet (*Beta vulgaris*)	Wu et al. (2014)
Potato (*Solanum tuberosum*)	Pino (2013), Evers et al. (2010)
Pepper (*Capsicum annum*)	De Britto et al. (2013), Sziderics et al. (2010)
Faba bean (*Vicia faba*)	Ali et al. (2013a, b)
Artichoke (*Cynara cardunculus*)	Zhang et al. (2011)

5.2.2.3 Glycine Betaine

Betaines are quaternary ammonium compounds of which glycine betaine (GB) is accumulated in the largest quantities and has the most important physiological functions in plant cells (Ladyman et al. 1980; Grumet and Hanson 1986; Ashraf and Harris 2004; Zhu et al. 2005; Flowers and Colmer 2008; Koyro et al. 2012; Singh et al. 2015). It is one of the most important osmoprotectants, able to mitigate the negative impacts of abiotic stresses (Türkan and Demiral 2009).

Table 5.4 Recent analytical studies on the function of glycine betaine in drought stress

Crop species	References
Wheat (*Triticum aestivum*)	Nikolaeva et al. (2015), Rybka and Nita (2015), Gupta and Thind (2015), Naeem et al. (2015), Malik et al. (2015), Gupta et al. (2014), Talat et al. (2013), Shahbaz et al. (2012), Iqbal et al. (2012)
Maize (*Zea mays*)	Di et al. (2015), Zhang et al. (2015), Kausar et al. (2014), He et al. (2013), Molazem et al. (2010)
Soybean (*Glycine max*)	Malekzadeh (2015), Rezaei et al. (2012), Celik and Atak (2012)
Rice (*Oryza sativa*)	Khan et al. (2015a, b), Abbasian et al. (2015), Tang et al. (2014), Niu et al. (2014), Pyngrope et al. (2013)
Barley (*Hordeum vulgare*)	Anbarasi et al. (2015), Taha et al. (2013)
Pea (*Pisum sativum* L.)	Nusrat et al. (2014)
Chickpea (*Cicer arietinum* L.)	Patel and Hemantaranjan (2012)
Canola (*Brassica napus* L.)	Khalid et al. (2015), Ashrafijou et al. (2010)
Arabidopsis (*Arabidopsis thaliana*)	Ogawa and Mitsuya (2012)
Sugar beet (*Beta vulgaris*)	Wu et al. (2014), Yousif et al. (2010)

GB has multiple functions in plant cells, in particular stabilizing the enzyme's quaternary structure and maintaining the membrane integrity under stress (Sakamoto and Murata 2000). GB is abundant in the cytoplasm and chloroplast of different—but probably not all—plant species (Leigh et al. 1981; Rhodes and Hanson 1993; Allard et al. 1998). It is particularly important for grasses and cereals (Grumet and Hanson 1986; Colmer et al. 1995; Islam et al. 2007). The most important recent contributions to the research of GB's effects appear in Table 5.4.

As mentioned, GB accumulates in the chloroplasts of some plant species in response to stress, but it was also reported in different kinds of microbes (Sakamoto and Murata 2002; Sawahel 2004; Ranganayakulu et al. 2013). This compound is associated with enhancing the water flow into cells to maintain the intracellular osmotic equilibrium and also with regulating the signal transduction cascade that is important for a stress response (Kumar et al. 2003; Ashraf and Foolad 2007; Ranganayakulu et al. 2013; Brestic et al. 2014). It was found that an increase in GB content, both in chloroplasts and in the cytosol, leads to a much higher increase in stress tolerance compared to the increase stimulated only in the cytosol (Sakamoto et al. 1998; Chen and Murata 2002; Park et al. 2004, 2007). GB contributes to osmoregulation directly (Wang et al. 2003) as well as indirectly by protecting the membrane stability, which is necessary for correct functions of channels and ion carriers (Ashraf and Foolad 2007). The protective role against reactive oxygen species has also been confirmed (Park et al. 2004; Quan et al. 2004; Einset et al. 2007). Extensive research has mostly confirmed that GB accumulation has positive effects on drought and salt tolerance in various crop plants (McNeil et al. 1999; Jagendorf and Takabe 2001; Sawahel 2004; Ranganayakulu et al. 2013) without having negative effects on growth and development (Alia et al. 1998; Sakamoto et al. 1998; Park et al. 2004).

5.2.2.4 Polyols

Polyols, sometime also called sugar alcohols, may have one of two structures: a cyclic structure, which describes the polyols myoinositol and pinitol, or a linear structure, which matches sorbitol, xylitol, mannitol, and ribitol (Tari et al. 2010). Sorbitol is widely distributed in plants. It is produced in parallel with sucrose during photosynthesis. It serves as translocation carbon and energy between sources and sinks (Jain et al. 2010). Mannitol, inositol, and sorbitol act as osmoregulators, enhancing plant tolerance to drought and other stresses (Williamson et al. 2002). Mannitol was found to be produced in high quantity, and its main function is probably quenching the hydroxyl radicals produced in electrochemical processes in plant cells (Mitoi et al. 2009; Gill and Tuteja 2010).

Inositols were shown to be important for normal growth and development, for biogenesis of membranes, and as a secondary messenger in signal transduction pathways (Loewus and Murthy 2000). Inositol and its derivatives (pinitol, galactinol, etc.) act as osmoprotectants and stress-related molecules (Taji et al. 2002). Inositol is also utilized to store carbohydrates in the cells (Bohnert et al. 1995).

Polyols contribute to the regulation of osmotic balance and general osmoregulation in plant cells. They provide significant protection of plant structures against indirect effects caused by drought and other stresses (Shen et al. 1997; Kanayama et al. 2006; Li et al. 2011). Several recent studies have shown that polyol molecules improve plant growth and may reduce the risk of damage caused by drought (Table 5.5).

5.2.2.5 Polyamines

Polyamines are small aliphatic organic molecules containing two or more primary amino groups. The major polyamines found in plants are putrescine (diamine), spermidine (triamine), and spermine (tetramine). They are engaged in different physiological processes such as cell division, growth, and differentiation (Wang et al. 2003; Alcázar et al. 2006, 2010; Kusano et al. 2007; Yamaguchi et al. 2007; Minguet et al. 2008). Polyamines in plant cells can exist in free or conjugated forms together with phenolic compounds or with proteins and nucleic acids. Their biological activity is associated with a polycationic nature; they are able to regulate the pH of cellular components (Groppa and Benavides 2008; Gill and Tuteja 2010). The specific functions of polyamines were identified in floral initiation and development (Gerats et al. 1988; Masgrau et al. 1997; Hanzawa et al. 2000; Panicot et al. 2002), root growth (Watson et al. 1998), fruit ripening (Mehta et al. 1997, 2002), and so forth.

Table 5.5 Recent analytical studies on the function of mannitol and sorbitol in crop and model plants

Crop species	References
Wheat (*Triticum aestivum*)	Khan et al. (2013), Al-Quraan et al. (2013), Iqbal et al. (2012), Sultan et al. (2012), Makhloufi et al. (2015), Ahmed et al. (2015), Melloul et al. (2014), Kamal et al. (2013)
Maize (*Zea mays*)	Rattan et al. (2012), Kaya et al. (2013), Nguyen et al. (2013), Bárzana et al. (2015), Thomson et al. (2014), Khazarin (2014), Ali and Ashraf (2011)
Rice (*Oryza sativa*)	Joseph et al. (2015), Theerakulpisut and Phongngarm (2013), Chutipaijit et al. (2012), Jnandabhiram and Sailen Prasad (2012), Jha et al. (2011), Chen et al. (2015), Theerakulpisut et al. (2013), Cha-um et al. (2013), Soren et al. (2010)
Barley (*Hordeum vulgare*)	Wu et al. (2013)
Sunflower (*Helianthus annuss*)	Andrade et al. (2013)
Arabidopsis (*Arabidopsis thaliana*)	Acosta-García et al. (2015), Wang et al. (2015), Ding et al. (2014), Chan et al. (2011), Ahn et al. (2011), Dai et al. (2011), Clauw et al. (2010), Liu et al. (2015), Aghdasi et al. (2011)
Tobacco (*Nicotiana tabacum*)	Kumar et al. (2013), Macaluso et al. (2007), Riahi and Ehsanpour (2013), Pospisilova et al. (2011)
Soybean (*Glycine max*)	Mohamed and Akladious (2014)
Flax (*Linum usitatissimum*)	Saker et al. (2014)
Tomato (*Solanum lycopersicum*)	Loukehaich et al. (2012), Al Hassan et al. (2015)
Sugar beet (*Beta vulgaris*)	Goudarzi et al. (2015), Wu et al. (2014, 2015)
Potato (*Solanum tuberosum*)	Kikuchi et al. (2015), Rahnama et al. (2011), Shi et al. (2015), Panta et al. (2014), Evers et al. (2010)
Pepper (*Capsicum annum*)	Sziderics et al. (2010)
Faba bean (*Vicia faba*)	Gao et al. (2013), Hanafy et al. (2013)
Common bean (*Phaseolus vulgaris*)	Fernandez-Aunión et al. (2010), Ramalho et al. (2014)

Heavy accumulation of polyamines under abiotic stress conditions has been found in many plant species (Evans and Malmberg 1989; Alcázar et al. 2006, 2010; Hussain et al. 2011). The protective role of polyamines against damage caused by extreme environmental stress was suggested (Liu et al. 2007; Hussain et al. 2011), including protective effects in conditions of water deficit (Nayyar and Chander 2004; Yamaguchi et al. 2007; Kubis et al. 2014; Singh et al. 2015).

The metabolism of polyamines is coupled with ethylene production, which can be important in responses to stress (Zapata et al. 2004). In addition to the osmoregulatory functions of polyamines, they are involved in stomatal closure (Liu et al. 2000).

Table 5.6 Recent analytical studies on the function of polyamines under drought in crops

Crop species	References
Wheat (*Triticum aestivum*)	Ibrahim (2014), Marcińska et al. (2013), Rana et al. (2013)
Maize (*Zea mays*)	Ludmerszki et al. (2014)
Common bean (*Phaseolus vulgaris*)	Abass and Mohamed (2011)
Rice (*Oryza sativa*)	Pandey and Shukla (2015), Maisura et al. (2014), Zahra et al. (2013), Summart et al. (2010)
Barley (*Hordeum vulgare*)	Fatehi et al. (2012)
Canola (*Brassica napus*)	Nasibi et al. (2014), Saadia et al. (2012)
Millet (*Setaria italica* L.)	Sudhakar et al. (2015)
Lettuce (*Lactuca sativa*)	Paradisone et al. (2015)
Sunflower (*Helianthus annuus*)	Andrade et al. (2013)
Arabidopsis (*Arabidopsis thaliana*)	Saibi et al. (2015), Ogawa and Mitsuya (2012), Missihoun et al. (2011), Watanabe et al. (2010)
Tobacco (*Nicotiana tabacum*)	Borgo et al. (2015), Ghahremani et al. (2014)
Flax (*Linum usitatissimum*)	Quéro et al. (2015)
Tomato (*Solanum lycopersicum*)	Montesinos-Pereira et al. (2014), Chaitali and Sengupta (2014), Sánchez-Rodríguez et al. (2010)
Potato (*Solanum tuberosum*)	Evers et al. (2010)
Pepper (*Capsicum annum*)	Yiu et al. (2012)

Moreover, polyamines are components of the antioxidant system, where they play the role of scavengers of free reactive oxygen radicals (Das and Misra 2004; Kuznetsov et al. 2007). The most recent studies on functions of polyamines in plant cells (Table 5.6) confirm the protective roles of polyamines in drought conditions.

5.2.2.6 Potassium and Other Inorganic Ions

The physiological function of inorganic ions in OA is thought to be mainly in osmoregulation, that is, decreasing the osmotic potential, with consequences for turgor maintenance, ability to open stomata, and so on. In conditions of drought, inorganic ions such as K^+, Na^+, and Cl^- contribute dominantly to the total osmotic potential in many species (Munns et al. 1979; Pugnaire et al. 1999). Potassium is broadly studied as a crop nutrient; however, numerous studies also assess the function of potassium in osmoregulation and osmoprotection (Table 5.7). Among all ions, potassium is the most important in the majority of field crops (Munns et al. 1979). For example, the contribution of K^+ to changes of the osmotic potential in wheat caused by drought was in the range of 40–80 % (Condon 1982; Morgan 1992). The contribution of individual osmolytes, however, depends on many factors, including the level of drought stress (Munns et al. 1979; Jones et al. 1981).

Table 5.7 Recent analytical studies on osmotic functions of potassium under drought stress

Crop species	References
Wheat (*Triticum aestivum*)	Li et al. (2015), Zia et al. (2014), Hanafy and Mohamed (2014), Rana et al. (2013), Al-Quraan et al. (2013), Aldesuquy et al. (2012), Dong et al. (2010), Nio (2009)
Maize (*Zea mays*)	Akbari et al. (2015), Chen et al. (2014), Estrada et al. (2013), Kojić et al. (2012), Ali et al. (2012a, b), Nawaz and Ashraf (2010)
Soybean (*Glycine max*)	Ali Rezaei et al. (2012)
Rice (*Oryza sativa*)	Joseph et al. (2015), Basu and Roychoudhury (2014), Bagheri et al. (2013, 2014), Pyngrope et al. (2013), Shehab et al. (2010)
Barley (*Hordeum vulgare*)	Good et al. (2010), Dong et al. (2010)
Bean (*Phaseolus vulgaris*)	Talaat (2015), Rosales et al. (2012), Fernandez-Aunión et al. (2010)
Brassica juncea	Fariduddin et al. (2015)
Canola (*Brassica napus* L.)	Leithy et al. (2015), Hosseini et al. (2014), El Habbasha and Mekki (2014)
Sugar beet (*Beta vulgaris* L.)	Wu et al. (2015)
Letuce (*Lactuca sativa)*	Eichholz et al. (2014)
Chickpea (*Cicer arietinum* L.)	Arefian et al. (2014), Hirich et al. (2014), Patel and Hemantaranjan (2012)
Arabidopsis (*Arabidopsis thaliana*)	Bhattacharyya et al. (2014), Zsigmond et al. (2012)
Tobacco (*Nicotiana tabacum*)	İşeri et al. (2013)

5.3 Methods and Experimental Approaches to Assess OA

The previous chapters have clearly shown that OA is a very complex issue. The physiological effects depend not only on the cumulative physical effects of all compounds, leading to a decreased water potential in plant tissues, but also on specific effects of all osmolytes. However, for screening purposes, there is a need for a fast, reliable method to assess the level of the overall OA, for example, to identify the species/genotypes with the capacity to produce more osmotically active compounds, without the need to analyze all key compounds contributing to OA. Moreover, the experimental design for testing biological material is also crucial. Therefore, in the next sections we present a brief summary of the main methods and propose possible sources of errors in OA studies.

5.3.1 *Methods to Estimate the Accumulation of Osmotically Active Compounds*

As previously summarized by Chandra Babu et al. (1999), several main methods exist to measure osmolyte accumulation in plants under water deficit:

1. The regression method
   Morgan (1992) used an estimation of osmolyte accumulation from the linear regressions between the relative water content (RWC) and osmotic potential (Ψ_s). The values used for this estimation can be derived from consecutive measurements of parameters during a progressive drought. The nonlinear shape of the relationship enables the active solute accumulation (because of the active accumulation of osmotically active compounds) to be recognized from the decrease in Ψ_s caused by the increase in concentration due to water loss from the tissues. Thus, the approach is based in principle on comparing the regression of the theoretical curve (without OA) with the actually measured regression. It enables the calculation of Ψ_s(non-OA) caused by the concentration effect, but not by OA:

   $$\Psi_{s,}(\text{non OA}) = \Psi_s(\text{initial}) \cdot [(\text{RWC}_i/100)/\text{RWC}_d/100],$$

   where Ψ_s(initial) is the initial osmotic potential in well-watered plants, RWC_i is the initial RWC in well-watered plants, and RWC_d is the RWC in drought-stressed plants. The level of OA (osmoregulatory effect of accumulated osmolytes) was calculated as OA = Ψ_s(non-OA) − Ψ_s(drought), where Ψ_s(drought) is an osmotic potential measured in drought-stressed samples.
2. The full turgor adjustment method
   Another method for estimation is based on recalculating the measured osmotic potential to the hypothetical status of full hydration, that is, the RWC at a level of 100 %. OA is estimated from the difference in the calculated OP values at a full hydrated state—$\Psi_s(100)$—between well-watered and water-deficit plants (Wilson et al. 1979; Ludlow et al. 1983, 1990). For this method, measurements of Ψ_s and RWC in nonstressed and drought-stressed conditions are needed. To make the estimation more correct, a correction for tissue apoplastic water is included in the formula: $\Psi_s(100) = \Psi_s\,[(\text{RWC}-B)/(100-B)]$, where B represents a proportion of bound water. For example, in rice a constant value for both well-watered and water-deficit leaves (18 %) was used (Turner et al. 1986).
3. The rehydration method
   The principle of this method is similar to the previous one: OA is estimated from the difference in Ψ_s values at a full hydrated state—$\Psi_s(100)$—between the well-watered and water-deficit plants. The difference lies in the fact that $\Psi_s(100)$ is not calculated, but detached samples (both in control and drought-stressed plants) are fully rehydrated before Ψ_s is measured (Turner and Jones 1980; Blum 1989).
4. Osmolyte accumulation effect can be estimated as the difference between the measured Ψ_s and the calculated Ψ_s , where the calculation of theoretical

Ψ_s(non-OA) is based on the concentration effect in a given leaf water content. Thus, OA is the net accumulation of solutes per unit of water in leaf tissue. The concentration effect on Ψ_s is the proportional decrease in leaf Ψ_s due only to the reduction in water content (WC) under water-deficit treatments:

Ψ_s(non-OA) = (WC_i/WC_d) · $\Psi_{s,I}$, where WC_i is the nonstressed water content, and WC_d is the water content in the stressed sample. The osmotic potential is again calculated as OA = Ψ_s(non-OA)–Ψ_s(drought), where Ψ_s(drought) is the osmotic potential measured in drought-stressed samples (Colmer et al. 1995; Ma et al. 2006; Ma and Turner 2006).

5.3.2 *Possible Sources of Errors in Assessment of OA*

All the methods proposed here to assess the OA are estimations, the preciseness of which is based on several assumptions. The most important assumption is that all inputs must be measured very precisely, as any incorrectness may lead to a very incorrect in estimation of the OA. It demands good practices in experimental design, sample handling, and measurement. Some possible sources of errors are highlighted below.

1. Impreciseness of the methods
   The regression method was found to be most comprehensive and had the best estimate of OA among the four methods; however, it is labor-intensive, as it requires frequent measurements of the water status (Chandra Babu et al. 1999; Zhang et al. 1999). The full turgor adjustment and rehydration methods are less demanding of time, labor, and plant tissues and more suitable for screening work; however, there are some risks of possible errors caused by incorrect corrections or losses of solutes during rehydration (Chandra Babu et al. 1999). Although these methods would not always provide the same numerical results, the proper use of any of them should be worthwhile for comparing the capacity for OA in different samples.
2. Handling plant material
   For all of the methods presented here, the fresh tissue must be taken and immediately measured while keeping it alive, in a fully functional state. In addition to preventing water loss, it is necessary to prevent overheating during handling. At least part of the work must be done a very short time after sampling (Nio 2009). Of the methods listed above, the rehydration method has the lowest demand for immediate work, as the only operation that must be done immediately is the rehydration of the detached samples. Samples can then be sealed and deep frozen, and the osmotic potential can be measured from sap extracted from the tissues after melting. If water loss is prevented well, the measurements can be done even a relatively long time after storage. This makes this method particularly interesting for screening purposes, as pointed out by Chandra Babu et al. (1999). However, to be correct, the rehydration should not significantly alter solute con-

centrations. Therefore, rehydration conditions (technique, length, temperature, etc.) for individual types of samples (different species) must be verified to reach full hydration without causing significant changes in osmolyte content by respiration or solute efflux.

3. Experimental design
   Although the capacity for OA is strongly inherited (Zhang et al. 1999), experiments performed in different conditions may bring very different results.
   Typically, numerous relevant studies have been realized in the field. Field conditions allow a gradual acclimation to water deficit and permit normal root development, and the results are clearly relevant for further applications in the field (Fereres et al. 1978). Although measurements of the accumulation of osmotically active compounds in the field will always be needed (at least in the last step), OA in the field may be influenced by any other biotic and abiotic factors, and so it is very difficult to control (Chandra Babu et al. 1999). Therefore, measurements in controlled conditions are also applied. Screening for OA expression may be useful under controlled conditions because photoperiod, temperature, water management, and nutrition can be regulated (Chandra Babu et al. 1999). The conditions can be adjusted to induce the maximum capacity for OA in (Glinka and Ludlow 1986). Also, the limited volume of soil water accessible to roots in small pots may influence the results (Fereres et al. 1978; Nio 2009).
   A specific tool for OA studies are the experiments in media with a low osmotic potential using additives such as polyethylene glycol (PEG). PEG is highly soluble in water and has low toxicity (Lawlor 1970), and in forms with a high molecular weight (e.g., PEG 6000), it does not permit the biological membranes (Bressan et al. 1981). PEG is suitable to induce a plant water deficit by decreasing the water potential of the medium (Lawlor 1970; Steuter et al. 1981). PEG can also be used to induce a water deficit in detached leaves in laboratory experiments. However, results on OA using PEG on detached leaves were not sufficiently consistent with laboratory or field experiments, and this method does not seem to be useful for fast screening for this trait (Nio 2009).

5.4 OA in Crop Improvement

Unlike wild species, in which stress survival is crucial, in crop plants, abiotic stress tolerance is measured by the ability to produce yield under adverse weather conditions. In general, it is expected that there is sufficient genetic variability for traits related to capacity to maintain the yield and quality of products under drought stress in crop plants. Technological progress has increased possibilities to mine genetic variability and improve knowledge about the mechanisms of abiotic stress tolerance (Dolferus 2014).

OA has previously been suggested as an efficient selection criterion useful in conventional breeding programs aimed at drought tolerance in main crops (Morgan

1983; Blum et al. 1983; Ludlow and Muchow 1990; Tangpremsri et al. 1991; Belhassen et al. 1995; Zhang et al. 1999). Genetic variability in the capacity for OA has been found in a number of species, which suggests that genes and molecular markers related to OA can be useful in breeding for crop drought tolerance (Belhassen et al. 1995; Nguyen et al. 1997; Zhang et al. 1999). In fact, molecular markers and quantitative trait loci (QTLs) associated with the accumulation of osmotically active compounds have been found in wheat (Van Deynze et al. 1995), rice (Lilley et al. 1996; Price and Courtois 1999), barley (Teulat et al. 1998), sunflower (Jamaux et al. 1997), and many other crops (Serraj and Sinclair 2002). On the other hand, there are some doubts whether the upregulation of osmoregulatory effects provided by the accumulation of osmotically active compounds can really have any positive effects on yield (Quisenberry et al. 1984; Grumet et al. 1987; Flower et al. 1990; Serraj and Sinclair 2002).

Manipulating stress-responsive genes to enhance the production of osmoprotectants is defined as one of the strategies to improve stress tolerance in plants (Reguera et al. 2012; Jain 2013). The effort to engineer plants upregulating the production of compatible osmoprotectants has taken almost two decades (LeRudulier et al. 1984; Singh et al. 2015). Numerous studies have shown that the enhanced synthesis of osmoprotectants such as proline and GB in transgenes led to higher drought tolerance (Holmström et al. 1996; Holmberg and Bülow 1998; Hare et al. 1998; Nuccio et al. 1999; Rathinasabapathi 2000). On the other hand, the role played by the accumulation of osmotically active compounds in osmoregulation and osmoprotection in crop science has become overemphasized although many studies in transgenic plants overaccumulating proline, GB, mannitol, or soluble sugars show only small improvements in stress tolerance (Blum 1996; Nguyen et al. 1997; Hare et al. 1998; Zhang et al. 1999).

Another approach, which can be used directly in plant improvement, is mining the existing genetic diversity using molecular markers for OA and/or accumulation of osmoprotective compounds. Genetic variability in dehydration tolerance characteristics was clearly identified in several crops and trees (Blum 1989, Gebre and Kuhns 1991, Morgan 1991, Tschaplinski et al. 1994, 1998). The genetic basis of osmolyte accumulation was already described in some species (Zhang et al. 1999). The QTLs for OA were identified in many species (e.g. Lilley et al. 1996; Bradshaw 1996; Saeed et al. 2011; Ghimire et al. 2012; Mu and Li 2013; Merewitz et al. 2014; Ali et al. 2015; Abdelraheem et al. 2015), although a direct association with yield improvements was not always clear (Lilley et al. 1996).

5.5 Conclusions and Future Perspectives in OA Research

OA has become an important topic in crop research and has kept the attention of scientists for decades. Whereas the early studies mostly focused on osmoregulatory functions provided by the accumulation of osmotically active compounds, the contribution of individual osmolytes to decrease the osmotic potential, and the osmoregulatory capacity of crop species and genotypes, recent studies have mostly targeted specific osmoprotective effects of compatible osmolytes. Research on drought stress tolerance mechanisms, including OA, has been driven onward by a wide range of "omics" technologies, such as genomics, metabolomics, phenomics, transcriptomics, or proteomics, which enable the genetic and molecular bases of observed phenomena to be uncovered. Investigation of the gene expression or regulation for realizing mechanisms of tolerance in particular resistant crops may help in translating information to other drought-sensitive crops. Considering the importance of compatible osmoprotective compounds in many plant physiological and biochemical functions under drought stresses, further research in the endogenous regulation of the metabolism of osmoprotective compounds may contribute to the next theoretical and practical applications in this field (Singh et al. 2015).

In addition to some success in conventional breeding using OA as a selection criterion, the most important progress has been made in engineering transgenic plants with drought stress tolerance by virtue of genes encoding osmoprotectants. However, the effect of the genes, which was apparent under controlled laboratory conditions, is often not present when introduced in field trials. This can be caused by the fact that testing protocols do not mimic the situation well in realistic field conditions, where plants experience multiple stresses and completely different environmental conditions, with fluctuating light, temperature, wind, and biotic factors. One of the reasons for the limited success of transgenes is the fact that abiotic stress tolerance in plants is controlled by many genes working in different stress response pathways (Vinocur and Altman 2005). However, the stress tolerance enhancement was provided by a single gene modification, which can be successful in particular conditions, but it can fail in many other environmental situations (Tayal et al. 2004; Ashraf and Akram 2009). Therefore, the combination of different strategies in future research, such as the multigenic approach, that is, incorporating more than one gene in transgenic plants, may bring real success. In this regard, the synthesis of genes encoding osmoprotectants needs to include being coexpressed with other stress-related genes such as ion transporters, transcription factors, and other functional genes. Combining these different approaches will result in enhanced stress tolerance, photosynthesis, and photoprotection and a more stable yield in transgenic plants. Moreover, research should also be undertaken in more realistic laboratory conditions that better mimic the real environment as well as in field conditions where the effects may be evaluated in terms of the yield, which is the ultimate target of improvement under drought (Khan et al. 2015a, 2015b).

Another promising tool is transcriptomics, which has yet to be efficiently used on this area. A paucity of notable studies was aimed at osmoprotectant-related gene expression under stress conditions (Barros et al. 2009; Chen et al. 2009; Kido et al. 2013). The studies should be extended to different crops to uncover the differences in the expression of genes related to OA and the accumulation of osmoprotectants under drought stress. The results can help to unravel the mechanism of stress tolerance. Moreover, it can be used in breeding programs and engineering of drought-tolerant crop varieties in a plant-specific manner.

Another important area is the need for new breeding approaches aimed at drought tolerance, for example, the application of stress-related QTLs and map-based cloning methods in association with new approaches such as microarray-based expression profiling of differential gene expression as was previously used in some cases (Salvi and Tuberosa 2005; Walia et al. 2007; Pandit et al. 2010). These methods can be combined with advanced "omics" techniques, such as next-generation sequencing, which can help to explore the function of genes in drought stress responses. Moreover, the development of high-throughput and precise techniques, such as multi-SNP analysis, SuperSAGE, PhyloCSF, and others, together with tools of high-throughput phenotyping can aid significantly improve future approaches for the development of abiotic stress-tolerant plants (Khan et al. 2015a, 2015b).

In summary, despite lengthy research and an enormous quantity of experimental results, the topic of OA is still quite attractive and provides numerous opportunities and promising ways for future research. To reach the ultimate goal, which is the increase in yield quantity, quality, and stability in conditions limited by water supply, the efficient use of modern tools and methods is needed. Moreover, the particular target of the research needs to be selected properly, to avoid impasses of research related to traits without any effect on crop yield.

References

Abass SM, Mohamed HI. Alleviation of adverse effects of drought stress on common bean (*Phaseolus vulgaris* L.) by exogenous application of hydrogen peroxide. Bangladesh J Bot. 2011;40(1):75–83.

Abbasian A, Mirshekari B, Vishekaei S, Naghi M, Rashidi V, Aminpanah H. Efecto de la aplicación foliar con metanol y presencia de distintas densidades de pasto dentado (*Echinochloa crusgalli*) sobre el desarrollo y rendimiento de arroz (*Oryza sativa* L.). Idesia Arica. 2015;33(2):69–75.

Abdelgawad ZA, Hathout TA, El-Khallal SM, Said EM, Al-Mokadem AZ. Accumulation of trehalose mediates salt adaptation in rice seedlings. Amer-Eur J Agr Environ Sci. 2014;14(12):1450–63.

Abdelraheem A, Hughs SE, Jones DC, Zhang J. Genetic analysis and quantitative trait locus mapping of PEG-induced osmotic stress tolerance in cotton. Plant Breeding. 2015;134(1):111–20.

Acosta-García G, Chapa-Oliver AM, Millán-Almaraz JR, Guevara-González RG, Cortez-Baheza E, Rangel-Cano RM, Torres-Pacheco I, et al. *CaLEA 73* gene from *Capsicum annuum* L. enhances drought and osmotic tolerance modulating transpiration rate in transgenic *Arabidopsis thaliana*. Can J Plant Sci. 2015;95(2):227–35.

Aghdasi M, Schluepmann H, Smeekens S. Characterization of *Arabidopsis* seedlings growth and development under trehalose feeding. J Cell Mol Res. 2011;2(1):1–9.

Ahmed M, Qadeer U, Ahmed ZI, Hassan FU. Improvement of wheat (*Triticum aestivum*) drought tolerance by seed priming with silicon. Arch Acker Pfl Boden. 2015

Ahn C, Park U, Park PB. Increased salt and drought tolerance by dononitol production in transgenic *Arabidopsis thaliana*. Biochem Biophys Res Commun. 2011;415:669–74.

Akbari MM, Mobasser HR, Ganjali HR. Influence of salt stress and variety on some characteristics of corn. Biol Forum. 2015;7(1):441.

Al Hassan M, Fuertes MM, Sánchez FJR, Vicente O, Boscaiu M. Effects of salt and water stress on plant growth and on accumulation of osmolytes and antioxidant compounds in cherry tomato. Notulae Botanicae Horti Agrobotanici Cluj-Napoca. 2015;43(1).

Alcázar R, Marco F, Cuevas JC, Patrón M, Ferrando A, Carrasco P, Tiburcio AF, Altabella T. Involvement of polyamines in plant response to abiotic stress. Biotechnol Lett. 2006;28: 1867–76.

Alcázar R, Altabella T, Marco F, Bortolotti C, Reymond M, Koncz C, Carrasco P, Tiburcio AF. Polyamines: molecules with regulatory functions in plant abiotic stress tolerance. Planta. 2010;231:1237–49.

Aldesuquy HS, Abo-Hamed SA, Abbas MA, Elhakem AH. Role of glycine betaine and salicylic acid in improving growth vigour and physiological aspects of droughted wheat cultivars. J Stress Physiol Biochem. 2012;8(1):149–71.

Aleksza D, Kovács H, Szabados L, Horváth GV. Regulation of protective proline synthesis in *Arabidopsis thaliana*. Adv Plant Breeding Biotechnol Tech. 2014.

Ali Rezaei M, Kaviani B, Masouleh AK. The effect of exogenous glycine betaine on yield of soybean *Glycine max* (L.) Merr. In two contrasting cultivars pershing and DPX under soil salinity stress. Plant Omics. 2012;5(2):87–93.

Ali Q, Ashraf M. Induction of drought tolerance in maize (*Zea mays* L.) due to exogenous application of trehalose: growth, photosynthesis, water relations and oxidative defence mechanism. J Agron Crop Sci. 2011;197(4):258–71.

Ali AAA, Ali AMM, Mohammed NK. ABA biosynthesis defective mutants reduce some free amino acids accumulation under drought stress in tomato leaves in comparison with *Arabidopsis* plants tissues. J Stress Physiol Biochem. 2012a;8(2):179–92.

Ali Q, Ashraf M, Anwar F, Al-Qurainy F. Trehalose-induced changes in seed oil composition and antioxidant potential of maize grown under drought stress. J Am Oil Chem Soc. 2012b;89(8):1485–93.

Ali MH, Siddiqui MH, Al-Wahibi MH, Basalah MO, Sakran AM, El-Zaidy M. Effect of proline and abscisic acid on growth and physiological performance of faba bean under water stress. Pak J Bot. 2013a;45(3):933–40.

Ali Q, Anwar F, Ashraf M, Saari N, Perveen R. Ameliorating effects of exogenously applied proline on seed composition, seed oil quality and oil antioxidant activity of maize (*Zea mays* L.) under drought stress. Int J Mol Sci. 2013b;14(1):818–35.

Ali A, Arshad M, Naqvi SS, Rasheed A, Sher H, Kazi AG, Mujeeb-Kazi A. Comparative assessment of synthetic-derived and conventional bread wheat advanced lines under osmotic stress and implications for molecular analysis. Plant Mol Biol Rep. 2015;22:1–11.

Alia HH, Chen THH, Murata N. Transformation with a gene for choline oxidase enhances the cold tolerance of *Arabidopsis* during germination and early growth. Plant Cell Environ. 1998;21:232–9.

Allard F, Houde M, Krol M, Ivanov A, Huner NPA, Sarhan F. Betaine improves freezing tolerance in wheat. Plant Cell Physiol. 1998;39:1194–202.

Al-Quraan NA, Sartawe FAB, Qaryouti MM. Characterization of γ-aminobutyric acid metabolism and oxidative damage in wheat (*Triticum aestivum* L.) seedlings under salt and osmotic stress. J Plant Physiol. 2013;170(11):1003–9.

Anbarasi G, Bhagavathi G, Vignesh R, Srinivasan M, Somasundaram ST. Effect of exogenous abscisic acid on growth and biochemical changes in the halophyte *Suaeda maritima*. J Microbiol Biotechnol Food Sci. 2015;4(5):442.

Andrade A, Vigliocco A, Alemano S, Llanes A, Abdala G. Comparative morpho-biochemical responses of sunflower lines sensitive and tolerant to water stress. Am J Plant Sci. 2013;4:12–32.

Anjum SA, Farooq M, Wang LC, Xue LL, Wang SG, Wang L, Chen M, et al. Gas exchange and chlorophyll synthesis of maize cultivars are enhanced by exogenously-applied glycine betaine under drought conditions. Plant Soil Environ. 2011a;57(7):326–31.

Anjum SA, Xie X, Wang LC, Saleem MF, Man C, Lei W. Morphological, physiological and biochemical responses of plants to drought stress. Afr J Agric Res. 2011b;6:2026–32.

Arefian M, Vessal S, Bagheri A. Biochemical changes in response to salinity in chickpea (*Cicer arietinum* L.) during early stages of seedling growth. J Animal Plant Sci. 2014;24(6):2014.

Ashraf M, Akram NA. Improving salinity tolerance of plants through conventional breeding and genetic engineering: an analytical comparison. Biotechnol Adv. 2009;27:744–52.

Ashraf M, Foolad MR. Roles of glycine betaine and proline in improving plant abiotic stress resistance. Environ Exp Bot. 2007;59:206–16.
Ashraf M, Harris PJC. Potential biochemical indicators of salinity tolerance in plants. Plant Sci. 2004;166:3–16.
Ashrafijou M, Sadat Noori SA, Izadi Darbandi A, Saghafi S. Effect of salinity and radiation on proline accumulation in seeds of canola (*Brassica napus* L.). Plant Soil Environ. 2010; 56(7):312–7.
Asthir B, Gulati A, Bains NS. Controlling water deficit by osmolytes and enzymes: enhancement of carbohydrate mobilization to overcome osmotic stress in wheat subjected to water deficit conditions. Afr J Biotechnol. 2014;13(20):2072.
Bagheri AA, Saadatmand S, Niknam V, Nejadsatari T, Babaeizad V. Effect of endophytic fungus, *Piriformospora indica*, on growth and activity of antioxidant enzymes of rice (*Oryza sativa* L.) under salinity stress. Int J Adv Biol BioMed Res. 2013;1(11):1337–50.
Bagheri AA, Saadatmand S, Niknam V, Nejadsatari T, Babaeizad V. Effects of *Piriformospora indica* on biochemical parameters of *Oryza sativa* under salt stress. Int J Biosci. 2014;4(4):24–32.
Balaguer L, Punaire FI, Martínez-Ferri E, Armas C, Valladares F, Manrique E. Ecophysiological significance of chlorophyll loss and reduced photochemical efficiency under extreme aridity in *Stipa tenacissima* L. Plant Soil. 2002;240:343–52.
Barlow EWR, Munns R, Scott NS, Reisner AH. Water potential, growth and polyribosome content of stressed wheat apex. J Exp Bot. 1977;28:909–16.
Barros PDS, Soares-Cavalcanti NM, Vieira-Mello GS, Wanderley-Nogueira AC, Calsa-Junior T, Benko-Iseppon AM. In silico evaluation of osmoprotectants in eucalyptus transcriptome. Lect Notes Comput Sci. 2009;5488:66–77.
Bárzana G, Aroca R, Ruiz-Lozano J. Localized and non-localized effects of arbuscular mycorrhizal symbiosis on accumulation of osmolytes and aquaporins and on antioxidant systems in maize plants subjected to total or partial root drying. Plant Cell Environ. 2015;38(8):108–25.
Basnayake J, Ludlow MM, Cooper M, Henzell RG. Genotypic variation of osmotic adjustment and desiccation tolerance in contrasting sorghum inbred lines. Field Crop Res. 1993;35:51–62.
Basu S, and Roychoudhury A. Expression profiling of abiotic stress-inducible genes in response to multiple stresses in rice (*Oryza sativa* L.) varieties with contrasting level of stress tolerance. BioMed Res Int. 2014.
Belhassen E, This D, Monneveux P. L'adaptation génétique face aux contraintes de sécheresse. Cah Agr. 1995;4:251–61.
Bhattacharyya D, Yu SM, Lee YH. Volatile compounds from *Alcaligenes faecalis* JBCS1294 confer salt tolerance in *Arabidopsis thaliana* through the auxin and gibberellin pathways and differential modulation of gene expression in root and shoot tissues. Plant Growth Regul. 2014;75(1):297–306.
Bîrsan A, Rotaru V, Jigău G, Nagacevschi T, Tofan E, Sîtari C. Influence of biological active substances on the proline content of soybean sorts with various resistances to drought. Soil Form Fact Proc Temp Zone. 2015;13(1):51–8.
Blum A. Plant breeding for stress environments. Boca Raton: CRC; 1988. p. 43–61.
Blum A. Osmotic adjustment and growth of barley genotypes under drought stress. Crop Sci. 1989;29:230–3.
Blum A. Crop responses to drought and the interpretation of adaptation. Plant Growth Regul. 1996;20:135–48.
Blum A, Mayer J, Gozlan G. Associations between plant production and some physiological components of drought resistance in wheat. Plant Cell Environ. 1983;6:219–25.
Blum A, Zhang J, Nguyen HT. Consistent differences among wheat cultivars in osmotic adjustment and their relationship to plant production. Field Crops Res. 1999;64:287–91.
Bohnert HJ, Jensen RG. Strategies for engineering water-stress tolerance in plants. Trends Biotechnol. 1996;14:89–97.
Bohnert HJ, Nelson DE, Jensen RG. Adaptations to environmental stresses. Plant Cell. 1995;7:1099–111.

Borgo L, Marur CJ, Vieira LGE. Effects of high proline accumulation on chloroplast and mitochondrial ultrastructure and on osmotic adjustment in tobacco plants. Acta Sci Agron. 2015;37(2):191–9.

Boukraâ D, Benabdelli K, Belabid L, Bennabi F. Effect of salinity on chickpea seed germination pre-treated with salicylic acid. Sci J Biol Sci. 2013;2(4):86–93.

Bowne JB, Erwin TA, Juttner J, Schnurbusch T, Langridge P, Bacic A, Roessner U. Drought responses of leaf tissues from wheat cultivars of differing drought tolerance at the metabolite level. Mol Plant. 2012;5(2):418–29.

Braatne JH, Hinckley TM, Stettler RF. Influence of soil water on the physiological and morphological components of plant water balance in *Populus trichocarpa*, *Pops deltoides* and their F1 hybrids. Tree Physiol. 1992;11:325–39.

Bradshaw HD. Molecular genetics of *Populus*. In: Stettler RF, Bradshaw HD, Heilman PE, Hinckley TM, editors. Biology of *Populus* and its implications for management and conservation. Ottawa: NRC Research Press, National Research Council of Canada; 1996. p. 183–99.

Bressan RA, Hasegawa PM, Handa AK. Resistance of cultured higher plant cells to polyethylene glycol-induced water stress. Plant Sci Lett. 1981;21:23–30.

Brestic M, Zivcak M. PSII fluorescence techniques for measurement of drought and high temperature stress signal in crop plants: protocols and applications. In: Molecular stress physiology of plants. Springer. 2013. pp 87-131.

Brestic M, Zivcak M, Olsovska K, Kalaji HM, Shao HB, Hakeem KR. Heat signaling and stress responses in photosynthesis. In: Brestic M, editor. Plant signaling: understanding the molecular cross-talk. New Delhi: Springer; 2014. p. 241–56.

Buermann W, Lintner BR, Koven CD, Angert A, Pinzon JE, Tucker CJ, et al. The changing carbon cycle at Mauna Loa Observatory. Proc Natl Acad Sci U S A. 2007;104:4249–54.

Celik O, Atak Ç. Evaluation of proline accumulation and Δ1-pyrroline-5-carboxylate synthetase (*P5CS*) gene expression during salinity stress in two soybean (*Glycine max* L. Merr.) varieties. Polish J Environ Stud. 2012;21:559–64.

Chaitali R, Sengupta DN. Effect of short term NaCl stress on cultivars of *S. lycopersicum*: a comparative biochemical approach. J Stress Physiol Biochem. 2014;10(1):112.

Chan Z, Grumet R, Loescher W. Global gene expression analysis of transgenic, mannitol-producing, and salt-tolerant *Arabidopsis thaliana* indicates widespread changes in abiotic and biotic stress-related genes. J Exp Bot. 2011;62(14):4787–803.

Chandra Babu R, Pathan MS, Blum A, Nguyen HT. Comparison of measurement methods of osmotic adjustment in rice cultivars. Crop Sci. 1999;39:150–8.

Charkazi F, Ramezanpour SS, Soltanloo H. Expression pattern of two sugar transporter genes (*SuT4* and *SuT5*) under salt stress in wheat. Plant Omics J. 2010;3:194–8.

Cha-Um S, Samphumphuang T, Kirdmanee C. Glycinebetaine alleviates water deficit stress in indica rice using proline accumulation, photosynthetic efficiencies, growth performances and yield attributes. Aust J Crop Sci. 2013;7(2):213–8.

Chaves MM, Oliveira MM. Mechanisms underlying plant resilience to water deficits: prospects for water-saving agriculture. J Exp Bot. 2004;55:2365–84.

Chaves MM, Pereira JS, Maroco J, Rodrigues ML, Ricardo CP, Osorio ML, Carvalho I, Faria T, Pinheiro C. How plants cope with water stress in the field. Photosynthesis and growth. Ann Bot. 2002;89:907–16.

Chaves MM, Flexas J, Pinheiro C. Photosynthesis under drought and salt stress: regulation mechanisms from whole plant to cell. Ann Bot. 2009;103:551–60.

Chen THH, Murata N. Enhancement of tolerance of abiotic stress by metabolic engineering of betaines and other compatible solutes. Curr Opin Plant Biol. 2002;5:250–7.

Chen S, Gollop N, Heuer B. Proteomic analysis of salt-stressed tomato (*Solanum lycopersicum*) seedlings: effect of genotype and exogenous application of glycinebetaine. J Exp Bot. 2009;60:2005–19.

Chen X, Song F, Liu F, Tian C, Liu S, Xu H, Zhu X. Effect of different arbuscular mycorrhizal fungi on growth and physiology of maize at ambient and low temperature regimes. Sci World J. 2014. 10.1155/2014/956141.

Chen YS, Lo SF, Sun PK, Lu CA, Ho THD, Yu SM. A late embryogenesis abundant protein HVA1 regulated by an inducible promoter enhances root growth and abiotic stress tolerance in rice without yield penalty. Plant Biotechnol J. 2015;13(1):105–16.

Chéour F, Kaddachi I, Achouri D, Bannour S, Zorgui L. Effects of water stress on relative water, chlorophylls and proline contents in barley (*Hordeum vulgare* L.) leaves. J Agr Vet Sci. 2014;7(6):13–6.

Chutipaijit S, Cha-Um S, Sompornpailin K. An evaluation of water deficit tolerance screening in pigmented indica rice genotypes. Pak J Bot. 2012;44(1):65–72.

Ciais P, Reichstein M, Viovy N, Granier A, Ogee J, Allard V, et al. Europe-wide reduction in primary productivity caused by the heat and drought in 2003. Nature. 2005;437:529–33.

Clauw P. Study of biomass production during drought in *Arabidopsis thaliana* (Doctoral dissertation, Faculty of Sciences Department of Plant Biotechnology and Genetics VIB Department of Plant Systems Biology, UGent). 2010.

Colmer TD, Epstein E, Dvorak J. Differential solute regulation in leaf blades of various ages in salt-sensitive wheat and a salt-tolerant wheat x *Lophopyrum elongatum* (Host) A. love amphiploid. Plant Physiol. 1995;108:1715–24.

Conde A, Silva P, Agasee A, Conde C, Gerós H. Mannitol transport and mannitol dehydrogenase activities are coordinated in *Olea japonica* under salt and osmotic stress. Plant Cell Physiol. 2011;52:1766–75.

Condon AG. Water relations and osmotic adjustment in wheat as influenced by water stress history and plant development. MSc Thesis. Sydney: Sydney University; 1982.

D'Souza Myrene R, Devaraj VR. Mercury-induced changes in growth and oxidative metabolism of field bean (*Dolichos lablab*). Res J Chem Environ. 2013;17(9):86–93.

Dai JR, Liu B, Feng DR, Liu HY, He YM, Qi KB, Wang JF, et al. MpAsr encodes an intrinsically unstructured protein and enhances osmotic tolerance in transgenic *Arabidopsis*. Plant Cell Rep. 2011;30(7):1219–30.

Das KC, Misra HP. Hydroxyl radical scavenging and singlet oxygen quenching properties of polyamines. Mol Cell Biochem. 2004;262:127–33.

Dawood MG, El-Awadi ME. Alleviation of salinity stress on *Vicia faba* L. plants via seed priming with melatonin. Acta Biol Colombiana. 2015;20(2):223–35.

De Britto JA, Roshan Sebastian S, Sheeba Gracelin DH. Effect of lead on malondialdehyde, superoxide dismutase, proline activity and chlorophyll content in *Capsicum annum*. Biores Bull. 2013;5(4):20–3.

De Ronde JA, Cress WA, Krüger GHJ, Strasser RJ, Van Staden J. Photosynthetic response of transgenic soybean plants, containing an *Arabidopsis P5CR* gene, during heat and drought stress. J Plant Physiol. 2004;161:1211–24.

Delauney AJ, Verma DPS. Proline biosynthesis and osmoregulation in plants. Plant J. 1993;4:215–23.

Deyanira QM, Estrada-Luna AA, Altamirano-Hernández J, Peña-Cabriales JJ, de Oca-Luna RM, Cabrera-Ponce JL. Use of trehalose metabolism as a biochemical marker in rice breeding. Mol Breeding. 2012;30(1):469–77.

Di H, Tian Y, Zu H, Meng X, Zeng X, Wang Z. Enhanced salinity tolerance in transgenic maize plants expressing a *BADH* gene from *Atriplex micrantha*. Euphytica. 2015;34:1–9.

Ding ZJ, Yan JY, Xu XY, Yu DQ, Li GX, Zhang SQ, Zheng SJ. Transcription factor WRKY46 regulates osmotic stress responses and stomatal movement independently in *Arabidopsis*. Plant J. 2014;79(1):13–27.

Dobra J, Motyka V, Dobrev P, Malbeck J, Prasil IT, Haisel D, Vankova R, et al. Comparison of hormonal responses to heat, drought and combined stress in tobacco plants with elevated proline content. J Plant Physiol. 2010;167(16):1360–70.

Dolferus R. To grow or not to grow: a stressful decision for plants. Plant Sci. 2014;229:247–61.

Dong J, Bowra S, Vincze E. The development and evaluation of single cell suspension from wheat and barley as a model system; a first step towards functional genomics application. BMC Plant Biol. 2010;10(1):239.

Duan B, Yang Y, Lu Y, Korpelainen H, Berninger F, Li C. Interactions between drought stress, ABA and genotypes in *Picea asperata*. J Exp Bot. 2007;58:3025–36.

Eichholz I, Förster N, Ulrichs C, Schreiner M, Huyskens-Keil S. Survey of bioactive metabolites in selected cultivars and varieties of *Lactuca sativa* L. under water stress. J Appl Bot Food Qual. 2014;87:265–73.

Einset J, Nielsen E, Connolly EL, Bones A, Sparstad T, Winge P, Zhu JK. Membrane-trafficking RabA4c involved in the effect of glycinebetaine on recovery from chilling stress in *Arabidopsis*. Physiol Plant. 2007;130:511–8.

El Habbasha SF, Mekki BB. Amelioration of the growth, yield and chemical constituents of canola plants grown under salinity stress condition by exogenous application of proline. Int J. 2014;2(11):501–8.

Engelbrecht BM, Comita LS, Condit R, Kursar TA, Tyree MT, Turner BL, Hubbell SP. Drought sensitivity shapes species distribution patterns in tropical forests. Nature. 2007;447:80–2.

Estrada B, Aroca R, Maathuis FJ, Barea JM, Ruiz-Lozano J. Arbuscular mycorrhizal fungi native from a Mediterranean saline area enhance maize tolerance to salinity through improved ion homeostasis. Plant Cell Environ. 2013;36(10):1771–82.

Evans PT, Malmberg RL. Do polyamines have roles in plant development? Annu Rev Plant Physiol Plant Mol Biol. 1989;40:235–69.

Evers D, Lefèvre I, Legay S, Lamoureux D, Hausman JF, Rosales ROG, Schafleitner R. Identification of drought-responsive compounds in potato through a combined transcriptomic and targeted metabolite approach. J Exp Bot. 2010;61(9):2327–43.

Farhad MS, Babak AM, Reza ZM, Hassan RSM, Afshin T. Response of proline, soluble sugars, photosynthetic pigments and antioxidant enzymes in potato (*Solanum tuberosum* L.) to different irrigation regimes in greenhouse condition. Aust J Crop Sci. 2011;5(1):55–60.

Fariduddin Q, Ahmed M, Mir BA, Yusuf M, Khan TA. 24-Epibrassinolide mitigates the adverse effects of manganese induced toxicity through improved antioxidant system and photosynthetic attributes in *Brassica juncea*. Environ Sci Pollut Res Int. 2015;22:1–11.

Fatehi F, Hosseinzadeh A, Alizadeh H, Brimavandi T, Struik PC. The proteome response of salt-resistant and salt-sensitive barley genotypes to long-term salinity stress. Mol Biol Rep. 2012;39(5):6387–97.

Fereres E, Acevedo E, Henderson DW, Hsiao TC. Seasonal changes in water potential and turgor maintenance in sorghum and maize under water stress. Physiol Plant. 1978;44:261–7.

Fernandez-Aunión C, Hamouda TB, Iglesias-Guerra F, Argandoña M, Reina-Bueno M, Nieto JJ, Vargas C, et al. Biosynthesis of compatible solutes in rhizobial strains isolated from *Phaseolus vulgaris* nodules in Tunisian fields. BMC Microbiol. 2010;10(1):192.

Filippou P, Bouchagier P, Skotti E, Fotopoulos V. Proline and reactive oxygen/nitrogen species metabolism is involved in the tolerant response of the invasive plant species *Ailanthus altissima* to drought and salinity. Environ Exp Bot. 2014;97:1–10.

Flexas J, Barbour MM, Brendel O, Cabrera HM, Carriquí M, Díaz-Espejo A, et al. Mesophyll diffusion conductance to CO_2: an unappreciated central player in photosynthesis. Plant Sci. 2012;193–194:70–84.

Flower DJ, Usha Rani A, Peacock JM. Influence of osmotic adjustment on the growth, stomatal conductance and light interception of contrasting sorghum lines in a harsh environment. Aust J Plant Physiol. 1990;17:91–105.

Flowers TJ, Colmer TD. Salinity tolerance in halophytes. New Phytol. 2008;179:945–63.

Foyer CH, Noctor G. Redox regulation in photosynthetic organisms: signaling, acclimation, and practical implications. Antioxid Redox Signal. 2009;11:861–905.

Gaikwad SA, Gaikwad DK, Chavan PD. Influence of water stress on free proline content in three different *Linum usitatissimum* varieties. World J Pharm Pharmaceut Sci. 2014;3(5):1528–33.

Galvani A. The challenge of the food sufficiency through salt tolerant crops. Rev Environ Sci Biotechnol. 2007;6:3–16.

Gao J, Wang N, Xu SS, Li Y, Wang Y, Wang GX. Exogenous application of trehalose induced H_2O_2 production and stomatal closure in *Vicia faba*. Biol Plantarum. 2013;57(2):380–4.

Gebre GM, Kuhns MR. Seasonal and clonal variations in drought tolerance of *Populus deltoides*. Canad J For Res. 1991;21:910–6.

Gebre G, Tschaplinski T. Role of osmotic adjustment in plant productivity: a summary report and review of current literature. Oak Ridge: Oak Ridge National Laboratory; 2000.

Gerats AGM, Kaye C, Collins C, Malmberg ML. Polyamine levels in *Petunia* genotypes with normal and abnormal floral morphologies. Plant Physiol. 1988;86:390–3.

Ghahremani M, Ghanati F, Bernard F, Azad T, Gholami M, Safari M. Ornithine-induced increase of proline and polyamines contents in tobacco cells under salinity conditions. Aust J Crop Sci. 2014;8:91–6.

Ghimire KH, Quiatchon LA, Vikram P, Swamy BM, Dixit S, Ahmed H, Kumar A, et al. Identification and mapping of a QTL (qDTY 1.1) with a consistent effect on grain yield under drought. Field Crop Res. 2012;131:88–96.

Gill SS, Tuteja N. Polyamines and abiotic stress tolerance in plants. Plant Signal Behav. 2010;51:26–33.

Glinka Z, Ludlow MM. Comparative osmotic adjustment to water deficit in Texas 671 and E 57. In: MA Foale, RG Henzell, PN Vance (eds.), Proc 2nd Aust Sorghum Conf Aust Inst of Agric Sci, Melbourne. 1986. pp 316–325.

Good AG, Stroeher VL, Muench DG. U.S. Patent No. 7,786,343. Washington, DC: U.S. Patent and Trademark Office, 2010.

Gosavi GU, Jadhav AS, Kale AA, Gadakh SR, Pawar BD, Chimote VP. Effect of heat stress on proline, chlorophyll content, heat shock proteins and antioxidant enzyme activity in sorghum (*Sorghum bicolor*) at seedlings stage. Indian J Biotechnol. 2014;13:356–63.

Goudarzi A, Jafari M, Safaie N, Jafari SM. Transgenic sugar beet expressing a bacterial mannitol-1-phosphate dehydrogenase (mtlD) gene shows enhanced resistance to fungal pathogens. Sugar Tech. 2015;1–12.

Groppa MD, Benavides MP. Polyamines and abiotic stress: recent advances. Amino Acids. 2008;34:35–45.

Grümberg BC, Urcelay C, Shroeder MA, Vargas-Gil S, Luna CM. The role of inoculum identity in drought stress mitigation by arbuscular mycorrhizal fungi in soybean. Biol Fert Soils. 2015;51(1):1–10.

Grumet R, Hanson AD. Genetic evidence for an osmoregulatory function of glycinebetaine accumulation in barley. Aust J Plant Physiol. 1986;13:353–64.

Grumet R, Albrechtsen RS, Handon AD. Growth and yield of barley isopopulations differing in solute potential. Crop Sci. 1987;27:991–5.

Gupta N, Thind SK, Bains NS. Glycine betaine application modifies biochemical attributes of osmotic adjustment in drought stressed wheat. Plant Growth Regul. 2014;72(3):221–8.

Gupta N, Thind S. Improving photosynthetic performance of bread wheat under field drought stress by foliar applied glycine betaine. J Agr Sci Technol. 2015;17(1):75–86.

Gurumoorthy P, Singh KG. Quantitative analysis of proteins and antioxidants during stress on *Oryza sativa*. J Global Biosci. 2014;3(2):552–61.

Hanafy MS, Mohamed HA. Effect of irradiation of wheat grains with fast neutrons on the grain yield and other characteristics of the plants. Appl Radiat Isotopes. 2014;86:71–8.

Hanafy MS, El-Banna A, Schumacher HM, Jacobsen HJ, Hassan FS. Enhanced tolerance to drought and salt stresses in transgenic faba bean (*Vicia faba* L.) plants by heterologous expression of the *PR10a* gene from potato. Plant Cell Rep. 2013;32(5):663–74.

Hanzawa Y, Takahashi T, Michael AJ, Burtin D, Long D, Pineiro M, Coupland G, Komeda Y. *ACAULIS5*, an *Arabidopsis* gene required for stem elongation, encodes a spermine synthase. EMBO J. 2000;19:4248–56.

Hare PD, Cress WA. Metabolic implications of stress-induced proline accumulation in plants. Plant Growth Regul. 1997;21:79–102.

Hare PD, Cress WA, Van Staden J. Dissecting the roles of osmolyte accumulation during stress. Plant Cell Environ. 1998;21:535–53.

He C, Zhang W, Gao Q, Yang A, Hu X, Zhang J. Enhancement of drought resistance and biomass by increasing the amount of glycine betaine in wheat seedlings. Euphytica. 2011;177(2):151–67.

He C, He Y, Liu Q, Liu T, Liu C, Wang L, Zhang J. Co-expression of genes *ApGSMT2* and *ApDMT2* for glycinebetaine synthesis in maize enhances the drought tolerance of plants. Mol Breeding. 2013;31(3):559–73.

Hirich A, El Omari H, Jacobsen SE, Lamaddalena N, Hamdy A, Ragab R, Choukr-Allah R, et al. Chickpea (*Cicer arietinum* L.) physiological, chemical and growth responses to irrigation with saline water. Aust J Crop Sci. 2014;8(5):646–54.

Holmberg N, Bülow L. Improving stress tolerance in plants by gene transfer. Trends Plant Sci. 1998;3:61–6.

Holmström KO, Mantyala E, Welin B, Mandal A, Palva TE, Tunnela O, Londsborough J. Drought tolerance in tobacco. Nature. 1996;379:683–4.

Hong Z, Lakkineni K, Zhang Z, Verma DPS. Removal of feedback inhibition of Δ1-pyrroline-5-carboxylate synthetase results in increased proline accumulation and protection of plants from osmotic stress. Plant Physiol. 2000;122:1129–36.

Hoque MA, Banu MNA, Nakamura Y, Shimoishi Y, Murata Y. Proline and glycine betaine enhance antioxidant defense and methyl glyoxal detoxification systems and reduce NaCl-induced damage in cultured tobacco cells. J Plant Physiol. 2008;165:813–24.

Hossain MA, Fujita M. Evidence for a role of exogenous glycine betaine and proline in antioxidant defense and methyl-glyoxal detoxification systems in mung bean seedlings under salt stress. Physiol Mol Biol Plants. 2010;16:19–29.

Hosseini SM, Hasanloo T, Mohammadi S. Physiological characteristics, antioxidant enzyme activities, and gene expression in 2 spring canola (*Brassica napus* L.) cultivars under drought stress conditions. Turk J Agr Forest. 2014;39:413–20.

Hsiao T, Xu LK. Sensitivity of growth of roots versus leaves to water stress: biophysical analysis and relation to water transport. J Exp Bot. 2000;51:1595–616.

Hsiao TC, Acevedo E, Fereres E, Henderson DW. Stress metabolism. Water stress, growth, and osmotic adjustment. Phil Trans R Soc Lond B. 1976;273:479–500.

Hsiao TC, O'Toole JC, Yambao EB, Turner NC. Influence of osmotic adjustment on leaf rolling and tissue death in rice (*Oryza sativa* L). Plant Physiol. 1984;75:338–41.

Hussain SS, Ali M, Ahmad M, Siddique KHM. Polyamines: natural and engineered abiotic and biotic stress tolerance in plants. Biotechnol Adv. 2011;29:300–11.

Ibrahim HM. Selenium pretreatment regulates the antioxidant defense system and reduces oxidative stress on drought-stressed wheat (*Triticum aestivum* L.) plants. Asian J Plant Sci. 2014;13(3):120–8.

Iqbal MJ, Rahman MU, Ashraf M, Sheikh MA, Jamil A. Trehalose expression in hexaploid wheat (*Triticum aestivum* L.) germplasm under drought stress. Pak J Life Socl Sci. 2012;10(2): 106–10.

İşeri ÖD, Körpe DA, Sahin FI, Haberal M. High salt induced oxidative damage and antioxidant response differs in *Nicotiana tabacum* L. and *Nicotiana rustica* L. cultivars. J Appl Biol Sci. 2013;7(1):15–23.

Islam S, Malik AI, Islam AKMR, Colmer TD. Salt tolerance in a *Hordeum marinum-Triticum aestivum* amphiploid, and its parents. J Exp Bot. 2007;58:1219–29.

Jagendorf AT, Takabe T. Inducers of glycine betaine synthesis in barley. Plant Physiol. 2001;127:1827–35.

Jain M. Emerging role of metabolic pathways in abiotic stress tolerance. J Plant Biochem Physiol. 2013;1:108.

Jain M, Tiwary S, Gadre R. Sorbitol-induced changes in various growth and biochemical parameters in maize. Plant Soil Environ. 2010;6:263–7.

Jamaux I, Steinmetz A, Belhassen E. Looking for molecular and physiological markers of osmotic adjustment in sunflower. New Phytol. 1997;137:117–27.

Javadian N, Karimzadeh G, Mahfoozi S, Ghanati F. Cold-induced changes of enzymes, proline, carbohydrates, and chlorophyll in wheat. Russ J Plant Physl. 2010;57(4):540–7.

Javadmanesh S, Rahmani F, Pourakbar L. UV-B radiation, soil salinity, drought stress and their concurrent effects on some physiological parameters in maize plant. Am Eurasian J Toxicol Sci. 2012;4(4):154–64.

Jha Y, Subramanian RB, Patel S. Combination of endophytic and rhizospheric plant growth promoting rhizobacteria in *Oryza sativa* shows higher accumulation of osmoprotectant against saline stress. Acta Physiol Plant. 2011;33(3):797–802.

Jnandabhiram C, Sailen Prasad B. Water stress effects on leaf growth and chlorophyll content but not the grain yield in traditional rice (*Oryza sativa* Linn.) genotypes of Assam, India. II. Protein and proline status in seedlings under PEG induced water stress. Amer J Plant Sci. 2012;3(7):971–80.

Jones XIM, Turner NC, Osmond CB. Mechanisms of drought resistance. In: Paleg LPG, Aspmall D, editors. The physiology and biochemistry of drought resistance in plants. Brisbane: Academic; 1981. p. 15–37.

Joseph EA, Radhakrishnan VV, Mohanan KV. A study on the accumulation of proline—an osmoprotectant amino acid under salt stress in some native rice cultivars of North Kerala, India. Universal J Agr Res. 2015;3(1):15–22.

Kadioglu A, Saruhan N, Saglam A, Terzi R, Acet T. Exogenous salicylic acid alleviates effects of long term drought stress and delays leaf rolling by inducing antioxidant system. Plant Growth Regul. 2011;64:27–37.

Kamal AHM, Cho K, Choi JS, Jin Y, Park CS, Lee JS, Woo SH. Patterns of protein expression in water-stressed wheat chloroplasts. Biol Plantarum. 2013;57(2):305–12.

Kamoshita A, Rodriguez R, Yamauchi A, Wade LJ. Response of rainfed-lowland rice genotypes to prolonged drought and rewatering during the vegetative stage. In: Fukai S, and Basnayake J (eds), Increased lowland rice production in the Mekong region. ACIAR Proceedings No 101, ACIAR, Canberra. 2001. pp 78–85.

Kanayama Y, Watanabe M, Moriguchi R, Deguchi M, Kanahama K, Yamaki S. Effects of low temperature and abscisic acid on the expression of the sorbitol-6-phosphate dehydrogenase gene in apple leaves. J Jpn Soc Hortic Sci. 2006;75:20–5.

Kausar N, Nawaz K, Hussain K, Bhatti KH, Siddiqi EH, Tallat A. Effect of exogenous applications of glycine betaine on growth and gaseous exchange attributes of two maize (*Zea mays* L.) cultivars under saline conditions. World Appl Sci J. 2014;29:1559–65.

Kaya C, Sonmez O, Aydemir S, Ashraf M, Dikilitas M. Exogenous application of mannitol and thiourea regulates plant growth and oxidative stress responses in salt-stressed maize (*Zea mays* L.). J Plant Interact. 2013;8(3):234–41.

Kerepesi I, Galiba G. Osmotic and salt stress-induced alteration in soluble carbohydrate content in wheat seedlings. Crop Sci. 2000;40:482–7.

Kesari R, Lasky JR, Villamor JG, Des Marais DL, Chen YJC, Liu TW, Verslues PE, et al. Intron-mediated alternative splicing of *Arabidopsis P5CS1* and its association with natural variation in proline and climate adaptation. Proc Natl Acad Sci U S A. 2012;109(23):9197–202.

Khalid A, Zafar ZU, Akram A, Hussain K, Manzoor H, Al-Qurainy F, Ashraf MA. Photosynthetic capacity of canola (*Brassica napus* L.) plants as affected by glycinebetaine under the salt stress. J Appl Bot. Food Qual. 2015;88(1).

Khan MA, Iqbal M, Jameel M, Nazeer W, Shakir S, Aslam MT, Iqbal B. Potentials of molecular based breeding to enhance drought tolerance in wheat (*Triticum aestivum* L.). Afr J Biotechnol. 2013;10(55):11340–4.

Khan MS, Ahmad D, Khan MA. Utilization of genes encoding osmoprotectants in transgenic plants for enhanced abiotic stress tolerance. Electron J Biotechnol. 2015a;18:257–66.

Khan SU, Khan A, Naveed S. Effect of exogenously applied kinetin and glycinebetaine on metabolic and yield attributes of rice (*Oryza sativa* L.) under drought stress. Emirates J Food Agr. 2015b;27(1):75.

Khavari-Nejad RA, Najafi F, Ranjbari M. The effects of GA3 application on growth, lipid peroxidation, antioxidant enzymes activities, and sugars levels of cadmium stressed tomato (*Lycopersicon esculentum* Mill. cv.CH) plants. Rom J Biol Plant Biol. 2013;58(1):51–60.

Khazarin HHV. Effects of deficit irrigation on soluble sugars, starch and proline in three corn hybrids. Indian J Sci Res. 2014;7(1):910–7.

Khomdram S, Barthakur S, Devi GS. Water stress inducible proline transporter from indica rice. Ann Agr Res. 2013;34(3):189–96.

Kido EA, Ferreira Neto JR, Silva RL, Belarmino LC, Bezerra Neto JP, Soares-Cavalcanti NM, et al. Expression dynamics and genome distribution of osmoprotectants in soybean: identifying important components to face abiotic stress. BMC Bioinf. 2013;14(1).
Kikuchi A, Huynh HD, Endo T, Watanabe K. Review of recent transgenic studies on abiotic stress tolerance and future molecular breeding in potato. Breeding Sci. 2015;65(1):85.
Kim TH, Bohmer M, Hu H, Nishimura N, Schroeder JI. Guard cell signal transduction network: advances in understanding abscisic acid, CO_2, and Ca^{2+} signaling. Annu Rev Plant Biol. 2010;61:561–91.
Kohzuma K, Cruz JA, Akashi K, Hoshiyasu S, Munekage YN, Yokota A, Kramer DM. The long-term responses of the photosynthetic proton circuit to drought. Plant Cell Environ. 2009;32:209–19.
Kojić D, Pajević S, Jovanović-Galović A, Purać J, Pamer E, Škondrić S, Grubor-Lajšić G, et al. Efficacy of natural aluminosilicates in moderating drought effects on the morphological and physiological parameters of maize plants (*Zea mays* L.). J Soil Sci Plant Nutr. 2012;12(1): 113–23.
Koyro HW, Ahmad P, Geissler N. Abiotic stress responses in plants: an overview. In: Ahmad P, Prasad MNV, editors. Environmental adaptations and stress tolerance of plants in the era of climate change. New York: Springer; 2012. p. 1–28.
Kubis J, Floryszak-Wieczorek J, Arasimowicz-Jelonek M. Polyamines induce adaptive responses in water deficit stressed cucumber roots. J Plant Res. 2014;127:151–8.
Kumar SG, Reddy AM, Sudhakar C. NaCl effects on proline metabolism in two high yielding genotypes of mulberry (*Morus alba* L.) with contrasting salt tolerance. Plant Sci. 2003;165: 1245–51.
Kumar D, Singh P, Yusuf MA, Upadhyaya CP, Roy SD, Hohn T, Sarin NB. The *Xerophyta viscosa* aldose reductase (*ALDRXV4*) confers enhanced drought and salinity tolerance to transgenic tobacco plants by scavenging methylglyoxal and reducing the membrane damage. Mol Biotechnol. 2013;54(2):292–303.
Kumar MN, Hsieh YF, Verslues PE. At14a-Like1 participates in membrane-associated mechanisms promoting growth during drought in *Arabidopsis thaliana*. Proc Natl Acad Sci U S A. 2015;112(33):10545–50.
Kusano T, Yamaguchi K, Berberich T, Takahashi Y. Advances in polyamine research in 2007. J Plant Res. 2007;120:345–50.
Kuznetsov V, Shorina M, Aronova E, Stetsenko L, Rakitin V, Shevyakova N. NaCl and ethylene-dependent cadaverine accumulation and its possible protective role in the adaptation of the common ice plant to salt stress. Plant Sci. 2007;172:363–70.
Ladyman JAR, Hitz WD, Hanson AD. Translocation and metabolism of glycinebetaine by barley plants in relation to water stress. Planta. 1980;150:191–6.
Lahuta LB, Dzik T. *D*-chiro-Inositol affects accumulation of raffinose family oligosaccharides in developing embryos of *Pisum sativum*. J Plant Physiol. 2011;168(4):352–8.
Lal A, Ku MSB, Edwards GE. Analysis of inhibition of photosynthesis due to water-stress in the C3 species *Hordeum vulgare* and *Vicia faba* —electron-transport, CO_2 fixation and carboxylation capacity. Photosynth Res. 1996;49:57–69.
Lawlor DW. Absorption of polyethylene glycols by plants and their effects on plant growth. New Phytol. 1970;69:501–13.
Lawlor DW, Cornic G. Photosynthetic carbon assimilation and associated metabolism in relation to water deficits in higher plants. Plant Cell Environ. 2002;25:275–94.
Le TN, Mcqueen-Mason SJ. Desiccation-tolerant plants in dry environments. Rev Environ Sci Biotechnol. 2006;5:269–79.
Leigh RA, Ahmad N, Wyn Jones RG. Assessment of glycinebetaine and proline compartmentation by analysis of isolated beet vacuoles. Planta. 1981;153:34–41.
Leithy SM, Leila BA, Abdallah EF, Gaballah MS. Response of *Canola* plants to antitranspirant levels and limited irrigation. Am Eurasian J Sustain Agr. 2015;9(4):83–7.
LeRudulier DL, Strom AR, Dandekar AM, Smith LT, Valentine RC. Molecular biology of osmoregulation. Science. 1984;224:1064–8.
Levitt J. Responses of plants to environmental stresses. II Water, radiation, salt and other stresses, 2nd edn. New York: Academic. 1980. pp 25-178.

Li F, Lei HJ, Zhao XJ, Tian RR, Li TH. Characterization of three sorbitol transporter genes in micropropagated apple plants grown under drought stress. Plant Mol Biol Rep. 2011;30:123–30.

Li X, Zhang SS, Ma JX, Guo GY, Zhang XY, Liu X, Bi CL. TaUBA, a UBA domain-containing protein in wheat (*Triticum aestivum* L.), is a negative regulator of salt and drought stress response in transgenic *Arabidopsis*. Plant Cell Rep. 2015;34(5):755–66.

Lilley JM, Ludlow MM. Expression of osmotic adjustment and dehydration tolerance in diverse rice lines. Field Crop Res. 1996;48:185–97.

Lilley JM, Ludlow MM, McCouch SR, O'Toole JC. Locating QTL for osmotic adjustment and dehydration tolerance in rice. J Exp Bot. 1996;47:1427–36.

Liu Q, Zhang Y, Shouyi C. Plant protein kinase genes induced by drought, high salt and cold stresses. Chin Sci Bull. 2000;45:1153–7.

Liu JH, Kitashiba H, Wang J, Ban Y, Moriguchi T. Polyamines and their ability to provide environmental stress tolerance to plants. Plant Biotechnol. 2007;24:117–26.

Liu C, Zhao L, Yu G. The dominant glutamic acid metabolic flux to produce gamma-amino butyric acid over proline in *Nicotiana tabacum* leaves under water stress relates to its significant role in antioxidant activity. J Integr Plant Biol. 2011;53:608–18.

Liu J, Zhang F, Zhou J, Chen F, Wang B, Xie X. Phytochrome B control of total leaf area and stomatal density affects drought tolerance in rice. Plant Mol Biol. 2012;78(3):289–300.

Liu R, Liu Y, Ye N, Zhu G, Chen M, Jia L, Zhang J, et al. AtDsPTP1 acts as a negative regulator in osmotic stress signalling during *Arabidopsis* seed germination and seedling establishment. J Exp Bot. 2015;66(5):1339–53.

Livingston DP, Hincha DK, Heyer AG. Fructan and its relationship to abiotic stress tolerance in plants. Cell Mol Life Sci. 2009;66:2007–23.

Loewus FA, Murthy PN. Myo-inositol metabolism in plants. Plant Sci. 2000;150:1–19.

Loukehaich R, Wang T, Ouyang B, Ziaf K, Li H, Zhang J, Ye Z, et al. SpUSP, an annexin-interacting universal stress protein, enhances drought tolerance in tomato. J Exp Bot. 2012;63(15):5593–606.

Ludlow MM, Muchow RC. A critical evaluation of traits for improving crop yields in water-limited environments. Adv Agron. 1990;43:107–53.

Ludlow MM, Chu ACP, Clements RJ, Kerslake RG. Adaptation of species *Centrosema* to water stress. Aust J Plant Physiol. 1983;10:119–30.

Ludlow MM, Santamaria JM, Fukai S. Contribution of osmotic adjustment to grain yield in *Sorghum bicolor* (L.) Moench under water-limited conditions. II Water stress after anthesis. Aust J Agr Res. 1990;41:67–77.

Ludmerszki E, Paldi K, Racz I, Szigeti Z, Rudnoy SZ. The promising role of exogenous s-methylmethionine in agriculture, in the case of maize cultivation. Appl Ecol Environ Res. 2014;12(3):777–85.

Lui J, Zhu JK. Proline accumulation and salt-stress-induced gene expression in a salt-hypersensitive mutant of *Arabidopsis*. Plant Physiol. 1997;114:591–6.

Ma Q, Turner DW. Osmotic adjustment segregates with and is positively related to seed yield in F3 lines of crosses between *Brassica napus* and *B. juncea* subjected to water deficit. Aust J Exp Agric. 2006;46:1621–7.

Ma Q, Niknam SR, Turner DW. Responses of osmotic adjustment and seed yield of *Brassica napus* and *B. juncea* to soil water deficit at different growth stages. Aust J Agric Res. 2006; 57:221–6.

Macaluso L, Bianco R, Rieger M. Mannitol-producing tobacco exposed to varying water, light, temperature and paraquat levels. J Hortic Sci Biotech. 2007;82(6):979–85.

Maevskaya SN, Nikolaeva MK. Response of antioxidant and osmoprotective systems of wheat seedlings to drought and rehydration. Russ J Plant Physiol. 2013;60(3):343–50.

Maggio A, Miyazaki S, Veronese P, Fujita T, Ibeas JI, Damsz B, Narasimhan ML, Hasegawa PM, Joly RJ, Bressan RA. Does proline accumulation play an active role in stress-induced growth reduction? Plant J. 2002;31:699–712.

Maisura CM, Lubis I, Junaedinand A, Ehara H. Some physiological character responses of rice under drought conditions in a paddy system. J Int Soc Southeast Asian Agric Sci. 2014;20(1):104–14.

Makhloufi E, Yousfi FE, Pirrello J, Bernadac A, Ghorbel A, Bouzayen M. TdERF1, an ethylene response factor associated with dehydration responses in durum wheat (*Triticum turgidum* L. subsp durum). Plant Signal Behav. 2015;10(10):66.

Malekzadeh P. Influence of exogenous application of glycinebetaine on antioxidative system and growth of salt-stressed soybean seedlings (*Glycine max* L.). Physiol Mol Biol Plant. 2015;21(2):225–32.

Malik S, Ashraf M, Arshad M, Malik TA. Effect of ascorbic acid application on physiology of wheat under drought stress. Pak J Agr Sci. 2015;52(1):209–17.

Manivannan P. Triazole induced tolerance to drought in *Helianthus annuus* L. Sunflower. 2015.

Marcińska I, Czyczyło-Mysza I, Skrzypek E, Grzesiak MT, Janowiak F, Filek M, Grzesiak S, et al. Alleviation of osmotic stress effects by exogenous application of salicylic or abscisic acid on wheat seedlings. Int J Mol Sci. 2013;14(7):13171–93.

Maroco JP, Rodrigues ML, Lopes C, Chaves MM. Limitations to leaf photosynthesis in grapevine under drought: metabolic and modelling approaches. Funct Plant Biol. 2002;29:1–9.

Masgrau C, Altabella T, Farrás R, Flores D, Thompson AJ, Besford RT, Tiburcio AF. Inducible over expression of oat arginine decarboxylase in transgenic tobacco plants. Plant J. 1997;11:465–73.

Matyssek R, Tang AC, Boyer JS. Plants can grow on internal water. Plant Cell Environ. 1991;14:925–30.

McNeil SD, Nuccio ML, Hanson AD. Betaines and related osmoprotectants. Targets for metabolic engineering of stress resistance. Plant Physiol. 1999;120:945–9.

Mehta RA, Handa A, Li N, Mattoo AK. Ripening activated expression of S-adenosylmethionine decarboxylase increases polyamine levels and influences ripening in transgenic tomato fruits. Plant Physiol. 1997;114:41–4.

Mehta RA, Cassol T, Li N, Ali N, Handa AK, Mattoo AK. Engineered polyamine accumulation in tomato enhances phytonutrient content, juice quality, and vine life. Nat Biotechnol. 2002;20:613–8.

Melkonian JJ, Wolfe J, Steponkus PL. Determination of the volumetric modulus of elasticity of wheat leaves by pressure-volume relations and the effect of drought conditioning. Crop Sci. 1982;22(1):116–23.

Melloul M, Iraqi D, El Alaoui M, Erba G, Alaoui S, Ibriz M, Elfahime E. Identification of differentially expressed genes by cDNA-AFLP technique in response to drought stress in *Triticum durum*. Food Technol Biotech. 2014;52(4):479–88.

Merewitz E, Belanger F, Warnke S, Huang B, Bonos S. Quantitative trait loci associated with drought tolerance in creeping bentgrass. Crop Sci. 2014;54(5):2314–24.

Minguet EG, Vera-Sirera F, Marina A, Carbonell J, Blazquez MA. Evolutionary diversification in polyamine biosynthesis. Mol Biol Evol. 2008;25:2119–28.

Missihoun TD, Schmitz J, Klug R, Kirch HH, Bartels D. Betaine aldehyde dehydrogenase genes from *Arabidopsis* with different sub-cellular localization affect stress responses. Planta. 2011;233(2):369–82.

Mitoi EN, Holobiuc I, Blindu R. The effect of mannitol on antioxidative enzymes *in vitro* long term cultures of *Dianthus tenuifolius* and *Dianthus spiculifolius*. Rom J Biol Plant Biol. 2009;54:25–30.

Mohamed HI, Akladious SA. Influence of garlic extract on enzymatic and non enzymatic antioxidants in soybean plants (*Glycine max*) grown under drought stress. Life Sci J. 2014;11(3s):46–58.

Molazem D, Qurbanov EM, Dunyamaliyev SA. Role of proline, Na and chlorophyll content in salt tolerance of corn (*Zea mays* L.). Am Eurasian J Agr Environ Sci. 2010;9:319–24.

Montesinos-Pereira D, Barrameda-Medina Y, Romero L, Ruiz JM, Sánchez-Rodríguez E. Genotype differences in the metabolism of proline and polyamines under moderate drought in tomato plants. Plant Biol. 2014;16(6):1050–7.

Morgan JM. Differences in osmoregulation between wheat genotypes. Nature. 1977;270:234–5.

Morgan JM. Osmoregulation as a selection criterion for drought tolerance in wheat. Aust J Agr Res. 1983;34:607–14.

Morgan JM. Osmoregulation and water stress in higher plants. Annu Rev Plant Physiol. 1984;35:299–319.

Morgan JM. A gene controlling differences in osmoregulation in wheat. Aust J Plant Physiol. 1991;18:249–57.

Morgan JM. Osmotic components and properties associated with genotypic differences in osmoregulation in wheat. Aust J Plant Physiol. 1992;19:67–76.

Morgan JM. Growth and yield of wheat lines with differing osmoregulative capacity at high soil water deficit in seasons of varying evaporative demand. Field Crop Res. 1995;40:143–52.

Morgan JM, Condon AG. Water use, grain yield, and osmoregulation in wheat. Aust J Plant Physiol. 1986;13:523–32.

Mu P, Li Z. Correlation analysis and QTL mapping of osmotic potential in japonica rice under upland and lowland conditions. Can J Plant Sci. 2013;93(5):785–92.

Munns R. Why measure osmotic adjustment? Aust J Plant Physiol. 1988;15:717–26.

Munns R, Tester M. Mechanisms of salinity tolerance. Annu Rev Plant Biol. 2008;59:651–81.

Munns R, Weir R. Contribution of sugars to osmotic adjustment in elongating and expanded zones of wheat leaves during moderate water deficits at two light levels. Aust J Plant Physiol. 1981;8:93–105.

Munns R, Brady CJ, Barlow EWR. Solute accumulation in the apex and leaves of wheat during water stress. Aust J Plant Physiol. 1979;6:379–89.

Murakeozy EP, Nagy Z, Duhaze C, Bouchereau A, Tuba Z. Seasonal changes in the levels of compatible osmolytes in three halophytic species of inland saline vegetation in Hungary. J Plant Physiol. 2003;160:395–401.

Naeem MK, Ahmed M, Noreen S, Shah MKN, Iqbal MS. Estimation of genetic components for plant growth and physiological traits of wheat (*Triticum aestivum* L.) under normal and stress conditions. SAARC J Agr. 2015;13(1):90–8.

Nasibi F, Kalantari KM, Barand A. Effect of seed pre-treatment with L-arginine on improvement of seedling growth and alleviation of oxidative damage in canola plants subjected to salt stress. Iran J Plant Physiol. 2014;5(1):1217–24.

Nawaz K, Ashraf M. Exogenous application of glycinebetaine modulates activities of antioxidants in maize plants subjected to salt stress. J Agron Crop Sci. 2010;196(1):28–37.

Nayyar H, Chander S. Protective effects of polyamines against oxidative stress induced by water and cold stress in chickpea. J Agron Crop Sci. 2004;190:355–65.

Nguyen HT, Babu RC, Blum A. Breeding for drought resistance in rice: physiology and molecular genetics considerations. Crop Sci. 1997;37:1426–34.

Nguyen TX, Nguyen T, Alameldin H, Goheen B, Loescher W, Sticklen M. Transgene Pyramiding of the HVA1 and mtlD in T3 maize (*Zea mays* L.) plants confers drought and salt tolerance, along with an increase in crop biomass. Int J Agron. 2013.

Nikolaeva MK, Maevskaya SN, Voronin PY. Activities of antioxidant and osmoprotective systems and photosynthetic gas exchange in maize seedlings under drought conditions. Russ J Plant Physiol. 2015;62(3):314–21.

Nio SA. Osmotic adjustment and solutes in leaves of wheat (*Triticum aestivum* L.) during. 2009.

Niu X, Bressan RA, Hasegawa PM, Pardo JM. Ion homeostasis in NaCl stress environments. Plant Physiol. 1995;109:735–42.

Niu X, Xiong F, Liu J, Sui Y, Zeng Z, Lu BR, Liu Y. Co-expression of ApGSMT and ApDMT promotes biosynthesis of glycine betaine in rice (*Oryza sativa* L.) and enhances salt and cold tolerance. Environ Exp Bot. 2014;104:16–25.

Nounjana N, Nghiab PT, Theerakulpisuta P. Exogenous proline and trehalose promote recovery of rice seedlings from salt-stress and differentially modulate antioxidant enzymes and expression of related genes. J Plant Physiol. 2012;169:596–604.

Nuccio ML, Rhodes D, Mcneil SD, Hanson AD. Metabolic engineering of plants for osmotic stress resistance. Curr Opin Plant Biol. 1999;2:128–34.

Nusrat N, Shahbaz M, Perveen S. Modulation in growth, photosynthetic efficiency, activity of antioxidants and mineral ions by foliar application of glycinebetaine on pea (*Pisum sativum* L.) under salt stress. Acta Physiol Plantarum. 2014;36(11):2985–98.

Ogawa S, Mitsuya S. S-methylmethionine is involved in the salinity tolerance of *Arabidopsis thaliana* plants at germination and early growth stages. Physiol Plantarum. 2012;144(1):13–9.

Ogren WL. Photorespiration: pathways, regulation, and modification. Annu Rev Plant Physiol. 1984;35:415–42.

Ort DR. When there is too much light. Plant Physiol. 2001;125:29–32.

Osakabe Y, Yamaguchi-Shinozaki K, Shinozaki K, Phan Tran LS. ABA control of plant macroelement membrane transport systems in response to water deficit and high salinity. New Phytol. 2014;202:35–49.

Pandey V, Shukla A. Acclimation and tolerance strategies of rice (*Oryza sativa* L.) under drought stress. Rice Sci. 2015;22(3):1.

Pandit A, Rai V, Bal S, Sinha S, Kumar V, Chauhan M, et al. Combining QTL mapping and transcriptome profiling of bulked RILs for identification of functional polymorphism for salt tolerance genes in rice (*Oryza sativa* L.). Mol Genet Genomics. 2010;284:121–36.

Panicot M, Masgrau C, Borrell A, Cordeiro A, Tiburcio AF, Altabella T. Effects of putrescine accumulation in tobacco transgenic plants with different expression of oat arginine decarboxylases. Physiol Plantarum. 2002;114:281–7.

Panta A, Panis B, Ynouye C, Swennen R, Roca W. Development of a PVS2 droplet vitrification method for potato cryopreservation. Cryo-Lett. 2014;35(3):255–66.

Paradisone V, Barrameda-Medina Y, Montesinos-Pereira D, Romero L, Esposito S, Ruiz JM. Roles of some nitrogenous compounds protectors in the resistance to zinc toxicity in *Lactuca sativa* cv. *Phillipus* and *Brassica oleracea* cv. Bronco Acta Physiol Plantarum. 2015;37(7):1–8.

Parihar P, Singh S, Singh R, Singh VP, Prasad SM. Effect of salinity stress on plants and its tolerance strategies: a review. Environ Sci Pollut Res. 2014;22(6):4056–75.

Park EJ, Jeknic Z, Sakamoto A, Denoma J, Yuwansiri R, Murata N, Chen THH. Genetic engineering of glycinebetaine synthesis in tomato protects seeds, plants, and flowers from chilling damage. Plant J. 2004;40:474–87.

Park EJ, Jeknic Z, Pino MT, Murata N, Chen THH. Glycinebetaine accumulation in chloroplasts is more effective than that in cytosol in protecting transgenic tomato plants against abiotic stress. Plant Cell Environ. 2007;30:994–1005.

Parry M, Andraloje PJ, Khan S, Lea PJ, Keys A. Rubisco activity: effect of drought stress. Ann Bot. 2002;89:833–39.

Passioura J. The drought environment: physical, biological and agricultural perspectives. J Exp Bot. 2007;58(2):113–7.

Patel PK, Hemantaranjan A. Salicylic acid induced alteration in dry matter partitioning, antioxidant defence system and yield in chickpea (*Cicer arietinum* L.) under drought stress. Asian J Crop Sci. 2012;4(3):86–102.

Peng Z, Lu Q, Verma DPS. Reciprocal regulation of D1-pyrroline-5-carboxylate synthetase and proline dehydrogenase genes control levels during and after osmotic stress in plants. Mol Gen Genet. 1996;253:334–41.

Pilon-Smits E, Ebskamp M, Paul MJ, Jeuken M, Weisbeek PJ, Smeekens S. Improved performance of transgenic fructan-accumulating tobacco under drought stress. Plant Physiol. 1995;107:125–30.

Pino MT. Enhanced *in vitro* drought tolerance of *Solanum tuberosum* and *Solanum commersonii* plants overexpressing the *ScCBF1* gene = *Solanum tuberosum* y *Solanum commersonii*, sobreexpresando el gen *ScCBF1*, mejoraron su tolerancia a sequía en condiciones de cultivo *in vitro*. 2013.

Pino MT, Ávila A, Molina A, Jeknic Z, Chen TH. Enhanced *in vitro* drought tolerance of *Solanum tuberosum* and *Solanum commersonii* plants overexpressing the *ScCBF1* gene. Cien Inv Agr. 2013;40:171–84.

Pintó-Marijuan M, Munné-Bosch S. Ecophysiology of invasive plants: osmotic adjustment and antioxidants. Trends Plant Sci. 2013;18(12):660–6.

Pospisilova J, Haisel D, Vankova R. Responses of transgenic tobacco plants with increased proline content to drought and/or heat stress. Am J Plant Sci. 2011;2(3):318.

Prasertsak A, Fukai S. Nitrogen availability and water stress interaction on rice growth and yield. Field Crop Res. 1997;52:249–60.

Price A, Courtois B. Mapping QTLs associated with drought resistance in rice: progress, problems and prospects. Plant Growth Regul. 1999;29:123–33.

Pugnaire FI, Serrano L, Pardos J. Constraints by water stress on plant growth. In: Passarakli M, editor. Handbook of plant and crop stress. 2nd ed. New York: Marcel Dekker; 1999.

Pyngrope S, Bhoomika K, Dubey RS. Reactive oxygen species, ascorbate–glutathione pool, and enzymes of their metabolism in drought-sensitive and tolerant indica rice (*Oryza sativa* L.) seedlings subjected to progressing levels of water deficit. Protoplasma. 2013;250(2):585–600.

Quan R, Shang M, Zhang H, Zhao Y, Zhang J. Improved chilling tolerance by transformation with *betA* gene for the enhancement of glycinebetaine synthesis in maize. Plant Sci. 2004;166:141–9.

Quéro A, Fliniaux O, Elboutachfaiti R, Petit E, Guillot X, Hawkins S, Mesnard F et al. β-Aminobutyric acid increases drought tolerance and reorganizes solute content and water homeostasis in flax (*Linum usitatissimum*). Metabolomics. 2015, pp 1–13.

Quisenberry JE, Cartwright GB, Mcmichael BL. Genetic relationship between turgor maintenance and growth in cotton germplasm. Crop Sci. 1984;24:479–82.

Rabert GA, Manivannan P, Somasundaram R, Panneerselvam R. Triazole compounds alters the antioxidant and osmoprotectant status in drought stressed *Helianthus annuus* L. plants. Emirates J Food Agr. 2014; 26(3): 265

Rahnama H, Vakilian H, Fahimi H, Ghareyazie B. Enhanced salt stress tolerance in transgenic potato plants (*Solanum tuberosum* L.) expressing a bacterial *mtlD* gene. Acta Physiol Plantarum. 2011;33(4):1521–32.

Ramalho JC, Zlatev ZS, Leitão AE, Pais IP, Fortunato AS, Lidon FC. Moderate water stress causes different stomatal and non-stomatal changes in the photosynthetic functioning of *Phaseolus vulgaris* L. genotypes. Plant Biol. 2014;16(1):133–46.

Rana RM, Rehman SU, Ahmed J, Bilal M. A comprehensive overview of recent advances in drought stress tolerance research in wheat (*Triticum aestivum* L.). Asian J Agric Biol. 2013;1(1):9–37.

Ranganayakulu GS, Veeranagamallaiah G, Sudhakar C. Effect of salt stress on osmolyte accumulation in two groundnut cultivars (*Arachis hypogaea* L.) with contrasting salt tolerance. Afr J Plant Sci. 2013;12:586–92.

Rathinasabapathi B. Metabolic engineering for stress tolerance: installing osmoprotectant synthesis pathways. Ann Bot. 2000;86:709–16.

Rattan A, Kapoor N, Bhardwaj R. Role of brassinosteroids in osmolytes accumulation under salinity stress in *Zea mays* plants. Int J Sci Res. 2012;3(9):1822–7.

Reguera M, Peleg Z, Blumwald E. Targeting metabolic pathways for genetic engineering abiotic stress-tolerance in crops. Biochim Biophys Acta. 2012;1819:186–94.

Rezaei MA, Kaviani B, Jahanshahi H. Application of exogenous glycine betaine on some growth traits of soybean (*Glycine max* L.) cv. DPX in drought stress conditions. Sci Res Essays. 2012;7:432–6.

Rhodes D, Hanson AD. Quaternary ammonium and tertiary sulfonium compounds in higher plants. Annu Rev Plant Physiol Plant Mol Biol. 1993;44:357–84.

Rhodes D, Samaras Y. Genetic control of osmoregulation in plants. In: Strange K, editor. Cellular and molecular physiology of cell volume regulation. Boca Raton: CRC; 1994. p. 347–61.

Riahi M, Ehsanpour AA. Responses of transgenic tobacco (*Nicotiana plambaginifolia*) overexpressing *P5CS* gene underin vitrosalt stress. Prog Biol Sci. 2013;2(2):76–84.

Rivero RM, Kojima M, Gepstein A, Sakakibara H, Mittler R, Gepstein S, Blumwald E. Delayed leaf senescence induces extreme drought tolerance in a flowering plant. Proc Natl Acad Sci U S A. 2007;104:19631–6.

Rosales MA, Ocampo E, Rodríguez-Valentín R, Olvera-Carrillo Y, Acosta-Gallegos J, Covarrubias AA. Physiological analysis of common bean (*Phaseolus vulgaris* L.) cultivars uncovers characteristics related to terminal drought resistance. Plant Physiol Biochem. 2012;56:24–34.

Rybka K, Nita Z. Physiological requirements for wheat ideotypes in response to drought threat. Acta Physiol Plantarum. 2015;37(5):1–13.

Saadia M, Jamil A, Akram NA, Ashraf M. A study of proline metabolism in canola (*Brassica napus* L.) seedlings under salt stress. Molecules. 2012;17(5):5803–15.

Saeed M, Guo W, Ullah I, Tabbasam N, Zafar Y, Zhang T. QTL mapping for physiology, yield and plant architecture traits in cotton (*Gossypium hirsutum* L.) grown under well-watered versus water-stress conditions. Electron J Biotechnol. 2011;14(3):3.

Saibi W, Feki K, Mahmoud RB, Brini F. Durum wheat dehydrin (DHN-5) confers salinity tolerance to transgenic *Arabidopsis* plants through the regulation of proline metabolism and ROS scavenging system. Planta. 2015;242(5):1187–94.

Sakamoto A, Murata N. Genetic engineering of glycinebetaine synthesis in plants: current status and implications for enhancement of stress tolerance. J Exp Bot. 2000;51:81–8.

Sakamoto A, Murata N. The role of glycine betaine in the protection of plants from stress: clues from transgenic plants. Plant Cell Environ. 2002;25:163–71.

Sakamoto A, Murata A, Murata N. Metabolic engineering of rice leading to biosynthesis of glycinebetaine and tolerance to salt and cold. Plant Mol Biol. 1998;38:1011–9.

Saker MM, Tawfik RS, Sammour RH. *In vitro* selection for drought tolerant line of *Linum usitatissimum* L. Cv. Sakha2. AOBR. 2014;1(1):41–51.

Salvi S, Tuberosa R. To clone or not to clone plant QTLs: present and future challenges. Trends Plant Sci. 2005;10:297–304.

Sánchez-Rodríguez E, Rubio-Wilhelmi M, Cervilla LM, Blasco B, Rios JJ, Rosales MA, Ruiz JM, et al. Genotypic differences in some physiological parameters symptomatic for oxidative stress under moderate drought in tomato plants. Plant Sci. 2010;178(1):30–40.

Saradhi P, Alia P, Arora S, Prasad KV. Proline accumulates in plants exposed to UV radiation and protects them against UV induced peroxidation. Biochem Biophys Res Commun. 1995;209:1–5.

Sarafraz-Ardakani MR, Khavari-Nejad RA, Moradi F, Najafi F. Abscisic acid and cytokinin-induced osmotic and antioxidant regulation in two drought-tolerant and drought-sensitive cultivars of wheat during grain filling under water deficit in field conditions. Not Sci Biol. 2014;6(3):354–62.

Sawahel W. Improved performance of transgenic glycine betaine-accumulating rice plants under drought stress. Biol Plant. 2004;47:39–44.

Seki M, Umezawa T, Urano K, Shinozaki K. Regulatory metabolic networks in drought stress responses. Curr Opin Plant Biol. 2007;10:296–302.

Serraj R, Sinclair TR. Osmolyte accumulation: can it really help increase crop yield under drought conditions? Plant Cell Environ. 2002;25:333–41.

Setter TL. Analysis of constituents for phenotyping drought tolerance in crop improvement. Front Physiol. 2012;3:180.

Shahbaz M, Masood Y, Perveen S, Ashraf M. Is foliar-applied glycinebetaine effective in mitigating the adverse effects of drought stress on wheat (*Triticum aestivum* L.)? J App Bot Food Qual. 2012;84(2):192.

Sharma SS, Dietz KJ. The significance of amino acids and amino acid-derived molecules in plant responses and adaptation to heavy metal stress. J Exp Bot. 2006;57:711–26.

Sharp RE, Davies WJ. Solute regulation and growth by roots and shoots of waterstressed maize plants. Planta. 1979;147:43–9.

Sharp RE, Hsiao TC, Silk WK. Growth of the maize primary root at low water potentials. II. Role of growth and deposition of hexose and potassium in osmotic adjustment. Plant Physiol. 1990;93:1337–46.

Shehab GG, Ahmed OK, El-Beltagi HS. Effects of various chemical agents for alleviation of drought stress in rice plants (*Oryza sativa* L.). Not Bot Horti Agr Cluj-Napoca. 2010;38(1):139–48.

Shen B, Jensen RG, Bohnert HJ. Increased resistance to oxidative stress in transgenic plants by targeting mannitol biosynthesis to chloroplasts. Plant Physiol. 1997;113:1177–83.

Shi S, Fan M, Iwama K, Li F, Zhang Z, Jia L. Physiological basis of drought tolerance in potato grown under long-term water deficiency. Int J Plant Prod. 2015;9(2).

Silvente S, Sobolev AP, Lara M. Metabolite adjustments in drought tolerant and sensitive soybean genotypes in response to water stress. PLoS One. 2012;7(6).

Sinclair TR. Model analysis of plant traits leading to prolonged crop survival during severe drought. Field Crop Res. 2000;68:211–7.

Sinclair TR, Muchow RC. System analysis of plant traits to increase grain yield on limited water supplies. Agron J. 2001;93:263–70.

Singh M, Kumar J, Singh S, Singh VP, Prasad SM. Roles of osmoprotectants in improving salinity and drought tolerance in plants: a review. Rev Environ Sci Bio Technol. 2015;14(3):407–26.

Siripornadulsil S, Train S, Verma DPS, Sayre RT. Molecular mechanisms of proline-mediated tolerance to toxic heavy metals in transgenic microalgae. Plant Cell. 2002;14:2837–47.

Slabbert MM, Krüger GHJ. Antioxidant enzyme activity, proline accumulation, leaf area and cell membrane stability in water stressed *Amaranthus* leaves. S Afr J Bot. 2014;95:123–8.

Smirnoff N, Cumbes QJ. Hydroxyl radical scavenging activity of compatible solutes. Phytochemistry. 1989;28:1057–60.

Somboonwatthanaku I, Dorling S, Leung S, McManus MT. Proline biosynthetic gene expression in tissue cultures of rice (*Oryza sativa* L.) in response to saline treatment. Plant Cell Tiss Organ Cult. 2010;103(3):369–76.

Soren KR, Ali K, Tyagi V, Tyagi A. Recent advances in molecular breeding of drought tolerance in rice (*Oryza sativa* L.). Indian J Biotechnol. 2010;9:233–51.

Steuter AA, Mozafar A, Goodin JR. Water potential of aqueous polyethylene glycol. Plant Physiol. 1981;67:64–7.

Sudhakar C, Veeranagamallaiah G, Nareshkumar A, Sudhakarbab O, Sivakumar M, Pandurangaiah M, Kiranmai K, Lokesh U. Polyamine metabolism influences antioxidant defense mechanism in foxtail millet (*Setaria italica* L.) cultivars with different salinity tolerance. Plant Cell Reports. 2015;34:141–56.

Sultan MARF, Hui L, Yang LJ, Xian ZH. Assessment of drought tolerance of some *Triticum* L. species through physiological indices. Czech J Genet Plant Breed. 2012;48(4):178–84.

Summart J, Thanonkeo P, Panichajakul S, Prathepha P, McManus MT. Effect of salt stress on growth, inorganic ion and proline accumulation in Thai aromatic rice, Khao Dawk Mali 105, callus culture. Afr J Biotechnol. 2010;9(2).

Székely G, Abraham E, Cseplo A, Rigo G, Zsigmond L, Csiszar J, Ayaydin F, Strizhov N, Jasik J, Schmelzer E, Koncz C, Szabados L. Duplicated *P5CS* genes of *Arabidopsis* play distinct roles in stress regulation and developmental control of proline biosynthesis. Plant J. 2008;53:11–28.

Sziderics AH, Oufir M, Trognitz F, Kopecky D, Matušíková I, Hausman JF, Wilhelm E. Organ-specific defence strategies of pepper (*Capsicum annuum* L.) during early phase of water deficit. Plant Cell Rep. 2010;29(3):295–305.

Szoke A, Miao GH, Hong Z, Verma DPS. Subcellular location of Δ1-pyrroline-5-carboxylate reductase in root/nodule and leaf of soybean. Plant Physiol. 1992;99:1642–9.

Tabaeizadeh Z. Drought-induced responses in plant cells. In: Jeon, KW (ed.), International Review of Cytology. 1998. pp. 193–247.

Taha MH, Abdallah MM, Abd El-Monem AA. Improving salinity tolerance and yield of barley (*Horde vulgare* L.) plants using arginine. Int J Acad Res. 2013;5(5):156–99.

Taji T, Ohsumi C, Iuchi S, Seki M, Kasuga M, Kobayashi M, Yamaguchi-Shinozaki K, Shinozaki K. Important roles of drought- and cold-inducible genes for galactinol synthase in stress tolerance in *Arabidopsis thaliana*. Plant J. 2002;29:417–26.

Talaat NB. Effective microorganisms improve growth performance and modulate the ROS-scavenging system in common bean (*Phaseolus vulgaris* L.) plants exposed to salinity stress. J Plant Growth Regul. 2015;34(1):35–46.

Talaat NB, Ghoniem AE, Abdelhamid MT, Shawky BT. Effective microorganisms improve growth performance, alter nutrients acquisition and induce compatible solutes accumulation in common bean (*Phaseolus vulgaris* L.) plants subjected to salinity stress. Plant Growth Regul. 2015;75(1):281–95.

Talat A, Nawaz K, Hussian K, Bhatti KH, Siddiqi EH, Khalid A, Sharif MU, et al. Foliar application of proline for salt tolerance of two wheat (*Triticum aestivum* L.) cultivars. World Appl Sci J. 2013;22(4):547–54.

Tan W, Blake TJ, Boyle TJB. Drought tolerance in faster- and slower-growing black spruce (*Picea mariana*) progenies: I. Stomatal and gas exchange response to osmotic stress. Physiol Plant. 1992;85(4):645–51.

Tang W, Sun J, Liu J, Liu F, Yan J, Gou X, Liu Y, et al. RNAi-directed downregulation of *betaine aldehyde dehydrogenase 1 (OsBADH1)* results in decreased stress tolerance and increased oxidative markers without affecting glycine betaine biosynthesis in rice (*Oryza sativa*). Plant Mol Biol. 2014;4(86):443–54.

Tangpremsri T, Fukai S, Fischer KS, Henzell RG. Genotypic variation in osmotic adjustment in grain sorghum. 2. Relation with some growth attributes. Aust J Agr Res. 1991;42:759–67.

Tari I, Kiss G, Deér AK, Csiszár J, Erdei L, Gallé Á, Gémes K, Horváth F, Poór P, Szepesi Á, Simon LM. Salicylic acid increased aldose reductase activity and sorbitol accumulation in tomato plants under salt stress. Biol Plant. 2010;54:677–83.

Tayal D, Srivastava PS, Bansal KC. Transgenic crops for abiotic stress tolerance. In: Srivastava PS, Narula A, Srivastava S, editors. Plant biotechnology and molecular markers. New Delhi: Anamaya; 2004. p. 346–65.

Teulat B, This D, Khairallah M, Borries C, Ragot C, Sourdille P, Leroy P, Monneveux P, Charrier A. Several QTLs involved in osmotic-adjustment trait variation in barley (*Hordeum vulgare* L.). Theor Appl Genet. 1998;96:688–98.

Tezara W, Mitchell VJ, Driscoll SD, Lawlor DW. Water stress inhibits plant photosynthesis by decreasing coupling factor and ATP. Nature. 1999;401:914–7.

Theerakulpisut P, Gunnula W. Exogenous sorbitol and trehalose mitigated salt stress damage in salt-sensitive but not salt tolerance rice seedlings. Asian J Crop Sci. 2012;4:165–70.

Theerakulpisut P, Phongngarm S. Alleviation of adverse effects of salt stress on rice seedlings by exogenous trehalose. Asian J Crop Sci. 2013;5(4):405–15.

Thomson JA, Mundree SG, Shepherd DM, Rybicki EP. The use of African indigenous genes in the development of transgenic maize tolerant to drought and resistant to maize streak virus. In: Wambugu F, Kamanga D, editors. Biotechnology in Africa. New York: Springer; 2014. p. 135–55.

Todaka D, Nakashima K, Shinozaki K, Yamaguchi-Shinozaki K. Toward understanding transcriptional regulatory networks in abiotic stress responses and tolerance in rice. Rice. 2012;5(1):1–9.

Tookalloo MR. Evaluation of the effect of foliar absorption of glycine betaine on physiological characteristics of rapeseed cultivars (*Brassica napus*). 2011.

Tschaplinski TJ, Tuskan GA, Gunderson CA. Water-stress tolerance of black cottonwood and eastern cottonwood clones and four of their hybrid progeny. I. Growth, water relations and gas exchange. Canad J For Res. 1994;24:346–71.

Tschaplinski TJ, Tuskan GA, Gebre GM, Todd DE. Drought resistance of two hybrid Populus clones grown in large-scale plantation. Tree Physiol. 1998;18:653–8.

Türkan I, Demiral T. Recent developments in understanding salinity tolerance. Environ Exp Bot. 2009;67:2–9.

Turner NC, Jones MM. Turgor maintenance by osmotic adjustment: a review and evaluation. In: Turner NC, Kramer PJ, editors. Adaptation of plants to water and high temperature stress. New York: Wiley; 1980. p. 87–103.

Turner NC, Begg JE, Tonnet ML. Osmotic adjustment of sorghum and sunflower crops on response to water deficits and its influence on the water potential at which stomata close. Aust J Plant Physiol. 1978;5:597–608.

Turner NC, O'Toole JC, Cruz RT, Yambao EB, Ahmad S, Namuco OS, Dingkuhn M. Responses of seven diverse rice cultivars to water deficits. II. Osmotic adjustment, leaf elasticity, leaf extension, leaf death, stomatal conductance and photosynthesis. Field Crop Res. 1986;13:273–86.

Tyagi A, Sairam RK. Physiology and molecular biology of salinity stress tolerance in plants. Curr Sci. 2004;86(3):407–20.

Valliyodan B, Nguyen HT. Understanding regulatory networks and engineering for enhanced drought tolerance in plants. Curr Opin Plant Biol. 2006;9:189–95.

Van Deynze AE, Nelson JC, Iglesias ES, Harrington SE, Braga DP, McCouch SR, Sorrels ME. Comparative mapping in grasses. Wheat relationships. Mol Gen Genet. 1995;248:744–54.

Vanková R, Dobrá J, Štorchová H. Recovery from drought stress in tobacco: an active process associated with the reversal of senescence in some plant parts and the sacrifice of others. Plant Signal Behav. 2012;7(1):19–21.

Vardharajula S, Ali SZ, Grover M, Reddy G, Bandi V. Drought-tolerant plant growth promoting *Bacillus* spp.: effect on growth, osmolytes, and antioxidant status of maize under drought stress. J Plant Interact. 2011;1:1–14.

Verbruggen N, Hermans C. Proline accumulation in plants: a review. Amino Acids. 2008;35: 753–9.

Vinocur B, Altman A. Recent advances in engineering plant tolerance to abiotic stress: achievements and limitations. Curr Opin Biotechnol. 2005;16:123–32.

Voetberg GS, Sharp RE. Growth of the maize primary root at low water potentials. III. Role of increased proline deposition in osmotic adjustment. Plant Physiol. 1991;96:1125–30.

Walia H, Wilson C, Zeng L, Ismail AM, Condamine P, Close TJ. Genome-wide transcriptional analysis of salinity stressed japonica and indica rice genotypes during panicle initiation stage. Plant Mol Biol. 2007;63:609–23.

Wang HY, Huang YC, Chen SF, Yeh KW. Molecular cloning, characterization and gene expression of a water deficiency and chilling induced proteinase inhibitor I gene family from sweet potato (*Ipomoea batatas* Lam.) leaves. Plant Sci. 2003;165:191–203.

Wang CL, Zhang SC, Qi SD, Zheng CC, Wu CA. Delayed germination of *Arabidopsis* seeds under chilling stress by overexpressing an abiotic stress inducible *GhTPS11*. Gene. 2015;575(2 Pt 1):206–12.

Wani SH, Singh NB, Haribhushan A, Mir JI. Compatible solute engineering in plants for abiotic stress tolerance-role of glycine betaine. Curr Genomics. 2013;14:157–65.

Watanabe S, Nakagawa A, Izumi S, Shimada H, Sakamoto A. RNA interference-mediated suppression of xanthine dehydrogenase reveals the role of purine metabolism in drought tolerance in *Arabidopsis*. FEBS Lett. 2010;584(6):1181–6.

Watson MB, Emory KK, Piatak RM, Malmberg RL. Arginine decarboxylase (polyamine synthesis) mutants of *Arabidopsis thaliana* exhibit altered root growth. Plant J. 1998;13:231.

Wei GP, Yang LF, Zhu YL, Chen G. Changes in oxidative damage, antioxidant enzyme activities and polyamine contents in leaves of grafted and non-grafted eggplant seedlings under stress by excess of calcium nitrate. Sci Hortic. 2009;120:443–51.

Westgate ME, Boyer JS. Osmotic adjustment and the inhibition of leaf, root, stem and silk growth at low water potentials in maize. Planta. 1985;164:540–9.

Williamson JD, Jennings DB, Guo WW, Pharr DM, Ehrenshaft M. Sugar alcohols, salt stress and fungal resistance: polyols: multifunctional plant protection? J Am Soc Hortic Sci. 2002;127:467–73.

Wilson JR, Fisher MJ, Schulze E-D, Dolby GR, Ludlow MM. Comparison between pressure-volume and dewpoint-hygrometry techniques for determining the water relations characteristics of grass and legume leaves. Oecologia. 1979;41:77–88.

Wu D, Cai S, Chen M, Ye L, Chen Z, Zhang H, Zhang G, et al. Tissue metabolic responses to salt stress in wild and cultivated barley. PLoS One. 2013;8(1).

Wu GQ, Wang CM, Su YY, Zhang JJ, Feng RJ, Liang N. Assessment of drought tolerance in seedlings of sugar beet (*Beta vulgaris* L.) cultivars using inorganic and organic solutes accumulation criteria. Soil Sci Plant Nutr. 2014;60(4):565–76.

Wu GQ, Feng RJ, Liang N, Yuan HJ, Sun WB. Sodium chloride stimulates growth and alleviates sorbitol-induced osmotic stress in sugar beet seedlings. Plant Growth Regul. 2015;75(1):307–16.

Yamaguchi K, Takahashi Y, Berberich T, Imai A, Takahashi T, Michael AJ, Kusano T. A protective role for the polyamine spermine against drought stress in *Arabidopsis*. Biochem Biophys Res Commun. 2007;352:486–90.

Yiu JC, Tseng MJ, Liu CW, Kuo CT. Modulation of NaCl stress in *Capsicum annuum* L. seedlings by catechin. Sci Hortic. 2012;134:200–9.

Yousif BS, Liu LY, Nguyen NT, Masaoka Y, Saneoka H. Comparative studies in salinity tolerance between New Zealand spinach (*Tetragonia tetragonioides*) and chard (*Beta vulgaris*) to salt stress. Agr J. 2010;5(1):19–24.

Zadehbagheri M, Azarpanah A, Javanmardi S. Proline metabolite transport an efficient approach in corn yield improvement as response to drought conditions. Nature. 2014;566:596.

Zahra ARF, De Costa DM, De Costa WAJM. Identification of differentially-expressed genes in response to salt stress in the salt-tolerant Sri Lankan rice variety At354. J Nat Sci F Sri Lanka. 2013;41:2.

Zapata PJ, Serrano M, Pretel MT, Amoros A, Botella MA. Polyamines and ethylene changes during germination of different plant species under salinity. Plant Sci. 2004;167:781–8.

Zhang J, Nguyen HT, Blum A. Genetic analysis of osmotic adjustment in crop plants. J Exp Bot. 1999;50:291–302.

Zhang M, Chen Q, Shen S. Physiological responses of two Jerusalem artichoke cultivars to drought stress induced by polyethylene glycol. Acta Physiol Plantarum. 2011;33(2):313–8.

Zhang LX, Lai JH, Liang ZS, Ashraf M. Interactive effects of sudden and gradual drought stress and foliar-applied glycine betaine on growth, water relations, osmolyte accumulation and antioxidant defence system in two maize cultivars differing in drought tolerance. J Agron Crop Sci. 2014;200:425–33.

Zhang LX, Zheng P, Ruan Z, Tian L, Ashraf M. Nitric oxide accumulation and glycinebetaine metabolism in two osmotically stressed maize cultivars supplied with different nitrogen forms. Biol Plantarum. 2015;59(1):183–6.

Zhao X, Xu M, Wei R, Liu Y. Expression of OsCAS (calcium-sensing receptor) in an *Arabidopsis* mutant increases drought tolerance. PLoS One. 2015;10(6).

Zhong T, Zhang L, Sun S, Zeng H, Han L. Effect of localized reduction of gibberellins in different tobacco organs on drought stress tolerance and recovery. Plant Biotechnol Rep. 2014;8(5): 399–408.

Zhu JK. Plant salt tolerance. Trends Plant Sci. 2001;6:66–71.

Zhu X, Gong H, Chen G, Wang S, Zhang C. Different solute levels in two spring wheat cultivars induced by progressive field water stress at different developmental stages. J Arid Environ. 2005;62:1–14.

Zia MA, Ashraf M, Akram NA, Hussain M. Heat-induced regulation of antioxidant defense system and nutrient accumulation in hexaploid bread wheat (*Triticum aestivum* l.). Pak J Bot. 2014;46(3):957–71.

Zivcak M, Brestic M, Olsovska K, Slamka P. Performance index as a sensitive indicator of water stress in *Triticum aestivum*. Plant Soil Environ. 2008;54:133–9.

Zivcak M, Repkova J, Olsovska K, Brestic M. Osmotic adjustment in winter wheat varieties and its importance as a mechanism of drought tolerance. Cereal Res Commun. 2009;37:569–72.

Zivcak M, Brestic M, Balatova Z, Drevenakova P, Olsovska K, Kalaji HM, Allakhverdiev SI. Photosynthetic electron transport and specific photoprotective responses in wheat leaves under drought stress. Photosynth Res. 2013;117(1–3):529–46.

Zivcak M, Kalaji HM, Shao HB, Olsovska K, Brestic M. Photosynthetic proton and electron transport in wheat leaves under prolonged moderate drought stress. J Photoch Photobiol B. 2014;137:107–15.

Zsigmond L, Szepesi Á, Tari I, Rigó G, Király A, Szabados L. Overexpression of the mitochondrial PPR40 gene improves salt tolerance in *Arabidopsis*. Plant Sci. 2012;182:87–93.

Chapter 6
Interplay Among Glutathione, Salicylic Acid, and Ethylene to Combat Environmental Stress

Sharmila Chattopadhyay

6.1 Introduction

Glutathione (GSH) is nearly ubiquitous nonprotein tripeptide thiol compound found in both prokaryotes and eukaryotes (Kunert and Foyer 1993; Rennenberg 1997) except for some organisms that use other thiol cofactors (Li et al. 2003). One of the most abundant thiols, GSH is gradually gaining importance and becoming a molecule of interest because of its role in plant resistance and/or tolerance to environmental stress conditions. It has a vast array of functions in both plant and animal systems, which include the detoxification mechanisms through conjugation reactions, thiol transfer, scavenging of free radicals, and metabolism of various exogenous and endogenous compounds. The role of GSH in plant stress defense has long been known (Dron et al. 1988; Wingate et al. 1988), in addition to its substantial role in stress tolerance and antioxidant signaling (Foyer et al. 1997; Foyer and Noctor 2005). Recent studies proposed a potential contribution of GSH in the induced defense signaling network in conjunction with other established signaling molecules (Gomez et al. 2004; Ghanta et al. 2011a, b; Han et al. 2013a, b). A substantial number of reports suggested the significant involvement of GSH in both abiotic (Kocsy et al. 2000; Singla-Pareek et al. 2003; Gomez et al. 2004; Kumar et al. 2009) and biotic stress management (Glazebrook and Ausubel 1994; Glazebrook et al. 1997; Parisy et al. 2007; Mhamdi et al. 2010; Ghanta et al. 2011a, b). In this chapter, we explored the involvement of GSH with other established signaling molecules to provide the stress tolerance potential in γ–ECS overexpressed transgenic tobacco.

S. Chattopadhyay (✉)
Plant Biology Lab, CSIR-Indian Institute of Chemical Biology, Kolkata 700 032, India
e-mail: sharmila@iicb.res.in

M.A. Hossain et al. (eds.), *Drought Stress Tolerance in Plants, Vol 1*,
DOI 10.1007/978-3-319-28899-4_6

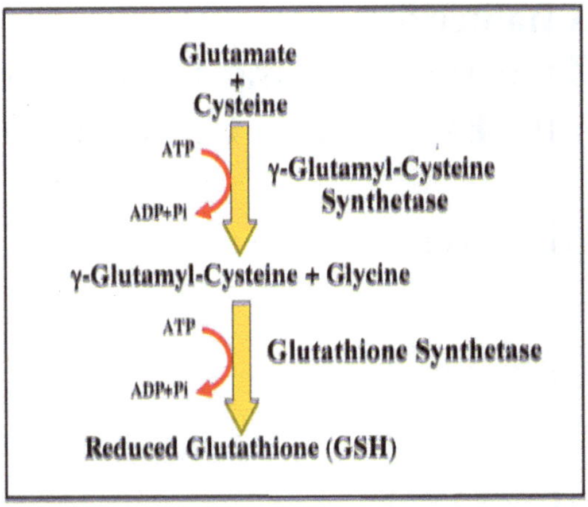

Fig. 6.1 Biosynthetic pathway of glutathione

6.2 Biosynthesis of GSH

The biosynthesis of GSH is well established and takes place in two ATP-dependent steps in all organisms, using two enzymes studied to date, except for *Streptococcus agalactiae,* where GSH is synthesized in a single step. This is a pseudopeptide, as the amino acid glutamine is bonded to cysteine through the side-chain carboxyl group rather than through its α-carbon carboxyl group; hence, its biosynthesis does not take place through the classical protein synthesis pathway. In the first step, *γ*-glutamylcysteine (*γ-EC*) is synthesized from L-glutamate and L-cysteine catalyzed by the *γ*-glutamylcysteine synthetase (*γ-ECS*) enzyme (Fig. 6.1). The second step, which involves the formation of GSH by the addition of glycine to the C-terminal end of γ-EC, is catalyzed by glutathione synthetase (GS) (Meister 1988). Both of the biosynthetic enzymes are found in the cytosol and plastids of the roots and leaves of plants (Hell and Bergmann 1988, 1990 Rüegsegger and Brunold 1993). The GSH biosynthesis has recently been analyzed in photosynthetic bacteria (*Synechocystis* spp.) (Musgrave et al. 2013), and its positive regulation by Lys R family regulators in *Sinorhizobium meliloti* has also been reported (Lu et al. 2013). The enzymes of GSH biosynthesis have also been analyzed by immunolocalization in leaves of *Arabidopsis thaliana.* These were observed to be localized in stroma (Preuss et al. 2014).

6.3 Role of GSH in Abiotic Stress

Climate change and abiotic stress adversely affect agriculture and crop production. When a plant is subjected to abiotic stress, a number of genes are turned on, resulting in increased levels of several metabolites and proteins, some of which may be responsible for conferring a certain degree of protection to these stresses. The major abiotic stresses in plants affecting the normal growth and development are salinity, drought, and fluctuation of temperature under natural conditions.

6.3.1 Role of GSH Against Salinity

One of the simplest ways to get an idea about the role of GSH in stress response is to measure the total GSH in stressed plants. A threefold increase in cysteine and GSH content was observed in wild-type *Brassica napus* exposed to salt stress, but not in the transgenic plants overexpressing vacuolar Na+/H+ antiporter (Ruiz and Blumwald 2002). Total GSH increased as a result of salt treatment in groundnut cell lines (Jain et al. 2002). A higher GSH concentration in a salt-resistant rice variety in comparison to its susceptible counterpart further supports the role of GSH in salt tolerance (Vaidyanathan et al. 2003). Salt increased both the GSH content and the GSH/oxidized GSH (GSSG) ratio associated with the salt-induced upregulation of γ-ECS protein in oxidative stress–tolerant *Lycopersicon pennellii* but not in *Lycopersicon esculentum* (Mittova et al. 2003). Maintenance of a higher GSH/GSSG ratio in transgenic tobacco overexpressing glyoxalase showed an improved salt tolerance potential (Yadav et al. 2005). In another study, recycling of GSH was thought to be the possible mechanism of tolerance to salinity stress in transgenic rice overexpressing glyoxalase II (Singla-Pareek et al. 2008). The overexpression of tobacco *NtGST/GPX* gene (encoding a bifunctional enzyme with both glutathione-s-transferase and glutathione peroxidase activity) in a transgenic tobacco plant had improved salt and chilling tolerances because of enhanced reactive oxygen species (ROS) scavenging and membrane damage. The differential modulation of both glutathione reductase and GSH in plants has been widely implicated for the significance of these two enigmatic antioxidants as major components of plant defense operations (Gill et al. 2013). In addition, the combined effect of GSH and ascorbic acid in drought and salt stress in GSTU17 knockout mutants of *A. thaliana* has been reported (Chen et al. 2012). The effect on GSH homeostasis under salinity is also reported for *Phaseolus vulgaris* (Kaymakanov et al. 2010) and *Gossypium hirsutum* (Kumari Vinodhana et al. 2013). A higher level of salt tolerance *in vivo* in transgenic *A. thaliana* plants has also been reported after GST overexpression (Qi et al. 2010; Chan and Lam 2014). Similarly, the importance of several enzymes of the ascorbate–GSH cycle has been shown during faba bean–*Rhizobium* symbiotic combination in root defense and adaptation against salt stress conditions (Oufdou et al. 2014).

6.3.2 Role of GSH in Drought Stress

GSH status and its relationship to protein synthesis during water deficit and subsequent rehydration have been examined in the drought-tolerant moss, *Tortula ruralis* (Dhindsa 1987). Total GSH was increased as a result of a water deficit in sunflower seedlings and detached poplar leaves (Sgherri and Navari-Izzo 1995; Morabito and Guerrier 2000). The accumulation of GSH in drought-stressed cowpea leaves is also reported (Cruz de Carvalho et al. 2010). The effect on ascorbate–GSH metabolism following water stress is also reported on *Andropyron cristatum* leaves. Likewise, the combined effect of GSH and ascorbic acid in drought and salt stress in GSTU17 knockout mutants of *A. thaliana* has been reported (Chen et al. 2012). The regulation of ROS-mediated abscisic acid (ABA) signaling by GSH in guard cells of *A. thaliana* during drought tolerance has also been reported (Munemasa et al. 2013).

6.3.3 Role of GSH in Metal Detoxification

The role of GSH as a metal chelator and its specific application in the phytoextraction of toxic metals have been reviewed (Seth et al. 2012). GSH is considered a key player in metal-induced oxidative stress management (Jozefczak et al. 2012).

6.3.4 Role of GSH Against Chilling Stress

According to previous reports, GSH has a significant role in chilling stress. In the past three decades, Anderson et al. (1992) reported that a greater total GSH content was observed in spruce during the winter, while a greater total GSH was observed in chilling-tolerant maize genotypes than in sensitive lines in cool spring periods (Leipner et al. 1999). A higher accumulation of γ-ECS transcript in cold-stressed maize was also observed (Gomez et al. 2004). The role of *GPX* in salt and chilling stress has also been reported in *Panax ginseng* (Kim et al. 2014).

6.3.5 Role of GSH in Heat Stress

GSH is also known to play a role in heat stress; a higher total GSH content was found to be associated with heat stress in wheat and maize (Nieto-Sotelo and Ho 1986; Dash and Mohanty 2002). The involvement of GSH in heat shock in rice has also been reported (Chao et al. 2009). The effect on ascorbate–GSH redox enzymes caused by heat shock has been noted as well (Locato et al. 2009). A similar effect on the plant antioxidative system following cadmium and heat stress has also been reported for rice (Zhao et al. 2009).

Together, GSH, which is an almost ubiquitous molecule, fulfills various important roles in plant functioning, making it an effective biomolecule (Kumar et al. 2010). The reduced form of GSH is considered to protect the cell from oxidative damage based on its redox buffering action and abundance in the cell (Ogawa 2005). Today considerable interest has been given to the functions of GSH because of its unique structural properties, abundance broad redox potential, and wide distribution in most living organisms (May et al. 1998; Meister 1988). GSH has also been suggested as an attractive target for engineering stress tolerance in plants because of its multiple roles in plant defenses against both biotic and abiotic stresses (Foyer et al. 1997; Kumar et al. 2009; Ghanta et al. 2011b). This tripeptide thiol is an important metabolite with a broad spectrum of functions, and its homeostasis is essential to maintain cellular redox poise and effective responses to stress in plants (Noctor et al. 2012). In addition to other reports, a cross talk of GSH with nitric oxide, carbon monoxide, and ROS has been reported in recalcitrant seeds of *Baccaurea ramiflora* (Bai et al. 2012). In concert, a growing body of evidence supports the notion that interplay between GSH and various established signaling components leads to the establishment of stress tolerance potential *in planta*.

6.4 Development of Transgenic Tobacco Lines, viz. *NtGp* Exhibiting Enhanced GSH Content

Tobacco (*Nicotiana tabacum* cv. Xanthi) and *Lycopersicon esculentum* seeds were sown on MS medium (Murashige and Skoog 1962) containing 3 % sucrose and 0.8 % agar and allowed to germinate at 22 ± 1 °C (16 h of 150 μE m^{-2}/s light and 8 h darkness).

Plasmid construction and plant transformation: *LeECS* (γ-ECS from *L. esculentum*) was PCR-cloned from 2-week-old seedlings by RT-PCR. For that, the 1571-base-pair full-length DNA sequence of *LeECS* was amplified using gene-specific primers; the PCR product was cloned into *pGEM-T* Easy vector (Promega, USA) and then finally subcloned into the binary vector *pBI*121 to obtain *NtGB* lines. In order to obtain *NtGp* lines, the *LeECS* construct was subcloned between SphI and SalI in *pJIT* 117 comprising the transit peptide to translocate the gene of interest to the chloroplast and finally into the binary vector *pCAMBIA* 2301 at the KpnI site, giving rise to the final construct (Russell and Sambrook 2001). These recombinant plasmids were introduced in *Agrobacterium tumefaciens* LBA4404 and used for transformation of tobacco leaf discs independently. The regenerated shoots were maintained on MS medium supplemented with required phytohormones and antibiotics. Finally, K^R, PCR-positive T0 transgenics plants were transferred to greenhouse conditions to obtain T2 generation for further analysis.

6.4.1 *HPLC analysis of* NtGp *Lines to Estimate GSH, SA, and ACC Content*

To estimate the GSH, SA, and ACC=1-aminocyclopropane-1-carboxylate (ACC) content of *NtGp* lines, we used the high-performance liquid chromatography (HPLC) system.

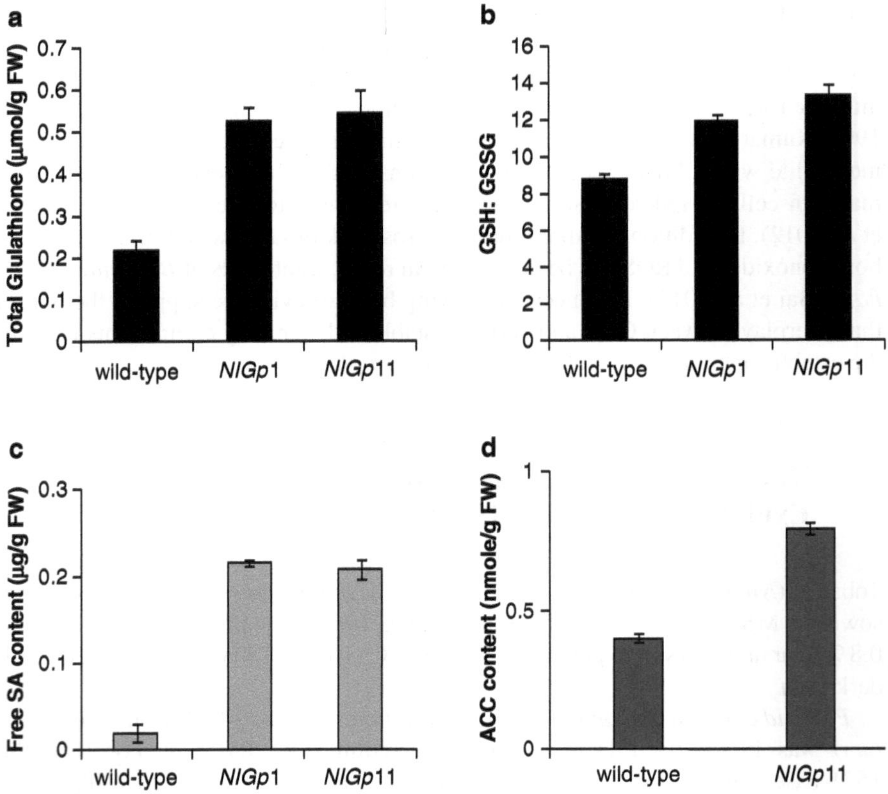

Fig. 6.2 Biochemical characterization of *NtGp* lines

6.4.1.1 Determination of GSH Content and GSH:GSSG Ratio

Estimation of GSH and its ratio with GSSG was analyzed from mature tobacco leaves of *NtGp* lines by HPLC using a 515-HPLC pump (Waters, USA), 2475 fluorescence detector, using AccQ Tag column (Waters, USA) (Tsakraklides et al. 2002) (Fig. 6.2a, b).

6.4.1.2 Determination of SA Content

SA was quantified by HPLC with a fluorescence detector, using Symmetry C-18 reverse-phase column (5 um, 4.6 × 250 mm) at an excitation wavelength of 254 nm and an emission wavelength of 395 nm (Freeman et al. 2005) (Fig. 6.2c).

6.4.1.3 Determination of ACC Content

Estimation of ACC was performed by the o-phthaldialdehyde precolumn derivatization method. The HPLC analysis was conducted using a 515 HPLC pump with a 2475 fluorescence detector as mentioned above, at a flow rate of 0.6 mL/min. An AccQ-Tag (3.9 mm × 150 mm) column with an excitation wavelength of 325 nm and an emission wavelength of 465 nm was used (Ghanta et al. 2014) (Fig. 6.2d).

6.5 Documentation of Drought and Salt Tolerance of *NtGp* Lines

The seeds of both wild-type and T2 generation of *NtGp* lines were grown in ½ MS media in a 16 h light/8 h dark cycle at 22 °C. For drought and salinity stress treatment, 2-week-old seedlings were further transferred to ½ MS media supplemented with 200 mM mannitol and NaCl for an additional 72 h and the leaves were collected for biochemical analysis (proline and chlorophyll estimation), RNA, and protein isolation as well as for drought and salinity stress analysis.

Stress analysis: The germination percentage of both wild-type and *NtGp*11 was measured daily after sowing it in ½ MS medium with different concentrations of mannitol and NaCl. The proline content was measured in 500 mg wild-type and *NtGp*11 leaf tissue treated with 200 mM of mannitol and NaCl after 72 h. In addition, leaf discs, approximately 1 cm in diameter, were detached from healthy, fully expanded leaves of wild-type and *NtGp*11 of the same age. The discs were floated in solutions of 200 mM of mannitol for 72 h, and then 100 mg of leaf tissue was immersed in 80 % acetone for 48 h to extract the chlorophyll (Fig. 6.3a, b).

6.5.1 Transcript Analysis Under Stress Condition

In response to drought and salinity stress, γ-ECS was found upregulated at the transcript level in *NtGp* lines (Fig. 6.4a). The expression level of all six abiotic stress–induced genes, including *osmotin*, *HSP*70, *ascorbate peroxidase* (*APX*), *glutathione-S-transferase* (*GST*), and *glutathione peroxidase* (*EIGC*08), was higher in the *NtGp* plant than in the wild-type plant in response to a treatment of 72 h 200 mM mannitol and NaCl (Fig. 6.5a). This clearly illustrated the effect of an enhanced GSH level on the elevated expression of abiotic stress–related genes in response to drought stress in the *NtGp* line.

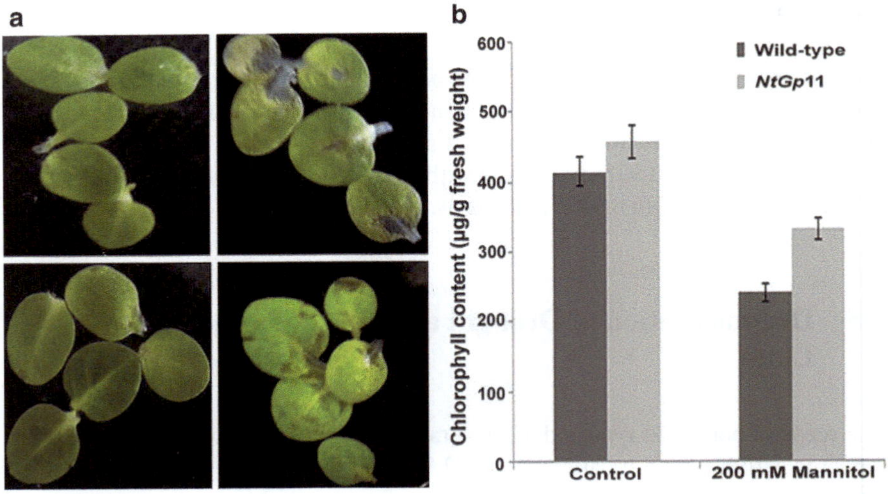

Fig. 6.3 Effect of 200 mM mannitol on the (**a**) morphology and (**b**) chlorophyll content of the detached leaves of wild-type and *NtGp11* plants after 72 h

Fig. 6.4 Expression level of *γ-ECS* in *NtGp*11 plant after 72 h of 200 mM mannitol and NaCl treatment by (**a**) semiquantitative real-time polymerase chain reaction and (**b**) Western blotting

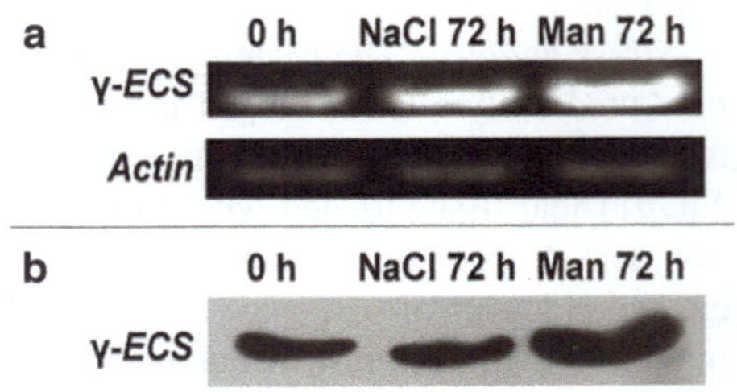

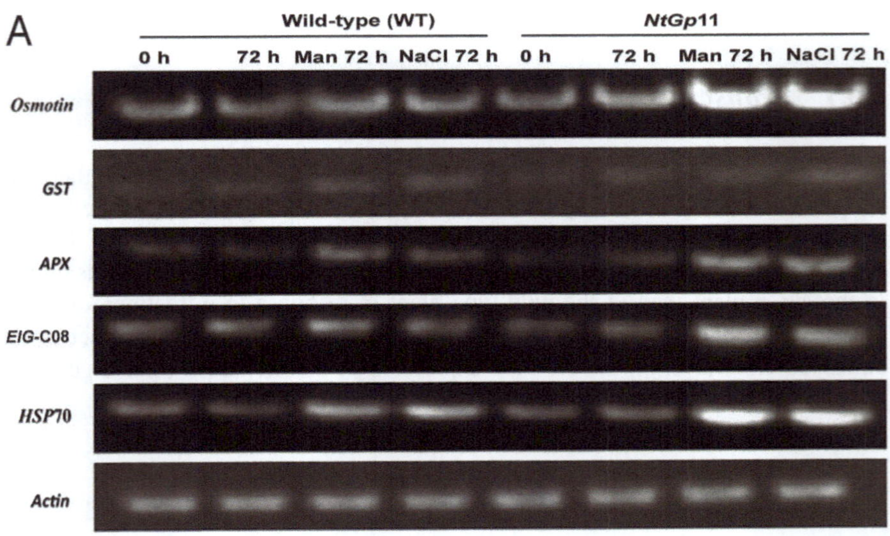

Fig. 6.5 (**a**) Expression level of stress-related genes in response to 200 mM mannitol and NaCl treatment in wild-type and *NtGp*11 plants after 72 h

6.5.2 Western Blot Analysis of γ-ECS

The accumulation of *γ-ECS* was noted more under drought stress condition in *NtGp* line than the salinity stress–treated plants (Fig. 6.4a), and this also corroborates with the *γ-ECS* expression through Western blot analysis (Fig. 6.4b).

6.6 Transcriptomic Analysis of *NtGp* Lines

Characterization of the expressed sequence tag (EST) sequences: In order to identify the differences in gene expression as a result of enhanced GSH content *in NtGp* lines, subtracted cDNA libraries were constructed. For this purpose, wild-type leaves were used as driver and *NtGp*11 leaves were used as tester. A total of 260 clones from the SSH cDNA library were screened for upregulated genes.

Transcript analysis of the SSH library identified the induction of many genes already known to be involved in the defense/stress pathways. Interestingly, many of these genes employ SA in the defense mechanism, thus confirming that SA is indispensable for GSH for its action in defense. Among the genes upregulated as a result of an enhanced GSH level, 11 ESTs matched with PR1a in the NCBI database, which corroborated with the previous study where PR1a, the well-known marker gene of SA-mediated pathway, was induced due to *γ-ECS* overexpression as well as by GSH feeding (Gomez et al. 2004). Systemic acquired resistance (SAR) is a form of inducible resistance that is triggered in systemic healthy tissues of locally infected plants (Vlot et al. 2008). Six ESTs were related to SAR8.2. The SAR8.2 is one of the gene families whose expression is induced in tobacco during SAR induction and contains at least 12 members (Alexander et al. 1992). Further, the phylloplanin gene represented by two ESTs had an enhanced expression level in *NtGp*11 as compared to the wild type. Phylloplanins are defensive proteins known to inhibit the biotrophic pathogen *Peronospora tabacina* (Shepherd et al. 2005; Shepherd and Wagner 2007). In addition, our present investigation significantly indicates a role of GSH in inducing an ET biosynthetic pathway. Among them, ACC oxidase, which is known to catalyze the conversion of ACC to ET, a key step in the ET biosynthesis, was induced in this forward SSH library, thus suggesting a probable induction of ET biosynthesis under enhanced GSH conditions. We got a clue about the probable ET–GSH interplay from our previous work, where the enhanced expression of ACC oxidase and PR4 was noted in *NtGB*9 and *NtGB*19 lines (Ghanta et al. 2011b). Thus, we next targeted exploring any probable cross talk of GSH with other signal molecules, and results demonstrate the induction of ET biosynthesis by GSH. It may be assumed that GSH induces the biosynthesis of ET by modulating the expression of ACC oxidase, which subsequently leads to an induction of the ET signaling pathway. This observation can be further corroborated with the previous reports that ET and SA may act synergistically to induce stress-related gene expression, although the nature of the SA–ET interaction is complex and less understood (Kunkel and Brooks 2002; De Vos et al. 2006; Pieterse et al. 2009).

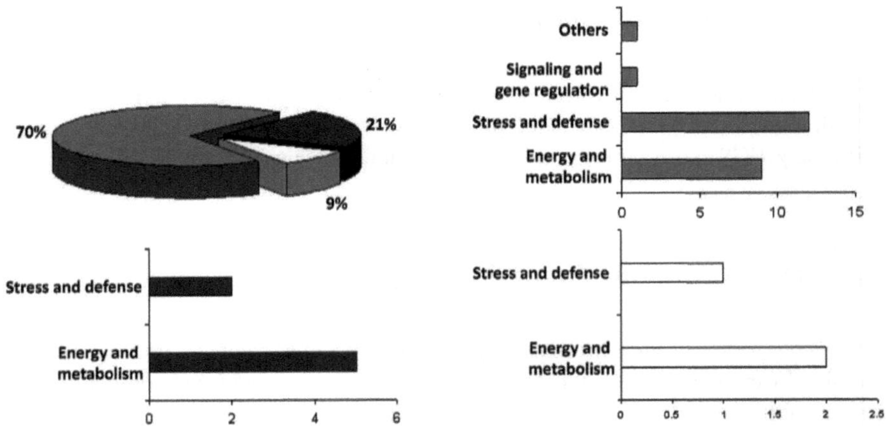

Fig. 6.6 Functional classification of identified proteins from chloroplast-enriched fraction of wild type and *NtGp11*. (**a**) Pie chart of proteins identified with increased accumulation (*gray*), decreased accumulation (*black*), and newly detected (*white*). (**b**) Bar graph showing functional categorization of proteins with increased accumulation. (**c**) Bar graph showing functional categorization of proteins with decreased accumulation. (**d**) Bar graph showing functional categorization of newly detected proteins

6.7 Comparative Proteomic Profiling *NtGp* Line Along with the Wild Type

Identification of differentially accumulated proteins from chloroplast-enriched fraction of NtGp11: Since the transgene was directed into chloroplasts, a comparative profiling of chloroplast-enriched protein fractions of the wild type and *NtGp* line was performed by two-dimensional gel electrophoresis (2-DE). Image analysis revealed an average of about 156 protein spots in the wild type and 210 protein spots in *NtGp*11, of which 125 spots could be matched to all gels. The overall mean coefficient of variation for the intensity of the spots marked was determined to be 38.91. The data were subjected to principal component analysis and results revealed that PC1 and PC2 explained 92 % and 5 % of variance, respectively (Fig. 6.6).

It is well known that the technological advances in proteomics will allow researchers to obtain high-quality proteomics data for complementary studies with transcriptome. However, one of the major limitations of 2-DE is that high abundant proteins are identified preferentially. Our proteome profile revealed that the major category was that of energy and metabolism class, accounting for 49 % of the proteins identified. Some of the proteins related to energy and metabolism, namely chloroplast ATP synthase, ribulose-1,5-bisphosphate carboxylase large subunit, ribulose bisphosphate carboxylase activase, and chlorophyll a/b-binding protein type III, were increased in accumulation. This may be because GSH plays some role in stabilizing the enzymes of the Calvin cycle (Foyer and Halliwell 1976). In addition, ACC synthase, a rate-limiting enzyme in ET biosynthesis, has been found to be induced (newly detected) in *NtGp*11, which corresponds with our transcript analysis data. The next-most abundant category was that of stress and defense, comprising about 42 % of the identified proteins. We identified HSP70 in our analysis, which are a major class of chaperones involved in protein-folding and organelle transport and play an important role in biotic and abiotic stress responses (Boston et al. 1996). Putative serine/threonine protein kinase was also identified, which represents a class of proteins that appear to act as a "central processor unit," accepting input information from receptors that sense varying environmental conditions and stress factors and converting it into appropriate outputs such as changes in metabolism, gene expression, and cell growth and division (Hardie 1999). The myb proteins are key factors in regulatory networks controlling development, metabolism, and responses to biotic and abiotic stresses (Dubos et al. 2010) and are known to be induced by the exogenous application of SA (Datta et al. 2013). The myb protein was also increased in accumulation in the present study.

6.8 Comparative Proteomics Analysis of *NtGp* Lines Under Stress Condition

*NtGp*11 was noted with a better drought tolerance potential than salinity, a higher expression of *γ-ECS*, as well as higher abiotic stress–induced genes, and so further comparative proteomic profiling was performed with *NtGp*11 in response to drought stress. The protein samples for 2-DE analyses were isolated from drought stress–treated wild-type and *NtGp*11 leaf samples. The differentially accumulated spots as identified by comparative protein profiling were further identified. The number of resolved spots from drought stress–treated wild type and *NtGp*11 were approximately 246 and 272, respectively. A total of 122 spots matched successfully, and 144 spots were found to be altered in intensity between the drought stress–treated wild type and *NtGp*11. The overall mean coefficient of variation for intensity of the spots matched was determined to be 38.7. Spots showing a statistically significant increase or decrease in response to stress treatment were excised from gels of the drought stress–treated wild-type and *NtGp*11 samples and were digested with trypsin and identified using MALDI TOF-TOF MS/MS. Among the 43 identified differentially accumulated protein spots, 58.13 % and 23.25 % spots were found up-accumulated [i.e., the spot intensity in *NtGp*11 is higher than in the wild type (control) by twofold or more] and down-accumulated (i.e., the spot intensity in *NtGp*11 is lower than in the wild type by twofold or more), respectively, in response to drought stress. Also identified were 9.3 % uniquely induced protein spots in response to drought stress from stress-treated *NtGp*11 plants and 9.3 % protein spots that were restricted only to wild-type plants (Fig. 6.7). Several proteins remained unidentified because the number and/or intensity of the fragment ions obtained by MS/MS was insufficient for a significant hit. For further confirmation on the role of

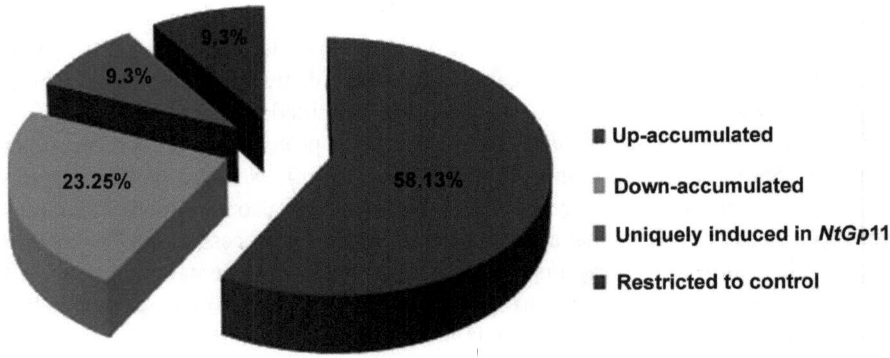

Fig. 6.7 Pie chart of differentially accumulated protein spots in response to 200 mM mannitol treatment in wild-type and *NtGp*11 plant after 72 h

GSH in mitigating drought stress, the identified differentially accumulated proteins in response to stress condition were distributed according to their functional categories. These proteins placed mostly into the categories of stress and defense, energy metabolism, carbon metabolism, gene and protein regulation, hypothetical, and unnamed proteins. Among the up-accumulated proteins in response to drought stress treatment in *NtGp*11, about 36 %, 24 %, and 20 % of the proteins were placed in the category of stress and defense, carbon metabolism, and energy metabolism, respectively. About 50 % of the down-accumulated proteins against stress condition in *NtGp*11 were related to carbon metabolism. Among uniquely induced proteins in response to stress in *NtGp*11, 50 % and 25 % of them were grouped under the energy metabolism and hormonal response categories, respectively. Several spots appeared more than once, which might be the result of different posttranslational modifications of proteins.

Among the *up-accumulated* proteins in response to drought stress in *NtGp*11, spot nos. 4 and 13 had identified to HSP70-like proteins and spot no. 12 identified to GSH peroxidise, which also showed a correlation with its gene expression in response to drought stress. HSP70 as stress-inducible genes respond to environmental abiotic stresses such as high/low temperature, drought, and high salinity. The sHSPs play a major role in decreasing the intracellular level of ROS, thereby protecting PSII function during stress. Spot no. 16 had identified to thioredoxin peroxidise, which together with glutathione peroxidase might play dual and distinctive roles in ROS homeostasis, acting as a general scavenger and specifically relaying the ROS signal as an oxidative signal transducer in drought stress signaling. Spot no. 10 was identified as chalcone synthase, which is a key enzyme of the flavonoid/isoflavonoid biosynthesis pathway. Flavonoids act as an antioxidant, which helps in neutralizing the effect of overproduced ROS in plant cell after drought stress. Spot no. 18 was identified as elicitor inducible protein EIG-J7, which was reported to be expressed in response to a variety of stress conditions, including wounding, drought stress, viral infection, and SA treatment. Spot no. 14 had the homology with ACC oxidase, which is an important enzyme of the ethylene biosynthetic pathway and it can be induced by stress. Spot no. 39 had identified to heme oxygenase I, which earlier was reported to play a protective role in plants against oxidative stress.

Among the *down-accumulated* proteins in *NtGp*11 in response to drought, spot nos. 24, 25, and 33 were identified as chlorophyll a/b binding protein type I, Photosystem I assembly protein Ycf3, and light harvesting chlorophyll a/b binding protein, respectively. Several studies have postulated that these genes were down-regulated in drought-tolerant plants in response to stress. Downregulation of these proteins revealed that there has been a reduction in the photosynthetic rate after stress treatment, which might indicate that the *NtGp*11 plant curtailed some of its energy expenditure from normal metabolism and utilized it for the development of resistance against drought stress. Spot nos. 22 and 38 were identified as maturase and homeobox transcription factor KN3. Changes in the expression level of these proteins indicated that the development of tolerance against abiotic stress requires several changes at the gene and protein regulation levels. These results clearly suggested that γ-ECS was upregulated at the transcription level by drought and salinity stress. Overexpression of γ-ECS resulted in enhanced GSH content in *NtGp*11 plants, ultimately enhancing the tolerance to drought stress by enhancing the germination rate, water recovery rate, chlorophyll, and proline content compared with the wild-type plant, which demonstrated that GSH may be a positive regulator of drought tolerance. Drought stress strongly induced *γ-ECS* activity in *NtGp*11, helped in enhancing the GSH content in the plant cell, which in turn upregulated the activities of stress and defense, energy metabolism, carbon metabolism, gene regulatory proteins and enzymes, which ultimately helped in minimizing the effect of drought in the *NtGp*11 plant. Taken together, we propose an interplay of GSH with SA and ET in mitigating drought stress. However, many further studies are necessary to identify the direct GSH targets in response to drought stress.

6.9 Conclusion and Future Perspectives

Plants must survive in this hostile environment as they encounter a wide range of stress throughout their life. To overcome this adverse situation as well as to carry out their normal life cycle, they have a variety of signaling molecules. The well-known members today are SA, JA, ET, ABA, and ROS, which plants use to combat various environmental stress conditions. In the past two decades, GSH has gradually gained in importance and become a molecule of interest to a number of researchers, especially in the field of environmental stress management. Although the role of GSH in plant defense has long been known, a dearth of information still exists, however, regarding the mechanism of how GSH takes part in this complex scenario. We have explored the interplay of GSH with various signaling molecules such as SA, ET, and ABA, using the genetic engineering approach to develop and establish transgenic tobacco overexpressing *γ-ECS* constitutively, with enhanced GSH content and abiotic/biotic stress tolerance potential. Transcriptomic as well as proteomics profiling identified the genes and proteins related to SA and ET and involved in providing the stress tolerance potential. Taken together, it is more than justified to say that, GSH, a dynamic biomolecule, is an active participant in the plant defense signaling network primarily through its interaction with SA and ET. This study will open up prospects for future investigations on GSH in plant stress management.

Acknowledgment The author is extremely thankful to the director of CSIR-IICB, Kolkata, for providing the necessary facilities.

References

Alexander D, Stinson J, Pear J, Glascock C, Ward E, Goodman RM, Ryals J. A new multigene family inducible by tobacco mosaic virus or salicylic acid in tobacco. Mol Plant Microbe Interact. 1992;5:513–5.

Anderson JV, Chevone BI, Hess JL. Seasonal variation in the antioxidant system of eastern white pine needles. Plant Physiol. 1992;98:501–8.

Bai XG, Chen JH, Kong XX, Todd CD, Yang YP, Hu XY, et al. Carbon monoxide enhances the chilling tolerance of recalcitrant *Baccaurea ramiflora* seeds via nitric oxide-mediated glutathione homeostasis. Free Radic Biol Med. 2012;15:710–20.

Boston RS, Vittanen PV, Vierling E. Molecular chaperones and protein folding in plants. Plant Mol Biol. 1996;32:191–222.

Chan C, Lam H-M. A putative lambda class glutathione-S-transferase enhances plant survival under salinity stress. Plant Cell Physiol. 2014;55(3):570–9.

Chao YY, Hsu YT, Kao CH. Involvement of glutathione in heat shock—and hydrogen peroxide—induced cadmium tolerance of rice (*Oryza sativa* L.) seedlings. Plant and Soil. 2009;318:37–45.

Chen JH, Jiang HW, Hsieh EJ, Chen HY, Chien CT, Hsieh HL, Lin TP. Drought and salt stress tolerance of an *Arabidopsis* glutathione S-transferase U17 knockout mutant are attributed to the combined effect of glutathione and abscisic acid. Plant Physiol. 2012;158:340–51.

Cruz de Carvalho MH, Brunet J, Bazin J, Kranner I, d' Arcy-Lameta A, Zuily-Fodil Y, Contour-Ansel D. Homoglutathione synthetase and glutathione synthetase in drought-stressed cowpea leaves: expression patterns and accumulation of low-molecular weight thiols. J Plant Physiol. 2010;167:480–7.

Dash S, Mohanty N. Response of seedlings to heat-stress in cultivars of wheat: growth temperature dependent differential modulation of photosystem 1 and 2 activity, and foliar antioxidant defense capacity. J Plant Physiol. 2002;159:49–59.

Datta R, Sinha R, Chattopadhyay S. Changes in leaf proteome profile of *Arabidopsis thaliana* in response to salicylic acid. J Biosci. 2013;38:317–28.

De Vos M, Denekamp M, Dicke M, Vuylsteke M, Van Loon L, Smeekens SC, Pieterse CM. The *Arabidopsis thaliana* transcription factor AtMYB102 functions in defense against the insect herbivore *Pieris rapae*. Plant Signal Behavior. 2006;1:305–11.

Dhindsa RS. Glutathione status and protein synthesis during drought and subsequent rehydration in *Tortula ruralis*. Plant Physiol. 1987;83:816–9.

Dron M, Clouse SD, Dixon RA, Lawton MA, Lamb CJ. Glutathione and fungal elicitor regulation of a plant defense gene promoter in electroporated protoplasts. Proc Natl Acad Sci U S A. 1988;85:6738–42.

Dubos C, Stracke R, Grotewold E, Weisshaar B, Martin C, Lepiniec L. MYB transcription factor in *Arabidopsis*. Trends Plant Sci. 2010;15:573–81.

Foyer CH, Halliwell B. The presence of glutathione and glutathione reductase in chloroplasts: a proposed role in ascorbic acid metabolism. Planta. 1976;133:21–5.

Foyer CH, Noctor G. Redox homeostasis and antioxidant signaling: a metabolic interface between stress perception and physiological responses. Plant Cell. 2005;17:1866–75.

Foyer CH, Lopez-Delgado H, Dat JF, Schott IM. Hydrogen peroxide- and glutathione-associated mechanisms of acclamatory stress tolerance and signalling. Physiol Plant. 1997;100:241–54.

Freeman JL, Garcia D, Kim D, Hopf A, Salt DE. Constitutively elevated salicylic acid signals glutathione-mediated nickel tolerance in *Thlaspi* nickel hyperaccumulators. Plant Physiol. 2005;137:1082–91.

Ghanta S, Bhattacharyya D, Chattopadhyay S. Glutathione signaling acts through NPR1-dependent SA-mediated pathway to mitigate biotic stress. Plant Signal Behav. 2011a;6:1–3.

Ghanta S, Bhattacharyya D, Sinha R, Banerjee A, Chattopadhyay S. *Nicotiana tabacum* overexpressing γ-ECS exhibits biotic stress tolerance likely through NPR1-dependent salicylic acid mediated pathway. Planta. 2011b;233:895–910.

Ghanta S, Datta R, Bhattacharyya D, Sinha R, Kumar D, Hazra S, Mazumdar AB, Chattopadhyay S. Multistep involvement of glutathione with salicylic acid and ethylene to combat environmental stress. J Plant Physiol. 2014;171:940–50.

Gill SS, Anjum NA, Hasanuzzaman M, Gill R, Trivedi DK, Ahmad I, Pereira E, Tuteja N. Glutathione and glutathione reductase: A boon in disguise for plant abiotic stress defense operations. Plant Physiol Biochem. 2013;70:204–12.

Glazebrook J, Ausubel FM. Isolation of phytoalexin-deficient mutants of *Arabidopsis thaliana* and characterization of their interactions with bacterial pathogens. Proc Natl Acad Sci U S A. 1994;91:8955–9.

Glazebrook J, Zook M, Merrit F, Kagan I, Rogers EE, Crute IR, Holub EB, Hammerschmidt R, Ausubel FM. Phytoalexin-deficient mutants of *Arabidopsis* reveal that PAD4 encodes a regulatory factor and that four PAD genes contribute to downy mildew resistance. Genetics. 1997;146:381–92.

Gomez LD, Noctor G, Knight MR, Foyer CH. Regulation of calcium signaling and gene expression by glutathione. J Exp Bot. 2004;55:1851–9.

Han Y, Chaouch S, Mhamdi A, Queval G, Zechmann B, Noctor G. Functional analysis of *Arabidopsis* mutants points to novel roles for glutathione in coupling H_2O_2 to activation of salicylic acid accumulation and signaling. Antioxid Redox Signal. 2013a;18:2106–21.

Han Y, Mhamdi A, Chaouch S, Noctor G. Regulation of basal and oxidative stress-triggered jasmonic acid-related gene expression by glutathione. Plant Cell Environ. 2013b;36:1135–46.

Hardie DG. Plant protein serine/threonine kinases: classification and function. Annu Rev Plant Physiol Plant Mol Biol. 1999;50:97–131.

Hell R, Bergmann L. Glutathione synthetase in tobacco suspension cultures; catalytic properties and localisation. Physiol Plant. 1988;72:70–6.

Hell R, Bergmann L. γ-Glutamylcysteine synthetase in higher plants: catalytic properties and subcellular localisation. Planta. 1990;180:603–12.

Jain M, Choudhary D, Kale RK, Bhalla-Sarin N. Salt- and glyphosate-induced increase in glyoxalase I activity in cell lines of groundnut (*Arachis hypogaea*). Physiol Plant. 2002;114:499–505.

Jozefczak M, Remans T, Vangronsveld J, Cuypers A. Glutathione is a key player in metal-induced oxidative stress defenses. Int J Mol Sci. 2012;13:3145–75.

Kaymakanov M, Lyubenova L, Schroder P, Stoeva N. Salt stress and glutathione-dependent enzyme activities in bean (*Phaseolus vulgaris* L.). Genet Appl Plant Physiol. 2010;36:55–9.

Kim YJ, Jang MG, Noh HY, Lee HJ, Sukweenadhi J, Kim JH, Kim SY, Kwon WS, Yang DC. Molecular characterisation of two glutathione peroxidase genes of *Panax ginseng* and their expression analysis under environmental stresses. Gene. 2014;535:33–41.

Kocsy G, Szalai G, Vagujfalvi A, Stehli L, Orosz G, Galiba G. Genetic study of glutathione accumulation during cold hardening in wheat. Planta. 2000;210:295–301.

Kumar A, Chakraborty A, Ghanta S, Chattopadhyay S. *Agrobacterium*-mediated genetic transformation of mint with *E. coli* glutathione synthetase gene. Plant Cell Tiss Organ Cult. 2009;96:117–26.

Kumar B, Singla-Pareek SL, Sopory SK. Glutathione homeostasis: crucial for abiotic stress tolerance in plants. In: Pareek A, Sopory SK, Bohnert HJ, Govindjee, editors. Abiotic stress adaptation in plants. The Hague: Springer; 2010. p. 263–82.

Kumari Vinodhana N, Gunasekaran M, Vindhiyavarman P. Genetic studies of variability, correlation and path coefficient analysis in cotton genotypes. Int J Pure Appl Biosci. 2013;1(5):6–10.

Kunert KJ, Foyer CH. Thiol/disulphide exchange in plants. In: De Kok LJ, Stulen I, Rennenberg H, Brunhold C, Rausen W, editors. Sulfur nutrition and assimilation in higher plants. Regulatory, agricultural and environmental aspects. The Hague: SPB Academic; 1993. p. 139–51.

Kunkel BN, Brooks DM. Cross talk between signaling pathways in pathogen defense. Curr Opin Plant Biol. 2002;5:325–31.

Leipner J, Fracheboud Y, Stamp P. Effect of growing season on the photosynthetic apparatus and leaf antioxidative defenses in two maize genotypes of different chilling tolerance. Environ Exp Bot. 1999;42:129–39.

Li H, Xu H, Graham DE, White RH. Glutathione synthetase homologs encode alpha-L-glutamate ligases for methanogenic coenzyme F420 and tetrahydrosarcinapterin biosyntheses. Proc Natl Acad Sci U S A. 2003;100:9785–90.

Locato V, de Pinto MC, De Gara L. Different involvement of the mitochondrial, plastidial and cytosolic ascorbate-glutathione redox enzymes in heat shock responses. Physiol Plant. 2009;135:296–306.

Lu D, Tang G, Wang D, Luo L. The *Sinorhizobium meliloti* LysR family transcriptional factors LsrB is involved in regulation of glutathione biosynthesis. Acta Biochim Biophys Sin. 2013;45:882–8.

May MJ, Vernoux T, Sánchez-Fernández R, Van Montagu M, Inzé D. Evidence for posttranscriptional activation of gamma-glutamylcysteine synthetase during plant stress responses. Proc Natl Acad Sci U S A. 1998;95:12049–54.

Meister A. Glutathione metabolism and its selective modification. J Biol Chem. 1988;263:17205–8.

Mhamdi A, Hager J, Chaouch S, Queval G, Han Y, Taconnat L, Saindrenan P, Gouia H, Issakidis-Bourguet E, Renou JP, Noctor G. *Arabidopsis* glutathione reductase 1 plays a crucial role in leaf responses to intracellular H_2O_2 and in ensuring appropriate gene expression through both salicylic acid and jasmonic acid signaling pathways. Plant Physiol. 2010;153:1144–60.

Mittova V, Theodoulou FL, Kiddle G, Gomez L, Volokita M, Tal M, Foyer CH, Guy M. Co-ordinate induction of glutathione biosynthesis and glutathione-metabolizing enzymes is correlated with salt tolerance in tomato. FEBS Lett. 2003;554:417–21.

Morabito D, Guerrier G. The free oxygen radical scavenging enzymes and redox status in roots and leaves of in response to osmotic stress, desiccation and rehydration. J Plant Physiol. 2000;157:74–80.

Munemasa S, Muroyama D, Nagahasi H, Nakamura Y, Mori IC, Murata Y. Regulation of reactive oxygen species- mediated abscisic acid signalling in guard cells and drought tolerance by glutathione. Front Plant Sci. 2013;20:472.

Murashige T, Skoog F. A revised medium for rapid growth and bioassays with tobacco tissue cultures. Physiol Plant. 1962;15:473–97.

Musgrave WB, Yi H, Kline D, Cameron JC, Wignes J, Dey S, Pakrasi HB, Jez JM. Probing the origin of glutathione biosynthesis through biochemical analysis of glutamate-cysteine ligase and glutathione synthetase from a model photosynthetic prokaryote. Biochem J. 2013;450:63–72.

Nieto-Sotelo J, Ho TD. Effect of heat shock on the metabolism of glutathione in maize roots. Plant Physiol. 1986;82:1031–5.

Noctor G, Mhamdi A, Chaouch S, Han Y, Neukermans J, Marquez-Garcia B, Queval G, Foyer CH. Glutathione in plants: an integrated overview. Plant Cell Environ. 2012;35:454–84.

Ogawa K. Glutathione-associated regulation of plant growth and stress responses. Antioxid Redox Signal. 2005;7:973–81.

Oufdou K, Benidire L, Lyubenova L, Daoui K, Fatemi ZEA, Schroder P. Enzymes of the glutathione-ascorbate cycle in leaves and roots of rhizobia inoculated faba bean plants (*Vicia faba* L.) under salinity stress. Eur J Soil Biol. 2014;60:98–103.
Parisy V, Poinssot B, Owsianowski L, Buchala A, Glazebrook J, Mauch F. Identification of PAD2 as a gamma-glutamylcysteine synthetase highlights the importance of glutathione in disease resistance of *Arabidopsis*. Plant J. 2007;49:159–72.
Pieterse CMJ, Leon-Reyes A, Van der Ent S, Van Wees SCM. Networking by small-molecules hormones in plant immunity. Nat Chem Biol. 2009;5:308–16.
Preuss ML, Cameron JC, Berg RH, Jez JM. Immunolocalisation of glutathione biosynthesis enzymes in *Arabidopsis thaliana*. Plant physiol Biochem. 2014;75:9–13.
Qi YC, Liu WQ, Qiu LY, Zhang SM, Ma L, Zhang H. Overexpression of glutathione S-transferase gene increases salt tolerance of *Arabidopsis*. Russ J Plant Physiol. 2010;57:233–40.
Rennenberg H. Molecular approaches to glutathione biosynthesis. In: Cram WJ, DeKok LJ, Stulen I, Brunold C, Rennenberg H, editors. Sulphur metabolism in higher plants. Leiden: Backhuys; 1997. p. 59–70.
Rüegsegger A, Brunold C. Localization of ã-glutamylcysteine synthetase and glutathione synthetase activity in maize seedlings. Plant Physiol. 1993;101:561–6.
Ruiz JM, Blumwald E. Salinity-induced glutathione synthesis in *Brassica napus*. Planta. 2002;214:965–9.
Russell D, Sambrook J. Molecular cloning: a laboratory manual. New York: Cold Spring Harbor Laboratory Press; 2001.
Seth CS, Reman T, Keunen E, Jozefczak M, Gielen H, Opdenakker K, Weywns N, Vangronsveld J, Cuypers A. Phytoextraction of toxic metals: a central role for glutathione. Plant Cell Environ. 2012;35:334–46.
Sgherri CLM, Navari-Izzo F. Sunflower seedlings subjected to increasing water deficit stress: oxidative stress and defence mechanisms. Physiol Plant. 1995;93:25–30.
Shepherd RW, Wagner GJ. Phylloplane proteins: emerging defenses at the aerial frontline. Trends Plant Sci. 2007;12:51–6.
Shepherd RW, Bass WT, Houtz RL, Wagner GJ. Phylloplanins of tobacco are defensive proteins deployed on aerial surfaces by short glandular trichomes. Plant Cell. 2005;17:1851–61.
Singla-Pareek SL, Reddy MK, Sopory SK. Genetic engineering of the glyoxalase pathway in tobacco leads to enhanced salinity tolerance. Proc Natl Acad Sci U S A. 2003;100:14672–7.
Singla-Pareek SL, Yadav SK, Pareek A, Reddy MK, Sopory SK. Enhancing salt tolerance in a crop plant by overexpression of glyoxalase II. Transgenic Res. 2008;17:171–80.
Tsakraklides G, Martin M, Chalam R, Tarczynski MC, Schmidt A, Leustek T. Sulfate reduction is increased in transgenic *Arabidopsis thaliana* expressing 5'- adenylylsulfate reductase from *Pseudomonas aeruginosa*. Plant J. 2002;32:879–89.
Vaidyanathan H, Sivakumar P, Chakrabarty R, Thomas G. Scavenging of reactive oxygen species in NaCl-stressed rice (*Oryza sativa* L.)—differential response in salt-tolerant and sensitive varieties. Plant Sci. 2003;165:1411–8.
Vlot AC, Klessig DF, Park SW. Systemic acquired resistance: the elusive signal(s). Curr Opin Plant Biol. 2008;11:436–42.
Wingate VMP, Lawton MA, Lamb CJ. Glutathione causes a massive and selective induction of plant defense genes. Plant Physiol. 1988;87:206–10.
Yadav SK, Singla-Pareek SL, Reddy MK, Sopory SK. Transgenic tobacco plants overexpressing glyoxalase enzymes resist an increase in methylglyoxal and maintain higher reduced glutathione levels under salinity stress. FEBS Lett. 2005;579:6265–71.
Zhao FY, Liu W, Zhang SY. Different responses of plant growth and antioxidant system to the combination of cadmium and heat stress in transgenic and nontransgenic rice. J Integr Plant Biol. 2009;51:942–50.

Chapter 7
Function of Heat-Shock Proteins in Drought Tolerance Regulation of Plants

Sruthy Maria Augustine

7.1 Introduction

Global warming and associated climate change is predicted to cause longer spells of drought than normal. One of the major environmental factors that limits plant growth and productivity worldwide is water stress (Valliyodan and Nguyen 2006; Flexas et al. 2006). Plants, because of their sessile nature, are unable to flee from unfavorable environmental conditions. Plants interact with several environmental factors, i.e., biotic and abiotic stress factors, all affecting plant growth and development, which consequently become the limiting factor for food production and sustainability (Seki et al. 2003; Mu et al. 2013). Both abiotic and biotic stresses affect the survival rate, biomass production, and crop yield (Agarwal et al. 2006). Abiotic stresses such as drought, cold, salinity, chemical pollutants, and heat stresses trigger damage to the plants by disrupting cellular structures and impairing major physiological functions (Larcher 2003). Almost all plants have the ability for stress tolerance, but its extent varies from species to species (Chaitanya et al. 2003). To cope with abiotic stress, plants adopt a series of molecular, physiological, and cellular responses.

Drought is one of the major abiotic stresses and is considered as the most catastrophic environmental stress leads to reduced plant productivity more than any other environmental stress (Lambers et al. 2008). Drought stress is an ever-growing problem, and it is one of the major limitations for crop production (Jones and Bradley 1992). Water is becoming an increasingly scarce and precious commodity. Currently, the agricultural sector uses 75 % of the total global consumption of water (Molden 2007). Climate change will probably reduce the available water further and that increasing the need for drought-tolerant crops (Hamdy et al. 2003). Besides the complexity of drought itself, the plant's responses to drought are complex

S.M. Augustine (✉)
Sugarcane Breeding Institute (ICAR), Coimbatore, India
e-mail: sruthymaria@gmail.com

M.A. Hossain et al. (eds.), *Drought Stress Tolerance in Plants, Vol 1*,
DOI 10.1007/978-3-319-28899-4_7

and acquire different mechanisms when they experience drought (Levitt 1980; Jones 2004). In drought stress, plants exhibit morphological, biochemical, physiological, and molecular changes (Greenway and Munns 1980; Hasegawa et al. 2000; Dubouzet et al. 2003). When plants are exposed to mild stress for the first time, it enables them to perform better when exposed to severe stress for the second time.

Drought stress has an effect on different developmental stages of the plant, i.e., germination, maturation, and senescence (Gyoergyey et al. 1991; Leprince et al. 1993; Hall 1993; Gagliardi et al. 1995; Kermode 1997). The developmental and physiological processes of the plants are negatively affected by drought stress. Stress becomes one of the major constraints of plant adaptation when it occurs at key developmental stages such as reproduction (Hall 2001). In a water deficit condition, plants experience stress at the cellular level and to protect themselves, they change their metabolism. The level of drought tolerance varies in plant species, even between genotypes of the same species, and it can be analyzed by the growth and productivity of the plant under drought stress conditions. Drought tolerance is a complex process that differs according to the severity of stress, age of the plant, and the water use efficiency. Moreover, drought stress may occur along with other abiotic or biotic stresses leading to a multi-stress environment to the plant.

Any change in the optimal growth conditions is perceived by plants as stress and transduced in the form of signals involving protein phosphorylation and/or dephosphorylation, calcium sensing, protein degradation, etc. In plants, one of the major problems caused by abiotic stress is protein dysfunction. Heat-shock proteins are a group of proteins that are expressed at high levels when exposed to stress, but are also present in cells under normal environmental conditions. As these proteins were first found in cells that were exposed to high temperature, they were named as heat-shock proteins. Hsp family in plants is larger compared to other kingdoms probably as a result of adaptation towards a wide range of stresses. The increase in expression of Hsps is transcriptionally regulated mostly by heat-shock factors (HSF). Heat-shock proteins are also called as molecular chaperones, which bind and stabilize proteins at various intermediate stages of its formation and helps in folding, assembly, degradation, and translocation across membranes. The mechanisms that are involved in adaptation, tolerance, and resistance to water stress are still elusive. The genes that are involved in the protection and repair mechanism, e.g., Hsps are not characterized fully and their functional role remains unknown.

The gene expression pattern, proteomics, and transcriptomics studies have identified the regulation and activation of several drought stress-related transcripts and proteins, which are generally classified into two major groups.

The first group (*functional proteins*) includes proteins that probably function in stress tolerance. They are protection factors such as LEA (Late Embryogenesis Abundant) proteins, chaperones and lipid transfer proteins, proteins involved in repair and protection from damage, that include proteinases, protease inhibitors, plant defense-related proteins, detoxification enzymes, and proteins involved in synthesis of osmoprotectants (proline, glycine betaine, and sugars). This group also includes proteins that have a role in cellular metabolic processes such as carbohydrate metabolism, fatty acid metabolism, secondary metabolism, proteins regulated by plant hormones (ABA, auxin, and JA), RNA-binding proteins, biosynthesis of plant hormones (ABA, ethylene, IAA, and JA), cellular structure, and organization-related proteins such as arabinogalactan proteins, senescence-related proteins, cytochrome P450, alcohol dehydrogenase, aldehyde dehydrogenase, reproduction development-related proteins such as pollen coat-like protein and respiration-related proteins such as flavin-containing monooxygenase.

The second group (*regulatory proteins*) is involved in regulation of signal transduction and transcription as part of drought response. These are transcription factors of multiple gene families such as DREB, ERF, zinc finger, WRKY, MYB, MYC, HD-ZIP, bZIP, and NAC families. Among the regulatory proteins, protein kinases such as MAPK, CDPK, S6K, and PRK can be found. This group also includes protein phosphatases such as PP2C, PI turnover-related proteins such as PLC, PLD, PIP5K, DGK, and PAP, and calmodulin-binding protein and Ca^{2+}-binding proteins. Understanding the mechanisms by which how the plants recognize environmental signals and transmit such signals to the cellular machinery to activate responses is a fundamental issue in crop improvement and is vital for the continued development of breeding and transgenic strategies to improve stress tolerance in crop plants.

7.1.1 *Mechanism of Drought Tolerance*

Plants have adopted different ways to adapt to drought stress namely drought escape, drought avoidance, and drought tolerance strategies. Plants that have the ability to escape drought show a rapid phenological development and high degree of developmental plasticity and being able to complete their life cycle before physiological water deficit occurs (Blum 1988). In drought adaptation strategies, i.e., by avoiding drought (growth stopped at the dry season), developing structures that help for the conservation of water or by increasing the water use efficiency (Price et al. 2002). Drought tolerance is defined as the relative capacity of plants to maintain functional growth under low leaf water status. Drought-tolerant plants have developed mechanisms to avoid water loss, stabilize or repair damaged proteins (Hsps, LEAs), and by the antioxidant mechanism.

7.2 Heat-Shock Protein Families

The Italian Scientist R. Ritossa observed a "new puffing pattern" in the gene expression of the chromosomes of *Drosophila melanogaster* after exposure to heat. This marked the beginning of the discovery of heat-shock proteins. Later on, these proteins were identified and named as heat-shock protein (HSP) (Tissieres et al. 1974). Because of its heat-shock response characteristics, i.e., increased expression under heat stress, researchers started studying the relationship of the synthesis of these proteins with stress tolerance. Heat-shock proteins were identified in almost all organisms (Bharti and Nover 2002). All the identified Hsps have a carboxylic terminal called heat-shock domain (Helm et al. 1993) and the molecular weights ranging from 10 to 200 kDa. Hsps are also referred to as chaperones where they involved in the induction of signaling mechanism during heat stress (Schoffl et al. 1999). The induced expression of heat-shock genes are mediated by the activation of heat-shock transcription factors (HSFs) that bind to the heat-shock elements in the promoter region (Sorger and Nelson 1989). The gene coding for HSF has been isolated and characterized from Drosophila, yeast, mice, chicken, humans (Wu 1995), tomato (Scharf et al. 1990), maize (Gagliardi et al. 1995), *Arabidopsis* (Hubel and Schoffl 1994), and soybean (Czarnecka-Verner et al. 2000).

Heat-shock proteins of archaea have been classified on the basis of their approximate molecular weight into: (1) Heat-shock proteins 100 kDa, i.e., Hsp100, (2) Hsp90, (3) Hsp70, (4) Hsp60, and small heat-shock proteins (sHsps), where the molecular weight ranges from 15 to 42 kDa (Trent 1996). The sHsps are usually occurring as a complex of small subunits and the molecular weight ranges from 200 to 800 kDa (Kim et al. 1998).

The abbreviations of Hsps names of bacteria differ from those in eukaryotic cells except for sHSPs and the nomenclature is given below:

Escherichia coli	Eukaryotic cell
ClpB	Hsp100
HtpG	Hsp90
Dnak	Hsp70
GroEL	Hsp60

Heat-shock proteins in mammals are not different from those of bacteria except for the presence of Hsp33 in the bacterial system (Schlesinger 1990). Mammalian Hsps were grouped into five families (Kregel 2002) and are listed in Table 7.1.

Table 7.1 Families of Hsps in mammalian system, cellular location, and proposed functions (Kregel 2002)

HSP families	Cellular location	Proposed functions
Hsp27 (sHSP)	Cytosol, nucleus	Microfilament stabilization, antiapoptotic
Hsp60	Mitochondria	Refolds proteins and prevent aggregation of denatured proteins, proapoptotic
Hsp70 family:		Antiapoptotic
Hsp72(Hsp70)	Cytosol, nucleus	Protein folding, cytoprotection
Hsp73(Hsc70)	Cytosol, nucleus	Molecular chaperones
Hsp75(mHsp70)	Mitochondria	Molecular chaperones
Hsp78(GRP78)	Endoplasmic reticulum	Cytoprotection, molecular chaperones
Hsp90	Cytosol, endoplasmic reticulum, nucleus	Regulation of steroid hormone receptors, protein translocation
Hsp110/104	Cytosol	Protein folding

In plants, Hsps are conservatively characterized as five principal classes based on their approximate molecular weight and they are: (1) Hsp100, (2) Hsp90, (3) Hsp70, (4) Hsp60, and (5) small heat-shock proteins (sHsps) (Table 7.2) (Schlesinger 1990; Schoffl et al. 1999; Kotak et al. 2007). In addition to major families, other proteins are also reported with chaperone functions, for example, protein disulfide isomerase and calnexin/calreticulin, which assist in protein folding in the endoplasmic reticulum (ER). Molecular Hsps/chaperones are located in both the cytoplasm and organelles, such as the nucleus, mitochondria, chloroplasts, and ER. Although the role of Hsps/chaperones has been revealed profoundly in other organisms, very little is known in plants. Here, we discuss the role of heat-shock proteins in plants along with the major recent findings describing its relation to abiotic stress responses (Table 7.2 footnote). Hsps are one of the important classes of stress-responsive proteins that play a key role in direct stress tolerance. In addition, Hsps also act via cross talk with other signaling pathways and function cooperatively with other components to decrease cellular damage.

The general aspects of Hsps have been discussed in detail in several previous reviews (Waters et al. 1996; Boston et al. 1996; Vierling 1991; Bukau and Horwich 1998; Hartl 1996; Frydman 2001; Buchner 1999; Morimoto 1998; Ranson et al. 1998; Miernyk 1997).

7.3 Role of Heat-Shock Proteins

The protein function is determined by its formation and folding into three-dimensional structures (Levitt et al. 1997). Fifty-percent of principle amino acid sequence is required for the formation of three-dimensional structure (Dobson et al. 1998). That is where the function of Hsps in the folding of other proteins is important. Hsps are usually cytosolic proteins and are induced by heat or any other stress at any stage of plant growth that has a major function in various intracellular processes. Maintaining the proper conformation of the protein and preventing the aggregation of non-native proteins are the key steps for survival of the cell under stress. In normal cellular processes, heat-shock proteins (Hsps) are responsible for protein folding, translocation and assembly, degradation, stabilize proteins and membranes, and during stress condition, it can assist in protein refolding. High temperature and other stresses make it more difficult for proteins to form proper tertiary structures and cause unfolding of some already structured proteins. When left uncorrected, they may form aggregates and cause the death of the cell. Hsps are induced rapidly at high levels to deal with this type of problem (Wang et al. 2004). They can play a crucial role in protecting plants against stress by reestablishing normal protein conformation and thus cellular homeostasis. Moreover, Hsps are able to protect the cells from injury and can also facilitate recovery and survival after return to normal growth conditions (Morimoto 1998).

Heat-shock proteins are also expressed under normal condition at moderate or low level because of their essential roles in protein maintenance, such as proper

Table 7.2 Five major classes of plant Hsps/molecular chaperones and their subfamilies, including specific examples for direct involvement of Hsps/molecular chaperones in plant tolerance to stress (Wang et al. 2004)

Classes	Representative members	Intracellular localization	Major functions	References
HSP70 Subfamily:			Preventing aggregation, assisting refolding, protein import and translocation, signal transduction, and transcriptional activation	Boston et al. 1996; Vierling 1991; Morimoto 1998.
Dank	Hsp/Hsc70	Cytosol		
	Hsp70	Chloroplast, mitochondria		
	Bip[a]	Endoplasmic reticulum		
HSP110/SSE	Hsp91	Cytosol		
Chaperonin/ HSP60 Subfamily:			Folding and assisting refolding	Boston et al. 1996; Hartl 1996; Morimoto 1998.
Group I	Cpn60[b]	Chloroplast, mitochondria		
Group II	CCt[c]	Cytosol		
HSP90	Hsp90		Facilitating maturation of signaling molecules, genetic buffering	Boston et al. 1996; Young et al. 2001a, b; Krishna and Gloor 2001.
	AtHsp90-1	Cytosol		
	AtHsp90-5	Chloroplast		
	AtHsp90-6	Mitochondria		
	AtHsp90-7	Endoplasmic reticulum		
HSP100/Clp Subfamily:	Hsp100[d]		Disaggregation, unfolding	Schirmer et al. 1996; Goloubinoff et al. 1999
Class I:	ClpB, ClpA/C			
	ClpD	Cytosol, mitochondria		
Class II:	ClpM, ClpN	Chloroplast		
	ClpX, ClpY	Chloroplast		
sHSP Subfamily:			Preventing aggregation, stabilizing non-native proteins	Waters et al. 1996; Boston et al. 1996; Vierling 1991
I	Hsp17.6	Cytosol		
II	Hsp17.9	Cytosol		
III	Hsp21	Chloroplast		
	Hsp26.2[e]			
IV	Hsp22	Endoplasmic reticulum		
V	Hsp23[e]	Mitochondria		
VI	Hsp22.3	Membrane		

[a]Enhanced accumulation of BiP in *Nicotiana tabacum* protoplast and transgenic plants conferred tolerance to water stress (Alvim et al. 2001a, b)

[b]Deletion of LEN1 (Cpn60b) triggered cell death in Arabidopsis (Ishikawa et al. 2003)

[c]CCTa from the mangrove plant *Bruguiera sexangula* enhanced the salt and osmotic stress tolerance of *Escherichia coli* transformants (Yamada et al. 2002)

[d]Hsp100 functional complementation of the temperature-sensitive yeast Hsp104 mutant cells was shown using AtHsp101 and gmhps101 cDNAs (Lee et al. 1994; Agarwal et al. 2001)

[e]*Zea mays* mitochondrial sHsp improved mitochondrial electron transport during salt stress, mainly by protection of the NADH: ubiquinone oxidoreductase activity (Complex I), but it failed to protect enzymes associated with Complex II (Hamilton and Heckathorn 2001). A mutant of the chloroplast sHsp of *Agrostis stolonifera* grass, sHsp26.2, with a point mutation that generated a premature stop-codon (sHsp26.2m) was isolated from a heat-sensitive variant; protein product of the mutant was not accumulated upon heat stress (Wang and Luthe 2003)

folding of newly synthesized protein. These are found in all organisms and have highly conserved sequences and tertiary structures. Hsps have a role in maintaining the membrane integrity during stress (Tsvetkova et al. 2002). The correlation between synthesis and accumulation of Hsps and heat tolerance suggests but does not prove that these are related. Binding of ATP and its hydrolysis are the essential steps for the chaperone activity of Hsp proteins both in vitro and in vivo (Mayer and Bukau 2005). One hypothesis is that HspP70 participates in ATP-dependent protein unfolding or assembly/disassembly reactions, and they prevent the protein denaturation during stress (Pelham 1986). Heat-shock proteins are involved in altering the biochemical processes necessary for drought adaptation (Iba 2002). Based on the previous studies, the roles of Hsps can be categorized into three: (1) refold denatured proteins; (2) participation in the finalization of newly synthesized proteins; (3) removal of protein aggregation (Trent 1996).

7.3.1 Heat-Shock Protein 70

Heat-shock protein 70 (Hsp70) proteins are one of the large families of highly conserved molecular chaperones and are extensively found in almost all organisms (Boorstein et al. 1994). This protein varies from 68 to 110 kDa. Hsp70 family proteins protect proteins under heat stress. In addition, it helps in the transport of proteins into mitochondria and to the endoplasmic reticulum. The sequence similarity between bacteria and eukaryotic Hsps is approximately 50 %, suggesting its crucial role in living organisms (Boorstein et al. 1994; Boston et al. 1996). Besides their known functions like prevention of aggregation and refolding of non-native proteins in stress condition, Hsp70 proteins also have critical roles in housekeeping activities under normal conditions. In addition to the stress-inducible Hsp70s, some Hsp70 homologs are also called heat-shock cognate 70 (HSC70). HSC70 are constitutively expressed in the cytosol, and the role is to stabilize nascent proteins that are being released from ribosomes, preventing the misfolding and aggregation of partially synthesized polypeptide chains before the end of protein expression (Boston et al. 1996; Fink 1999).

In *Arabidopsis*, the Hsp70 family has 18 members, 9 in the cytosol, 4 in the ER, 3 in the chloroplast, and 2 in the mitochondrion (Lin et al. 2001; Sung et al. 2001). Most of the Hsp70 proteins in plants share similar structures compared to other higher eukaryotes. The Hsp70 proteins play a major role in a wide range of processes, including the folding and assembly of newly synthesized proteins, membrane translocation of organellar and secretory proteins, refolding of misfolded and aggregated proteins, and control of the activity of regulatory proteins (Bukau et al. 2000; Hartl and Hayer-Hartl 2002; Young et al. 2003). The Hsp70 proteins achieved the cellular functions through (1) the amplification and diversification of *Hsp70* genes in evolution that helps the generation of specialized Hsp70 chaperones, (2) Hsp70 chaperones selectively recruited the co-chaperones to fulfill specific cellular functions and (3) Association of Hsp70s with other chaperone systems to extent their activity spectrum (Mayer and Bukau 2005). Hsp70 proteins along with their co-chaperones and cooperating chaperones constitute a complex network of folding machines.

7.3.1.1 Protein Folding Process Assisted by Hsp70

The function of Hsp70 in the folding of non-native proteins can be divided into three categories: preventing aggregation of the polypeptide, help to fold properly to the native state, and solubilization and refolding of aggregated proteins (Mayer and Bukau 2005). Hsp70 proteins, together with one of the co-chaperones of the J-domain protein (JDP) family, prevent the non-native proteins aggregation through association with the hydrophobic patches of substrate molecules that shields them from any intermolecular interactions. The crucial role of JDPs is to mediate the ATP hydrolysis-dependent locking of substrates into the binding cavity of Hsp70 proteins, and this is very essential for almost all chaperone activities of Hsp70 proteins (Laufen et al. 1998; Kelley 1999). Hsp70 chaperone also helps in non-native folding intermediates to fold to the native state (Mayer and Bukau 2005). The mechanism behind Hsp70-chaperones role in the folding of non-native substrates is still unclear.

7.3.2 Heat-Shock Protein 60

Heat-shock protein 60 (Hsp60) was the first molecular chaperone to be identified (Vierling 1991), and it is also known as chaperonin 60 (cpn60). The term chaperonins (Hsp60) was first suggested (Hemmingsen et al. 1988) to describe one of the molecular chaperone classes that are evolutionarily homologous to *E. coli* GroEL, a class of molecular chaperones found in prokaryotes and in the plastids and mitochondria of eukaryotes (Boston et al. 1996; Hartl 1996). Hsp60 also plays a major role in ATP-dependent protein folding like Hsp70. The Hsp60 and Hsp70 families share some overlapping function (Hartl 1996), but their structures and mechanisms are distinct. Hsp60 has an important role in assisting plastid proteins such as Rubisco (Wang et al. 2004). Previous studies stated that Hsp60 might participate in folding and aggregation of many proteins that were transported to organelles such as chloroplasts and mitochondria (Lubben et al. 1989). Hsp60 prevents the protein aggregation by binding to them after their transcription and before folding (Parsell and Lindquist 1993). Functionally, plant chaperonins are limited and the stromal chaperones (Hsp70 and Hsp60) are involved in attaining functional conformation of newly imported proteins to the chloroplast (Jackson-Constan et al. 2001). Characterization of the functions of plant chaperonins is still scanty, but it is widely agreed that they are important in facilitating plastid proteins such as Rubisco (Boston et al. 1996; Hemmingsen et al. 1988)

7.3.3 Heat-Shock Protein 90

Heat-shock protein 90 (Hsp90) is distinct from many other molecular chaperones, and the protein varies from 82 to 96 kDa. Most of the known substrates for Hsp90 to date are signal-transduction proteins such as signaling kinases and steroid hormone receptors (Young et al. 2001a, b). Hsp90 requires ATP for its functions and is one of the most abundant proteins in the cell, i.e., 1–2 % of total cellular protein (Frydman 2001). The major role of Hsp90 is to manage protein folding (Frydman 2001; Buchner 1999), but it also plays a key role in signal-transduction networks, cell-cycle control, protein degradation, and protein trafficking (Young et al. 2001a, b; Richter and Buchner 2001; Pratt et al. 2001). Hsp90 proteins are widely expressed in most organisms and these are ATP-dependent molecular chaperones (Pearl and Prodromou 2006). Hsp90 proteins usually interact with moderately well-folded proteins that are involved in signal-transduction pathways and transcription regulation instead of binding a wide spectrum of unfolded proteins (Majoul et al. 2003; Zhao et al. 2005). One of the major functions of Hsp90 is to help in the formation of large protein complexes involving multiple co-chaperones, including Hsp70 and Hsp40 that indicates a close collaboration between different chaperone families (Zhao et al. 2005). Furthermore, it might also have a role in morphological evolution and stress adaptation in Drosophila and *Arabidopsis* (Rutherford and Lindquist 1998; Queitsch et al. 2002). Imai et al. (2003) reported that Hsp90 interacts with the 26S proteasome and plays a crucial role in its assembly and maintenance. Hsp90 also plays a major role in modulating the cellular signals, i.e., regulating the activity of glucocorticoid receptor activity (Pratt et al. 2004).

Hsp90 genes that are localized at cytosol, ER, and plastid have been isolated from several plant species and that shares 63–71 % amino acid similarities with yeast and animal origin Hsp90 gene (Krishna and Gloor 2001). Hsp90 chaperones are constitutively expressed in both prokaryotes and eukaryotes; in addition, their expression increases in response to stress in most organisms. In *Arabidopsis*, Hsp90 expression is developmentally regulated and responds to cold, heat, heavy metals, salt stress, phyto-hormones, and light and dark transitions (Krishna and Gloor 2001; Milioni and Hatzopoulos 1997). Rutherford and Lindquist (1998) stated that Hsp90 also assists the functions of mutated proteins that participate in the signaling pathways of development and morphogenesis.

7.3.4 Heat-Shock Protein 100/Clp Family

Heat-shock protein 100 (Hsp100) is another class of ATP-dependent molecular chaperones. The Hsp100 family is the member of the larger superfamily AAA ATPase with diverse functional properties (Schirmer et al. 1996; Neuwald et al. 1999). The uniqueness of Hsp100 family is their ability to solubilize aggregated proteins and involvement in protein degradation (Horwich 1995; Boston et al. 1996).

The removal of polypeptides that are nonfunctional and potentially harmful resulting from degradation, misfolding, or aggregation is crucial for maintaining the cellular homeostasis. Unlike other Hsp100 proteins, some members are not involved in protein degradation, i.e., Hsp101, Hsp104, and ClpB. One of the cytoplasmic members of this class is necessary for thermal stress tolerance but not necessary for the germination and growth of the plant under normal environmental conditions (Queitsch et al. 2000; Hong and Vierling 2001). The functions of Hsp100 family protein are not restricted to thermal stress tolerance but also involve housekeeping functions, which are important for chloroplast development (Lee et al. 2006). Gurley (2000) suggested that Hsp100 family proteins involve and facilitate the normal situation of the plant after severe stress.

Hsp100 plays an important role in plant survival in case of severe heat stress (Hong and Vierling 2000a, b), but it is absent in Drosophila and some vertebrates that depend on Hsp70 and other Hsps to prevent aggregation following refolding under severe heat stress (Gurley 2000; Xu et al. 2012). In addition to its role in stress tolerance, it also plays a major role under normal growth condition (Tonsor et al. 2008).

7.3.5 *Small Heat-Shock Proteins*

Small heat-shock proteins (sHsps) are ubiquitous molecular chaperones and the molecular weights ranging from 12 to 42 kDa. These are heat inducible, exhibit chaperone activity in vitro, and thermo protection in vivo. The common structure of this protein is an alpha-crystallin domain containing 80–100 amino acid residues present in the C-terminal region (Seo et al. 2006). One of the distinctive functions of this class of protein is the degradation of proteins that have inappropriate folding. Heat-induced dissociation of sHsp oligomers may reveal the hydrophobic patches inside the oligomeric interface that results in binding and stabilization of denatured proteins (Van Montfort et al. 2001). One of the characteristic proteins is the sHsp ubiquitin and its molecular weight is 8.5 kDa with its bound enzymes (Ferguson et al. 1990).

sHsps do not have a known protein folding role, but they associate with Hsp70 and help in the repair of denatured proteins (Lee and Vierling 2000). Another distinctive feature of sHsps is that their activity is independent from ATP (Miernyk 1999). sHsps are not able to refold non-native proteins, but they prevent the irreversible unfolding or wrong protein aggregation by binding to partially folded or denatured substrate proteins (Sun et al. 2002). Previous studies showed that under in vitro conditions sHsp 18.1 isolated from *Pisum sativum*, and the sHsp 16.6 from *Synechocystis* spp. binds to unfolded proteins and allows further refolding by Hsp70/Hsp100 complexes (Mogk et al. 2003). Moreover, small heat-shock proteins play an important role in membrane quality control and thereby maintaining the membrane integrity under stress conditions (Nakamoto and Vıgh 2007).

7.4 Phenomena of Induction of Hsps in Plants During Drought Stress

The major function that we know for Hsp is protein folding and trafficking of signaling proteins (Vierling 1991; Pratt et al. 2001; Wang et al. 2004). In the cell, Hsps are ubiquitous and found in all subcellular locations. Hsps in higher plants were first observed in tobacco and soybean cell culture (Barnett et al. 1980). The expression of Hsp occurs in different phases of the plant life cycle (pollen development, seed development, and germination), and it is suggested that they might play a major role in stress (drought, heat, and salinity) tolerance (Almoguera and Jordano 1992; DeRocher and Vierling 1994; Wehmeyer et al. 1996). The abundance and diversification of Hsps in plant reflect the extensive role of Hsps in stress tolerance mechanism. The expression of Hsps under stress is intense, rapid, and transient suggesting that it is an emergency response of the plant to the stress. Key et al. (1981) reported that when soybean was treated at 40 °C for four hours, ten new proteins identified but they disappeared when treated at 28 °C for 3 h. The gene expression analysis in rice showed that expression of Hsp87 was high after 2 h of heat shock and stable even after 4 h of heat treatment. In addition, Hsp90 was also induced in the presence of other stresses such as drought, salinity, and cold (Pareek et al. 1998).

Liu et al. (2006) reported that Hsp90 is involved in the abiotic stress tolerance in plants, especially salinity, desiccation, high pH, and temperature. The synthesis of Hsp both qualitatively and quantitatively was dependent on cell or tissue type and/or the degree of differentiation and development. Studies also reported the presence of Hsps in plants when subjected to two or more stresses at the same time. Under natural environmental condition, heat stress is usually accompanied by drought, salinity, high radiation, or other abiotic/biotic stress, but the studies of this kind are scanty. Wallner et al. (1982) suggested that heat and drought stresses are often correlated in plants. Heat stress is often coexisting with the drought stress; the expression of Hsp genes under drought stress is not surprising for the same reason. One such study was conducted in cotton plant (*Gossypium hirsutum*) under irrigated and nonirrigated condition, where most of the growth parameters decreased to 80–85 % (Burke et al. 1985). The sHsp 17.4 expression levels were higher in drought-tolerant *Arabidopsis* seeds than in the non-tolerant seed. This suggests that sHsp has a protective role in drought stress (Wehmeyer and Vierling 2000).

The Hsps that accumulate in response to heat stress or other abiotic stress such as drought or salinity are developmentally regulated (Vierling 1991). Heikkila et al. (1984) reported a relationship between ABA, drought stress, and the expression of Hsp70 genes. In *Erianthus arundinaceus,* a highly drought-tolerant plant (Augustine et al. 2014), Hsp70 was expressed sevenfold higher under drought-stressed condition compared to the irrigated plants (Augustine et al. 2015). Up-regulation of some of the Hsps is only during the photosynthetic acclimation, e.g., Hsc70-1 and Hsc70-3 (Vasquez-Robinet et al. 2008). In pine trees, the mitochondrial Hsps, i.e., mitochondrial Hsp70 and sHsp have a major role in high intensity drought stress and the expression of these genes increased at each cycle of severe stress (Vasquez-Robinet et al. 2008), which might be due to the increased antioxidant proteins in the mitochondria during severe drought stress. Moreover, they reported that a homolog of *Arabidopsis* Hsp90-7 was up-regulated in pine trees after rehydration. Kuznetsov et al. (1999) reported that adaptation of cotton plant to drought stress was along with resistance to high temperature. The induction of Hsp was reported in both laboratory and field grown plants when subjected to drought stress (Prasad 1997; Augustine et al. 2014).

Hu et al. (2009) analyzed the expression profiling of heat-stressed rice and compared with the data for drought, cold, and salt stresses. From the study, they concluded that Hsps and HSFs are probably involved in the cross talk of different stress signal networks. In general, Hsps and HSFs are induced by heat, drought, cold, and salinity. Previous studies reported that there is a relationship between heat stress and osmotic stress as both stress signaling pathways leading to the expression of Hsps (Dat et al. 1998; Lee et al. 2000).

7.5 Transgenic Plants Overexpressing Hsps and Drought Stress Tolerance in Plants

Global climate change causes a major threat in water availability in the past and still further in a larger amount of land (Hamdy et al. 2003), and this leads to an increased need for drought-tolerant crops. The variation in the expression of stress-responsive genes in the tolerant and sensitive plants revealed that tolerance is conferred by the mechanism that is genetically encoded (Bray 1993). One of the central themes in plant stress studies is the identification and isolation of the stress-responsive genes. Stress-induced gene expression can be broadly categorized into three groups: (1) genes encoding proteins with known enzymatic or structural functions, (2) proteins with yet unknown functions, and (3) regulatory proteins. Previously, transgenics for abiotic stress tolerance was developed by using genes that are responsible for the modification of a single metabolite that would ultimately lead to increased tolerance to drought or salt stress. In the last century, genetic improvement of grass germplasms was contributed by conventional breeding (Humphreys 1999). Now, the technology is available to make pinpoint genetic changes by using direct gene transfer methods like biolistic transformation and agrobacterium-mediated transformation in grass species (Sticklen and Kenna 1998; Wang and Ge 2006). Although it is well established that transgenic plants overexpressing heat-shock proteins show enhanced tolerance to heat stress, little is known for drought stress tolerance. In case of rice, the incorporation of *Hsp* genes provide increased heat tolerance. For example, thermotolerance was improved in the transgenic rice by overexpressing *Arabidopsis Hsp101* gene (Katiyar-Agarwal et al. 2003). Overexpression of a rice chloroplast *sHsp* (*OsHsp26*) gene improves tolerance to heat and oxidative stress in *E. coli* (Lee et al. 2000), and in rice the overexpression of sHsp17.7 enhanced both heat tolerance and UV-B resistance (Murakami et al. 2004) (Table 7.3).

Table 7.3 Genetic engineering for stress tolerance in plants by employing HSF-Hsp (Hsc) regulon proteins (modified from Grover et al. 2013)

Protein	Source sps	Host sps	Promoter	Comments	References
Engineering Hsp levels					
HSP16.9	*Zea mays*	Tobacco	CaMV35S	Overexpression resulted enhanced seed germination rate, root length, and antioxidant enzyme activity under heat stress (40 °C; 9 h)	Sun et al. (2012)
HSP17.5	*Nelumbo nucifera*	*A. thaliana*	CaMV35S	Transgenic plants had higher seed germination vigor and improved basal thermotolerance (44 °C; 60 and 70 min)	Zhou et al. (2012)
HSP17.8	*Rosa chinensis*	*A .thaliana*	CaMV35S	Overexpressed plants showed higher level of basal thermotolerance (45 °C; 2 h), lower relative electrolyte leakage and more proline content	Jiang et al. (2009)
MT-sHSP	Tomato	Tobacco	CaMV35S	Overexpression of the sHSP increased thermotolerance in tobacco plant	Sanmiya et al. (2004)
HSP21	*Solanum lycopersicum*	*S. lycopersicum*	CaMV35S	Overexpression protected PSII from temperature-dependent oxidative stress; early accumulation of carotenoids noted (40, 47, 50 °C; 2 h and then exposed to high light)	Sharir et al. (2005)
HSP22	*Z. mays*	*A. thaliana*	CaMV35S	Overexpression resulted in increased thermotolerance (42 °C; 30 min)	Rhoads et al. (2005)
HSP23.6	*Tomato*	*Tobacco*	CaMV35S	Overexpression confers thermotolerance	Sanmiya et al. (2004)
HSP26	*Gossypium arboreum*	*Gossypium hirsutum*	CaMV35S	Overexpression resulted enhanced drought tolerance	Maqbool et al. (2009)
HSP26	*Saccharomyces cerevisiae*	*A. thaliana*	CaMV35S	Greater tolerance to heat stress (45 °C; 16 h), increased free proline content and elevated expression of proline biosynthetic pathway genes in transgenic plants	Xue et al. (2010)
HSP26	*Oryza sativa*	*Festuca arundinacea*	CaMV35S	Reduction of electrolyte leakage and accumulation of thiobarbituric acid reactive substance and higher PSII activity on exposure to heat stress (42 °C; 24 h) in transgenics	Kim et al. (2012)

(continued)

Table 7.3 (continued)

Protein	Source sps	Host sps	Promoter	Comments	References
DnaK/HSP70	*Aphanothece halophytica*	Tobacco	CaMV35S	Transgenics were more tolerant to heat stress (40 °C) during germination (seeds were stressed for 40, 45, 50, 55 °C; 6 h in dark) and early growth.	Ono (2001)
BIP/HSP70	Soybean	Tobacco	CaMV35S	Overexpression conferred tolerance to water stress and a reduction of relative water content to 65 %	
DnaK/HSP70	*Aphanothece halophytica*	Tobacco, Rice	CaMV35S	Transgenics showed higher activity of ascorbate peroxidase and catalase. Rice transgenics exhibited enhanced activities of Calvin cycle enzymes and greater tolerance to heat stress (33–37 °C; 16 h)	Uchida et al. (2008)
HSP70	*Trichoderma harzianum*	*A. thaliana*	CaMV35S	Transgenics showed enhanced tolerance to heat stress (45 °C; 5 h); increased transcript levels of SOS1 and APX1 noted	Montero-Barrientos (2010)
HSP70-1	*N. tabacum*	Tobacco	CaMV35S	Overexpressing seedlings grew better after heat stress; trans-protein prevented fragmentation and degradation of nuclear DNA during heat stress (45 °C; 2 h)	Cho and Choi (2009)
HSP70-1 (Hsc70-3)	*N. tabacum*	Tobacco	CaMV35S	Overexpression conferred drought-stress tolerance in 3-week-old tobacco. In addition it suggested a role in regulating the water flux in seedlings	Cho and Hong (2006)
HSP70 (mitochondrial)	*O. sativa*	Rice	CaMV35S	Overexpression manifested enhanced tolerance to heat stress (48 °C; 15 min); programmed cell death was suppressed in overexpression lines	Qi et al. (2011)
HSP70	*Chrysanthemum morifolium*	*A. thaliana*	CaMV35S	Overexpression enhances tolerance to heat, drought, and salinity	Song et al. (2014)
HSP70	*E. arundinaceus*	Sugarcane (*S. officinarum*)	Port Ubi 2.3	Higher level of tolerance to drought (stayed green for more than 10 days) and salinity (germination at 300 mM of NaCl)	Augustine et al. (2015)
Hsc70-1	*A. thaliana*	*A. thaliana*	CaMV35S	Overexpression resulted in increased thermotolerance	Sung and Guy (2003)
HSP100	*A. thaliana*	*A. thaliana*	Maize Ubi 1	Transformants constitutively expressed HSP 100; tolerated sudden shifts to heat; antisense/co-suppression lines had diminished acquired thermotolerance (45 °C; 2 h with or without a conditioning pre-treatment at 38 °C; 90 min)	Queitsch et al. (2000)

Protein	Source sps	Host sps	Promoter	Comments	References
HSP100	*A .thaliana*	Rice	CaMV35S	Transformants expressing HSP100 evidenced better growth in the recovery phase after heat stress (45 °C; 3 h, 47 °C; 2 h and 50 °C; 40 min)	Katiyar-Agarwal et al. (2003)
HSP100	*O. sativa*	Tobacco	CaMV35S	Transgenics survived heat stress (50 °C; 50 min) relatively better than non-transgenic control plants	Chang et al. (2007)
Engineering HSF levels					
HSFA1	*Glycine max*	*G. max*	CaMV35S	Enhanced heat stress (45 °C; 3 h) tolerance through activation of GmHSP70 in transgenics	Zhu et al. (2006)
HSFA2	*A. thaliana*	*A. thaliana*	CaMV35S	Transgenics developed enhanced basal (45 °C; 90 min) and acquired (37 °C; 1 h; 22 °C; 3 h: 45 °C; 90 min) thermotolerance; dominant negative mutants displayed reduced heat stress tolerance	Ogawa et al. (2007)
HSFA3	*A .thaliana*	*A. thaliana*	CaMV35S	Transgenics showed enhanced heat stress (46 °C; 45 min) tolerance; induction of many heat stress associated genes	Yoshida et al. (2008)
HSF7	*O. sativa*	*A. thaliana*	CaMV35S	Higher basal thermotolerance (42 °C; 16 h) in transgenics; transcription of some HSF target genes was enhanced	Liu et al. (2009)
HSF1	*Boea hygrometrica*	*A. thaliana*, Tobacco	CaMV35S	Overexpression resulted in increased basal (48 °C; 2 h 30 min for tobacco and 45 °C; 1 h for *Arabidopsis*) and acquired (40 °C; 3 h: 50 °C; 2 h for tobacco and 37 °C; 1 h: 22 °C; 3 h: 45 °C; 90 min for *Arabidopsis*) thermotolerance via regulation of genes involved in stress protection and mitotic cell cycle	Zhu et al. (2009)
HSFA2	*Lilium longiflorum*	*A. thaliana*	CaMV35S	Overexpression resulted in the activation of genes like HSP101, HSP70, HSP25.3, and APX2, thereby enhancing heat stress (45 °C; 1 h) tolerance in transgenic plants	Xin et al. (2010)
HSFA7	*O. sativa*	*O. sativa*	CaMV35S	Enhanced adaptation to high salinity or drought stress in transgenic rice	Liu and Charng (2012)

7.6 Summary and Conclusions

Drought is one of the major threats facing agriculture in recent times. Climate change will significantly affect the sustainable water content in the coming decades. Drought stress leads to the induced expression of stress-responsive genes, e.g., Hsps and increased abundance of their protein. These stress-responsive proteins help the plant to adapt to the stressful environment through physiological, biochemical, and molecular pathways. Most of the stresses induce proteins that are specific to the stress. However, some proteins are common and expressed in more than one type of stress. Cold, drought, salinity, and heat lead to the expression of some common proteins. Many of the stress-responsive genes have been isolated and characterized.

Hsps play a major role in stress tolerance mechanism and cellular protein homeostasis, but we don't yet know how Hsps act as regulatory molecules, participate in stress signaling mechanism and activation of stress-responsive genes. Recent study showed that introduction of single Hsp70 gene is sufficient for providing stress (drought and salinity) tolerance in crop plant (Augustine et al. 2015). Nevertheless, further experiments are necessary to determine the regulatory role of Hsp. Hsps are thought to be expressed in both normal and stressful conditions. The molecular mechanisms underlying these processes and the role of Hsps in protecting plant from drought stress is still needed to be elucidated. Until now, the research mainly focuses on the identification of changes in the expression of Hsps under stress. Moreover, in several cases, the induced expression of Hsps under stress is mostly based on in vitro experiments because of the lack of availability of mutants for the specific Hsp. Future research should focus mainly on the generation of mutant plants for specific Hsps (Hong and Vierling 2000a, b; Hong et al. 2003). In addition, the response of the plant to more than one stress simultaneously with changes in the expression of Hsps is also important. A detailed study regarding the cross talk between Hsps and other stress response mechanisms in plants will provide a more in-depth understanding of acquired stress tolerance in plants. Moreover, functional identification of the Hsps under drought stress will be of immense help in producing stress-tolerant plants using transgenic technology.

References

Agarwal M, et al. *Arabidopsis thaliana* Hsp100 proteins: kith and kin. Cell Stress Chaperones. 2001;6:219–24.

Agarwal PK, Agarwal P, Reddy MK, Sopory SK. Role of DREB transcription factors in abiotic and biotic stress tolerance in plants. Plant Cell Rep. 2006;25:1263–74.

Almoguera C, Jordano J. Developmental and environmental concurrent expression of sunflower dry-seed-stored low-molecular-weight heat-shock protein and Lea mRNAs. Plant Mol Biol. 1992;19:781–92.

Alvim FC, Carolino SM, Cascardo JC, Nunes CC, Martinez CA, Otoni WC, Fontes EP. Enhanced accumulation of BiP in transgenic plants confers tolerance to water stress. Plant Physiol. 2001a;126:1042–54.

Alvim FC, et al. Enhanced accumulation of BiP in transgenic plants confers tolerance to water stress. Plant Physiol. 2001b;126:1042–54.

Augustine SM, Syamaladevi DP, Premachandran MN, Ravichandran V, Subramonian N. Physiological and molecular insights to drought responsiveness in *Erianthus* spp. Sugar Tech. 2014. doi:10.1007/s12355-014-0312-7.

Augustine SM, Ashwin NJ, Syamaladevi DP, Appunu C, Chakravarthi M, Ravichandran V. *Erianthus arundinaceus* HSP70 (EaHSP70) overexpression increases drought and salinity tolerance in sugarcane (*Saccharum* spp. hybrid). Plant Sci. 2015;232:23–34.

Barnett T, Altschuler M, McDaniel CN, Mascarenhas JP. Heat shock induced proteins in plant cells. Dev Genet. 1980;1:331–40.

Bharti K, Nover L. Heat stress-induced signaling. In: Scheel D, Wasternack C, editors. Plant Signal Transduction: *Frontiers in Molecular Biology*. Oxford: Oxford University Press; 2002. p. 74–115.

Blum A. Plant breeding for stress environments. Boca Raton: CRC Press; 1988. 223 pages.

Boorstein WR, Ziegelhoffer T, Craig EA. Molecular evolution of the HSP70 multi gene family. J Mol Evol. 1994;38(1):1–17.

Boston RS, et al. Molecular chaperones and protein folding in plants. Plant Mol Biol. 1996;32:191–222.

Bray EA. Molecular response to water deficit. Plant Physiol. 1993;103:1035–40.

Buchner J. Hsp90 & Co.—a holding for folding. Trends Biochem Sci. 1999;24:136–41.

Bukau B, Horwich AL. The Hsp70 and Hsp60 chaperone machines. Cell. 1998;92:351–66.

Bukau B, Deuerling E, Pfund C, Craig EA. Getting newly synthesized proteins into shape. Cell. 2000;101:119–22.

Burke JJ, Hatfield JL, Klein RR, Mullet JE. Accumulation of heat shock proteins in field grown cotton. Plant Physiol. 1985;78:394–8.

Chaitanya KV, Sundar D, Jutur PP, Ramachandrareddy A. Water stress effects on photosynthesis in different mulberry cultivars. Plant Growth Regul. 2003;40:75–80.

Chang CC, Huang PS, Lin HR, Lu CH. Transactivation of protein expression by rice HSP101 in planta and using Hsp101 as a selection marker for transformation. Plant Cell Physiol. 2007;48(8):1098–107.

Cho EK, Choi YJ. A nuclear-localized HSP70 confers thermo protective activity and drought-stress tolerance on plants. Biotechnol Lett. 2009;31:597–606.

Cho EK, Hong CB. Over-expression of tobacco NtHSP70-1 contributes to drought-stress tolerance in plants. Plant Cell Rep. 2006;25(4):349–58.

Czarnecka-Verner E, Yuan CX, Scharf KD, Englich G, Gurley WB. Plants contain a novel multi-member class of heat shock factors without transcriptional activator potential. Plant Mol Biol. 2000;43(4):459–71.

Dat JF, Foyer CH, Scott IM. Changes in salicylic acid and antioxidants during induction of thermotolerance in mustard seedlings. Plant Physiol. 1998;118:1455–61.

DeRocher AE, Vierling E. Developmental control of small heat shock protein expression during pea seed maturation. Plant J. 1994;5:93–102.

Dobson CM, Sali A, Karplus M. Protein folding: a perspective from theory and experiment. Angew Chem Int. 1998;37:868–93.

Dubouzet JG, Sakuma Y, Ito Y, Kasuga M, Dubouzet EG, Miura S, Seki M, Shinozaki K, Yamaguchi-Shinozaki K. OsDREB genes in rice, *Oryza sativa* L., encode transcription activators that function in drought-, high salt and cold responsive gene expression. Plant J. 2003;33:751–63.

Ferguson DL, Guikema JA, Paulsen GM. Ubiquitin pool modulation and protein degradation in wheat roots during high temperature stress. Plant Physiol. 1990;92:740–6.

Fink AL. Chaperone-mediated protein folding. Physiol Rev. 1999;79(2):425–49.

Flexas J, Bota J, Galmés J, Medrano H, Ribas-Carbó M. Keeping a positive carbon balance under adverse conditions: responses of photosynthesis and respiration to water stress. Physiol Plant. 2006;127:343–52.

Frydman J. Folding of newly translated proteins *in vivo*: the role of molecular chaperones. Annu Rev Biochem. 2001;70:603–47.

Gagliardi D, Breton C, Chaboud A, Vergne P, Dumas C. Expression of heat shock factors and heat shock protein 70 genes during maize pollen development. Plant Mol Biol. 1995;29:841–56.

Goloubinoff P, et al. Sequential mechanism of solubilization and refolding of stable protein aggregates by a bichaperone network. Proc Natl Acad Sci. 1999;96:13732–7.

Greenway H, Munns R. Mechanisms of salt tolerance in non-halophytes. Annu Rev Plant Physiol. 1980;312:149–90.

Grover A, Mittal D, Negi M, Lavania D. Generating high temperature tolerant transgenic plants: achievements and challenges. Plant Sci. 2013;205–206:38–47.

Gurley WB. HSP101: a key component for the acquisition of thermotolerance in plants. Plant Cell. 2000;12:457–60.

Gyoergyey J, Gartner A, Nemet K, Magyar Z, Hirt H, Heberle-Bors E, Dudits D. Alfalfa heat-shock genes are differentially expressed during somatic embryogenesis. Plant Mol Biol. 1991;16:999–1007.

Hall EA. Is dehydration tolerance relevant to genotypic differences in leaf senescence and crop adaptation to dry environments? In: Close TJ, Bray EA, editors. Plant responses to cellular dehydration during environmental stress. American Society of Plant Physiologists, Rockville: MD pp; 1993. p. 1–13.

Hall A. Crop developmental responses to temperature, photoperiod, and light quality. In: Hall AE, editor. Crop response to environment. Boca Raton: CRC; 2001. p. 83–7.

Hamdy A, Ragab R, Scarascia-Mugnozza E. Coping with water scarcity: water saving and increasing water productivity. Irrig Drain. 2003;52:3–20.

Hamilton EWIII, Heckathorn SA. Mitochondrial adaptations to NaCl. Complex I is protected by anti-oxidants and small heat shock proteins, whereas complex II is protected by proline and betaine. Plant Physiol. 2001;126:1266–74.

Hartl FU. Molecular chaperones in cellular protein folding. Nature. 1996;381:571–80.

Hartl FU, Hayer-Hartl M. Molecular chaperones in the cytosol: from nascent chain to folded protein. Science. 2002;295:1852–8.

Hasegawa PM, Bressan RA, Zhu J-K, Bohnert HJ. Plant cellular and molecular responses to high salinity. Annu Rev Plant Physiol Plant Mol Biol. 2000;51:463–99.

Heikkila JJ, Papp JET, Schultz GA, Bewley JD. Induction of heat shock messenger RNA in maize mesocotyls by water stress, abscisic acid, and wounding. Plant Physiol. 1984;76:270–4.

Helm KW, Lafayete PR, Nago RT, Key JL, Vierling E. Localization of small heat shock proteins to the higher plant endomembrane system. Mol Cell Biol. 1993;13:238–47.

Hemmingsen SM, et al. Homologous plant and bacterial proteins chaperone oligomeric protein assembly. Nature. 1988;26:330–4.

Hong SW, Vierling E. Mutants of Arabidopsis thaliana defective in the acquisition of tolerance to high temperature stress. Proc Natl Acad Sci. 2000a;97:4392–7.

Hong SW, Vierling E. Mutants of *Arabidopsis thaliana* defective in the acquisition of tolerance to high temperature stress. Proc Natl Acad Sci. 2000b;97:4392–7.

Hong SW, Vierling E. Hsp101 is necessary for heat tolerance but dispensable for development and germination in the absence of stress. Plant J. 2001;27:25–35.

Hong S-W, Lee U, Vierling E. Arabidopsis *hot* mutants define multiple functions required for acclimation to high temperatures. Plant Physiol. 2003;132:757–67.

Horwich AL. Resurrection or destruction? Curr Biol. 1995;5:455–8.

Hu W, Hu G, Han B. Genome-wide survey and expression profiling of heat shock proteins and heat shock factors revealed overlapped and stress specific response under abiotic stresses in rice. Plant Sci. 2009;176:583–90.

Hubel A, Schoffl F. Arabidopsis Heat shock factor: isolation and characterization of the gene and the recombinant protein. Plant Mol Biol. 1994;26:353–62.

Humphreys MO. The contribution of conventional plant breeding to forage crop improvement, in Proceedings of the 18th International Grassland Congress. Canada: Saskatoon; 1999.

Iba K. Acclimation response to temperature stress in higher plants: approaches of gene engineering for temperature tolerance. Annu Rev Plant Biol. 2002;53:225–45.

Imai J, Maruya M, Yashiroda H, Yahara I, Tanaka K. The molecular chaperone Hsp90 plays a role in the assembly and maintenance of the 26S proteasome. EMBO J. 2003;22:3557–67.
Ishikawa A, et al. Deletion of a chaperonin 60 b gene leads to cell death in the Arabidopsis lesion initiation 1 mutant. Plant Cell Physiol. 2003;44:255–61.
Jackson-Constan D, Akita M, Keegstra K. Molecular chaperones involved in chloroplast protein import. Biochim Biophys Acta. 2001;1541:102–13.
Jiang C, et al. A cytosolic class I small heat shock protein, RcHSP17.8, of *Rosa chinensis* confers resistance to a variety of stresses to *Escherichia coli*, yeast and *Arabidopsis thaliana*. Plant Cell Environ. 2009;32:1046–59.
Jones PD, Bradley RS. Climatic variations over last 500 years. In: Bradley RS, Jones PD, editors. Climate since A.D. 1500. London: Routledge; 1992. p. 665.
Jones H. What is water use efficiency? In: Bacon MA, editor. Water use efficiency in plant biology. Oxford: Wiley-Blackwell; 2004. p. 27–41.
Katiyar-Agarwal S, Agarwal M, Grover A. Heat-tolerant basmati rice engineered by over-expression of hsp101. Plant Mol Biol. 2003;51:677–86.
Kelley WL. Molecular chaperones: how J domains turn on Hsp70s. Curr Biol. 1999;9:R305–8.
Kermode AR. Approaches to elucidate the basis of desiccation tolerance in seeds. Seeds Sci Res. 1997;7:75–95.
Key JL, Lin CY, Chen YM. Heat shock proteins of higher plants. Proc Natl Acad Sci U S A. 1981;78:3526–30.
Kim KK, Yakota H, Santoso S, Lerner D, Kim R, Kim SH. Purification, crystallization and preliminary X-ray crystallographic data analysis of a small heat shock protein homolog from *Methanococcus jannaschii*, a hyperthermophile. J Struct Biol. 1998;121:76–80.
Kim KH, et al. Overexpression of a chloroplast-localized small heat shock protein OsHSP26 confers enhanced tolerance against oxidative and heat stresses in tall fescue. Biotechnol Lett. 2012;34:371–7.
Kotak S, Larkindale J, Lee U, von Koskull-Doring P, Vierling E, Scharf KD. Complexity of the heat stress response in plants. Curr Opin Plant Biol. 2007;10:310–6.
Kregel KC. Heat shock proteins: modifying factors in physiological stress responses and acquired thermotolerance. J Appl Physiol. 2002;92:2177–86.
Krishna P, Gloor G. The Hsp90 family of proteins in *Arabidopsis thaliana*. Cell Stress Chaperones. 2001;6:238–46.
Kuznetsov VV, Rakitin V, Zholkevich VN. Effects of preliminary heat-shock treatment on accumulation of osmolytes and drought resistance in cotton plants during water deficiency. Physiol Plant. 1999;107:399–406.
Lambers H, Chapin FS, Pons TL. Plant physiological ecology. 2nd ed. New York: Springer; 2008.
Larcher W. Physiological plant ecology. Ecophysiology and stress physiology of functional groups. 4th ed. Berlin: Springer; 2003. p. 504.
Laufen T, Zuber U, Buchberger A, Bukau B. DnaJ proteins. In: Fink AL, Goto Y, editors. Molecular chaperones in proteins: structure, function and mode of action. New York: Marcel Dekker; 1998. p. 241–74.
Lee GJ, Vierling E. A small heat shock protein cooperates with heat shock protein 70 systems to reactivate a heat-denatured protein. Plant Physiol. 2000;122:189–97.
Lee YRJ, et al. A soybean 101-kD heat shock protein complements a yeast HSP 104 deletion mutant in acquiring thermotolerance. Plant Cell. 1994;6:1889–97.
Lee BH, Won SH, Lee HS, et al. Expression of the chloroplast-localized small heat shock protein by oxidative stress in rice. Gene. 2000;245(2):283–90.
Lee U, Rioflorido I, Hong SW, Larkindale J, Waters ER, Vierling E. The Arabidopsis ClpB/ Hsp100 family of proteins: chaperones for stress and chloroplast development. Plant J. 2006;49:115–27.
Leprince O, Hendry GAF, McKersie BD. The mechanisms of desiccation tolerance in seeds. Seed Sci Res. 1993;3:231–46.
Levitt J. Responses of plants to environmental stress: chilling, freezing and high temperature stresses. 2nd ed. New York: Academic; 1980.

Levitt M, Gerstein M, Huang E, Subbiah S, Tsai J. Protein folding: the endgame. Annu Rev Biochem. 1997;66:549–79.

Lin BL, Wang JS, Liu HC, Chen RW, Meyer Y, Barakat A, Delseny M. Genomic analysis of the Hsp70 superfamily in *Arabidopsis thaliana*. Cell Stress Chaperones. 2001;6:201–8.

Liu HC, Charng YY. Acquired thermotolerance independent of heat shock factor A1 (HsfA1), the master regulator of the heat stress response. Plant Signal Behav. 2012;7:547–50.

Liu D, Zhang X, Cheng Y, Takano T, Liu S. RHsp90 gene expression in response to several environmental stresses in rice (*Oryza sativa* L.). Plant Physiol Biochem. 2006;44:380–6.

Liu JG, et al. OsHSF7 gene in rice, *Oryza sativa* L., encodes a transcription factor that functions as a high temperature receptive and responsive factor. BMB Rep. 2009;42:16–21.

Lubben TH, Donaldson GK, Viitanen PV, Gatenby AA. Several proteins imported into chloroplasts form stable complexes with the GroEL-related chloroplast molecular chaperone. Plant Cell. 1989;1(1223):1230.

Majoul T, Bancel E, Triboi E, Ben Hamida J, Branlard G. Proteomic analysis of the effect of heat stress on hexaploid wheat grain: characterization of heat-responsive proteins from total endosperm. Proteomics. 2003;3:175–83.

Maqbool A, Abbas W, Rao AQ, Irfan M, Zahur M, Bakhsh A, Riazuddin S, Husnain T. *Gossypium arboreum* GHSP26 enhances drought tolerance in *Gossypium Hirsutum*. Amer Inst Chem Engrs. 2009;26(1):21–5.

Mayer MP, Bukau B. Hsp70 chaperones: cellular functions and molecular mechanism. Cell Mol Life Sci. 2005;62:670–84.

Miernyk JA. The 70kDa stress-related proteins as molecular chaperones. Trends Plant Sci. 1997;2:180–7.

Miernyk JA. Protein folding in the plant cell. Plant Physiol. 1999;121:695–703.

Milioni D, Hatzopoulos P. Genomic organization of Hsp90 gene family in *Arabidopsis*. Plant Mol Biol. 1997;35:955–61.

Mogk A, Schlieker C, Friedrich KL, Schonfeld HJ, Vierling E, Bukau B. Refolding of substrates bound to small Hsps relies on a disaggregation reaction mediated most efficiently by ClpB/DnaK. J Biol Chem. 2003;278:31033–42.

Molden D, editor. London, UK: Earthscan; Colombo. Sri Lanka: International Water Management Institute (IWMI); 2007. Water for food, water for life: A comprehensive assessment of water management in agriculture.

Montero-Barrientos M. Transgenic expression of the *Trichoderma harzianum* hsp70 gene increases *Arabidopsis* resistance to heat and other abiotic stresses. J Plant Physiol. 2010;167:659–65.

Morimoto RI. Regulation of the heat shook transcriptional response: cross talk between a family of heat shock factors, molecular chaperones, and negative regulators. Genes Dev. 1998;12:3788–96.

Mu C, Zhang S, Yu G, Chen N, Li X. Overexpression of small heat shock protein LimHSP16.45 in Arabidopsis enhances tolerance to abiotic stresses. PLos One. 2013;8(12):e82264. doi:10.1371/journal.pone.0082264.

Murakami T, Matsuba S, Funatsuki H, et al. Overexpression of a small heat shock protein, sHSP17.7, confers both heat tolerance and UV-B resistance to rice plants. Mol Breed. 2004;13(2):165–75.

Nakamoto H, Vıgh L. The small heat shock proteins and their clients. Cell Mol Life Sci. 2007;64:294–306.

Neuwald AF, Aravind L, Spouge JL, Koonin EV. AAA+: A class of chaperone-like ATPases associated with assembly, operation, and disassembly of protein complexes. Genome Res. 1999;9:27–43.

Ogawa D, Yamaguchi K, Nishiuchi T. High-level overexpression of the *Arabidopsis* HsfA2 gene confers not only increased thermotolerance but also salt/osmotic stress tolerance and enhanced callus growth. J Exp Bot. 2007;58:3373–83.

Ono K. Overexpression of DnaK from a halotolerant cyanobacterium *Aphanothece halophytica* enhances the high temperature tolerance of tobacco during germination and early growth. Plant Sci. 2001;160:455–61.

Pareek A, Singla SL, Grover A. Plant Hsp90 family with special reference to rice. J Biosci. 1998;23:361–7.
Parsell PA, Lindquist S. The function of heat-shock proteins in stress tolerance degradation and reactivation of damaged proteins. Annu Rev Genet. 1993;27:437–96.
Pearl LH, Prodromou C. Structure and mechanism of the hsp90 molecular chaperone machinery. Annu Rev Biochem. 2006;75:271–94.
Pelham H. Speculations on the major heat shock and glucose regulated proteins. *Cell* 46: 959-961. Plant Cell. 1986;6:1889–97.
Prasad MVN. Plant Ecophysiology. New York, NY: John Wiley and Sons; 1997. ISBN 0-47-13157-1.
Pratt WB, et al. Hsp90-binding immunophilins in plants: the protein movers. Trends Plant Sci. 2001;6:54–5.
Pratt WB, Galigniana MD, Harrell JM, Deranco DB. Role of hsp90 and the hsp90-binding immunophilins in signaling protein movement. Cell Signal. 2004;16:857–72.
Price AH, Cairns JE, Horton P, Jones HG, Griffiths H. Linking drought-resistance mechanisms to drought avoidance in upland rice using a QTL approach: progress and new opportunities to integrate stomatal and mesophyll responses. J Exp Bot. 2002;53:989–1004.
Qi Y, et al. Over-expression of mitochondrial heat shock protein 70 suppresses programmed cell death in rice. FEBS Lett. 2011;585:231–9.
Queitsch C, Hong SW, Vierling E, Lindquist S. Heat shock protein 101 plays a crucial role in thermotolerance in *Arabidopsis*. Plant Cell. 2000;12:479–92.
Queitsch C, Sangster TA, Lindquist S. Hsp90 as a capacitor of phenotypic variation. Nature. 2002;417:618–24.
Ranson NA, et al. Chaperonins. Biochem J. 1998;333:233–42.
Rhoads DM, White SJ, Zhou Y, Muralidharan M, Elthon TE. Altered gene expression in plants with constitutive expression of a mitochondrial small heat shock protein suggests the involvement of retrograde regulation in the heat stress response. Physiol Plant. 2005;123:435–44.
Richter K, Buchner J. Hsp90: chaperoning signal transduction. J Cell Physiol. 2001;188:281–90.
Rutherford SL, Lindquist S. Hsp90 as a capacitor for morphological evolution. Nature. 1998;396:336–42.
Sanmiya K, Suzuki K, Egawa Y, Shono M. Mitochondrial small heat-shock protein enhances thermotolerance in tobacco plants. FEBS Lett. 2004;557:265–8.
Scharf KD, Rose S, Zott W, Schoffl F, Nover L. Three tomato genes code for heat stress transcription factors with a region of remarkable homology to the DNA-binding domain of the yeast HSF. EMBO J. 1990;9:4495–501.
Schirmer EC, Glover JR, Singer MA, Lindquist SL. Hsp100/Clp proteins: A common mechanism explains diverse functions. Trends Biochem Sci. 1996;21:289–96.
Schlesinger MJ. Heat shock proteins. J Biol Chem. 1990;265:12111–4.
Schoffl F, Prandl R, Reindl A. Molecular responses to heat stress. In: Shinozaki K, Yamaguchi-Shinozaki K, editors. Molecular responses to cold, drought, heat and salt stress in higher plants. Austin: R.G. Landes; 1999. p. 81–98.
Seki M, Kamei A, Satou M, Sakurai T, Fujita M, Oono Y, Yamaguch-Shinozaki K, Shinozaki K. Transcriptome analysis in abiotic stress conditions in higher plants. Topics Curr Genet. 2003;4:271–95.
Seo JS, Lee YM, Park HG, Lee JS. The inter tidal copepod Tigriopus japonicas small heat shock protein 20 gene (Hsp20) enhances thermotolerance of transformed *Escherichia coli*. Biochem Biophys Res Co. 2006;340:901–8.
Sharir NI, Isaacson T, Lurie S, Weiss D. Dual role for tomato heat shock protein 21: protecting photosystem II from oxidative stress and promoting color changes during fruit maturation. Plant Cell. 2005;17:1829–38.
Song A, Zhu X, Chen F, et al. A Chrysanthemum heat shock protein confers tolerance to abiotic stress. Int J Mol Sci. 2014;15(3):5063–78.
Sorger PK, Nelson HCM. Trimerization of a yeast transcriptional activator via a coiled-coil motif. Cell. 1989;59:807–13.

Sticklen MB, Kenna MP. Turf grass Biotechnology. Cell and Molecular Approaches to Turf grass Improvement. Chelsea: Ann Arbor Press; 1998.

Sun W, Motangu MV, Verbruggen N. Small heat shock proteins and stress tolerance in plants. Biochim Biophys Acta. 2002;1577:1–9.

Sun L, et al. ZmHSP16.9, a cytosolic class I small heat shock protein in maize (*Zea mays*), confers heat tolerance in transgenic tobacco. Plant Cell Rep. 2012;31:1473–84.

Sung DY, Guy CL. Physiological and molecular assessment of altered expression of Hsc70-1 in Arabidopsis. Evidence for pleiotropic consequences. Plant Physiol. 2003;132:979–87.

Sung DY, Vierling E, Guy CL. Comprehensive expression profile analysis of the Arabidopsis Hsp70 gene family. Plant Physiol. 2001;126:789–800.

Tissieres A, Mitchell HK, Tracy UM. Protein synthesis in salivary glands of D. Melanogaster Relation to chromosome puffs. J Mol Biol. 1974;84:389–98.

Tonsor SJ, Scott C, Boumaza I, Liss TR, Brodsky JL, Vierling E. Heat shock protein 101 effects in A. thaliana: genetic variation, fitness and pleiotropy in controlled temperature conditions. Mol Ecol. 2008;17:1614–26.

Trent JD. A review of acquired thermotolerance, heat-shock proteins and molecular chaperones in Archaea. FEMS Microbiol Rev. 1996;18:249–58.

Tsvetkova NM, Horvath I, Torok Z, et al. Small heat-shock proteins regulate membrane lipid polymorphism. Proc Natl Acad Sci U S A. 2002;99:13504–9.

Uchida A, et al. Overexpression of DnaK chaperone from a halotolerant cyanobacterium *Aphanothece halophytica* increases seed yield in rice and tobacco. Plant Biotechnol. 2008;25:141–50.

Valliyodan B, Nguyen HT. Understanding regulatory networks and engineering for enhanced drought tolerance in plants. Curr Opin Plant Biol. 2006;9:189–95.

van Montfort RL, Basha E, Friedrich KL, Slingsby C, Vierling E. Crystal structure and assembly of a eukaryotic small heat shock protein. Nat Struct Biol. 2001;8:1025–30.

Vasquez-Robinet C, Mane SP, Ulanov AV, Watkinson JI, Stromberg VK, De Koeyer D, Schafleitner R, Willmot DB, Bonierbale M, Bohnert HJ, Grene R. Physiological and molecular adaptations to drought in Andean potato genotypes. J Exp Bot. 2008;59(8):2109–23.

Vierling E. The roles of heat shock proteins in plants. Annu Rev Plant Physiol Plant Mol Biol. 1991;42:579–620.

Wallner SJ, Becwar MR, Butler JD. Measurement of turfgrass heat tolerance *in vitro*. J Am Soc Hortic Sci. 1982;107:608–13.

Wang ZY, Ge Y. Invited review: recent advances in genetic transformation of forage and turf grasses. Vitro Cell Dev Biol Plant. 2006;42(1):1–18.

Wang D, Luthe DS. Heat sensitivity in a bentgrass variant. Failure to accumulate a chloroplast heat shock protein isoform implicated in heat tolerance. Plant Physiol. 2003;133:319–27.

Wang W, Vinocur B, Shoseyov O, Altman A. Role of plant heat-shock proteins and molecular chaperones in the abiotic stress response. Trends Plant Sci. 2004;9:244–52.

Waters ER, et al. Evolution, structure and function of the small heat shock proteins in plants. J Exp Bot. 1996;47:325–38.

Wehmeyer N, Vierling E. The expression of small heat shock proteins in seeds responds to discrete developmental signals and suggest a general protective role in desiccation tolerance. Plant Physiol. 2000;122:1099–108.

Wehmeyer N, Hernandez LD, Flinkelstein RR, Vierling E. Synthesis of small heat-shock proteins is part of the developmental program of late seed maturation. Plant Physiol. 1996;112:747–57.

Wu C. Heat stress transcription factors. Annu Rev Cell Biol. 1995;11:441–6.

Xin H, et al. Cloning and characterization of HsfA2 from Lily (*Lilium longiflorum*). Plant Cell Rep. 2010;29:875–85.

Xu X, Sarbeng EB, Vorvis C, Kumar DP, Zhou L, Liu Q. Unique peptide substrate binding properties of 110-kDa heat-shock protein (Hsp110) determine its distinct chaperone activity. J Biol Chem. 2012;287:5661–72.

Xue Y, et al. Over-expression of heat shock protein gene hsp26 in *Arabidopsis thaliana* enhances heat tolerance. Biol Plant. 2010;54(1):105–11.

Yamada A, et al. The role of plant CCTa in salt- and osmotic stress tolerance. Plant Cell Physiol. 2002;43:1043–8.

Yoshida T, Sakuma Y, Todaka D, Maruyama K, Qin F, Mizoi J, Kidokoro S, Fujita Y, Shinozaki K, Yamaguchi-Shinozaki K. Functional analysis of an *Arabidopsis* heat-shock transcription factor HsfA3 in the transcriptional cascade downstream of the DREB2A stress-regulatory system. Biochem Biophys Res Commun. 2008;368:515–21.

Young JC, et al. Hsp90: a specialized but essential protein folding tool. J Cell Biol. 2001a;154:267–73.

Young JC, Moarefi I, Hartl U. Hsp90: a specialized but essential protein-folding tool. J Cell Biol. 2001b;154:267–73.

Young JC, Hoogenraad NJ, Hartl FU. Molecular chaperones Hsp90 and Hsp70 deliver pre-proteins to the mitochondrial import receptor Tom70. Cell. 2003;112:41–50.

Zhao R, Davey M, Hsu YC, Kaplanek P, Tong A, Parsons AB, Krogan N, Cagney G, Mai D, Greenblatt J, Boone C, Emili A, Houry WA. Navigating the chaperone network: an integrative map of physical and genetic interactions mediated by the hsp90 chaperone. Cell. 2005;120:715–27.

Zhou Y, et al. NnHSP17.5 a cytosolic class II small heat shock protein gene from *Nelumbo nucifera*, contributes to seed germination vigor and seedling thermotolerance in transgenic *Arabidopsis*. Plant Cell Rep. 2012;31:379–89.

Zhu B, et al. Identification and characterization of a novel heat shock transcription factor gene, GmHsfA1, in soybeans (*Glycine max*). J Plant Res. 2006;119:247–56.

Zhu Y, et al. Ectopic over-expression of BhHsf1 a heat shock factor from the resurrection plant *Boea hygrometrica*, leads to increased thermotolerance and retarded growth in transgenic *Arabidopsis* and tobacco. Plant Mol Biol. 2009;71:451–67.

Chapter 8
Ascorbate–Glutathione Cycle: Controlling the Redox Environment for Drought Tolerance

Lyuben Zagorchev, Denitsa Teofanova, and Mariela Odjakova

8.1 Introduction

8.1.1 Reactive Oxygen Species vs. Antioxidants

Reactive oxygen species (ROS) are continually produced in all aerobic life forms and plants do not make an exception. While generally regarded as detrimental by-products that should be detoxified immediately, recently ROS were also established as important signalling molecules that may be produced on purpose at least in some cases. Extensive review of the production, detoxification, and function of ROS in plants was recently provided (Apel and Hirt 2004). The balance between ROS production and scavenging is usually regarded as the thin line that would define whether a plant cell will live or die (Foyer and Noctor 2005b) and the excessive ROS production, accompanied by ineffective or insufficient ROS scavenging is generally regarded as a condition of oxidative stress (Mittler 2002).

Hydrogen peroxide (H_2O_2) and superoxide anion (O_2^-) are probably the most abundant ROS in aerobic organisms and consequently the scavenging mechanisms, involved in their detoxification are the best studied. The sites and sources of production and the respective detoxifying mechanisms in plants were summarized by Mittler (Mittler 2002). The main subcellular ROS producing compartments are chloroplasts, mitochondria and peroxisomes, although a significant amount of O_2^- is also produced in the apoplast by NADPH-dependent oxidases (reviewed by Marino et al. 2012) in various processes, associated with abiotic and biotic stress responses, developmental processes, growth etc. The deleterious effect of ROS is attributed to

L. Zagorchev (✉) • D. Teofanova • M. Odjakova
Department of Biochemistry, Faculty of Biology, Sofia University,
8 Dragan Tsankov Blvd, Sofia 1164, Bulgaria
e-mail: lzagortchev@yahoo.co.uk; denyrt@yahoo.com; modjakova@biofac.uni-sofia.bg

M.A. Hossain et al. (eds.), *Drought Stress Tolerance in Plants, Vol 1*,
DOI 10.1007/978-3-319-28899-4_8

the direct oxidation of cellular macromolecules including but not restricted to DNA, protein thiols and membrane lipids (Mittler 2002). The role of ROS was recently revised and the term oxidative stress, expanded to oxidative signalling (Foyer and Noctor 2005a). Generally, the ROS scavenging mechanisms are subdivided into nonenzymatic and enzymatic (Foyer and Noctor 2005b; Mittler 2002), summarized in Table 8.1. The redox state of some nonenzymatic antioxidant couples such as glutathione/glutathione disulphide and ascorbate/ dehydroascorbate defines the redox environment to a great extent and is decisive for the fate of the cell (Kranner et al. 2006; Foyer and Noctor 2005b, 2011; Birtić et al. 2011).

Obviously, ROS and antioxidants are two important players in the cellular homeostasis, either in normal conditions or during oxidative stress. It seems, however, that ROS production is an inevitable consequence of all types of abiotic and biotic stresses such as salt (Miller et al. 2010), drought (Yang et al. 2015; Miller et al. 2010), cold (Theocharis et al. 2012), heavy metals (Hossain et al. 2012), pathogen attacks (Scheler et al. 2013), and physical damages (Suzuki and Mittler 2012) summarized by Sharma (Sharma et al. 2012). Hence ROS signalling and ROS scavenging are considered important factors in the stress response and stress tolerance of crop plants (Gill and Tuteja 2010) and the ability of some plant species or plant cultivars to successfully neutralize ROS might be discriminative over close relatives in their ability to withstand harsh environmental conditions (Yang et al. 2015). Reactive oxygen species production and scavenging in relation to drought stress response and tolerance are the focus of the current overview.

The NADPH oxidase-mediated mechanism of ROS production is regarded as a signalling rather than a harmful event. Most plant NADPH oxidases are membrane-bound or cell-wall-associated enzymes. Several studies showed the involvement of this mechanism in ROS-mediated activation of antioxidant mechanisms or compatible solute accumulation during water deficit in maize (*Zea mays*) (Voothuluru and Sharp 2012), *Pluchea indica* (Chang et al. 2012) and Arabidopsis (Ben Rejeb et al. 2015). Clearly, this is a widespread mechanism for signalling in conditions of abiotic stress and it may be related to the drought stress tolerance of particular cultivars. However, this is not the classical case of oxidative burst, followed by an increase in the antioxidant defense and will not be extensively discussed in this chapter.

The main sites of ROS production in plants are the chloroplasts, peroxisomes and mitochondria (Miller et al. 2010). Asada (Asada 2006) established the reaction centers of both photosystems (PSI and PSII) as the major sources of ROS in plants. Water deficit-induced stomatal closure is the main defense mechanism, allowing reduced water losses due to transpiration especially in warm arid climate, but this also leads to the restriction of the CO_2 uptake, enhanced photorespiration and shifting of the photosynthetic machinery to excessive production of singlet oxygen by PSII and H_2O_2 by PSI. Furthermore the same cascade of events leads to overproduction of H_2O_2 in the peroxisomes by glycolate oxidase (Noctor et al. 2014; Cruz de Carvalho 2008). Although not directly, but drought induces ROS production in the mitochondrial electron transport chains (Noctor et al. 2014). C4 and CAM plants are not good only in reducing photorespiration and/or water losses, but also in

Table 8.1 Plant nonenzymatic and enzymatic antioxidative systems. Multiorganelle distribution is used for antioxidants, found in cytosol, chloroplasts, mitochondria, and peroxisomes and in some cases nucleus (as for glutathione)

Antioxidant	Localization	Primary ROS	References
Nonenzymatic			
Glutathione	Multiorganelle distribution	H_2O_2	Noctor et al. (2012), Foyer and Noctor (2011), Kranner et al. (2006)
Ascorbate	Multiorganelle distribution	H_2O_2	Foyer and Noctor (2011), Szarka et al. (2012)
α-tocopherol	Membranes	Lipid peroxyl radical	Szarka et al. (2012)
Flavonoids	Vacuole, Cell wall	H_2O_2	Agati et al. (2012)
Sugars and sugar alcohols: galactinol, mannitol, raffinose, and others	Vacuole, Chloroplasts	Hydroxyl and superoxide radicals	Van den Ende and Peshev (2013)
Proline	Cytosol	Hydroxyl and superoxide radicals	Shevyakova et al. (2009)
Carotenoids	Chloroplasts	Singlet molecular oxygen and peroxyl radicals	Stahl and Sies (2003)
Enzymatic			
Superoxide dismutase	Multiorganelle distribution	Superoxide radical	Mittler (2002)
Catalase	Peroxisomes	H_2O_2	Mhamdi et al. (2012)
Ascorbate peroxidase[a]	Multiorganelle distribution	H_2O_2	Caverzan et al. (2012)
Glutathione peroxidase[a]	Multiorganelle distribution	H_2O_2, lipid peroxyl radicals	Gaber et al. (2012)
Thioredoxin peroxidase[a]	Chloroplasts, mitochondria, cytosol	H_2O_2	Mittler (2002)
Monodehydroascorbate reductase, Dehydroascorbate reductase, and Glutathione reductase	Multiorganelle distribution	MDHA, DHA and GSSG[b]	Foyer and Noctor (2011), Szarka et al. (2012)

[a]A variety of other peroxidases exists in plants, but is not included here for the conciseness of the list
[b]Please note that MDHAR, DHAR, and GR are AsA and GSH-regenerating enzymes, rather than antioxidant enzymes per se. However, their involvement in the antioxidant defense is so profound, that they are treated as antioxidant enzymes in the predominant literature

reducing ROS caused damages. This means neither that limiting photorespiration is sufficient trait, conferring drought tolerance nor that C4 or CAM plants don't over-produce ROS in water deprived conditions. Efficient antioxidative capacity is also needed to discriminate between drought-tolerant and sensitive lines as shown in maize (Yang et al. 2015).

8.1.2 *Definition and Significance of Ascorbate–Glutathione Pathway*

The most important of the nonenzymatic antioxidants comprise the so called ascorbate–glutathione-α-tocopherol triad (Szarka et al. 2012). Considering that tocopherol is lipid-soluble and destined for the specific role of membrane protection (Marquardt et al. 2013), ascorbate (AsA) and glutathione (GSH) remain the major, water-soluble, redox-active molecules. It is not exaggerated that this couple was called the "heart of the redox hub" (Foyer and Noctor 2011), and it is a matter not only of efficiency but also of concentrations. Both ascorbate and glutathione are presented in the millimolar range (Foyer and Noctor 2005a) and their concentration could be further increased in folds in response to different stimuli. The abundance of GSH makes it, along with its oxidized form, glutathione disulphide, the main intracellular redox couple and thus, a significant redox buffer (Kranner et al. 2006).

Since both ascorbate and glutathione were long known as potent antioxidants, it was not until 1976 when Foyer and Halliwell (1976) proposed the first, and still valid, scheme for interconnection) between them. According to this founding work, an electron flow from NADPH through GSH and AsA ultimately leads to reduction of H_2O_2 by ascorbate peroxidase (APX; EC 1.11.1.11). Thus monodehydroascorbate (MDHA) is formed and two moles of MDHA disproportionate spontaneously to AsA and dehydroascorbate (DHA) at a significant rate (Potters et al. 2002). The oxidized forms monodehydroascorbate, dehydroascorbate, and glutathione disulphide are produced, that should be recycled by monodehydroascorbate reductase (MDHAR; EC 1.6.5.4, NAD(P)H as electron donor), dehydroascorbate reductase (DHAR; 1.8.5.1, GSH as electron donor), and glutathione reductase (GR; 1.8.1.7, NADPH as electron donor).

From a biochemical point of view NADPH/$NADP^+$, 2 GSH/GSSG and AsA/DHA makes a perfect chain of electron donors and acceptors (Fig. 8.1), based on the standard half reduction potential, but AsA and GSH, though interconnected, are also independent players (Foyer and Noctor 2011). Dehydroascorbate for example could be reduced to AsA in a GSH-independent mechanism, by thioredoxin reductase (May et al. 1997). It was even proposed, that GSH-dependent DHA reduction in the classical AsA–GSH cycle is more important in optimal conditions, while in stressful environment the GSH-independent DHA reduction is prevalent (Gallie 2013). Glutathione is also capable of ROS scavenging directly through glutathione peroxidase (GPX; 1.11.1.9) and plays a variety of functions, ranging from protein

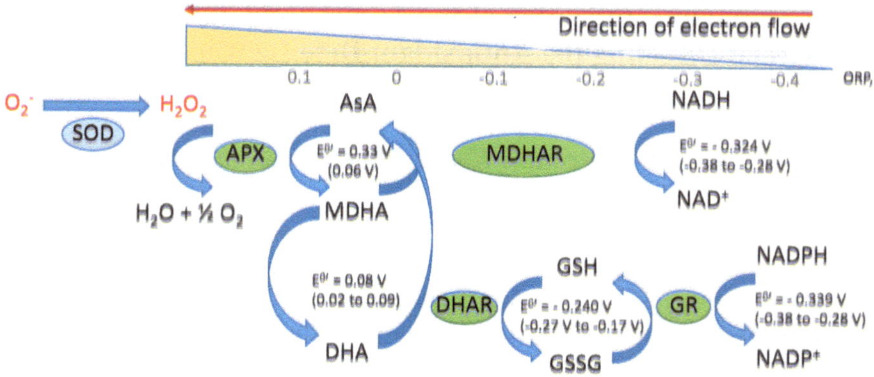

Fig. 8.1 Schematic representation of the ascorbate–glutathione cycle. The enzymes of the AsA–GSH cycle are represented by green circles, while other enzymes in the blue circle. The standard half-cell reduction potential E^{0}' for the redox couples NADPH/NADP$^+$, NADH/NAD$^+$, GSSG/2GSH, DHA/AsA, and MDHA/AsA are shown. Values of E^{0}' are according to (Tinoco et al. 2014). Please note that the actual reduction potential depends on the relative concentrations and differs depending on physiological conditions and within different compartments of the cell. The figures in brackets represent the actual reduction potential of the respective redox couple under physiological conditions. For references, see (Kranner et al. 2006; Potters et al. 2002; Noctor 2006; Schafer and Buettner 2001). The direction of the electron flow from more negative toward more positive reduction potential (ORP, V) is represented on the top

thiol protection and xenobiotics compartmentalization to chelation of heavy metals and transcriptional control (Noctor et al. 2012; Zagorchev et al. 2013). It is also evident as in ascorbate-deficient *vtc1* Arabidopsis mutants, that GSH may compensate the AsA deficiency by accumulating in higher concentrations than in the wild-type plants in response to drought stress (Niu et al. 2013). These mutants, however, were not able to keep the growth rates of wild-type plants, even at optimal conditions (Veljovic-Jovanovic et al. 2001), suggesting that the role of AsA is not restricted to ROS scavenging in the stress response.

What is the main function of the AsA–GSH pathway in plant life? First of all, considering the ROS scavenging properties, it is regarded as a central mechanism conferring stress tolerance in plants. A book, edited by Anjum et al. (2010) was published recently and is fully dedicated to this subject. However, it does not deal strictly with the classical succession of events—electron flow from NADPH through GSH and AsA eventually reduces H_2O_2, derived either directly from electron transport chain or from dismutation of superoxide by superoxide dismutases (Polle 2001). In a broader sense the components of the AsA–GSH pathway are involved in a variety of processes, ranging from signalling and differentiation to resistance to a spectrum of stress factors, either abiotic or biotic (for reference, see (Anjum et al. 2010). In the current chapter, we will discuss the role of the enzymes involved in the AsA–GSH cycle in drought stress tolerance of plants, although the mechanisms, underlying such resistance are common and usually confer tolerance to multiple stresses.

8.2 Compartmentalization of the Ascorbate: Glutathione Cycle: Directing the Antioxidant Defense where It Is Needed

8.2.1 Thylakoids Protection Keeps Photosynthesis Running

The chloroplast photosystems seem to be the most pronounced ROS source and the primary site of ROS overproduction in plants during drought stress. It is not surprising that the AsA–GSH cycle was first defined (Foyer and Halliwell 1976) and best studied as a protective mechanism for the photosynthetic machinery in these organelles. As plant catalases are almost exclusively peroxisomal proteins (Mhamdi et al. 2012), the role of AsA–GSH cycle in H_2O_2 scavenging is of particular importance in chloroplasts. The maintenance of active photosynthesis under water deficit defines not simply the ability of plants to survive short periods of unfavorable conditions, but to sustain growth and development in the overall water depleted environment.

Plant chloroplasts APXs are presented both in the stroma, the thylakoid lumen and in thylakoids, but depending on the plant species might be encoded either by a single gene (tobacco, *Nicotiana tabacum*), with following alternative splicing of the product, or by multiple genes (Fig. 8.2) as in Arabidopsis (*Arabidopsis thaliana*), rice (*Oryza sativa*) and tomato (*Solanum lycopersicum*) with differential regulation (Caverzan et al. 2012). The expression of either stromal or thylakoid-bound APX isoenzyme is controlled in a tissue dependent manner in the first case (Ishikawa and Shigeoka 2008). The comparative contribution to drought tolerance of the thylakoid or stromal isoenzyme seems to be different as shown in wheat (*Triticum aestivum*) where the transcript levels of the thylakoid APX increased in drought-tolerant genotype, compared to the sensitive one (Sečenji et al. 2010). Overexpression of the thylakoid-bound APX in rapeseed (*Brassica napus*) also leads to increased drought and salt tolerance (Wang et al. 2013a). The heterologous expression of either stromal APX (from *Suaeda salsa* into Arabidopsis) (Li et al. 2012b) or thylakoid APX (from *Jatropha curcas* into tobacco) (Liu et al. 2013) increased salt tolerance of the transgenic plants. It is possible that the thylakoid APXs are more important in abiotic stress conditions, but clearly both isoenzymes would have a positive effect.

Dehydroascorbate reductases from plant chloroplasts were isolated and characterized (Shimaoka et al. 2000; Foyer and Halliwell 1977) in spinach (*Spinacia oleracea*) comparatively a long time ago and its catalytic mechanism was revealed, showing a much higher specific activity and higher substrate affinity than other characterized DHA reducing enzymes (Shimaoka et al. 2003). Monodehydroascorbate reductase was also cloned from spinach chloroplasts and expressed as a recombinant protein (Sano et al. 2005). Assuming the importance of chloroplasts APXs (and its increased activity in response to abiotic stresses) in ROS scavenging, the effective regeneration of AsA, to which DHAR contributes substantially, is an invaluable feature of tolerant plants as decrease in AsA concentration would ultimately lead to inactivation and degradation of APX (Caverzan et al. 2012). Unlike

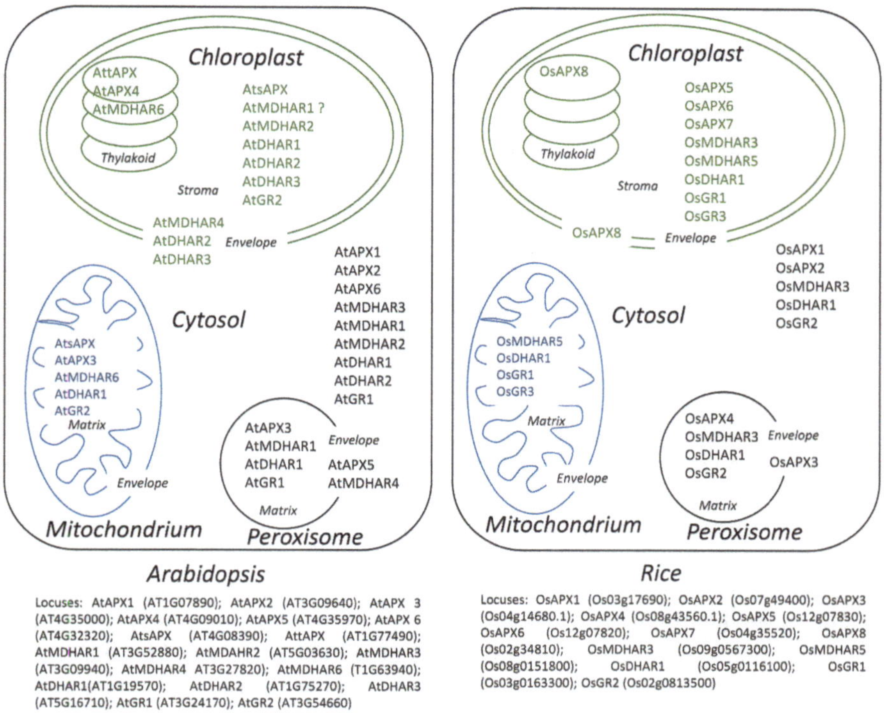

Fig. 8.2 Schematic representation of the distribution of the AsA–GSH cycle isoenzymes between chloroplasts, mitochondria, peroxisomes, and cytosol in Arabidopsis and rice. Gene products are roughly situated in the respective compartment, but the precise localization of some isoenzymes is not reflected (either thylakoid membrane or thylakoid lumen). Most of the cell organelles are omitted for the clearness of the picture. Locuses of the particular genes are included for reference purposes. Database used are The Arabidopsis Information Resource (Phoenix Bioinformatics Corporation) and (National Bioresource Project: Rice 2000)

mitochondria, chloroplasts need to import AsA (Foyer 2015). Ascorbate is transported from the cytosol to the chloroplast stroma through a recently characterized transporter, AtPHT4:4, in the envelope membrane (Miyaji et al. 2015). Its further movement to the thylakoid lumen is proposed to be dependent on slow diffusion (Foyer 2015), which is by any mean, unfavorable in case of ROS overproduction.

The presence of GSH and GR in plant chloroplasts and their link to AsA metabolism was first proposed by Foyer and Halliwell back in 1976, thus defining the AsA–GSH cycle (Foyer and Halliwell 1976) and GR was further characterized as a key regulator of the GSH-to-GSSG ratio (Schaedle and Bassham 1977). In the latter study GSH was proposed to act mainly as protein thiol-protective molecule in oxidative conditions. Gamble and Burke (Gamble and Burke 1984) studied the effect of water deficit on the chloroplast GR activity in winter wheat, showing an overall increase, that is most pronounced when expressed as a function of the chlorophyll content. The subcellular distribution of GR activity was studied in pea (*Pisum sativum*) leaves and it was shown, that over 77 % of the total activity is concentrated in the chloroplasts, with 20 % in the cytosol and 3 % in mitochondria (Edwards et al. 1990). This is not surprising, considering that light-dependent reactions of the photosynthetic apparatus are the major source of NADPH, and vice versa, GR is linked to the protection of photosystems as decreased activity is associated with increased H_2O_2 production and higher sensitivity of PSII (Ding et al. 2012), conferring higher sensitivity to environmental stresses. Glutathione may act as an electron donor for DHAR in the regeneration of AsA, but may also be directly involved in H_2O_2 scavenging through the action of glutathione peroxidases (GPX). The involvement of chloroplasts GPXs in stress tolerance was also shown (Zhai et al. 2013).

A recent study by Cao (Cao et al. 2015) on PEG-treated tomato plants might be exemplary for the AsA–GSH cycle enzyme activity in drought-sensitive plants, showing simultaneous and gradual increase in ROS production and decrease of chlorophyll concentration and overall photosynthetic parameters. PEG-treated plants showed an increase in SOD, APX, and MDHAR up to the third day, accompanied by decrease in DHAR and GR activities. In prolonged stress, up to 12 days all the five enzymes showed decrease in activity. Thus, it seems, that in drought-sensitive plants the chloroplasts antioxidant system depends mainly on AsA for H_2O_2 scavenging through APX with subsequent immediate reduction of MDA by MDHAR and no involvement of GSH-dependent DHA reduction. This mechanism is however not efficient for even comparatively short periods and lead to the gradual decline of photosynthetic capacity. A partially positive effect of Si-treatment was observed, largely linked to higher GR and DHAR activities (Cao et al. 2015). Double-transformed tobacco plants, overexpressing GR and DHAR also showed improved abiotic stress tolerance (Le Martret et al. 2011).

8.2.2 Mitochondria

Mitochondria are a well-known source of ROS production, though in smaller amounts compared to chloroplasts and peroxisomes (Foyer and Noctor 2005b). The major sites of ROS production are complex I and complex III of the electron transport chain where overreduction of the ubiquinone pool result in electron-pass to O_2 and formation of O_2^- . The initial step of detoxification involves the action of a Mn-SOD that further generates H_2O_2. In conditions of severe drought stress the production of ATP in the chloroplasts is diminished. Thus, increased respiration in mitochondria compensates the ATP deficiency, simultaneously leading to overproduction of ROS (Miller et al. 2010).

Mitochondrial AsA–GSH cycle was first defined in pea leaves along with the peroxisomal one (Jimenez et al. 1997) and its function in leaf senescence was proposed (Jiménez et al. 1998). Collected data suggest that mitochondrial APX is membrane-bound, while MDHAR, DHAR, and GR are proposed to be matrix proteins (Chew et al. 2003). Arabidopsis studies showed that at least some of the AsA–

GSH cycle enzymes are dual-targeted to chloroplasts and mitochondria (Fig. 8.2) (Chew et al. 2003), predetermined by multiple transcription starts, as shown for MDHAR (Obara et al. 2002). This might not be a globally valid concept as Mittova et al. (2004) suggested, that chloroplast and mitochondrial APXs may be encoded by different genes and Teixeira (Teixeira et al. 2006) suggested an exclusively mitochondrion-located APX in rice. Morgante et al. (2009) summarized the data on dual-targeted proteins in Arabidopsis and rice and showed that at least some of the APX, MDHAR, and GR have been found both in chloroplasts and mitochondria. Ascorbate peroxidases might display both exclusively mitochondrial and exclusively chloroplasts isoenzymes. However, substantial prediction differences, depending on the proteomics tool (Predotar, Target P or iPSORT) and lack of experimental confirmation hamper the proper analysis.

Plant mitochondria serve a central role during water deficit. Their function includes, but is not restricted to ATP synthesis, glycine oxidation with concomitant matrix NADH reoxidation without energy conservation, regulation of proline concentrations and control of redox environment (Atkin and Macherel 2009). Bartoli et al. (2004) reported that leaf mitochondria are the main target of oxidative damage in stressed wheat. It should be noted, that plant mitochondria possess alternative oxidase (AOX; EC:1.10.3.11), an efficient mechanism to prevent overaccumulation of ubiquinol thus uncoupling electron flow toward complex III and reducing ROS production (Blokhina and Fagerstedt 2010). Therefore upregulation of mitochondrial AOX is crucial for survival in drought conditions (Bartoli et al. 2005). Moreover AsA–GSH cycle independent regeneration of AsA from DHA also exist (Szarka et al. 2007). Plant mitochondria are able to import DHA through glucose transporter and to reduce it by succinate dehydrogenase (complex II), thus effectively maintaining the AsA pool. The relative share of both mechanisms for DHA reduction was further studied (Szarka 2013), showing that under certain conditions either succinate dehydrogenase or the GSH-dependent process may prevail.

8.2.3 *Peroxisomes*

Plant peroxisomes are implicated in a variety of metabolic pathways, including photorespiration, fatty acid oxidation and the glyoxylate cycle. Furthermore, plant peroxisomes are also well-known center of H_2O_2 production and degradation. The contemporary opinion defines peroxisomes also as an important source of O_2^- and nitric oxide (NO·) production, and all three of them are regarded as signalling molecules, able to permeate through the peroxisomal membrane and transfer the signal into the cytosol (Corpas et al. 2001). Peroxisomes overproduce hydrogen peroxide in conditions of drought stress, mainly by metabolizing the chloroplasts-derived glycolate (Noctor et al. 2014). Next, H_2O_2 could be dismutated by catalase or APX of the AsA–GSH cycle, or to serve as intracellular signalling molecule, when transported to the cytosol (Corpas et al. 2001; Neill et al. 2002).

Evidences for the presence of AsA–GSH cycle enzymes in plant peroxisomes were published for pea leaves (Jimenez et al. 1997) and their role in leaf senescence was proposed (Jiménez et al. 1998). As summarized by Corpas (Corpas et al. 2001) GR and DHAR are presented in the peroxisomal matrix, where they utilize the intra-organelle NADPH, GSH, and AsA pools to dismutate H_2O_2, while peroxisomal APX and MDHAR are membrane-bound on the cytosolic site. During leaf senescence APX and MDHAR activities decreased significantly, while little or no apparent change was observed in GR and DHAR activities (Jiménez et al. 1998). While the AsA/DHA ratio was not affected, a significant shift toward GSSG was also observed. A sixfold increase in H_2O_2 concentration in peroxisomes was accompanied with little decrease of the total water-soluble concentration. On the other hand, stress factors, such as high salt concentrations did not lead to an actual increase in peroxisomal H_2O_2 concentration (Mittova et al. 2003), supposedly due to diffusion through porines.

A pea leaf peroxisomal GR was first purified and characterized only recently (Romero-Puertas et al. 2006). Its specific activity was found to be up to threefold higher than the specific activity, reported for other GR isozymes. In response to Cd stress, the activity of peroxisomal isozyme was increased almost two times with no respective increase of the protein band or increase in the overall GR activity. In opposite, peroxisomal GR activity in both salt-sensitive tomato (*Lycopersicon esculentum*) and salt-tolerant *L. pennellii* diminished in response to salt stress (Mittova et al. 2003). Later a proteomics study established, that GR1 in Arabidopsis is dually targeted to the cytosol and peroxisomes (Kataya and Reumann 2010), and its organelle import may depend on the current needs and H_2O_2 production.

Peroxisomal MDHAR isozymes are well characterized in a number of plant species. A good summary of the available data and a phylogenetic tree was presented by Leterrier et al. (2005) and Lisenbee et al. (2005). It seems that the different MDHAR isozymes are presented in both peroxisomal membrane and matrix. Peroxisomal MDHARs are also responsive to salt stress (Mittova et al. 2003) with an increase and overall higher activity in the salt-tolerant wild tomato species. To summarize, it is evident, that some of the AsA–GSH peroxisomal enzymes are differentially affected by different abiotic stress factors, independently of the whole cell extract activity. The responsiveness of the peroxisomal antioxidative system may rely on MDHA (through MDHAR) or GSSG (through GR) recycling independently rather than on electron flow through the AsA–GSH pathway.

Successful cloning and expression of putative membrane-bound peroxisomal APX was achieved relatively long time ago in Arabidopsis and its stress activation was shown (Zhang et al. 1997). Certain subfractions of the ER were identified as a sorting compartment for constitutively expressed peroxisomal APX in Arabidopsis (Lisenbee et al. 2003) The cytosolic orientation of the active site of peroxisomal APXs was proven even earlier (Yamaguchi et al. 1995). Currently, two APX (APX3 and APX5) were identified as peroxisomal in Arabidopsis. As summarized and further concluded by Narendra et al. (2006), the presence of APX in the peroxisomal membrane increase the stress tolerance to various factors, but did not affect substantially the performance of plants in normal conditions.

8.2.4 Cytosol

Although the enzymes of the AsA–GSH cycle are mostly compartmentalized between chloroplasts, mitochondria and peroxisomes, where the main biochemical sources of ROS occur, they could be also detected in the cytosol as a general ROS scavenging mechanism. Hydrogen peroxide, generated in other compartments diffuses in the cytosol, but it is also produced in stress conditions by cell-wall peroxidases and from O_2^-, generated by a plasma membrane and cell-wall-bound NADPH oxidases and enters the cell by facilitated diffusion through peroxiporins, serving an important signalling function (Neill et al. 2002). The cytosolic AsA and GSH pools are directly involved in H_2O_2 dismutation through the AsA–GSH cycle.

Early evidence of a cytosolic APX, involved in drought stress response was published for pea (Mittler and Zilinskas 1994). Two cytosolic APXs were identified in Arabidopsis (Fig. 8.2), a constitutive (APX1) and stress-inducible one (APX2), and their role in the cross-protection of chloroplasts was shown (Davletova et al. 2005). Rizhsky et al. (2002) showed differential response to heat stress (enhanced expression of cytosolic APX and CAT) and drought stress (leading role for glutathione peroxidase) in tobacco. Rice cytosolic APX knockout mutants showed increased sensitivity to drought, as well as salt and cold stresses (Zhang et al. 2013). Transgenic tobacco plants, overexpressing cytosolic APX alone (Shrivastava et al. 2015) or cytosolic APX and Cu/Zn-SOD (Faize et al. 2011) performed better under water deficit conditions and also showed improved photosynthetic rate, confirming the role of cytosolic ROS scavenging in chloroplast protection.

Cytosolic DHAR and MDHAR were also purified and characterized in various plant species (summarized in Eltayeb et al. 2007) and it was shown that their overexpression in transgenic plants also confers drought tolerance (Eltayeb et al. 2006, 2007, 2011). Tomato peroxisomal matrix MDHAR (SlMDHAR3), highly homologous to *Arabidopsis thaliana* AtMDHAR1 was shown to manifest peroxisomal and cytosolic localization (Gest et al. 2013). Increase in cytosolic GR transcript levels in pea (Stevens et al. 1997) and transcript levels and activity in cowpea (*Vigna unguiculata*) (Contour-Ansel et al. 2006) was detected under drought stress with delayed response of drought-tolerant compared to drought-sensitive cultivars of the latter (Contour-Ansel et al. 2006). The expression levels of cytosolic GR were shown to be highly increased under drought conditions in a genome-wide study, compared to no change in chloroplasts GR (Tahmasebi et al. 2012), suggesting that GSH regeneration in the cytosol is of higher importance in a stressful environment.

The AsA/(M)DHA and GSH/GSSG pools are of crucial importance to the ROS scavenging machinery in most if not all cell compartments. Enhanced ROS production during drought stress determines higher demands for effective detoxifying capacity, directly involving either or both of these antioxidants, thus altering the redox state toward more unfavorable figures. Increased expression and/or activity of the (M)DHA (DHAR and MDHAR) and GSSG-reducing enzyme (GR) are inevitable characteristics of the drought stress response. The relative contribution may be different in the different compartments and, as shown, many of the AsA–GSH cycle enzymes are dually targeted, thus allowing higher plasticity of the stress response. The better understanding of the drought tolerance phenomena requires a compartment-wide enzyme characterization rather than the overall activity measurements that are rarely provided.

8.3 The Molecular Basis of Antioxidant Defense Activation

8.3.1 Hormonal Triggering

The hormonal regulation of drought stress response is primarily provided by abscisic acid (ABA), the dominant stress-related plant hormone. ABA is a terpenoid phytohormone, regulator of many developmental and functional processes as stomatal aperture, hydraulic conductivity, seed dormancy, etc.), which leads to changes in gene expression and in adaptive physiological responses. Drought and salt stresses for example leads to ABA-induced gene expression, followed by stomatal closure, lower water loss and restriction of cellular growth, occurring promptly upon the sensing of stress. Osmotic stress induces the gene transcription of the ABA-biosynthetic pathway enzymes, most probably by Ca^{2+}-induced phosphorylation cascade, simultaneously inhibits ABA degradation and thus mediate responses, related to regulation of the water balance inside the cells and ensuring the long-distance signalling from the primary site of osmotic sensing (roots) to the main site of water evaporation control (leaves) (Zhang et al. 2006). ABA-induced genes encoded dehydrins associated proteins, ROS-detoxifying enzymes, transporters, regulatory proteins such as transcription factors, etc. and represses mainly proteins associated with growth, ribosomal function, plasma membrane, and chloroplast proteins (Wang et al. 2013b). Simultaneously a drop in cytokinin levels, hormones with opposite to ABA functions was reported under drought conditions (Carvalho et al. 2015). Expression profiles and transcriptome analyses indicated a high percentage of genes, significantly regulated by ABA and involved primarily in the metabolic response to drought, salinity, and to a lesser extent to cold (Wang et al. 2013b).

A recent review (Kao 2014) of the role of H_2O_2 summarized that ABA increases the gene expression and activity of the AsA–GSH cycle enzymes (APX and GR1) in leaves but also increase H_2O_2 levels in maize embryos, seedlings, and leaves. H_2O_2 production induced by ABA was first observed in guard cells and subsequently reported in maize seedlings exposed to water stress. In rice it was shown that ABA treatment increases H_2O_2 in roots and leaves with simultaneous increase in APX and GR activities in rice roots which suggest that ABA-induced H_2O_2 accumulation could be a signal for increasing these enzyme activities (Kao 2014).

Drought decreases cytokinin levels leading to increased shoot responses to ABA and to stomatal closure. Cytokinin and ABA stress-induced changes promote early leaf senescence and abscission, thus decreasing water loss. SARK-IPT (isopentenyltransferase under the senescence-associated-receptor protein-kinase promoter) transgenic tobacco plants in conditions of drought stress showed better water loss control efficiency and enhanced drought tolerance together with expression of ROS metabolism genes, especially AsA–GSH cycle genes (Rivero et al. 2007). Among the transcriptional control and stomatal closure, ABA affects the production of ROS, acting as second messengers and the stable H_2O_2 subsequently induces the accumulation of various hormones as jasmonic acid (JA), salicylic acid (SA), and ethylene (ET). The complex network of interaction between ROS and enzymatic oxidant scavengers provides the response to abiotic stresses (Karuppanapandian et al. 2011).

Salicylic acid is a phenolic plant growth regulator, with several physiological and biochemical functions and in particular in response to abiotic stresses affecting the antioxidative systems and components of the AsA–GSH cycle. Accumulation of endogenous SA under drought stress was reported for Arabidopsis, conferring higher tolerance (Okuma et al. 2014). Treatment with exogenous SA is generally shown to increase the antioxidative response. In drought stressed mustard seedlings it was shown that the activities of MDHAR, DHAR, GR were enhanced with a concomitant decrease in H_2O_2. This suggests that SA helps the plants to become more tolerant to oxidative damage and supports their enzymatic antioxidant defense in drought stress response (Alam et al. 2013). The transcript levels of eight AsA–GSH cycle-related genes were investigated in wheat seedlings under drought stress conditions as well. It was found that the applied exogenous SA leads to increased transcription of the *GR* and *MDHAR* genes, while the transcription of *DHAR* showed time-dependent manner. This allows the presumption that SA regulates the expression of mentioned above genes differentially and could cause GSH and AsA redox state changes under drought stress. The effect of SA showed enhancement of drought tolerance related to its influence on the transcription of the genes encoding the enzymes from AsA–GSH cycle (Kang et al. 2013). Salicylic acid induced drought tolerance, mediated by increased antioxidant capacity was also reported for maize (Saruhan et al. 2012), bean (*Phaseolus vulgaris*) and tomato (Senaratna et al. 2000). This concept was, however argued by Németh et al. (2002), suggesting that exogenous SA increase the polyamines levels and confers cold resistance, but causes drought sensitivity in maize and wheat. In support of this it was shown that excessive accumulation of SA may trigger a PCD (Programmed Cell Death) pathway and SA under accumulating transgenic Arabidopsis plants showed lower decrease of the GSH/GSSG ratio, compared to wild types in NaCl-induced osmotic stress (Borsani et al. 2001).

Jasmonates (JA and derivatives) have a significant role in abiotic stress response (drought, salt, etc.) through activation of the defensive protein encoding genes and regulation of plant growth and development. They may have antagonistic to SA signalling regulation (SA can suppress the JA-dependent response) or they can share signalling pathways (Wasternack 2007).

The levels of JA and methyl jasmonate (MeJA) increase during stress. There are evidences for expression changes of jasmonate-responsive genes (JRGs) under drought stress and in particular their induction in this case. The transcript levels and activities of APX, GR, DHAR, and MDHAR increased under water stress and that presents the assumption that the accumulation of JA is involved in the regulation of

AsA and GSH metabolism under stress conditions (Shan and Liang 2010). JA has similar to ABA features and functions, such as stomatal closure, inhibited plant growth, plant senescence, and stress response involvement. They can act dependently or independently with each other. It is possible that increased JA levels could subsequently enhance ABA levels and afterwards to result in plant drought tolerance. This was reported in apple (*Malus domestica*) and barley (*Hordeum vulgare*) (Shan and Liang 2010). A study on JA and ABA-deficient Arabidopsis plants (Brossa et al. 2011) showed that the interconnection between the two hormonal control pathways is essential in drought stress tolerance, but JA rather than ABA is primarily involved in the control of the AsA and GSH metabolism. Meanwhile the GSH redox status, controlled by GR could exert control on the H_2O_2-mediated SA and JA-pathway-dependent gene expression (Mhamdi et al. 2010).

Salicylic acid and JA-responsive *cis*-elements, but also ET-responsive elements were identified in several ROS-responsive genes as *Apx1*, in addition to other control elements which suggest a broader role for these hormone-mediated responses (Miller et al. 2010). The subfamily of ethylene response factors (ERFs) are *cis*-elements-binding proteins that were shown to enhance plant tolerance to multiple stresses and playing a role in abiotic stress responses. An osmotic- and oxidative-stress ERF-expressing transgenic tobacco showed better drought and salinity adaptation, increased expression of ROS-detoxifying enzymes (primary SOD), and decreased ROS accumulation (Wu et al. 2008). Another ET-responsive factor, the Arabidopsis AtERF98 was also shown to trigger the transcriptional activation of AsA synthesis (Zhang et al. 2012).

Brassinosteroids (BRs) are plant hormones that take part in various physiological and biochemical processes like development, stem elongation, vascular differentiation, root inhibition, plant immunity, induction of ethylene biosynthesis, regulation of gene expression, nucleic acid and protein synthesis, photosynthesis, etc. (Yang et al. 2011). They also have an important role in plant protection and response to different abiotic stresses, including drought, through induction of the AsA–GSH cycle (Jin et al. 2015). It was shown that BRs increase the ABA concentration in tomato plants under drought stress, but it is not clear if the drought tolerance was caused by this elevation or other factors. The activity of APX was also increased and it was speculated that BRs induces the biosynthesis of endogenous ABA thus causing upregulation of antioxidant system (Yuan et al. 2010). The induced by BRs drought stress resistance and their effect on antioxidant system was also investigated in blue mustard (*Chorispora bungeana*) (Li et al. 2012c). The enzyme (APX and GR) activities increased significantly in plants under drought stress after application of BRs compared with drought stress alone. This clearly shows that BRs enhance the antioxidative system and improve the plants resistance to drought stress. All the above mentioned could mean that BRs possess regulation function or the increased enzyme activities could be a result of *de novo* synthesis and/or activation of the enzymes, mediated through transcription and/or translation of specific genes (Li et al. 2012c).

Hormones and especially ABA play significant role in plant response to abiotic stress, including drought. They indirectly influence the production of ROS, in chloroplasts, but also in peroxisomes and mitochondria and simultaneously control the expression of the AsA–GSH cycle genes and other defensive mechanisms. The action of different hormones is often interconnected and ABA, along with SA and JA are primary involved in plant response to water deficit. Higher endogenous levels or exogenous addition are associated with higher activity and expression levels of the AsA–GSH cycle enzymes.

8.3.2 Gene Expression Control

Drought stress-induced genes protect cells from water losses, protect intracellular macromolecules from dehydration-induced damage and have also a role in the regulation of genes for signal transduction during plant response. These gene products are divided into two groups. The first group consists of proteins with function in stress tolerance (porins), enzymes osmoprotectants biosynthetic pathways, proteins protecting macromolecules and membranes (LEA protein, etc.), proteases for protein turnover and detoxification enzymes (including AsA–GSH cycle enzymes). The second group contains protein factors for regulation of signal transduction and gene expression that probably function in stress response (protein kinases, transcription factors, PLC, and 14-3-3 proteins) (Hadiarto and Tran 2011).

Multiple genes are induced or repressed by abiotic stresses. The resulting products play role in stress response and in establishing plant stress tolerance and these genes may find application in genetic engineering of transgenic plants with induced stress tolerance. Different microarray technologies were used for analyzing gene expression profiles of plants under abiotic stresses and to identify stress-induced genes (Nakajima et al. 2002; Seki et al. 2001). A striking similarity of the functional and regulatory proteins was reported for Arabidopsis and rice. They were related to similar stress responses at the molecular level as well (Shinozaki and Yamaguchi-Shinozaki 2007).

A study on ABA and PEG 6000—induced expression of the AsA–GSH cycle genes in maize showed differential response (Liu et al. 2012). A total of 17 genes were studied (8 *Apx*, 4 *Mdhar*, 3 *Dhar* and 2 *Gr*). Two cytosolic APXs (*Apx1.1* and *Apx1.2*), one cytosolic MDHAR (*Mdhar1*), one cytosolic DHAR (*DharR2*), and one cytosolic GR (*Gr2*) showed increased relative transcript levels in both ABA and PEG—treated plants in the first three hours with diminishing abundance at 6 and 12 h. Additionally *Mdhar2* (cytosolic), *Dhar1* (chloroplasts), *Dhar3* (mitochondrial) and *Gr1* (cytosolic) showed increased transcript levels under PEG-treatment, suggesting that most of the drought-responsive AsA–GSH cycle genes are under ABA-dependent induction control, but ABA-independent pathway is also present.

Genes from the AsA–GSH cycle showed differential transcript abundance in drought tolerant, compared to drought-sensitive wheat cultivars (Sečenji et al. 2010). All six APXs showed increased transcription in the tolerant, compared to the sensitive one, with stromal (sAPX I and sAPX II) and thylakoid (tAPX) expressing greater differences than cytosolic and mitochondrial APXs). Similarly cytosolic (cGR) and chloroplasts (chlGR) GRs were drought-inducible in the tolerant cultivar, cytosolic (cDHAR), and chloroplastic (chlDHAR) DHARs showed constitutively higher and drought-inducible expression, while no MDHAR expression (except for the cytosolic MDHAR II in the first stages of the stress response) proved indicative for drought tolerance.

In one of the ABA-dependent pathways dehydration inducible genes do not require protein biosynthesis for their expression and contain potential ABREs (*cis*-acting DNA elements). cDNAs for ABRE and G-box-binding proteins represents a large gene family and have a region adjacent to a Leu-zipper motif (bZIP). Their core motif determines binding specificity of bZIP proteins, but it is still unknown how ABA activates them for initiation of gene transcription. There are other *cis*-acting elements in ABA-responsive gene expression. For example, ABA- and VP1-dependent expression of the maize *C1* gene (encodes MYB-related transcription factor) is regulated from the Sph box and GTGTC motifs regulate which subsequently controls the anthocyanin biosynthesis during seed development. *VP1* encodes a transcriptional activator cooperating with bZIP proteins. There is a similar to the VP1 protein in Arabidopsis (AB13).

There are two *cis*-acting ABRE motifs for the control of ABA-responsive expression of the Arabidopsis *rd29B* gene and two bZIP transcription factors (AREB/ABF) binding to them. The AREB/ABF proteins require an ABA-mediated signal for their activation through their ABA-dependent phosphorylation. Overexpression of ABF3 or AREB2/ABF4 in transgenic Arabidopsis showed reduced transcription and increased drought tolerance. On the other hand expression of the mutated and phosphorylated form of AREB1 leads to induction of ABA-responsive genes without exogenous ABA application which leads to the conclusion that mutations of transcription factors could enhance drought tolerance (Shinozaki and Yamaguchi-Shinozaki 2007).

In another ABA-dependent pathway biosynthesis of protein factors is required for the expression of drought-inducible genes (e.g. *rd22* gene in Arabidopsis with 67-bp region of the promoter needed for the expression and containing DNA-binding protein motifs such as MYC (*rd22BP1* is MYC homolog) and MYB (the *Atmyb2* gene that encodes a MYB-related protein induced by dehydration stress), but no ABREs). Concerning the ATMYB2 protein, in the above example, it was reported that it binds to MYBRS in the *rd22* 67-bp promoter region and may act in cooperation with RD22BP1 protein as a transcription factor for gene expression. It was revealed that target genes of MYC/MYB in overexpressing transgenic plants to be alcohol dehydrogenase and ABA- or JA-inducible genes. It was shown that bZIP transcription factors from rice, maize, and Arabidopsis bind to G-box-like sequences and are also involved in the ABA-dependent pathway. In such a pathway, many stresses and ABA-inducible genes encode transcription factors which play a role in the regulation of ABA-inducible genes for drought stress response (Shinozaki and Yamaguchi-Shinozaki 1997, 2007).

There are genes induced by drought (for example *rd29A* (*lti78* and *cor78*), *kinl*, *cor6.6* (*kin2*), and *cor47* (*rdl7*) in Arabidopsis) for which expression is suggested that no ABA is required but they respond to exogenous ABA. For them both ABA-independent and ABA-dependent regulation systems function during drought stress gene expression. The most significant for regulation of the induction of *rd29A* under drought conditions is a 9-bp conserved DRE (drought-responsive elements) (including C-repeat with CCGAC core motif) that is linked to ABA-independent pathway and there is an ABRE in the *rd29A* promoter for ABA-responsive expression. For the interaction with DRE motifs there are several protein factors (DRE/C-repeat-binding proteins with conserved DNA-binding motif). Such DNA-binding motifs were also reported in EREBP and AP2 proteins involved in ethylene-responsive gene expression and floral morphogenesis, respectively. These elements are for example *CBF/DREB1* and *DREB2* and the overexpression of *CBF/DREB1* in transgenic plants was shown to increase stress tolerance to freezing, drought, and salt stresses. The *DREB2* genes are drought-induced and may activate other genes involved in drought stress tolerance, but their overexpression do not contribute to enhanced stress tolerance, which indicates that there are proteins that activate DREB2 posttranslationally. The overexpression of mentioned above genes shown similar effects both in rice and Arabidopsis, which suggests similar transcription factors function in abiotic stress tolerance between dicotyledonous and monocotyledonous plants (Shinozaki and Yamaguchi-Shinozaki 2007).

There is also suggestion for another ABA-independent pathway for the drought stress response expression of genes such *rd19* and *rd21* (for thiol proteases) and *erdl* (for Clp protease regulatory subunit). The reported *cis*-acting elements in *erdl* promoter are involved both in stress-induced and senescence-activated gene expression. There are DNA-binding proteins which interact with these *cis*-elements called NAC transcription factors. The drought-inducible gene that encodes NAC factor is *rd26* and its expression could be induced not only by stress but also by ABA and JA. It was reported that typical ABA-inducible genes such as *lea*, *rd*, *erd*, *cor*, and *kin* are not target genes of RD26, whereas many JA-inducible genes are target genes of RD26 which suggests its significance as a mediator between ABA and JA signalling during stress responses (Shinozaki and Yamaguchi-Shinozaki 1997, 2007)

It is proven that different stress signals are transmitted separately in plants to activate DRE-dependent transcription of the *rd29A/cor78* gene and because of these *trans*-acting elements that regulate DRE-dependent gene expression are important. For the function of the DREB1A and DREB2A proteins as *trans*-acting factors were studied through transient expression in Arabidopsis leaf protoplasts and overexpression in transgenic Arabidopsis plants and their different functions were reported as DREB1A-related proteins function in the DRE/C-repeat-dependent expression of *rd29A* during cold stress while the DREB2A-related proteins are linked in its expression during drought and salt stress. All that data evidenced that these two protein families function as *trans*-acting factors in two separate signal transduction pathways (Liu et al. 1998).

A number of studies indicated that genes of the AsA–GSH cycle are regulated in both ABA and ABA-independent pathways. ABRE-like sequence was reported in the 5′-flanking region of field mustard (*Brassica campestris*) *Grl* gene (Lee et al. 2002). Previously, two ABRE motifs were identified in the 5′-flanking region of the

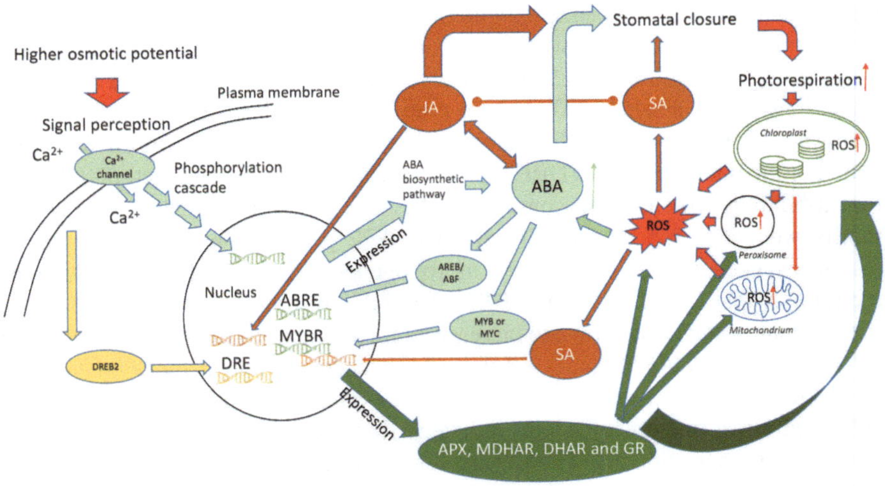

Fig. 8.3 Schematic representation of the hormonal and transcriptional control of the AsA/GSH cycle enzymes. The ABA (abscisic acid) -responsive pathway is shown in pale green. Higher osmotic potential (either water deficit or high salinity) is sensed, most possibly by osmotic-sensitive plasma membrane Ca^{2+} channels. Subsequent phosphorylation cascade leads to transcriptional activation of ABA-biosynthetic pathway enzymes and increased ABA concentrations. This ensures the long-distance signalling and stomatal closure in order to minimize transpiration and water losses. The simultaneous increase in photorespiration (*red*) leads to overproduction of ROS in chloroplasts, peroxisomes, and mitochondria (see subtitle 1), which in turns further increase ABA accumulation. Two ABA-dependent (*pale green*)—AREB/ABRE and MYB/MYBR and an ABA-independent (*yellow*)—DREB2/DRE transcription activation systems (see subtitle 3.2) lead to the expression of the AsA/GSH cycle enzymes (dark green)—APXs, MDHARs, DHARs, and GRs among other stress-responsive genes and they function in ROS scavenging and AsA/DHA and GSH/GSSG redox state maintenance in the cytosol or are targeted to other compartments (see subtitles 2.1–2.4). Two other plant hormones—JA (jasmonic acid) and SA salicylic acid) are also induced by ROS and/or ABA and are further responsible for stomatal closure and AsA/GSH cycle gene expression (*brown*)

rice *Gr2* gene (Kaminaka et al. 1998). Rice *Gr3*, co-localized to chloroplasts and mitochondria is under ABA-dependent (ABRE element) and ABA-independent (C-repeat/DRE-like element) transcriptional control (Wu et al. 2013). A grey mangrove (*Avicennia marina*) peroxisomal APX is also under ABRE-controlled expression, among Myc cold/freeze/dehydration *cis*-acting element and ERE element (Kavitha et al. 2008). A comprehensive study on *cis*-acting regulatory elements of the rice AsA–GSH cycle genes was recently published (Pandey et al. 2015). The frequency of control elements, involved in drought stress response showed that APXs are mainly under ABA-control, ABRE, but predominantly MYB-binding site (MBS), GRs showed a similar gene promoter distribution and no ABA-independent drought regulation pathway at all, MDHARs are predominantly ABRE-controlled and DHARs showed the strongest ABA-independent C-repeat/DRE-controlled transcription (Fig. 8.3).

8.4 Ascorbate–Glutathione Cycle and Drought Tolerance

8.4.1 Genetic Diversity of AsA–GSH Cycle Enzymes

The AsA–GSH cycle enzymes in plants are all encoded by multigene families. APXs belong to class-I of the superfamily of bacterial, fungal, and plant peroxidases and are proved to be with prokaryotic origin. They are found mainly in higher plants, but also in algae and some cyanobacteria, insects, and ascorbate-rich animal tissues. APX isoforms differ in molecular weight, optimal pH, stability, substrate specificity, localization, and the level of response to specific stress conditions. For APXs there are two evolutionary models proposed. The first one presumed that genetic diversity is due to interlocus recombination or gene conversion with subsequent natural selection and the other is based on the assumption that genes are created by repeated duplication of existing ones, some of which function for a long time, while others are removed or cease to function. It is also suggested that APXs have a common ancestral origin, which is supported by the presence of conserved regions in the ion-binding area of all their isoforms. It is reported that the APXs evolution started with diversification of the cytosolic and chloroplast forms, and it is shown that these in the same compartments from different plants are more closely related as well as these are in the same species from different evolutionary branches. It is also determined that tAPX and sAPX share high similarity except of the region responsible for membrane-spanning in the thylakoidal isoform, which consequently presumes that they are encoded by a single gene with alternative splicing. Still in some species it is possible that the diversification of tAPX and sAPX happened earlier in the evolution. Cytosolic and peroxisomal APXs respectively, are proved to be highly homologous within many higher plant species. This data leads to the suggestion that the enzymes from the same compartment have common genetic origin (Dąbrowska et al. 2007).

The phylogenetic analyses indicated also that different APX isoforms originated molecular evolutionary complex process of gene duplications and their structural organization also reflects this process. It is proposed that cytosolic and peroxisomal isoforms diverged early from chloroplasts ones. Two features are found to identify higher plant chloroplastic isoforms—(K-[ND]-I-[ETK]-E-W-P) motif near the active site and (E-T-K-Y-T-[KE]-[DNTE]-G-PG-[ANEK]-[PA]-G-G-Q-S) motif near the heme-binding site, while generally additional single amino-acid changes represent the differences among the isoforms. Still these differences clearly distinguish chloroplastic from nonchloroplastic isoforms (cytosolic and peroxisome membrane-bound) which represent the dichotomous divergence of APXs phylogeny. This is used for evidencing that the nonchloroplastic isoforms were generated by duplication events of a single nonchloroplastic ancestral gene. There is also a new APX group of isoforms close to the peroxisomal isoforms, suggesting a common origin to these branches but found only in a few plants. The comparison of the structural organization of APX genes additionally supports the theory of a common origin of each isoform arising from a common ancestor for all APX genes—isoforms in the same cellular compartment have a similar gene structure. The evolutionary pathway was proposed to begin with a duplication event that generated the ancestral genes encoding the chloroplastic and the nonchloroplastic isoforms, from a common ancestral APX gene most likely before the divergence between Viridiplantae and Euglenazoa and subsequent particular evolutionary path for each group. Another duplication event was proposed for diversification of stromal and thylakoid-bound isoforms of the chloroplastic APXs. The possibility of alternative splicing of the genes for the chloroplastic proteins recently on the evolution of a certain lineage of eudicots was presumed and supported by the absence of alternative splicing of the corresponding genes in rice, *Arabidopsis*, and *Chlamydomonas*. As it concerns the diversification within the nonchloroplastic APX branch, it was suggested to occur after the divergence of Euglenazoa and Viridiplantae possibly from cytosolic form as the nonchloroplastic isoform in Euglena is cytosolic. Most probably the peroxisomal and new APX isoforms arise from duplication of the ancestral APX gene, creating one gene for cytosolic and a second one for peroxisomal forms. Another duplication event separated peroxisomal from new APX isoforms (Teixeira et al. 2004).

Glutathione reductases are encoded by a small gene family and different numbers of genes encoding GRs were described from different plant genomes (e.g. two for *Arabidopsis thaliana*, *Nicotiana tabacum* and *Pisum sativum*, and three for *Oryza sativa* and *Populus trichocarpa*). GRs are also compartmentalized differently in cells—plastid, mitochondria, cytosol, glyoxysome, or peroxisome. The phylogenetic analysis of GR isoforms revealed the same pattern as for the APXs—there was a clear divergence between cytosolic and chloroplastic isoforms both divided into monocot and dicot subgroups. Two smaller groups were found to present in the monocot subgroup with no isoform found in the dicot genomes (Wu et al. 2013). The comparison of GR genes in rice and Arabidopsis, as the representatives of mono- and dicotyledonous plants, was also performed and the phylogenetic relationship between GR gene families showed the presence of homologous and orthologous group and revealed evolutionary conserved pyridine nucleotide-disulphide oxidoreductases class-I active site among the GR protein (Trivedi et al. 2013).

To study the phylogenetic relationships, genetic diversity, and evolutionary history of MDHARs, an investigation of their isoforms in the moss *Physcomitrella* was performed. The MDHAR enzyme activity is found both within the plants and animals. They also have cytosolic isoforms and isoforms positioned in different cellular compartments. The genetic diversity of available plant MDHARs showed the identity and conservation of the FAD/NAD-binding sites and clear division to three subfamilies of monophyletic origin—cytosolic, peroxisomal and chloroplast/mitochondrial based mainly on compartmental targeting motifs. It was confirmed that the genes for the MDHAR isoforms in *Physcomitrella* share a common ancestor with cytosolic MDHAR genes from higher plants (they are similar to cytosolic isoforms except for organelle-targeting motifs). The phylogenetic analysis showed that cytosolic MDHAR genes' ancestor evolved before the divergence of bryophytes. Even within the cytosolic clade of the higher plants, the bryophytes MDHARs form a separated group which confirms that these are not orthologs with a conserved function but the gene duplications have occurred after diversification and speciation. In addition the higher plant chloroplast/mitochondrial, membrane-bound peroxisomal and cytosolic MDHARs are divided into separate phylogenetic branches which suggest that two duplication and diversification events lead to the separation of these subfamilies. This data in addition to the amino-acid and gene structural organization indicated that the chloroplast/mitochondrial MDHARs diverged early in evolution and that the membrane-bound peroxisomal MDHARs evolved from a more recent duplication event. It was also speculated that MDHARs and APXs families followed a similar evolutionary path (Lunde et al. 2006).

DHAR enzymes were studied in *Eucalyptus* and three clusters encoding DHAR isoforms were identified. In the same manner as the other AsA–GSH cycle enzymes DHAR branching showed phylogenetic division with regards to the cell compartment—two of the clusters consisted from cytosolic isoforms and one—chloroplastic. All three of them were grouped with particular previously described organelle isoforms and also the compartment-targeting motifs were highly similar to previously described ones (Teixeira et al. 2005).

To summarize, a clear, compartment-dependent divergence of the genes, encoding AsA–GSH cycle enzymes was established in most of the studied plants and the subcompartmental division occurred early in the evolution. Cytosolic and peroxisomal isoforms are clearly different from the mitochondrial/chloroplast clade. At least some of the isoenzymes are products of alternative splicing, originating from a single gene. Dual-targeting is also common, especially for the mitochondria/chloroplast gene products. The pattern of phylogenetic relationship is similar for all four of the AsA–GSH cycle enzymes.

8.4.2 *AsA–GSH Cycle in Drought-Tolerant vs. Drought-Sensitive Cultivars*

The application of biochemical markers for fast screening of multiple cultivars for stress tolerance is an attractive, though not decisive approach in contemporary agriculture. The integration of genomics approaches, used in marker-assisted selection with the metabolic and physiological data analyses ensures a better understanding of the complex response of plants to drought and more successful improvement of drought tolerance in crops (Mir et al. 2012). Oxidative stress and the effectiveness of the antioxidative systems, including the AsA–GSH cycle, are often used as stress tolerance markers (Raheleh et al. 2012; Abbas et al. 2014; Devi et al. 2012; Zhang et al. 2014).

The overall value of antioxidants as stress tolerance markers is however questioned. As drought and other abiotic stresses cause enhanced ROS production with detrimental effect on biological membranes, proteins, and DNA (Noctor et al. 2014), more effective antioxidant systems might provide higher drought or other stress resistance. Higher tolerance may be achieved due to better protection of photosynthetic machinery (Shamim et al. 2013), reduced levels of membrane lipid peroxidation (Abbas et al. 2014) or successful maintenance of the cellular redox state (Marquez-Garcia et al. 2015). This could be achieved either by an increase in the antioxidant defense (Saruhan et al. 2012) or due to constitutively higher antioxidant activity in the drought-tolerant plant (Türkan et al. 2005).

However, conclusions on drought tolerance, based solely on the antioxidative status should be drowned with great caution. The response is differential both in concentration and time manner. For example, in an experiment, screening sugarcane (*Saccharum officinarum*) cultivars for drought tolerance, the sensitive one showed increased activity of antioxidant enzymes, compared to the tolerant one at lower water deficit levels (Cia et al. 2012). The higher antioxidative capacity of the drought-tolerant cultivar is evident only at higher water deficit levels. Similarly, no apparent differences or higher antioxidant activity of sensitive cultivars was observed after short-term drought treatment, with clear differences only after long-term stress (Boaretto et al. 2014). Assuming that drought stress would eventually lead to enhanced ROS production, we could define a tolerant plant as this one that will activate its antioxidative machinery after longer exposition or higher levels of stress. Thus, increased activity of antioxidants at certain conditions may be associated with susceptibility rather than tolerance and misinterpretation of data.

The use of the AsA–GSH cycle enzymes as biochemical markers for drought tolerance involves not only changes in the enzymatic activity, but also isoenzyme profiles. Drought-inducible APXs were cloned from finger millet (*Eleusine coracana*) (Bhatt et al. 2013) and peanut (*Arachis hypogaea*) (Shrivastava et al. 2015). Both are associated with better drought tolerance. Another important point here is that the performance of the AsA–GSH cycle enzymes alone is rarely credited as drought tolerance marker. Such experiments typically involve GSH/GSSG and AsA/DHA concentrations and ratios (Pyngrope et al. 2013), activity of other antioxidant enzymes—catalase, guaiacol peroxidases, and superoxide dismutases (Cia et al. 2012), ROS (typically H_2O_2, more rarely O_2^-) and levels of lipid peroxidation (by MDA concentration) (Abbas et al. 2014), glutathione-S-transferases (Le Martret et al. 2011) and osmoprotectants (Abbas et al. 2014). A major advantage of this approach is the possibility to measure all of the above by common and inexpensive laboratory equipment—a UV-VIS spectrophotometer.

Examples of comparative analyses of the AsA–GSH cycle in drought-tolerant and drought-sensitive cultivars of crop plants are summarized in Table 8.2. Several key conclusions could be made here. The methodology for drought stress treatment varies significantly. Different concentrations of PEG-6000 (Pyngrope et al. 2013; Sekmen et al. 2014) or irrigation withholding (Huseynova 2012; Marok et al. 2013; Chugh et al. 2013) are the preferred methods, but they are not directly comparable. Unlike other abiotic stress treatments where the concentration of the stressor is the easily controllable variable, in drought stress treatment results may vary significantly depending on the experimental system. Therefore, a unified measure unit—osmotic potential or water availability should be used, or an internal physiological criterion—relative water content (RWC), having also in mind that keeping the water inside the plant is a drought tolerance characteristic. The stress prolongation also varies significantly among the cited studies—immediate (24–72 h) (Pyngrope et al.

Table 8.2 Selected recent (2012–2015) reports on experiments, comparing drought-tolerant and drought-sensitive cultivars of various crop plants

	AsA–GSH cycle		Reference
Crop plant	Sensitive	Tolerant	
Oryza sativa	GSH/GSSG ratio decreased	GSH/GSSG ratio stable	Pyngrope et al. (2013)
	AsA/DHA ratio slightly decreased	AsA/DHA ratio stable	
	APX not changed	APX increased	
	MDHAR increased	MDHAR increased	
	DHAR slight decrease	DHAR slight increase	
	GR slight decrease	GR slight increase	
Triticum aestivum	APX increase	APX increase/decrease depending on the developmental stage	Huseynova (2012)
	GR increase/decrease depending on the developmental stage	GR increase	
Triticum durum	APX increase/decrease depending on the developmental stage	APX increase	
	GR slight decrease	GR increase	
Gossypium hirsutum	APX twofold increase	APX slight increase	Sekmen et al. (2014)
	GR slight increase	GR slight increase	
Hordeum vulgare	GSH and GSSG increase	No significant change in GSH and GSSG	Marok et al. (2013)
	AsA and DHA increase	AsA and DHA decrease	
Zea mays	GSH decrease	GSH increase up to 21 days	Chugh et al. (2013)
	AsA decrease	AsA increase up to 21 days	
	GR decrease	GR increase up to 21 days	
	APX not changed	APX increase up to 21 days	
Saccharum officinarum	GR increase	GR decrease in mild, increase in severe stress	Boaretto et al. (2014)
Arachis hypogaea	APX decrease	APX increase up to 15 days	Padmavathi and Rao (2014)
	GR decrease	GR showed little change	
Triticum aestivum	AsA decrease in leaves and roots	AsA decrease in leaves and roots	Singh et al. (2012)
	APX decrease in leaves and roots	APX increase in leaves and roots	
	GR decrease in leaves and roots	GR increase in leaves and roots	

2013), short term (about 15–20 days) (Padmavathi and Rao 2014) to long term (40–150 days) (Singh et al. 2012; Chugh et al. 2013). The overall response is also highly dependent on the developmental stage (Huseynova 2012). So, could the AsA–GSH cycle be used as drought stress tolerance marker? This concept is already questioned as a useful approach for antioxidant enzyme activity for salt tolerance breeding (Fan et al. 2014) and no dramatic changes in the redox status during drought stress (unless sublethal levels) were reported (Noctor et al. 2014). This, however, does not mean that the AsA–GSH cycle does not participate in drought stress response and numerous examples improved drought tolerance by overexpression of the enzymes in transgenic plants exist (see subtitle 5).

8.4.3 Surviving in Harsh Environment: Lessons from Xerophytes

Xerophytes are a diverse group of flowering plants, displaying the remarkable feature of surviving in environmental conditions with little availability of liquid water. A number of physiological and anatomical adaptations allow them to (1) increase water uptake; (2) store large amount of water; (3) reduce water losses. The drought stress is often combined with high light intensity, extremely high (deserts) or extremely low (high mountains) temperature, which impose further metabolic challenge to cope with multiple stresses.

Xerophytes, like any other plant, would also undergo concomitant oxidative stress in conditions of low water availability. For example caper (*Capparis ovata*) seedlings, subjected to PEG-induced osmotic potential of -0.81 MPa displayed almost three times increase in MDA concentration after 14 days, accompanied with an 8 % decrease in RWC (Ozkur et al. 2009). Lipid peroxidation in *Gypsophila aucheri* after 14 days drought treatment was also increased, although at a lower rate—with 9.6 % (Esen et al. 2012). In a comparative study of four xerophytic and two sensitive *Caragana* species, lipid peroxidation decreased or did not show changes in tolerant with significant increase in sensitive species, while H_2O_2 concentrations increased in all but one after 48 days of drought treatment (Kang et al. 2012).

It is obvious in the above mentioned examples that xerophytic plants produce more ROS in drought conditions, therefore a more effective antioxidant strategy is needed to sustain photosynthetic capacity and membrane integrity. In all cases, antioxidant enzymes were studied and showed a remarkable increase during stress. In *Capparis ovata* both APX and GR showed a threefold increase in activity, accompanied by an increase in SOD, POX and CAT activities (Ozkur et al. 2009). The same set of five enzymes showed increased activity during salt (100 and 300 mM NaCl treatment) and drought stress in *Gypsophila aucheri* with APX showing the most pronounce and probably drought-specific change (Esen et al. 2012). Ascorbate peroxidase and GR activities were also significantly higher in xerophytic *Caragana* species compared to sensitive ones (Kang et al. 2012). A proteomics study also showed a sevenfold increase in the amount of DHAR in wild watermelon (*Citrullus lanatus*) in the condition of water deficit after only 3 days treatment (Yoshimura et al. 2008).

Clearly, enzymes of the AsA–GSH cycle were shown to participate in drought stress response of xerophytic plants. In addition, it should be noted that antioxidant enzymes are also a potent defense mechanism in extreme halophytes, where water unavailability is comparable to extreme drought. This was shown for APX (and to a lesser extent GR) in seepweed *Suaeda salsa* (Cai-Hong et al. 2005), APX, GR, MDHAR and DHAR in a tolerant, compared to sensitive European searocket (*Cakile maritima*) accession (Amor et al. 2006), APX, MDHAR and DHAR (with little decrease in GR activity) in *Salsola crassa* (Yildiztugay et al. 2014), subjected to gradually increasing NaCl concentrations (up to 1.5 M). These are just a few examples from studies on halophytes, responding to drought. Not surprising, but some, if not all, halophytes also perform well in water deficit conditions as shown for Umari keerai (*Salicornia brachiata*) (Parida and Jha 2013) and AsA–GSH promoted H_2O_2 scavenging is one of several mechanisms along with compatible solute accumulation, conferring tolerance. In all of the examples above it seems that the AsA-dependent ROS scavenging through APX is the predominant event, supported by continuous reduction of DHA.

8.4.4 Ascorbate–Glutathione Cycle during the Desiccation: Rehydration Cycle of Resurrection Plants

The extreme level of drought tolerance is observed in the peculiar group of poikilohydric plants, also known as resurrection plants. Unlike all other groups that undergo dehydration–rehydration cycles at particular developmental stages—seeds and pollen, resurrection plant are characterized by the ability to lose up to 95 % of their water content in vegetative tissues—leaves and roots. They can survive for several years in this completely desiccated condition and to return to normal physiological state once water availability is restored (Hartung et al. 1998; Scott 2000). While all lichens and most bryophytes are considered desiccation tolerant (Heber and Lüttge 2011), this remarkable feature is restricted to few vascular plants (Porembski 2011). Excluding ferns the number of resurrection plant species is roughly estimated at around 300, belonging both to monocots and dicots (Porembski 2011). Besides the indisputable ecological interest, resurrection angiosperms are also an attractive model to study desiccation tolerance in plants (Ingram and Bartels 1996; Hoekstra et al. 2001) and are considered an important gene pool for drought-tolerant transgenic plant production (Liu et al. 2009; Iturriaga et al. 1992).

Desiccation tolerance in resurrection plants is like any other trait, a complex aggregate of genetic and metabolic strategies. An important question is whether resurrection plants are just better in being drought tolerant than any other plants or they possess some unique feature. Recent transcriptome studies under drought, desiccation and subsequent rehydration of the "trendy" *Craterostigma plantagineum* and *Haberlea rhodopensis* revealed that both possibilities might be true (Gechev et al. 2012). These include, but is not restricted to expression of dehydration-specific proteins such as dehydrins (Rorat 2006) or LEA proteins (Liu et al. 2009), accumulation of oligosaccharides (Peters et al. 2007) and adjustment of the photosynthetic efficiency (Gashi et al. 2013; Farrant 2000). Last, but not least, the successful revival of a resurrection plant may be strongly dependent on its antioxidant status (Kranner et al. 2002; Sgherri et al. 2004) which in turns suggest a role for the enzymes of the ascorbate–glutathione pathway.

One of the first studies on the antioxidant systems in resurrection plants date back to the 1990s. In *Sporobolus staphianus* it was shown that both GSH and AsA tend to decrease in dehydrated plants. Simultaneously the activities of GR and DHAR increased approximately two fold while the activity of ascorbate peroxidase decreased (Sgherri et al. 1994). Navari-Izzo (Navari-Izzo et al. 1997) studied the alteration in glutathione concentrations and glutathione-related enzymes during dehydration of Queensland rock violet (*Boea hygroscopica*). The initial stages of dehydration, up to 80 % RWC were characterized by GSH depletion and a shift toward the oxidized state—GSSG. Further dehydration lead to normalization of the GSSG percentage, followed by a trigger for a substantial increase in GSH concentration—a twofold raise at around 50 % RWC. Such increase is related to ROS scavenging but also to the protein thiols' protective role of GSH. The initial increase in GSH concentration might be crucial for the revival of resurrection plants as its concentration gradually decreases after the initial raise as shown in *Myrothamnus flabellifolia* (Kranner et al. 2002). Both ascorbate and GSH depletion correlated with lower survival rates, but *Myrothamnus* seems not to accumulate ascorbate during dehydration. In both cases, no significant change in GR activity was observed in the first days of dehydration. Sherwin and Farrant (Sherwin and Farrant 1998) showed that the role of glutathione reductase during dehydration of resurrection plants may be species-specific as *Craterostigma wilmsii* showed overall higher GR activity with a transient increase in 50 % RWC, while *Xerophyta viscosa* showed dramatically lower GR activity with a peak at the completely dry stage. Interestingly the ascorbate peroxidase assay in these plants showed an opposite pattern—higher activity with a peak at 50 % RWC in *Xerophyta* and a decrease of the overall lower activity in *Craterostigma*. In phoenix flower (*Ramonda nathaliae*) both enzymes showed increased activity at 80 % RWC, followed by decrease even bellow initial levels at 50 % RWC (Jovanović et al. 2011).

Recently an increase in GSH concentration during dehydration was observed in *Boea hygrometrica* (*Jiang et al.* 2007), *Ramonda serbica* (Sgherri et al. 2004) and *Haberlea rhodopensis* (Djilianov et al. 2011). The later study also provided an important comparison with non-resurrection, closely related plant and showed that the ability to accumulate GSH rather than the ability to recycle GSSG might make the difference at least in some desiccation tolerant plants. In *Ramonda serbica* (Sgherri et al. 2004) the peaks both in total AsA and GSH concentrations were detected in 4.2 % RWC.

What happens during rehydration, or revival of the resurrection plant? Studies in lichens showed significant increase in reactive oxygen species production during rehydration (Weissman et al. 2005). However, this was not attributed to photosynthesis. In all studied resurrection plants the GSH+GSSG concentration declined during rehydration with little or no change in GSSG percentage (Djilianov et al. 2011; Jiang et al. 2007; Kranner et al. 2002; Sgherri et al. 1994). Simultaneously the

AsA concentration increased in *Sporobolus* (Sgherri et al. 1994) and *Myrothamnus* (Kranner et al. 2002), accompanied by an increase in the APX activity in the latter. All these data suggest that during dehydration, resurrection plants are more concerned to protect protein thiols in a GSH-dependent mechanism rather than using the ascorbate–glutathione pathway as a ROS scavenging machinery. Later, during rehydration, the probable ROS production is accompanied by increase in AsA concentration and normalization of GSH levels to meet the new requirements.

Despite their indisputable attractiveness as model organisms to study desiccation tolerance in plants, studies on resurrection plants are still scarce and not satisfactory. Few of the published research are extended into multiple aspects of the biochemical events during dehydration and rehydration. Moreover, it seems that there are no established model resurrection plants—most of the working groups prefer to focus on locally available species. With few exceptions, there are no closely related, non-resurrection plant species as controls and most of the studies were performed at different RWC percentage or different timeframes, thus not allowing a direct comparison.

8.5 Engineering the Ascorbate–Glutathione Cycle for Better Drought Tolerance

Contemporary approaches toward drought-tolerant crop plants breeding include marker-assisted selection and genetic engineering. High throughput methods are needed to ensure fast and efficient identification of molecular markers for drought tolerance and subsequent breeding of cultivars with enhanced traits. Both approaches result in high number of journal articles though the practical application is currently facing substantial constraints (Xu and Crouch 2008). This phenomenon is mainly attributed to the complex nature of the abiotic stress tolerance trait (Wang et al. 2003) that makes the efficient engineering of stress tolerant cultivars difficult to achieve. Currently a number of molecular markers for marker-assisted selection of a variety of crop plants as maize (Ribaut and Ragot 2007), pearl millet (Serraj et al. 2005), wheat (Fleury et al. 2010), rice (Siangliw et al. 2007) etc. exist. The application of QTLs (quantitative trait loci) in MAS (and genome-wide selection) gives promising results in a variety of crops (Mir et al. 2012). The further identification of candidate genes for drought tolerance, however allows for a more precise engineering of drought tolerance.

Wang (Wang et al. 2003) proposed several groups of genes and gene products as potential targets for drought, salinity and extreme temperature adaptation (as plant response to different abiotic stresses often share similar metabolic pathways) in crop plants. These include transcription factors, genes, involved in compatible solute biosynthetic pathways and ion transport, late embryogenesis abundant proteins (LEA) and heat shock proteins (HSP) encoding genes and antioxidants and detoxification genes. The members of the AsA–GSH cycle belong to the last group. The latter was recently proposed as an efficient biotechnological target for improving of

Table 8.3 A selection of examples, involving homologous or heterologous expression of the isoenzymes of the AsA–GSH cycle in transgenic plants, conferring drought tolerance. In notes, co-expression of the AsA–GSH cycle enzyme and another antioxidative enzyme, and/or tolerance to additional abiotic stress are reflected

Host plant	Source plant	Enzyme	Compartment	Note	Reference
Nicotiana tabacum	*Pisum sativum*	APX	cytosolic	Co-expression with Cu/Zn-SOD	Faize et al. (2011)
Nicotiana tabacum	*Arabidopsis thaliana*	DHAR	cytosolic	Ozone tolerance	Eltayeb et al. (2006)
Nicotiana tabacum	*Arabidopsis thaliana*	MDHAR	cytosolic	Ozone and salt tolerance	Eltayeb et al. (2007)
Ipomoea batatus	*Pisum sativum*	APX	chloroplasts	Co-expression with Cu/Zn-SOD	Lu et al. (2010)
Brassica napus	*Brassica napus*	APX	Thylakoid-bound	Salt-stress tolerance	Wang et al. (2013a)
Solanum tuberosum	*Arabidopsis thaliana*	DHAR	cytosolic	Herbicide and salt-stress tolerance	Eltayeb et al. (2011)
Lycopersicon esculentum	*Lycopersicon esculentum*	MDHAR	chloroplasts	Salt-stress tolerance	Li et al. (2012a)
Nicotiana tabacum	*Salicornia brachiata*	APX	peroxisomes	Salt-stress tolerance	Singh et al. (2014)
Nicotiana tabacum	*Arabidopsis thaliana*	APX	chloroplasts	Salt-stress tolerance	Badawi et al. (2004)
Arabidopsis thaliana	*Puccinellia tenuiflora*	APX	peroxisomes	Multiple abiotic stresses tolerance	Guan et al. (2015)
Arabidopsis thaliana	*Thellungiella salsuginea*	APX	cytosolic	Salt-stress tolerance	Li et al. (2015)

plant salt tolerance (Ashraf 2009). Assuming the similarity of plant response to the osmotic component of salt and drought stress and the role of AsA–GSH pathway in drought tolerance as summarized in this chapter, it is not surprising that overexpression and/or heterologous expression of one or several enzymes of the cycle may also contribute significantly to the drought adaptation of transgenic plants. Recent papers on this subject are summarized in Table 8.3.

There are several open questions that remain partially unsolved when referring to drought stress tolerance: (1) Whether transgenic expression of any AsA–GSH isoenzymes confer tolerance; (2) Whether transgenic expression in any compartment confers tolerance; (3) Whether co-expression of two or more AsA–GSH isoenzymes (or other antioxidant enzyme) improve drought tolerance; (4) Whether the expression of isoenzyme from drought or desiccation tolerant plant species improve drought tolerance. An important aspect of the experiments, involving transgenesis of the AsA–GSH cycle enzymes is the impact on multiple stress tolerance. In many, if not all, of the cited papers improved drought adaptation is accompanied by improved salt tolerance. Therefore transgenic plants, overexpressing some of these enzymes may be also drought tolerant, but not tested. The list of successful genetically modified drought-tolerant plants might be much longer.

It seems that overexpression of any of the AsA–GSH cycle enzymes might have a positive effect on drought adaptation, but APX is often the enzyme of choice. However, in some cases it was shown that some isoenzymes might work while others not. For example, overexpression of DHAR, but not MDHAR in transgenic tobacco conferred tolerance to aluminum stress (Yin et al. 2010). In transgenic Arabidopsis plants two different rice cytosolic APXs affected differentially stress tolerance under high salinity (Lu et al. 2007). Whether this is valid for drought tolerance should be established in the future. Overexpression of either cytosolic, chloroplastic, or peroxisomal isoenzymes has a positive effect on drought stress tolerance. Transgenic plants, expressing mitochondrial isoenzymes are not presented in Table 8.3, but there are reports on positive effect on freezing tolerance in plants, expressing mitochondrial MDHAR (Shin et al. 2014).

In two cases APX was co-expressed with Cu/Zn-SOD either in the cytosol (Faize et al. 2011) or the chloroplasts (Lu et al. 2010). Co-expression of MDHAR and DHAR was also shown to be more effective than a single transgene in freezing tolerance in rice (Shin et al. 2013). Overall, it seems that two is more than one, or in other words, the cooperative action of several antioxidant enzymes would give better results in terms of stress tolerance. Although in several cases the genetic source is a halophyte (Singh et al. 2014; Guan et al. 2015; Li et al. 2015), homologous overexpression (Li et al. 2012a; Wang et al. 2013a) and heterologous expression (all other cases in Table 8.3) from non-tolerant plant species proved to be equally beneficial for drought adaptation of the transgenic plants. No reports on transgenic expression of AsA–GSH cycle isoenzyme from xerophytes were found.

8.6 Conclusion and Future Perspectives

The ascorbate–glutathione cycle represent the central antioxidant mechanism in plants, primarily responsible for the detoxification of H_2O_2 mainly through ascorbate peroxidase in the major cellular compartments. The maintenance of favorable redox status of the major, water-soluble redox couples—2GSH/GSSG and AsA/DHA as provided by glutathione reductase, monodehydroascorbate reductase and dehydroascorbate reductase is of crucial importance for the survival strategy in conditions of drought stress. Constitutively higher or induced activity and expression of these enzymes is often associated with better adaptation to water deficit and consequently used as a biochemical marker for stress tolerance although its importance is questioned.

Several important aspects of the role of the ascorbate glutathione cycle in drought tolerance remain elusive. First of all, substantial differences in the activity and gene expression levels of the AsA–GSH enzymes were observed depending on the developmental stage, the severity and prolongation of the stress treatment. The diverse compartmentalization also complicates the identification of tolerance-associated isoenzymes. Systemic approaches toward exhaustive characterization of all AsA–GSH isoenzymes in conditions of water deficit and in multiple cultivars, differing in tolerance are still scarce and incomplete.

The AsA–GSH cycle enzymes are also an attractive target for genetic modification to acquire cultivars with improved tolerance. It seems that overexpression of either of these enzymes has a positive effect on multiple stress tolerance, and the co-expression of two or more of the AsA–GSH cycle enzymes leads to better results compared to a single transgene. Drought tolerance in plants is, however, a complex trait, with multiple molecular players in which the regulation and the effectiveness of the AsA–GSH cycle is an important, but not decisive part.

References

Abbas S, Ahmad S, Sabir S, Shah A. Detection of drought tolerant sugarcane genotypes (*Saccharum officinarum*) using lipid peroxidation, antioxidant activity, glycine-betaine and proline contents. J Soil Sci Plant Nutr. 2014;14(1):233–43.

Agati G, Azzarello E, Pollastri S, Tattini M. Flavonoids as antioxidants in plants: location and functional significance. Plant Sci. 2012;196:67–76.

Alam MM, Hasanuzzaman M, Nahar K, Fujita M. Exogenous salicylic acid ameliorates short-term drought stress in mustard (*Brassica juncea* L.) seedlings by up-regulating the antioxidant defense and glyoxalase system. Aust J Crop Sci. 2013;7(7):1053–63.

Amor NB, Jiménez A, Megdiche W, Lundqvist M, Sevilla F, Abdelly C. Response of antioxidant systems to NaCl stress in the halophyte Cakile maritima. Physiol Plant. 2006;126(3):446–57.

Anjum NA, Umar S, Chan M-T. Ascorbate-glutathione pathway and stress tolerance in plants. 2010. Springer Science & Business Media.

Apel K, Hirt H. Reactive oxygen species: metabolism, oxidative stress, and signal transduction. Annu Rev Plant Biol. 2004;55:373–99.

Asada K. Production and scavenging of reactive oxygen species in chloroplasts and their functions. Plant Physiol. 2006;141(2):391–6.

Ashraf M. Biotechnological approach of improving plant salt tolerance using antioxidants as markers. Biotechnol Adv. 2009;27(1):84–93.

Atkin OK, Macherel D. The crucial role of plant mitochondria in orchestrating drought tolerance. Ann Bot. 2009;103(4):581–97.

Badawi GH, Kawano N, Yamauchi Y, Shimada E, Sasaki R, Kubo A, Tanaka K. Over-expression of ascorbate peroxidase in tobacco chloroplasts enhances the tolerance to salt stress and water deficit. Physiol Plant. 2004;121(2):231–8.

Bartoli CG, Gómez F, Martínez DE, Guiamet JJ. Mitochondria are the main target for oxidative damage in leaves of wheat (*Triticum aestivum* L.). J Exp Bot. 2004;55(403):1663–9.

Bartoli CG, Gomez F, Gergoff G, Guiamét JJ, Puntarulo S. Up-regulation of the mitochondrial alternative oxidase pathway enhances photosynthetic electron transport under drought conditions. J Exp Bot. 2005;56(415):1269–76.

Ben Rejeb K, Vos LD, Le Disquet I, Leprince AS, Bordenave M, Maldiney R, Jdey A, Abdelly C, Savouré A. Hydrogen peroxide produced by NADPH oxidases increases proline accumulation during salt or mannitol stress in Arabidopsis thaliana. 2015. New Phytologist.

Bhatt D, Saxena SC, Jain S, Dobriyal AK, Majee M, Arora S. Cloning, expression and functional validation of drought inducible ascorbate peroxidase (*Ec-apx1*) from *Eleusine coracana*. Mol Biol Rep. 2013;40(2):1155–65.

Birtić S, Colville L, Pritchard HW, Pearce SR, Kranner I. Mathematically combined half-cell reduction potentials of low-molecular-weight thiols as markers of seed ageing. Free Radic Res. 2011;45(9):1093–102.

Cao BL, Ma Q, Zhao Q, Wang L, Xu K. Effects of silicon on absorbed light allocation, antioxidant enzymes and ultrastructure of chloroplasts in tomato leaves under simulated drought stress. Sci Hortic. 2015;194:53–62.

Blokhina O, Fagerstedt KV. Reactive oxygen species and nitric oxide in plant mitochondria: origin and redundant regulatory systems. Physiol Plant. 2010;138(4):447–62.

Boaretto LF, Carvalho G, Borgo L, Creste S, Landell MG, Mazzafera P, Azevedo RA. Water stress reveals differential antioxidant responses of tolerant and non-tolerant sugarcane genotypes. Plant Physiol Biochem. 2014;74:165–75.

Borsani O, Valpuesta V, Botella MA. Evidence for a role of salicylic acid in the oxidative damage generated by NaCl and osmotic stress in Arabidopsis seedlings. Plant Physiol. 2001;126(3):1024–30.

Brossa R, López-Carbonell M, Jubany-Marí T, Alegre L. Interplay between abscisic acid and jasmonic acid and its role in water-oxidative stress in wild-type, ABA-deficient, JA-deficient, and ascorbate-deficient Arabidopsis plants. J Plant Growth Regul. 2011;30(3):322–33.

Cai-Hong P, Su-Jun Z, Zhi-Zhong G, Bao-Shan W. NaCl treatment markedly enhances H_2O_2-scavenging system in leaves of halophyte Suaeda salsa. Physiol Plant. 2005;125(4):490–9.

Carvalho LC, Vidigal P, Amâncio S. Oxidative stress homeostasis in grapevine (*Vitis vinifera* L.). Front Environ Sci. 2015;3:20–5.

Caverzan A, Passaia G, Rosa SB, Ribeiro CW, Lazzarotto F, Margis-Pinheiro M. Plant responses to stresses: role of ascorbate peroxidase in the antioxidant protection. Genet Mol Biol. 2012;35(4):1011–9.

Chang M-L, Liao L-J, Lin J-H, Liu Z-H, Hsu Y-T, Lee T-M. Modulation of antioxidant defense system and NADPH oxidase in *Pluchea indica* leaves by water deficit stress. Bot Stud. 2012;53(1):45–54.

Chew O, Whelan J, Millar AH. Molecular definition of the ascorbate-glutathione cycle in Arabidopsis mitochondria reveals dual targeting of antioxidant defenses in plants. J Biol Chem. 2003;278(47):46869–77.

Chugh V, Kaur N, Grewal M, Gupta AK. Differential antioxidative response of tolerant and sensitive maize (*Zea mays* L.) genotypes to drought stress at reproductive stage. Indian J Biochem Biophys. 2013;50(2):150–8.

Cia M, Guimarães A, Medici L, Chabregas S, Azevedo R. Antioxidant responses to water deficit by drought-tolerant and-sensitive sugarcane varieties. Ann Appl Biol. 2012;161(3):313–24.

Contour-Ansel D, Torres-Franklin ML, DE CARVALHO MHC, D'Arcy-Lameta A, ZUILY-FODIL Y. Glutathione reductase in leaves of cowpea: cloning of two cDNAs, expression and enzymatic activity under progressive drought stress, desiccation and abscisic acid treatment. Ann Bot. 2006;98(6):1279–87.

Corpas FJ, Barroso JB, del Río LA. Peroxisomes as a source of reactive oxygen species and nitric oxide signal molecules in plant cells. Trends Plant Sci. 2001;6(4):145–50.

Cruz de Carvalho MH. Drought stress and reactive oxygen species: production, scavenging and signaling. Plant Signal Behav. 2008;3(3):156–65.

Dąbrowska G, Kata A, Goc A, Szechyńska-Hebda M, Skrzypek E. Characteristics of the plant ascorbate peroxidase family. Acta Biol Cracow Ser Bot. 2007;49:7–17.

Davletova S, Rizhsky L, Liang H, Shengqiang Z, Oliver DJ, Coutu J, Shulaev V, Schlauch K, Mittler R. Cytosolic ascorbate peroxidase 1 is a central component of the reactive oxygen gene network of Arabidopsis. Plant Cell. 2005;17(1):268–81.

Devi R, Kaur N, Gupta AK. Potential of antioxidant enzymes in depicting drought tolerance of wheat (Triticum aestivum L.). Indian J Biochem Biophys. 2012;49(4):257–65.

Ding S, Lei M, Lu Q, Zhang A, Yin Y, Wen X, Zhang L, Lu C. Enhanced sensitivity and characterization of photosystem II in transgenic tobacco plants with decreased chloroplast glutathione reductase under chilling stress. Biochimica et Biophysica Acta (BBA)-Bioenergetics. 2012; 1817(11):1979–91.

Djilianov D, Ivanov S, Moyankova D, Miteva L, Kirova E, Alexieva V, Joudi M, Peshev D, Van den Ende W. Sugar ratios, glutathione redox status and phenols in the resurrection species *Haberlea rhodopensis* and the closely related non-resurrection species *Chirita eberhardtii*. Plant Biol. 2011;13(5):767–76.

Edwards EA, Rawsthorne S, Mullineaux PM. Subcellular distribution of multiple forms of glutathione reductase in leaves of pea (*Pisum sativum* L.). Planta. 1990;180(2):278–84.

Eltayeb AE, Kawano N, Badawi GH, Kaminaka H, Sanekata T, Morishima I, Shibahara T, Inanaga S, Tanaka K. Enhanced tolerance to ozone and drought stresses in transgenic tobacco overexpressing dehydroascorbate reductase in cytosol. Physiol Plant. 2006;127(1):57–65.

Eltayeb AE, Kawano N, Badawi GH, Kaminaka H, Sanekata T, Shibahara T, Inanaga S, Tanaka K. Overexpression of monodehydroascorbate reductase in transgenic tobacco confers enhanced tolerance to ozone, salt and polyethylene glycol stresses. Planta. 2007;225(5):1255–64.

Eltayeb AE, Yamamoto S, Habora MEE, Yin L, Tsujimoto H, Tanaka K. Transgenic potato overexpressing Arabidopsis cytosolic AtDHAR1 showed higher tolerance to herbicide, drought and salt stresses. Breeding Sci. 2011;61(1):3–10.

Esen AHS, Özgur R, Uzilday B, Tanyolac ZÖ, Dinc A. The response of the xerophytic plant Gypsophila aucheri to salt and drought stresses: the role of the antioxidant defence system. Turk J Bot. 2012;36:697–706.

Faize M, Burgos L, Faize L, Piqueras A, Nicolas E, Barba-Espin G, Clemente-Moreno M, Alcobendas R, Artlip T, Hernandez J. Involvement of cytosolic ascorbate peroxidase and Cu/Zn-superoxide dismutase for improved tolerance against drought stress. Journal of Experimental Botany. J Exp Bot. 2011;62(8):2599–613.

Fan Y, Zhu M, Shabala S, Li C, Johnson P, Zhou M. Antioxidant Activity in Salt-Stressed Barley Leaves: Evaluating Time-and Age-Dependence and Suitability for the Use as a Biochemical Marker in Breeding Programs. J Agronomy Crop Sci. 2014;200(4):261–72.

Farrant JM. A comparison of mechanisms of desiccation tolerance among three angiosperm resurrection plant species. Plant Ecol. 2000;151(1):29–39.

Fleury D, Jefferies S, Kuchel H, Langridge P. Genetic and genomic tools to improve drought tolerance in wheat. J Exp Bot. 2010;61(12):3211–22.

Foyer CH. Redox homeostasis: Opening up ascorbate transport. Nat Plants. 2015;1(1).

Foyer CH, Halliwell B. The presence of glutathione and glutathione reductase in chloroplasts: a proposed role in ascorbic acid metabolism. Planta. 1976;133(1):21–5.

Foyer CH, Halliwell B. Purification and properties of dehydroascorbate reductase from spinach leaves. Phytochemistry. 1977;16(9):1347–50.

Foyer CH, Noctor G. Oxidant and antioxidant signalling in plants: a re-evaluation of the concept of oxidative stress in a physiological context. Plant Cell Environ. 2005a;28(8):1056–71.

Foyer CH, Noctor G. Redox homeostasis and antioxidant signaling: a metabolic interface between stress perception and physiological responses. Plant Cell. 2005b;17(7):1866–75.

Foyer CH, Noctor G. Ascorbate and glutathione: the heart of the redox hub. Plant Physiol. 2011;155(1):2–18.

Gaber A, Ogata T, Maruta T, Yoshimura K, Tamoi M, Shigeoka S. The involvement of Arabidopsis glutathione peroxidase 8 in the suppression of oxidative damage in the nucleus and cytosol. Plant Cell Physiol. 2012;53(9):1596–606.

Gallie DR. The role of L-ascorbic acid recycling in responding to environmental stress and in promoting plant growth. J Exp Bot. 2013;64(2):433–43.

Gamble PE, Burke JJ. Effect of water stress on the chloroplast antioxidant system I. Alterations in glutathione reductase activity. Plant Physiol. 1984;76(3):615–21.

Gashi B, Babani F, Kongjika E. Chlorophyll fluorescence imaging of photosynthetic activity and pigment contents of the resurrection plants Ramonda serbica and Ramonda nathaliae during dehydration and rehydration. Physiol Mol Biol Plants. 2013;19(3):333–41.

Gechev TS, Dinakar C, Benina M, Toneva V, Bartels D. Molecular mechanisms of desiccation tolerance in resurrection plants. Cell Mol Life Sci. 2012;69(19):3175–86.

Gest N, Garchery C, Gautier H, Jiménez A, Stevens R. Light-dependent regulation of ascorbate in tomato by a monodehydroascorbate reductase localized in peroxisomes and the cytosol. Plant Biotechnol J. 2013;11(3):344–54.

Gill SS, Tuteja N. Reactive oxygen species and antioxidant machinery in abiotic stress tolerance in crop plants. Plant Physiol Biochem. 2010;48(12):909–30.

Guan Q, Wang Z, Wang X, Takano T, Liu S. A peroxisomal APX from Puccinellia tenuiflora improves the abiotic stress tolerance of transgenic *Arabidopsis thaliana* through decreasing of H_2O_2 accumulation. J Plant Physiol. 2015;175:183–91.

Hadiarto T, Tran L-SP. Progress studies of drought-responsive genes in rice. Plant Cell Rep. 2011;30(3):297–310.

Hartung W, Schiller P, Dietz K-J. Physiology of poikilohydric plants. In: Progress in botany. Springer; 1998. p. 299–327.

Heber U, Lüttge U. Lichens and bryophytes: light stress and photoinhibition in desiccation/rehydration cycles–mechanisms of photoprotection. In: Plant Desiccation Tolerance. Springer; 2011. p. 121–37.

Hoekstra FA, Golovina EA, Buitink J. Mechanisms of plant desiccation tolerance. Trends Plant Sci. 2001;6(9):431–8.

Hossain MA, Piyatida P, da Silva JAT, Fujita M. Molecular mechanism of heavy metal toxicity and tolerance in plants: central role of glutathione in detoxification of reactive oxygen species and methylglyoxal and in heavy metal chelation. J Bot. 2012;2012.

Huseynova IM. Photosynthetic characteristics and enzymatic antioxidant capacity of leaves from wheat cultivars exposed to drought. Biochimica et Biophysica Acta (BBA)-Bioenergetics. 2012;1817(8):1516–23.

Ingram J, Bartels D. The molecular basis of dehydration tolerance in plants. Annu Rev Plant Biol. 1996;47(1):377–403.

Ishikawa T, Shigeoka S. Recent advances in ascorbate biosynthesis and the physiological significance of ascorbate peroxidase in photosynthesizing organisms. Biosci Biotechnol Biochem. 2008;72(5):1143–54.

Iturriaga G, Schneider K, Salamini F, Bartels D. Expression of desiccation-related proteins from the resurrection plant *Craterostigma plantagineum* in transgenic tobacco. Plant Mol Biol. 1992;20(3):555–8.

Jiang G, Wang Z, Shang H, Yang W, Hu Z, Phillips J, Deng X. Proteome analysis of leaves from the resurrection plant Boea hygrometrica in response to dehydration and rehydration. Planta. 2007;225(6):1405–20.

Jimenez A, Hernandez JA, del Rio LA, Sevilla F. Evidence for the presence of the ascorbate-glutathione cycle in mitochondria and peroxisomes of pea leaves. Plant Physiol. 1997;114(1):275–84.

Jiménez A, Hernández JA, Pastori G, del Río LA, Sevilla F. Role of the ascorbate-glutathione cycle of mitochondria and peroxisomes in the senescence of pea leaves. Plant Physiol. 1998;118(4):1327–35.

Jin SH, Li XQ, Wang GG, Zhu XT. Brassinosteroids alleviate high-temperature injury in Ficus concinna seedlings via maintaining higher antioxidant defence and glyoxalase systems. AoB Plants. 2015;7:9–25.

Jovanović Ž, Rakić T, Stevanović B, Radović S. Characterization of oxidative and antioxidative events during dehydration and rehydration of resurrection plant *Ramonda nathaliae*. Plant Growth Regul. 2011;64(3):231–40.

Kaminaka H, Morita S, Nakajima M, Masumura T, Tanaka K. Gene cloning and expression of cytosolic glutathione reductase in rice (*Oryza sativa* L.). Plant Cell Physiol. 1998;39(12):1269–80.

Kang H-M, Chen K, Bai J, Wang G. Antioxidative system's responses in the leaves of six Caragana species during drought stress and recovery. Acta Physiologiae Plantarum. 2012;34(6):2145–54.

Kang G, Li G, Liu G, Xu W, Peng X, Wang C, Zhu Y, Guo T. Exogenous salicylic acid enhances wheat drought tolerance by influence on the expression of genes related to ascorbate-glutathione cycle. Biologia Plantarum. 2013;57(4):718–24.

Kao CH. Role of hydrogen peroxide in rice plants. Crop Environ Bioinf. 2014;11(1):1–10.

Karuppanapandian T, Moon J-C, Kim C, Manoharan K, Kim W. Reactive oxygen species in plants: their generation, signal transduction, and scavenging mechanisms. Aust J Crop Sci. 2011;5(6):709–25.

Kataya AR, Reumann S. Arabidopsis glutathione reductase 1 is dually targeted to peroxisomes and the cytosol. Plant Signal Behav. 2010;5(2):171–5.

Kavitha K, Venkataraman G, Parida A. An oxidative and salinity stress induced peroxisomal ascorbate peroxidase from *Avicennia marina*: molecular and functional characterization. Plant Physiol Biochem. 2008;46(8):794–804.

Kranner I, Beckett RP, Wornik S, Zorn M, Pfeifhofer HW. Revival of a resurrection plant correlates with its antioxidant status. Plant J. 2002;31(1):13–24.

Kranner I, Birtić S, Anderson KM, Pritchard HW. Glutathione half-cell reduction potential: a universal stress marker and modulator of programmed cell death? Free Radical Biol Med. 2006;40(12):2155–65.

Le Martret B, Poage M, Shiel K, Nugent GD, Dix PJ. Tobacco chloroplast transformants expressing genes encoding dehydroascorbate reductase, glutathione reductase, and glutathione-S-transferase, exhibit altered anti-oxidant metabolism and improved abiotic stress tolerance. Plant Biotechnol J. 2011;9(6):661–73.

Lee H, Won S-H, Lee B-H, Park H-D, Chung W-I, Jo J. Genomic cloning and characterization of glutathione reductase gene from *Brassica campestris* var. *Pekinensis* Mol Cells. 2002;13(2): 245–51.

Leterrier M, Corpas FJ, Barroso JB, Sandalio LM, Luis A. Peroxisomal monodehydroascorbate reductase. Genomic clone characterization and functional analysis under environmental stress conditions. Plant Physiol. 2005;138(4):2111–23.

Li F, Wu Q, Duan M, Dong X, Li B, Meng Q. Transgenic tomato plants overexpressing chloroplastic monodehydroascorbate reductase are resistant to salt-and PEG-induced osmotic stress. Photosynthetica. 2012a;50(1):120–8.

Li K, Pang C-H, Ding F, Sui N, Feng Z-T, Wang B-S. Overexpression of Suaeda salsa stroma ascorbate peroxidase in Arabidopsis chloroplasts enhances salt tolerance of plants. South African J Bot. 2012b;78:235–45.

Li Y, Liu Y, Xu X, Jin M, An L, Zhang H. Effect of 24-epibrassinolide on drought stress-induced changes in *Chorispora bungeana*. Biologia Plantarum. 2012c;56(1):192–6.

Li Z, Li J, Li H, Shi Z, Zhang G. Overexpression of TsApx1 from Thellungiella salsuginea improves abiotic stress tolerance in transgenic *Arabidopsis thaliana*. Biologia Plantarum. 2015;59(3):497–506.

Lisenbee CS, Heinze M, Trelease RN. Peroxisomal ascorbate peroxidase resides within a subdomain of rough endoplasmic reticulum in wild-type Arabidopsis cells. Plant Physiol. 2003;132(2):870–82.

Lisenbee CS, Lingard MJ, Trelease RN. Arabidopsis peroxisomes possess functionally redundant membrane and matrix isoforms of monodehydroascorbate reductase. Plant J. 2005;43(6): 900–14.

Liu Q, Kasuga M, Sakuma Y, Abe H, Miura S, Yamaguchi-Shinozaki K, Shinozaki K. Two transcription factors, DREB1 and DREB2, with an EREBP/AP2 DNA binding domain separate two cellular signal transduction pathways in drought-and low-temperature-responsive gene expression, respectively, in Arabidopsis. Plant Cell. 1998;10(8):1391–406.

Liu X, Wang Z, Wang L, Wu R, Phillips J, Deng X. LEA 4 group genes from the resurrection plant Boea hygrometrica confer dehydration tolerance in transgenic tobacco. Plant Sci. 2009;176(1):90–8.

Liu Y-J, Yuan Y, Liu Y-Y, Liu Y, Fu J-J, Zheng J, Wang G-Y. Gene families of maize glutathione–ascorbate redox cycle respond differently to abiotic stresses. J Plant Physiol. 2012;169(2): 183–92.

Liu Z, Bao H, Cai J, Han J, Zhou L. A novel thylakoid ascorbate peroxidase from Jatrophacurcas enhances salt tolerance in transgenic tobacco. Int J Mol Sci. 2013;15(1):171–85.

Lu Z, Liu D, Liu S. Two rice cytosolic ascorbate peroxidases differentially improve salt tolerance in transgenic Arabidopsis. Plant Cell Rep. 2007;26(10):1909–17.

Lu Y-Y, Deng X-P, Kwak S-S. Over expression of CuZn superoxide dismutase (CuZn SOD) and ascorbate peroxidase (APX) in transgenic sweet potato enhances tolerance and recovery from drought stress. Afr J Biotechnol. 2010;9(49):8378–91.

Lunde C, Baumann U, Shirley NJ, Drew DP, Fincher GB. Gene structure and expression pattern analysis of three monodehydroascorbate reductase (Mdhar) genes in Physcomitrella patens: implications for the evolution of the MDHAR family in plants. Plant Mol Biol. 2006;60(2): 259–75.

Marino D, Dunand C, Puppo A, Pauly N. A burst of plant NADPH oxidases. Trends Plant Sci. 2012;17(1):9–15.

Marok MA, Tarrago L, Ksas B, Henri P, Abrous-Belbachir O, Havaux M, Rey P. A drought-sensitive barley variety displays oxidative stress and strongly increased contents in low-molecular weight antioxidant compounds during water deficit compared to a tolerant variety. J Plant Physiol. 2013;170(7):633–45.

Marquardt D, Williams JA, Kučerka N, Atkinson J, Wassall SR, Katsaras J, Harroun TA. Tocopherol activity correlates with its location in a membrane: a new perspective on the antioxidant vitamin E. J Am Chem Soc. 2013;135(20):7523–33.

Marquez-Garcia B, Shaw D, Cooper JW, Karpinska B, Quain MD, Makgopa EM, Kunert K, Foyer CH. Redox markers for drought-induced nodule senescence, a process occurring after drought-induced senescence of the lowest leaves in soybean (*Glycine max*). Ann Bot. 2015;116(4): 497–510.

May JM, Mendiratta S, Hill KE, Burk RF. Reduction of dehydroascorbate to ascorbate by the selenoenzyme thioredoxin reductase. J Biol Chem. 1997;272(36):22607–10.

Mhamdi A, Hager J, Chaouch S, Queval G, Han Y, Taconnat L, Saindrenan P, Gouia H, Issakidis-Bourguet E, Renou J-P. Arabidopsis GLUTATHIONE REDUCTASE1 plays a crucial role in leaf responses to intracellular hydrogen peroxide and in ensuring appropriate gene expression through both salicylic acid and jasmonic acid signaling pathways. Plant Physiol. 2010;153(3):1144–60.

Mhamdi A, Noctor G, Baker A. Plant catalases: peroxisomal redox guardians. Arch Biochem Biophys. 2012;525(2):181–94.

Miller G, Suzuki N, Ciftci-Yilmaz S, Mittler R. Reactive oxygen species homeostasis and signalling during drought and salinity stresses. Plant Cell Environ. 2010;33(4):453–67.

Mir RR, Zaman-Allah M, Sreenivasulu N, Trethowan R, Varshney RK. Integrated genomics, physiology and breeding approaches for improving drought tolerance in crops. Theor Appl Genet. 2012;125(4):625–45.

Mittler R. Oxidative stress, antioxidants and stress tolerance. Trends Plant Sci. 2002;7(9):405–10.

Mittler R, Zilinskas BA. Regulation of pea cytosolic ascorbate peroxidase and other antioxidant enzymes during the progression of drought stress and following recovery from drought. Plant J. 1994;5(3):397–405.

Mittova V, Tal M, Volokita M, Guy M. Up-regulation of the leaf mitochondrial and peroxisomal antioxidative systems in response to salt-induced oxidative stress in the wild salt-tolerant tomato species *Lycopersicon pennellii*. Plant Cell Environ. 2003;26(6):845–56.

Mittova V, Theodoulou F, Kiddle G, Volokita M, Tal M, Foyer C, Guy M. Comparison of mitochondrial ascorbate peroxidase in the cultivated tomato, *Lycopersicon esculentum*, and its wild, salt-tolerant relative, *L. pennellii*–a role for matrix isoforms in protection against oxidative damage. Plant Cell Environ. 2004;27(2):237–50.

Miyaji T, Kuromori T, Takeuchi Y, Yamaji N, Yokosho K, Shimazawa A, Sugimoto E, Omote H, Ma JF, Shinozaki K. AtPHT4; 4 is a chloroplast-localized ascorbate transporter in Arabidopsis. Nat Commun. 2015. doi:10.1038/ncomms6928.

Morgante CV, Rodrigues RA, Marbach PA, Borgonovi CM, Moura DS, Silva-Filho MC. Conservation of dual-targeted proteins in Arabidopsis and rice points to a similar pattern of gene-family evolution. Mol Genet Genomics. 2009;281(5):525–38.

Nakajima M, Enju A, Sakurai T, Satou M, Akiyama K, Taji T, Yamaguchi-Shinozaki K, Carninci P, Kawai J, Hayashizaki Y. Monitoring the expression proles of 7000 Arabidopsis genes under drought, cold and high-salinity stresses using a full-length cDNA microarray. Plant J. 2002;31(3):279292.

Narendra S, Venkataramani S, Shen G, Wang J, Pasapula V, Lin Y, Kornyeyev D, Holaday AS, Zhang H. The Arabidopsis ascorbate peroxidase 3 is a peroxisomal membrane-bound antioxidant enzyme and is dispensable for Arabidopsis growth and development. J Exp Bot. 2006;57(12):3033–42.

National BioResource Project: Rice. Oryzabase. 2000. Accessed 7 Sept 2015.

Navari-Izzo F, Meneguzzo S, Loggini B, Vazzana C, Sgherri C. The role of the glutathione system during dehydration of *Boea hygroscopica*. Physiol Plant. 1997;99(1):23–30.

Neill S, Desikan R, Hancock J. Hydrogen peroxide signalling. Curr Opin Plant Biol. 2002;5(5):388–95.

Német M, Janda T, Horváth E, Páldi E, Szalai G. Exogenous salicylic acid increases polyamine content but may decrease drought tolerance in maize. Plant Sci. 2002;162(4):569–74.

Niu Y, Wang Y, Li P, Zhang F, Liu H, Zheng G. Drought stress induces oxidative stress and the antioxidant defense system in ascorbate-deficient vtc1 mutants of *Arabidopsis thaliana*. Acta Physiologiae Plantarum. 2013;35(4):1189–200.

Noctor G. Metabolic signalling in defence and stress: the central roles of soluble redox couples. Plant Cell Environ. 2006;29(3):409–25.

Noctor G, Mhamdi A, Chaouch S, Han Y, Neukermans J, Marquez-Garcia B, Queval G, Foyer CH. Glutathione in plants: an integrated overview. Plant Cell Environ. 2012;35(2):454–84.

Noctor G, Mhamdi A, Foyer CH. The roles of reactive oxygen metabolism in drought: not so cut and dried. Plant Physiol. 2014;164(4):1636–48.

Obara K, Sumi K, Fukuda H. The use of multiple transcription starts causes the dual targeting of Arabidopsis putative monodehydroascorbate reductase to both mitochondria and chloroplasts. Plant Cell Physiol. 2002;43(7):697–705.

Okuma E, Nozawa R, Murata Y, Miura K. Accumulation of endogenous salicylic acid confers drought tolerance to Arabidopsis. Plant Signal Behav. 2014;9(3), e28085.

Ozkur O, Ozdemir F, Bor M, Turkan I. Physiochemical and antioxidant responses of the perennial xerophyte Capparis ovata Desf. to drought. Environ Exp Bot. 2009;66(3):487–92.

Padmavathi T, Rao DM. Oxidative stress damage and antioxidant activity in 4 cultivars of peanut (*Arachis hypogaea* L.) under drought stress. Trends Biosci. 2014;7(23):3802–10.

Pandey S, Subramanaym Reddy C, Yaqoob U, Negi Y, Arora S. In silico analysis of cis acting regulatory elements CAREs in upstream regions of ascorbate glutathione pathway genes from Oryza sativa. Biochem Physiol. 2015;4(159):2.

Parida AK, Jha B. Physiological and biochemical responses reveal the drought tolerance efficacy of the halophyte Salicornia brachiata. J Plant Growth Regul. 2013;32(2):342–52.

Peters S, Mundree SG, Thomson JA, Farrant JM, Keller F. Protection mechanisms in the resurrection plant Xerophyta viscosa (Baker): both sucrose and raffinose family oligosaccharides (RFOs) accumulate in leaves in response to water deficit. J Exp Bot. 2007;58(8):1947–56.

Phoenix Bioinformatics Corporation The Arabidopsis Information Resource (TAIR). www.arabidopsis.org. Accessed 7 Sept 2015.

Polle A. Dissecting the superoxide dismutase-ascorbate-glutathione-pathway in chloroplasts by metabolic modeling. Computer simulations as a step towards flux analysis. Plant Physiol. 2001;126(1):445–62.

Porembski S. Evolution, diversity, and habitats of poikilohydrous vascular plants. In: Plant desiccation tolerance. Springer; 2011. p. 139–56.

Potters G, De Gara L, Asard H, Horemans N. Ascorbate and glutathione: guardians of the cell cycle, partners in crime? Plant Physiol Biochem. 2002;40(6):537–48.

Pyngrope S, Bhoomika K, Dubey R. Reactive oxygen species, ascorbate–glutathione pool, and enzymes of their metabolism in drought-sensitive and tolerant indica rice (*Oryza sativa* L.) seedlings subjected to progressing levels of water deficit. Protoplasma. 2013;250(2):585–600.

Raheleh R, Ramazanali K, Ali G. Use of biochemical indices and antioxidant enzymes as a screening technique for drought tolerance in Chickpea genotypes (*Cicer arietinum* L.). Afr J Agr Res. 2012;7(39):5372–80.

Ribaut J-M, Ragot M. Marker-assisted selection to improve drought adaptation in maize: the back-cross approach, perspectives, limitations, and alternatives. J Exp Bot. 2007;58(2):351–60.

Rivero RM, Kojima M, Gepstein A, Sakakibara H, Mittler R, Gepstein S, Blumwald E. Delayed leaf senescence induces extreme drought tolerance in a flowering plant. Proc Natl Acad Sci. 2007;104(49):19631–6.

Rizhsky L, Liang H, Mittler R. The combined effect of drought stress and heat shock on gene expression in tobacco. Plant Physiol. 2002;130(3):1143–51.

Romero-Puertas MC, Corpas FJ, Sandalio LM, Leterrier M, Rodríguez-Serrano M, Del Río LA, Palma JM. Glutathione reductase from pea leaves: response to abiotic stress and characterization of the peroxisomal isozyme. New Phytol. 2006;170(1):43–52.

Rorat T. Plant dehydrins—tissue location, structure and function. Cell Mol Biol Lett. 2006;11(4):536–56.

Sano S, Tao S, Endo Y, Inaba T, Hossain MA, MIYAKE C, Matsuo M, Aoki H, Asada K, Saito K. Purification and cDNA cloning of chloroplastic monodehydroascorbate reductase from spinach. Biosci Biotechnol Biochem. 2005;69(4):762–72.

Saruhan N, Saglam A, Kadioglu A. Salicylic acid pretreatment induces drought tolerance and delays leaf rolling by inducing antioxidant systems in maize genotypes. Acta Physiologiae Plantarum. 2012;34(1):97–106.

Schaedle M, Bassham JA. Chloroplast glutathione reductase. Plant Physiol. 1977;59(5):1011–2.

Schafer FQ, Buettner GR. Redox environment of the cell as viewed through the redox state of the glutathione disulfide/glutathione couple. Free Radical Biol Med. 2001;30(11):1191–212.

Scheler C, Durner J, Astier J. Nitric oxide and reactive oxygen species in plant biotic interactions. Curr Opin Plant Biol. 2013;16(4):534–9.

Scott P. Resurrection plants and the secrets of eternal leaf. Ann Bot. 2000;85(2):159–66.

Sečenji M, Hideg É, Bebes A, Györgyey J. Transcriptional differences in gene families of the ascorbate–glutathione cycle in wheat during mild water deficit. Plant Cell Rep. 2010;29(1):37–50.

Seki M, Narusaka M, Abe H, Kasuga M, Yamaguchi-Shinozaki K, Carninci P, Hayashizaki Y, Shinozaki K. Monitoring the expression pattern of 1300 Arabidopsis genes under drought and cold stresses by using a full-length cDNA microarray. Plant Cell. 2001;13(1):61–72.

Sekmen AH, Ozgur R, Uzilday B, Turkan I. Reactive oxygen species scavenging capacities of cotton (*Gossypium hirsutum*) cultivars under combined drought and heat induced oxidative stress. Environ Exp Bot. 2014;99:141–9.

Senaratna T, Touchell D, Bunn E, Dixon K. Acetyl salicylic acid (Aspirin) and salicylic acid induce multiple stress tolerance in bean and tomato plants. Plant Growth Regul. 2000;30(2):157–61.

Serraj R, Tom Hash C, Rizvi SMH, Sharma A, Yadav RS, Bidinger FR. Recent advances in marker-assisted selection for drought tolerance in pearl millet. Plant Prod Sci. 2005;8(3):334–7.

Sgherri CLM, Loggini B, Puliga S, Navari-Izzo F. Antioxidant system in *Sporobolus stapfianus*: changes in response to desiccation and rehydration. Phytochemistry. 1994;35(3):561–5.

Sgherri C, Stevanovic B, Navari-Izzo F. Role of phenolics in the antioxidative status of the resurrection plant Ramonda serbica during dehydration and rehydration. Physiol Plant. 2004;122(4):478–85.

Shamim F, Johnson GN, NAQVI SS, Waheed A. Higher antioxidant capacity protects photosynthetic activities as revealed by Chl a fluorescence in drought tolerant tomato genotypes. Pak J Bot. 2013;45(5):1631–42.

Shan C, Liang Z. Jasmonic acid regulates ascorbate and glutathione metabolism in Agropyron cristatum leaves under water stress. Plant Sci. 2010;178(2):130–9.

Sharma P, Jha AB, Dubey RS, Pessarakli M. Reactive oxygen species, oxidative damage, and antioxidative defense mechanism in plants under stressful conditions. J Bot. 2012;2012.

Sherwin HW, Farrant JM. Protection mechanisms against excess light in the resurrection plants Craterostigma wilmsii and Xerophyta viscosa. Plant Growth Regul. 1998;24(3):203–10.

Shevyakova N, Bakulina E, Kuznetsov VV. Proline antioxidant role in the common ice plant subjected to salinity and paraquat treatment inducing oxidative stress. Russ J Plant Physiol. 2009;56(5):663–9.

Shimaoka T, Yokota A, Miyake C. Purification and characterization of chloroplast dehydroascorbate reductase from spinach leaves. Plant Cell Physiol. 2000;41(10):1110–8.

Shimaoka T, Miyake C, Yokota A. Mechanism of the reaction catalyzed by dehydroascorbate reductase from spinach chloroplasts. Eur J Biochem. 2003;270(5):921–8.

Shin S-Y, Kim M-H, Kim Y-H, Park H-M, Yoon H-S. Co-expression of monodehydroascorbate reductase and dehydroascorbate reductase from *Brassica rapa* effectively confers tolerance to freezing-induced oxidative stress. Mol Cells. 2013;36(4):304–15.

Shin S, Kim Y, Kim I, Kim Y, Park H, Yoon H. The expression of BrMDHAR gene in chloroplasts and mitochondria enhances tolerance to freezing stress in *Arabidopsis thaliana*. Biologia Plantarum. 2014;58(3):456–68.

Shinozaki K, Yamaguchi-Shinozaki K. Gene expression and signal transduction in water-stress response. Plant Physiol. 1997;115(2):327.

Shinozaki K, Yamaguchi-Shinozaki K. Gene networks involved in drought stress response and tolerance. J Exp Bot. 2007;58(2):221–7.

Shrivastava DC, Kisku AV, Saxena M, Deswal R, Sarin NB. Stress inducible cytosolic ascorbate peroxidase (Ahcapx) from *Arachis hypogaea* cell lines confers salinity and drought stress tolerance in transgenic tobacco. Nucleus. 2015;58(1):3–13.

Siangliw JL, Jongdee B, Pantuwan G, Toojinda T. Developing KDML105 backcross introgression lines using marker-assisted selection for QTLs associated with drought tolerance in rice. Sci Asia. 2007;33:207–14.

Singh S, Gupta A, Kaur N. Differential responses of antioxidative defence system to long-term field drought in wheat (*Triticum aestivum* L.) genotypes differing in drought tolerance. J Agron Crop Sci. 2012;198(3):185–95.

Singh N, Mishra A, Jha B. Over-expression of the peroxisomal ascorbate peroxidase (SbpAPX) gene cloned from halophyte *Salicornia brachiata* confers salt and drought stress tolerance in transgenic tobacco. Marine Biotechnol. 2014;16(3):321–32.

Stahl W, Sies H. Antioxidant activity of carotenoids. Mol Aspects Med. 2003;24(6):345–51.

Stevens RG, Creissen GP, Mullineaux PM. Cloning and characterisation of a cytosolic glutathione reductase cDNA from pea (*Pisum sativum* L.) and its expression in response to stress. Plant Mol Biol. 1997;35(5):641–54.

Suzuki N, Mittler R. Reactive oxygen species-dependent wound responses in animals and plants. Free Radical Biol Med. 2012;53(12):2269–76.

Szarka A. Quantitative data on the contribution of GSH and Complex II dependent ascorbate recycling in plant mitochondria. Acta Physiologiae Plantarum. 2013;35(11):3245–50.

Szarka A, Horemans N, Kovács Z, Gróf P, Mayer M, Bánhegyi G. Dehydroascorbate reduction in plant mitochondria is coupled to the respiratory electron transfer chain. Physiol Plant. 2007;129(1):225–32.

Szarka A, Tomasskovics B, Bánhegyi G. The ascorbate-glutathione-α-tocopherol triad in abiotic stress response. Int J Mol Sci. 2012;13(4):4458–83.

Tahmasebi A, Aram F, Ebrahimi M, Mohammadi-Dehcheshmeh M. Ebrahimie E (2012) Genome-wide analysis of cytosolic and chloroplastic isoforms of glutathione reductase in plant cells. Springerplus. 2012;1:30.

Teixeira FK, Menezes-Benavente L, Margis R, Margis-Pinheiro M. Analysis of the molecular evolutionary history of the ascorbate peroxidase gene family: inferences from the rice genome. J Mol Evol. 2004;59(6):761–70.

Teixeira FK, Menezes-Benavente L, Galvão VC, Margis-Pinheiro M. Multigene families encode the major enzymes of antioxidant metabolism in *Eucalyptus grandis* L. Genet Mol Biol. 2005;28(3):529–38.

Teixeira FK, Menezes-Benavente L, Galvão VC, Margis R, Margis-Pinheiro M. Rice ascorbate peroxidase gene family encodes functionally diverse isoforms localized in different subcellular compartments. Planta. 2006;224(2):300–14.

Theocharis A, Clément C, Barka EA. Physiological and molecular changes in plants grown at low temperatures. Planta. 2012;235(6):1091–105.

Tinoco I, Sauer K, Wang JC, Puglisi JD, Harbison G, Rovnyak D. Physical chemistry: principles and applications in biological sciences. 5th ed. 2014. Pearson.

Trivedi DK, Gill SS, Yadav S, Tuteja N. Genome-wide analysis of glutathione reductase (GR) genes from rice and Arabidopsis. Plant Signal Behav. 2013;8(2), e23021.

Türkan İ, Bor M, Özdemir F, Koca H. Differential responses of lipid peroxidation and antioxidants in the leaves of drought-tolerant *P. acutifolius* Gray and drought-sensitive *P. vulgaris* L. subjected to polyethylene glycol mediated water stress. Plant Sci. 2005;168(1):223–31.

Van den Ende W, Peshev D. Sugars as antioxidants in plants. In: Crop improvement under adverse conditions. Springer; 2013. p. 285–307.

Veljovic-Jovanovic SD, Pignocchi C, Noctor G, Foyer CH. Low ascorbic acid in the vtc-1 mutant of Arabidopsis is associated with decreased growth and intracellular redistribution of the antioxidant system. Plant Physiol. 2001;127(2):426–35.

Voothuluru P, Sharp RE. Apoplastic hydrogen peroxide in the growth zone of the maize primary root under water stress. I. Increased levels are specific to the apical region of growth maintenance. J Exp Bot. 2012. doi:10.1093/jxb/ers277.

Wang W, Vinocur B, Altman A. Plant responses to drought, salinity and extreme temperatures: towards genetic engineering for stress tolerance. Planta. 2003;218(1):1–14.

Wang J, Fan Z, Liu Z, Xiang J, Chai L, Li X, Yang Y. Thylakoid-bound ascorbate peroxidase increases resistance to salt stress and drought in Brassica napus. Afr J Biotechnol. 2013a;10(41):8039–45.

Wang Y, Tao X, Tang X-M, Xiao L, J-l S, Yan X-F, Li D, Deng H-Y, Ma X-R. Comparative transcriptome analysis of tomato (*Solanum lycopersicum*) in response to exogenous abscisic acid. BMC Genomics. 2013b;14(1):841.

Wasternack C. Jasmonates: an update on biosynthesis, signal transduction and action in plant stress response, growth and development. Ann Bot. 2007;100(4):681–97.

Weissman L, Garty J, Hochman A. Rehydration of the lichen *Ramalina lacera* results in production of reactive oxygen species and nitric oxide and a decrease in antioxidants. Appl Environ Microbiol. 2005;71(4):2121–9.

Wu L, Zhang Z, Zhang H, Wang X-C, Huang R. Transcriptional modulation of ethylene response factor protein JERF3 in the oxidative stress response enhances tolerance of tobacco seedlings to salt, drought, and freezing. Plant Physiol. 2008;148(4):1953–63.

Wu T-M, Lin W-R, Kao Y-T, Hsu Y-T, Yeh C-H, Hong C-Y, Kao CH. Identification and characterization of a novel chloroplast/mitochondria co-localized glutathione reductase 3 involved in salt stress response in rice. Plant Mol Biol. 2013;83(4-5):379–90.

Xu Y, Crouch JH. Marker-assisted selection in plant breeding: from publications to practice. Crop Sci. 2008;48(2):391–407.

Yamaguchi K, Mori H, Nishimura M. A novel isoenzyme of ascorbate peroxidase localized on glyoxysomal and leaf peroxisomal membranes in pumpkin. Plant Cell Physiol. 1995;36:1157–62.

Yang C-J, Zhang C, Lu Y-N, Jin J-Q, Wang X-L. The mechanisms of brassinosteroids' action: from signal transduction to plant development. Mol Plant. 2011;4(4):588–600.

Yang L, Fountain J, Ni X, Ji P, Lee R, Scully B, Kemerait R, Guo B. Maize sensitivity to drought stress is associated with differential responses to reactive oxygen species. Plant and Animal Genome Conference, 2015.

Yildiztugay E, Ozfidan-Konakci C, Kucukoduk M. The role of antioxidant responses on the tolerance range of extreme halophyte *Salsola crassa* grown under toxic salt concentrations. Ecotoxicol Environ Saf. 2014;110:21–30.

Yin L, Wang S, Eltayeb AE, Uddin MI, Yamamoto Y, Tsuji W, Takeuchi Y, Tanaka K. Overexpression of dehydroascorbate reductase, but not monodehydroascorbate reductase, confers tolerance to aluminum stress in transgenic tobacco. Planta. 2010;231(3):609–21.

Yoshimura K, Masuda A, Kuwano M, Yokota A, Akashi K. Programmed proteome response for drought avoidance/tolerance in the root of a C3 xerophyte (wild watermelon) under water deficits. Plant Cell Physiol. 2008;49(2):226–41.

Yuan G-F, Jia C-G, Li Z, Sun B, Zhang L-P, Liu N, Wang Q-M. Effect of brassinosteroids on drought resistance and abscisic acid concentration in tomato under water stress. Sci Hortic. 2010;126(2):103–8.

Zagorchev L, Seal CE, Kranner I, Odjakova M. A central role for thiols in plant tolerance to abiotic stress. Int J Mol Sci. 2013;14(4):7405–32.

Zhai C-Z, Zhao L, Yin L-J, Chen M, Wang Q-Y, Li L-C, Xu Z-S, Ma Y-Z. Two wheat glutathione peroxidase genes whose products are located in chloroplasts improve salt and H_2O_2 tolerances in Arabidopsis. PLoS One. 2013;8(10).

Zhang H, Wang J, Allen RD, Goodman HM. Cloning and expression of an Arabidopsis gene encoding a putative peroxisomal ascorbate peroxidase. Plant Mol Biol. 1997;34(6):967–71.

Zhang J, Jia W, Yang J, Ismail AM. Role of ABA in integrating plant responses to drought and salt stresses. Field Crop Res. 2006;97(1):111–9.

Zhang Z, Wang J, Zhang R, Huang R. The ethylene response factor AtERF98 enhances tolerance to salt through the transcriptional activation of ascorbic acid synthesis in Arabidopsis. Plant J. 2012;71(2):273–87.

Zhang Z, Zhang Q, Wu J, Zheng X, Zheng S, Sun X, Qiu Q, Lu T. Gene knockout study reveals that cytosolic ascorbate peroxidase 2 (OsAPX2) plays a critical role in growth and reproduction in rice under drought, salt and cold stresses. PLoS One. 2013;8(2), e57472.

Zhang LX, Lai JH, Liang ZS, Ashraf M. Interactive effects of sudden and gradual drought stress and foliar-applied glycinebetaine on growth, water relations, osmolyte accumulation and antioxidant defence system in two maize cultivars differing in drought tolerance. J Agron Crop Sci. 2014;200(6):425–33.

Chapter 9
Sulfur Metabolism and Drought Stress Tolerance in Plants

Walid Abuelsoud, Felix Hirschmann, and Jutta Papenbrock

9.1 Introduction

Drought stress is one of the major environmental limitations with tremendous effects on the plant growth and development (Harb et al. 2010; Song et al. 2012). Drought stress causes a decrease in the crop productivity and nearly 28 % of the world's soil surface is too dry for regular crop yields (Ambrosone et al. 2013; Bray 2002). Drought-tolerant plants have developed a huge spectrum of morphological, physiological, and metabolic adaptations to a shortage of water. A better understanding of these mechanisms might help develop crop plants with a higher drought stress tolerance. During the last years the important role of sulfur-containing compounds during defense against biotic stress has been investigated in some detail (Rausch and Wachter 2005). Recently, a role of sulfur and sulfur-containing compounds in abiotic stress defenses has also been postulated (Chan et al. 2013). However, so far the role of sulfur-containing compounds in stress tolerance is not well understood. In this review, we summarize the most important aspects in sulfur metabolism from our point of view that might play significant roles in drought stress responses. A complex balancing act is required to coordinate primary and secondary sulfur metabolism during the drought stress response (Fig. 9.1).

W. Abuelsoud
Institute of Botany, Leibniz University Hannover,
Herrenhäuserstr. 2, Hannover D-30419, Germany

Botany Department, Faculty of Science, Cairo University, Giza 12613, Egypt

F. Hirschmann • J. Papenbrock (✉)
Institute of Botany, Leibniz University Hannover,
Herrenhäuserstr. 2, Hannover D-30419, Germany
e-mail: Jutta.Papenbrock@botanik.uni-hannover.de

M.A. Hossain et al. (eds.), *Drought Stress Tolerance in Plants, Vol 1*,
DOI 10.1007/978-3-319-28899-4_9

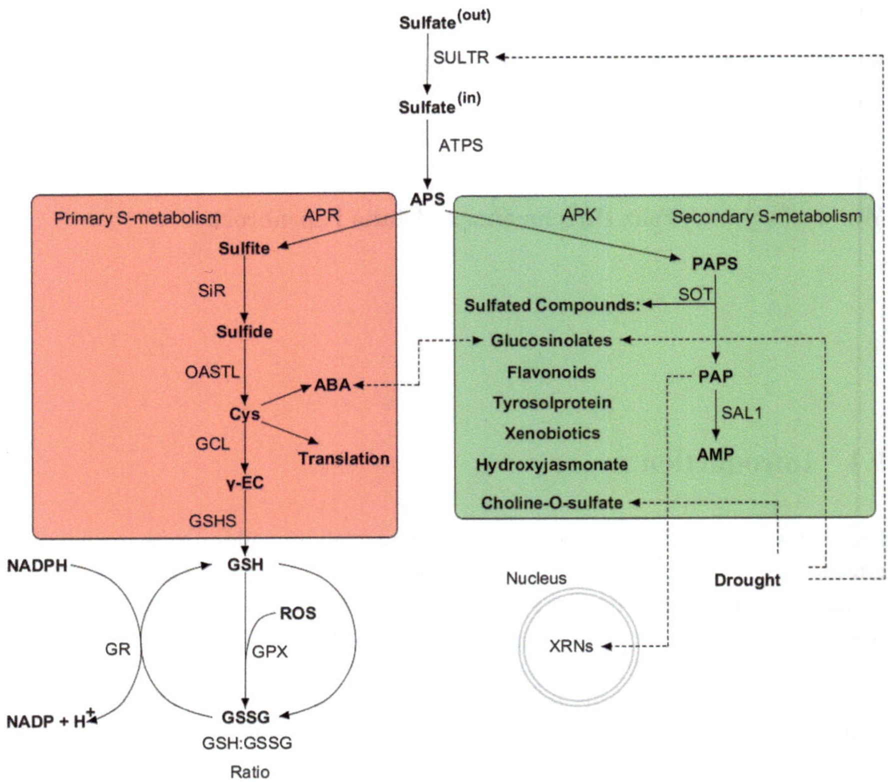

Fig. 9.1 Schematic overview of plant sulfur metabolism and effects of drought. Sulfate is taken up by sulfate transporters (SULTR) and activated by adenosine triphosphate (ATP) to form adenosine phosphosulfate (APS). APS can either be reduced to sulfite as part of the primary sulfur metabolism (*red box*), or phosphorylated to 5′-phosphoadenosine 3′-phosphosulfate (PAPS) as part of the secondary sulfur metabolism (*green box*). Impacts of drought and regulatory effects of key enzymes during drought stress are indicated by *broken lines*. Cosubstrates and products are not displayed. *ATPS* ATP sulfurylase, *APR* APS reductase, *APK* APS kinase, *SiR* sulfite reductase, *OASTL* O-acetylserine lyase, *Cys* cysteine, *ABA* abscisic acid, *GCL* Glutamate cysteine lyase, *γ-EC* γ-glutamyl cysteine, *GSH* reduced glutathione, *ROS* reactive oxygen species, *GSSG* glutathione disulfide, *GSHS* GSH synthetase, *GR* glutathione reductase, *GPX* glutathione peroxidase, *PAP* 5′-phosphoadenosine phosphate, *SOT* sulfotransferase, *AMP* adenosine monophosphate, *XRNs* 5′-3′exoribonucleases

9.2 Uptake of Sulfur

9.2.1 Uptake of Sulfur and Sulfate Transporters

Interestingly, sulfate is the only macronutrient that increases in the xylem sap during drought stress treatments. Other macronutrients, such as nitrate and phosphate, are not affected. These results indicate that sulfur partitioning is regulated in a different way in comparison to nitrate and phosphate (Ernst et al. 2010). Sulfate is taken up by sulfate transporters from the soil. There are four groups of sulfate transporters abbreviated SULTR with the respective number within a group. So far the knowledge about the expression of sulfate transporters during stress responses is only limited (Gallardo et al. 2014). In group 1, high affinity transporters expressed in the roots are found. Group 2, low affinity transporters, enables loading and unloading of sulfate to and from the xylem and the phloem. Transporter proteins in group 3 are responsible for transporting sulfate over plastid membranes. Proteins localized in the tonoplast enable the export of stored sulfate out of the vacuoles and are summarized in group 4. With respect to abiotic stress, the up-regulation of the plastidal *SULTR3;1* gene in roots of several plant species, such as *Arabidopsis* and *Medicago*, subjected to drought and salt stress is of particular interest. Remarkably, the expression of *AtSULTR3;1* is enhanced by abscisic acid (ABA) and is strongly required for cysteine synthesis (Cao et al. 2014). Cysteine plays a major role in the defense against abiotic stress, because it serves as a precursor for glutathione biosynthesis and also as a sulfur donor for the sulfuration of molybdenum cofactor (Moco), acting as a cofactor in its sulfurylated form for the last reaction step of ABA biosynthesis. Therefore, ABA biosynthesis and sulfur metabolism probably interplay during abiotic stress reactions to ensure sufficient cysteine for ABA production (Cao et al. 2013). Interestingly, another gene in group 3 *AtSULTR3;4* is co-expressed with *AtSULTR3;1* in roots in response to drought. In *AtSULTR3;1* and *AtSULTR3;4* mutants, the concentration of ABA is reduced also supporting a role of these sulfate transporters in ABA production.

It was found that the microRNA395 is up-regulated in response to drought stress in rice. In part, the flux of sulfur from roots to shoots is controlled by this particular microRNA which limits the expression of *AtSULTR2;2* to xylem parenchyma and thereby enhancing sulfate translocation to aerial parts (Kawashima et al. 2011). Therefore, microRNA395 might be involved in maintaining the flux of sulfur towards aerial parts during abiotic stress conditions. Recently, more microRNAs have been analyzed for their involvement in abiotic stresses and sulfur deficiency. Experiments presented added new miRNA players in a complex network of gene expression regulation in plant response to a wide array of abiotic stresses; however, none of them was influenced by sulfur deficiency (Barciszewska-Pacak et al. 2015).

9.2.2 *Higher Sulfur Demand During Drought Stress*

The flux of sulfate is increased during the stress response in relation to other ions, like nitrate or phosphate, reflecting a higher demand of sulfate in source organs during drought (Ernst et al. 2010). In addition, sulfate from the xylem acts as a chemical signal for ABA-dependent stomatal closure in leaves during early stages of water stress when ABA biosynthesis is restricted to leaves (Ernst et al. 2010). Vice versa, an effect of ABA on cysteine biosynthesis was observed (Barroso et al.

1999). However, although it was shown that *O*-acetylserine (thiol) lyase (OAS-TL) expression is regulated by ABA, typical ABA- or dehydration-responsive elements in any of the regulatory key genes for sulfur assimilation were not found (Urano et al. 2009).

9.3 Role of Abscisic Acid Under Drought Stress

Under biotic and abiotic stresses, ABA acts as an endogenous stress signal, which triggers plant defenses (Adie et al. 2007; Kowitcharoen et al. 2015; Mehrotra et al. 2014; Melotto et al. 2006). ABA plays versatile physiological and regulatory roles during plant development, including bud dormancy, seed maturation and root development. ABA controls stomatal movement, thus allowing decreased transpiration under drought or salinity stresses (Ye et al. 2012).

9.3.1 Biosynthesis of ABA and Drought Stress

ABA is a 15 carbon isoprenoid plant hormone, which is synthesized in plastids from carotenoids within almost all plant cells that contain plastids (Fig. 9.2). Zeaxanthin is subjected to epioxidation by zeaxanthin epoxidase (ZEP) into violaxanthin within the plastids. Then violaxanthin is isomerized to 9′-*cis*-neoxanthin, which is then cleaved by 9-*cis*-epoxycarotenoid dioxygenase (NCED) to form the 15-carbon xanthoxal, a natural growth inhibitor. Xanthoxal is then oxidized first into abscisic aldehyde and then into ABA. The later step is catalyzed by ABA aldehyde oxidase (AAO). ZEP and NCED activities seem to be the rate limiting steps in ABA biosynthesis (Xue et al. 2015; Ye et al. 2012).

ZEP is encoded by a locus called *ABA1* in *Arabidopsis*. *aba1* mutants were shown to contain lowered levels of ABA and lower osmotic stress tolerance. Upon osmotic stress, *ABA1* transcripts increased in wild type *Arabidopsis* (Leon-Kloosterziel et al. 1996; Xiong et al. 2002). Moreover, *vp2*, *vp5*, *vp7*, and *vp9* mutants of maize have blocked synthesis of carotenoids and were found to contain lower levels of ABA compared to wild type and grains exhibit precocious germination (Neill et al. 1986). This confirms that ABA is synthesized through carotenoids rather than directly as a small molecule.

Sulfate transporter SULTR3 was found to be preferentially expressed in leaves and was shown to act in sulfate transport into chloroplasts. *Arabidopsis* mutants, *sultr3;1-2* and *sultr3;1-4*, have decreased rates of sulfate transport into the chloroplast and they were shown to contain lower amounts of ABA compared to wild type under non-stressed conditions. The normal levels of ABA could be restored in these mutants by expression of a 35S-SULTR3;1 construct (Cao et al. 2013, 2014). These results emphasized the importance of sulfate metabolism in chloroplasts for ABA biosynthesis. The connection between sulfate metabolism in chloroplasts and ABA

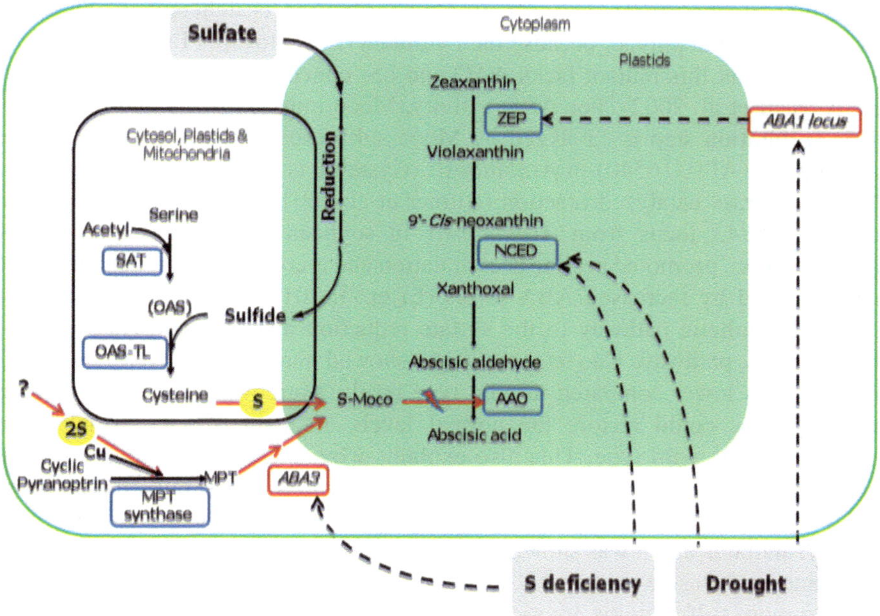

Fig. 9.2 Schematic overview of interaction between sulfur (S) metabolism and abscisic acid (ABA) biosynthesis. Sulfur supply in form of sulfate produces sulfide as a by-product of its reduction pathway, this sulfide fuels the biosynthesis of cysteine which acts as a vehicle for S to activate molybdenum cofactor (Moco) into S-Moco. S-Moco is the active cofactor necessary for the activity of abscisic aldehyde oxidase (AAO) enzyme that catalyzes the last step in ABA biosynthesis. Two more S atoms from an yet unknown donor (?), in addition to one copper atom, are required for the biosynthesis of metal-containing pterin (MPT), the direct precursor of Moco. Drought activates the biosynthesis of ABA through up-regulation of zeaxanthin epoxidase (ZEP) and 9-*cis*-epoxycarotenoid dioxygenase (NCED) expression, two key enzymes in ABA biosynthesis. Under S deficiency, the expression of *NCED* and *ABA3*, the gene responsible for Moco biosynthesis, is activated. *OAS-TL O*-acetylserine (thiol) lyase, *SAT* serine acetyl transferase

biosynthesis started to reveal when normal levels of ABA could be restored in *sultr3,1* knockout *Arabidopsis* seedlings upon feeding with cysteine (Cao et al. 2014), thus highlighting the role of cysteine in ABA biosynthesis. The precursor of cysteine *O*-acetylserine (OAS) is biosynthesized in the cytosol, mitochondria, and plastids through the activity of serine acetyl transferase (SAT), which transfers acetyl groups to serine, thus OAS. OAS in turn is converted to cysteine by the activity OAS-TL, which transfers sulfide to OAS. The sulfide is synthesized in the plastid through sulfate reduction pathway (Feldman-Salit et al. 2009). Hence, the availability of sulfides is of pivotal importance for cysteine biosynthesis, which in turn is necessary for ABA biosynthesis.

AAO catalyzes the last step of ABA biosynthesis and it is one of the four molybdenum cofactors containing enzymes in plants and its activity was correlated with the level of ABA (Schwarz and Mendel 2006; Szepesi et al. 2009;

Zdunek-Zastocka 2010). The tomato *flacca* mutation is deficient in ABA and exhibits wilty phenotype, because their stomata resist closure (Tal et al. 1979). It was found that this mutant lacks AAO and has concomitant decreased levels of ABA (Sagi et al. 2002). For AAO activity, Moco must be sulfurated (S-Moco). This sulfuration step is catalyzed by Moco sulfurase, which is encoded by the *ABA3* gene (AT1G16540) in *Arabidopsis*. Cysteine is the donor of sulfur in the Moco sulfurase catalyzed reaction (Bittner et al. 2001). Constitutive expression of *LOS5*/*ABA3* locus from *Arabidopsis* in soybean alleviated wilting under drought stress, promoted proline accumulation and increased antioxidant enzymes accompanied by increased ABA levels (Li et al. 2013). These results link the ABA biosynthetic pathway to the sulfate reduction and the following cysteine biosynthesis pathway. Cao et al. (2014) showed that AAO activity in *sultr3* mutants was lower compared to wild type levels. Moreover, exogenous feeding with cysteine could restore the normal levels of AAO activity, which were observed in the wild type. This also explains why *sultr3* mutants could restore normal ABA levels after cysteine feeding. Under low sulfur supply, the levels of ABA3 and NCED3 transcripts increased in both wild type *Arabidopsis* and *sultr3;1* mutants and it was shown that the promoters of both genes contain sulfur deficiency responsive elements.

Two more sulfur atoms are required for the synthesis of ABA. In the cytosol, metal-containing pterin (MPT) synthase produces MPT, the direct precursor of Moco, by transferring two sulfur atoms and one copper atom to cyclic pyranopterin monophosphate. The donor of these two sulfur atoms is not yet known in plants (Bittner and Mendel 2010). The *vp10* mutant of maize and *cnx1* mutants of *Arabidopsis*, lacking the CNX1 enzyme that catalyzes that last step in the MPT biosynthesis, were shown to be deficient of AAO activity and to have a lower ABA and indole acetic acid (IAA) content compared to the wild type (Porch et al. 2006).

Drought imposes effects indirectly through affecting steps in ABA biosynthesis. Xiong et al. (2002) showed that osmotic stress enhanced the transcription level of *LOS6*/*ABA1*, encoding ZEP in *Arabidopsis*, which is involved in an early step in ABA biosynthesis. Furthermore, they demonstrated that ABA exhibits a positive feedback regulation in *LOS6*/*ABA1* expression. The oxidative cleavage of neoxanthin by NCED is a rate limiting step in ABA biosynthesis. It was found that levels of ZEP and NCED expression are increased by drought and salt stress (Xiong et al. 2002). Constitutive expression of NCED from tomato in *Petunia* under the control of the stress-inducible promoter *rd29A* improved the plant's tolerance to drought (Estrada-Melo et al. 2015). The dependence of AAO on Moco cofactor that must be sulfurated in order to be effective in activation of AAO highlights the importance of sulfur metabolism for plants to cope with drought and other abiotic stresses through ABA-dependent mechanisms. Moreover, knockout of sulfur transferase SULTR3 affects ABA levels (Cao et al. 2014; Gallardo et al. 2014). These results indicate a sufficient sulfur supply, and coordinated sulfur metabolism is crucial for plants to cope with different stresses.

9.3.2 Role of ABA in Drought Stress Signal Cascades

Drought is sensed by the root system and leads to both hydraulic and chemical signals that start in the root system and extend to shoot. Drought also triggers ABA biosynthesis in roots and in leaf chloroplasts and finally results in passive and active (ABA-induced) closure of stomata (Comstock 2002; Tombesi et al. 2015). However, the contribution of each factor is not well characterized. ABA biosynthesis under drought stress imposes its effect through changes in gene expression as well as through a complex network of signaling. Unlike the pathway of ABA biosynthesis, which is well identified, the mechanisms by which ABA regulates different physiological and developmental processes are beginning to unfold through genetic analyses accompanied by physiological analyses of *Arabidopsis* and other important crop plants with sequenced genomes (Munemasa et al. 2013; Ye et al. 2012). Gonzalez-Guzman et al. (2012) and Kuhn et al. (2006) identified in a screen of a 35S::cDNA library of *Arabidopsis* an ABA insensitive mutant that could germinate on a medium containing 100 μM ABA, a concentration high enough to completely inhibit the germination of wild type seeds. This mutant overexpressed *AtPP2CA*, encoding protein phosphatase type 2C (PP2C). PP2C is a group of protein phosphatases, the subgroup A is involved in the ABA signaling pathway. They act as negative regulators through binding to SNF-1 related protein kinases and inactivating them through dephosphorylation at the Ser/Thr residues in their activation loop (Umezawa et al. 2009). Similar results have been obtained by Gosti et al. (1999) and Rubio et al. (2009) in *Arabidopsis*, by Zhang et al. (2014) in *Artemisia*, and by You et al. (2014) in rice. These results clearly demonstrate that PP2C acts as negative regulator of ABA signaling. The PP2Cs, ABI1 and ABI2, are key players in ABA signal transduction in *Arabidopsis* (Kuhn et al. 2006; Yoshida et al. 2006). They negatively regulate the ABA signaling at early steps of the pathway.

Pyrabactin is a synthetic sulfonamide that has been shown to act as an agonist of ABA (Melcher et al. 2010) and was shown to act through *PYRABACTIN RESISTANCE* (*PYR1*), a START-containing protein (PYR/PYL), known to be involved in intercellular lipid transport and metabolism as well as signal transduction (Park et al. 2009). Furthermore, ABA was shown to bind to PYR1 and upon binding the ABA/PYR1 complex can bind PP2Cs group A and inhibits their enzymatic activity. This demonstrates that PYR/PYLs are the terminal ABA acceptors in the ABA signaling pathway (Ma et al. 2009) and they are essential for the ABA-induced stomatal closure under drought stress as it has been shown in tomato (Gonzalez-Guzman et al. 2012) and *Arabidopsis* (Wang et al. 2013). Similar results have been shown in rice, that OsPP108 (group PP2C) is inducible under ABA, salt, and drought treatments. Constitutive expression of OsPP108 in transgenic *Arabidopsis* renders them highly insensitive to high levels of exogenous ABA. This adds to the evidence that PP2C group act as negative regulators of ABA signaling (Singh et al. 2015). PP2Cs are known to interact with SNF1 (Sucrose Non-Fermenting kinase 1)-related protein kinases OST1/SnRK2.6/SnRK2E that act as positive ABA regulators. OST1 functions as a positive regulator of stomatal closure by activating SLAC, the ion channel, and inhibits KAT1, the cation channel, in guard cell membrane (Raghavendra et al. 2010). Upon the interaction between PP2C and OST1/SnRK2.6/SnRK2E kinases, they lose their kinase activity, due to dephosphorylation by PP2C at Ser/Thr residues in their active sites (Umezawa et al. 2009; Vlad et al. 2009).

H_2O_2 is produced in tissues under stress conditions. It acts as a secondary messenger of ABA signaling. ABI1 (a PP2C group member) has been shown to be reversibly inhibited by H_2O_2, which was produced due to stress (Meinhard and Grill 2001). Under low water potential, ABA activates NADH oxidase to produce H_2O_2. Thus, H_2O_2 generation during ABA signaling seems to inactivate the negative regulator of the ABA response. Hence, inactivation of PP2C activities can be directly or indirectly imposed by ABA upon drought stress (Meinhard and Grill 2001; Sridharamurthy et al. 2014).

9.4 Role of Sulfur-Containing Metabolites During Drought Stress

9.4.1 Glutathione and Its Precursor Cysteine

Glutathione is a major antioxidant molecule in eukaryotic systems that is involved in maintaining the redox homeostasis of the cell, detoxification of xenobiotics as well as regulation of cell cycle transition from G to S phases (Anjum et al. 2012; Cairns et al. 2006; Pasternak et al. 2008). It is widely spread in the plant kingdom and in all tissues. All plant families contain glutathione except Fabaceae, which contain homoglutathione instead (a homologue of glutathione, which partially or completely replaces glutathione) (Colville et al. 2015; de Carvalho et al. 2010).

Glutathione is synthesized in the cytoplasm and chloroplast through the catalysis of two enzymes requiring ATP. γ-glutamylcysteine synthetase (GSH1 or γ-ECS) catalyzes the synthesis of γ-glutamylcysteine and it exists exclusively in chloroplasts and was demonstrated to be the rate limiting step. Glutathione synthetase (GSH2 or GSHS) adds glycine to γ-glutamylcysteine to create glutathione. GSH2 exists both in the cytoplasm and in the chloroplast (Pasternak et al. 2008).

Glutathione transporters are localized in the membranes of different cellular organelles, controlling the shuffling of glutathione among them, in order to keep cellular homeostasis, especially during biotic and abiotic stresses. However, the molecular identity of these transporters remains unknown to a large extent (Bachhawat et al. 2013; Bogs et al. 2003; Zechmann et al. 2014; Zhang et al. 2004). The yeast (*Saccharomyces cerevisiae*) mutant *hgt1*, that is deficient of glutathione transporter, was able to grow normally on a medium containing glutathione as the sole sulfur source, after its transformation with the *OsGT1* gene from rice (Zhang et al. 2004). Similarly, *BjGT1* from *Brassica juncea* was able to complement the *hgt1* mutant (Bogs et al. 2003). *BjGT1* is expressed in the leaves, to lower extend in stems and not in roots. The level of *BjGT1l* transcripts increased upon exposure to

Cd stress, indicating its role in cellular glutathione redistribution under heavy metals stress (Bogs et al. 2003). Three plastidal thiol transporters have been identified in *Arabidopsis*. Knockout mutants of these transporters showed GSH deficiency, heavy metal sensitivity, and hypersensitivity to *Phytophthora* infection. These transporters are homologues to *PfCRT* that was shown to have a glutathione transport function in *Plasmodium falciparum* (Maughan et al. 2010; Patzewitz et al. 2013). Subcellular compartmentation of glutathione is essential for proper growth under normal growth conditions. Maintaining the glutathione level in mitochondria is very essential for normal phenotypic growth in *Arabidopsis*. In the *Arabidopsis thaliana* mutant *pad2-1*, deficient of glutamylcysteine synthetase, the mitochondrial glutathione levels remained similar to the wild type, but they decreased in other cellular compartments (Zechmann et al. 2008). The *rml1* mutant contained lower glutathione content in all cellular compartments, including mitochondria. *rml1* showed a dwarf root system and small shoots and leaves, while *pad2-1* showed a normal phenotype under non-stressed conditions, thus reflecting the importance of maintaining glutathione levels in mitochondria for proper growth (Cheng et al. 1995; Zechmann and Müller 2010).

GSH1 knockout mutants of *Arabidopsis* that are deficient of postembryonic γ-glutamylcysteine synthetase fail to develop embryos. This can be rescued by external supplementation with glutathione (Cairns et al. 2006).

9.4.2 Role of Glutathione and Its Precursor Cysteine in Drought Stress

Drought affects the biosynthesis of glutathione at different pathways, either the enzymes of glutathione biosynthesis per se or the biosynthesis of its precursor molecules, notably cysteine. Drought affects glutathione biosynthesis enzymes at levels of expression and posttranslational modification. The expression of γ-glutamylcysteine synthetase (γ-GCS) from *Vigna radiata* does not change under conditions of progressive drought. However, the enzymatic activity is altered during drought and recovery periods, thus indicating a posttranslational regulation of Vrγ-GCS. These changes correlated with changes in H_2O_2 and lipid peroxidation levels, linking the glutathione biosynthesis process to the signals of drought stress (Sengupta et al. 2012). Similarly, the γ-glutamylcysteine synthetase activity increased under drought in *Arabidopsis* (May et al. 1998), *Nicotiana* (Kumar et al. 2014), and in *Phragmites communis* under drought and saline conditions (Chen et al. 2003a, b).

The expression of *Brassica rapa* BrECS1 and BrECS2 in transgenic rice, under the control of the stress-inducible promoter *Rab21*, increased the germination rate under salt conditions, led to a better glutathione redox state, enhanced growth and lowered oxidative stress and generally enhanced tolerance to abiotic stresses (Bae et al. 2013). Impaired expression of sulfite reductase, a key enzyme in the synthesis of cysteine, which is the precursor of glutathione biosynthesis, by RNAi caused early leaf senescence in tomato and it has been found to contain higher levels of sulfite and sulfate and lower levels of glutathione (Yarmolinsky et al. 2014).

Upon drought stress, *Vitis vinifera* leaves expressed higher levels of SERTA2.1, one of the four serine acetyl transferases found in *Vitis* and that is part of the cysteine synthase complex. SERTA2.1 is localized in the cytosol and plastids. High light and temperature did not affect the expression of this isoform, thus indicating drought-specific sulfur metabolism changes in *Vitis* (Tavares et al. 2015).

The interplay between glutathione content and ABA signaling in guard cells and the consequent stomatal closure has been demonstrated (Akter et al. 2010; Munemasa et al. 2013; Okuma et al. 2011). Glutathione has been shown to negatively regulate ABA-induced guard cell movement. The *Arabidopsis cad2-1* mutant that is deficient in γ-GCS enzyme and the application of 1-chloro-2,4-dinitrobenzene (CDNB), a chemical that decreases GSH content, to wild type *Arabidopsis* enhanced the ABA-induced stomatal closure. Restoring levels of GSH by external application of glutathione monoethyl ether restored the phenotype of wild type *Arabidopsis*. The *cad2-1* mutant showed increased ABA-induced ROS accumulation in the protoplast and enhanced ROS activation of Ca^{++} permeable channels.

The different organelles differ in their content of glutathione under normal non-stressed conditions. The mitochondria contain the highest concentration of glutathione, ranging from 8.7 to 15.1 mM in young and old leaf apexes of *Arabidopsis*, respectively. The second highest concentration has been found in nuclei (5.7–9.5 mM) followed by peroxisomes (2.6–4.8 mM) and cytosol (2.8–4.5 mM) (Koffler et al. 2013). The changes in glutathione content under drought stress were shown to be organelle- and stage-specific. The *Arabidopsis* mutant *pad2-1* (glutathione-deficient) and *vtc2-1* (ascorbate-deficient) both showed a drop in the glutathione, but not in the ascorbate content in mitochondria and nuclei. This occurs at early stages of drought when soil water content dropped, but drought was not yet measurable in leaves. At later stages of drought, the content of glutathione and ascorbate in chloroplasts and the peroxisome was correlated to the high accumulation of H_2O_2 in vacuoles. These results may indicate that glutathione could function as a drought stress signal from root to shoot system at early stages of drought (Koffler et al. 2014).

9.5 Role of Sulfated Compounds During Drought Stress

9.5.1 Glucosinolates

Glucosinolates (Gls) are major secondary compounds found in 16 families within the order Brassicales, including *A. thaliana* and economically important crop plants such as *B. napus* (rapeseed), *B. rapa* (Chinese cabbage, Chinese mustard, bok choy), and *B. oleracea* (broccoli, cauliflower, kale). Glucosinolates consist of a β-D-glucopyranose moiety, which is linked via a sulfur atom to a (*Z*)-*N*-hydroximino sulfate ester (Fig. 9.3). They are divided into three groups according to their amino

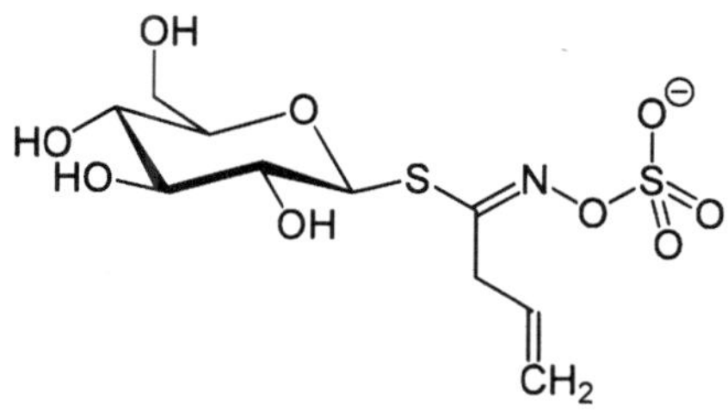

Fig. 9.3 Sinigrin as an exemplary structure of a glucosinolate. Sinigrin is a common aliphatic, methionine derived glucosinolate found in Brassicaceae

acid precursor: aliphatic Gls, derived from Ala, Leu, Ile, Val, and Met; indolic Gls, derived from Trp; and aromatic Gl derived either from Phe or Tyr. The biosynthesis can be divided into three stages: In the first stage, precursor amino acid side chains (only in case of Met and Phe) are elongated, by the addition of one or more methylene groups. In the second stage, the precursor amino acid is transferred into the core Gl. In the third stage, secondary modifications of the amino acid side chain take place (Sønderby et al. 2010). Due to the side chain elongation and the secondary modifications, there are over 200 different Gl structures reported in the plant order Brassicales (Clarke 2010). The Gl composition and the quantity depend on the species, the plant organ, and developmental stage (Kliebenstein et al. 2001). Intact Gls are stored in the vacuole and are non-toxic to cells. However, when the cell tissue is damaged, Gls are hydrolyzed by thioglucosidases called myrosinases, leading to biologically active compounds such as thiocyanates, isothiocyanates, and nitriles. These breakdown products of Gls function as defense compounds against insects, pathogens, and herbivores (Agrawal and Kurashige 2003; Hopkins et al. 2009; Manici et al. 1997; Rask et al. 2000; Tierens et al. 2001). Gl-derived isothiocyanates, for example, had antimicrobial activity against 7 out of 9 tested fungi and all out of 4 bacteria. *Pseudomonas syringae* was the most sensitive, as 28 μM isothiocyanates led to 50 % growth inhibition (Tierens et al. 2001). Furthermore, Gl breakdown products are interesting compounds for plant breeding as their consumption might reduce the risk of heart disease and carcinogenesis (Traka and Mithen 2009). Biosynthesis and functionality have been reviewed previously by Ishida et al. (2014) and Sønderby et al. (2010).

9.5.2 *Influence of Drought on Gl Biosynthesis*

As the function of Gls in response to biotic stress is overall agreed, it is still not yet clear what role Gls play in abiotic stress, including drought stress. The effect of drought on Gl content has been studied in several species, though with inconsistent results. Numerous studies agreed that drought stress elevates the Gl content, whereat especially the aliphatic Gl content increased, while the indolic Gls decreased or were not affected. These effects were shown on *A. thaliana* (Mewis et al. 2012), *B. napus* (Jensen et al. 1996), *B. oleracea* (Radovich et al. 2005), *B. oleracea* var.

italica (Paschold et al. 2000), *B. rapa* ssp. *rapifera* (Zhang et al. 2008), *B. juncea* (Tong et al. 2014), *B. carinata* (Schreiner et al. 2009), *N. officinale* (Gardner 2002) and suggested for *R. nasturtium-aquaticum* (Gershenzon 1984). For example, in *B. juncea* leaves, a reduction of soil water content from 18 to 6 % for 21 days led to an increase of aliphatic Gls from 13.6 mg g^{-1} dry weight to 16.7 mg g^{-1}, while the indolic Gl content remained unaffected (Tong et al. 2014). In contrary, studies on *B. oleracea* (Robbins et al. 2005), *B. oleracea* var. *italica* (Khan et al. 2010, 2011), *Boechera holboellii* (Haugen et al. 2008), and *Alliaria petiolata* (Gutbrodt et al. 2011) came to the conclusion that Gls were not affected by drought as there were no changes or even reduction of Gl content upon drought stress. For example, in *B. oleracea* var. *italica*, a reduction of soil water content from 70–75 to 35–40 % for 14 days only led to an insignificant reduction of aliphatic Gls from 0.42 μmol g^{-1} dry weight to 0.34 μmol g^{-1} dry weight, while indolic Gl content decreased from 2.39 μmol g^{-1} dry weight to 0.92 μmol g^{-1} dry weight (Khan et al. 2010).

A possible explanation for the contradictory results of the studies listed above could be the application of different stress levels. In case of *B. napus*, seed Gl content only increased after the water potential was below −1.4 MPa for approximately 10 days, but not under mild but prolonged stress (Jensen et al. 1996). Khan et al. (2011) argued similarly when discussing their observation of total Gl decrease upon drought stress in *B. oleracea* var. *italica* (reduction from 2.81 μmol g^{-1} dry weight to 1.26 μmol g^{-1} dry weight after 2 weeks of drought stress). It has to be considered, too, that the majority of the Gls in treated and control *B. oleracea* var. *italica* plants were indolic (85 % and 73 %, respectively) and a decrease of indolic Gls upon drought was also observed in studies arguing towards an effect of drought on Gl synthesis (Mewis et al. 2012; Schreiner et al. 2009; Tong et al. 2014). Additional possible explanation given for the contradicting results was genotypically different responses to drought stress and different vegetative stages of the plants (Khan et al. 2011; Schreiner et al. 2009).

Further questions remain about the potential functions of Gls in relation to drought stress. Are the observed changes in Gl content a specific answer to drought stress or is it an unspecific consequence of stress in general? The results of Schreiner et al. (2009) indicated an inverse correlation of leaf Gl content with the relative water content (RWC) of the leaves in *B. carinata*. For example, when the RWC dropped under 75 %, for every 10 % drop in RWC, the 2-propenyl Gl linearly increased by 1.7 mg g^{-1} dry weight. Hence, it was concluded that the Gl metabolism is linked to the soil water content. A correlation of the plant's water status with the total Gl was also reported for *B. napus* (Jensen et al. 1996). Here the Gl content increased at low turgor by 1.49 μmol g^{-1} dry weight per day of drought stress when the RWC was less than 82 %. Supported by findings that salt stress in *B. oleracea* var. *italica* also increased Gl concentrations (López-Berenguer et al. 2008), it was therefore suggested that Gls are involved in the process of osmotic adjustment (Schreiner et al. 2009). In *A. thaliana*, drought stress also increased the concentrations of *MYB29*, *MYB76*, and *MAM1* (Mewis et al. 2012), which are transcription factors responsible for the up-regulation of aliphatic and downregulation of indolic Gls (Gigolashvili et al. 2008; Textor et al. 2007). These findings indicate that the increased Gl contents were due to a specific response and not a consequence of shifts in other connected metabolic pathways.

Though several studies have shown an effect of drought on Gl content, the function that Gls might play on a physiological level as a stress response remains unclear. Several studies indicated a connection of drought-induced Gl accumulation to ABA formation. In a topsoil drying experiment with *B. juncea*, where one part of the roots was under drought stress, while the other was well watered, an increase of aliphatic Gls and ABA was only measured in the stressed part of the roots. Additionally, aliphatic Gl and ABA concentrations increased in the leaves, indicating a connected function in root-to-shoot signaling (Tong et al. 2014). Furthermore, it was shown that the application of ABA to *B. campestris* ssp. *oleifera* led to an increase of Gls in the seeds, especially the aliphatic 2-hydroxy-3-butenyl Gl (Bano et al. 2009). Further interactions between Gls and the plant hormone network were indicated by cytochrome P450 CYP79F1 and CYP79F2 RNAi mutant analyses in *A. thaliana* (Chen et al. 2012b). CYP79F1 and CYP79F2 catalyze the formation of short- and long-chained aliphatic Gls from chain-elongated methionines (Chen et al. 2003a, b; Reintanz et al. 2001). The perturbation of aliphatic Gl synthesis by RNAi of CYP79F1 and CYP79F2 led to a decrease of aliphatic and increase of indolic Gls. Furthermore, a cross-talk between Gl and hormone metabolism was shown. The mutants showed increased levels of ABA and cytokinins, while jasmonic acid, salicylic acid and IAA decreased (Chen et al. 2012b). Water stressed *A. thaliana* plants showed downregulation of *PR1*, which is a gene associated with the salicylic acid pathway, and decreased levels of 4-methoxyindol-3-ylmethyl Gl led to the conclusion that the salicylic acid signaling pathway is a key element in regulating 4-methoxyindol-3-ylmethyl Gl production from the indol-3-ylmethyl Gl precursor (Mewis et al. 2012). Further results showing a connection of Gls and the plant's hormone network (Chen et al. 2012b; Malitsky et al. 2008; Morant et al. 2010) support the suggestion of Gls playing a part in drought stress response.

The shifts in Gl content have been explained by a shift from primary to secondary metabolism upon drought stress (Schreiner et al. 2009; Tong et al. 2014), based on the Growth-Differentiation Balance hypothesis by Herms and Mattson (1992). According to this hypothesis, growth-related processes and differentiation-related processes compete for common substrates and energy. Hence, any environmental factor decreasing the growth rate to a degree lower than photosynthesis leads to an increase of the resource pool for secondary metabolites. Therefore, the increase of aliphatic Gls would be a rather unspecific result of stress instead of a specific answer to drought. This is supported by the findings that Gl content in general is also affected by other abiotic stresses such as temperature and UV radiation (Schonhof et al. 2007a) and increased CO_2 levels (Schreiner et al. 2006; Schonhof et al. 2007b). It was also suggested that Gls increase as a response to oxidative stress (Schreiner et al. 2009), due to reported shifts within the aliphatic Gls towards methylsulfinyl and alkenyl Gls upon drought stress in *B. oleracea* (Radovich et al. 2005) and *B. rapa* ssp. *rapifera* (Zhang et al. 2008). This was supported by the RNAi *CYP79F1* and *CYP79F2* mutants, which proteomic and metabolomic data also indicated increased oxidative stress (Chen et al. 2012b). Oxidative stress is caused by several kinds of abiotic stress, such as radiation, temperature, heavy metals and drought, hence, the changes in Gl contents would rather indicate an unspecific response to stress in general.

Overall, there are several studies indicating an effect of drought on Gl contents. Whether or not Gls play an active role in response to drought and what function they have on a molecular level still remains unclear. Hence, suggested functions such as Gls playing part in osmotic regulation, interconnection with hormone signaling or serving as antioxidants, yet still need more experimental evidence.

9.5.3 Accumulation of More Sulfated Compounds During Drought Stress

Beside Gls, there are several other sulfated compounds in plants. Most of them are sulfated by a protein family called sulfotransferases (SOTs) (EC 2.8.2.-). They catalyze the transfer of a sulfuryl group from the ubiquitous donor 3′-phosphoadenosine 5′-phosphosulfate to numerous compounds, such Gls, flavonoids, brassinosteroids and choline (Hirschmann et al. 2014). There is only little information about drought-associated functions of sulfated compounds available. However, it was shown in halophytic *Limonium* species that choline-*O*-sulfate accumulates at high levels during stress (Hanson et al. 1991). Also the respective SOT showed an increased activity under high salinity conditions (Rivoal and Hanson 1994). The choline-*O*-sulfate acts as an osmoprotectant and is suggested to contribute to adaption to rough environmental conditions. In comparison to other osmoprotectants that accumulate in halophytes, such as glycine betaine, choline-*O*-sulfate was advantageous in sulfate-containing soils (Hanson et al. 1994). It was suggested that choline-*O*-sulfate was preferred due to sulfate detoxification properties (Hanson et al. 1994).

Sulfated polysaccharides were found in the cell walls of all marine algae to date (Arad and Levy-Ontman 2010), but their specific functions remain unclear. Aquino et al. (2005, 2011) also identified sulfated polysaccharides in marine angiosperms and could show a correlation between salinity and sulfated polysaccharides in the cell walls. On the other hand, there were no sulfated polysaccharides detected in *O. sativa* after salt treatment. Therefore, it was concluded that the presence of sulfated polysaccharides in the cell wall is an adaption to high salt environments. As a possible function, it was speculated that the sulfated polysaccharides increase the Donnan potential, thus facilitating the ion transport. Surprisingly, Dantas-Santos et al. (2012) also detected sulfated polysaccharides in the fresh water plants *Eichhornia crassipes*, *Hydrocotyle bonariensis* and *Nymphaea ampla*, which suggests further factors for sulfated polysaccharide production in plants.

Otherwise there are only indications for drought-related functions of sulfated compounds. The *A. thaliana* SOT12, which sulfates several substrates such as flavonones, brassinosteroids, salicylic acid and xenobiotics (Chen et al. 2015; Hirschmann et al. 2014), was induced by salt stress, osmotic stress, and ABA (Baek et al. 2010). Additionally, a *sot12* knockout mutant showed hypersensitivity to salt stress and ABA in seed germination (Baek et al. 2010). Due to the broad specificity of AtSOT12, it is difficult to make conclusions regarding its function in abiotic stress. It was suggested that the sulfation detoxifies salicylic acid (Baek et al. 2010), which was shown to increase the creation of reactive oxygen species during salt and osmotic stress (Borsani et al. 2001). In *O. sativa*, 35 *SOTs* we identified and ten of those showed an up-regulation upon drought stress, though in different developmental stages (Chen et al. 2012a). However, due to the unknown substrates of these SOTs, no further conclusions about the functions can be made.

The by-product of SOT catalyzed sulfation reactions, 3′-phosphoadenosine 5′-phosphate (PAP), is suggested to function as a retrograde drought signal from the chloroplast to the nucleus. In *A. thaliana*, a 30-fold increase of PAP was observed under drought conditions (Estavillo et al. 2011). The PAP content is regulated in the chloroplasts by the adenosine bisphosphate phosphatase SAL1, which dephosphorylates PAP to adenosine monophosphate (Quintero et al. 1996). Consequently, a loss of function mutation of SAL1 led to an increase of PAP, but also to a 50 % higher drought tolerance. The targeting of SAL1 to the nucleus of *sal1* knockout mutants led to the complete complementation of PAP and drought tolerance to wild type levels (Estavillo et al. 2011). It was suggested that PAP moves into the nucleus, possibly transported by a PAPS/PAP chloroplastic antiporter (Gigolashvili et al. 2012), and there inhibits the RNA-degrading activity of 5′-3′exoribonucleases (XRNs), as it was previously shown in yeast (Dichtl et al. 1997). The XRNs degrade uncapped RNAs, such as excised hairpin loops that form part of precursor *miRNA* (Kastenmayer and Green 2000). Furthermore, *Arabidopsis xrn* and *sal1* mutants show similar morphological and molecular phenotypes and the *sal1* mutants accumulate XRN substrates. The inhibition of XRNs is assumed to lead to the prevention of posttranscriptional gene silencing of stress response genes (Estavillo et al. 2011).

9.6 Conclusion

An increasing demand for sulfate during metabolic adaptation reactions during drought stress reflects specific roles of sulfur-containing compounds. Not only sulfur-containing mass products such as osmolytes and osmoprotectants are formed but products of sulfur-containing compounds are involved in (retrograde) signaling pathways, like PAP. ABA biosynthesis and sulfur metabolism interplay during abiotic stress reactions to ensure sufficient cysteine for ABA production. Overexpression of *AtSULTR3;1* and/or *AtSULTR3;4* might increase the concentration of cysteine and ABA, and therefore increase drought tolerance in these plants lines. Since induced expression of key genes in sulfur assimilation during drought stress was observed more analysis is needed to identify ABA- and drought-responsive elements within these genes. Finally, the role of Gl in drought stress responses is not clear because concentrations of aliphatic Gl are related to ABA formation whereas indole and aromatic Gl decreased during drought stress. Probably, the breakdown products of Gl need to be analyzed in detail before a conclusive role of Gl can be described. A better knowledge of all global metabolic cross-links is a prerequisite for a successful improvement of drought tolerance, a multi-gene trait, in plants, either by conventional breeding or biotechnological means.

Acknowledgement The cooperation was funded by the German Research Foundation (PA 764/10-1) and the Egyptian Government (fellowship to Dr. Abu El-Soud).

References

Adie BA, Perez-Perez J, Perez-Perez MM, Godoy M, Sanchez-Serrano J, Schmelz EA, Solano R. ABA is an essential signal for plant resistance to pathogens affecting JA biosynthesis and the activation of defenses in *Arabidopsis*. Plant Cell. 2007;19(5):1665–81. doi:10.1105/tpc.106.048041.

Agrawal AA, Kurashige NS. A role for isothiocyanates in plant resistance against the specialist herbivore *Pieris rapae*. J Chem Ecol. 2003;29(6):1403–15.

Akter N, Sobahan MA, Hossain MA, Uraji M, Nakumara Y, Mori IC, Murata Y. The involvement of intracellular glutathione in methyl jasmonate signaling in *Arabidopsis* guard cells. Biosci Biotechnol Biochem. 2010;74(12):2504–6. doi:10.1271/bbb.100513.

Ambrosone A, di Giacomo M, Leone A, Grillo MS, Costa A. Identification of early induced genes upon water deficit in potato cell cultures by cDNA-AFLP. J Plant Res. 2013;126:169–78.

Anjum NA, Ahmad I, Mohmood I, Pacheco M, Duarte AC, Pereira E, Umar S, Ahmad A, Khan NA, Iqbal M, Prasad M. Modulation of glutathione and its related enzymes in plants' responses to toxic metals and metalloids—a review. Environ Exp Bot. 2012;75:307–24. doi:10.1016/j.envexpbot.2011.07.002.

Aquino RS, Landeira-Fernandez AM, Valente AP, Andrade LR, Mourão PAS. Occurrence of sulfated galactans in marine angiosperms: evolutionary implications. Glycobiology. 2005; 15(1):11–20. doi:10.1093/glycob/cwh138.

Aquino RS, Grativol C, Mourão PAS. Rising from the sea: correlations between sulfated polysaccharides and salinity in plants. PLoS One. 2011;6(4):e18862. doi:10.1371/journal.pone.0018862.

Arad SM, Levy-Ontman O. Red microalgal cell-wall polysaccharides: biotechnological aspects. Curr Opin Biotechnol. 2010;21(3):358–64. doi:10.1016/j.copbio.2010.02.008.

Bachhawat AK, Thakur A, Kaur J, Zulkifli M. Glutathione transporters. Biochim Biophys Acta Gen Subj. 2013;1830(5):3154–64. doi:10.1016/j.bbagen.2012.11.018.

Bae M, Kim Y, Kim I, Choe Y, Lee E, Kim Y, Park H, Yoon H. Transgenic rice overexpressing the *Brassica juncea* gamma-glutamylcysteine synthetase gene enhances tolerance to abiotic stress and improves grain yield under paddy field conditions. Mol Breed. 2013;31(4):931–45. doi:10.1007/s11032-013-9846-8.

Baek D, Pathange P, Chung J, Jiang J, Gao L, Oikawa A, Hirai MY, Saito K, Pare PW, Shi H. A stress-inducible sulphotransferase sulphonates salicylic acid and confers pathogen resistance in *Arabidopsis*. Plant Cell Environ. 2010;33(8):1383–92. doi:10.1111/j.1365-3040.2010.02156.x.

Bano A, Khan N, Faizanullah. Nutritive values of *Brassica campestris* L. Oil as affected by growth regulator treatments. J Chem Soc Pak. 2009;31(5):819–22.

Barciszewska-Pacak M, Milanowska K, Knop K, Bielewicz D, Nuc P, Plewka P, Pacak AM, Vazquez F, Karlowski W, Jarmolowski A, Szweykowska-Kulinska Z. *Arabidopsis* microRNA expression regulation in a wide range of abiotic stress responses. Front Plant Sci. 2015;6:410. doi:10.3389/fpls.2015.00410.

Barroso C, Romero LC, Cejudo FJ, Vega JM, Gotor C. Salt-specific regulation of the cytosolic O-acetylserine(thiol)lyase gene from *Arabidopsis thaliana* is dependent on abscisic acid. Plant Mol Biol. 1999;40(4):729–36.

Bittner F, Mendel RR. Cell biology of molybdenum. In: Hell R, Mendel RR, editors. Cell biology of metals and nutrients. Heidelberg: Springer; 2010. p. 119–43.

Bittner F, Oreb M, Mendel RR. ABA3 is a molybdenum cofactor sulfurase required for activation of aldehyde oxidase and xanthine dehydrogenase in *Arabidopsis thaliana*. J Biol Chem. 2001;276(44):40381–4. doi:10.1074/jbc.C100472200.

Bogs J, Bourbouloux A, Cagnac O, Wachter A, Rausch T, Delrot S. Functional characterization and expression analysis of a glutathione transporter, BjGT1, from *Brassica juncea*: evidence for regulation by heavy metal exposure. Plant Cell Environ. 2003;26(10):1703–11. doi:10.1046/j.1365-3040.2003.01088.x.

Borsani O, Valpuesta V, Botella MA. Evidence for a role of salicylic acid in the oxidative damage generated by NaCl and osmotic stress in *Arabidopsis* seedlings. Plant Physiol. 2001;126(3): 1024–30.

Bray EA. Classification of genes differentially expressed during water-deficit stress in *Arabidopsis thaliana*: an analysis using microarray and differential expression data. Ann Bot. 2002;89: 803–11.

Cairns NG, Pasternak M, Wachter A, Cobbett CS, Meyer AJ. Maturation of *Arabidopsis* seeds is dependent on glutathione biosynthesis within the embryo. Plant Physiol. 2006;141(2):446–55. doi:10.1104/pp.106.077982.

Cao M, Wang Z, Wirtz M, Hell R, Oliver DJ, Xiang C. SULTR3;1 is a chloroplast-localized sulfate transporter in *Arabidopsis thaliana*. Plant J. 2013;73(4):607–16. doi:10.1111/tpj.12059.

Cao M, Wang Z, Zhao Q, Mao J, Speiser A, Wirtz M, Hell R, Zhu J, Xiang C. Sulfate availability affects ABA levels and germination response to ABA and salt stress in *Arabidopsis thaliana*. Plant J. 2014;77(4):604–15. doi:10.1111/tpj.12407.

Chan KX, Wirtz M, Phua SY, Estavillo GM, Pogson BJ. Balancing metabolites in drought: the sulfur assimilation conundrum. Trends Plant Sci. 2013;18(1):18–29.

Chen K, Gong H, Chen G, Wang S, Zhang C. Up-regulation of glutathione metabolism and changes in redox status involved in adaptation of reed (*Phragmites communis*) ecotypes to drought-prone and saline habitats. J Plant Physiol. 2003a;160(3):293–301. doi:10.1078/0176-1617-00927.

Chen S, Glawischnig E, Jørgensen K, Naur P, Jørgensen B, Olsen C, Hansen CH, Rasmussen H, Pickett JA, Halkier BA. CYP79F1 and CYP79F2 have distinct functions in the biosynthesis of aliphatic glucosinolates in *Arabidopsis*. Plant J. 2003b;33(5):923–37.

Chen R, Jiang Y, Dong J, Zhang X, Xiao H, Xu Z, Gao X. Genome-wide analysis and environmental response profiling of SOT family genes in rice (*Oryza sativa*). Genes Genomics. 2012a;34(5):549–60. doi:10.1007/s13258-012-0053-5.

Chen Y, Pang Q, He Y, Zhu N, Branstrom I, Yan X, Chen S. Proteomics and metabolomics of *Arabidopsis* responses to perturbation of glucosinolate biosynthesis. Mol Plant. 2012b;5(5):1138–50. doi:10.1093/mp/sss034.

Chen J, Gao L, Baek D, Liu C, Ruan Y, Shi H. Detoxification function of the *Arabidopsis* sulphotransferase AtSOT12 by sulphonation of xenobiotics. Plant Cell Environ. 2015;38:1673–82. doi:10.1111/pce.12525.

Cheng JC, Seeley KA, Sung ZR. RML1 and RML2, *Arabidopsis* genes required for cell proliferation at the root tip. Plant Physiol. 1995;107(2):365–76.

Clarke DB. Glucosinolates, structures and analysis in food. Anal Methods. 2010;2(4):310–25. doi:10.1039/b9ay00280d.

Colville L, Sáez B, Clara M, Lewis GP, Kranner I. The distribution of glutathione and homoglutathione in leaf, root and seed tissue of 73 species across the three sub-families of the *Leguminosae*. Phytochemistry. 2015;115:175–83. doi:10.1016/j.phytochem.2015.01.011.

Comstock JP. Hydraulic and chemical signalling in the control of stomatal conductance and transpiration. J Exp Bot. 2002;53(367):195–200. doi:10.1093/jexbot/53.367.195.

Dantas-Santos N, Gomes DL, Costa LS, Cordeiro SL, Costa MSSP, Trindade ES, Franco CRS, Scortecci KC, Leite EL, Rocha HAO. Freshwater plants synthesize sulfated polysaccharides: heterogalactans from water hyacinth (*Eicchornia crassipes*). Int J Mol Sci. 2012;13(1):961–76. doi:10.3390/ijms13010961.

de Carvalho C, Maria H, Brunet J, Bazin J, Kranner I, Ad A-L, Zuily-Fodil Y, Contour-Ansel D. Homoglutathione synthetase and glutathione synthetase in drought-stressed cowpea leaves:

expression patterns and accumulation of low-molecular-weight thiols. J Plant Physiol. 2010;167(6):480–7. doi:10.1016/j.jplph.2009.10.023.

Dichtl B, Stevens A, Tollervey D. Lithium toxicity in yeast is due to the inhibition of RNA processing enzymes. EMBO J. 1997;16(23):7184–95. doi:10.1093/emboj/16.23.7184.

Ernst L, Goodger JQD, Alvarez S, Marsh EL, Berla B, Lockhart E, Jung J, Li P, Bohnert HJ, Schachtman DP. Sulphate as a xylem-borne chemical signal precedes the expression of ABA biosynthetic genes in maize roots. J Exp Bot. 2010;61(12):3395–405. doi:10.1093/jxb/erq160.

Estavillo GM, Crisp PA, Pornsiriwong W, Wirtz M, Collinge D, Carrie C, Giraud E, Whelan J, David P, Javot H, Brearley C, Hell R, Marin E, Pogson BJ. Evidence for a SAL1-PAP chloroplast retrograde pathway that functions in drought and high light signaling in *Arabidopsis*. Plant Cell. 2011;23(11):3992–4012. doi:10.1105/tpc.111.091033.

Estrada-Melo AC, Chao, Reid MS, Jiang C. Overexpression of an ABA biosynthesis gene using a stress-inducible promoter enhances drought resistance in *petunia*. Hortic Res. 2015;2:15013. doi:10.1038/hortres.2015.13.

Feldman-Salit A, Wirtz M, Hell R, Wade RC. A mechanistic model of the cysteine synthase complex. J Mol Biol. 2009;386(1):37–59. doi:10.1016/j.jmb.2008.08.075.

Gallardo K, Courty P, Le Signor C, Wipf D, Vernoud V. Sulfate transporters in the plant's response to drought and salinity: regulation and possible functions. Front Plant Sci. 2014;5:580. doi:10.3389/fpls.2014.00580.

Gardner G. The center for plants and human health: an interdisciplinary approach. NABC report on foods for health: integrating agriculture, medicine and food for future health. In: Eaglesham A, Carlson C, Hardy RWF, editors. National Agricultural Biotechnology Council. Ithaca: USG; 2002. p. 299–308.

Gershenzon J. Changes in level of plant secondary metabolites under water and nutrient stress. In: Timmermann BN et al., editors. Phytochemical adaptations to stress. New York: Plenum; 1984. p. 273–321.

Gigolashvili T, Engqvist M, Yatusevich R, Müller C, Flügge U. HAG2/MYB76 and HAG3/MYB29 exert a specific and coordinated control on the regulation of aliphatic glucosinolate biosynthesis in *Arabidopsis* thaliana. New Phytol. 2008;177(3):627–42. doi:10.1111/j.1469-8137.2007.02295.x.

Gigolashvili T, Geier M, Ashykhmina N, Frerigmann H, Wulfert S, Krueger S, Mugford SG, Kopriva S, Haferkamp I, Flügge UI. The Arabidopsis thylakoid ADP/ATP carrier TAAC has an additional role in supplying plastidic phosphoadenosine 5'-phosphosulfate to the cytosol. Plant Cell. 2012;24(10):4187–204. doi:10.1105/tpc.112.101964.

Gonzalez-Guzman M, Pizzio GA, Antoni R, Vera-Sirera F, Merilo E, Bassel GW, Fernández MA, Holdsworth MJ, Perez-Amador MA, Kollist H, Rodriguez PL. *Arabidopsis* PYR/PYL/RCAR receptors play a major role in quantitative regulation of stomatal aperture and transcriptional response to abscisic acid. Plant Cell. 2012;24(6):2483–96. doi:10.1105/tpc.112.098574.

Gosti F, Beaudoin N, Serizet C, Webb AA, Vartanian N, Giraudat J. ABI1 protein phosphatase 2C is a negative regulator of abscisic acid signaling. Plant Cell. 1999;11:1987–90.

Gutbrodt B, Mody K, Dorn S. Drought changes plant chemistry and causes contrasting responses in lepidopteran herbivores. Oikos. 2011;120(11):1732–40. doi:10.1111/j.1600-0706.2011.19558.x.

Hanson AD, Rathinasabapathi B, Chamberlin B, Gage DA. Comparative physiological evidence that β-alanine betaine and choline-O-sulfate act as compatible osmolytes in halophytic *Limonium* species. Plant Physiol. 1991;97(3):1199–205.

Hanson AD, Rathinasabapathi B, Rivoal J, Burnet M, Dillon MO, Gage DA. Osmoprotective compounds in the *Plumbaginaceae*: a natural experiment in metabolic engineering of stress tolerance. Proc Natl Acad Sci U S A. 1994;91(1):306–10.

Harb A, Krishnan M, Ambavaram MMR, Pereira A. Molecular and physiological analysis of drought stress in Arabidopsis reveals early response leading to acclimation in plant growth. Plant Physiol. 2010;154:1254–71.

Haugen R, Steffes L, Wolf J, Brown P, Matzner S, Siemens DH. Evolution of drought tolerance and defense: dependence of tradeoffs on mechanism, environment and defense switching. Oikos. 2008;117(2):231–44. doi:10.1111/j.2007.0030-1299.16111.x.

Herms DA, Mattson WJ. The dilemma of plants: to grow or defend. Q Rev Biol. 1992;67(3):283–335.

Hirschmann F, Krause F, Papenbrock J. The multi-protein family of sulfotransferases in plants: composition, occurrence, substrate specificity, and functions. Front Plant Sci. 2014;5:556. doi:10.3389/fpls.2014.00556.

Hopkins RJ, van Dam NM, van Loon JJA. Role of glucosinolates in insect-plant relationships and multitrophic interactions. Annu Rev Entomol. 2009;54:57–83. doi:10.1146/annurev.ento.54.110807.090623.

Ishida M, Hara M, Fukino N, Kakizaki T, Morimitsu Y. Glucosinolate metabolism, functionality and breeding for the improvement of *Brassicaceae* vegetables. Breed Sci. 2014;64(1):48–59. doi:10.1270/jsbbs.64.48.

Jensen CR, Mogensen VO, Mortensen G, Fieldsend JK, Milford G, Andersen MN, Thage JH. Seed glucosinolate, oil and protein contents of field-grown rape (*Brassica napus* L.) affected by soil drying and evaporative demand. Field Crop Res. 1996;47(2–3):93–105. doi:10.1016/0378-4290(96)00026-3.

Kastenmayer JP, Green PJ. Novel features of the XRN-family in *Arabidopsis*: evidence that AtXRN4, one of several orthologs of nuclear Xrn2p/Rat1p, functions in the cytoplasm. Proc Natl Acad Sci U S A. 2000;97(25):13985–90. doi:10.1073/pnas.97.25.13985.

Kawashima CG, Matthewman CA, Huang S, Lee B, Yoshimoto N, Koprivova A, Rubio-Somoza I, Todesco M, Rathjen T, Saito K, Takahashi H, Dalmay T, Kopriva S. Interplay of SLIM1 and miR395 in the regulation of sulfate assimilation in *Arabidopsis*. Plant J. 2011;66(5):863–76. doi:10.1111/j.1365-313X.2011.04547.x.

Khan MA, Ulrichs C, Mewis I. Influence of water stress on the glucosinolate profile of *Brassica oleracea* var. *italica* and the performance of *Brevicoryne brassicae* and *Myzus persicae*. Entomol Exp Appl. 2010;137(3):229–36. doi:10.1111/j.1570-7458.2010.01059.x.

Khan MAM, Ulrichs C, Mewis I. Water stress alters aphid-induced glucosinolate response in *Brassica oleracea* var. *italica* differently. Chemoecology. 2011;21(4):235–42. doi:10.1007/s00049-011-0084-4.

Kliebenstein DJ, Kroymann J, Brown P, Figuth A, Pedersen D, Gershenzon J, Mitchell-Olds T. Genetic control of natural variation in *Arabidopsis* glucosinolate accumulation. Plant Physiol. 2001;126(2):811–25. doi:10.1104/pp.126.2.811.

Koffler BE, Bloem E, Zellnig G, Zechmann B. High resolution imaging of subcellular glutathione concentrations by quantitative immunoelectron microscopy in different leaf areas of *Arabidopsis*. Micron. 2013;45:119–28. doi:10.1016/j.micron.2012.11.006.

Koffler BE, Luschin-Ebengreuth N, Stabentheiner E, Müller M, Zechmann B. Compartment specific response of antioxidants to drought stress in *Arabidopsis*. Plant Sci. 2014;227:133–44. doi:10.1016/j.plantsci.2014.08.002.

Kowitcharoen L, Wongs-Aree C, Setha S, Komkhuntod R, Srilaong V, Kondo S. Changes in abscisic acid and antioxidant activity in sugar apples under drought conditions. Sci Hortic. 2015;193:1–6. doi:10.1016/j.scienta.2015.06.043.

Kuhn JM, Boisson-Dernier A, Dizon MB, Maktabi MH, Schroeder JI. The protein phosphatase AtPP2CA negatively regulates abscisic acid signal transduction in *Arabidopsis*, and effects of *abh1* on AtPP2CA mRNA. Plant Physiol. 2006;140(1):127–39. doi:10.1104/pp.105.070318.

Kumar D, Datta R, Sinha R, Ghosh A, Chattopadhyay S. Proteomic profiling of γ-ECS overexpressed transgenic *Nicotiana* in response to drought stress. Plant Signal Behav. 2014;9(8), e29246. doi:10.4161/psb.29246.

Leon-Kloosterziel KM, Gil MA, Ruijs GJ, Jacobsen SE, Olszewski NE, Schwartz SH, Zeevaart JA, Koornneef M. Isolation and characterization of abscisic acid-deficient *Arabidopsis* mutants at two new loci. Plant J. 1996;10(4):655–61. doi:10.1046/j.1365-313X.1996.10040655.x.

Li Y, Zhang J, Zhang J, Hao L, Hua J, Duan L, Zhang M, Li Z. Expression of an Arabidopsis molybdenum cofactor sulphurase gene in soybean enhances drought tolerance and increases yield under field conditions. Plant Biotechnol J. 2013;11(6):747–58. doi:10.1111/pbi.12066.

López-Berenguer C, Martínez-Ballesta MC, García-Viguera C, Carvajal M. Leaf water balance mediated by aquaporins under salt stress and associated glucosinolate synthesis in broccoli. Plant Sci. 2008;174(3):321–8. doi:10.1016/j.plantsci.2007.11.012.

Ma Y, Szostkiewicz I, Korte A, Moes D, Yang Y, Christmann A, Grill E. Regulators of PP2C phosphatase activity function as abscisic acid sensors. Science. 2009;324:1064–8. doi:10.1126/science.1172408.

Malitsky S, Blum E, Less H, Venger I, Elbaz M, Morin S, Eshed Y, Aharoni A. The transcript and metabolite networks affected by the two clades of *Arabidopsis* glucosinolate biosynthesis regulators. Plant Physiol. 2008;148(4):2021–49. doi:10.1104/pp.108.124784.

Manici LM, Lazzeri L, Palmieri S. *In vitro* fungitoxic activity of some glucosinolates and their enzyme-derived products toward plant pathogenic fungi. J Agric Food Chem. 1997;45(7): 2768–73.

Maughan SC, Pasternak M, Cairns N, Kiddle G, Brach T, Jarvis R, Haas F, Nieuwland J, Lim B, Muller C, Salcedo-Sora E, Kruse C, Orsel M, Hell R, Miller AJ, Bray P, Foyer CH, Murray JAH, Meyer AJ, Cobbett CS. Plant homologs of the *Plasmodium falciparum* chloroquine-resistance transporter, PfCRT, are required for glutathione homeostasis and stress responses. Proc Natl Acad Sci U S A. 2010;107(5):2331–6. doi:10.1073/pnas.0913689107.

May MJ, Vernoux T, Sánchez-Fernández R, van Montagu M, Inzé D. Evidence for posttranscriptional activation of γ-glutamylcysteine synthetase during plant stress responses. Proc Natl Acad Sci U S A. 1998;95:12049–54.

Mehrotra R, Bhalothia P, Bansal P, Basantani MK, Bharti V, Mehrotra S. Abscisic acid and abiotic stress tolerance—different tiers of regulation. J Plant Physiol. 2014;171(7):486–96. doi:10.1016/j.jplph.2013.12.007.

Meinhard M, Grill E. Hydrogen peroxide is a regulator of ABI1, a protein phosphatase 2C from *Arabidopsis*. FEBS Lett. 2001;508(3):443–6. doi:10.1016/S0014-5793(01)03106-4.

Melcher K, Xu Y, Ng L, Zhou XE, Soon F, Chinnusamy V, Suino-Powell KM, Kovach A, Tham FS, Cutler SR, Li J, Yong E, Zhu J, Xu HE. Identification and mechanism of ABA receptor antagonism. Nat Struct Mol Biol. 2010;17(9):1102–8. doi:10.1038/nsmb.1887.

Melotto M, Underwood W, Koczan J, Nomura K, He SY. Plant stomata function in innate immunity against bacterial invasion. Cell. 2006;126(5):969–80. doi:10.1016/j.cell.2006.06.054.

Mewis I, Khan MA, Glawischnig E, Schreiner M, Ulrichs C. Water stress and aphid feeding differentially influence metabolite composition in *Arabidopsis thaliana* (L.). PLoS One. 2012;7(11), e48661. doi:10.1371/journal.pone.0048661.

Morant M, Ekstrøm C, Ulvskov P, Kristensen C, Rudemo M, Olsen CE, Hansen J, Jørgensen K, Jørgensen B, Møller BL, Bak S. Metabolomic, transcriptional, hormonal, and signaling cross-talk in superroot2. Mol Plant. 2010;3(1):192–211. doi:10.1093/mp/ssp098.

Munemasa S, Muroyama D, Nagahashi H, Nakumara Y, Mori IC, Murata Y. Regulation of reactive oxygen species-mediated abscisic acid signaling in guard cells and drought tolerance by glutathione. Front Plant Sci. 2013;4:472. doi:10.3389/fpls.2013.00472.

Neill SJ, Horgan R, Parry AD. The carotenoid and abscisic acid content of viviparous kernels and seedlings of *Zea mays* L. Planta. 1986;169(1):87–96. doi:10.1007/BF01369779.

Okuma E, Jahan MS, Munemasa S, Hossain MA, Muroyama D, Islam MM, Ogawa K, Watanabe-Sugimoto M, Nakumara Y, Shimoishi Y, Mori IC, Murata Y. Negative regulation of abscisic acid-induced stomatal closure by glutathione in *Arabidopsis*. J Plant Physiol. 2011;168(17): 2048–55. doi:10.1016/j.jplph.2011.06.002.

Park S, Fung P, Nishimura N, Jensen DR, Fujii H, Zhao Y, Lumba S, Santiago J, Rodrigues A, Chow TF, Alfred SE, Bonetta D, Finkelstein R, Provart NJ, Desveaux D, Rodriguez PL, McCourt P, Zhu J, Schroeder JI, Volkman BF, Cutler SR. Abscisic acid inhibits type 2C protein phosphatases via the PYR/PYL family of START proteins. Science. 2009;324:1068–71. doi:10.1126/science.1173041.

Paschold PJ, Kleber J, Adam ST, Bognar A, Tauscher B. Einfluss von Bewässerung und N-Düngung auf Ertrag und Sulforaphangehalt von Brokkoli (*Brassica oleracea*). Karlsruhe: Deutsche Gesellschaft für Qualitätsforschung; 2000. p. 57–66.

Pasternak M, Lim B, Wirtz M, Hell R, Cobbett CS, Meyer AJ. Restricting glutathione biosynthesis to the cytosol is sufficient for normal plant development. Plant J. 2008;53(6):999–1012. doi:10.1111/j.1365-313X.2007.03389.x.

Patzewitz E, Salcedo-Sora JE, Wong EH, Sethia S, Stocks PA, Maughan SC, Murray JA, Krishna S, Bray PG, Ward SA, Müller S. Glutathione transport: a new role for PfCRT in chloroquine resistance. Antioxid Redox Signal. 2013;19(7):683–95. doi:10.1089/ars.2012.4625.

Porch TG, Tseung C, Schmelz EA, Settles AM. The maize *Viviparous10/Viviparous13* locus encodes the *Cnx1* gene required for molybdenum cofactor biosynthesis. Plant J. 2006;45(2):250–63. doi:10.1111/j.1365-313X.2005.02621.x.

Quintero FJ, Garciadeblás B, Rodríguez-Navarro A. The *SAL1* gene of *Arabidopsis*, encoding an enzyme with 3′(2′),5′-bisphosphate nucleotidase and inositol polyphosphate 1-phosphatase activities, increases salt tolerance in yeast. Plant Cell. 1996;8(3):529–37.

Radovich TJ, Kleinhenz MD, Streeter JG. Irrigation timing relative to head development influences yield components, sugar levels, and glucosinolate concentrations in cabbage. J Am Soc Hortic Sci. 2005;130(6):943–9.

Raghavendra AS, Gonugunta VK, Christmann A, Grill E. ABA perception and signalling. Trends Plant Sci. 2010;15(7):395–401. doi:10.1016/j.tplants.2010.04.006.

Rask L, Andréasson E, Ekbom B, Eriksson S, Pontoppidan B, Meijer J. Myrosinase: gene family evolution and herbivore defense in *Brassicaceae*. Plant Mol Biol. 2000;42(1):93–113.

Rausch T, Wachter A. Sulfur metabolism: a versatile platform for launching defence operations. Trends Plant Sci. 2005;10:503–9.

Reintanz B, Lehnen M, Reichelt M, Gershenzon J, Kowalczyk M, Sandberg G, Godde M, Uhl R, Palme K. Bus, a bushy *Arabidopsis CYP79F1* knockout mutant with abolished synthesis of short-chain aliphatic glucosinolates. Plant Cell. 2001;13(2):351–67.

Rivoal J, Hanson AD. Choline-O-sulfate biosynthesis in plants. Identification and partial characterization of a salinity-inducible choline sulfotransferase from species of *Limonium* (Plumbaginaceae). Plant Physiol. 1994;106(3):1187–93.

Robbins RJ, Keck A, Banuelos G, Finley JW. Cultivation conditions and selenium fertilization alter the phenolic profile, glucosinolate, and sulforaphane content of broccoli. J Med Food. 2005;8(2):204–14. doi:10.1089/jmf.2005.8.204.

Rubio S, Rodrigues A, Saez A, Dizon MB, Galle A, Kim T, Santiago J, Flexas J, Schroeder JI, Rodriguez PL. Triple loss of function of protein phosphatases type 2C leads to partial constitutive response to endogenous abscisic acid. Plant Physiol. 2009;150(3):1345–55. doi:10.1104/pp.109.137174.

Sagi M, Scazzocchio C, Fluhr R. The absence of molybdenum cofactor sulfuration is the primary cause of the *flacca* phenotype in tomato plants. Plant J. 2002;31(3):305–17. doi:10.1046/j.1365-313X.2002.01363.x.

Schonhof I, Kläring HP, Krumbein A, Claußen W, Schreiner M. Effect of temperature increase under low radiation conditions on phytochemicals and ascorbic acid in greenhouse grown broccoli. Agr Ecosyst Environ. 2007a;119(1–2):103–11. doi:10.1016/j.agee.2006.06.018.

Schonhof I, Kläring H, Krumbein A, Schreiner M. Interaction between atmospheric CO_2 and glucosinolates in broccoli. J Chem Ecol. 2007b;33(1):105–14. doi:10.1007/s10886-006-9202-0.

Schreiner MC, Peters PJ, Krumbein AB. Glucosinolates in mixed-packaged mini broccoli and mini cauliflower under modified atmosphere. J Agric Food Chem. 2006;54(6):2218–22. doi:10.1021/jf0525636.

Schreiner M, Beyene B, Krumbein A, Stützel H. Ontogenetic changes of 2-propenyl and 3-indolylmethyl glucosinolates in *Brassica carinata* leaves as affected by water supply. J Agric Food Chem. 2009;57(16):7259–63. doi:10.1021/jf901076h.

Schwarz G, Mendel RR. Molybdenum cofactor biosynthesis and molybdenum enzymes. Annu Rev Plant Biol. 2006;57:623–47. doi:10.1146/annurev.arplant.57.032905.105437.

Sengupta D, Ramesh G, Mudalkar S, Kumar KRR, Kirti PB, Reddy AR. Molecular cloning and characterization of γ-glutamyl cysteine synthetase (VrγECS) from roots of *Vigna radiata* (L.) Wilczek under progressive drought stress and recovery. Plant Mol Biol Report. 2012;30(4):894–903. doi:10.1007/s11105-011-0398-y.

Singh A, Jha SK, Bagri J, Pandey GK, Zhang J. ABA inducible rice protein phosphatase 2C confers ABA insensitivity and abiotic stress tolerance in *Arabidopsis*. PLoS One. 2015;10(4), e0125168. doi:10.1371/journal.pone.0125168.

Sønderby IE, Geu-Flores F, Halkier BA. Biosynthesis of glucosinolates-gene discovery and beyond. Trends Plant Sci. 2010;15(5):283–90. doi:10.1016/j.tplants.2010.02.005.

Song Y, Wang Z, Bo W, Ren Y, Zhang Z, Zhang D. Transcriptional profiling by cDNA-AFLP analysis showed differential transcript abundance in response to water stress in *Populus hopeiensis*. BMC Genomics. 2012;13:286.

Sridharamurthy M, Kovach A, Zhao Y, Zhu J, Xu HE, Swaminathan K, Melcher K. H_2O_2 inhibits ABA-signaling protein phosphatase HAB1. PLoS One. 2014;9(12), e113643. doi:10.1371/journal.pone.0113643.

Szepesi Á, Csiszár J, Gémes K, Horváth E, Horváth F, Simon ML, Tari I. Salicylic acid improves acclimation to salt stress by stimulating abscisic aldehyde oxidase activity and abscisic acid accumulation, and increases Na^+ content in leaves without toxicity symptoms in *Solanum lycopersicum* L. J Plant Physiol. 2009;166(9):914–25. doi:10.1016/j.jplph.2008.11.012.

Tal M, Imber D, Erez A, Epstein E. Abnormal stomatal behavior and hormonal imbalance in flacca, a wilty mutant of tomato: V. Effect of abscisic acid on indoleacetic acid metabolism and ethylene evolution. Plant Physiol. 1979;63(6):1044–8.

Tavares S, Wirtz M, Beier MP, Bogs J, Hell R, Amâncio S. Characterization of the serine acetyltransferase gene family of *Vitis vinifera* uncovers differences in regulation of OAS synthesis in woody plants. Front Plant Sci. 2015;6:74. doi:10.3389/fpls.2015.00074.

Textor S, de Kraker J, Hause B, Gershenzon J, Tokuhisa JG. MAM3 catalyzes the formation of all aliphatic glucosinolate chain lengths in *Arabidopsis*. Plant Physiol. 2007;144(1):60–71. doi:10.1104/pp.106.091579.

Tierens KF, Thomma BP, Brouwer M, Schmidt J, Kistner K, Porzel A, Mauch-Mani B, Cammue BP, Broekaert WF. Study of the role of antimicrobial glucosinolate-derived isothiocyanates in resistance of *Arabidopsis* to microbial pathogens. Plant Physiol. 2001;125(4):1688–99.

Tombesi S, Nardini A, Frioni T, Soccolini M, Zadra C, Farinelli D, Poni S, Palliotti A. Stomatal closure is induced by hydraulic signals and maintained by ABA in drought-stressed grapevine. Sci Rep. 2015;5:12449. doi:10.1038/srep12449.

Tong Y, Gabriel-Neumann E, Ngwene B, Krumbein A, George E, Platz S, Rohn S, Schreiner M. Topsoil drying combined with increased sulfur supply leads to enhanced aliphatic glucosinolates in *Brassica juncea* leaves and roots. Food Chem. 2014;152:190–6. doi:10.1016/j.foodchem.2013.11.099.

Traka M, Mithen R. Glucosinolates, isothiocyanates and human health. Phytochem Rev. 2009;8(1):269–82. doi:10.1007/s11101-008-9103-7.

Umezawa T, Sugiyama N, Mizoguchi M, Hayashi S, Myouga F, Yamaguchi-Shinozaki K, Ishihama Y, Hirayama T, Shinozaki K. Type 2C protein phosphatases directly regulate abscisic acid-activated protein kinases in *Arabidopsis*. Proc Natl Acad Sci U S A. 2009;106(41):17588–93. doi:10.1073/pnas.0907095106.

Urano K, Maruyama K, Ogata Y, Morishita Y, Takeda M, Sakurai N, Suzuki H, Saito K, Shibata D, Kobayashi M, Yamaguchi-Shinozaki K, Shinozaki K. Characterization of the ABA-regulated global responses to dehydration in *Arabidopsis* by metabolomics. Plant J. 2009;57(6):1065–78. doi:10.1111/j.1365-313X.2008.03748.x.

Vlad F, Rubio S, Rodrigues A, Sirichandra C, Belin C, Robert N, Leung J, Rodriguez PL, Lauriere C, Merlot S. Protein phosphatases 2C regulate the activation of the Snf1-related kinase OST1 by abscisic acid in *Arabidopsis*. Plant Cell. 2009;21(10):3170–84. doi:10.1105/tpc.109.069179.

Wang Y, Chen Z, Zhang B, Hills A, Blatt MR. PYR/PYL/RCAR abscisic acid receptors regulate K^+ and Cl^- channels through reactive oxygen species-mediated activation of Ca^{2+} channels at the plasma membrane of intact *Arabidopsis* guard cells. Plant Physiol. 2013;163(2):566–77. doi:10.1104/pp.113.219758.

Xiong L, Lee H, Ishitani M, Zhu J. Regulation of osmotic stress-responsive gene expression by the *LOS6/ABA1* Locus in *Arabidopsis*. J Biol Chem. 2002;277(10):8588–96. doi:10.1074/jbc.M109275200.

Xue J, Wang S, Zhang P, Zhu F, Ren X, Liu C, Zhang X. On the role of physiological substances, abscisic acid and its biosynthetic genes in seed maturation and dormancy of tree peony (*Paeonia ostii* 'Feng Dan'). Sci Hortic. 2015;182:92–101. doi:10.1016/j.scienta.2014.11.021.

Yarmolinsky D, Brychkova G, Kurmanbayeva A, Bekturova A, Ventura Y, Khozin-Goldberg I, Eppel A, Fluhr R, Sagi M. Impairment in sulfite reductase leads to early leaf senescence in tomato plants. Plant Physiol. 2014;165(4):1505–20. doi:10.1104/pp.114.241356.

Ye N, Jia L, Zhang J. ABA signal in rice under stress conditions. Rice (N Y). 2012;5(1):1. doi:10.1186/1939-8433-5-1.

Yoshida T, Nishimura N, Kitahata N, Kuromori T, Ito T, Asami T, Shinozaki K, and Hirayama T. *ABA-hypersensitive germination3* encodes a protein phosphatase 2C (AtPP2CA) that strongly regulates abscisic acid signaling during germination among *Arabidopsis* protein phosphatase 2Cs. Plant Physiol. 2006;140(1):115–26. doi:10.1104/pp.105.070128.

You J, Zong W, Hu H, Li X, Xiao J, Xiong L. A stress-responsive NAC1-regulated protein phosphatase gene rice protein phosphatase18 modulates drought and oxidative stress tolerance through abscisic acid-independent reactive oxygen species scavenging in rice. Plant Physiol. 2014;166(4):2100–14. doi:10.1104/pp.114.251116.

Zdunek-Zastocka E. The activity pattern and gene expression profile of aldehyde oxidase during the development of *Pisum sativum* seeds. Plant Sci. 2010;179(5):543–8. doi:10.1016/j.plantsci.2010.08.005.

Zechmann B, Müller M. Subcellular compartmentation of glutathione in dicotyledonous plants. Protoplasma. 2010;246(1–4):15–24. doi:10.1007/s00709-010-0111-2.

Zechmann B, Mauch F, Sticher L, Müller M. Subcellular immunocytochemical analysis detects the highest concentrations of glutathione in mitochondria and not in plastids. J Exp Bot. 2008;59(14):4017–27. doi:10.1093/jxb/ern243.

Zechmann B, Mauch F, Sticher L, Müller M. Compartment-specific importance of glutathione during abiotic and biotic stress. Front Plant Sci. 2014;5:566. doi:10.3389/fpls.2014.00566.

Zhang M, Bourbouloux A, Cagnac O, Srikanth CV, Rentsch D, Bachhawat AK, Delrot S. A novel family of transporters mediating the transport of glutathione derivatives in plants. Plant Physiol. 2004;134(1):482–91. doi:10.1104/pp.103.030940.

Zhang H, Schonhof I, Krumbein A, Gutezeit B, Li L, Stützel H, Schreiner M. Water supply and growing season influence glucosinolate concentration and composition in turnip root (*Brassica rapa* ssp. *rapifera* L.). J Plant Nutr Soil Sci. 2008;171(2):255–65. doi:10.1002/jpln.200700079.

Zhang F, Fu X, Lv Z, Shen Q, Yan T, Jiang W, Wang G, Sun X, Tang K. Type 2C phosphatase 1 of *Artemisia annua* L. is a negative regulator of ABA signaling. Biomed Res Int. 2014;2014(2):1–9. doi:10.1155/2014/521794.

Chapter 10
Effects of Elevated Carbon Dioxide and Drought Stress on Agricultural Crops

Jong Ahn Chun, Sanai Li, and Qingguo Wang

10.1 Introduction

The atmospheric CO_2 concentration has increased exponentially from about 280 ppm at the beginning of the industrial revolution to about 380 ppm today, and is expected to double preindustrial levels during this century (Keeling and Whorf 2001). The increase in atmospheric CO_2 concentrations may contribute to climate change including changes in precipitation patterns and evapotranspiration (Kruijt et al. 2008; Long et al. 2004; Schneider 2001). This climate change may increase in the risks of drought in many areas (Bates et al. 2008).

Seasonal variability in rainfall is one of the crucial factors contributing to variations of crop yields (Hu and Buyanovsky 2003). Approximately 40 % of the world land surface was covered by arid and semiarid areas, where drought stress is a main limiting factor for the conventional rain-fed agriculture (Gamo 1999). In some areas of the world, water supply is already a limiting factor for agricultural production (Penning de Vries et al. 1995). Climate change and variability will impose significant impacts on agricultural productivity by altering precipitation pattern, rising temperature, and carbon dioxide.

Climate change would influence the hydrological cycle and water resource availability, suggesting that it has an impact on crop productivity (Evans 1996). Climate change can accelerate the hydrological cycle through altering rainfall, evapotranspiration, and the intensity and frequency of extreme climate events such as floods and droughts (Watson et al. 1996). Under future climate, the potential and actual evapotranspiration possibly increase by the rising temperature (Riedo et al. 2001). The agricultural production is likely to be greatly impacted by a decrease in soil moisture and an increase in the possible extreme events such as droughts and floods

J.A. Chun (✉) • S. Li • Q. Wang
Climate Change Research Team, Department of Climate Research, APEC Climate Center, 12 Centum 7-ro, Haeundae-gu, Busan 48058, Republic of Korea
e-mail: jachun@apcc21.org

M.A. Hossain et al. (eds.), *Drought Stress Tolerance in Plants, Vol 1*,
DOI 10.1007/978-3-319-28899-4_10

caused by combined effects of rising CO_2 concentrations and temperatures (Chiotti and Johnston 1995). It is therefore important to know how drought and elevated CO_2 will affect crop growth, development, water use, and productivity.

There is continued interest in how agricultural crops will respond to future CO_2, since CO_2 is an essential substrate for photosynthesis and limits the rate of photosynthesis in many crops at current conditions. Generally, plants sense and respond to elevated CO_2 through increased photosynthesis and reduced stomatal conductance. All other effects are derived from these two fundamental responses (Long et al. 2004). Elevated CO_2 stimulates photosynthesis and reduces the opening of plant stomata, contributing to a decrease in plant transpiration. As a result, plants growing in elevated CO_2 conditions will improve water use efficiency (WUE, the ratio of rate of carbon assimilation to the rate of transpiration).

There are two main plants categorized into C_3, C_4, or C_3–C_4 intermediate plants according to the spatial distribution of pathways of CO_2 fixation within leaf tissues, and as crassulacean acid metabolism (CAM) plants with a temporal distribution (Freschi and Mercier 2012). C_3 plants represent over 95 % of the Earth plant species, mainly growing in cool and wet climate areas. C_4 and CAM plants occur in hot and dry climatic conditions. Elevated CO_2 concentrations will, in general, lead to increased photosynthesis and decreased transpiration in C_3 plants. Agricultural crops with a C_3 photosynthetic pathway often exhibit greater assimilation responses than those with a C_4 pathway due to differences in CO_2 use during photosynthetic procedures (Amthor 1995; Rogers et al. 1997).

It is widely known that drought is the single most critical threat to world food security. Because the world's water supply is limiting, future food demand for rapidly increasing population pressures is likely to further aggravate the effects of drought (Somerville and Briscoe 2001). Under water stress conditions, photosynthesis decreases through direct effects, as the decreased CO_2 availability caused by diffusion limitations through the stomata and the mesophyll (Flexas et al. 2004, 2007; Warren 2008) or the alterations of photosynthetic metabolism (Lawlor and Cornic 2002). These water stress conditions can arise as secondary effects, namely oxidative stress, and feedback regulation by end-product accumulation (Nikinmaa et al. 2013).

The purpose of this review is to provide: (1) an overview of physiological processes including photosynthesis and transpiration of agricultural crops under elevated CO_2 and drought stress and (2) summary of recent research on those crop responses to elevated CO_2 and drought stress based on field experiments and crop modeling studies.

10.2 Physiological Processes Under Elevated CO_2 and Drought Stress

10.2.1 Photosynthesis

Two key processes occur in photosynthesis: light-dependent reactions and light-independent (or dark) reactions. In the former reactions, light energy is converted into adenosine triphosphate (ATP) and nicotinamide adenine dinucleotide phosphate hydrogen (NADPH), and O_2 is released. In the latter reactions, the enzyme Ribulose-1,5-bisphosphate carboxylase/oxygenase (RuBisCO) captures atmospheric CO_2 and releases three-carbon sugars by utilizing ATP and NADPH.

CO_2 and soil water considerably influence the process of photosynthesis in most plants by altering stomatal regulation, the ultrastructure of the organelles, concentration of various pigments and metabolites. A great number of research found that the plant photosynthetic rates were greatly enhanced under elevated CO_2 in the short-term (Radmer and Kok 1977; Witter 1979), and these increases were likely to be more moderate due to various feedback responses and constraints in the long-term (Kramer 1981). There was a significant and marked increase in photosynthesis of C_3 plants (Norby et al. 1999; Ainsworth and Long 2005), but there were significant differences between species and cultivars. In C_3 plants, the maximum carboxylation rate and the maximum rate of electron transport were also significantly reduced at elevated CO_2. There was a significant increase in photosynthesis of C_4 crops, as an indirect effect resulting through the mitigation of drought stress due to reduced stomatal conductance (Ghannoum et al. 2000). Increases in photosynthesis in sorghum and maize were associated with improved water status or were limited to periods of low rainfall where drought stress was likely ameliorated at elevated CO_2 (Leakey et al. 2004; Kimball 2006).

Photosynthetic responses to water stress are highly complex. These effects vary according to the intensity and duration of progression of the water stress as well as with the leaf age and the plant species and at different time scales in relation to plant development (Lawlor and Cornic 2002; Flexas et al. 2004). Both stomatal and non-stomatal limitations to photosynthesis are important. Photosynthesis acclimation under drought indirectly affects photosynthesis. This acclimation will help to maintain plant water status and therefore photosynthesis. Osmotic compounds that build up in response to water stress will lead to restoration of cellular homeostasis and detoxification.

10.2.2 Stomatal Conductance

The regulation of leaf stomatal conductance is a key phenomenon in plants for photosynthesis and transpiration (Medici et al. 2007). One of the most consistent responses of plants to elevated CO_2 is a reduction in stomatal conductance (Ainsworth and Long 2005). However, the responses are significantly different among species and cultivars. As an exception, Ellsworth (1999) reported that *Pinus taeda* guard cells appear to be insensitive to elevated CO_2. The decrease in stomatal conductance may be largely determined by stomatal aperture rather than density. Ainsworth and Rogers (2007) found that a decrease in the density is statistically insignificant through a meta-analysis of stomatal density responses to elevated CO_2.

Guard cells sense intercellular CO_2 rather than at the leaf surface. Stomatal conductance responses to elevated CO_2 may vary according to the duration of plants grown in elevated CO_2. In the short term, stomatal aperture generally decreases in response to high CO_2. In the long-term, stomatal conduction may acclimate to elevated CO_2. Ball et al. (1987) reported that stomatal conductance would decrease in response to elevated CO_2. Medlyn et al. (2001) found that stomatal conductance only in water-stressed *Phillyrea angustifolia* was acclimated to elevated CO_2 in six tree species. However, there is little evidence that stomatal conductance independently acclimates to elevated CO_2 for *Lolium perenne* grown at 600 μmol mol^{-1} (Leakey et al. 2006a, b).

The magnitude of the effect of elevated CO_2 on stomatal conductance varies considerably with environmental factors (Medlyn et al. 2001; Leakey et al. 2006a, b). There is generally a smaller effect of elevated CO_2 on stomatal conductance under water stress (Leakey et al. 2006a, b). For example, there was no significant change in stomatal conductance at elevated CO_2 in *Liquidambar styraciflua* when vapor pressure deficit was high (Herrick et al. 2004). For long-term water stress, stomatal conductance will be much less reduced in elevated CO_2 compared to ambient conditions (Leakey et al. 2006a, b). A small decline in stomatal conductance may have protective effects against water stress, by less transpiration rate and improving plant water use efficiency.

Under water-stress conditions, the first response of plants is the stomatal closure to prevent the water loss due to transpiration to maintain the photosynthesis at low water availability (Pan et al. 2011). The stomata closure under water stress generally occurs due to decreased leaf turgor or water potential and low humidity atmosphere along with root-generated chemical signals (Chaves et al. 2009). The stomata closure is caused mainly by the action of a plant hormone, abscisic acid (ABA). High ABA level can cause an increase in cytosolic Ca^{2+} and activation of plasma membrane-localized anion channels (Kohler and Blatt 2002). This causes potassium efflux, guard cell depolarization, loss of guard cell volume and turgor, high water production, and finally the stomata closure (Wang et al. 2012).

10.2.3 Rubisco Activity and Content

Rubisco is usually fully active and carbamylated at current CO_2 under steady-state high light conditions (von Caemmerer and Quick 2000). Under elevated CO_2 conditions, photosynthesis increases; there is an increasing demand for ATP and control of photosynthesis shifts from being limited by Rubisco to being limited by the capacity for ribulose-1,5-bisphosphate (RuBP) regeneration (Farquhar et al. 1980; von Caemmerer and Quick 2000). Reductions in the ATP:ADP ratio lead to a reduction in activase activity. The reductions in Rubisco activation state have been reported under elevated CO_2 (Cen and Sage 2005).

One of the most prominent effects of water stress is the stomata closure, which leads to a lower concentration of intercellular CO_2, which in turn causes deactivation of Rubisco (Mumm et al. 2011). Medrano et al. (1997) observed that water deficit conditions reduced the initial and total Rubisco activity, but it did not decrease the overall amount of Rubisco per unit of leaf area in subterranean clover (*Trifolium subterraneum*). Marques and Arrabica (1995) reported that Rubisco activity in *Setaria sphacelota* declined slightly under moderate water stress, but substantially under severe water stress. Using transgenic tobacco plants, Gunasekera and Berkowitz (1993) showed that a 68 % decrease in Rubisco activity did not hamper photosynthesis under water-limited regimes. They concluded that drought stress may affect any of the steps involved in the regeneration of RuBP rather than Rubisco itself.

10.3 Effects of Elevated CO_2 and Drought Stress on Crops

It is widely known that elevated CO_2 concentrations contribute to the increases of crop photosynthetic exchange rates (CER) and yield by decreasing photorespiration. This response of C_3 plants to elevated atmospheric CO_2 is higher than that of C_4 plants (Sage and Monson 1999). Increases in the growth of C_3 plants under doubled atmospheric CO_2 concentrations are approximately 40–45 %, while the growth of C_4 plants under doubled atmospheric CO_2 concentrations increases by 10–20 % (Ghannoum et al. 2000).

The water relations for most plants exhibit improved under-elevated CO_2, and showed less transpiration by inducing the partial stomatal closure. Studies have shown that elevated CO_2 reduces transpiration for both C_3 (Allen et al. 1994; Prior et al. 1991) and C_4 (Chaudhuri et al. 1986) plants. Using stem flow gauges under elevated CO_2, Dugas et al. (1997) reported the reduction in whole-plant transpiration for both soybean (C_3) and sorghum (C_4) crops.

The reduction in transpiration under elevated CO_2, coupled with increased photosynthesis, can contribute to increase in WUE (Baker et al. 1990; Sionit et al. 1984). Kimball and Idso (1983) analyzed 46 observations for transpiration and over 500 observations for economic yield, and suggested a doubling of WUE for a doubling of CO_2 concentrations. Under elevated CO_2, C_4 plants show a smaller response to elevated CO_2 than C_3 plants. However, both C_3 and C_4 plants show reduced transpiration. These results indicate that WUE should be primarily controlled by transpiration in C_4 plants, whereas both photosynthesis and transpiration are important in C_3 plants (Acock and Allen 1985).

Obviously, water-stressed plants have lower relative water content than non-stressed ones. For example, exposure of wheat and rice plants to drought stress substantially decreased the leaf water potential and transpiration rate (Siddique et al. 2001). Nerd and Nobel (1991) suggested that during drought stress, total water contents of *Opuntia ficusindica* cladode were decreased by 57 %. In another study on *Hibiscus rosasinensis*, transpiration, stomatal conductance, and WUE were declined under drought stress (Egilla et al. 2005). Abbate et al. (2004) reported that under limited water supply, WUE of wheat was greater than in well-watered conditions due to stomatal closure to reduce the transpiration under water stress conditions. Lazaridou and Koutroubas (2004) concluded that WUE of clover (*Trifolium alexandrinum*) was increased due to decreased transpiration rates and leaf area. In studies on *Artemisia tridentata* (DeLucia and Heckathorn 1989) and *Medicago sativa* (Lazaridou et al. 2003), drought stress increased WUE mainly due to a decrease in stomatal conductance with increasing water deficit.

Given the fact that elevated CO_2 can reduce transpiration, it has been suggested that this might partially ameliorate the effects of drought (Bazzaz 1990) and allow plants to maintain increased photosynthesis. This has frequently been observed (Acock and Allen 1985; Sionit et al. 1981; Prior et al. 1991). It has been suggested that under elevated CO_2 whole-plant water use may be differentially affected as a result of leaf area index (LAI) or plant size, although instantaneous WUE is increased. Allen (1994) reported that higher LAI could counter balance the reduction in water use. Jones et al. (1985) showed that increase in WUE was greater for plants with a lower LAI than higher LAI.

Elevated CO_2 intends to increase photosynthesis through raising the CO_2 gradient between the atmosphere and the inside of leaves, and consequently improve its conversion into carbohydrates (Rosenzweig and Hillel 1998). The impacts of elevated CO_2 on crop yield may vary among different experimental studies due to differences in experimental methods and its corresponding environmental conditions. The free-air CO_2 enrichment (FACE) showed that crop yield of C_3 plants such as rice, wheat, cotton, and sorghum increased by about 17–20 % at 550 ppm (Long et al. 2004; Ainsworth and Long 2005). On the other hand, the glasshouse and growth chamber experiments showed an 18–23 % increase in crop yield (Amthor 2001; Tubiello et al. 2007), and the response of crops to elevated CO_2 is slightly higher than the FACE results. Under elevated CO_2, increases in the number of grains per plant and the harvest index lead to an increase in crop yield (Wu et al. 2004). However, the CO_2 fertilization effect may be limited by some severe environmental stress, such as temperature, rooting volume, light, nutrient, and drought (Batts et al. 1997; Arp 1991; Kramer 1981).

The impacts of drought on crop depend on the magnitude of water stress and the developmental stages (Sau and Mínguez 2000). The negative impacts of drought are more severe during some moisture-sensitive phenological stages (Nesmith and Ritchie 1992). In the early growth stages, extreme water stress can postpone sowing of crop and affect seed germination (Hu and Buyanovsky 2003). From emergence to double ridge stages, drought stress can significantly affect the leaf expansion of crops (Acevedo et al. 1971). The leaf expansion rate of wheat is expected to be greatly reduced when the extractable soil water is smaller than 50 % (Meyer and Green 1980, 1981). During the pre-anthesis stage, the number of kernels per spike of wheat can be greatly reduced by drought stress (Fischer 1980). This result can be explained by considering that the number of kernels per spike largely contributed to grain yield particularly under drought conditions (García del Moral et al. 2003). Shpiler and Blum (1991) found that the grain yield of wheat showed the most sensitivity to moisture deficit during double ridge to anthesis stages due to the substantial effect of water deficit on both spikelet number and kernels per spike. However, van Herwaarden et al. (1998) reported that the grain yield of wheat was mostly impacted by the moisture deficit after anthesis. The different conclusions may be resulted from the differences in crop varieties, field management, and climatic conditions. In addition, crop development can also be accelerated by soil moisture deficit during anthesis (Simane et al. 1993). During the grain filling period, grain weight can be greatly decreased by drought stress mainly through accelerating senescence rates and shortening growth duration (Hochman 1982). These results suggest that efficacious adaptation strategies can be provided by focusing on the most moisture-sensitive stages.

10.4 Interactive Effects of Elevated CO_2 and Drought Stress on Crops

The interaction of elevated CO_2 and water on crop growth has been studied. The water use of C_4 crops under elevated CO_2 decreases by reducing stomatal conductance without an increase in photosynthesis (Morison 1993; Leakey et al. 2006a, b; Long et al. 2006). Loomis and Lafitte (1987) reported that large changes in the supplies of CO_2 and water little affected corn growth. An increase in WUE was found regardless of water supply (Surano and Shinn 1984). Prior et al. (2010) reported that elevated CO_2 significantly increases WUE, suggesting better soil moisture conservation at elevated CO_2.

In an outdoor growth chamber study conducted by Chun et al. (2011), some points (denoted as "breaking points") from high to low rates of soil water uptake were observed in the bottom depth (between 0.625 and 0.85 m from the surface), indicating a decrease in water availability. The breaking points were earlier under ambient CO_2 than under elevated CO_2, suggesting that the depletion of the easily available water occurred later under elevated CO_2 than under ambient CO_2.

The effects of elevated atmospheric CO_2 concentrations on plants under drought are complex. Plants reduce transpiration by closing stomata, but this substantially reduces photosynthetic rates. However, elevated CO_2 enhances photosynthetic rates in C_3 plants. If the photosynthesis-stimulating effect of elevated CO_2 is greater than the reduction in photosynthesis from drought-induced stomatal closure, the overall effects of CO_2 and water stress will be positive. Otherwise, the overall effects will be negative. Morgan et al. (2004) observed that the relative photosynthetic benefits of elevated CO_2 are generally greater in more arid environments in large-scale studies. Numerous studies have shown that increasing CO_2 may benefit photosynthesis and survival during droughts of moderate duration, while the negative effects may overwhelm the benefits of elevated CO_2 where droughts become more severe. Elevated CO_2 caused a smaller reduction in evapotranspiration under water stress and different species have different responses to elevated CO_2 under water stress conditions. Reddy et al. (2000) found that there was no reduction in evapotranspiration for cotton; however, Hunsaker et al. (2000) reported 4 % reduction in evapotranspiration for wheat.

Elevated CO_2 can alleviate drought stress and improve crop yields by improvement of water use efficiency under higher CO_2 concentrations (Allen et al. 1998; Makino and Mae 1999; Maroco et al. 1999). In the Free-air CO_2 enrichment (FACE), there is a 7 % increase in water use efficiency at 550 ppm of CO_2 concentrations Hunsaker et al. (1996). Similarly, Allen (1991) found that there is a 10 % reduction in crop canopy water use under doubled CO_2. In contrast, Yoshimoto et al. (2005)

reported that in a FACE experiment, there is a 19 % increase in WUE of rice at 587 ppm of CO_2 concentrations. The response of crop water use to elevated CO_2 depends on crop species and environmental conditions. For example, a doubled CO_2 can lead to a decrease in evapotranspiration (ET) of rice at 26 °C, while it increased in ET at 29.5 °C (Horie et al. 2000).

Drought stress has a great impact on the magnitude of CO_2 fertilization effect of a crop. Some experimental results found that there were higher increases in growth and yield of wheat in response to elevated CO_2 under drought stress conditions than under high soil moisture (Gifford 1979; Chaudhuri et al. 1990; Samarakoon et al. 1995). However, other research on wheat showed that there were greater CO_2 fertilization effects under optimal soil water conditions than in water deficit conditions (Kramer 1981; Kimball 1983; Poorter 1998; Wu and Wang 2000). Similarly, Smith et al. (2000) found that in dry year CO_2, fertilization effect has no beneficial impacts on desert shrub growth under severe water deficit conditions (Acevedo et al. 1991). These results imply that sufficient soil moisture is an important factor in maintaining stomata opening and improving CO_2 conductance (Loomis and Amthor 1996).

10.5 Applications of Crop Models

There have been many studies on investigation of the impact of water on crops using various crop models. For example, Yang et al. (2009) modified the leaf area module of a soil–plant–atmosphere continuum corn simulation model (MaizeSim) to better simulate leaf area of corn crops at different water status and reported that the modified model improved the simulation of leaf area. Katerji et al. (2013) investigated the impacts of water stress on productivity, evapotranspiration, and water use efficiency of corn and tomato crops using the FAO AquaCrop model (a crop water productivity model). They concluded that the model can be a useful tool for research purposes to enhance the water use efficiency and to manage irrigation practices.

Crop models have been widely used to simulate the response of crops to elevated CO_2. Tubiello et al. (2007) compared the simulated response of crop yield to elevated CO_2 from the DSSAT-CERES which is widely used for cereal grains, Environmental Policy Integrated Climate (EPIC), and Agro-Ecological Zones (AEZ) models. The results showed that at 550 ppm of CO_2 concentrations, the yields of C_3 crops increased by 10–19 %, while yields of C_4 crops only increased by 4–8 %. The magnitude of CO_2 fertilization effect is close to the reported value by Long et al. (2006) for FACE experiments. However, the results simulated from CERES (Boote and Pickering 1994) and EPIC/Cropping Systems Simulation Model (CropSyst) (Tubiello et al. 2000) showed a 25 % increase in C_3 crop yield for a doubling of CO_2. The effects of climate change with combined CO_2 fertilization on potential crop yield (e.g., Tubiello and Ewert 2002) and water use (Asseng et al. 2004) have been investigated using crop models. However, the long-term and large-scale CO_2 fertilization effect still remains uncertain. The uncertainties in land use change scenarios under future climate conditions may contribute to this uncertainty (Levy et al. 2004).

The interactions of water and CO_2 not only affect the crop growth and yield, but also crop development. The results from FACE experiments showed that the crop developmental rate can be accelerated by the water and CO_2 interactions; however, many crop models may not be able to accurately capture these interactions due to the ignorance of CO_2-related canopy temperature (Ewert et al. 2002; Tubiello et al. 1999). The effects of water and CO_2 interactions on canopy temperature were included in the DEMETER crop model, and Kartschall et al. (1995) reported that the simulated values of phenology, growth, and yields are in good agreement with the observed values.

Under dryland conditions, grain yield was highly related with evapotranspiration (Sadras and Angus 2006). The different effects of drought stress on crops were reported at each phenological period (Andresen et al. 1989). From emergence to anthesis, leaf area expansion can be greatly affected by water deficit (Acevedo et al. 2002). Eitzinger et al. (2003) found that during the grain filling stage, crop yield was most sensitive to drought stress, whereas Chipanshi et al. (1999) showed the flowering and heading periods were most sensitive stages to drought stress. The difference in environmental conditions and parameterization of drought stress for crop modeling may contribute to this discrepancy.

Even though lots of crop models have been developed and evaluated as discussed in this section, the models still need to be improved to adequately address phenology with respect to water stress. A stomatal control and transpiration models were incorporated into the photosynthesis model initially proposed by Farquhar et al. (1980) to address stomatal limitations to CO_2 assimilation (Ball et al. 1987). This approach is generally considered as one of the most popular approaches for coupled models of stomatal control and photosynthesis. However, there are still controversies on the use of crop models that resulted from complexity, testability, and parameterization (Timlin et al. 2008). It is suggested that more robust and realistic parameters should be provided to address these controversies.

10.6 Summary and Conclusions

Increasing CO_2 may change precipitation patterns and evapotranspiration, implying increases in the risks of drought in many areas. The impacts of elevated CO_2 and drought stress on growth and development of crops were discussed in the previous sections. The different responses of CO_2 have been reported according to the spatial distribution of pathways of CO_2 fixation within leaf tissues. The response of C_4 plants to elevated atmospheric CO_2 is lower than that of C_3 plants. Elevated CO_2 reduces transpiration for both C_3 and C_4 plants. These results indicate that WUE should be primarily controlled by transpiration in C_4 plants, while both photosynthesis and transpiration are important in C_3 plants. Numerous literatures suggest that crops will use less water under high atmospheric CO_2 in the future than at present.

The use of crop models has been used for assessment of the impacts of elevated CO_2 and drought stress on crop growth and development. However, many crop models still need to be improved to adequately address phenology with respect to water stress. In addition, there are still controversies on the use of crop models that resulted from complexity, testability, and parameterization, suggesting that more robust and realistic parameters should be provided to address these controversies. It is concluded that crop models can be a useful tool to quantify the impacts of elevated CO_2 and drought stress and to assess agricultural management practices. This review can provide a better understanding of the interactive effects of elevated CO_2 and drought stress on crop growth and development.

References

Abbate PE, Dardanellib JL, Cantareroc MG, Maturanoc M, Melchiorid RJM, Sueroa EE. Climatic and water availability effects on water-use efficiency in wheat. Crop Sci. 2004;44:474–83.

Acevedo E, Hsiao TC, Henderson DW. Immediate and subsequent growth responses of maize leaves to changes in water status. Plant Physiol. 1971;48:631–6.

Acevedo E, Harris H, Cooper PJM. Crop architecture and water use efficiency in Mediterranean environments. In: Harris H, Cooper PJM, Pala M, editors. Soil and crop management for improved water use efficiency in rainfed areas. Ankara: ICARDA; 1991. p. 106–18.

Acevedo E, Silva P, Silva H. Wheat growth and physiology. In: Curtis BC, Rajaram S, Gómez Macpherson H, editors. Bread wheat improvement and production. Rome: FAO plant production and protection series; 2002. p. 567.

Acock B, Allen Jr LH. Crop responses to elevated carbon dioxide concentrations. In: Strain BR, Cure JD, editors. Direct effects of increasing carbon dioxide on vegetation. Washington, DC: DOE/ER-0238, Office of Energy Research, U.S. Dept. of Energy; 1985. p. 317–46.

Ainsworth EA, Long SP. What have we learned from 15 years of free-air CO_2 enrichment (FACE)? A meta-analysis of the responses of photosynthesis, canopy properties and plant production to rising CO_2. New Phytol. 2005;165:351–72.

Ainsworth EA, Rogers A. The response of photosynthesis and stomatal conductance to rising CO_2: mechanisms and environmental interactions. Plant Cell Environ. 2007;30:258–70.

Allen Jr LH. Effects of increasing carbon dioxide levels and climate change on plant growth, evapotranspiration, and water resources. Proceedings of a colloquium on managing water resources in the west under conditions of climatic uncertainty. Washington, DC: National Research Council, National Academy Press; 1991. p. 101–47.

Allen LH. Carbon dioxide increase: direct impact on crops and indirect effects mediated through anticipated climatic changes. In: Boote KJ, Sinclair TR, Bennett JM, editors. Physiology and determination of crop yield. Madison, WI: ASA, CSSA, SSSA; 1994. p. 425–59.

Allen Jr LH, Valle RR, Mishoe JW, Jones JW. Soybean leaf gas-exchange responses to carbon dioxide and water stress. Agron J. 1994;86:625–36.

Allen Jr LH, Valle RR, Jones JW, Jones PH. Soybean leaf water potential responses to carbon dioxide and drought. Agron J. 1998;90:375–83.

Amthor JS. Terrestrial higher-plant response to increasing atmospheric CO_2 in relation to the global carbon cycle. Glob Chang Biol. 1995;1:243–74.

Amthor JS. Effects of atmospheric CO_2 concentration on wheat yield: review of results from experiments using various approaches to control CO_2 concentration. Field Crop Res. 2001;73:1–34.

Andresen JA, Dale RF, Fletcher JJ, Preckel PV. Prediction of county-level corn yields using an energy-crop growth index. J Climate. 1989;2:48–56.

Arp WJ. Effects of source-sink relations on photosynthetic acclimation to elevated CO_2. Plant Cell Environ. 1991;14:869–75.

Asseng S, Jamieson PD, Kimball BA, Pinter Jr PJ, Sayred K, Bowden JW, Howden SM. Simulated wheat growth affected by rising temperature, increased water deficit and elevated atmospheric CO_2. Field Crop Res. 2004;85:85–102.
Baker JY, Allen Jr LH, Boote KJ. Growth and yield responses of rice to carbon dioxide concentration. J Agric Sci. 1990;115:313–20.
Ball JT, Woodrow IE, Berry JA. A model predicting stomatal conductance and its contribution to the control of photosynthesis under different environmental conditions. In: Biggens J, editor. Progress in photosynthesis research. Dordrecht: Martinus-Nijhoff; 1987. p. 221–4.
Bates BC, Kundzewicz ZW, Wu S, Palutikof JP, editors. Climate change and water. Technical paper of the intergovernmental panel on climate change. Geneva: IPCC Secretariat; 2008.
Batts GR, Morison JIL, Ellis RH, Hadley P, Wheeler TR. Effects of CO_2 and temperature on growth and yield of crops of winter wheat over several seasons. Eur J Agron. 1997;7:43–52.
Bazzaz FA. The response of natural ecosystems to the rising global CO_2 levels. Annu Rev Ecol Syst. 1990;21:167–96.
Boote KJ, Pickering NB. Modeling photosynthesis of row crop canopies. Hort. Science 1994;29:1423–1434.
Cen Y-P, Sage RF. The regulation of rubisco activity in response to variation in temperature and atmospheric CO^+ partial pressure in sweet potato. Plant Physiol. 2005;139:979–90.
Chaudhuri UN, Burnett RB, Kirkham MB, Kanemasu ET. Effect of carbon dioxide on sorghum yield, root growth, and water use. Agr For Meteorol. 1986;37:109–22.
Chaudhuri UN, Kirkham MB, Kanemasu ET. Carbon dioxide and wheat level effects on yield and water use of winter wheat. Agron J. 1990;82:637–41.
Chaves MM, Flexas J, Pinheiro C. Photosynthesis under drought and salt stress: regulation mechanisms from whole plant to cell. Ann Bot. 2009;103:551–60.
Chiotti QP, Johnston T. Extending the boundaries of climate change research: a discussion on agriculture. J Rural Stud. 1995;11:335–50.
Chipanshi AC, Ripley EA, Lawford RG. Large-scale simulation of wheat yield in semi-arid environments using a crop-growth model. Agric Syst. 1999;59:57–66.
Chun JA, Wang Q, Timlin D, Fleisher D, Reddy VR. Effect of elevated carbon dioxide and water stress on Gas exchange and water Use efficiency in corn. Agric For Meteorol. 2011;151:378–84.
DeLucia EH, Heckathorn SA. The effect of soil drought on water-use efficiency in a contrasting Great Basin desert and Sierran montane species. Plant Cell Environ. 1989;12:935–40.
Dugas WA, Prior SA, Rogers HH. Transpiration from sorghum and soybean growing under ambient and elevated CO_2 conditions. Agric For Meteorol. 1997;83:37–48.
Egilla JN, Davies Jr FT, Boutton TW. Drought stress influences leaf water content, photosynthesis, and water-use efficiency of *Hibiscus rosa-sinensis* at three potassium concentrations. Photosynthetica. 2005;43:135–40.
Eitzinger J, Štastná M, Žalud Z, Dubrovský M. A simulation study of the effect of soil water balance and water stress on winter wheat production under different climate change scenarios. Agric Water Manage. 2003;61:195–217.
Ellsworth DS. CO_2 enrichment in a maturing pine forest: are CO_2 exchange and water status in the canopy affected? Plant Cell Environ. 1999;22:461–72.
Evans TE. The effects of changes in the world hydrological cycle on availability of water resources. In: Bazzaz F, Sombroek W, editors. Global climate change and agricultural production. Chichester: Wiley; 1996. p. 248.
Ewert F, Rodriguez D, Jamieson P, Semenov MA, Mitchell RAC, Goudriaan J, Porter JR, Kimball BA, Pinter Jr PJ, Manderscheid R, Weigel HJ, Fangmeier A, Fereres E, Villalobos F. Effects of elevated CO_2 and drought on wheat: testing crop simulation models for different experimental and climatic conditions. Agric Ecosyst Environ. 2002;93:249–66.
Farquhar GD, von Caemmerer S, Berry JA. A biochemical model of photosynthetic CO_2 assimilation in leaves of C3 species. Planta. 1980;149:78–90.
Fischer RA. Influence of water stress on crop yield in semi arid regions. In: Turner NC, Kramer P, editors. Adaptation of plants to water and high temperature stress. New York: Willey; 1980. p. 323–40.

Flexas J, Bota J, Loreto F, Cornic G, Sharkey TD. Diffusive and metabolic limitations to photosynthesis under drought and salinity in C3 plants. Plant Biol. 2004;6:269–79.

Flexas J, Diaz-Espejo A, Galmés J, Kaldenhoff R, Medrano H, Ribas-Carbo M. Rapid variations of mesophyll conductance in response to changes in CO_2 concentration around leaves. Plant Cell Environ. 2007;30:1284–98.

Freschi L, Mercier H. Connecting environmental stimuli and crassulacean acid metabolism expression: phytohormones and other signaling molecules. Prog Bot. 2012;73:231–55.

Gamo M. Classification of arid regions by climate and vegetation. J Arid Land Stud. 1999;1:9–17.

García del Moral LF, Rharrabti Y, Villegas D, Royo C. Evaluation of grain yield and its components in durum wheat under Mediterranean conditions: an ontogenic approach. Agron J. 2003;95:266–74.

Ghannoum O, von Caemmerer S, Ziska LH, Conroy JP. The growth response of C4 plants to rising atmospheric CO_2 partial pressure: a reassessment. Plant Cell Environ. 2000;23:931–42.

Gifford RM. Growth and yield of CO_2 enriched wheat under water-limited conditions. Aust J Plant Physiol. 1979;6:367–78.

Gunasekera D, Berkowitz GA. Use of transgenic plants with ribulose-1,5-bisphosphate carboxylase/oxygenase antisense DNA to evaluate the rate limitation of photosynthesis under water stress. Plant Physiol. 1993;103:629–35.

Herrick JD, Maherali H, Thomas RB. Reduced stomatal conductance in sweetgum (Liquidambar styraciflua) sustained over long-term CO_2 enrichment. New Phytol. 2004;162:387–96.

Hochman ZVI. Effect of water stress with phasic development on yield of wheat grown in a semi-arid environment. Field Crop Res. 1982;5:55–67.

Horie T, Baker JT, Nakagawa H, Matsui T, Kim HY. Crop ecosystem responses to climatic change: rice. In: Reddy KR, Hodges HF, editors. Climate change and global crop productivity. New York: CAB International; 2000. p. 81–106.

Hu Q, Buyanovsky G. Climate effects on corn yield in Missouri. J Appl Meteorol. 2003;42:1626–35.

Hunsaker DJ, Kimball BA, Pinter Jr PJ, LaMorte RL, Wall GW. Effects of CO_2 enrichment and irrigation on soil water balance evapotranspiration of wheat grown under open-air field conditions. Trans ASAE. 1996;4:1345–55.

Hunsaker DJ, Kimball BA, Pinter PJ, Wall GW, LaMorte RL, Adamsen FJ, Leavitt SW, Thompson TL, Matthias AD, Brooks TJ. CO_2 enrichment and soil nitrogen effects on wheat evapotranspiration and water use efficiency. Agric For Meteorol. 2000;104:85–105.

Jones P, Allen Jr LH, Jones JW, Valle R. Photosynthesis and transpiration responses of soybean canopies to short- and long-term CO_2 treatments. Agron J. 1985;77:119–26.

Kartschall T, Grossman S, Pinter Jr PJ, Garcia RL, Kimball BA, Wall GW, Hunsaker DJ, LaMorte RL. A simulation of phenology, growth, carbon dioxide exchange and yields under ambient atmosphere and free-air carbon dioxide enrichment (FACE) Maricopa, Arizona, for wheat. J Biogeogr. 1995;22:2467–77.

Katerji N, Campi P, Mastrorilli M. Productivity, evapotranspiration, and water use efficiency of corn and tomato crops simulated by AquaCrop under contrasting water stress conditions in the Mediterranean region. Agric Water Manage. 2013;130:14–26.

Keeling CD, Whorf TP. Trends: A compendium of data on global change, Atmospheric CO_2 records from sites in the SIO air sampling network. Oak Ridge, TN: Carbon Dioxide Information Analysis Center, Oak Ridge National Laboratory, U.S. Dept. Energy; 2001. p. 14–21.

Kimball BA. Carbon dioxide and agricultural yield: an assemblage and analysis of 430 prior observations. Agron J. 1983;75:779–88.

Kimball BA. The effects of free-air CO_2 enrichment of cotton, wheat and sorghum. In: Nösberger J, Long SP, Norby RJ, Stitt M, Hendrey GR, Blum H, editors. Managed ecosystems and CO_2, Case studies, processes and perspectives. Heidelberg: Springer; 2006. p. 47–70.

Kimball BA, Idso SB. Increasing atmospheric CO_2: effects on crop yield, water use, and climate. Agric Water Manage. 1983;7:55–72.

Kohler B, Blatt MR. Protein phosphorylation activates the guard cell Ca^{2+} channel and is a prerequisite for gating by abscisic acid. Plant J. 2002;32:185–94.

Kramer PJ. Carbon dioxide concentration, photosynthesis and dry matter production. BioScience. 1981;31:29–33.

Kruijt B, Witte J-PM, Jacos CMJ, Kroon T. Effects of rising atmospheric CO_2 on evapotranspiration and soil moisture: a practical approach for the Netherlands. J Hydrol. 2008;349:257–67.

Lawlor DW, Cornic G. Photosynthetic carbon assimilation and associated metabolism in relation to water deficits in higher plants. Plant Cell Environ. 2002;25:275–94.

Lazaridou M, Kirilov A, Noitsakis B, Todorov N, Katerov I. The effect of water deficit on yield and water use efficiency of lucerne. Optimal forage systems for animal production and the environment. Proceedings of the 12th Symposium of the European Grassland Federation, Pleven, Bulgaria; 2003. 26–28 May 2003.

Lazaridou M, Koutroubas SD. Drought effect on water use efficiency of berseem clover at various growth stages. New directions for a diverse planet. Proceedings of the 4th International Crop Science Congress Brisbane, Australia; 2004. 26 Sept–1 Oct 2004.

Leakey ADB, Bernacchi CJ, Dohleman FG, Ort DR, Long SP. Will photosynthesis of maize (Zea mays) in the US Corn Belt increase in future CO_2 rich atmospheres? An analysis of diurnal courses of CO_2 uptake under free-air concentration enrichment (FACE). Glob Chang Biol. 2004;10:951–62.

Leakey ADB, Bernacchi CJ, Ort DR, Long SP. Long-term growth of soybean at elevated CO_2 does not cause acclimation of stomatal conductance under fully open-air conditions. Plant Cell Environ. 2006a;29:1794–800.

Leakey ADB, Uribelarrea M, Ainsworth EA, Naide SL, Rogers A, Ort DR, Long SP. Photosynthesis, productivity, and yield of maize are not affected by open-air elevation of CO_2 concentration in the absence of drought. Plant Physiol. 2006b;55:591–628.

Levy PE, Cannel MGR, Friend AD. Modelling the impact of future changes in climate, CO_2 concentration and land use on natural ecosystems and the terrestrial carbon sink. Glob Environ Change. 2004;14:21–30.

Long SP, Ainsworth EA, Rogers A, Ort DR. Rising atmospheric carbon dioxide: plants FACE the future. Annu Rev Plant Biol. 2004;55:591–628.

Long SP, Ainsworth EA, Leakey ADB, Nosberger J, Ort DR. Food for thought: lower-than-expected crop yield stimulation with rising CO_2 concentrations. Science. 2006;312:1918–21.

Loomis RS, Amthor JS. Limits of yield revisited. In: Reynolds MP, Rajaram S, McNab A, editors. Increasing yield potential in wheat: breaking the barriers DF. Mexico: CIMMYT; 1996. p. 76–89.

Loomis RS, Lafitte HR. The carbon economy of a maize crop exposed to elevated CO_2 concentrations and water stress, as determined from elemental analyses. Field Crop Res. 1987;17:63–74.

Makino A, Mae T. Photosynthesis and plant growth at elevated levels of CO_2. Plant Cell Physiol. 1999;40:999–1006.

Maroco JP, Edwards GE, Ku MSB. Photosynthetic acclimation of maize to growth under elevated levels of carbon dioxide. Planta. 1999;210:115–25.

Marques da Silva J, Arrabica MC. Effect of water stress on Rubisco activity of Setaria sphacelota. In: Mathis P, editor. Photosynthesis: from light to biosphere, vol. 4. Dordrecht: Kluwer; 1995. p. 545–8.

Medici LO, Azevedo RA, Canellas LP, et al. Stomatal conductance of maize under water and nitrogen deficits. Pesq Agrop Brasileira. 2007;42:599–601.

Medlyn BE, Barton CVM, Broadmeadow MSJ, et al. Stomatal conductance of forest species after long-term exposure to elevated CO_2 concentration: a synthesis. New Phytol. 2001;149:247–64.

Medrano H, Parry MAJ, Socias X, Lawlor DW. Longterm water stress inactivates rubisco in subterranean clover. Ann Appl Biol. 1997;131:491–501.
Meyer WS, Green GC. Water use by wheat and plant indicators of available soil water. Agron J. 1980;72:253–7.
Meyer WS, Green GC. Plant indicators of wheat and soybean crop water stress. Irrig Sci. 1981;2:167–76.
Morgan JA, Pataki DE, Körner C, et al. Water relations in grassland and desert ecosystems exposed to elevated atmospheric CO_2. Oecologia. 2004;140:11–25.
Morison JIL. Response of plants to CO_2 under water limited conditions. Vegetatio. 1993;104(105):193–209.
Mumm P, Wolf T, Fromm J, et al. Cell type-specific regulation of ion channels within the maize stomatal complex. Plant Cell Physiol. 2011;52:1365–75.
Nerd A, Nobel PS. Effects of drought on water relations and nonstructural carbohydrates in cladodes of Opuntia ficus-indica. Physiol Plant. 1991;81:495–500.
Nesmith DS, Ritchie JT. Effects of soil water deficits during tassel emergence on development and yield component of maize (*Zea mays* L.). Field Crop Res. 1992;28:251–6.
Nikinmaa E, Hölttä T, Hari P, Kolari P, Mäkelä A, Sevanto S, Vesala T. Assimilate transport in phloem sets conditions for leaf gas exchange. Plant Cell Environ. 2013;36:655–69.
Norby RJ, Wullschleger SD, Gunderson CA, Johnson DW, Ceulemans R. Tree responses to rising CO_2 in field experiments: implications for the future forest. Plant Cell Environ. 1999;22:683–714.
Pan Y, Birdsey RA, Fang J, Houghton R, Kauppi PE, Kurz WA, Phillips OL, Shvidenko A, Lewis SL, Canadell JG, et al. A large and persistent carbon sink in the world's forests. Science. 2011;333:988–93.
Penning de Vries FWT, Keulen van H, Rabbinge R. Natural resources and the limits of food production. In: Bouma J et al., editors. Ecoregional approaches for sustainable land use and food production. Dordrecht: Kluwer Academic Press; 1995.
Poorter H. Do slow-growing species and nutrient-stressed plants respond relatively strongly to elevated CO_2? Glob Chang Biol. 1998;4:693–7.
Prior SA, Rogers HH, Sionit N, Patterson RP. Effects of elevated atmospheric CO_2, on water relations of soya bean. Agr Ecosyst Environ. 1991;35:13–25.
Prior SA, Runion GB, Rogers HH, Arriaga FJ. Elevated atmospheric carbon dioxide effects on soybean and sorghum gas exchange in conventional and no-tillage systems. J Environ Qual. 2010;39:596–608.
Radmer R, Kok B. Photosynthesis: limited yields, unlimited dreams. BioScience. 1977;27:599–605.
Reddy KR, Hodeges HF, Kimball BA. Crop ecosystem response to climatic change: cotton. In: Reddy KR, Hodges HF, editors. Climate change and global crop productivity. New York: CABI; 2000. p. 161–87.
Riedo M, Gyalistras D, Fuhrer J. Pasture responses to elevated temperature and doubled CO_2 concentration: assessing the spatial pattern across an alpine landscape. Clim Res. 2001;17:19–31.
Rogers HH, Runion GB, Krupa SV, Prior SA. In: Allen Jr LH, Kirkham MB, Olszyk DM, Whitman CE, editors. Advances in carbon dioxide effects research, Plant responses to atmospheric CO_2 enrichment: Implications in root-soil-microbe interactions. Madison, WI: ASA Special Publication No. 61. ASA, CSSA, and SSSA; 1997. p. 1–34.
Rosenzweig C, Hillel D. Climate change and the global harvest. New York: Oxford University Press; 1998. p. 324.
Sadras VO, Angus JF. Benchmarking water use efficiency of rainfed wheat in dry environments. Aust J Agr Res. 2006;57:847–56.
Sage RF, Monson RK. C4 plat biology. San Diego: Academic; 1999.
Samarakoon AB, Müller WJ, Gifford RM. Transpiration and leaf area under elevated CO_2: effects of soil water status and genotype in wheat. Aust J Plant Physiol. 1995;22:33–44.

Sau F, Mínguez MI. Adaptation of indeterminate faba beans to weather and management under a Mediterranean climate. Field Crop Res. 2000;66:81–99.

Schneider SH. What is 'dangerous' climate change. Nature. 2001;411:17–9.

Shpiler L, Blum A. Heat tolerance to yield and its components in different wheat cultivars. Euphytica. 1991;51:257–263.

Siddique MRB, Hamid A, Islam MS. Drought stress effects on water relations of wheat. Bot Bull Acad Sin. 2001;41:35–9.

Simane B, Peacock JM, Struik PC. Differences in development and growth rate among drought-resistant and susceptible cultivars of durum wheat (*Triticum turgidum L. var. durum*). Plant Soil. 1993;157:155–166.

Sionit N, Strain BR, Hellmers H, Kramer PJ. Effects of atmospheric CO_2 concentrations and water stress on water relations of wheat. Bot Gaz. 1981;142:191–6.

Sionit N, Rogers HH, Bingham GE, Strain BR. Photosynthesis and stomatal conductance with CO_2-enrichment of container- and field-grown soybeans. Agron J. 1984;76:447–51.

Smith SD, Huxman TE, Zitzer SF, Charlet TN, Housman DC, Coleman JS, Fenstermaker LK, Seemann JR. Nowak RS, Elevated CO_2 increases productivity and invasive species success in an arid ecosystem. Nature. 2000;408;79–81.

Somerville C, Briscoe J. Genetic engineering and water. Science. 2001;292:2217.

Surano KA, Shinn JH. CO_2 and water stress effects on yield, water-use efficiency and photosynthate partitioning in field grown corn. UCRL-90771. Livermore: Lawrence Livermore National Laboratory; 1984.

Timlin D, Bunce J, Fleisher D, et al. Simulation of the effects of limited water on photosynthesis and transpiration in field crops: can we advance our modeling approaches? In: Ahuja L et al., editors. Response of crops to limited water: understanding and modeling water stress effects on plant growth processes. Madison, WI: Advances in Agricultural Systems Modeling Ser. 1. ASA, CSSA, SSSA; 2008.

Tubiello FN, Ewert F. Simulating the effects of elevated CO_2 on crops: approaches and applications for climate change. Eur J Agron. 2002;18:57–74.

Tubiello FN, Rosenzweig C, Kimball BA, Pinter Jr PJ, Wall GW, Hunsaker DJ, Lamorte RL, Garcia RL. Testing CERES-wheat with FACE data: CO_2 and water interactions. Agron J. 1999;91:1856–65.

Tubiello FN, Donatelli M, Rosenzweig C, Stockle CO. Effects of climate change and elevated CO_2 on cropping systems: model predictions at two Italian locations. Eur J Agron. 2000;13:179–89.

Tubiello FN, Amthor JA, Boote K, Donatelli M, Easterling WE, Fisher G, Gifford R, Howden M, Reilly J, Rosenzweig C. Crop response to elevated CO_2 and world food supply. Eur J Agron. 2007;26:215–23.

van Herwaarden AF, Farquhar GD, Angus JF, Richards RA, Howe GN. 'Haying-off ', the negative grain yield response of dryland wheat to nitrogen fertilizer. I.Biomass, grain yield, and water use. Aust. J. Agr. Res. 1998;49:1067–1082.

von Caemmerer S, Quick PW. Rubisco, physiology in vivo. In: Leegood RC, Sharkey TD, von Caemmerer S, editors. In photosynthesis: physiology and metabolism. Dordrecht: Kluwer Academic; 2000. p. 85–113.

Wang W-H, Yi X-Q, Han A-D, et al. Calcium-sensing receptor regulates stomatal closure through hydrogen peroxide and nitric oxide in response to extracellular calcium in Arabidopsis. J Exp Bot. 2012;63:177–90.

Warren CR. Soil water deficits decrease the internal conductance to CO_2 transfer but atmospheric water deficits do not. J Exp Bot. 2008;59:327–34.

Watson RT, Zinyowera MC, Moss RH. Climate Change 1995: Impacts, Adaptation and Mitigation of Climate Change. Cambridge University Press, Cambridge; 1996.

Witter SH. Future technological advances in agriculture and their impact on the regulatory environment. BioScience. 1979;29:603–10.

Wu DX Wang GX. Interaction of CO_2 enrichment and drought on growth, water use, and yield of broad bean (*Vicia faba*). Environ. Exp. Bot. 2000;43:131–139.

Wu DX, Wang GX, Bai YF, Liao JX. Effects of elevated CO_2 concentration on growth, water use, yield and grain quality of wheat under two soil water levels. Agr. Ecosyst. Environ. 2004;104:493–507.

Yang Y, Timlin DJ, Fleisher DH, Kim S-H, Quebedeaux B, Reddy VR. Simulating leaf area of corn plants at contrasting water status. Agric For Meteorol. 2009;149:1161–7.

Yoshimoto M, Oue H, Kobayashi K. Energy balance and water use efficiency of rice canopies under free-air CO_2 enrichment. Agric. Forest Meteorol. 2005;133;226–246.

Chapter 11
Drought Stress Tolerance in Relation to Polyamine Metabolism in Plants

Miren Sequera-Mutiozabal, Antonio F. Tiburcio, and Rubén Alcázar

11.1 Introduction

The levels of polyamines (PAs, mainly putrescine, spermidine, and spermine) fluctuate in response to a diversity of abiotic and biotic stresses. Early experiments several decades ago, already reported the accumulation of putrescine in response to potassium starvation, oxidative stress, UV treatment, drought and osmotic stresses in different plant species. Correlational studies and the use of PA biosynthesis inhibitors suggested the implication of different PAs in abiotic stress tolerance. However, it was not until the discovery of genes encoding PA biosynthetic enzymes in different plant species that genetic manipulation of the PA pathway becomes feasible. Overexpression and loss-of-function approaches during the last decades support the conclusions provided by early polyamine researchers, pioneers of an area of research with promising practical applications. Current research by contemporary researchers in the field aims at the identification of PA mechanisms of action. This seems a challenging task given that PAs exert their functions through complex interactions with metabolic networks and diverse-signaling pathways. The overall picture currently involves ROS and NO signaling, ABA cross talk, GABA biosynthesis and interactions with primary metabolism, although this is only part of it. In this book chapter, we focus on the role of PAs during drought stress and currently known PA mechanisms of action contributing to desiccation tolerance.

M. Sequera-Mutiozabal • A.F. Tiburcio • R. Alcázar (✉)
Department of Natural Products, Plant Biology and Soil Science, Faculty of Pharmacy, University of Barcelona, Avda Joan XXIII s/n 08028, Barcelona, Spain
e-mail: ralcazar@ub.edu

M.A. Hossain et al. (eds.), *Drought Stress Tolerance in Plants, Vol 1*,
DOI 10.1007/978-3-319-28899-4_11

11.2 Polyamine Metabolism and Plant Tolerance to Abiotic Stress

Global climate change is expected to intensify the frequency and severity of drought and flooding events in many regions worldwide, severely affecting crop production (Pottosin and Shabala 2014). As sessile organisms, plants are exposed to several environmental stresses and have evolved diverse strategies to face life-threatening situations (Berberich et al. 2015). In consequence, plant stress physiology has been pointed out towards dissection of genetic elements involved in stress tolerance. In regard to this topic, polyamines (PAs) are essential molecules because they participate in abiotic and biotic stress responses in plants (Alcázar and Tiburcio 2014).

PAs are organic polycations having variable hydrocarbon chains and two or more primary amino groups (Takahashi and Kakehi 2010). The structure and chemistry of the most abundant PAs in plants diamine putrescine (Put), triamine spermidine (Spd), and tetraamine spermine (Spm) were elucidated in the late 1920s. It was revealed that they are nitrogen-containing compounds of low molecular weight (Alcázar et al. 2010a). Later on, it was shown that thermospermine (T-Spm), an isomer of spermine, was also present in higher plants (Moschou et al. 2008b), and this is not a minor (qualitatively) PA (Takano et al. 2012).

In the context of individual PAs, Put is important as a precursor for the biosynthesis of higher molecular weight PAs. According to a study using transgenic plants with altered PA levels, Put levels must exceed certain threshold to enhance the synthesis of Spd and Spm under stress, such synthesis being necessary for recovery from stress conditions (Capell et al. 2004). In addition, Put oxidation in plants produces 4-aminobutanal, which spontaneously cyclizes to Δ^1-pyrroline and can be further converted to γ-aminobutyric acid (GABA) (Petřivalský et al. 2007), a reaction common in animals and plants (Cona et al. 2006). GABA is an important metabolite which levels tend to be altered during stress, although its function is currently unknown (Shelp et al. 2012). Spd is a higher PA that is essential during embryogenesis in *Arabidopsis* (Imai et al. 2004b), its conjugates are implicated in protection against pathogens, detoxifying phenolic compounds, and/or serving as a reserve of PAs that becomes available in actively proliferating tissues (Takahashi and Kakehi 2010). No requirement for Spm under normal growth conditions has been demonstrated in an *Arabidopsis* mutants which cannot produce this PA (Imai et al. 2004a). However, this mutant showed sensitivity to drought and salt stress (Yamaguchi et al. 2007). To date, evidence suggests that Spm plays versatile roles in stress response (Takahashi and Kakehi 2010). Regarding to Spm isomers, an *Arabidopsis* mutant deficient in T-SPM synthesis has been isolated that displays reduced stem elongation (Kim et al. 2014). Indeed, T-Spm modifies the expression of auxin-related genes required for vascular tissue differentiation (Tong et al. 2014). Although the potential role of T-Spm in biotic stress protection has been shown (Sagor et al. 2012; Marina et al. 2013), its role in abiotic stress needs further investigations.

11.2.1 The Polyamine Biosynthetic Pathway

Metabolic studies indicate that the intracellular levels of PAs in plants are mostly regulated by anabolic and catabolic processes, as well as by their conjugation to hydroxycinnamic acids (Alcázar et al. 2010a). PA biosynthesis begins with Put formation. This PA is derived either directly from ornithine (Orn) by ornithine decarboxylase (ODC) or from arginine (Arg) through several steps catalyzed by arginine decarboxylase (ADC), which produces agmatine (Agm); agmatine iminohydrolase (AIH), and N-carbamoylputrescine amidohydrolase (CPA). In contrast to animals and fungi, in which ODC is the first and rate-limiting enzyme in the synthesis of PAs, plants typically use ADC (Takahashi and Kakehi 2010). In *Arabidopsis*, Put content is modulated by the expression of two gene isoforms encoding ADC (*ADC1* and *ADC2*), with contrasting expression patterns depending on the nature of the activation signal (Alcázar et al. 2010b). Spd, Spm, and T-Spm are synthesized by aminopropyltransferases (APT), which transfer aminopropyl residues to amine acceptors Put or Spd, producing Spd, Spm, or its isomers (e.g., T-Spm), respectively.

The donor of the aminopropyl groups is decarboxylated S-adenosylmethionine (dcSAM), which is formed by decarboxylation of S-adenosylmethionine (SAM), through an enzymatic reaction catalyzed by SAM decarboxylase (*SAMDC*). The APTs donating aminopropyl residues to Put or Spd for production of Spd or Spm are spermidine synthase (SPDS) and spermine synthase (SPMS). SDPS from *Arabidopsis thaliana* are encoded by two gene paralogs (*SPDS1* and *SPDS2*). *SPMS* and thermospermine synthase (ACL5) are single genes in *Arabidopsis* (Kakehi et al. 2008).

11.2.2 Polyamine Catabolism and Back-Conversion Pathways: PAOs

Endogenous PA levels mostly depend on the dynamic balance between *de novo* biosynthesis and catabolism (Fig. 11.1). PA oxidation is catalyzed by two types of amine oxidases (AO), copper-containing amine oxidases (CuAO), and flavin-containing polyamine oxidases (PAO), which carry covalently bound FAD as cofactor (Cona et al. 2006; Angelini et al. 2010). The reactions catalyzed by both type of enzymes contribute to several physiological processes through their reaction products (i.e., aminoaldehydes, 1,3-diaminopropane (DAP), and hydrogen peroxide) (Šebela et al. 2001; Cona et al. 2006; Angelini et al. 2010; Fincato et al. 2011).

CuAO are frequently found in dicots (Cona et al. 2006), and are homodimeric enzymes which exhibit high affinity for the oxidation of the primary amino groups of Put and cadaverine (Cad), and lower affinity for Spd and Spm (Moschou et al. 2008a). CuAOs participates in PA terminal catabolism in the apoplast and peroxisomes (Planas-Portell et al. 2013). *Arabidopsis* carries ten putative CuAO-encoding genes,

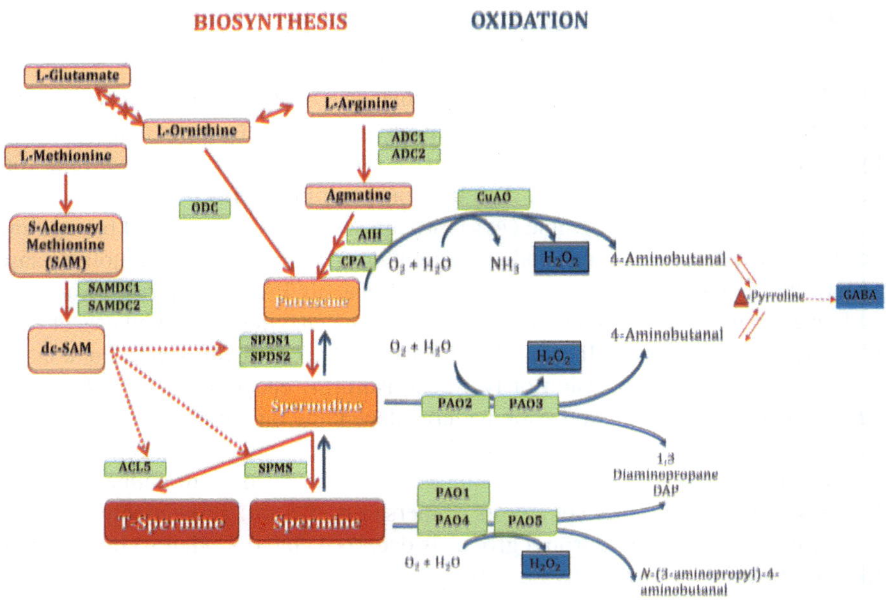

Fig. 11.1 Polyamine metabolism in the model species *Arabidopsis thaliana*. *ADC* arginine decarboxylase, *AIH* agmatine iminohydrolase, *CPA* N-carbamoylputrescine amidohydrolase, *ODC* ornithine decarboxylase, *dc-SAM* decarboxylated SAM, *ACL5 ACAULIS* 5, *PAO* polyamine oxidase

four of which (*ATAO1* and *AtCuAO1–3*) have been characterized. *CuAO* genes are differentially modulated during development, wounding, and treatment with hormones or elicitors. CuAO proteins also differ in their localization, with AtCuAO1 and AtAO1 being apoplastic, whereas AtCuAO2 and AtCuAO3 are peroxisomal enzymes (Møller and McPherson 1998; Reumann et al. 2009; Planas-Portell et al. 2013), all of them involved in terminal catabolism of Put and Spd.

PAO catalyze the oxidation of Spd, Spm, and/or acetylated derivatives at their secondary amino groups (Federico et al. 1996; Tavladoraki et al. 2012). They are classified into one of two families depending on whether they terminally oxidize PAs, as is the case on *Zea mays* (Federico et al. 1996; Tavladoraki et al. 1998; Moschou et al. 2008a) or catalyze PA back-conversion as in *Arabidopsis* or *Oryza sativa* (Tun et al. 2006; Moschou et al. 2008c; Takahashi et al. 2010; Fincato et al. 2011; Ono et al. 2012). PAOs of the first family oxidize the carbon at the *endo* side of the N^4 of Spd and Spm, producing 4-aminobutanal and N-(3-aminopropyl)-4-aminobutanal, respectively (Moschou et al. 2012). PAOs catalyzing PA back-conversion oxidize the carbon at the *exo* side of the N^4 of Spd, Spm, or T-Spm (and/or their acetylated derivatives) producing Put and Spd, respectively (Angelini et al. 2010). The most well-characterized plant PAOs involved in PA back-conversion are from *Arabidopsis*. This plant carries five PAO-encoding genes (*AtPAO1–5*) (Takahashi et al. 2010; Fincato et al. 2011). Tissue- and organ-specific expression studies of various *AtPAO* genes have shown some overlapping patterns but also important differences. This, together with their contrasted substrate specificity, suggests a functional diversity of *AtPAO* genes (Takahashi et al. 2010). The different subcellular localizations of AtPAO proteins also support the view that AtPAO2–4 are localized in peroxisomes, whereas AtPAO1 and AtPAO5 are predicted to be cytosolic. Thus, PA catabolism in the *Arabidopsis*apoplast is mediated predominantly by CuAO. Recently, it has been demonstrated that PA back-conversion mediated by PAO and terminal catabolism mediated by CuAO are co-localized in peroxisomes of *Arabidopsis* (Planas-Portell et al. 2013). Put level at certain points is able to inhibit peroxisomal PAO enzymes, which goes in favor of Spd or Put terminal degradation. A model was proposed where PA homeostasis is maintained by a tight coordination between both catabolic enzyme machineries (Planas-Portell et al. 2013). Thus, Spm and T-Spm homeostasis relies in part on the activity of PAO enzymes (Fig. 11.1).

Essentially PA pools are dynamic, changing over time, and PAs also undergo rapid interconversion in what is called the "polyamine cycle". Furthermore, catabolism of higher PAs may also lead to lower PA-releasing H_2O_2 and Put formation (Pál et al. 2015), which has been proposed as a mechanism to induce plant stress tolerance (Tavladoraki et al. 2012) and has been related to signal transduction during stress, affecting different cellular compartments (Ahou et al. 2014; Andronis et al. 2014; Liu et al. 2014; Tong et al. 2014). Therefore, PA catabolism by AO is not merely a PA degradation mechanism, but an important component of PA-signaling pathway, and an emerging field of PA research.

11.2.3 Interactions of PAs with Primary Metabolism

PAs are major sinks of assimilated nitrogen due to their intracellular high concentration (Moschou et al. 2012). However, many studies suggest the interaction of PAs with primary metabolism during plant development and stress (Walden et al. 1997; Forde and Lea 2007; Mattoo et al. 2010; Minocha et al. 2014). A unique feature of plant PA metabolism is that Put, proline (Pro), and GABA are all synthesized from a common substrate: glutamate (Glu), a hub molecule of nitrogen metabolism (Mattoo et al. 2010). Glu signaling in plants presents different spatiotemporal components in a complex way (Forde and Lea 2007; Forde 2014), impacting in all cases amino acid metabolism, which is connected at several levels with carbon mobilization pathways. In this sense, GABA is also a molecule for which the function is still unknown, but has been related to nitrogen and carbon metabolism. At the metabolic level, GABA transamination and further oxidation yields succinic acid, which enters directly into the Krebs cycle (Rea et al. 2004). Hence, a connection of GABA with carbon metabolism exists and has been suggested to coordinate C:N balance (Bouché et al. 2003; Bouché and Fromm 2004) via Glu receptors (Kang and Turano 2003). PA and GABA accumulation has been reported in both control and stress conditions (Shelp et al. 2012). Similarly, metabolic interactions have been reported between PAs and sugars such as glucose and sucrose (Handa and Mattoo 2010). This is interesting because sugar signaling is emerging as an important element in the plant's stress response (Van den Ende 2014).

Apart from known metabolic connections described, PAs have been involved in the biosynthesis of other metabolites not metabolically connected. To mention few examples, it has recently been reported that Put acts as buffer and osmolite that induces proline (Pro) production, leading to maintenance of leaf water status under stress conditions (Kotakis et al. 2014). Microarray analyses of Put overexpressor *Arabidopsis* plants revealed both the up- and down-regulation of stress-responsive, hormone and signaling-related genes, involved in the biosynthesis of auxin, ethylene (ET), ABA, gibberellin, and salicylic acid (SA). Furthermore, genes for auxin transport, genes coding for auxin-responsive proteins, ET and ABA-responsive transcriptional factors, and also jasmonate (JA)-induced proteins (Marco et al. 2011) were identified. Overexpression of Spd synthase up-regulated the expression of various putative stress-related genes in chilling-stressed transgenic *Arabidopsis* compared with the wild type. These genes putatively encode transcription factors, calmodulin-related protein and stress-protective proteins, such as *RD29A* (Kasukabe et al. 2004). *Arabidopsis* plants with increased Spm levels showed altered expression of genes involved in the biosynthesis of JA, ABA, and SA, receptor-like kinases, mitogen-activated protein kinases and genes with a role in calcium regulation (Marco et al. 2011). In tobacco, Spm accumulation caused up-regulation of transcripts for anti-oxidative enzymes, especially those induced by abiotic stresses, such as salt, cold, or acidic stress (Wi et al., 2006). In agreement with this, it was demonstrated that T-Spm modifies the expression of auxin-related genes (Tong et al. 2014).

The identification of PA-regulated downstream targets and the discovery of connections between PA and other stress-responsive molecules (mostly related to primary metabolism) have opened new possibilities to investigate the function of individual PAs at molecular level (Pál et al. 2015). However, further studies are needed to elucidate the PA-signaling pathways (Shi and Chan 2014).

11.3 Implications of the PA Pathway During the Drought Stress Response

Drought stress is a tremendous limitation for plant growth and hence, crop productivity. Upon severe environmental stresses such as drought, plants activate physiological, metabolic and defense systems to survive and sustain growth. PA accumulation has been associated with plant tolerance during water stress in different species (Liu et al. 2007; Groppa and Benavides 2008; Alcázar et al. 2010a; Takahashi and Kakehi 2010; Minocha et al. 2014; Shi and Chan 2014).

11.3.1 Changes in Primary Metabolism Induced By Drought

Reduction of photosynthetic activity, accumulation of organic acids and osmolytes, and changes in carbohydrate metabolism are typical, physiological, and biochemical responses to stress (Valliyodan and Nguyen 2006; Xiong et al. 2002). Plant features associated with tolerance mechanisms are multigenic, and thus, difficult to elucidate. Omics approaches and gene expression studies have identified the activation and regulation of several stress-related transcripts and proteins, generally classified into two major groups. One group is involved in signaling cascades and in transcriptional control, whereas the other in membrane protection such as osmoprotectants or antioxidants (Valliyodan and Nguyen 2006). The first group is constituted by transcriptions factors (TFs), which are essential components in the abiotic stress signal transduction of ABA-dependent and ABA-independent pathways. TFs also function in hubs with other proteins partners, and inside dynamic networks serving as interacting nodes between different pathways (Lindemose et al. 2013). TFs that contribute to water stress signaling include basic leucine zipper (bZIP type) proteins, APETALA 2/ethylene-responsive element-binding factor (AP2/ERF type), NAM/ATAF1/CUC2 (NAC type), or MYB domain-containing proteins (MYB type) (Hadiarto and Tran 2011; Lindemose et al. 2013; Todaka et al. 2015). The synthesis of osmoprotectants or compatible solutes is one another mechanism by which plants adapt to water deficit (Valliyodan and Nguyen 2006). Redox metabolism and associated signaling also participate in tolerance against abiotic stress (Munné-bosch et al. 2013).

The most significant changes regarding primary metabolism after water stress are related to carbon mobilization and reallocation. One of the first things occurring during dehydration is that cell division and expansion are severely inhibited (Xiong et al. 2002). The oxidative stress, generated as secondary effect, may cause damage to the photosynthetic apparatus (Ort 2001). This is in part because carbon uptake is further reduced due to the concomitant or earlier inhibition of growth (Chaves and Oliveira 2004), and also because it may allow plants to divert energy resources to generate protective molecules (Xiong et al. 2002). However, it is generally accepted that the decrease in photosynthetic rate is primarily due to marked stomatal closure, mainly mediated by ABA (Chaves et al. 2002).

Regarding sugar molecules, the signaling role is not fully understood. Nonetheless, there is evidence of cross-talk mechanisms between hexoses and ABA. An increase in acid invertase activity was observed in leaves of dehydrated maize plants, in accordance with a rapid accumulation of glucose and fructose, which was highly correlated with a xylem-located ABA increase (Trouverie et al. 2003). Conversely, in stressed *Arabidopsis*seedlings, glucose promoted the transcription of several genes in ABA biosynthesis (Cheng et al. 2002). In any case, sugars being transported into the xylem of dehydrated plants, or an abrupt increase of sugars in the apoplast of guard cells under high light or stress, are mechanisms that likely exert an important influence on ABA-mediated stomatal dynamics (Wilkinson and Davies 2002). Lately, trehalose has emerged as redox-signaling molecule with a proposed role on stress (Luo et al. 2008; O'Hara et al. 2013; Krasensky et al. 2014). Trehalose degradation and glucose production have been associated with drought tolerance (Van Houtte et al. 2013). Raffinose oligosaccharides (RFOs) (Van den Ende 2013; Elsayed et al. 2014) are also important molecules during drought. RFOs accumulation occurs during drought stress, and they function as osmolites maintaining cell turgor (Bartels and Sunkar 2005; Nakabayashi et al. 2014) and antioxidants that alleviate the accumulation of ROS under stress conditions (Nishizawa et al. 2008; Van den Ende and Valluru 2008; Bolouri-Moghaddam et al. 2010; Nakabayashi et al. 2014). Sugars or hydrocarbons are also involved in the control of the expression of different genes related to lipid and nitrogen metabolism (Price et al. 2004; Weisman et al. 2010), both altered by drought stress.

Regarding to nitrogen metabolism, amino acids or metabolites related to their metabolism are the most altered molecules by drought stress (Kalamaki et al. 2009; Mao et al. 2010; Nakabayashi et al. 2014), in some cases to cope with oxidative stress (Lehmann et al. 2009), or to promote the functioning of central metabolic pathways implied in redox balance (Araújo et al. 2010; Obata et al. 2011). On the other hand, the homeostasis of reactive nitrogen species (RNS) like nitric oxide (NO) is a key-signaling element during stress including drought (Qiao and Fan 2008; Tanou et al. 2009; Zhao et al. 2009; Filippou et al. 2011; Fan et al. 2013; Ziogas et al. 2013), in part because of its relation with stomatal closure (García-Mata and Lamattina 2003; Neill et al. 2008) and lateral root induction (Sun et al. 2015).

11.3.2 Evidences Supporting a Role for PAs in Drought Tolerance

During the last decade, different approaches have been undertaken to generate plants tolerant to abiotic stress. Genetic engineering of PA pathway or exogenous application of PAs has been reported. From these approaches, the main conclusion is that PAs provide tolerance against water stress at different levels and in different species of agronomical interest such as maize, rice, cacao, or wheat (Bae et al. 2008; Mao et al. 2010; An et al. 2012; Agudelo-Romero et al. 2014; Hatmi et al. 2014). This might be because their increased concentration gives in some way protection to cells, or because PA degradation products are signal molecules that trigger defense signaling.

11.3.2.1 Genetic Engineering Approaches for the Manipulation of PA Homeostasis

By the use of transgenics approaches, it has been possible to determine that increases in the levels of major PAs in plants are related to drought stress tolerance. Constitutive *ADC* expression in rice provided a continuous supply of Put, which was necessary to synthesize sufficient Spd and Spm after SAMdc induction during drought stress.

This mechanism was determinant to provide stress tolerance in transgenic rice (Capell et al. 2004). On the other hand, overexpression of *SPDS* gene in *Arabidopsis* generated almost two-fold increases of free Spd levels which were associated with tolerance to drought, and induction of TF involved in drought protection (Kasukabe et al. 2004). To distinguish the role of Put from that of Spd and Spm, transgenic rice was generated overexpressing *ADC* or *SAMdc*. The ADC overexpressors showed a marked increase in all PAs and drought-tolerant phenotypes. SAMdc overexpressors exhibited a similar behavior to the wild-type plants although the recovery was significantly more robust. Authors assigned an immediate protective effect to Put and protective effect to Spm, which was associated to recovery (Peremarti et al. 2009)

Indirect stimulation of PA biosynthesis has also been reported to lead to drought tolerance. Overexpression of a TF of the MYBs family (R2R3 type-MYB gene) stimulated the expression of *ADC* gene during several stresses including drought, thus increasing PA levels. The mutants were tolerant to water deprivation in part by modulation of PA metabolism (Sun et al. 2014)

11.3.2.2 Exogenous Application of PAs and Use of PA Biosynthesis Inhibitors

Generation of plants completely depleted of PAs is not possible. However, by the use of plants defective in the production of at least one major PA, followed by exogenous application (rescue), it has been possible to understand more about the physiological functions of PAs in plants. *Arabidopsis*mutants defective in the production of Spm (*acl5/spms*) were sensitive to drought stress, this phenotype was cured by Spm pretreatment but not by addition of Put or Spd, which led to the notion that the induction of PA biosynthetic pathway after drought stress is necessary to overcome the stressful condition by means of Spm production, assigning a possible protective role for this higher PA (Yamaguchi et al. 2007). Nonetheless, pretreatment after exogenous addition of Put, Spd, and Spm improved drought tolerance in Bermudagrass (Shi et al. 2013). Proteomic approaches indicated an enhancement of enzymes involved on ROS metabolism, proline, and sugar content after pretreatment with higher PAs specially Spm, in the same manner enhancement of proteins related to electron transport and energy pathways was observed after all kinds of pretreatments (Shi et al. 2013), establishing that improvement of abiotic stress plant tolerance can be achieved by exogenous PAaddition (Table 11.1).

11.4 PA Modes of Action During Drought Stress Tolerance: ABA, NO, and ROS

11.4.1 Importance of ABA in the Drought Stress Response

Water availability is an essential feature to plant physiology. In that sense, ABA is extremely important (Desikan et al. 2004). In brief, water stress induces gene expression towards ABA biosynthesis, with a concomitant accumulation and

Table 11.1 Genetic engineering of PA metabolism and its effects on drought stress tolerance

Species	Stimulus/genetic approach	Drought phenotype	Remarks	Citation
Oryza sativa	Constitutive generation of Put	Tolerant	Put pools are determinant to produce enough Spd and Spm to present drought tolerance	Capell et al. (2004)
Arabidopsis thaliana	Over-production of Spd	Tolerant	Spd was associated with expression of TF involved on drought protection	Kasukabe et al. (2004)
Arabidopsis thaliana	Mutants defective in Spm production	Sensitive	Drought sensitivity was rescued only after Spm exogenous addition.	Yamaguchi et al. (2007)
Arabidopsis thaliana	Exogenous addition of Spm	Tolerant	Spm has a protective role on drought stress	Yamaguchi et al. (2007)
Arabidopsis thaliana	Overexpression of Glutamate-synthase (*SlNAGS1*)	Tolerant	Accumulation of PA precursors (e.g., Orn, Arg, and Glu) is necessary to drought stress alleviation	Kalamaki et al. (2009)
Oryza sativa	Overexpression of *ADC*	Tolerant	Put exerts a direct protective effect during stress	Peremarti et al. (2009)
Oryza sativa	Overexpression of *SAMDC*	NP	Spm might have a protective role during stress recovery	Peremarti et al. (2009)
Arabidopsis thaliana	Overexpression of *ADC2*	Tolerant	Put production by ADC2 is involved on drought stress tolerance	Alcázar et al. (2010b)
Egyptian cotton	Overexpression of SAMDC	Tolerant	Spm accumulation led to drought tolerance of cotton transgenics varieties	Momtaz et al. (2010)
Arabidopsis thaliana	T-DNA mutants for *ADC, SPDS, SPM*	NP	PAs back-conversion pathway is involved on drought response rather than SPM terminal oxidation	Alcázar et al. (2011)
Arabidopsis thaliana	Expression of *ADC* under stress-inducible promoter	Tolerant	Put generation is stress-inducible and correlates with induction of stress-responsive genes	Alet et al. (2011)
Solanum lycopersicum	Overexpression of *SAMDC*	Tolerant	Spd and Spm enhancement lead to tolerance against biotic and abiotic stress on the whole plant	Hazarika and Rajam (2011)
Arabidopsis thaliana	Overexpression of *ADC/SAMDC*	NP	Transcriptomic profiling revealed Put and Spm induce ABA biosynthetic genes	Marco et al. (2011)
Arabidopsis thaliana	Overexpression of *PtADC* on *adc* mutant	Tolerant	ADC inhibitor D-Arginine reversed drought-tolerant phenotype and ROS scavenging properties of PtADC	Wang et al. (2011)
Arabidopsis thaliana	T-DNA mutants for *CuAO1*	NP	Put oxidation by CuAO1 is involved on ABA-mediated stomatal closure	Wimalasekera et al. (2011)

(continued)

Table 11.1 (continued)

Species	Stimulus/genetic approach	Drought phenotype	Remarks	Citation
Cynodon dactilon	Exogenous addition of Put, Spd, and Spm	Tolerant	Proteins involved with ROS balance were stimulated by higher PAs. All treatments stimulated energy-related pathways	Shi et al. (2013)
Nicotiana tabacum	Overexpression of MYB type TF	Tolerant	Overexpression of MYB type TF induce *ADC* gene after drought stress maintaining increase PAs	Sun et al. (2014)
White clover	Spm exogenous addition	Tolerant	Protective Spm effect may be due to enhancement of sugar metabolism and dehydrin biosynthesis	Li et al. (2015)

redistribution of this plant hormone in the guard cells surrounding the stomata. This triggers highly interactive signaling cascades, where a consequent movement of ions across the membrane, leads to release of water and turgor loss of guard cells, causing stomatal closure. ABA and its role after water deprivation have been extensively studied and reviewed (Bray 1997; Desikan et al. 2004; Cutler et al. 2010).

The role of ABA role during drought is unquestionable, but later on it was discovered that ABA was not the only molecule in stomatal signaling. In the early 2000s, it was demonstrated that ABA treatment of *Arabidopsis* guard cells induced an oxidative burst by H_2O_2 that resulted in stomatal closure (Pei et al. 2000). Soon, it was observed the same mechanism in other species thus establishing that the generation of ROS was essential for ABA-induced stomatal closure (Desikan et al. 2004). Sources for the generation of this ROS were proposed. Early experiments using tobacco epidermal cells led to the notion that flavin-containing enzymes such as peroxidases or AOs (e.g., PAOs) were responsible for H_2O_2 production (Allan and Fluhr 1997). However, further investigations suggested that NADPH oxidase-mediated H_2O_2 release is required for ABA-mediated stomatal closure (Kwak et al. 2003). During the same period, NO was also found essential for stomatal dynamics. Reports on pea, wheat, and maize provided evidence that NO is a signaling component in ABA-mediated stomatal closure (Neill et al. 2002) after drought imposition (García-Mata and Lamattina 2003) through a feedback mechanism in which NO synthesis is required for ABA-induced stomatal closure and ABA enhances NO synthesis in guard cells (Neill et al. 2002). Recent reports have also demonstrated that *S*-nitrosylation by NO may also modulate ABA-mediated signaling, exerting a negative regulation on an ABA-dependent kinase (OST1), which increases the complexity of ABA-NO interaction referred to stomatal dynamics (Wang et al. 2015).

Sugar metabolism is also related to ABA induction, glucose is necessary to induce ABA synthesis (Cheng et al. 2002) and sucrose is produced after ABA induction during drought stress (Trouverie et al. 2003). Obviously, ABA is a hub molecule for stress signaling as PAs are, thus, the interaction between both is of relevance during drought stress.

11.4.2 Modulation of PA Metabolism By ABA: Involvement in Stomatal Aperture

In *Arabidopsis*, drought-mediated induction of PA biosynthetic enzymes ADC, SPDS, and SPMS and subsequent increases in PAs levels is an ABA-dependent response (Alcázar et al. 2006). A similar trend was observed in *ADC* expression in response to salt stress (Urano et al. 2004) and cold stress (Cuevas et al. 2008). Conversely, in *Lotus tenuis* Put exerts a positive regulation on *NCED* (an essential ABA biosynthetic gene) promoting ABA formation in response to drought, while the levels of higher PAs did not show alteration (Espasandin et al. 2014). Feedback mechanism between Put and ABA cannot be ruled out. Indeed, transcriptomic analyses of transgenic *Arabidopsis* plants accumulating Spm by *SAMDC1* or SPMS overexpression showed significant increases in the expression of important ABA biosynthetic genes (e.g., *NCED*) followed by ABA accumulation, and up-regulation of genes associated with water deprivation and defense response including drought-related TFs (Marco et al. 2011). Interestingly, *SPMS* is an ABA-inducible gene (Rambla et al. 2010), which might be linked with a role for Spm during water stress (Yamaguchi et al. 2007). The connection between PAs and ABA is not trivial, and some overlapping functions between Put and Spm regarding to ABA have been suggested (Minocha et al. 2014).

11.4.3 Interactions with ROS and NO Signaling: Are PAOs the Main Drivers in the PA-Abiotic Stress Response?

Previous observations relate PAs to ROS through H_2O_2 via their catabolism pathway; nonetheless, their relationship appears to be more complex. H_2O_2 derived from PA degradation not always leads to the same effect. Transgenic tobacco plants over-expressing *ZmPAO* showed low apoplastic Spd and Spm levels under control conditions (due to their degradation), as well as lower levels of ROS because of an enhancement of anti-oxidative machinery (Moschou et al. 2008a). Surprisingly, these transgenic plants were sensitive to oxidative stress compared with the wild type, leading to the notion that PAs itself were key elements of the oxidative response (Moschou et al. 2008a). Another aspect is that it is becoming increasingly clear that ROS derived from PA oxidation is necessary to trigger stress responsiveness. In this sense, AOs like PAO enzymes are emerging modulators of ROS-related signaling, instead of being only PAs catabolism enzymes. A recent study demonstrated that Spd oxidation by *PAO3* is required for balanced respiration, proposing that this peroxisomal PAO is a key element for balancing superoxide/hydrogen peroxide production. Spd homeostasis by *PAO3* was demonstrated to be involved in ROS production other than H_2O_2, and the ratio ($O_2^{\bullet -}/H_2O_2$) showed to be an important signal for transcriptional induction (Andronis et al. 2014).

At present, PA researchers recognize that NO biosynthesis and PA metabolism are tightly related. Spd and Spm are able to promote NO biosynthesis in *Arabidopsis* seedlings, root tip, and primary leaves (Tun et al. 2006). A recent report in citrus seedlings provided evidence that there is a tissue-specific modulation of *PAO* expression, when oxidative/nitrosative stress is imposed as pretreatment. In both cases, pretreated plants showed salinity tolerance, suggesting that PAs may represent a molecular link between oxidative and nitrosative signaling (Tanou et al. 2012). The same authors further demonstrated that PAs are able to reprogram oxidative and nitrosative status as well as the proteome of salt-stressed plants (Tanou et al. 2014) suggesting a feedback mechanism between these three elements.

A significant reduction in NO release was observed in *Arabidopsis cuao1* suggesting that this AO might be involved in NO biosynthesis induced by PAs. Interestingly, these mutants were insensitive to exogenous addition of ABA (Wimalasekera et al. 2011). In grape, it has been demonstrated that ABA is an upstream signal for induction of PA exodus to the apoplast, and subsequent activation of PA catabolic pathway. H_2O_2 derived from PAO-mediated oxidation is a signal for stomatal closure (Konstantinos et al. 2010). Moreover, tomato seedlings under drought stress showed that significant increases of higher PAs stimulated ABA biosynthesis, which was positively correlated with increased PAO activity (Zhang and Huang 2013). PAO enzymes seem to be part of the signaling mechanism triggering plant defense. However, further investigations are necessary to understand the nature of these effects.

A recent work reported that ABA-mediated stomatal closure involves a coordinated mechanism in which ABA downstream targets were not just H_2O_2 and NO (Xie et al. 2014). In this report, the novelty relies in the involvement of hydrogen gas (H_2), which seems to be highly induced by ABA and is a subsequent stimulator of signaling by H_2O_2 and NO. The cascade involves activation of an outward/rectifying K^+ channel of the guard cell (GORK) in an ROS/RNS-dependent manner (Xie et al. 2014). Interestingly, the connection between ROS and PA signals for activation of plasma membrane cation channels have been established in *Arabidopsis* (Pottosin et al. 2012; Pottosin and Shabala 2014).

ABA dynamics are complex; nonetheless, multiple lines of evidence discussed above suggest that there is a biologically active interplay between these molecules and the triad H_2O_2-NO-PAs, especially induced after stress signals.

11.5 Conclusion

Polyamine pathway engineering enables the development of crops tolerant to different types of abiotic stresses, those including drought, salinity, freezing and oxidative stress. The discovery of PA back-conversion opens new possibilities for the accumulation of Spm and Spd, for which the biosynthesis is tightly regulated. Future research in PA transport and PA exchange between plants and soil microbes will enable the development of crops with improved PA contents by root uptake.

Acknowledgements Research has been supported by the Spanish Ministerio de Ciencia e Innovación (BIO2011-29683 and CSD2007-00036) and the Generalitat de Catalunya (SGR2009-1060) to A.F.T. Current research by R.A. is supported by the Ramón y Cajal Program (RYC-2011-07847) of the Ministerio de Ciencia e Innovación (Spain), the BFU2013-41337-P grant of the Programa Estatal de Fomento de la Investigación Científica y Técnica de Excelencia (Ministerio de Economía y Competitividad, Spain) and the Marie Curie Career Integration Grant (DISEASENVIRON, PCIG10-GA-2011-303568) of the European Union. A.F.T. and R.A. are members of the Grup de Recerca Consolidat 2014 SGR-920 of the Generalitat de Catalunya, Spain.

References

Agudelo-Romero P, Ali K, Choi YH, Sousa L, Verpoorte R, Tiburcio AF, Fortes AM. Perturbation of polyamine catabolism affects grape ripening of *Vitis vinifera* cv. Trincadeira. Plant Physiol Biochem. 2014;74:141–155.

Ahou A, Martignago D, Alabdallah O, et al. A plant spermine oxidase/dehydrogenase regulated by the proteasome and polyamines. J Exp Bot. 2014;65:1585–603.

Alcázar R, Tiburcio AF. Plant polyamines in stress and development: an emerging area of research in plant sciences. Front Plant Sci. 2014;5:1–2.

Alcázar R, Cuevas JC, Patron M, Altabella T, Tiburcio AF. Abscisic acid modulates polyamine metabolism under water stress in *Arabidopsis thaliana*. Physiol Plant. 2006;128:448–55.

Alcázar R, Altabella T, Marco F, Bortolotti C, Reymond M, Koncz C, Carrasco P, Tiburcio AF. Polyamines: molecules with regulatory functions in plant abiotic stress tolerance. Planta. 2010a;231:1237–49.

Alcázar R, Planas-Portell J, Saxena T, Zarza X, Bortolotti C, Cuevas J, Bitrián M, Tiburcio AF, Altabella T. Putrescine accumulation confers drought tolerance in transgenic *Arabidopsis* plants over-expressing the homologous *Arginine decarboxylase* 2 gene. Plant Physiol Biochem. 2010b;48:547–52.

Alcázar R, Bitrián M, Bartels D, Koncz C, Altabella T, Tiburcio AF. Polyamine metabolic canalization in response to drought stress in Arabidopsis and the resurrection plant *Craterostigma plantagineum*. Plant Signal Behav. 2011;6:243–50.

Alet AI, Sanchez DH, Cuevas JC, et al. Putrescine accumulation in Arabidopsis thaliana transgenic lines enhances tolerance to dehydration and freezing stress. Plant Signal Behav. 2011;6:278–86.

Allan A, Fluhr R. Two distinct sources of elicited reactive oxygen species in tobacco epidermal cells. Plant Cell. 1997;9:1559–72.

An ZF, Li CY, Zhang LX, Alva AK. Role of polyamines and phospholipase D in maize (*Zea mays* L.) response to drought stress. S Afr J Bot. 2012;83:145–50.

Andronis EA, Moschou PN, Toumi I, Roubelakis-Angelakis KA. Peroxisomal polyamine oxidase and NADPH-oxidase cross-talk for ROS homeostasis which affects respiration rate in *Arabidopsis thaliana*. Front Plant Sci. 2014;5(132):1–10.

Angelini R, Cona A, Federico R, Fincato P, Tavladoraki P, Tisi A. Plant amine oxidases 'on the move': an update. Plant Physiol Biochem. 2010;48:560–4.

Araújo WL, Ishizaki K, Nunes-Nesi A, et al. Identification of the 2-hydroxyglutarate and isovaleryl-CoA dehydrogenases as alternative electron donors linking lysine catabolism to the electron transport chain of Arabidopsis mitochondria. Plant Cell. 2010;22:1549–63.

Bae H, Kim SH, Kim MS, Sicher RC, Lary D, Strem MD, Natarajan S, Bailey BA. The drought response of *Theobroma cacao* (cacao) and the regulation of genes involved in polyamine biosynthesis by drought and other stresses. Plant Physiol Biochem. 2008;46:174–88.

Bartels D, Sunkar R. Drought and salt tolerance in plants. Crit Rev Plant Sci. 2005;24:23–58.

Berberich T, Sagor GHM, Kusano T. Polyamines in plant stress response. In: Kusano T, Suzuki H, editors. Polyamines. Tokio: Springer; 2015. p. 155–68.

Bolouri-Moghaddam MR, Le Roy K, Xiang L, Rolland F, Van den Ende W. Sugar signalling and antioxidant network connections in plant cells. FEBS J. 2010;277:2022–37.

Bouché N, Fromm H. GABA in plants: just a metabolite? Trends Plant Sci. 2004;9:110–5.

Bouché N, Lacombe B, Fromm H. GABA signaling: a conserved and ubiquitous mechanism. Trends Cell Biol. 2003;13:607–10.

Bray BA. Plant responses to water deficit. Trends Plant Sci. 1997;2:48–54.

Capell T, Bassie L, Christou P. Modulation of the polyamine biosynthetic pathway in transgenic rice confers tolerance to drought stress. Proc Natl Acad Sci U S A. 2004;101:9909–14.

Chaves MM, Oliveira MM. Mechanisms underlying plant resilience to water deficits: prospects for water-saving agriculture. J Exp Bot. 2004;55:2365–84.

Chaves MM, Pereira JS, Maroco J, Rodrigues ML, Ricardo CPP, Osório ML, Carvalho I, Faria T, Pinheiro C. How plants cope with water stress in the field? Photosynthesis and growth. Ann Bot. 2002;89:907–16.

Cheng W-H, Endo A, Zhou L, et al. A unique short-chain dehydrogenase/reductase in Arabidopsis glucose signaling and abscisic acid biosynthesis and functions. Plant Cell. 2002;14:2723–43.

Cona A, Rea G, Angelini R, Federico R, Tavladoraki P. Functions of amine oxidases in plant development and defence. Trends Plant Sci. 2006;11:80–8.

Cuevas JC, López-Cobollo R, Alcázar R, Zarza X, Koncz C, Altabella T, Salinas J, Tiburcio AF, Ferrando A. Putrescine is involved in Arabidopsis freezing tolerance and cold acclimation by regulating abscisic acid levels in response to low temperature. Plant Physiol. 2008;148:1094–105.

Cutler SR, Rodriguez PL, Finkelstein RR, Abrams SR. Abscisic acid: emergence of a core signaling network. 2010.

Desikan R, Cheung MK, Bright J, Henson D, Hancock JT, Neill SJ. ABA, hydrogen peroxide and nitric oxide signalling in stomatal guard cells. J Exp Bot. 2004;55:205–12.

Elsayed AI, Rafudeen MS, Golldack D. Physiological aspects of raffinose family oligosaccharides in plants: protection against abiotic stress. Plant Biol. 2014;16:1–8.

Espasandin FD, Maiale SJ, Calzadilla P, Ruiz OA, Sansberro PA. Transcriptional regulation of *9-cis*-epoxycarotenoid dioxygenase (NCED) gene by putrescine accumulation positively modulates ABA synthesis and drought tolerance in Lotus tenuis plants. Plant Physiol Biochem. 2014;76:29–35.

Fan HF, Du CX, Guo SR. Nitric oxide enhances salt tolerance in cucumber seedlings by regulating free polyamine content. Environ Exp Bot. 2013;86:52–9.

Federico R, Ercolini L, Laurenzi M, Angelini R. Oxidation of acetylpolyamines by maize polyamine oxidase. Phytochemistry. 1996;43:339–41.

Filippou P, Antoniou C, Fotopoulos V. Effect of drought and rewatering on the cellular status and antioxidant response of Medicago truncatula plants. Plant Signal Behav. 2011;6:270–7.

Fincato P, Moschou PN, Spedaletti V, Tavazza R, Angelini R, Federico R, Roubelakis-Angelakis KA, Tavladoraki P. Functional diversity inside the Arabidopsis polyamine oxidase gene family. J Exp Bot. 2011;62:1155–68.

Forde BG. Glutamate signalling in roots. J Exp Bot. 2014;65:779–87.

Forde BG, Lea PJ. Glutamate in plants: metabolism, regulation, and signalling. J Exp Bot. 2007; 58:2339–58.

García-Mata C, Lamattina L. 2003. Abscisic acid, nitric oxide and stomatal closure—is nitrate reductase one of the missing links? Trends Plant Sci. 2003;8(1):20–6.

Groppa MD, Benavides MP. Polyamines and abiotic stress: recent advances. Amino Acids. 2008;34:35–45.

Hadiarto T, Tran LSP. Progress studies of drought-responsive genes in rice. Plant Cell Rep. 2011;30:297–310.

Handa AK, Mattoo AK. Differential and functional interactions emphasize the multiple roles of polyamines in plants. Plant Physiol Biochem. 2010;48:540–6.

Hatmi S, Gruau C, Trotel-Aziz P, Villaume S, Rabenoelina F, Baillieul F, Eullaffroy P, Clement C, Ferchichi A, Aziz A. Drought stress tolerance in grapevine involves activation of polyamine

oxidation contributing to improved immune response and low susceptibility to Botrytis cinerea. J Exp Bot. 2014;66:775–87.
Hazarika P, Rajam MV. Biotic and abiotic stress tolerance in transgenic tomatoes by constitutive expression of S-adenosylmethionine decarboxylase gene. Physiol Mol Biol Plants. 2011;17:115–28.
Imai A, Akiyama T, Kato T, Sato S, Tabata S, Yamamoto KT, Takahashi T. Spermine is not essential for survival of Arabidopsis. FEBS Lett. 2004a;556:148–52.
Imai A, Matsuyama T, Hanzawa Y, Akiyama T, Tamaoki M. Spermidine Synthase Genes Are Essential for Survival of Arabidopsis. Plant Physiol. 2004b;135:1565–73.
Kakehi J-I, Kuwashiro Y, Niitsu M, Takahashi T. Thermospermine is Required for Stem Elongation in Arabidopsis thaliana. Plant Cell Physiol. 2008;49:1342–9.
Kalamaki MS, Alexandrou D, Lazari D, Merkouropoulos G, Fotopoulos V, Pateraki I, Aggelis A, Carrillo-López A, Rubio-Cabetas MJ, Kanellis AK. Over-expression of a tomato *N*-acetyl-L-glutamate synthase gene (*SlNAGS1*) in *Arabidopsis thaliana* results in high ornithine levels and increased tolerance in salt and drought stresses. J Exp Bot. 2009;60:1859–71.
Kang J, Turano FJ. The putative glutamate receptor 1.1 (AtGLR1.1) functions as a regulator of carbon and nitrogen metabolism in Arabidopsis thaliana. Proc Natl Acad Sci U S A. 2003;100:6872–7.
Kasukabe Y, He LX, Nada K, Misawa S, Ihara I, Tachibana S. Overexpression of spermidine synthase enhances tolerance to multiple environmental stresses and up-regulates the expression of various stress regulated genes in transgenic *Arabidopsis thaliana*. Plant Cell Physiol. 2004;45:712–22.
Kim DW, Watanabe K, Murayama C, Izawa S, Niitsu M, Michael AJ, Berberich T, Kusano T. Polyamine oxidase 5 regulates *Arabidopsis thaliana* growth through a thermospermine oxidase activity. Plant Physiol. 2014;165:1575–90.
Konstantinos PA, Imene T, Panagiotis MN, Roubelakis-Angelakis KA. ABA-dependent amine oxidases-derived H_2O_2 affects stomata conductance. Plant Signal Behav. 2010;5:1153–6.
Kotakis C, Theodoropoulou E, Tassis K, Oustamanolakis C, Ioannidis NE, Kotzabasis K. Putrescine, a fast-acting switch for tolerance against osmotic stress. J Plant Physiol. 2014;171:48–51.
Krasensky J, Broyart C, Rabanal FA, Jonak C. The redox-sensitive chloroplast trehalose-6-phosphate phosphatase AtTPPD regulates salt stress tolerance. Antioxid Redox Signal. 2014;21:1–16.
Kwak JM, Mori IC, Pei Z-M, Leonhardt N, Torres MA, Dangl JL, Bloom RE, Bodde S, Jones JDG, Schroeder JI. NADPH oxidase *AtrbohD* and *AtrbohF* genes function in ROS-dependent ABA signaling in Arabidopsis. EMBO J. 2003;22:2623–33.
Lehmann M, Schwarzländer M, Obata T, et al. The metabolic response of Arabidopsis roots to oxidative stress is distinct from that of heterotrophic cells in culture and highlights a complex relationship between the levels of transcripts, metabolites, and flux. Mol Plant. 2009;2:390–406.
Li Z, Jing W, Peng Y, Zhang XQ, Ma X, Huang LK, Yan Y. Spermine alleviates drought stress in white clover with different resistance by influencing carbohydrate metabolism and dehydrins synthesis. PLoS One. 2015;10, e0120708 (1–16).
Lindemose S, O'Shea C, Jensen MK, Skriver K. Structure, function and networks of transcription factors involved in abiotic stress responses. Int J Mol Sci. 2013;14:5842–78.
Liu JH, Kitashiba H, Wang J, Ban Y, Moriguchi T. Polyamines and their ability to provide environmental stress tolerance to plants. Plant Biotechnol. 2007;24:117–26.
Liu T, Dobashi H, Kim DW, Sagor GHM, Niitsu M, Berberich T, Kusano T. Arabidopsis mutant plants with diverse defects in polyamine metabolism show unequal sensitivity to exogenous cadaverine probably based on their spermine content. Physiol Mol Biol Plants. 2014;20:151–9.
Luo Y, Li WM, Wang W. Trehalose: protector of antioxidant enzymes or reactive oxygen species scavenger under heat stress? Environ Exp Bot. 2008;63:378–84.

Mao X, Zhang H, Tian S, Chang X, Jing R. TaSnRK2.4, an SNF1-type serine/threonine protein kinase of wheat (*Triticum aestivum* L.), confers enhanced multi stress tolerance in Arabidopsis. J Exp Bot. 2010;61:683–96.

Marco F, Alcázar R, Tiburcio AF, Carrasco P. Interactions between polyamines and abiotic stress pathway responses unraveled by transcriptome analysis of polyamine overproducers. OMICS. 2011;15:775–81.

Marina M, Sirera FV, Rambla JL, Gonzalez ME, Blázquez MA, Carbonell J, Pieckenstain FL, Ruiz OA. Thermospermine catabolism increases *Arabidopsis thaliana* resistance to *Pseudomonas viridiflava*. J Exp Bot. 2013;64:1393–402.

Mattoo AK, Minocha SC, Minocha R, Handa AK. Polyamines and cellular metabolism in plants: transgenic approaches reveal different responses to diamine putrescine versus higher polyamines spermidine and spermine. Amino Acids. 2010;38:405–13.

Minocha R, Majumdar R, Minocha SC. Polyamines and abiotic stress in plants: a complex relationship. Front Plant Sci. 2014;5:175 (11–17).

Møller SG, McPherson MJ. Developmental expression and biochemical analysis of the Arabidopsis atao1 gene encoding an H_2O_2-generating diamine oxidase. Plant J. 1998;13:781–91.

Momtaz OA, Hussein EM, Fahmy EM, Ahmed SE. Expression of S-adenosyl methionine decarboxylase gene for polyamine accumulation in Egyptian cotton Giza 88 and Giza 90. GM Crops. 2010;1:257–66.

Moschou PN, Paschalidis KA, Delis ID, Andriopoulou AH, Lagiotis GD, Yakoumakis DI, Roubelakis-Angelakis KA. Spermidine exodus and oxidation in the apoplast induced by abiotic stress is responsible for H_2O_2 signatures that direct tolerance responses in tobacco. Plant Cell. 2008a;20:1708–24.

Moschou PN, Paschalidis KA, Roubelakis-Angelakis KA. Plant polyamine catabolism: the state of the art. Plant Signal Behav. 2008b;3:1061–6.

Moschou PN, Sanmartin M, Andriopoulou AH, Rojo E, Sanchez-Serrano JJ, Roubelakis-Angelakis KA. Bridging the gap between plant and mammalian polyamine catabolism: a novel peroxisomal polyamine oxidase responsible for a full back-conversion pathway in Arabidopsis. Plant Physiol. 2008c;147:1845–57.

Moschou PN, Wu J, Cona A, Tavladoraki P, Angelini R, Roubelakis-Angelakis KA. The polyamines and their catabolic products are significant players in the turnover of nitrogenous molecules in plants. J Exp Bot. 2012;63:5003–15.

Munné-Bosch S, Queval G, Foyer CH. 2013. The impact of global change factors on redox signaling underpinning stress tolerance. Plant Physiol. 2013;161(1):5-19.

Nakabayashi R, Yonekura-Sakakibara K, Urano K, et al. Enhancement of oxidative and drought tolerance in Arabidopsis by overaccumulation of antioxidant flavonoids. Plant J. 2014;77:367–79.

Neill SJ, Desikan R, Clarke A, Hancock JT. Nitric oxide is a novel component of abscisic acid signaling in stomatal guard cells. Plant Physiol. 2002;128:13–6.

Neill S, Barros R, Bright J, Desikan R, Hancock J, Harrison J, Morris P, Ribeiro D, Wilson I. Nitric oxide, stomatal closure, and abiotic stress. J Exp Bot. 2008;59:165–76.

Nishizawa A, Yabuta Y, Shigeoka S. Galactinol and raffinose constitute a novel function to protect plants from oxidative damage. Plant Physiol. 2008;147:1251–63.

O'Hara LE, Paul MJ, Wingler A. How do sugars regulate plant growth and development? new insight into the role of trehalose-6-phosphate. Mol Plant. 2013;6:261–74.

Obata T, Matthes A, Koszior S, Lehmann M, Araújo WL, Bock R, Sweetlove LJ, Fernie AR. Alteration of mitochondrial protein complexes in relation to metabolic regulation under short-term oxidative stress in Arabidopsis seedlings. Phytochemistry. 2011;72:1081–91.

Ono Y, Kim DW, Watanabe K, Sasaki A, Niitsu M, Berberich T, Kusano T, Takahashi Y. Constitutively and highly expressed *Oryza sativa* polyamine oxidases localize in peroxisomes and catalyze polyamine back conversion. Amino Acids. 2012;42:867–76.

Ort R. When there is too much light. Plant Physiol. 2001;125:29–32.

Pál M, Szalai G, Janda T. Speculation: polyamines are important in abiotic stress signaling. Plant Sci. 2015;237:16–23.

Pei ZM, Murata Y, Benning G, Thomine S, Klüsener B, Allen GJ, Grill E, Schroeder JI. Calcium channels activated by hydrogen peroxide mediate abscisic acid signalling in guard cells. Nature. 2000;406:731–4.

Peremarti A, Bassie L, Christou P, Capell T. Spermine facilitates recovery from drought but does not confer drought tolerance in transgenic rice plants expressing *Datura stramonium* *S*-adenosylmethionine decarboxylase. Plant Mol Biol. 2009;70:253–64.

Petřivalský M, Brauner F, Luhová L, Gagneul D, Šebela M. Aminoaldehyde dehydrogenase activity during wound healing of mechanically injured pea seedlings. J Plant Physiol. 2007;164:1410–8.

Planas-Portell J, Gallart M, Tiburcio AF, Altabella T. Copper-containing amine oxidases contribute to terminal polyamine oxidation in peroxisomes and apoplast of *Arabidopsis thaliana*. BMC Plant Biol. 2013;13:109 (1–13).

Pottosin I, Shabala S. Polyamines control of cation transport across plant membranes: implications for ion homeostasis and abiotic stress signaling. Front Plant Sci. 2014;5:154 (1–16).

Pottosin I, Velarde-Buendía A-M, Zepeda-Jazo I, Dobrovinskaya O, Shabala S. Synergism between polyamines and ROS in the induction of Ca^{2+} and K^{+} fluxes in roots. Plant Signal Behav. 2012;7:1084–7.

Price J, Laxmi A, St Martin SK, Jang J-C. Global transcription profiling reveals multiple sugar signal transduction mechanisms in Arabidopsis. Plant Cell. 2004;16:2128–50.

Qiao W, Fan L-M. Nitric oxide signaling in plant responses to abiotic stresses. J Integr Plant Biol. 2008;50:1238–46.

Rambla JL, Vera-Sirera F, Blázquez MA, Carbonell J, Granell A. Quantitation of biogenic tetraamines in *Arabidopsis thaliana*. Anal Biochem. 2010;397:208–11.

Rea G, de Pinto MC, Tavazza R, Biondi S, Gobbi V, Ferrante P, De Gara L, Federico R, Angelini R, Tavladoraki P. Ectopic expression of maize polyamine oxidase and pea copper amine oxidase in the cell wall of tobacco plants. Plant Physiol. 2004;134:1414–26.

Reumann S, Quan S, Aung K, et al. In-depth proteome analysis of Arabidopsis leaf peroxisomes combined with in vivo subcellular targeting verification indicates novel metabolic and regulatory functions of peroxisomes. Plant Physiol. 2009;150:125–43.

Sagor GHM, Takahashi H, Niitsu M, Takahashi Y, Berberich T, Kusano T. Exogenous thermospermine has an activity to induce a subset of the defense genes and restrict cucumber mosaic virus multiplication in *Arabidopsis thaliana*. Plant Cell Rep. 2012;31:1227–32.

Šebela M, Radová A, Angelini R, Tavladoraki P, Frébort I, Peč P. FAD-containing polyamine oxidases: a timely challenge for researchers in biochemistry and physiology of plants. Plant Sci. 2001;160:197–207.

Shelp BJ, Bozzo GG, Trobacher CP, Zarei A, Deyman KL, Brikis CJ. Hypothesis/review: contribution of putrescine to 4-aminobutyrate (GABA) production in response to abiotic stress. Plant Sci. 2012;193–194:130–5.

Shi H, Chan Z. Improvement of plant abiotic stress tolerance through modulation of the polyamine pathway. J Integr Plant Biol. 2014;56:114–21.

Shi H, Ye T, Chan Z. Comparative proteomic and physiological analyses reveal the protective effect of exogenous polyamines in the bermudagrass (*Cynodon dactylon*) response to salt and drought stresses. J Proteome Res. 2013;12:4951–64.

Sun P, Zhu X, Huang X, Liu J-H. Overexpression of a stress-responsive MYB transcription factor of Poncirus trifoliata confers enhanced dehydration tolerance and increases polyamine biosynthesis. Plant Physiol Biochem. 2014;78:71–9.

Sun H, Li J, Song W, Tao J, Huang S, Chen S, Hou M, Xu G, Zhang Y. Nitric oxide generated by nitrate reductase increases nitrogen uptake capacity by inducing lateral root formation and inorganic nitrogen uptake under partial nitrate nutrition in rice. J Exp Bot. 2015;66:2449–59.

Takahashi T, Kakehi JI. Polyamines: ubiquitous polycations with unique roles in growth and stress responses. Ann Bot. 2010;105:1–6.

Takahashi Y, Cong R, Sagor GHM, Niitsu M, Berberich T, Kusano T. Characterization of five polyamine oxidase isoforms in *Arabidopsis thaliana*. Plant Cell Rep. 2010;29:955–65.

Takano A, Kakehi JI, Takahashi T. Thermospermine is not a minor polyamine in the plant kingdom. Plant Cell Physiol. 2012;53:606–16.

Tanou G, Molassiotis A, Diamantidis G. Hydrogen peroxide- and nitric oxide-induced systemic antioxidant prime-like activity under NaCl-stress and stress-free conditions in citrus plants. J Plant Physiol. 2009;166:1904–13.

Tanou G, Filippou P, Belghazi M, Job D, Diamantidis G, Fotopoulos V, Molassiotis A. Oxidative and nitrosative-based signaling and associated post-translational modifications orchestrate the acclimation of citrus plants to salinity stress. Plant J. 2012;72:585–99.

Tanou G, Ziogas V, Belghazi M, Christou A, Filippou P, Job D, Fotopoulos V, Molassiotis A. Polyamines reprogram oxidative and nitrosative status and the proteome of citrus plants exposed to salinity stress. Plant Cell Environ. 2014;37:864–85.

Tavladoraki P, Schininà ME, Cecconi F, Di Agostino S, Manera F, Rea G, Mariottini P, Federico R, Angelini R. Maize polyamine oxidase: primary structure from protein and cDNA sequencing. FEBS Lett. 1998;426:62–6.

Tavladoraki P, Cona A, Federico R, Tempera G, Viceconte N, Saccoccio S, Battaglia V, Toninello A, Agostinelli E. Polyamine catabolism: target for antiproliferative therapies in animals and stress tolerance strategies in plants. Amino Acids. 2012;42:411–26.

Todaka D, Shinozaki K, Yamaguchi-Shinozaki K. Recent advances in the dissection of drought-stress regulatory networks and strategies for development of drought-tolerant transgenic rice plants. Front Plant Sci. 2015;6:1–20.

Tong W, Yoshimoto K, Kakehi JI, Motose H, Niitsu M, Takahashi T. Thermospermine modulates expression of auxin-related genes in Arabidopsis. Front Plant Sci. 2014;5:94 (1–10).

Trouverie J, Thévenot C, Rocher JP, Sotta B, Prioul J. The role of abscisic acid in the response of a specific vacuolar invertase to water stress in the adult maize leaf. J Exp Bot. 2003;54:2177–86.

Tun NN, Santa-Catarina C, Begum T, Silveira V, Handro W, Segal Floh EI, Scherer GFE. Polyamines induce rapid biosynthesis of nitric oxide (NO) in *Arabidopsis thaliana* seedlings. Plant Cell Physiol. 2006;47:346–54.

Urano K, Yoshiba Y, Nanjo T, Ito T, Yamaguchi-Shinozaki K, Shinozaki K. Arabidopsis stress-inducible gene for arginine decarboxylase AtADC2 is required for accumulation of putrescine in salt tolerance. Biochem Biophys Res Commun. 2004;313:369–75.

Valliyodan B, Nguyen HT. Understanding regulatory networks and engineering for enhanced drought tolerance in plants. Curr Opin Plant Biol. 2006;9:189–95.

Van den Ende W. Multifunctional fructans and raffinose family oligosaccharides. Front Plant Sci. 2013;4:247 (1–11).

Van den Ende W. Sugars take a central position in plant growth, development, and stress responses. A focus on apical dominance. Front Plant Sci. 2014;5:1–3.

Van den Ende W, Valluru R. Sucrose, sucrosyl oligosaccharides, and oxidative stress: scavenging and salvaging? J Exp Bot. 2008;60:9–18.

Van Houtte H, Vandesteene L, López-Galvis L, et al. Overexpression of the trehalase gene AtTRE1 leads to increased drought stress tolerance in Arabidopsis and is involved in abscisic acid-induced stomatal closure. Plant Physiol. 2013;161:1158–71.

Walden R, Cordeiro A, Tiburcio AF. Polyamines: small molecules triggering pathways in plant growth and development. Plant Physiol. 1997;113:1009–13.

Wang J, Sun P-P, Chen C-L, Wang Y, Fu X-Z, Liu J-H. An arginine decarboxylase gene PtADC from Poncirus trifoliata confers abiotic stress tolerance and promotes primary root growth in Arabidopsis. J Exp Bot. 2011;62:2899–914.

Wang P, Du Y, Hou Y-J, Zhao Y, Hsu C-C, Yuan F, Zhu X, Tao WA, Song C-P, Zhu J-K. Nitric oxide negatively regulates abscisic acid signaling in guard cells by S-nitrosylation of OST1. Proc Natl Acad Sci. 2015;112:613–8.

Weisman D, Alkio M, Colón-Carmona A. Transcriptional responses to polycyclic aromatic hydrocarbon-induced stress in *Arabidopsis thaliana* reveal the involvement of hormone and defense signaling pathways. BMC Plant Biol. 2010;10:59.

Wilkinson S, Davies WJ. ABA-based chemical signalling: the co-ordination of responses to stress in plants. Plant Cell Environ. 2002;25:195–210.

Wimalasekera R, Villar C, Begum T, Scherer GFE. COPPER AMINE OXIDASE1 (CuAO1) of Arabidopsis thaliana contributes to abscisic acid-and polyamine-induced nitric oxide biosynthesis and abscisic acid signal transduction. Mol Plant. 2011;4:663–78.

Xie Y, Mao Y, Zhang W, Lai D, Wang Q, Shen W. Reactive oxygen species-dependent nitric oxide production contributes to hydrogen-promoted stomatal closure in Arabidopsis. Plant Physiol. 2014;165:759–73.

Xiong L, Schumaker KS, Zhu JK. Cell signaling during cold, drought, and salt stress. Plant Cell. 2002;14:S165.

Yamaguchi K, Takahashi Y, Berberich T, Imai A, Takahashi T, Michael AJ, Kusano T. A protective role for the polyamine spermine against drought stress in Arabidopsis. Biochem Biophys Res Commun. 2007;352:486–90.

Zhang C, Huang Z. Effects of endogenous abscisic acid, jasmonic acid, polyamines, and polyamine oxidase activity in tomato seedlings under drought stress. Sci Hortic. 2013;159:172–7.

Zhao M-G, Chen L, Zhang L-L, Zhang W-H. Nitric reductase-dependent nitric oxide production is involved in cold acclimation and freezing tolerance in Arabidopsis. Plant Physiol. 2009; 151:755–67.

Ziogas V, Tanou G, Filippou P, Diamantidis G, Vasilakakis M, Fotopoulos V, Molassiotis A. Nitrosative responses in citrus plants exposed to six abiotic stress conditions. Plant Physiol Biochem. 2013;68:118–26.

Chapter 12
Plant–Rhizobacteria Interaction and Drought Stress Tolerance in Plants

Cohen Ana Carmen, Piccoli Patricia, Bottini Rubén, and Salomon María Victoria

12.1 Introduction

By 2050 the world's population will reach 9.1 billion and food production will be insufficient (Tomlinson 2013). This has been a source of debate and worry for many ages, and it was estimated that the farmers will produce 50–70 % more food to feed population (FAO, Food and Agriculture Organization). In an effort to adjust to the exponential trends of growth population, the farmers all over the world have been using chemical fertilizers and synthetic herbicides and pesticides to increase crop yield. In addition to the chemical advances, high-yield crops have been also developed and introduced. However, while the goal of conventional agriculture is to maximize yields, biodiversity and environmental health are usually not preserved (Phalan et al. 2011). The uses indiscriminately of these products have increased the emission greenhouse gases, key factor in climate factor (Eisenhauer et al. 2012). Also, they are very expensive and can harm the environment if are not used correctly causing continuous environmental degradation. The soil, water systems around the fields, and rhizosphere microorganisms are polluted by chemical products, thus the interaction between host plants and bacteria may be impacted as well (Compant et al. 2010; Eisenhauer et al. 2012).

On the next century, it has been predicted that global climate will change drastically and a range of parameters will be affected in this moving environment (Houghton et al. 2001). It's widely accepted that climatic change produces an increase in atmospheric CO_2 concentration and in temperature (IPCC, Climate Change, 2007); in some area, it is expected to decrease soil water content, with the

C.A. Carmen (✉) • P. Patricia • B. Rubén • S.M. Victoria
Laboratorio de Bioquímica Vegetal, Instituto de Biología Agrícola de Mendoza, Consejo Nacional de Investigaciones Científicas y Técnicas-Universidad Nacional de Cuyo, Almirante Brown 500, Chacras de Coria M5528AHB, Argentina
e-mail: acohen@mendoza-conicet.gob.ar

M.A. Hossain et al. (eds.), *Drought Stress Tolerance in Plants, Vol 1*,
DOI 10.1007/978-3-319-28899-4_12

concomitant increase of droughts. Environmental deterioration has become a great problem to sustainable crop production because it causes a significant loss of plant productivity (Munns 2005), and it is becoming more severe and widespread. It's estimated that drought covers approximately 41 % of earth's land surface (Reynolds et al. 2007). Higher plants are in permanent relationship with their environment, and they are strongly dependent on its effects.

Environmental stresses cause significant loss of plant productivity. Abiotic stresses, e.g., light UV, temperature, drought, soil salinity, air pollution, and mechanical damage directly affect crop production (Vickers et al. 2009). Drought is one of the main environmental factors that negatively affect plant growth and development, and it restricts agricultural productivity (Boyer 1982). In plants, drought is associated with other stresses, for example, osmotic stress produced by cellular dehydration, which diminishes cell expansion (Bartels and Sunkar 2005). Moreover, drought, in turn, brought some other significant problems as high vapor pressure, increase of soil salinity, and diminish nutrient availability and mechanical impedance to root growth (Wilkinson and Davies 2010). According to the intensity and duration of drought, the plants respond to differential water status through a series of molecular, cellular, and physiological events (Chaves et al. 2003, 2009).

One of the best known effects of water restriction in plants is by increment of ABA biosynthesis and/or decrease in its catabolism (Bray 2002; Tardieu et al. 2010). This produces modifications in physiological and genes expression (Seki et al. 2002; Shinozaki and Yamaguchi-Shinozaki 2007). ABA has a great impact on stomatal closure to reducing water loss and limit of gas exchange reducing transpiration and photosynthesis (Liu et al. 2005; Xu et al. 2008; Zhang et al. 2009a, b). ABA increases in leaf by increasing ABA transportation (via xylem) since the roots are in contact with the dry soil (Schachtman and Goodger 2008).

A water deficit causes a decrease in water potential and turgor loss, stomatal closure, and disruption of membrane integrity along with protein denaturation. Stomatal closure in response to water deficit causes a decline in the rate of photosynthesis (Chen and Murata 2008). Also, inhibition of photosynthesis enzymes and photosystem II activity can occur (Guóth et al. 2009; Corrêa de Souza et al. 2013). As consequence, drought produces a reduction in biomass accumulation and in plant yield (Vile et al. 2012).

Elevated temperature and drought due to climate change might induce changes in plant physiology and root exudation. This combined effect might modify the composition, abundance, and activity of plant-associated microorganism communities which depend on the exudates of these roots for their survival (Whipps 1990). The bacterial communities which live in the rhizosphere are also able to colonize plant roots. They have beneficial effects on plant growth and on crop yield and/or quality and they are known like PGPR (plant growth-promoting rhizobacteria, Kloepper and Schroth 1978; Kloepper et al. 1991). Numerous authors informed that PGPR positively affected plant growth subjected to drought stress and plant-associated microorganisms are important factors that influence the response of plants to climate change (Compant et al. 2010).

PGPR can improve plant growth by direct and indirect mechanism. The latter has been found in most PGPR strains by inhibiting the growth of plant pathogens (Glick and Bashan 1997; Persello-Cartieaux et al. 2003; Bashan and de-Bashan 2005; 2010). Among PGPR genuses that are biological control agents are cited: *Agrobacterium, Bacillus, Burkholderia, Pseudomonas*, and *Streptomyces.* They suppress plant disease by inducing systemic resistance or by antibiotics production, and synthesis of hydrogen cyanide HCN (Bano and Musarrat 2003; Bashan and de-Bashan 2010; Tan et al. 2013).

Among the direct mechanisms, we may mention plant growth regulators (PGR) production (Arshad and Frankenberger 1993; Costacurta and Vanderleyden 1995; Glick 1995; Bastián et al. 1998; Bloemberg and Lugtenberg 2001; Bottini et al. 2004; Cohen et al. 2008; Piccoli et al. 2011, Cohen et al. 2015a, b), fixation of atmospheric nitrogen (Boddey and Dobereiner 1995), phosphate and minerals solubilization, and siderophore production (Barea et al. 1976; Kloepper et al. 1989; Glick 1995; De Freitas et al. 1997; Rodriguez and Fraga 1999; Richardson 2001; Chen et al. 2006; Rodriguez and Fraga 1999; Ayyadurai et al. 2007; Hu et al. 2009; Jha et al. 2009; Sharma et al. 2011). In addition to these promotion mechanisms, there are many researches that show the use of PGPR not only increases plant's growth under ideal condition but also increases plant's resistance to damaging effects of environmental stresses like drought (Mayak et al. 2004a; Cohen et al. 2015a, b), salinity (Egamberdieva 2008; Mayak et al. 2004b; Zahir et al. 2004; Kaymak et al. 2009; Tank and Saraf 2010; Ahmad et al. 2011), nutrient deficiency, and heavy metal contamination (Chanway and Holl 1994). These mechanisms include PGR production (that will be discussed in the next section), lowering of stress-induced ethylene (Glick et al. 2007; Zahir et al. 2009), production of exopolysaccharides, regulating nutrient uptake, and enhancing the activity of antioxidant enzymes (Glick et al. 2007; Sandhya et al. 2009). In addition, some PGPRs are used to remediate and rehabilitate non-fertile and contaminated land into fertile ones (Glick 2010).

In the recent years, there is a significant interest in eco-friendly and sustainable agriculture. Therefore, one strategy could be the use of PGPR as inoculants for biofertilization, phytostimulation, and biocontrol (Lugtenberg and Kamilova 2009; Babalola 2010). Frequently, the researchers use a bacterial consortium (co-inoculation with more than one strain) to obtain mayor results, so the bioinoculants are effective to protect the plant from disease under yield condition, to increase nutrient availability for plants (N, P, K, Ca, Mg, Fe, and Mn) (Freitas et al. 2007; Dursun et al. 2010) and to participate in carbon, nitrogen, sulfur, and phosphorous cycling. In this way, the use of PGPR in the form of biofertilizers is an effective supportive strategy to provide crop nutrition due to high price and contamination environmental concerns about the chemical fertilizers (Cakmakci et al. 2006). Furthermore, these bioinoculants contribute to the development of sustainable agriculture under stressed conditions (Glick et al. 2007; Dodd and Pérez-Alfocea 2012; Berg et al. 2013). Also, with the rise of organic agriculture, the demand of PGPR biofertilizers has been increasing. These are a promising solution for sustainable, environmentally friendly agriculture (Tsavkelova et al. 2006). Furthermore, *Azospirillum* and *Pseudomonas* strains have capacity to degrade glyphosate in maize plants growing in field to minimize the persistence of xenobiotic compound in the environment (Travaglia et al. 2015). Biofertilizer containing efficient PGPR may improve crop production, reduce agrochemical use, and support eco-friendly sustainable food production.

12.2 Plant Growth Regulators

The PGR are a group of naturally occurring, organic substances involve in physiological process such as growth, differentiation, and development, as well as responses against both biotic and abiotic stresses (Schmelz et al. 2003; Davies 2010). As it was mentioned, several bacteria also have the ability to produce different PGRs, and it has been proposed as one of the main direct mechanisms by which bacteria exert the benefit for the plant, increasing its growth and yield (Piccoli and Bottini 2013). In the literature, there are many studies that provide evidence regarding how PGPRs exhibit their capacity to improve plant development due to the influence of PGR production.

12.2.1 Auxins

Auxins are recognized as the most active plant growth stimulators, mainly indole-3-acetic acid (IAA), implicated in cell division and enlargement, root initiation, tissue differentiation, and tropistic responses (Davies 2010). IAA biosynthesis is widespread among plant-associated bacteria since it has been suggested that more than 80 % of isolates are able to produce IAA via different biosynthetic pathways (Patten and Glick 1996; Spaepen et al. 2007); and it is quantitatively produced in high amounts by a diverse set of both symbiotic and free living bacteria (Tsavkelova et al. 2005; Karadeniz et al. 2006). It has been shown for PGPR such as *Azospirillum*, *Pseudomonas*, and *Bradyrhizobium* that the positive effect on root proliferation found in plants is correlated with bacterial IAA (Vessey 2003; Spaepen et al. 2007). Also, it was related to the increase in branches number and oil content (Asghar et al. 2002). Furthermore, it was suggested that the enhancement of root growth and development via bacterial IAA results in a greater root surface area which allows the plant to access more nutrients from soil (Vessey 2003), and in an augment of root exudation which stimulates bacterial colonization (Lambrecht et al. 2000). The effect of IAA in plants has been related with cytokinins, which are other PGR involved in aspects as cell division, leaf senescence, apical dominance, nutrient mobilization, and chloroplast differentiation (Sakakibara 2010) that bacteria have the ability to synthesize. Various bacterial organisms are reported to synthesize them (Akiyoshi et al. 1987; Taller and Wong 1989; García de Salamone et al. 2001; Karadeniz et al. 2006). Regarding to the effect in plants, Arkhipova et al. (2005, 2007) found that *Bacillus subtilis* produces cytokinins and that may be involved in increase its content in lettuce and enhance the plant growth under drought conditions; and Noel et al. (1996) reported that *Rhizobium* inoculation promote root growth of canola and lettuce probably via IAA and cytokinins. Besides these antecedents, the implication of cytokinins produced by bacteria on growth plants requires other studies in order to fully elucidate its role.

12.2.2 Ethylene

The gas ethylene is a hydrocarbon PGR that controls processes such as germination, senescence of organs, fruit ripening and abscission, as well as stress responses to biotic and abiotic stress (Nehring and Ecker 2010). Although some bacteria produce ethylene (Mansouri and Bunch 1989; Weingart and Völksch 1997), the main studied modulating effect of PGPR on its concentration in plants, is the production of the enzyme 1-aminocyclopropane-1-carboxylate (ACC) deaminase which was found in a diversity of bacteria (Vessey 2003; Glick 2005). This enzyme cleaves the immediate precursor molecule of ethylene in plants (the ACC), so bacteria that produce this enzyme promote plant growth by lowering the level of ethylene in the plant, thereby allows the plant to resist efficiently a wide variety of environmental stresses such as salt and heavy metal toxicity (Burd et al. 1998; Glick 2005; Mayak et al. 2004b; Siddikee et al. 2011).

12.2.3 Gibberellins

Also, it has been reported that bacteria are able to synthesize the terpenic PGR gibberellins (GAs) and ABA already mentioned above. Gibberellins are hormonal diterpenes involved in several processes in higher plants, including seed germination, stem, leaves and root growth, floral induction, and flower and fruit growth (Sponsel and Hedden 2010). Although the production by bacteria is known from several years ago (Bottini et al. 1989), little is known regarding to GA synthesis pathway in these microorganisms. However, the capacity to produce them has been reported in bacteria such as *Azospirillum lipoferum* and *A. brasilense* (Bottini et al. 1989; Janzen et al. 1992), *Acetobacter diazotrophicus*, *Herbaspirillum seropedicae* (Bastián et al. 1998), *Bacillus subtilis* and *B. licheniformis, B. cereus* (Gutiérrez-Mañero et al. 2001), *Burkholderia* sp. (Joo et al. 2009), and *Pseudomonas fluorescens* (Salomon et al. 2014). The inoculation of GA-producing bacteria showed positive effects on growth of diverse plants, which was evidenced in a number of studies extensively revised by Bottini et al. (2004) and Piccoli and Bottini (2013); the GAs produced by bacteria has been related to promote root and shoot growth in rice (Yanni et al. 2001), reversal of dwarfism in rice and maize seedlings and black alder (Lucangeli and Bottini 1996, 1997; Gutiérrez-Mañero et al. 2001), increase total carbohydrate accumulation (Bastián et al. 1999) and enhance growth of *Pinus pinea* (Probanza et al. 2002) and pepper (Joo et al. 2004).

12.2.4 Abscisic Acid

ABA (subject of analysis in this chapter) is a sesquiterpene stress signaling hormone that becomes elevated in plants under water stress because it is directly involved in regulation of stomatal aperture-closure (Davies and Zhang 1991; Jiang and Zhang 2002). ABA production has been reported in the well-characterized PGPR *Azospirillum* as well as *Bacillus* and *Pseudomonas* (Karadeniz et al. 2006; Forchetti et al. 2007; Cohen et al. 2008, 2009; Salomon et al. 2014). The ABA-producing bacteria have been related to increases in ABA content in the plant thereby helping them to cope with drought, which is analyzed and extended in the following section.

12.3 Role of ABA Produced by PGPR in Plants Submitted to Drought Stress

Climate change and other anthropogenic factors have exacerbated the droughts (frequency and severity, IPCC 2014). Drought stress is considered one of the major detrimental limiting the growth events, nutrient uptake and metabolism, and the crop productions worldwide, especially in areas where irrigation is an inevitable aid to agriculture (Boyer 1982; Engelbrecht et al. 2007; Lambers et al. 2008; Farooq et al. 2009; Li et al. 2009). Agricultural drought is the lack of moisture required for normal plant growth and development to complete the life cycle (Manivannan et al. 2008). Areas that are currently major producers could deal drastic yield decreased caused by lack of water. A continuous shortfall in precipitation coupled with higher evapo-transpiration demand leads to drought (Mishra and Cherkauer 2010). Plants have the capacity to perceive environmental stress signals and rapidly regulate their physiology and metabolism to cope them (Jiang and Zhang 2003; Seki et al. 2007). That means morphological adaptation and responses at biochemical and genetic levels including maintenance of water-use efficiency, net carbon gain, and osmotic adjustment (Bohnert et al. 2006; Farooq et al. 2009). In water deficits, increase ABA biosynthesis and/or ABA deactivation (Bray 2002; Ren et al. 2007; Huang et al. 2008), triggering downstream responses that confer drought tolerance to plants, or prepare the plant to resist water loss. Physiological responses to drought include stomatal closure (Zhang and Outlaw 2001), decrease in photosynthetic activity and vegetative shoot growth, modification in cell wall elasticity, imbalance between production of reactive oxygen species and antioxidant defense (Hu et al. 2006) and generation of toxic metabolites (Ahuja et al. 2010).

ABA has long been recognized as an endogenous messenger that module several physiological processes controlling plant response to biotic and abiotic factors. Different abiotic stress-inducible genes are controlled by ABA, however, others are ABA independent (Yamaguchi-Shinozaki and Shinozaki 2005). In addition, ABA regulates a variety of plant processes, including seed development, modulation of growth and development, fruit ripening, and responses to environmental stress thereby improving the plant water uptake capacity (De Smet et al. 2006; Zhang et al. 2009a, b). ABA plays an important role in the regulation of stomatal closure during drought and in the decrease stomatal conductance (g_s, Tardieu and Davies 1992, Davies et al. 2005) to regulate the balance between water loss (Davies and Zhang 1991), CO_2 uptake, and assimilation. The increase in leaf ABA is crucial for the decrease in mesophyll conductance (g_m) under drought conditions (Mizokami et al. 2015). ABA sprayed on leaves promotes growth in *Ilex paraguariensis* plants by alleviating diurnal water stress and dry matter accumulation in stem and leaves (Sansberro et al. 2004) and ABA induces leaf growth in maize by augmenting water movement in the plant because of increased tissue hydraulic conductivity (Tardieu et al. 2010). In tomato, ABA overproduction enhanced transpiration efficiency and root hydraulic conductivity, thereby affecting leaf expansion through improvements in water status (Thompson et al. 2007). Also, ABA treatments increase yield under moderate water restriction in field-grown wheat and carbohydrates redistribution (Travaglia et al. 2012); also ABA enhance the transport of photo-assimilates from leaves and stem to developing grains, without that modify the quality (Travaglia et al. 2007, 2010). In grape, ABA enhances fruit yield (Quiroga et al. 2009) and increases sugar transport and promotes carbon allocation toward sink organs involved in plant survival (roots and fruits; Moreno et al. 2011; Murcia et al. 2015).

ABA is found in organism form different kingdoms from higher plants, bryophytes, algae, fungi, and bacteria (Zeevaart 1999; Takezawa et al. 2011), and it can modulate physiological functions of various organisms. More recently, this hormone has been identified as an endogenous pro-inflammatory cytokine in human's granulocytes (Bruzzone et al. 2007; Bassaganya-Riera et al. 2011). In plants, ABA biosynthesis is formed by cleavage of C40 carotenoids involving an oxidative cleavage to give rise xanthoxin, followed by two-step conversion to ABA (Nambara and Marion-Poll 2005). Reports on ABA production by PGPR was detected by radioimmunoassay (Kolb and Martin 1985; Belimov et al. 2001); latter ABA has been characterized with more accuracy by full scan mass spectrometry in chemically defined growth cultures of *A. brasilense* Sp 245, *Arthrobacter koreensis, and B. licheniformis* (Perrig et al. 2007; Cohen et al. 2008; Sgroy et al. 2009; Piccoli et al. 2011). *A. brasilense* Sp 245 increase ABA production in culture medium plus NaCl (Cohen et al. 2008). Moreover, *Corynebacterium* sp. converts ABA to dehydrovomifoliol in vitro and possessed vomifoliol dehydrogenase activity (Hasegawa et al. 1984). *Rhodococcus* sp. P1Y and *Novosphingobium* sp. P6W can metabolize ABA in vitro as a sole carbon and energy source using ABA-supplemented medium (Belimov et al. 2014). The ABA biosynthetic pathway in bacteria has not been elucidated yet. It was proposed that bacterial-synthesized ABA is a product of carotenoid metabolism (Marasco and Schmidt-Dannert 2008), and we have some evidences that the gene CtrZ, responsible of the synthesis of xanthoxine in plants, is expressed in *P. fluorescens* Rt6M10 cultures (Domínguez et al. 2011).

One possibility to increase crop growth and tolerance stress conditions is to use PGPR as bioinoculants (Barea et al 2005; Azcón and Barea 2010; Calvo et al. 2014). It is necessary that PGPR can colonize the plant root and survive in rhizosphere

(Normander and Prosser 2000). Among the PGPR more studied appear *Azospirillum*, *Azotobacter*, *Acetobacter*, *Bacillus*, *Burkholderia*, *Paenibacillus*, *Pseudomonas*, *Rhizobium*, and *Serratia* (Glick 1995; Kloepper et al. 1989; Bashan et al. 2014; Cohen et al. 2015b). However, the plant response to stress is regulated by signaling molecules that may be generated by the plant or its associated microbial populations (Marasco et al. 2012; Bakker et al. 2014).

It is accepted that PGPR such as *Azospirillum* sp. are very effective in enhancing the ability of plants to become established and to cope with stress situation such as drought, salt, and nutrient limitation (Azcón and Barea 2010; Creus et al. 1997, 1998, 2004; Bashan and de-Bashan 2010; Cohen et al. 2015a, b). We will focus on results found with *Azospirillum* sp. in our group and by other researchers related to PGR. In a study with maize plants, we show that *A. lipoferum* increase ABA levels and reverse the effects of inhibitors of ABA and GA synthesis (fluridone and prohexadione-Ca, respectively). In well-watered (WW) plants, fluridone application decreases the ABA levels and it affects growth (shorter plants) than those submitted to a period of water stress. These ones, don't control water loss efficiently, which in turn reduces cell turgidity, decreases growth, and as a consequence reduces shoot and root dry weight. However, *A. lipoferum* application reverses growth parameters at the level of control (WW) in fluridone-plants, suggesting that might supply the plant with ABA as to cover the deficit produced by fluridone. This also affects the relative water content (RWC) in both, WW and D-stressed plants, and *A. lipoferum* reverse this effect (Cohen et al. 2009). These results are confirmed in *Arabidopsis* mutant *aba*2-1 (defective in ABA biosynthesis), and wild-type Col-0 plants (Cohen et al. 2015a, b).

Recently, we analyzed the effects of *Azospirillum brasilense* Sp 245 strain (Sp 245) on *Arabidopsis thaliana* plants and their relation with ABA and we proposed a model to explain the beneficial bacteria effects (see Fig. 12.1).

In in vitro grown system, Sp 245 colonizes the roots and rosettes of *Arabidopsis* wild-type Col-0 (Col-0) and on *aba*2-1 plants. The root architecture in plants inoculated (Col-0+Sp 245 and *aba*2-1+Sp 245) is modified by increasing the main root length, the number of lateral roots (LR) that also are longer and roots fresh weight (FW) than the non-inoculated (Col-0 and *aba*2-1), while Col-0 plants present fewer LR than *aba*2-1 mutants. This effect had been observed in our preliminary experiments where plant-fluridone increased the LR length and number (Cohen et al. 2007), something that was observed previously by Deak and Malamy (2005). Thus, ABA-signaling pathway could participate in this response. *Azospirillum* sp. produces both IAA (Crozier et al. 1988; Bashan and de-Bashan 2010) and GAs (Bottini et al. 1989, 2004). So, IAA and GA_3 *Azospirillum*-produced taking part in the signaling cascade that changes the root architecture it could compensate the effect of the ABA lessening LR development. The root system of pot-grown Col-0+Sp 245 plants were higher than Col-0. It favored exploration of the whole soil volume of the pot, so increasing the ability to obtain water from the soil under water stress according to the RWC_{soil}. Also, different PGPR produce small amounts of IAA increasing considerably the development of roots, plant growth, and its crop productivity (maize, rice, sorghum, potato, canola, among the most cited, reviewed by Kloepper et al. (1989).

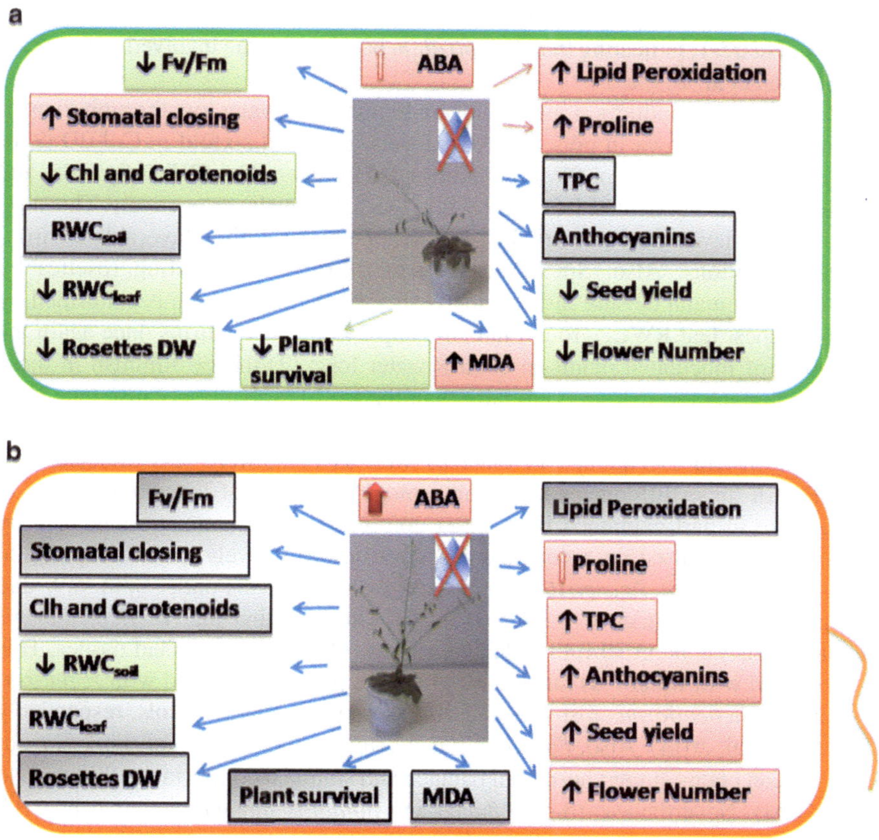

Fig. 12.1 Morpho-physiological and biochemical changes in (**a**) Col-0D and (**b**) Col-0D+Sp 245 *Arabidopsis* plants against drought (D). In (**a**) Col-0D plants, the maximum photochemical efficiency of photosystem II (F_v/F_m), chlorophylls (Chl), carotenoids levels, RWC_{leaf}, rosette DW (dry weight), flower number, plant survival, and seed yield were decreased (*green box*), compared with water plants (Col-0 W) while the abscisic acid (ABA), stomatal closing, lipid peroxidation, and proline were increased (*red box*). The total phenolic compounds (TPC) and anthocyanins levels were not modified significantly (*gray box*). Whereas that in (**b**) Col-0D+Sp 245, ABA, and proline increased much more than (**a**), TPC and anthocyanins levels, flowers number, and seed yield were augmented (*red box*). However, these plants were not modified the stomatal closing, F_v/F_m, Chl and carotenoids levels, RWC_{leaf}, rosettes DW, lipid peroxidation, and plant survival compared to Col-0W (*gray box*)

In the rosettes, Sp 245 increase leaf area (LA) and FW as consequence of root branching that improves the active area in water and nutrient uptake. ABA is well known to be essential in plant responses to D. In in vitro *system*, Sp 245 augments the plants ABA levels in rosettes of Col-0 and *aba*2-1 plants, but more markedly in *aba*2-1. The *aba*2-1 treatment have 63 % less ABA as compared to Col-0 plants, while *aba*2-1+Sp 245 plants show higher ABA levels than Col-0; this was the first report showing the effects of Sp 245 in *aba*2-1 mutants confirming that endophytic

Sp 245 produces ABA per se and/or increases the plant biosynthesis of ABA in both Col-0 and *aba*2-1, indicating that *Azospirillum* has the enzyme involved in this reaction (Cohen et al. 2015a, b).

During D, ABA induces stomatal closure to minimize water loss through transpiration. In plants grown in pots, the ABA levels increased sixfold in Col-0D compared to those under Col-0WW conditions. However, Col-0WW+Sp 245 and Col-0D+Sp 245 augmented 1.5- and 18-fold the ABA levels, respectively. This augment in ABA levels may prepare the plant to cope better with unfavorable environmental conditions. Also, Sp 245 delay water losses after cutting rosettes by controlling stomatal closure through increasing ABA levels which is confirmed by (1) the highest leaf RWC_{leaf} found in Col-0+Sp 245 plants confirms once again the capability of inoculated plants to control water loss; (2) the *gs* diminish in these plants that reach the wilting point later than the Col-0 probably because they have more ABA than the non-inoculated (even though Col-0+Sp 245 plants had greater LA). Col-0D+Sp 245 plants, in plate as well as pots, were less affected than Col-0D. All these differences in the physiologic response of inoculated plants to drought are in part explained by a better control of stomata closure mediated by ABA. Drought caused an accentuated increase in ABA levels when compared to those WW; however, *Azospirillum* increased the ABA levels under both WW and drought conditions (Cohen et al. 2015a, b). Other strains like *B. licheniformis* increased ABA content 70-fold and *P. fluorescens* 40-fold in grape leaf tissues. This is correlated with water loss rate assay, where the plants inoculated with *P. fluorescens* and *B. licheniformis* lost 4 % and 10 % less water than controls, respectively (Salomon et al. 2014). Also, *Paenibacillus polymyma* and *Rhizobium tropici* inoculated in bean plants altered hormonal balance and stomatal conductance (Figueiredo et al. 2008). Nevertheless, tomato *flacca* and *notabilis* mutants deficient in ABA inoculated with *Rhodococcus* sp. P1Y and *Novosphingobium* sp. P6W decreased root and/or leaf ABA concentrations (Belimov et al. 2014).

The increment in chlorophyll and enhanced photosynthesis is a well-known response of plants to inoculation with several PGPR (Deka Boruah and Dileep Kumar 2002; Bashan et al. 2006; Zhang et al. 2008). Sp 245 increases photosynthetic (total chlorophylls and carotenoids) and photoprotective pigments (total phenolic compounds, TPC and anthocyanins). However, photoprotector pigments are incremented only under drought conditions (Cohen et al. 2015a, b). These latter compounds are related with stress conditions in grapevine with Solar UV-B and ABA treatments (Berli et al. 2010, 2011), and the first have photoprotective role due to radiation filtering and/or ROS quenching through the powerful antioxidative capacity (Kumar 2011; Sperdouli and Moustakas 2012). Additionally, phenolic compounds may also enhance protection against oxidative stress, as they possess chemical structures capable of scavenging free radicals (Blokhina et al. 2002; Berli et al. 2010). ABA applications in wheat and grapevines plant treated whit ABA present higher content of carotenoids (Travaglia et al. 2007, 2009, 2010; Berli et al. 2010) indicating that inoculation with *A. brasilense* Sp 245 may be involved in this process. Also, under drought stress, wheat plants inoculated with *Azospirillum* showed an enhanced osmotic adjustment that maintains cell turgor, so preventing degenerative processes

(Creus et al. 2004). ABA application to maize hybrids DK390 increases the ability to maintain the water status and to combat oxidative stress via antioxidant enzymes (Corrêa de Souza et al 2014). Drought stress induced changes in lipid peroxidation that can be quantified by malondialdehyde (MDA) levels. Under drought conditions, Col-0 plants had elevated MDA levels (high lipid peroxidation), while Col-0+Sp 245 and Col-0+ABA were similar maintaining at lower levels their values (less damage). This indicates that these plants are protected against the adverse effects of oxidative stress and demonstrate the efficiency of both *A. brasilense* Sp 245 and ABA to induce antioxidative defense mechanisms. Also, Col-0+Sp 245 increased the concentration of proline compared to Col-0 or Col-0+ABA (Cohen et al. 2015a, b). Proline contributes to osmotic adjustment during stress allowing the plant to obtain water even with very low soil water potentials, and it protects the structure of membranes during extreme dehydration (Meloni et al. 2001). *Azospirillum* enhance osmotic adjustment in wheat plants under drought stress preventing degenerative processes (Mayak et al. 2004a, b). These results indicate the efficiency of the Sp 245 and ABA to induce antioxidative defense mechanisms in plants. Proline synthesis also was reported with *Burkholderia* sp., *Arthrobacter* sp., and *Bacillus* sp. in stressed plants (Dodd and Pérez-Alfocea 2012).

The other mechanism that increase drought tolerance was found with *Paenibacillus polymyxa* in *Arabidopsis thaliana* plants and where augmented mRNA transcriptions of a drought-response gene ERD15 (EARLY RESPONSE TO DEHYDRATION 15, Timmusk and Wagner 1999). *Pseudomonas mendocina* inoculated in lettuce plants increase phosphatase activity in roots and proline accumulation in leaves (Kohler et al. 2008). *Pseudomonas* sp. inoculated in maize plants increased solutes and modified antioxidants status in drought conditions (Sandhya et al. 2010). *Pseudomonas* sp. inoculated in *Ociumum basilicum* improves plant growth, as well as auxin and protein contents under drought stress conditions (Heidari et al. 2011). *Pseudomonas aeruginosa* GGRJ21 strain elicits water stress tolerance in mug bean plants by accelerating the accumulation of antioxidant enzymes, cell osmolytes, and consistently advancing the up-regulation of stress-responsive genes: dehydration-responsive element-binding protein (DREB2A), catalase, and dehydrin in PGPR-treated plants (Sarma and Saikia 2014).

Azospirillum strains improve plant–water relationships and cell wall elasticity with higher seed yield in sorghum and wheat (Sarig et al 1988; Creus et al. 1997, 1998, 2004). It is also observed in *Bacillus thuringiensis* that it is able to increase plant water uptake in *Retama sphaerocarpa* (Marulanda et al. 2006). As was explained before, Col-0+Sp 245 enlarged the root system so improve root exploration, suggesting an increased ability to obtain water from the soil under water stress with higher RWC_{soil} (Cohen et al. 2015a, b). Such ability may be related to the presence of aquaporins, as it was seen in *Azospirillum*-inoculated barley seedlings where a higher root expression of aquaporin gene was detected (Zawoznik et al. 2011). On the other hand, Dardanelli et al. (2008) reported that *A. brasilense* promote root branching in bean seedling roots and increased secretion of flavonoids and lipochitooligosaccharides. Also, *Proteus penneri* Pp1, *Pseudomonas aeruginosa* Pa_2, and *Alcaligenes faecalis* AF3 increase exopolysaccharides (EPS) in maize plants. The EPS produced by bacteria protect the microbes against inhospitable conditions and enable their survival.

Water stress significantly decreases inflorescences and flowers number, rosettes and inflorescence dry weight (DW), seed yield, and plant survival. Col-0+Sp 245 plants (pot-grown) increased different parameters evaluated as both vegetative and reproductive stages in WW and drought conditions. The Col-0WW+Sp 245 increased the inflorescences DW and seed yield; however, flowers number, seed yield per plant, and their survival rate are increased in Col-0D+Sp245 alleviating in part the effect of drought (Cohen et al. 2015a, b). A recent study demonstrates that Sp 245 increase maximum photochemical efficiency of photosystem II (F_v/F_m) in Col-0D+Sp 245 plants whereas F_v/F_m decreased in Col-0D plants (results unpublished). In *Arabidopsis*, the growth increases and photosynthesis was attributed to the emission of volatile compounds by *B. subtilis* strain GB03, and there are evidence that it is modulated by ABA signaling (Zhang et al. 2008).

Col-0D+Sp 245 affect the whole life cycle of the plant, accelerating its growth rate and shortening its vegetative period, providing evidence of the relevant effect of the bacteria in agricultural production (Cohen et al. 2015a, b). Poupin et al. (2013) reported similar results with *Burkholderia phytofirmans* PsJN in *Arabidopsis* plants. However, other bacterium strain, i.e., GB03, delayed flowering in *Arabidopsis* (Xie et al. 2009; Bresson et al. 2013). Thereby, flowering time depend in part on the soil bacteria strains or the PGPR strains applied to the plant.

The findings of the work allow us to postulate a model to explain the physiological, biochemical, and morphological changes that allow inoculated plants tolerate drought condition. In other words, the Sp 245 inoculation on *A. thaliana* under drought conditions has an impact on the increase of F_v/F_m, plant biomass, and seed yield production (Fig. 12.1). The enhancement in root surface, RWC_{leaf} and ABA levels induced by Sp 245 was correlated with a higher sensitivity of inoculated plants to close stomata when experiencing D. However, the increase in photosynthetic and photoprotective compounds and the decrease in RWC_{soil} and MDA levels suggest that *Azospirillum* enhances plant tolerance to drought and seed yield by additional biochemical mechanisms that include phytohormones production, comprising ABA, and also augmenting osmoprotectors compounds like proline and also photoprotective pigments.

12.4 Conclusion and Future Perspective

The productivity of important agricultural crops is drastically reduced when they experience stress induced by both biotic and abiotic factors. Climate change is anticipated to further reduce water availability for agriculture in near future. It is evident from the above explained the capacity of PGPR to help agricultural yield to increase their tolerance and adaptation to drought conditions. It is relevant to suggest that PGPR alone or with ABA applications could help improve agricultural production in crops in which the irrigation is not possible if they have drought. Also, as useful tools to increase crop yield in an efficient and ecological way substitute the old practices from the green revolution with a new agriculture based on biotechnological technique.

Acknowledgments This chapter was supported by Fondo para la Investigación Científica y Tecnológica (FONCYT, PICT 2008 1666 to R. Bottini and PICT 2007 02190 to P. Piccoli), Consejo Nacional de Investigaciones Científicas y Tecnológicas (CONICET, PIP 2008 to P. Piccoli), and Secretaría de Ciencia y Técnica-Universidad Nacional de Cuyo to A. Cohen, R. Bottini and P. Piccoli. A. Cohen, R. Bottini and P. Piccoli are fellows of CONICET.

References

Ahmad M, Zahir ZA, Asghar N, Asghar M. Inducing salt tolerance in mung bean through coinoculation with rhizobia and plant-growth promoting rhizobacteria containing 1-aminocyclopropane-1-carboxylate deaminase. Can J Microbiol. 2011;57:578–89.

Ahuja I, de Vos RC, Bones AM, Hall RD. Plant molecular stress responses face climate change. Trends Plant Sci. 2010;15(12):664–74.

Akiyoshi DE, Regier AD, Gordon MP. Cytokinin production by *Agrobacterium* and *Pseudomonas* spp. J Bacteriol. 1987;169:135–40.

Arkhipova TN, Veselov SU, Melentiev AI, Martynenko EV, Kudoyarova GR. Ability of bacterium *Bacillus subtilis* to produce cytokinins and to influence the growth and endogenous hormone content of lettuce plants. Plant Soil. 2005;272:201–9.

Arkhipova TN, Prinsen E, Veselov SU, Martynenko EV, Martynenko EV, Kudoyarova GR. Cytokinin producing bacteria enhance plant growth in drying soil. Plant Soil. 2007;292:305–15.

Arshad M, Frankenberger Jr WT. Microbial production of plant growth regulators. In: Metting Jr FB, editor. Soil microbial ecology. New York: Marcel and Dekker; 1993. p. 307–47.

Asghar H, Zahir Z, Arshad M, Khaliq A. Relationship between *in vitro* production of auxins by rhizobacteria and their growth-promoting activities in *Brassica juncea* L. Biol Fert Soils. 2002;35:231–7.

Ayyadurai N, Ravindra Naik P, Sakthivel N. Functional characterization of antagonistic fluorescent pseudomonads associated with rhizospheric soil of rice (*Oryza sativa* L.). J Microbiol Biotechnol. 2007;17:919–92.

Azcón R, Barea JM. Mycorrhizosphere interactions for legume improvement. In: Khanf MS, Zaidi A, Musarrar J, editors. Microbes for legume improvement. Vienna: Springer; 2010. p. 237–71.

Babalola OO. Beneficial bacteria of agricultural importance. Biotechnol Lett. 2010;32:1559–70.

Bakker MG, Schlatter DC, Otto-Hanson L, Kinkel LL. Diffuse symbioses: roles of plant-plant, plant-microbe and microbe-microbe interactions in structuring the soil microbiome. Mol Ecol. 2014;23:1571–83.

Bano N, Musarrat J. Characterization of a new *Pseudomonas aeruginosa* strain NJ-15 as a potential biocontrol agent. Curr Microbiol. 2003;46(5):324–8.

Barea J, Navarro M, Montoya E. Production of plant growth regulators by rhizosphere phosphate-solubilizing bacteria. J Appl Bacteriol. 1976;40:129–34.

Barea J-M, Pozo MJ, Azcón R, Azcón-Aguilar C. Microbial cooperation in the rhizosphere. J Exp Bot. 2005;56:1761–78.

Bartels D, Sunkar R. Drought and salt tolerance in plants. Crit Rev Plant Sci. 2005;24(1):23–58.

Bashan Y, de-Bashan L. Plant Growth-Promoting. In: D. Hillel (Editor-in-Chief), Encyclopedia of soils in the environment, Vol. 1. Oxford: Elsevier; 2005, p. 103–15.

Bashan Y, Bustillos JJ, Leyva LA, Hernandez J-P, Bacilio M. Increase in auxiliary photoprotective photosynthetic pigments in wheat seedlings induced by *Azospirillum brasilense*. Biol Fertil Soils. 2006;42:279–285.

Bashan Y, de-Bashan L. How the plant growth-promoting bacterium *Azospirillum* promotes plant growth. A critical assessment. Adv Agron. 2010;108:77–136.

Bashan Y, de-Bashan LE, Prabhu SR, Hernandez J-P. Advances in plant growth-promoting bacterial inoculants technology: Formulations and practical perspectives (1998-2013). Plant Soil. 2014;378(1-2):1–33.

Bassaganya-Riera J, Guri AJ, Lu P, Climent M, Carbo A, Sobral BW, Horne WT, Lewis SN, Bevan DR, Hontecillas R. Abscisic acid regulates inflammation via ligand-binding domain-independent activation of peroxisome proliferator-activated receptor gamma. J Biol Chem. 2011;286(4):2504–16.

Bastián F, Cohen AC, Piccoli P, Luna V, Baraldi R, Bottini R. Production of indole-3-acetic acid and gibberellins A_1 and A_3 by *Acetobacter diazotrophicus* and *Herbaspirillum seropedicae* in chemically-defined culture media. Plant Growth Regul. 1998;24:7–11.

Bastián F, Rapparini F, Baraldi R, Piccoli P, Bottini R. Inoculation with *Acetobacter diazotrophicus* increases glucose and fructose content in shoots of *Sorghum bicolor* (L.) Moench. Symbiosis 1999;27:147–156.

Belimov AA, Safronova VI, Sergeyeva TA, Egorova TN, Matveyeva VA, Tsyganov VE, Borisov AY, Tikhonovich IA, Kluge C, Preisfeld A, Dietz KJ, Stepanok VV. Characterization of plant growth promoting rhizobacteria isolated from polluted soils and containing 1-aminocyclopropane-1-carboxylate deaminase. Can J Microbiol. 2001;47:242–52.

Belimov AA, Dodd IC, Safronova VI, Dumova VA, Shaposhnikov AI, Ladatko AG, Davies WJ. Abscisic acid metabolizing rhizobacteria decrease ABA concentrations *in planta* and alter plant growth. Plant Physiol Biochem. 2014;74:84–91.

Berg G, Alavi M, Schmidt CS, Zachow C, Egamberdieva D, Kamilova F, Lugtenberg B. Biocontrol and osmoprotection for plants under salinated conditions. In: de Bruijn FJ, editor. Molecular microbial ecology of the rhizosphere. Hoboken: Wiley-Blackwell; 2013. p. 561–73.

Berli FJ, Moreno D, Piccoli P, Hespanhol-Viana L, Silva MF, Bressan-Smith R, Cavagnaro JB, Bottini R. Abscisic acid is involved in the response of grape (*Vitis vinifera* L.) cv. Malbec leaf tissues to ultraviolet-B radiation by enhancing ultraviolet-absorbing compounds, antioxidant enzymes and membrane sterols. Plant Cell Environ. 2010;33:1–10.

Berli FJ, Fanzone M, Piccoli P, Bottini R. Solar UV-B and ABA are involved in phenol metabolism of *Vitis vinifera* L. increasing biosynthesis of berry skin polyphenols. J Agric Food Chem. 2011;59:4874–84.

Bloemberg GV, Lugtenberg BJ. Molecular basis of plant growth promotion and biocontrol by rhizobacteria. Review. Curr Opin Plant Biol. 2001;4(4):343–50.

Blokhina O, Virolainen E, Fagerstedt KV. Antioxidants, oxidative damage and oxygen deprivation stress: a review. Ann Bot. 2002;91:179–94.

Boddey RM, Dobereiner J. Nitrogen fixation associated with grasses and cereals: recent progress and perspectives for the future. Fert Res. 1995;42:241–50.

Bohnert HJ, Gong Q, Li P, Ma S. Unraveling a biotic stress tolerance mechanism-getting genomics going. Plant Biol. 2006;9:180–8.

Bottini R, Fulchieri M, Pearce D, Pharis RP. Identification of gibberellins A_1, A_3 and iso-A_3 in cultures of *Azospirillum lipoferum*. Plant Physiol. 1989;90:45–7.

Bottini R, Cassán F, Piccoli P. Gibberellin production by bacteria and its involvement in plant growth promotion and yield increase. Appl Microbiol Biotechnol. 2004;65:497–503.

Boyer JS. Plant productivity and environment. Science. 1982;218:443–8.

Bray EA. Abscisic acid regulation of gene expression during water-deficit stress in the era of the *Arabidopsis* genome. Plant Cell Environ. 2002;25:153–61.

Bresson J, Varoquaux F, Bontpart T, Touraine B, Vile D. The PGPR strain *Phyllobacterium brassicacearum* STM196 induces a reproductive delay and physiological changes that result in improved drought tolerance in *Arabidopsis*. New Phytol. 2013;200:558–69.

Bruzzone S, Moreschi I, Usai C, Guida L, Damonte G, Salis A, Scarfì S, Millo E, De Flora A, Zocchi E. Abscisic acid is an endogenous cytokine in human granulocytes with cyclic ADP-ribose as second messenger. Proc Natl Acad Sci U S A. 2007;104(14):5759–64.

Burd GI, Dixon DG, Glick BR. A plant growth promoting bacterium that decreases nickel toxicity in plant seedlings. Appl Environ Microbiol. 1998;64:3663–8.

Cakmakci R, Dönmez F, Aydın A, Sahin F. Growth promotion of plants by plant growth-promoting rhizobacteria under greenhouse and two different field soil conditions. Soil Biol Biochem. 2006;38:1482–7.

Calvo P, Nelson L, Kloepper JW. Agricultural uses of plant biostimulants. Plant Soil. 2014;383(1-2):3–41.

Chanway CP, Holl FB. Ecological growth response specificity of two Douglas-fir ecotypes inoculated with coexistent beneficial bacteria. Can J Bot. 1994;72:582–6.

Chaves MM, Maroco JP, Pereira JS. Understanding plant responses to drought-from genes to the whole plant. Funct Plant Biol. 2003;30:239–64.

Chaves MM, Felxas J, Pnheiro C. Photosynthesis under drought and salt stress: regulation mechanisms from whole plant to cell. Ann Bot. 2009;103:551–60.

Chen TH, Murata N. Glycinebetaine: an effective protectant against abiotic stress in plants. Trends Plant Sci. 2008;13(9):499–505.

Chen YP, Rekha PD, Arun AB, Shen FT, Lai WA, Young CC. Phosphate solubilizing bacteria from subtropical soil and their tricalcium phosphate solubilizing abilities. Appl Soil Ecol. 2006;34(1):33–41.

Cohen AC, Pontin M, Bottini R, Piccoli P. *Azospirillum brasilense* and ABA improve growth in *Arabidopsis thaliana*, 19th International plant growth substances association meeting. México: Puerto Vallarta; 2007.

Cohen AC, Bottini R, Piccoli P. *Azospirillum brasilense* Sp 245 produces ABA in chemically-defined culture medium and increases ABA content in Arabidopsis plants. Plant Growth Regul. 2008;54:97–103.

Cohen AC, Travaglia C, Bottini R, Piccoli P. Participation of abscisic acid and gibberellins produced by entophytic *Azospirillum* in the alleviation of drought effects in maize. Botany. 2009;87:455–62.

Cohen AC, Bottini R, Pontin M, Berli FJ, Moreno D, Boccanlandro H, Travaglia CN, Piccoli PN. *Azospirillum brasilense* ameliorates the response of *Arabidopsis thaliana* to drought mainly via enhancement of ABA levels. Physiol Plant. 2015a;153(1):79–90.

Cohen AC, Bottini R, Piccoli P. Role of abscisic acid producing PGPR in sustainable agriculture. In: Maheshwari DK, editor. Bacterial metabolites in sustainable agro-ecosystem. New York: Springer; 2015b (in press).

Compant E, van der Heijden MGA, Sessitsch A. Climate change effects on beneficial-plant microorganism interactions. FEMS Microbiol Ecol. 2010;73:197–214.

Corrêa de Souza T, Magalhães PC, de Castro EM, Pereira de Albuquerque PM, Marabesi MA. The influence of ABA on water relation, photosynthesis parameters, and chlorophyll fluorescence under drought conditions in two maize hybrids with contrasting drought resistance. Acta Physiol Plant. 2013;35:515–27.

Corrêa de Souza T, Magalhães PC, de Castro EM, Carneiro NP, Padilha FA, Gomes Júnior CC. ABA application to maize hybrids contrasting for drought tolerance: changes in water parameters and in antioxidant enzyme activity. Plant Growth Regul. 2014;73(3):205–17.

Costacurta A, Vanderleyden J. Synthesis of phytohormones by plant-associated bacteria. Crit Rev Microbiol. 1995;21(1):1–18.

Creus C, Sueldo R, Barassi C. Shoot growth and water status in *Azospirillum*-inoculated wheat seedlings grown under osmotic and salt stresses. Plant Physiol Biochem. 1997;35:939–44.

Creus CM, Sueldo RJ, Barassi CA. Water relations in *Azospirillum*-inoculated wheat seedlings under osmotic stress. Can J Bot. 1998;76:238–44.

Creus CM, Sueldo RJ, Barassi CA. Water relations and yield in *Azospirillum* inoculated wheat exposed to drought in the field. Can J Bot. 2004;82:273–81.

Crozier A, Arruda P, Jasmim JM, Monteiro AM, Sandberg G. Analysis of indole-3-acetic acid and related indoles in culture medium from *Azospirillum lipoferum* and *Azospirillum brasilense*. Appl Environ Microbiol. 1988;54:2833–7.

Dardanelli MS, Fernandez de Cordoba FJ, Espuny MR, Rodriguez Carvajal MA, Soria Diaz ME, Gil Serrano A, Okon Y, Megias M. Effect of *Azospirillum brasilense* coinoculated with

Rhizobiumon Phaseolus vulgaris flavonoids and Nod factor production under salt stress. Soil Biol Biochem. 2008;40:2713–21.

Davies P. The plant hormones; their nature occurrence and function. In: Davies P, editor. Plant Hormones. Biosynthesis, Signal Transduction, Action. Dordrecht: Kluwer Academic Press; 2005;1–15.

Davies P. The plant hormone: their nature, occurrence and functions. In: Davies PJ, editor. Plant hormones: biosynthesis, signal transduction, action. Dordrecht: Springer; 2010. p. 1–15.

Davies WJ, Zhang JH. Root signals and the regulation of growth and development of plants in drying soil. Annu Rev Plant Physiol. 1991;42:55–76.

De Freitas JR, Banerjee MR, Germida JJ. Phosphate solubilizing rhizobacteria enhance the growth and yield but not phosphorus uptake of canola (*Brassica napus* L.). Biol Fertil Soil. 1997;24:358–64.

De Smet I, Zhang H, Inze D, Beeckman T. A novel role for abscisic acid emerges from underground. Trends Plant Sci. 2006;11:434–9.

Deak KI, Malamy J. Osmotic regulation of root system architecture. Plant J. 2005;43:17–28.

Deka Boruah HP, Dileep Kumar BS. Plant disease suppression and growth promotion by *florescent Pseudomonas* strain. Folia Microbiol. 2002;47(2):137–43.

Dodd IC, Pérez-Alfocea F. Microbial amelioration of crop salinity stress. J Exp Bot. 2012;63:3415–28.

Domínguez JE, García Lampasona S, Cohen AC, Salomon MV, Piccoli P. Estudio del gen CrtZ, intermediario clave en la vía del Ácido Abscísico en Pseudomonas fluorescens, aisladas de raíces de plantas de vid. Tucumán: XVII Congr. SAMIGE; 2011.

Dursun A, Turan M, Ekinci M, Gunes A, Ataoglu N, Esringü A, Yildirim E. Effects of boron fertilizer on tomato, pepper, and cucumber yields and chemical composition. Commun Soil Sci Plant Anal. 2010;41:1576–93.

Egamberdieva D. Plant growth promoting properties of rhizobacteria isolated from wheat and pea grown in loamy sand soil. Turk J Biol. 2008;32(1):9–15.

Eisenhauer N, Cesarz S, Koller R, Worm K, Reich PB. Global change belowground: impacts of elevated CO_2, nitrogen, and summer drought on soil food webs and biodiversity. Global Change Biol. 2012;18:435–47.

Engelbrecht BMJ, Comita LS, Condit R, Kursar TA, Tyree MT, Turner BL, Hubbell SP. Drought sensitivity shapes species distribution patterns in tropical forests. Nature. 2007;447:80–2.

Farooq M, Wahid A, Kobayashi N, Fujita D, Basra SMA. Plant drought stress: effects, mechanisms and management. Agron Sustain Dev. 2009;29:185–212.

Figueiredo MVB, Burity HA, Martinez CR, Chanway CP. Alleviation of drought stress in common bean (*Phaseolus vulgaris* L.) by co-inoculation with *Paenibacillus polymyxa* and *Rhizobium tropici*. Appl Soil Ecol. 2008;40:182–8.

Forchetti G, Masciarelli O, Alemano S, Alvarez D, Abdala G. Endophytic bacteria in sunflower (*Helianthus annuus* L.): isolation, characterization, and production of jasmonates and abscisic acid in culture medium. Appl Microb Biotechnol. 2007;76:1145–52.

Freitas ADS, Vieira CL, Santos CERS, Stamford NP, Lyra MCCP. Caracterização de rizóbios isolados de Jacatupé cultivado em solo salino no Estado de Pernanbuco, Brazil. Bragantia. 2007;66:497–504.

García de Salamone IE, Hynes RK, Nelson LM. Cytokinin production by plant growth promoting rhizobacteria and selected mutants. Can J Microbiol. 2001;47:404–11.

Glick BR. The enhancement of plant growth by free-living bacteria. Can J Microbiol. 1995;41:109–17.

Glick BR, Bashan Y. Genetic manipulation of plant growth-promoting bacteria to enhance biocontrol of fungal phytopathogens. Biotechnol Adv. 1997;15:353–78.

Glick BR. Modulation of plant ethylene levels by the bacterial enzyme ACC deaminase. FEMS Microbiol Lett. 2005;251:1–7.

Glick BR, Cheng Z, Czarny J, Cheng Z, Duan J. Promotion of plant growth by ACC deaminase-producing soil bacteria. Eur J Plant Pathol. 2007;119:329–39.

Glick BR. Using soil bacteria to facilitate phytoremediation. Biotechnol Adv. 2010;28:367–74.

Guóth A, Tari I, Gallé A, Csiszár J, Pécsváradi A, Cseuz L, Erdei L. Comparison of the drought stress responses of tolerant and sensitive wheat cultivars during grain filling: changes in flag leaf photosynthetic activity, ABA levels, and grain yield. J Plant Growth Regul. 2009;28:167–76.

Gutiérrez-Mañero F, Ramos-Solano B, Probanza A, Mehouachi J, Tadeob F, Talon M. The plant-growth-promoting rhizobacteria *Bacillus pumilus* and *Bacillus licheniformis* produce high amounts of physiologically active gibberellins. Physiol Plant. 2001;111:206–11.

Hasegawa S, Poling SM, Maier VP, Bennett RD. Metabolism of abscisic acid: bacterial conversion to dehydrovomifoliol and vomifoliol dehydrogenase activity. Phytochem. 1984; 23(12):2769–2771.

Heidari M, Mousavinik SM, Golpayegani A. Plant growth promoting rhizobacteria (PGPR) effect on physiological parameters and mineral uptake in basil (*Ocium basilicum* L.) under water stress. ARPN J Agric Biol Sci. 2011;6(5):6–11.

Houghton JT, Ding Y, Griggs DJ, Noguer M, Vander Linden PJ, Xiaosu D (2001) Climate change 2001: the scientific basis. In: Houghton JT, Ding Y, Griggs DJ, Noguer M, Vander Linder PJ, Dai X, Maskell K, Johnson CA (eds). Cambridge: Cambridge University Press, p. 1–83.

Hu H, Dai M, Yao J, Xiao B, Li X, Zhang Q, Xiong L. Overexpressing a NAM, ATAF, and CUC (NAC) transcription factor enhances drought resistance and salt tolerance in rice. Proc Natl Acad Sci U S A. 2006;103:12987–92.

Hu J, Lin X, Wang J, Chu H, Yin R, Zhang J. Population size and specific potential of P-mineralizing and P-solubilizing bacteria under long-term P-deficiency fertilization in a sandy loam soil. Pedobiologia. 2009;53:49–58.

Huang D, Wu W, Abrams SR, Cutler AJ. The relationship of drought-related gene expression in *Arabidopsis thaliana* to hormonal and environmental factors. J Exp Bot. 2008;59:2991–3007.

IPCC (Intergovernmental Panel on Climate Change 2007): Climate Change 2007: The Physical Science Basis. Contribution of Working Group I to the Fourth Assessment Report of the Intergovernmental Panel on Climate Change. Solomon S, Qin D, Manning M, Chen Z, Marquis M, Averyt FB, Tignor M, Miller HL (eds). Cambridge University Press, Cambridge, United Kingdom and New York, NY, USA, 996 pp

IPCC (Intergovernmental Panel on Climate Change). Climate Change 2014: Synthesis Report. In: Core Writing Team, Pachauri RK, Meyer LA, editors. Contribution of Working Groups I, II and III to the Fifth Assessment Report of the Intergovernmental Panel on Climate Change. Geneva: IPCC; 2014. 151 p.

Janzen R, Rood S, Dormar J, McGill W. *Azospirillum brasilense* produces gibberellins in pure culture and chemically- medium and in co-culture on straw. Soil Biol Biochem. 1992;24:1061–4.

Jha BK, Pragash MG, Cletus J, Raman G, Sakthivel N. Simultaneous phosphate solubilization potential and antifungal activity of new fluorescent pseudomonad strains, *Pseudomonas aeruginosa*, *P. plecoglossicida* and *P. mosselii*. World J Microbiol Biotechnol. 2009;25:573–81.

Jiang M, Zhang J. Water stress-induced abscisic acid accumulation triggers the increased generation of reactive oxygen species and up-regulates the activities of antioxidant enzymes in maize leaves. J Exp Bot. 2002;53:2401–10.

Jiang M, Zhang J. Cross-talk between calcium and reactive oxygen species originated from NADPH oxidase in abscisic acid-induced antioxidant defense in leaves of maize seedlings. Plant Cell Environ. 2003;26:929–93.

Joo G-J, Kang S-M, Hamayun M, Kim S-K, Na C-I, Shin D-H, Lee I-J. *Burkholderia* sp. 11096BP as a newly isolated gibberellin producing bacterium. J Microbiol. 2009;47(2):167–71.

Joo G-J, Kim Y-M, Lee I-J, Song K-S, Rhee I-K. Growth promotion of red pepper plug seedlings and the production of gibberellins by *Bacillus cereus*, *Bacillus macroides* and *Bacillus pumilus*. Biotechnol Lett. 2004;26:487–91.

Karadeniz A, Topcuoğlu SF, İnan S. Auxin, gibberellin, cytokinin and abscisic acid production in some bacteria. World J Microbiol Biotechnol. 2006;22:1061–4.

Kaymak HC, Guvenc I, Yarali F, Donmez MF. The effects of bio-priming with PGPR on germination of radish (*Raphanus sativus* L.) seeds under saline conditions. Turk J Agric For. 2009;33(2):173–9.

Kloepper JW, Schroth M (1978) Plant growth promoting rhizobacteria on radishes. Prov. IV Int Conf Plant Pathol Bacteria, Angers 2:879–82.

Kloepper JW, Lifshitz R, Zablotowicz RM. Free living bacterial inocula for enhancing crop productivity. Trends Biotechnol. 1989;7:39–44.

Kloepper JW, Zablotowiez RM, Tipping EM, Lifshitz R. Inorganic plant growth promotion mediated by bacterial rhizosphere colonizer. In: Keister KL, Gregan PB, editors. The rhizosphere and plant growth. Dordrecht: Kluwer; 1991. p. 315–26.

Kohler J, Hernández JA, Caravaca F, Roldán A. Plant-growth promoting rhizobacteria and arbuscular mycorrhizal fungi modify alleviation biochemical mechanisms in water-stressed plants. Funct Plant Biol. 2008;35:141–51.

Kolb W, Martin P. Response of plant roots to inoculation with *Azospirillum brasilense* and to application of indole acetic acid. In: Klingmüller W, editor. *Azospirillum* III: genetics, physiology, ecology. Berlin: Springer; 1985. p. 215–21.

Kumar KR. Plant responses to water stress: Role of reactive oxygen species. Plant Signal Behav. 2011;6(1):1741–5.

Lambers H, Raven JA, Shaver GR, Smith SE. Plant nutrient-acquisition strategies change with soil age. Trends Ecol Evol. 2008;23(2):95–103.

Lambrecht M, Okon Y, VandeBroek A, Vanderleyden J. Indole-3-acetic acid: a reciprocal signalling molecule in bacteria–plant interactions. Trends Microbiol. 2000;8:298–300.

Li YP, Ye W, Wang M, Yan XD. Climate change and drought: a risk assessment of crop-yield impacts. Clim Res. 2009;39:31–46.

Liu F, Jensen CR, Shahanzari A, Andersen MN, Jacobse S-E. ABA regulated stomatal control and photosynthetic water use efficiency of potato (*Solanum tuberosum* L.) during progressive soil drying. Plant Sci. 2005;168(3):831–6.

Lucangeli C, Bottini R. Reversion of dwarfism in dwarf-1 maize (*Zea mays* L.) and dwarf-x rice (*Oryza sativa* L.) mutants by endophytic *Azospirillum* spp. Biocell 1996;20:223–228.

Lucangeli C, Bottini R. Effects of Azospirillum spp. on endogenous gibberellin content and growth of maize (*Zea mays* L.) treated with uniconazole. Symbiosis 1997;23:63–72.

Lugtenberg B, Kamilova F. Plant-growth-promoting rhizobacteria. Annu Rev Microbiol. 2009;63:54–6.

Manivannan P, Jaleel CA, Somasundaram R, Panneerselvam R. Osmoregulation and antioxidant metabolism in drought stressed *Helianthus annuus* under triadimefon drenching. C R Biol. 2008;331(6):418–25.

Mansouri S, Bunch AW. Bacterial ethylene synthesis from 2-0×0-4-thiobutyric acid and from methionine. J Gen Microbiol. 1989;135:2819–27.

Marasco EK, Schmidt-Dannert C. Exploring and accessing plant natural product biosynthesis in engineered microbial hosts. In: Kayser O, Quax WJ, editors. Medicinal plant biotechnology: from basic research to industrial applications. Weinheim: Wiley; 2007. p. 287–317.

Marasco R, Rolli E, Ettoumi B, Vigani G, Mapelli F, Borin S, et al. A Drought resistance-promoting microbiome is selected by root system under desert farming. PLoS One. 2012;7:48479.

Marulanda A, Barea JM, Azcón R. An indigenous drought-tolerant strain of *Glomus intraradices* associated with a native bacterium improves water transport and root development in *Retama sphaerocarpa*. Microb Ecol. 2006;52:670–8.

Mayak S, Tirosh T, Glick BR. Plant growth-promoting bacteria that confer resistance to water stress in tomatoes and pepper. Plant Sci. 2004a;166:525–30.

Mayak S, Tirosh T, Glick BR. Plant growth promoting bacteria confers resistance in tomato plants to salt stress. Plant Physiol Biochem. 2004b;42:565–72.

Meloni DA, Oliva MA, Ruiz HA, Martinez CA. Contribution of proline and inorganic solutes to osmotic adjustment in cotton under salt stress. J Plant Nutr. 2001;24:599–612.

Mishra V, Cherkauer KA. Retrospective droughts in the crop growing season: Implications to corn and soybean yield in the Midwestern United States. Agr Forest Meteorol. 2010;150(7–8):1030–45.

Mizokami Y, Noguchi K, Kojima M, Sakakibara H, Terashima I. Mesophyll conductance decreases in the wild type but not in an ABA-deficient mutant (*aba*1) of *Nicotiana plumbaginifolia* under drought conditions. Plant Cell Environ. 2015;38:388–98.

Moreno D, Berli FJ, Piccoli P, Bottini R. Gibberellins and abscisic acid promote carbon allocation in roots and berries of grape plants. J Plant Growth Regul. 2011;30:220–8.

Munns R. Genes and salt tolerance: bringing them together. New Phytol. 2005;167(3):645–63.

Murcia G, Pontin M, Reinoso H, Baraldi R, Bertazza G, Gómez-Talquenca S, Bottini R, Piccoli P. ABA and GA_3 increase carbon allocation in different organs of grapevine plants by inducing accumulation of non-structural carbohydrates in leaves, enhancement of phloem area and expression of sugar transporters. Physiol Plant. 2015. doi:10.1111/ppl.12390 (in press).

Nambara E, Marion-Poll A. Abscisic acid biosynthesis and catabolism. Annu Rev Plant Biol. 2005;56:165–85.

Nehring RB, Ecker JR. Ethylene responses in seedling growth and development. InPlant Hormones Springer Netherlands. 2010;1:358–376.

Noel ITC, Sheng C, Yost CK, Pharis RP, Hynes MF. *Rhizobium leguminosarum* as a plant I growth-promoting rhizobacterium: direct growth promotion of canola and lettuce. Can J Microbiol. 1996;42:279–83.

Normander B, Prosser JI. Bacterial origin and community composition in the barley phytosphere as a function of habitat and pre-sowing conditions. Appl Environ Microbiol. 2000;66(10):4372–7.

Patten CL, Glick BR. Bacterial biosynthesis of indole-3-acetic acid. Can J Microbiol. 1996;42:207–20.

Perrig D, Boiero L, Masciarelli O, Penna C, Cassán F, Luna V. Plant growth promoting compounds produced by two agronomically important strains of *Azospirillum brasilense*, and their implications for inoculant formulation. Appl Microbiol Biotechnol. 2007;75:1143–50.

Persello-Cartieaux F, Nussaume L, Robaglia C. Tales from the underground: molecular plant-rhizobacterial interactions. Plant Cell Environ. 2003;26:189–99.

Phalan B, Onial M, Balmford A, Green RE. Reconciling food production and biodiversity conservation: Land sharing and land sparing compared. Science. 2011;333:1289–91.

Piccoli P, Travaglia C, Cohen A, Sosa CP, Masuelli R, Bottini R. An endophytic bacterium isolated from roots of the halophyte *Prosopis strombulifera* produces ABA, IAA, gibberellins A_1 and A_3 and jasmonic acid in chemically-defined culture medium. Plant Growth Regul. 2011;64:207–10.

Piccoli P, Bottini R. Abiotic stress tolerance induced by endophytic PGPR (Chapter 3). In: Aroca R, editor. Progress in symbiotic endophytes, Soil biology book series. Dordrecht: Springer; 2013. p. 151–63.

Poupin MJ, Timmermann T, Vega A, Zuñiga A, González B. Effects of the plant growth-promoting bacterium *Burkholderia phytofirmans* PsJN throughout the life cycle of *Arabidopsis thaliana*. PLoS One. 2013;8, e69435.

Probanza A, Garcíia JAL, Palomino MR, Ramos B, Manero FJG. *Pinus pinea* L. seedling growth and bacterial rhizosphere structure after inoculation with PGPR Bacillus (*B. licheniformis* CECT 5106 and *B. pumilus* CECT 5105). Appl Soil Ecol. 2002;20:75–84.

Quiroga AM, Berli F, Moreno D, Cavagnaro JB, Bottini R. Abscisic acid sprays significantly increase yield per plant in vine yard grown wine grape (*Vitis vinifera* L.) cv. *Cabernet sauvignon* through increased berry set with no negative effects on anthocyanin content and total polyphenol index of both juice and wine. J Plant Growth Regul. 2009;28:28–35.

Ren H, Gao Z, Chen L, Wei K, Liu J, Fan Y, Davies WJ, Jia W, Zhang J. Dynamic analysis of ABA accumulation in relation to the rate of ABA catabolism in maize tissues under water deficit. J Exp Bot. 2007;58:211–9.

Reynolds JF, Stafford Smith DM, Lambin EF, Turner BL, Mortimore M, Batterbury SPJ, Downing TE, Dowlatabadi H, et al. Global desertification: building a science for dry land development. Science. 2007;316:847–51.

Richardson AE. Prospects for using soil microorganisms to improve the acquisition of phosphorus by plants. Aust J Plant Physiol. 2001;28:897–906.

Rodriguez H, Fraga R. Phosphate solubilizing bacteria and their role in plant growth promotion. Biotechnol Adv. 1999;17:319–39.

Sakakibara H. Cytokinin biosynthesis and metabolism. In: Davies PJ, editor. Plant hormones: biosynthesis, signal transduction, action. Dordrecht: Springer; 2010. p. 95–114.

Salomon MV, Bottini R, de Souza Filho GA, Cohen AC, Moreno D, Gil M, Piccoli P. Bacteria isolated from roots and rhizosphere of *Vitis vinifera* retard water losses, induce abscisic acid accumulation and synthesis of defense-related terpenes in in vitro cultured grapevine. Physiol Plant. 2014;151(4):359–74.

Sandhya V, Ali SZ, Grover M, Reddy G, Venkateswarlu B. Alleviation of drought stress effects in sunflower seedlings by the exopolysaccharides producing Pseudomonas putida strain GAP-P45. Biol Fertil Soils. 2009;1:46(1):17–26.

Sandhya V, Ali SZ, Grover M, Reddy G, Venkateswarlu B. Effect of plant growth promoting *Pseudomonas* spp. on compatible solutes, antioxidant status and plant growth of maize under drought stress. Plant Growth Regul. 2010;62(1):21–30.

Sansberro P, Mroginski L, Bottini R. Abscisic Acid promotes growth of *Ilex paraguariensis* plants by alleviating diurnal water stress. Plant Growth Regul. 2004;42:105–11.

Sarig S, Blum A, Okon Y. Improvement of water status and yield of field-grown grain sorghum (*Sorghum bicolor*) by inoculation with *Azospirillum brasilense*. J Agric Sci. 1988;110:271–7.

Sarma RK, Saikia R. Alleviation of drought stress in mung bean by strain *Pseudomonas aeruginosa* GGRJ21. Plant Soil. 2014;377:111–26.

Schachtman DP, Goodger JQD. Chemical root to shoot signaling under drought. Trends Plant Sci. 2008;13:281–7.

Schmelz EA, Engelberth J, Alborn HT, O'Donnell P, Sammons M, Toshima H, Tumlinson III JH. Simultaneous analysis of phytohormones, phytotoxins, and volatile organic compounds in plants. Proc Natl Acad Sci. 2003;100(18):10552–7.

Seki M, Narusaka M, Ishida J, Nanjo T, Fujita M, Oono Y, Kamiya A, Nakajima M, Enju A, Sakurai T, et al. Monitoring the expression profiles of ca 7000 Arabidopsis genes under drought, cold and high-salinity stresses using a full-length cDNA microarray. Plant J. 2002;31:279–92.

Seki M, Umezawa T, Urano K, Shinozaki K. Regulatory metabolic networks in drought stress responses. Curr Opin Plant Biol. 2007;10:296–302.

Sgroy V, Cassán F, Masciarelli O, Del Papa M, Lagares A, Luna V. Isolation and characterization of endophytic plant growth-promoting (PGPB) or stress homeostasis regulating (PSHB) bacteria associated to the halophyte *Prosopis strombulifera*. Appl Microbiol Biotechnol. 2009;85:371–81.

Sharma S, Kumar V, Tripathi RB. Isolation of phosphate solubilizing microorganism (PSMs) from soil. J Microbiol Biotechnol Res. 2011;1:90–5.

Shinozaki K, Yamaguchi-Shinozaki K. Gene networks involved in drought stress response and tolerance. J Exp Bot. 2007;58:221–7.

Siddikee MA, Glick BR, Chauhan PS, Yim WJ, Sa T. Enhancement of growth and salt tolerance of red pepper seedlings (*Capsicum annuum* L.) by regulating stress ethylene synthesis with halotolerant bacteria containing 1-aminocyclopropane-1-carboxylic acid deaminase activity. Plant Physiol Biochem. 2011;49:427–34.

Spaepen S, Vanderleyden J, Remans R. Indole-3-acetic acid in microbial and microorganism-plant signaling. FEMS Microbiol Rev. 2007;31:425–48.

Sperdouli I, Moustakas M. Interaction of proline, sugars, and anthocyanins during photosynthetic acclimation of *Arabidopsis thaliana* to drought stress. J Plant Physiol. 2012;169:577–85.

Sponsel VM, Hedden P. Gibberellin biosynthesis and inactivation. In: Davies PJ, editor. Plant hormones: biosynthesis, signal transduction, action. Dordrecht: Springer; 2010. p. 63–94.

Takezawa D, Komatsu K, Sakata Y. ABA in bryophytes: how a universal growth regulator in life became a plant hormone? J Plant Res. 2011;124:437–53.

Taller BG, Wong T-Y. Cytokinins in *Azotobacter vinelandii* culture medium. Appl Environ Microbiol. 1989;55(1):266–7.

Tan S, Jiang Y, Song S, Huang J, Ling N, Xu Y, Shen Q. Two *Bacillus amyloliquefaciens* strains isolated using the competitive tomato root enrichment method and their effects on suppressing *Ralstonia solanacearum* and promoting tomato plant growth. Crop Prot. 2013;43:134–40.

Tank N, Saraf M. Salinity resistant plant growth promoting rhizobacteria ameliorates sodium chloride stress on tomato plants. J Plant Interact. 2010;5:51–8.

Tardieu F, Davies WJ. Stomatal response to abscisic acid is a function of current plant water status. Plant Physiol. 1992;98:540–5.

Tardieu F, Parent B, Simonneau T. Control of leaf growth by abscisic acid: hydraulic or non-hydraulic processes? Plant Cell Environ. 2010;33:636–47.

Thompson AJ, Andrews J, Mulholland BJ, McKee JM, Hilton HW, Horridge JS, Farquhar GD, Smeeton RC, Smillie IR, Black CR, Taylor IB. Overproduction of abscisic acid in tomato increases transpiration efficiency and root hydraulic conductivity and influences leaf expansion. Plant Physiol. 2007;143:1905–17.

Timmusk S, Wagner GH. The plant-growth-promoting rhizobacterium *Paenibacillus polymyxa* induces changes in *Arabidopsis thaliana* gene expression: a possible connection between biotic and abiotic stress responses. Mol Plant Microbe Interact. 1999;12:951–9.

Tomlinson I. Doubling food production to feed the 9 billion: a critical perspective on a key discourse of food security in the UK. J Rural Stud. 2013;21:81–90.

Travaglia C, Cohen AC, Reinoso H, Castillo C, Bottini R. Exogenous abscisic acid increases carbohydrate accumulation and redistribution to the grains in wheat grown under field conditions of soil water restriction. J Plant Growth Regul. 2007;26:285–9.

Travaglia C, Reinoso H, Bottini R. Application of abscisic acid promotes yield in field-cultured soybean by enhancing production of carbohydrates and their allocation in seed. Crop Pasture Sci. 2009;60:1131–6.

Travaglia C, Reinoso H, Cohen AC, Luna C, Castillo C, Bottini R. Exogenous ABA increases yield in field-grown wheat with a moderate water restriction. J Plant Growth Regul. 2010;29:366–74.

Travaglia C, Balboa G, Espósito G, Reinoso H. ABA action on the production and redistribution of field-grown maize carbohydrates in semiarid regions. Plant Growth Regul. 2012;1:67(1):27–34.

Travaglia C, Masciarelli O, Fortuna J, Marchetti G, Cardozo P, Lucero M, et al. Towards sustainable maize production: Glyphosate detoxification by *Azospirillum* sp. and *Pseudomonas* sp. Crop Prot. 2015;77:102–9.

Tsavkelova EA, Cherdyntseva TA, Netrusov AI. Auxin production by bacteria associated with orchid roots. Microbiology. 2005;74(1):46–53.

Tsavkelova EA, Klimova SY, Cherdyntseva TA, Netrusov AI. Microbial producers of plant growth stimulators and their practical use: a review. Appl Biochem Microb. 2006;42:117–26.

Vessey JK. Plant growth promoting rhizobacteria as biofertilizers. Plant Soil. 2003;255:571–86.

Vickers CE, Gershenzon J, Lerdau MT, Loreto F. A unified mechanism of action for volatile isoprenoids in plant abiotic stress. Nat Chem Biol. 2009;1:5(5):283–91.

Vile D, Pervent M, Belluau M, Vasseur F, Bresson J, Muller B, Granier C, Simonneau T. *Arabidopsis* growth under prolonged high temperature and water deficit: independent or interactive effects? Plant Cell Environ. 2012;35:702–18.

Weingart H, Völksch B. Ethylene production by *Pseudomonas syringae* pathovars *in vitro* and *in planta*. Appl Environ Microbiol. 1997;63:156–61.

Whipps JM. Carbon economy. In: Lynch JM, editor. The rhizosphere. Wiley: Chichester; 1990. p. 59–97.

Wilkinson S, Davies WJ. Drought, ozone, ABA and ethylene: new insights from cell to plant to community. Plant Cell Environ. 2010;33:510–25.

Xie X, Zhang H, Paré PW. Sustained growth promotion in *Arabidopsis* with long-term exposure to the beneficial soil bacterium *Bacillus subtilis* (GB03). Plant Signal Behav. 2009;4:948–53.

Xu ZZ, Zhou GS, Wang YL, Han GX, Li YJ. Changes in chlorophyll l fluorescence in maize plants with imposed rapid dehydration at different leaf ages. J Plant Growth Regul. 2008;27:83–92.

Yamaguchi-Shinozaki K, Shinozaki K. Organization of cis-acting regulatory elements in osmotic- and cold-stress-responsive promoters. Trends Plant Sci. 2005;10:88–94.

Yanni YG, Rizk RY, El-Fattah FK, Squartini A, Corich V, Giacomini A, de Bruijn F, Rademaker J, Maya-Flores J, Ostrom P, Vega-Hernandez M. The beneficial plant growth-promoting association of *Rhizobium leguminosarum* bv. *trifolii* with rice roots. Functional Plant Biology. 2001;21:28(9):845–70.

Zahir ZA, Arshad M, Frankenberger Jr WT. Plant growth promoting rhizobacteria application and perspectives in agriculture. Adv Agron. 2004;81:96–168.

Zahir ZA, Ghani U, Naveed M, Nadeem SM, Asghar HN. Comparative effectiveness of *Pseudomonas* and *Serratia* sp. containing ACC-deaminase for improving growth and yield of wheat (*Triticum aestivum* L.) under salt-stressed conditions. Arch Microbiol. 2009;191:415–24.

Zawoznik MS, Ameneiros M, Benavides MP, Vázquez S, Groppa MD. Response to saline stress and aquaporin expression in *Azospirillum*-inoculated barley seedlings. Appl Microbiol Biotechnol. 2011;90:1389–97.

Zeevaart J. Abscisic acid metabolism and its regulation. In: Hooykaas P, Hall M, Libbenga K, editors. Biochemistry and molecular biology of plant hormones. Amsterdam: Elsevier Science; 1999. p. 189–207.

Zhang H, Xie X, Kim MS, Kornyeyev DA, Holaday S, Paré PW. Soil bacteria augment *Arabidopsis* photosynthesis by decreasing glucose sensing and abscisic acid levels in plant. Plant J. 2008;56:264–73.

Zhang M, Yuan B, Leng P. The role of ABA in triggering ethylene biosynthesis and ripening of tomato fruit. J Exp Bot. 2009a;60(6):1579–88.

Zhang SQ, Outlaw Jr WH. Abscisic acid introduced into the transpiration stream accumulates in the guard-cell apoplast and causes stomatal closure. Plant Cell Environ. 2001;24:1045–54.

Zhang Y, Zhu H, Zhang Q, Li M, Yan M, Wang R, Wang L, Welti R, Zhang W, Wang X. Phospholipase Dα1 and phosphatidic acid regulate NADPH oxidase activity and production of reactive oxygen species in ABA-mediated stomatal closure in *Arabidopsis*. Plant Cell. 2009b;21:2357–77.

Chapter 13
Signaling Role of ROS in Modulating Drought Stress Tolerance

Ana Laura Furlan, Eliana Bianucci, and Stella Castro

13.1 Introduction to Drought-Induced Oxidative Stress and ROS Signaling

For a long time, the symptoms associated with ROS accumulation were considered harmful to biomolecules and termed oxidative stress, indicating a negative misbalance between oxidant generation and antioxidant provision. Nowadays, the signaling function of ROS has been recognized as a fundamental principle in cellular communication, and the concept of the redox regulatory network of the cell has been developed as a central element in acclimation (Dietz 2008; Jacquot et al. 2013). As synthesized by Dietz (2014), six functional elements that co-operate in the redox regulatory network can be identified: *redox input elements* that feed electrons to the redox regulatory network, such as NADPH, ferredoxin, and glutathione (GSH); *redox transmitters* that transfer and distribute the electrons to downstream proteins (i.e., thioredoxin, glutaredoxin); *redox target proteins* that have redox-sensitive thiols controlled by transmitters and includes a vast number of proteins members of metabolic pathways, translation, and transcription; *redox buffers proteins* that are accumulated in a high amount in organelles, such as the Rubisco in chloroplasts that possess a high number of cysteins susceptible for oxidation; *redox sensors* that deliver information from ROS to the redox regulatory network and make the cross-talk with other signaling pathways; *final electron acceptors* are low-molecular-mass redox species such as ROS, reactive nitrogen species (RNS), reactive sulfur species (RSS), and reactive carbonyl species (RCS). It is interesting to note that the final electron acceptors include reactive and nonreactive molecules

A.L. Furlan (✉) • E. Bianucci • S. Castro
Departamento de Ciencias Naturales, Facultad de Ciencias Exactas, Físico-Químicas y Naturales, Universidad Nacional de Río Cuarto, Ruta 36, Km. 601, 5800 Río Cuarto, Córdoba, Argentina
e-mail: afurlan@exa.unrc.edu.ar

M.A. Hossain et al. (eds.), *Drought Stress Tolerance in Plants, Vol 1*,
DOI 10.1007/978-3-319-28899-4_13

that can be considered both toxic by-products and signals in the redox regulatory network. The activation of acclimation mechanisms in response to non-challenged tissues (or abiotic stresses, to distinguish from biotic stress) is termed systemic acquired acclimation (SAA) (Baxter et al. 2014). Drought is an environmental factor that limit CO_2 fixation and reduce the $NADP^+$ regeneration by the Calvin cycle, consequently, the over-reduction of the photosynthetic electron transport chain occurs, producing superoxide radicals and singlet oxygen in the chloroplasts (Shao et al. 2007, 2008). Abscisic acid (ABA) also increases the production of ROS, which serve as a signaling intermediate to promote stomatal closure (Yan et al. 2007) via the H_2O_2 generation and indirectly via the Mehler reaction and photorespiration (Cruz de Carvalho 2008). It is important to highlight that the activation of different kinds of all these sources of ROS determines different responses in an overall view. In this sense, temporal and spatial coordination of ROS signals will determine the response to a specific stimulus. Regarding the temporal coordination of ROS signals in plants, RBOHD-dependent long-distance signals play an important biological role in the SAA response of plants to heat or high light (Suzuki et al. 2013). A biphasic production of ROS comprising a primary phase that occurs within minutes and a secondary phase that occurs within hours/days was described by several authors (Soares et al. 2009; Kunihiro et al. 2011; Mittler et al. 2011). There is a general agreement in the explanation that the rapid burst of ROS is required for the second phase of ROS production responsible for the regulation of downstream pathways and plant acclimation to stress. Recent findings reveal that these two phases of the ROS burst are linked via a ROS wave that communicates the initial ROS burst in the local tissue to the systemic tissue via a cell-to-cell relay mechanism supporting the concept of SAA mentioned previously (Suzuki et al. 2013). In the other dimension, the spatial coordination of ROS signals in plants, some studies demonstrated differences in transcripts or metabolites related to redox state between different types of tissues and even among different leaves during SAA. In this sense, Gordon et al. (2012) reported that, in systemic leaves, transcript levels of ZAT10, and Redox Responsive Factor 1 (RRTF1) transcripts were accumulated in response to local high light treatment. As expected, a spatial–temporal interaction is possible as well, in this way SAA of plants to heat stress was correlated with activation of the ROS wave and transient accumulation of ABA in systemic tissues, and these responses were suppressed in a mutant lacking RBOHD or ABA (Suzuki et al. 2013). The mentioned results indicate that RBOHD-dependent ROS and ABA accumulation in terms of temporal and spatial interactions mediate SAA to heat stress (Suzuki et al. 2013). Non-toxic levels of ROS must be maintained to accomplish the role of ROS as signaling molecules. Thus, a balance between ROS production and ROS-scavenging pathways is essential (Mittler et al. 2004). In plants, NADPH oxidases, respiratory burst oxidase homologues (RBOHs), are key components in the network of ROS production (Suzuki et al. 2011). RBOH proteins produce superoxide ($O_2^{\cdot-}$) at the apoplast, which dismutates to H_2O_2 spontaneously or catalytically by the action of superoxide dismutase (SOD) (Lin et al. 2009). Besides, regulatory mechanisms of RBOH protein homologues make the activation of members of the family of proteins to become highly specific to respond to different stressors.

These regulatory mechanisms depend on well-known signaling components including protein phosphorylation, Ca^{2+}, calcium-dependent protein kinases (CDPKs), and phospholipase Dα1 (PLDα1) (Lin et al. 2009; Drerup et al. 2013; Dubiella et al. 2013). However, the current information shows that RBOH proteins are not the only source for ROS in plant cells. Other producing pathways for ROS are photosynthesis (via the electron transport chain and photosystems I and II), respiration (via the electron transport chain), glycolate oxidase, oxalate oxidase, xanthine oxidase, amine oxidase, excited chlorophyll, fatty acid oxidation, and peroxidases (Mittler 2002).

In this chapter, we discuss the recent advances in understanding the cellular signaling networks of plant acclimation to drought. The current knowledge on hormonal signal perception and transduction is integrated in the context of plant signaling networks under drought and the interaction of these molecules with other signals are discussed in the context of SAA.

13.2 Downstream Signaling Events: ROS Sensing

Plant cell can sense, transduce, and translate the ROS signals into appropriate cellular responses through the involvement of redox-sensitive proteins. Redox-sensitive proteins operate through reversible processes of oxidation/reduction switching "on" and "off" in a redox-dependent manner, constituting the candidates for ROS sensing. Cys S-glutathionylation and S-nitrosylation are just two of the possible reversible oxidative modifications that may be involved in redox signaling during drought (Colville and Kranner 2010). A specified mechanism underlying the sensing of ROS in plant cells has not been revealed yet; however, three mechanisms of ROS sensing were proposed: unidentified receptor proteins, redox-sensitive transcription factors and phosphatases (Huang et al. 2012). In this regard, numerous studies revealed that ROS interact with other signal transmission components, such as phytohormones, MAPK cascades, and calcium ions (Xia et al. 2015); despite this, the initial steps of ROS perception are still ignored.

Redox-response transcription factors act upstream to a cascade of other transcription factors. In this way, ROS accumulated in the cytosol can be detected in plants. Thus, in *Arabidopsis thaliana* the redox-sensitive protein NPR1 (non-expressor of pathogenesis-related gene 1) is involved in the transduction of redox changes induced by salicylic acid, and, also is a key regulator of SA-mediated suppression of signaling via jasmonic acid (Love et al. 2008; Leon-Reyes et al. 2009). In addition, ROS could produce a global shift in the main cellular redox buffers such as glutathione or thioredoxins and then be perceived by sensitive proteins that interact with these components. Specificity may be conferred by peroxidases that transmit oxidative signals to sensitive proteins such as transcription factors as revealed in the Gpx3 yeast protein, which after peroxide oxidation, activates the oxidized transcription factor, Yap1 (Delaunay et al. 2002). This has been proposed as a general mechanism of peroxide-based signaling taking into account several examples such as the oxidation of redox-sensitive green fluorescent protein (roGFP). These findings support the concept that some peroxidases have the capacity to act as H_2O_2-dependent protein thiol oxidases when they are in the proximity of oxidizable proteins (Gutscher et al. 2009). Another level of regulation was described by Benina et al. (2015) in *A. thaliana* lines expressing a FLAG-tagged ribosomal protein to immunopurify polysome-bound mRNAs before and after oxidative stress. These authors revealed that ROS-responsive transcripts are regulated both by the common mechanisms that control translation, including the presence of uAUGs and 5′-UTR length, but also potential specific mechanisms as two uncharacterized cis-elements enriched in the 5′-UTRs. This study showed a novel mechanism underlying ROS signaling in plant tissues. Mitogen-activated protein kinases (MAPKs) cascade can indirectly activate other transcription factors. It is well known that serine/threonine protein kinases sense ROS and activates some MAPKs by calcium. This protein kinase is also activated by phosphoinositide-dependent kinase-1 (PDK1) through phospholipase-C/D-phosphatidic-acid pathway. The expression of an *A. thaliana* gene (OXI1) encoding a serine/threonine kinase is induced in response to biotic (*Peronospora parasitica*) and abiotic (cold, osmotic, heat) H_2O_2-generating stresses.

Another aspect to consider when redox regulatory networks are involved is the role of cells being in a reductive state, thus generating a reductive signaling. In this case, thioredoxins are implicated in the regulation of enzyme activity in response to environmental changes in light conditions (Schürmann and Buchanan 2008). During pathogen challenge, a reductive signaling has been described for the NPR1-TGA transcription factor interaction in the regulation of the expression of pathogenesis-related (PR) genes (Després et al. 2003; Mou et al. 2003). Besides, in chloroplasts, several enzymes are activated by disulfide reduction. However, the detailed mechanisms that drive the cytosolic reductive signaling remain to be elucidated. In this regard, the protein folding mediated by protein disulfide isomerases oxidized by homologous endoplasmic reticulum oxyreductins can be inadequate in the insufficiently oxidizing conditions of the compartment (Sevier and Kaiser 2008). However, the importance of reductive stress in other subcellular compartments which maintain reducing environment remains to be determined. Two concepts of reductive stress can perhaps be distinguished. In the first, over-reduction of redox-active compounds would favor production of ROS. Such effects are well described in plants at the level of electron transport chains and reflect increased ROS production rates caused by over-reduction of auto-oxidizable compounds. The second type of reductive stress involving modifications of protein function through a drop in the redox potential of pyridine nucleotides, thioredoxins, or glutathione was suggested by Foyer and Noctor (2009).

After ROS sensing, downstream signaling events amplify the ROS signal and transduce the response to counteract the environmental constraint. Such mechanisms implicate Ca^+, calcium-binding proteins such as calmodulins, G-proteins, and phospholipids that mediate phosphatidic acid accumulation. All these mechanisms will be discussed in the following sections.

13.3 Achieving Specificity in ROS Signaling

ROS are relatively simple molecules, thus some unanswered questions are how: (a) ROS signals generated in a compartment or a particular cell are specific for a determined stimulus; (b) ROS increases act as specific signals to trigger an appropriate acclimation response; (c) ROS signal generated in a group of cells can be transferred to the entire plant and still being stress specific. Mittler et al. (2011) exposed several explanations: the first is that ROS activate the cellular signaling network of cells and together with ROS achieve the required specificity. These signals could be small peptides, hormones, lipids, cell wall fragments, and others. The second is that the ROS signal has specific oscillation patterns carrying a decoded message similar to calcium signals. Then, the different amplitude, frequency, and/or localization of the signal could be perceived and decoded to trigger specific gene expression patterns. A third possibility is that each cellular compartment or individual cell has its own set(s) of ROS receptors to decode ROS signals generated within it, which are then transferred by other networks such as calcium and/or protein phosphorylation. Finally, a combination of the different mechanisms described above is probable.

The first possibility described by Mittler et al. (2011) is in accordance with the mechanism proposed by Möller and Sweetlove (2010). The authors reported that oxidatively damaged proteins can originate oxidized peptides that can behave as specific and selective secondary ROS messengers to the nucleus. Evidences that support these statements are that mitochondria and other plant cell compartments contain oxidized proteins. Some of these oxidations are irreversible and can occur on the side-chain of Pro, His, Arg, Lys, and Thr-producing ketone or aldehyde derivatives (protein carbonyls) (Möller et al. 2007; Sweetlove and Møller 2009). Carbonylated proteins were found in chloroplasts, peroxisomes, and in a greater extent in mitochondria of wheat (*Triticum aestivum*) leaves during drought stress (Bartoli et al. 2004). Another aspect that contributes to achieve specificity is that depending on the ROS species their reactivity and capacity of generation of protein modifications will be different (Halliwell and Gutteridge 2007). Thus, oxidized peptides acting as secondary ROS messengers can regulate gene expression in a compartment, sub-compartment and ROS-species-specific manner. Proteolytically produced peptides can function as retrograde signals to coordinate mitochondrial and nuclear gene expression (Koppen and Langer 2007; Tatsuta 2009). One of the protein modifications triggered by ROS accumulation is the reversible redox modulation of Cys residues, and this modification can trigger the formation of disulfides with other protein thiol groups or soluble thiols such as glutathione, as well as production of more oxidized sulfur states (sulfenic, sulfinic, and sulfonic groups). Thiol modification is a canonical mechanism of oxidative signaling, for instance, in the bacterial oxyR and the yeast Yap1 systems (Delaunay et al. 2002). Other mechanisms are glutathionylation and the oxidation of methionine residues to sulfoxide forms (MetSO) that can be reversed by peptide methionine sulfoxide reductases (PMSR). Besides, different forms of protein oxidative damage can occur by reaction with lipid peroxidation products, or by conjugation with sugars (glycation) or their oxidation products (glycoxidation). Highly oxidized proteins, found in all cellular compartments, are generally assumed to lose their catalytic activity (Møller and Kristensen 2004; Davletova et al. 2005a). Although "damage" may be a useful term for oxidation-induced loss of function at the protein level, it is misleading when applied at levels of greater complexity (e.g., whole cells, tissues, or organisms). In accordance with this statement, the development and response to environmental stresses of *A. thaliana* and maize is controlled by mechanisms of protein oxidation (Johansson et al. 2004; Kurepa et al. 2008). Mano (2012) described other reactive species with signaling functions as reactive carbonyl species (RCS) which designates the α,β-unsaturated aldehydes and ketones that are derived from lipid peroxides (LOOH). In HepG2 cells, the activity of transcription factor Nrf2 (NF-E2-related factor 2) is dependent on its redox-sensitive inhibitor Keap1 (kelch-like ECH-associated protein 1). Under non-stressful conditions, the transcription factor Nrf2 is bound to Keap1. The formation of this complex triggers the ubiquitination of Nrf2 and subsequent proteasomal degradation (Botzen and Grune 2007). Under stress conditions, RCS levels rise and the exposure of Keap1 changes its conformation and becomes unable to bind Nrf2, resulting in the increased Nrf2 concentration. Further Nrf2 migrates into the nucleus and up-regulates the transcription of genes encoding antioxidant enzymes and other defensive proteins (Farmer and Davoine 2007; Kaspar et al. 2009; Kaspar and Jaiswal 2010). Interestingly, Nrf2 homologues were not found in the *A. thaliana* genome, although some genes show similarity to Keap1 and are considered to be involved in RCS signaling (Farmer and Davoine 2007). These data reveal that another actor in the redox regulatory network may be RCS. More evidences pointing in the direction of RCS as mediators in redox signaling are the enzymatic control of RCS production/elimination and the RCS ability to cross biological membranes and diffuse for relatively long distances. Also, recent studies show that RCS activate specific receptors (Forman et al. 2008; Yadav and Ramana 2013). The degradation and resynthesis of RCS-modified proteins contribute in the reversible aspect of RCS signaling (Forman et al. 2008). On the other hand, sulfur-containing molecules, particularly where the sulfur atoms are at higher oxidation states, can behave as reactive sulfur species or RSS (Giles and Jacob 2002). Hydrolysed nitrosothiols give rise to sulfenic acids which have important biological activity in signal transduction and metabolic regulation. RCS have the potential to be signaling molecules due to: (1) their level is tightly controlled in vivo; (2) they are sufficiently stable, small, and hydrophobic to diffuse across biological membranes; (3) they bind to specific receptors, triggering a chain of events within the cell; (4) the signaling effects triggered are reversible Semchyshyn (2014). A similar hypothesis could be described for RNS and RSS.

Compartmentation is another way to achieving specificity in redox signaling (Foyer and Noctor 2015). The different compartments of the cell are not all equally buffered; even inside organelles the distinct areas have different antioxidant capacities, as is the case with the highly reducing stroma in the chloroplasts versus the limited antioxidant capacity of the lumen. On the other hand, the cytosol has greater redox stability being important for signaling involving the gene expression because the cytosolic redox state directly influences that of the nucleus. In the opposite side, the apoplast is where the oxidative burst and many oxidant-requiring reactions occur. This is the site of production of hydroxyl radicals and other strong oxidants and antioxidants, such as SOD, ASC, and GSH are in low levels (Pignocchi and Foyer 2003). In this compartment, the presence of ascorbate oxidase suggests that ascorbate is in the oxidative state and glutathione is thought to be degraded here (Ohkamu-Ohtsu et al. 2007; Parsons and Fry 2012). The vacuole has similar redox characteristics to the apoplast, where the reduced states of antioxidants are in low levels and reductant enzymes are not present at quantifiable amounts according to the available information. Moreover, the vacuole accumulates GSSG probably from the cytosolic pool to diminish the excessive oxidation from the cytosol during stress (Noctor et al. 2013).

Understanding the mechanisms by which plants achieve the specificity in signals derived from ROS is of fundamental importance to describe the processes underlying stress acclimation responses and hence develop more stress-tolerant crops to improve production efficiency in a growing world population.

13.4 Interaction of ROS Signaling with Other Signaling Pathways

Signal transduction pathways are the link between the sensing mechanism and the genetic response. Among the intermediates in the process that links ROS with the gene expression that will end with the plant-adaptive response, there are molecules such as protein kinases, calcium, and hormones that traduce and amplify the signal.

13.4.1 Protein Kinases

Eukaryotic mitogen-activated protein kinase (MAPK) cascades transduce environmental and developmental cues into intracellular responses. In a general model, stimulated plasma membrane receptors activate MAP kinase kinase kinases (MAP3Ks; also called MAPKKKs or MEKKs) or MAP kinase kinase kinase kinases (MAP4Ks). Sequential phosphorylations ensue as MAP3Ks activate downstream MAP kinase kinases (MAP2Ks; also called MKKs or MEKs) that in turn activate MAPKs. Then, MAPKs target effector proteins, which include other kinases, enzymes, or transcription factors. The deactivation and regulation of MAPK activity are mediated by tyrosine and serine-/threonine-specific phosphatases (Suarez Rodriguez et al. 2010). It is well known that MAPKs mediate drought responses being ABA and ROS intermediates in the signaling pathway. In alfalfa, p44MKK4 (MAP kinase kinase) gene expression and kinase activity are activated under drought conditions in an ABA-independent manner (Jonak et al. 1996). The expression of AtMEKK1 and AtMPK3 in *A. thaliana* can be induced by drought (Mizoguchi et al. 1996). Under the same conditions, OsMSRMK2 and OsMAPK5 were activated in rice plants (Xiong and Yang 2003). In addition, the expression patterns of MaMAPK and ZmMPK3 suggested that activities of MAPKs are molecular mechanisms of drought tolerance in Malus and maize (Peng et al. 2006; Wang et al. 2010). The putative rice MAPKKK gene (DSM1) overexpressed in rice increased the tolerance to dehydration stress regulating early responses to drought stress by scavenging ROS (Ning et al. 2010). Moreover, some MAPKs constitute hubs for biotic and abiotic stress signaling as the cotton MAPK GhMPK16 which is functionally involved in pathogen resistance, drought tolerance, and ROS accumulation (Shi et al. 2011). In *A. thaliana*, exogenous H_2O_2 can activate MPK1 and MPK2 (Ortiz-Masia et al. 2007), MPK3 and MPK6 (Kovtun et al. 2000), MPK4 (Nakagami et al. 2006), and MPK7 (Doczi et al. 2007). The overexpression of the Nicotiana H_2O_2-dependent ANP1 (a homologue of NPK1) enhanced abiotic stress tolerance in transgenic tobacco (Kovtun et al. 2000) and maize (Shou et al. 2004a, b). Similarly, the overexpression of DSM1, a Raf-Like MAPKKK, increased tolerance to dehydration and oxidative stress at the seedling stage in rice (Ning et al. 2010). Transgenic tobacco overexpressing ZmMPK7, a maize MAPK gene, showed improved protection by peroxidases during osmotic stress (Zong et al. 2009). Taking together the literature reveals that a co-regulation and interaction of the MAP kinase pathway and ROS signaling within the cellular signaling framework exist. Thus, the understanding of MAP kinase as a hub in signaling under environmental adversity increased.

13.4.2 *Calcium*

An important second messenger released in response to a variety of biotic and abiotic stresses and mediator of stress-response reactions as well as of developmental processes is calcium (Ca^{2+}) (Steinhorst and Kudla 2013). Ca^{2+} has diverse functions in plants and its accumulation patterns are characterized by specific signatures that determine temporal and spatial features that can vary in terms of amplitude, frequency, and duration encoding information in response to particular stimuli (Dodd et al. 2010; Kudla et al. 2010). Ca^{2+} functions in concert with other important second messengers, being ROS one of the most significant. Taking into account, the generation of ROS in a controlled manner by NADPH oxidases and recent findings that point to connections between ROS and Ca^{2+} signaling pathways, a cell-to-cell communication and thereby long-distance transmission of signals in plants is suggested (Steinhorst and Kudla 2013). In this regard, a burst of ROS production mediated by RBOH proteins is initiated in plant cells in response to many different abiotic stresses. This ROS production is conducted by neighboring cells initiating a long-distance signal termed the ROS wave (Mittler et al. 2011; Gilroy et al. 2014). This process is integrated with the Ca^{2+} wave through the action of vacuolar ion channel two pore channel 1 (TPC1), a cation permeable channel, implicated in calcium-induced calcium release, which phosphorylates and activates the ROS-producing RBOHD by the Ca^{2+}-dependent protein kinase (CPK). This triggers the activation of SAA in plants exposed to local stimuli such as high light or heat (Suzuki et al. 2013). Convincing evidence for this model with regard to pathogen responses exist, however, is a challenge to investigate the overall significance of the Ca^{2+}-dependent phosphoregulation of ROS-producing NADPH oxidases for other physiological processes (Steinhorst and Kudla 2013). One example was described by Miller et al. (2009), in *A. thaliana* where local heat stress application induced systemic heat stress responses through ROS waves generated by the NADPH oxidase RBOHD activated by Ca^{2+}-regulated kinases. This mechanism was associated with the finding that cyclic nucleotide-gated ion channels (CNGCs) can facilitate Ca^{2+} influx in response to heat stress (Gao et al. 2012; Tunc-Ozdemir et al. 2013). Therefore, after exposition to local heat stress, it was suggested that CNGCs are responsible for Ca^{2+} influx and enable signal propagation through the activation of RBOHD, resulting in systemic heat stress tolerance. Another example of abiotic stress-mediated induction of calcium waves through NADPH oxidase activity was described in root tips of *A. thaliana* exposed to salt stress which resulted in the initiation of a Ca^{2+} wave that traveled and spread through the roots to the aerial parts of the plant at 2.4 cm/min (400 μm/s). An interesting finding was the fact that Ca^{2+} wave propagated through the root cortex and endodermis layers, demonstrating cell-type specificity. In addition, it was dependent on the function of the vacuolar ion channel TPC1 (Choi et al. 2014). Taking together the background information that reveals the increasing evidence of systemic acquired resistance to abiotic stresses, it may be suggested that similar mechanisms could be described in plants exposed to drought stress in the nearest future.

13.4.3 Hormones

Increasing evidence supports the concept that ROS are essential second messengers in hormone signaling that coordinately regulate plant development and stress tolerance (Xia et al. 2015). Hormones regulate plant development and stress tolerance through the hormone-dependent activation of ROS production, often through the activation of NADPH oxidases, which are encoded by RBOH genes in plant genomes (Sagi and Fluhr 2006). Particularly, in this section, the mode of action of the hormone ABA, which is essential in drought stress response, will be focused. An approach to examine the potential influence of ROS in the drought response was investigated by Noctor et al. (2014) who analyzed how many of the drought-induced genes were induced by different oxidative stress conditions. The authors extracted the responses of the drought-induced genes to different oxidative stresses from Genevestigator (Hruz et al. 2008). Data were available for externally supplied H_2O_2 (Davletova et al. 2005b), paraquat (which mainly stimulates light-dependent production of superoxide and H_2O_2 in the chloroplast), the flu mutant (excess singlet oxygen production in the chloroplast; Laloi et al. 2007), and the photorespiratory *cat2* mutant (excess H_2O_2 in the peroxisomes; Queval et al. 2012). The authors revealed that of the drought-associated genes that were induced by these three oxidative stresses, 57–72 % were also induced by ABA. These percentages were higher than the overall proportion of the drought-induced genes that were induced by ABA (173 of 375; 46 %). This was consistent with a close relationship between oxidative stress and ABA in ROS-dependent drought responses. ABA is a stress hormone that plays a general role in developmental processes as well as being a key regulator of plant responses to abiotic stresses (Xiong et al. 2002). In addition to the roles in ABA-induced stomatal closure, ROS production is also critical for ABA-mediated stress tolerance of seedlings. Drought stress or ABA treatment enhance ROS accumulation in maize, together with increased expression of genes encoding antioxidant enzymes and their activities (Jiang and Zhang 2002). Additionally, ROS scavengers block ABA-induced increases in antioxidant activities (Zhang et al. 2006). Besides, NADPH oxidase plays a role in ABA-induced ROS accumulation (Jiang and Zhang 2003; Zhang et al. 2006). In tomato, ABA induces ROS accumulation in chloroplasts; however, the DPI-dependent inhibition of NADPH oxidase or silencing of RBOH1 partially blocked ABA-induced ROS accumulation and associated increases in antioxidant enzymes (Zhou et al. 2014). In this way, Furlan et al. (2014) demonstrated the accumulation of AREB1 transcripts during the stress period, which was associated with ABA accumulation in peanut plants as described in previous studies (Furlan et al. 2012). Increased transcript levels of AREB1 in stressed peanut were also described by Hong et al. (2013). Interestingly, following rehydration, AREB1 transcript levels were lower than in control conditions. The transcript AREB1 can be down-regulated as consequence of declining ABA accumulation. In peanut, drought stress initiated ABA accumulation, which may trigger H_2O_2 production as demonstrated in Furlan et al. (2013). Closely associated with this finding, the up-regulation of ferritin, a protein involved in Fe sequestration, could be instrumental to prevent the formation of hydroxyl radicals in the presence of H_2O_2 under this stress condition (op den Camp et al. 2003). The marker transcript Fer1 is activated in response to excess light and is associated with oxidative stress (Oelze et al. 2012). In peanut nodules, increased expression of Fer1 was correlated with H_2O_2 accumulation, suggesting, for the time being, that Fer1 is a suitable marker of oxidative stress in this plant system as well. The protein RPR-10, which is homologous to RPR-10 from *Retama raetam*, is strongly expressed in peanut plants under drought stress (Pnueli et al. 2002; Luo et al. 2005). It has been suggested that RPR-10 acts as a dehydrin or chaperone similar to small heat-shock proteins based on the high number of polar residues per total number of side chains found in PR-10 (~40%) compared with dehydrin (~50%) (Pnueli et al. 2002). Structural or regulatory genes that are expressed in a stimulus-specific manner were used as marker transcripts to assess the possible involvement of signaling pathways (Fig. 13.1). Accepting RPR-10, AREB1, and Fer1 as powerful stress markers allowed describing efficient recovery after relief of the water deficit: accumulation of each transcript was reversed after rehydration (Furlan et al. 2014).

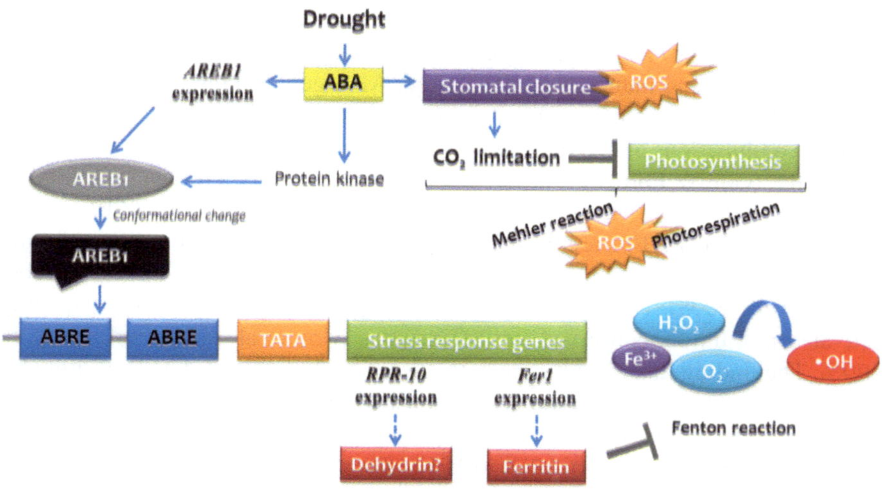

Fig. 13.1 Model of response of peanut plants to drought stress. ABA regulates the expression of many genes, the products of which may function in dehydration tolerance. ABA activates a protein kinase which, in turn, phosphorylates Ser–Thr residues sites in conserved regions of AREB1. The ABA-dependent multisite phosphorylation of AREB1 regulates its own activation and induces the transcription of stress-response genes (Furihata et al. 2006). In peanut, such genes are *RPR-10*, a hypothetical dehydrin, and *Fer1* which codes a ferritin, a protein that sequestrates Fe and diminishes Fenton reactions. ABA also promotes the stomatal closure in guard cells, a process mediated by ROS accumulation, and limits CO_2 availability decreasing photosynthesis and increasing ROS production. → Indicate positive interactions; ⊣ indicate inhibitory interactions

13.4.4 An Example of Interaction of ROS Signaling with Other Signaling Pathways

An interesting example that reveals the interaction between hormones, protein phosphatases, and ROS is the perception and transduction of ABA during drought stress. Current information indicates that the earliest events of ABA signal transduction occur via a module made up of proteins belonging to three protein classes: Pyracbactin Resistance/Pyracbactin resistance-like/Regulatory Component of ABA Receptor (PYR/PYL/RCARs) proposed to be the ABA receptors, Protein Phosphatase 2Cs (PP2Cs) which act as negative regulators, and SNF1-related protein kinase 2 s (SnRKs) which are positive regulators (Park et al. 2009; Umezawa et al. 2009). In the presence of ABA, the PYR/PYL/RCAR-PP2C complex formation inhibits PP2C activity (Fujii et al. 2009; Park et al. 2009; Santiago et al. 2009), thus activating SnRKs which target membrane proteins, ion channels, transcription factors, and trigger transcription of ABA-responsive genes (Sheard and Zheng 2009; Soon et al. 2012). The PYR/PYL/RCAR ABA receptors are a soluble ABA receptor family while PP2Cs are negative regulators of ABA signaling. ABA binding to PYR/PYL/RCARs induces a conformational change that exposes the interaction surface allowing for favorable binding of some PP2Cs (Cutler et al. 2010).

As positive regulators of ABA signaling are a group of plant-specific Ser/Thr kinases, SnRK2s, which phosphorylate basic-domain leucine zipper (bZIPs) transcription factors and induce gene expression. ABA gene expression requires multiple cis-elements (also called ABA-responsive elements; ABREs—PyACGTGG/TC), or combinations of an ABRE with a coupling element such as CE1, CE3, and DRE/CRT (Gomez-Porras et al. 2007; Zhang et al. 2005). Proteins that bind to ABRE are called ABRE-binding (AREB) or ABRE-binding factors (ABFs) (Uno et al. 2000). AREB/ABFs are transcription factors members of the bZIP subfamily with 13 members in *A. thaliana* (Yamaguchi-Shinozaki and Shinozaki 2006; Correa et al. 2008). AREB/ABF proteins are phosphorylated and consequently activated in multiple conserved RxxS/T regions (Uno et al. 2000; Furihata et al. 2006). The finding that MAPKs mediate ABA signaling in guard cells of drought-stressed plants emerged after the convergence of the knowledge that ROS mediate ABA signaling in guard cells (Zhang and Klessig 2001; Jammes et al. 2009) and components of the ABA-activated MAPKs are also activated by ROS (Desikan et al. 2004). Considering all the available information, the proposed model of action of ABA is that following ABA perception in guard cells, active SnRK2 kinases such as OST1 (released from inhibition by PYR/PYL/RCAR-mediated sequestration of PP2Cs) phosphorylate the NADPH oxidase RbohF, leading to ROS accumulation. ROS activate two MAPKs, MPK9 and MPK12, which function positively to regulate ABA-mediated stomatal closure. An intriguing aspect is that considering the family of NADPH oxidases, RBOHD mediates only pathogen and RBOHF only ABA signals (Mersmann et al. 2010). It is noteworthy that guard cells can differentiate between ROS generated by the different RBOHs and couple to distinct MAPK pathways; however, the signaling pathways are presently unclear. Additional ABA signal perception and transduction modules were proposed as potential ABA receptors including the G-protein-coupled receptor (GCPR)-type G proteins (GTG1 and GTG2) and ABA-binding protein (ABAR)/Mg-chelatase H subunit (CHLH)/Genomes uncoupled 5 (GUN5), but more research will be required to clarify these mechanisms.

13.5 Transcription Factors

As explained before, ABA is the hormone of stress and is responsible for the adaptive response of plants. Based on results of studies that showed that exogenous ABA triggers the expression of transcripts related to drought stress tolerance, an ABA-dependent pathway was described. This ABA-dependent signaling system is composed of (1) AREB/ABF (ABA-responsive element-binding protein/ABA-binding factor); and (2) MYC/MYB (Busk and Pagès 1998). However, some genes are not dependent on ABA, since some mutants lacking the ABA signal expressed the transcripts coding for genes involved in the stress response. In this case, the ABA-independent genes are: (1) the CBF/DREB (cold-binding factor/dehydration-responsive element binding); and (2) the NAC and ZF-HD (zinc-finger homeodomain) (Saibo et al. 2009).

In the ABA-dependent pathway, the bZIP transcription factors ABRE-binding protein (AREB)/ABRE-binding factor (ABF))can bind to ABRE and activate ABA-dependent gene expression (Uno et al. 2000). The activation of the AREB/ABF proteins has been shown to require an ABA-mediated signal, which is ABA-dependent phosphorylation (Furihata et al. 2006). Cloning and transgenic analysis of a DREB1-related transcription factor, CBF4 in *A. thaliana*, showed that regulation of DRE elements is also mediated by an ABA-dependent pathway. Genes of the CBF/DREB1 family are mainly induced by cold stress, but the drought-inducible gene CBF4 functions to provide crosstalk between DREB2 and CBF/DREB1 regulatory systems. CBF4 gene expression is up-regulated by drought and ABA, but not by cold stress. Other important transcriptional regulators, such as the MYC and MYB proteins, are known to function as activators in one of the ABA-dependent regulatory systems (Abe et al. 2003; Valliyodan and Nguyen 2006).

In the ABA-independent pathway, the dehydration-responsive element (DRE), a 9-bp conserved sequence, TACCGACAT, is an essential cis-element component of the promoter regions of drought- and cold-inducible genes (Yamaguchi-Shinozaki and Shinozaki 2006). Transcription factors from the ethylene-responsive factor (ERF)/APETALA2 (AP2) family that bind to these DRE/C-Repeats (CRT) elements were isolated and termed C-Repeat binding factors (CBF)/dehydration responsive element-binding factors (DREB) and their conserved DNA-binding motif is A/GCCGAC. Two identified DREB proteins namely DREB1 and DREB2 are involved in two separate signal transduction pathways under low temperature and dehydration, respectively. Despite this, in the *A. thaliana* genome, at least six DREB2 homologues other than DREB2A and DREB2B were described. In this organism, among the eight DREB2-type proteins, DREB2A and DREB2B are thought to be major transcription factors that function under drought and high-salinity stress conditions (Sakuma et al. 2002). For its activation, the DREB2A protein requires posttranslational modification, such as phosphorylation (Sakuma et al. 2006). The DREB2A transcriptional activation domain is between residues 254 and 335 in *A. thaliana* protoplasts, and deletion of a region between residues 136 and 165 transforms DREB2A to a constitutive active form. It is noteworthy that both DREB2A and DREB1A can bind to the DRE sequence, but the DNA-binding specificities of each to the neighboring sequences of the DRE core motif were slightly different; therefore, the downstream genes of each are partially different. The available information reveals that the stability of the DREB2A protein is important for its activation, and the activated DREB2A regulates drought stress-responsive gene expression, which enhances drought stress tolerance in plants (Sakuma et al. 2006).

The NAC and ZF-HD (zinc-finger homeodomain) is another family of transcription factors involved in plant developmental processes and stress responses to abiotic factors (Saibo et al. 2009). This family contains a highly conserved N-terminal DNA-binding domain and a diversified C-terminal domain (Hu et al. 2008). The names of the first three described TFs containing NAC domain, namely NAM (no apical meristem), ATAF1-2, and CUC2 (cup-shaped cotyledon), are the origin of the name of the family (NAC) (Aida et al. 1997). The cis-element of NAC TF [NAC-recognized sequence (NACRS)] was also identified in *A. thaliana* (Tran et al. 2004). NAC TFs increased the tolerance to drought stress in several species such as *A. thaliana*, *Oryza sativa*, and *Setaria italica* (Tran et al. 2007; Zheng et al. 2009; Puranik et al. 2011).

Among the different families of Zn-finger TFs, the C_2H_2-type was described as involved in several stress signaling pathways (Huang et al. 2007; Xu et al. 2008). Another class of Zn finger TFs that has been implicated in abiotic stress signaling is the ZF-HD, which is characterized by the presence of Zn-finger-like motifs upstream of a homeodomain (Windhovel et al. 2001). Interestingly, the co-expression of the stress-inducible ZFHD1 and NAC transcription factors enhances the early responsive to dehydration 1 (ERD1) gene expression in *A. thaliana* (Tran et al. 2007). Thus, revealing that these transcription factors can modulate the stress responses by interacting with other TFs. Figueiredo et al. (2012) revealed that seven novel TFs bind to the promoter and interplay to repress the expression of OsDREB1B, modulating the response to different abiotic stresses. Together with previous reports, the authors suggested that Zn-finger TFs may be a pivotal component in the regulation of DREB1/CBF genes in plants.

Because plants are exposed to complex environment, it is expected to find more signaling components involved in plant responses to abiotic stresses. A candidate is the heptahelical protein 1 (HHP1), a negative regulator in stresses. HHP1 is a negative regulator in ABA and osmotic signaling and is suggested to be a novel signaling component in the cross-talk between cold and osmotic signaling pathways in *A. thaliana* (Chen et al. 2010).

13.6 Antioxidant Systems as Redox Signal Transmitters

Plants possess an enzymatic and nonenzymatic antioxidant defense system that allow maintaining ROS in a low quantity and protect cells from oxidative damage. The subcellular localization and biochemical properties of antioxidant enzymes, their induction at the enzyme and gene expression level and the associated nonenzymatic scavengers make the antioxidant system an efficient mechanism to control ROS accumulation temporally and spatially (Shao et al 2008). As mentioned in previous sections ROS are important molecules functioning as signals acting in stress conditions, a process that requires the presence of redox-sensitive proteins that can be reversibly oxidized/reduced depending upon the cellular redox state (Shao et al. 2007). These redox-sensitive proteins can be oxidized by ROS directly or indirectly via nonenzymatic compounds, such as GSH or thioredoxins, which are major players in redox signaling when antioxidants are involved (Shao et al. 2008). Noctor et al. (2014) suggest that thiol-based enzymes may have antioxidative and signaling functions through changes in glutathione or TRX redox potentials with repercussions for sensitive target proteins, or by structural changes in the enzymes themselves and knock-on effects on their partners. Although the main role of heme-based antioxidative enzymes may be to antagonize ROS signaling by decreasing ROS concentrations, the enzyme increase may itself be an integral part of the signals that are generated (Tripathi et al. 2009). Examples in the literature revealed that an *A. thaliana* GPX can couple H_2O_2 reduction to oxidation of a transcription factor to allow the oxidation of components involved in ABA signaling during drought (Miao et al. 2006). Moreover, forward genetics suggests that CAT is an ROS-dependent activator of signals involved in cell death or autophagy, possibly through a secondary CAT-peroxidase reaction (Juul et al. 2010; Hackenberg et al. 2013). Furlan (2014) reported that in peanut plants exposed to drought stress and rehydration, the content of the thiol GSH did not show changes. However, the content of the oxidized form was reduced, although the effect may not have biological significance, since it contributed to only 1–3% of the total GSH. The lack in GSSG accumulation has been reported in other species experiencing drought which contrasted with other stresses where glutathione oxidation was evident (Noctor et al. 2014). A possible explanation may be that changes in thiol status were involved in redox signaling pathways and consequently, the activity of thiol-dependent antioxidant enzymes may have an indirect role in signaling (Noctor et al. 2014). Besides, in peanut, increased antioxidant enzyme activity revealed by transcript accumulation of genes coding for CuZn-SOD, GR, and GST and the specific activity of APX, CAT, GR, and GPX (Furlan et al. 2014) may be linked to ABA accumulation, since the antioxidant system was induced by external application of ABA as described by several authors (Zhou et al. 2005; Bright et al. 2006; Zhang et al. 2007; Lu et al. 2009). Further research is necessary to reveal the interaction of thiols and antioxidant enzymes with ABA in signaling responses to drought stress.

Acknowledgements The authors thank Secretaría de Ciencia y Técnica, Universidad Nacional de Río Cuarto for providing financial assistance for this research. A. Furlan has a postdoctoral fellowship from CONICET. E. Bianucci is a member of the research career from CONICET. S. Castro is titular professor in Universidad Nacional de Río Cuarto.

References

Abe H, Urao T, Ito T, Seki M, Shinozaki K, Yamaguchi-Shinozaki K. *Arabidopsis* AtMYC2 (bHLH) and At- MYB2 (MYB) function as transcriptional activators in abscisic acid signaling. Plant Cell. 2003;15:63–78.

Aida M, Ishida T, Fukaki H, Fujisawa H, Tasaka M. Genes involved in organ separation in *Arabidopsis*: an analysis of the cup-shaped cotyledon mutant. Plant Cell. 1997;9:841–57.

Bartoli CG, Gómez F, Martínez DE, Guiamet JJ. Mitochondria are the main target for oxidative damage in leaves of wheat (*Triticum aestivum* L.). J Exp Bot. 2004;55:1663–9.

Baxter A, Mittler R, Suzuki N. ROS as key players in plant stress signaling. J Exp Bot. 2014;65(5):1229–40. doi:10.1093/jxb/ert375.

Benina M, Ribeiro DM, Gechev TS, Mueller-Rober B, Schippers JHM. A cell type- specific view on the translation of mRNAs from ROS-responsive genes upon paraquat treatment of *Arabidopsis thaliana* leaves. Plant Cell Environ. 2015;38:349–63.

Botzen D, Grune T. Degradation of HNE-modified proteins—possible role of ubiquitin. Redox Rep. 2007;12:63–7.

Bright J, Desikan R, Tancock JT, Weir IS, Neill SJ. ABA-induced NO generation and stomatal closure in Arabidopsis are dependent on H_2O_2 synthesis. Plant J. 2006;45:113–22.

Busk PK, Pagès M. Regulation of abscisic acid-induced transcription. Plant Mol Biol. 1998;37: 425–35.

Chen CC, Liang CS, Kao AL, Yang CC. HHP1, a novel signalling component in the cross-talk between the cold and osmotic signalling pathways in *Arabidopsis*. J Exp Bot. 2010;61(12): 3305–20.

Choi WG, Toyota M, Kim SH, Hilleary R, Gilroy S. Salt stress-induced Ca^{2+} waves are associated with rapid, long-distance root-to-shoot signaling in plants. Proc Natl Acad Sci U S A. 2014;111: 6497–502.

Colville L, Kranner I. Desiccation tolerant plants as model systems to study redox regulation of protein thiols. Plant Growth Regul. 2010;62:241–55.

Correa LG, Riano-Pachon DM, Schrago CG, dos Santos RV, Mueller-Roeber B, Vincentz M. The role of bZIP transcription factors in green plant evolution: adaptive features emerging from four founder genes. PLoS One. 2008;3, e2944.

Cruz de Carvalho MH. Drought stress and reactive oxygen species. Production, scavenging and signaling. Plant Signal Behav. 2008;3:156–65.

Cutler SR, Rodriguez PL, Finkelstein RR, Abrams SR. Abscisic acid: emergence of a core signaling network. Annu Rev Plant Biol. 2010;61:651–79.

Davletova S, Rizhsky L, Liang H, Shengqiang Z, Oliver DJ, Coutu J, Shulaev V, Schlauch K, Mittler R. Cytosolic ascorbate peroxidase 1 is a central component of the reactive oxygen gene network of *Arabidopsis*. Plant Cell. 2005a;17:268–81.

Davletova S, Schlauch K, Coutu J, Mittler R. The zinc-finger protein Zat12 plays a central role in reactive oxygen and abiotic stress signaling in *Arabidopsis*. Plant Physiol. 2005b;139: 847–56.

Delaunay A, Pflieger D, Barrault MB, Vinh J, Toledano MB. A thiol peroxidase is an H_2O_2 receptor and redox-transducer in gene activation. Cell. 2002;111:471–81.

Desikan R, Cheung MK, Bright J, Henson D, Hancock JT, Neill SJ. ABA, hydrogen peroxide and nitric oxide signalling in stomatal guard cells. J Exp Bot. 2004;55:205–12.

Després C, Chubak C, Rochon A, Clark R, Bethune T, Desveaux D, Fobert PR. The *Arabidopsis* NPR1 disease resistance protein is a novel cofactor that confers redox regulation of DNA binding activity to the basis domain/leucine zipper transcription factor TGA1. Plant Cell. 2003;15:2181–91.

Dietz KJ. Redox signal integration: from stimulus to networks and genes. Physiol Plant. 2008;133: 459–68.

Dietz KJ. Redox regulation of transcription factors in plant stress acclimation and development. Antioxid Redox Signal. 2014;21(9):1356–72.

Doczi R, Brader G, Pettko-Szandtner A, Rajh I, Djamei A, Pitzschke A, Teige M, Hirt H. The *Arabidopsis* mitogen-activated protein kinase kinase MKK3 is upstream of group C mitogen-activated protein kinases and participates in pathogen signaling. Plant Cell. 2007;19:3266–79.

Dodd AN, Kudla J, Sanders D. The language of calcium signaling. Annu Rev Plant Biol. 2010;61: 593–620.

Drerup MM, Schlucking K, Hashimoto K, Manishankar P, Steinhorst L, Kuchitsu K, Kudla J. The calcineurin B-like calcium sensors CBL1 and CBL9 together with their interacting protein kinase CIPK26 regulate the *Arabidopsis* NADPH oxidase RBOHF. Mol Plant. 2013;6: 559–69.

Dubiella U, Seybold H, Durian G, Komander E, Lassig R, Witte CP, Schulze WX, Romeis T. Calcium-dependent protein kinase/NADPH oxidase activation circuit is required for rapid defense signal propagation. Proc Natl Acad Sci U S A. 2013;110:8744–9.

Farmer EE, Davoine C. Reactive electrophile species. Curr Opin Plant Biol. 2007;10:380–6.

Figueiredo DD, Barros PM, Cordeiro AM, Serra TS, Lourenc T, Chander S, Oliveira MM, Saibo NJM. Seven zinc-finger transcription factors are novel regulators of the stress responsive gene *OsDREB1B*. J Exp Bot. 2012;63(10):3643–56.

Forman HJ, Fukuto JM, Miller T, Zhang H, Rinna A, Levy S. The chemistry of cell signaling by reactive oxygen and nitrogen species and 4-hydroxynonenal. Arch Biochem Biophys. 2008;477(2):183–95.

Foyer CH, Noctor G. Redox regulation in photosynthetic organisms: signaling, acclimation, and practical implications. Antioxid Redox Signal. 2009;11(4):861–905.

Foyer CH, Noctor G. Stress-triggered redox signalling: what's in pROSpect? Plant Cell Environ. 2015. doi:10.1111/pce.12621.

Fujii H, Chinnusamy V, Rodrigues A, Rubio S, Antoni R, Park S-Y, Cutler SR, Sheen J, Rodriguez PL, Zhu JK. *In vitro* reconstitution of an abscisic acid signalling pathway. Nature. 2009;462:660–4.

Furihata T, Maruyama K, Fujita Y, Umezawa T, Yoshida R, Shinozaki K, Yamaguchi-Shinozaki K. Abscisic acid-dependent multisite phosphorylation regulates the activity of a transcription activator AREB1. Proc Natl Acad Sci U S A. 2006;103:1988–93.

Furlan A, Llanes A, Luna V, Castro S. Physiological and biochemical responses to drought stress and subsequent rehydration in the symbiotic association peanut—*Bradyrhizobium* sp. ISRN Agron. 2012;318083. doi:10.5402/2012/318083.

Furlan A, Llanes A, Luna V, Castro S. Abscisic acid mediation in hydrogen peroxide production in peanut under water stress. Biol Planta. 2013;57(3):555–8.

Furlan AL, Bianucci E, Tordable MC, Castro S, Dietz KJ. Antioxidant enzyme activities and gene expression patterns in peanut nodules during a drought and rehydration cycle. Funct Plant Biol. 2014;41:704–13.

Furlan A. Doctoral Thesis. Antioxidant role in the defense of peanut plants to abiotic stress. Córdoba, Argentina: Universidad Nacional de Río Cuarto; 2014.

Gao F, Han X, Wu J, Zheng S, Shang Z, Sun D, Zhou R, Li B. A heat-activated calcium-permeable channel -*Arabidopsis* cyclic nucleotide-gated ion channel 6- is involved in heat shock responses. Plant J. 2012;70:1056–69.

Giles GI, Jacob C. Reactive sulfur species: an emerging concept in oxidative stress. J Biol Chem. 2002;383:375–88.

Gilroy S, Suzuki N, Miller G, Choi WG, Toyota M, Devireddy AR, Mittler R. A tidal wave of signals: calcium and ROS at the forefront of rapid systemic signaling. Trends Plant Sci. 2014;19:623–30.

Gomez-Porras JL, Riano-Pachon DM, Dreyer I, Mayer JE, Mueller-Roeber B. Genome-wide analysis of ABA-responsive elements ABRE and CE3 reveals divergent patterns in *Arabidopsis* and rice. BMC Genomics. 2007;8:260.

Gordon MJ, Carmody M, Albrecht V, Pogson B. Systemic and local responses to repeated HL stress-induced retrograde signaling in *Arabidopsis*. Front Plant Sci. 2012;3:303.

Gutscher M, Sobotta MC, Wabnitz GH, Ballikaya S, Meyer AJ, Samstag Y, Dick TP. Proximity-based protein thiol oxidation by H_2O_2-scavenging peroxidases. J Biol Chem. 2009;284: 31532–40.

Hackenberg T, Juul T, Auzina A, Gwizdz S, Malolepszy A, Van Der Kelen K, Dam S, Bressendorff S, Lorentzen A, Roepstorff P, Nielsenf KL, Jørgensena JE, Hofiusd D, Van Breusegem F, Petersen M, Andersen SU. Catalase and no Catalase activity1 promote autophagy-dependent cell death in *Arabidopsis*. Plant Cell. 2013;25:4616–26.

Halliwell B, Gutteridge JMC. Free radicals in biology and medicine (4th edn). Oxford: Oxford University Press; 2007. p. 47.

Hong L, Hu B, Liu X, He CY, Yao Y, Li XL, Li L. Molecular cloning and expression analysis of a new stress-related AREB gene from *Arachis hypogaea*. Biol Planta. 2013;57(1):56–62.

Hruz T, Laule O, Szabo G, Wessendorp F, Bleuler S, Oertle L, Widmayer P, Gruissem W, Zimmermann P. Genevestigator v3: a reference expression database for the meta-analysis of transcriptomes. Adv Bioinformatics. 2008;2008:420747. doi:10.1155/2008/420747.

Hu H, You J, Fang Y, Zhu X, Qi Z, Xiong L. Characterization of transcription factor gene SNAC2 conferring cold and salt tolerance in rice. Plant Mol Biol. 2008;67:169–81.

Huang J, Yang X, Wang MM, Tang HJ, Ding LY, Shen Y, Zhang HS. A novel rice C_2H_2-type zinc finger protein lacking DLN-box/EAR-motif plays a role in salt tolerance. Biochim Biophys Acta. 2007;1769:220–7.

Huang GT, Ma SL, Bai LP, Zhang L, Ma H, Jia P, Liu J, Zhong M, Guo ZF. Signal transduction during cold, salt, and drought stresses in plants. Mol Biol Rep. 2012;39:969–87.

Jacquot JP, Dietz KJ, Rouhier N, Meux E, Lallement PA, Selles B, Hecker A. Redox regulation in plants: glutathione and "redoxin"-related families. In: Jakob U, Reichmann D, editors. Oxidative Stress and Redox Regulation. Dordrecht: Springer Science and Business Media; 2013. p. 213–31.

Jammes F, Song C, Shin D, Munemasa S, Takeda K, Gu D, Cho D, Lee S, Giordo R, Sritubtim S, Leonhardt N, Ellis BE, Murata Y, Kwak JM. MAP kinases MPK9 and MPK12 are preferentially expressed in guard cells and positively regulate ROS-mediated ABA signaling. Proc Natl Acad Sci U S A. 2009;106:20520–5.

Jiang M, Zhang J. Water stress-induced abscisic acid accumulation triggers the increased generation of reactive oxygen species and up-regulates the activities of antioxidant enzymes in maize leaves. J Exp Bot. 2002;53:2401–10.

Jiang M, Zhang J. Cross-talk between calcium and reactive oxygen species originated from NADPH oxidase in abscisic acid-induced antioxidant defence in leaves of maize seedlings. Plant Cell Environ. 2003;26:929–39.

Johansson E, Olsson O, Nystrom T. Progression and specificity of protein oxidation in the life cycle of *Arabidopsis thaliana*. J Biol Chem. 2004;279:22204–8.

Jonak C, Kiegerl S, Ligterink W, Barker PJ, Huskisson NS, Hirt H. Stress signaling in plants: a mitogen-activated protein kinase pathway is activated by cold and drought. Proc Natl Acad Sci U S A. 1996;93:11274–9.

Juul T, Malolepszy A, Dybkaer K, Kidmose R, Rasmussen JT, Andersen GR, Johnsen HE, Jørgensen JE, Andersen SU. The in vivo toxicity of hydroxyurea depends on its direct target catalase. J Biol Chem. 2010;285:21411–5.

Kaspar JW, Jaiswal AK. Antioxidant-induced phosphorylation of tyrosine 486 leads to rapid nuclear export of Bach1 that allows Nrf2 to bind to the antioxidant response element and activate defensive gene expression. J Biol Chem. 2010;285(1):153–62.

Kaspar JW, Niture SK, Jaiswal AK. Nrf2:INrf2 (Keap1) signaling in oxidative stress. Free Radic Biol Med. 2009;47(9):1304–9.

Koppen M, Langer T. Protein degradation within mitochondria: versatile activities of AAA proteases and other peptidases. Crit Rev Biochem Mol Biol. 2007;42:221–42.

Kovtun Y, Chiu WL, Tena G, Sheen J. Functional analysis of oxidative stress-activated mitogen-activated protein kinase cascade in plants. Proc Natl Acad Sci U S A. 2000;97:2940–5.

Kudla J, Batistic O, Hashimoto K. Calcium signals: the lead currency of plant information processing. Plant Cell. 2010;22:541–63.

Kunihiro S, Hiramatsu T, Kawano T. Involvement of salicylic acid signal transduction in aluminum-responsive oxidative burst in *Arabidopsis thaliana* cell suspension culture. Plant Signal Behav. 2011;6:611–6.

Kurepa J, Toh-E A, Smalle JA. 26S proteasome regulatory particle mutants have increased oxidative stress tolerance. Plant J. 2008;53:102–14.

Laloi C, Stachowiak M, Pers-Kamczyc E, Warzych E, Murgia I, Apel K. Cross-talk between singlet oxygen- and hydrogen peroxide- dependent signaling of stress responses in *Arabidopsis thaliana*. Proc Natl Acad Sci U S A. 2007;104:672–7.

Leon-Reyes A, Spoel SH, De Lange ES, Abe H, Kobayashi M, Tsuda S, Millenaar FF, Welschen RAM, Ritsema T, Pieterse CMJ. Ethylene modulates the role of non-expressor of pathogenesis-related genes 1 in cross talk between salicylate and jasmonate signaling. Plant Physiol. 2009;149:1797–809.

Lin F, Ding HD, Wang JX, Zhang H, Zhang AY, Zhang Y, Tan MP, Dong W, Jiang MY. Positive feedback regulation of maize NADPH oxidase by mitogen-activated protein kinase cascade in abscisic acid signalling. J Exp Bot. 2009;60:3221–38.

Love A, Geri C, Laird J, Yun B, Loake G, Sadanandom A, Milner J. An effector protein encoded by cauliflower mosaic virus inhibits SA-dependent defence responses in *Arabidopsis* via an NPR1-dependent mechanism. Comp Biochem Physiol A Mol Integr Physiol. 2008;150:193.

Lu S, Su W, Li H, Guo Z. Abscisic acid improves drought tolerance of triploid bermudagrass and involves H_2O_2- and NO-induced antioxidant enzyme activities. Plant Physiol Biochem. 2009;47:132–8.

Luo M, Liang XQ, Dang P, Holbrook CC, Bausher MG, Lee RD, Guo BZ. Microarray-based screening of differentially expressed genes in peanut in response to *Aspergillus parasiticus* infection and drought stress. Plant Sci. 2005;169:695–703.

Mano J. Reactive carbonyl species: Their production from lipid peroxides, action in environmental stress, and the detoxification mechanism. Plant Physiol Biochem. 2012;59(2012):90–7.

Mersmann S, Bourdais G, Rietz S, Robatzek S. Ethylene signaling regulates accumulation of the FLS2 receptor and is required for the oxidative burst contributing to plant immunity. Plant Physiol. 2010;154:391–400.

Miao Y, Lv D, Wang P, Wang XC, Chen J, Miao C, Song CP. An *Arabidopsis* glutathione peroxidase functions as both a redox transducer and a scavenger in abscisic acid and drought stress responses. Plant Cell. 2006;18:2749–66.

Miller G, Schlauch K, Tam R, Cortes D, Torres MA, Shulaev V, Dangl JL, Mittler R. 2009. The plant NADPH oxidase RBOHD mediates rapid systemic signaling in response to diverse stimuli. Sci Signal. 2:ra45. doi:10.1126/scisignal.2000448

Mittler R, Vanderauwera S, Gollery M, Van Breusegem F. Reactive oxygen gene network of plants. Trends Plant Sci. 2004;9(10):1360–85.

Mittler R, Vanderauwera S, Suzuki N, Miller G, Tognetti VB, Vandepoele K, Gollery M, Shulaev V, Van Breusegem F. ROS signaling: the new wave? Trends Plant Sci. 2011;16:300–9.

Mittler R. Oxidative stress, antioxidants and stress tolerance. Trends Plant Sci. 2002;7:405–10.

Mizoguchi T, Irie K, Hirayama T, Hayashida N, Yamaguchi-Shinozaki K, Matsumoto K, Shinozaki K. A gene encoding a mitogen-activated protein kinase kinase kinase is induced simultaneously with genes for a mitogen-activated protein kinase and an S6 ribosomal protein kinase by touch, cold and water stress in *Arabidopsis thaliana*. Proc Natl Acad Sci U S A. 1996;93:765–9.

Møller IM, Kristensen BK. Protein oxidation in plant mitochondria as a stress indicator. Photochem Photobiol Sci. 2004;3:730–5.

Möller IM, Jensen PE, Hansson A. Oxidative modifications to cellular components in plants. Annu Rev Plant Biol. 2007;51:459–81.

Möller IM, Sweetlove LJ. ROS signaling—specificity is required. Trends Plant Sci. 2010;15:370–4.

Mou Z, Fan W, Dong X. Inducers of plant systemic acquired resistance regulate NPR1 function through redox changes. Cell. 2003;27:935–44.

Nakagami H, Soukupová H, Schikora A, Zárský V, Hirt H. A mitogen-activated protein kinase kinase kinase mediates reactive oxygen species homeostasis in *Arabidopsis*. J Biol Chem. 2006;281:38697–704.

Ning J, Li X, Hicks LM, Xiong L. A Raf-like MAPKKK gene DSM1 mediates drought resistance through reactive oxygen species scavenging in rice. Plant Physiol. 2010;152:876–90.

Noctor G, Mhamdi A, Foyer CH. The roles of reactive oxygen metabolism in drought: not so cut and dried. Plant Physiol. 2014;164:1636–48.

Noctor G, Mhamdi A, Queval G, Foyer CH. Regulating the redox gatekeeper: vacuolar sequestration puts glutathione disulfide in its place. Plant Physiol. 2013;163:665–71.

Oelze ML, Vogel MO, Alsharafa K, Kahmann U, Viehhauser A, Maurino VG, Dietz KJ. Efficient acclimation of the chloroplast antioxidant defence of *Arabidopsis thaliana* leaves in response to a 10- or 100-fold light increment and the possible involvement of retrograde signals. J Exp Bot. 2012;63(3):1297–313.

Ohkamu-Ohtsu N, Radwan S, Peterson A, Zhao P, Badr AF, Xiang C, Oliver DJ. Characterization of the extracellular γ-glutamyltranspeptidases, GGT1 and GGT2, in *Arabidopsis*. Plant J. 2007;49:865–77.

op den Camp RGL, Przybyla D, Ochsenbein C, Laloi C, Kim C, Danon A, Wagner D, Hideg E, Göbel C, Feussner I, Nater M, Apel K. Rapid induction of distinct stress responses after the release of singlet oxygen in *Arabidopsis*. Plant Cell. 2003;15:2320–32.

Ortiz-Masia D, Perez-Amador MA, Carbonell J, Marcote MJ. Diverse stress signals activate the C1 subgroup MAP kinases of *Arabidopsis*. FEBS Lett. 2007;581:1834–40.

Park SY, Fung P, Nishimura N, Jensen DR, Fujii H, Zhao Y, Lumba S, Santiago J, Rodrigues A, Tsz-fung FC. Abscisic acid inhibits type 2C protein phosphatases via the PYR/PYL family of START proteins. Science. 2009;324:1068–71.

Parsons HT, Fry SC. Oxidation of dehydroascorbic acid and 2,3-diketogulonate under plant apoplastic conditions. Phytochemistry. 2012;75:41–9.

Peng LX, Gu LK, Zheng CC, Li DQ, Shu HR. Expression of MaMAPK gene in seedlings of *Malus* L. under water stress. Acta Biochim Biophys Sin (Shanghai). 2006;38:281–6.

Pignocchi C, Foyer CH. Apoplastic ascorbate metabolism and its role in the regulation of cell signalling. Curr Opin Plant Biol. 2003;6:379–89.

Pnueli L, Hallak-Herr E, Rozenberg M, Cohen M, Goloubinoff P, Kaplan A, Mittler R. Molecular and biochemical mechanisms associated with dormancy and drought tolerance in the desert legume *Retama raetam*. Plant J. 2002;31(3):319–30.

Puranik S, Bahadur RP, Srivastava PS, Prasad M. Molecular cloning and characterization of a membrane associated NAC family gene, SiNAC from foxtail millet [*Setaria italica* (L.) P. Beauv.]. Mol Biotechnol. 2011;49(2):138–50.

Queval G, Neukermans J, Vanderauwera S, Van Breusegem F, Noctor G. Day length is a key regulator of transcriptomic responses to both CO_2 and H_2O_2 in *Arabidopsis*. Plant Cell Environ. 2012;35:374–87.

Sagi M, Fluhr R. Production of reactive oxygen species by plant NADPH oxidases. Plant Physiol. 2006;141:336–40.

Saibo NJM, Lourenco T, Oliveira MM. Transcription factors and regulation of photosynthetic and related metabolism under environmental stresses. Ann Bot. 2009;103:609–23.

Sakuma Y, Liu Q, Dubouzet JG, Abe H, Shinozaki K, Yamaguchi-Shinozaki K. DNA-binding specificity of the ERF/AP2 domain of *Arabidopsis* DREBs, transcription factors involved in dehydration- and cold-inducible gene expression. Biochem Biophys Res Commun. 2002;290:998–1009.

Sakuma Y, Maruyama K, Osakabe Y, Qin F, Seki M, Shinozaki K, Yamaguchi-Shinozaki K. Functional analysis of an *Arabidopsis* transcription factor, DREB2A, involved in drought-responsive gene expression. Plant Cell. 2006;18:1292–309.

Santiago J, Rodrigues A, Saez A, Rubio S, Antoni R, Dupeux F, Park SY, Márquez JA, Cutler SR, Rodriguez PL. Modulation of drought resistance by the abscisic acid receptor PYL5 through inhibition of clade A PP2Cs. Plant J. 2009;60:575–88.

Schürmann P, Buchanan B. The ferredoxin-thioredoxin system of oxygenic photosynthesis. Antioxid Redox Signal. 2008;10:1235–73.

Semchyshyn HM. 2014. Reactive carbonyl species in vivo: generation and dual biological effects. ScientificWorldJournal. 2014;2014:417842. http://dx.doi.org/10.1155/2014/417842

Sevier CS, Kaiser CA. Ero1 and redox homeostasis in the endoplasmic reticulum. Biochim Biophys Acta. 2008;1783:549–56.

Shao HB, Chu LY, Lu ZH, Kang CM. Primary antioxidant free radical scavenging and redox signaling pathways in higher plant cells. Int J Biol Sci. 2008;4:8–14.

Shao HB, Chu LY, Wu G, Zhang JH, Lu ZH, Hu YC. Changes of some antioxidative physiological indices under soil water deficits among 10 wheat (*Triticum aestivum* L.) genotypes at tillering stage. Biointerfaces. 2007;54(2):143–9.

Sheard LB, Zheng N. Plant biology: signal advance for abscisic acid. Nature. 2009;462:575–6.

Shi J, Zhang L, An H, Wu C, Guo X. GhMPK16, a novel stress-responsive group D MAPK gene from cotton, is involved in disease resistance and drought sensitivity. BMC Mol Biol. 2011;12:22.

Shou H, Bordallo P, Fan J-B, Yeakley JM, Bibikova M, Sheen J, Wang K. Expression of an active tobacco mitogen-activated protein kinase kinase kinase enhances freezing tolerance in transgenic maize. Proc Natl Acad Sci U S A. 2004a;101:3298–303.

Shou H, Bordallo P, Wang K. b. Expression of the *Nicotiana* protein kinase (NPK1) enhanced drought tolerance in transgenic maize. J Exp Bot. 2004b;55:1013–9.

Soares NC, Francisco R, Vielba JM, Ricardo CP, Jackson PA. Associating wound-related changes in the apoplast proteome of *Medicago* with early steps in the ROS signal-transduction pathway. J Proteome Res. 2009;8:2298–309.

Soon F-F, Ng L-M, Zhou XE, West GM, Kovach A, Tan MHE, Suino-Powell KM, He Y, Xu Y, Chalmers MJ, Brunzelle JS, Zhang H, Yang H, Jiang H, Li J, Yong EL, Cutler S, Zhu JK, Griffin PR, Melcher K, Xu HE. Molecular mimicry regulates ABA signaling by SnRK2 kinases and PP2C phosphatases. Science. 2012;335:85–8.

Steinhorst L, Kudla J. Calcium and reactive oxygen species rule the waves of signaling. Plant Physiol. 2013;163:471–85.

Suarez Rodriguez MC, Petersen M, Mundy J. Mitogen-activated protein kinase signaling in plants. Annu Rev Plant Biol. 2010;61:621–49.

Suzuki N, Miller G, Morales J, Shulaev V, Torres MA, Mittler R. Respiratory burst oxidases: the engines of ROS signaling. Curr Opin Plant Biol. 2011;14:691–9.

Suzuki N, Miller G, Salazar C, Mondal HA, Shulaev E, Cortes DF, Shuman JL, Luo X, Shah J, Schlauch K. Temporal-spatial interaction between reactive oxygen species and abscisic acid regulates rapid systemic acclimation in plants. Plant Cell. 2013;25:3553–69.

Sweetlove LJ, Møller IM. Oxidation of proteins in plants—mechanisms and consequences. Adv Bot Res. 2009;52:1–23.

Tatsuta T. Protein quality control in mitochondria. J Biochem. 2009;146:455–61.

Tran L-SP, Nakashima K, Sakuma Y, Simpson SD, Fujita Y, Maruyama K, Fujita M, Seki M, Shinozaki K, Yamaguchi-Shinozaki K. Isolation and functional analysis of *Arabidopsis* stress-inducible NAC transcription factors that bind to a drought- responsive cis-element in the early responsive to dehydration stress 1 promoter. Plant Cell. 2004;16:2481–98.

Tran LS, Nakashima K, Sakuma Y, Osakabe Y, Qin F, Simpson SD, Maruyama K, Fujita Y, Shinozaki K, Yamaguchi-Shinozaki K. Co-expression of the stress- inducible zinc finger homeodomain ZFHD1 and NAC transcription factors enhances expression of the ERD1 gene in *Arabidopsis*. Plant J. 2007;49:46–63.

Tripathi V, Parasuraman B, Laxmi A, Chattopadhyay D. CIPK6, a CBL-interacting protein kinase is required for development and salt tolerance in plants. Plant J. 2009;58:778–90.

Tunc-Ozdemir M, Tang C, Ishka MR, Brown E, Groves NR, Myers CT, Rato C, Poulsen LR, McDowell S, Miller G, Mittler R, Harper JF. A cyclic nucleotide-gated channel (CNGC16) in pollen is critical for stress tolerance in pollen reproductive development. Plant Physiol. 2013;161:1010–20.

Umezawa T, Sugiyama N, Mizoguchi M, Hayash S, Myouga F, Yamaguchi-Shinozaki K, Ishihama Y, Hirayama T, Shinozaki K. Type 2C protein phosphatases directly regulate abscisic acid-activated protein kinases in *Arabidopsis*. Proc Natl Acad Sci U S A. 2009;106:17588–93.

Uno Y, Furihata T, Abe H, Yoshida R, Shinozaki K, Yamaguchi-Shinozaki K. *Arabidopsis* basic leucine zipper transcription factors involved in an abscisic acid- dependent signal transduction pathway under drought and high-salinity conditions. Proc Natl Acad Sci U S A. 2000;97:11632–7.

Valliyodan B, Nguyen HT. Understanding regulatory networks and engineering for enhanced drought tolerance in plants. Plant Biol. 2006;9(2):189–95.

Wang J, Ding H, Zhang A, Ma F, Cao J, Jiang M. A novel mitogen-activated protein kinase gene in maize (*Zea mays*), ZmMPK3, is involved in response to diverse environmental cues. J Integr Plant Biol. 2010;52:442–52.

Windhovel A, Hein I, Dabrowa R, Stockhaus J. Characterization of a novel class of plant homeodomain proteins that bind to the C4 phosphoenolpyruvate carboxylase gene of *Flaveria trinervia*. Plant Mol Biol. 2001;45:201–14.

Xia XJ, Zhou YH, Shi K, Zhou J, Foyer CH, Yu JQ. Interplay between reactive oxygen species and hormones in the control of plant development and stress tolerance. J Exp Bot. 2015;66(10):2839–56. doi:10.1093/jxb/erv089.

Xiong L, Yang Y. Disease resistance and abiotic stress tolerance in rice are inversely modulated by an abscisic acid-inducible mitogen-activated protein kinase. Plant Cell. 2003;15:745–59.

Xiong L, Schumaker KS, Zhu JK. Cell signaling during cold, drought, and salt stress. Plant Cell. 2002;14(Suppl):S165–83.

Xu DQ, Huang J, Guo SQ, Yang X, Bao YM, Tang HJ, Zhang HS. Overexpression of a TFIIIA-type zinc finger protein gene ZFP252 enhances drought and salt tolerance in rice (*Oryza sativa* L.). FEBS Lett. 2008;582:1037–43.

Yadav UCS Ramana KV. Regulation of NF-κB-induced inflammatory signaling by lipid peroxidation-derived aldehydes. Oxid Med Cell Longev. 2013;690545:11 p.

Yamaguchi-Shinozaki K, Shinozaki K. Transcriptional regulatory networks in cellular responses and tolerance to dehydration and cold stresses. Annu Rev Plant Biol. 2006;57:781–803.

Yan J, Tsuichihara N, Etoh T, Iwai S. Reactive oxygen species and nitric oxide are involved in ABA inhibition of stomatal opening plant. Plant Cell Environ. 2007;30:1320–5.

Zhang A, Jiang M, Zhang J, Ding H, Xu S, Hu X, Tan M. Nitric oxide induced by hydrogen peroxide mediates abscisic acid-induced activation of the mitogen-activated protein kinase cascade involved in antioxidant defense in maize leaves. New Phytol. 2007;175:36–50.

Zhang J, Jia W, Yang J, Ismail A. Role of ABA in integrating plant responses to drought and salt stresses. Field Crop Res. 2006;97:111–9.

Zhang S, Klessig DF. MAPK cascades in plant defense signaling. Trends Plant Sci. 2001;6:520–7.

Zhang W, Ruan J, Ho TH, You Y, Yu T, Quatrano RS. Cis-regulatory element based targeted gene finding: genome-wide identification of abscisic acid- and abiotic stress- responsive genes in *Arabidopsis thaliana*. Bioinformatics. 2005;21:3074–81.

Zheng X, Chen B, Lu G, Han B. Overexpression of a NAC transcription factor enhances rice drought and salt tolerance. Biochem Biophys Res Commun. 2009;379:985–9.

Zhou B, Guo Z, Xing J, Huang B. Nitric oxide is involved in abscisic acid induced antioxidant activities in *Stylosanthes guianensis*. J Exp Bot. 2005;56:3223–8.

Zhou J, Wang J, Li X, Xia XJ, Zhou YH, Shi K, Chen Z, Yu JQ. H_2O_2 mediates the crosstalk of brassinosteroid and abscisic acid in tomato responses to heat and oxidative stresses. J Exp Bot. 2014;65:4371–83.

Zong XJ, Li DP, Gu LK, Li DQ, Liu LX, Hu XL. Abscisic acid and hydrogen peroxide induce a novel maize group CMAP kinase gene, ZmMPK7, which is responsible for the removal of reactive oxygen species. Planta. 2009;229:485–95.

Chapter 14
Improving Crop Yield Under Drought Stress Through Physiological Breeding

Veena Pandey and Alok Shukla

14.1 Introduction

Drought stress is a major constraint to crop production worldwide and improving yield stability under drought is a major goal of recent studies so as to ensure food security. Crop yield severely reduces under drought stress and this might be attributed due to drought induced reduced stomatal conductance, reduction in CO_2 assimilation rates, photosynthetic pigments, small leaf size, reduced stems extension, disturbed plant water relations, reduced water-use efficiency, reduced activities of sucrose and starch synthesis enzymes and reduced assimilate partitioning, leading to a reduction in plant growth and productivity (Anjum et al. 2011). Recently, several plant traits that govern yield under drought stress have been identified and are being used for crop improvement practices, but still various bottlenecks are to be addressed. To facilitate the development of tolerant cultivars which can survive and give better yield under drought conditions, a thorough understanding of the various morphological, physiological and molecular characters that govern the yield under water stress condition is a prerequisite (Pandey and Shukla 2015). A major challenge for the genetic improvement of drought resistance is the lack of complete molecular basis for drought perception, signal transduction, and stress adaptation. Thus it's a challenge to unravel the complex mechanisms of drought resistance in crops through more intensive and integrative studies in order to find key functional components or machineries that can be used as tools for engineering drought-resistant crops (Hu and Xiong 2014). Breeders have made exciting progress in improving and developing drought-tolerant crops, but these still cannot meet the demands of food security in the face of an increasing world population, global

V. Pandey (✉) • A. Shukla
Department of Plant Physiology, College of Basic Sciences and Humanities,
G.B. Pant University of Agriculture and Technology, Pantnagar 263145, Uttarakhand, India
e-mail: veenapandey28@gmail.com

M.A. Hossain et al. (eds.), *Drought Stress Tolerance in Plants, Vol 1*,
DOI 10.1007/978-3-319-28899-4_14

warming, and a water shortage (Hu and Xiong 2014). Therefore, the aim of this article is to review complex drought associated physiological traits used in plant breeding for improved drought tolerance.

14.2 Breeding for Improved Drought Tolerance

Plant breeding aims to develop cultivars which fit well in specific environment and to develop production practices for high yield. The increase in crop yield under drought through breeding and selection remains the primary method of choice under stressed conditions. Breeders use a step-wise selection procedure to screen the best-performing genotypes and finally promising varieties is evaluated in farmers fields (Fischer et al. 2003). Breeders create new gene combinations and useful variability among genotypes by intercrossing parents that possess desirable characteristics or by introducing new germplasm from another breeding program (Bänziger et al. 2000). Progress in any breeding program is based on the amount of genetic variability available, effectiveness of selection and heritability of the trait (Graham et al. 2008). However, the lack of effective selection criteria and low heritability of grain yield (GY)under drought are the major reasons for the slow progress in breeding drought-tolerant varieties (Ouk et al. 2006).

New crop varieties with improved drought resistance (DR) have been developed through conventional breeding approaches in the past decade which adopts mainly a large-scale backcross strategy to develop new varieties with improved DR and high yield potential. Recent research has shown that varieties developed through direct selection for grain yield (GY) under drought-stress and non-stress conditions from progenies derived from crosses of drought-tolerant donors and high-yielding drought-susceptible varieties provide a yield advantage under drought (Venuprasad et al. 2008; Kumar et al. 2008) in addition to maintaining a high yield potential under non-stress conditions. For example, in rice, the International Rice Research Institute made 322 crosses between 3 elite recurrent and 163 donor varieties of diverse origins (Yu et al. 2003). The backcross progenies derived from these crosses that showed higher yields than the recurrent parents were selected in severe-drought lowland or upland nurseries (Lafitte et al. 2006). Similarly, Shanghai Agrobiological Gene Center developed new rice varieties with improved DR and high yield by performing large scale crossing and backcrossing using elite rice varieties as recurrent parents and upland drought-resistant rice as donors (Hu and Xiong, 2014). In recent years, a series of water-saving and drought-resistant crop varieties have been developed and released by breeders.

Selecting for improved phenotype has some limitations especially when interest is focused on more complex physiological traits. A more accurate way of selection would be at the genetic level where markers linked to the gene(s) or quantitative trait loci (QTLs) underlying the trait can be screened for. A prerequisite for genotypic selection is the establishment of associations between traits of interest and genetic markers. Understanding the genetic control of physiological traits and the linkage of these physiological characteristics to molecular markers on chromosomes, and ultimately the gene(s) underlying the trait is the future of plant breeding (Graham et al. 2008). Molecular breeding approaches such as marker-assisted backcrossing, marker-assisted recurrent selection and genome-wide selection have also been suggested to be integrated in crop improvement strategies to develop drought-tolerant cultivars that will enhance food security in the context of a changing and more variable climate (Mir et al. 2012). Recently identification of QTLs underlying desirable phenotypic traits becomes crucial for molecular breeding approaches. Thereafter, transfer of the identified QTL underlying drought resistance-associated traits is done, followed by evaluation of the effect of the transferred QTL on plant phenotype (Kosová et al. 2014).

14.3 Physiological Traits Used in Breeding Programmes

Physiology forms the basis of proper phenotyping and thus a full understanding of physiology is needed to design the traits targeted by various breeding approaches. The use of physiological traits in a breeding programme, either by direct selection or through a surrogate such as molecular markers, depends on their relative genetic correlation with yield, extent of genetic variation, heritability and genotype × environment interactions. A large number of studies have identified several physiological traits, whose presence is associated with plant adaptability to drought-prone environments (Cattivelli et al. 2008). The early escape from drought stress, through the manipulation of plant phenology, is the most commonly exploited genetic strategy used to ensure relatively stable yields under terminal drought conditions (Richards 1991). Traits such as small plant size, reduced leaf area, early maturity and prolonged stomatal closure lead to a reduced total seasonal evapo-transpiration (Fischer and Wood 1979; Karamanos and Papatheohari 1999). Osmotic adjustment, accumulation and remobilization of stem reserves, superior photosynthesis, heat- and desiccation-tolerant enzymes, etc. are other important physiological traits under drought environment. However, it is important to establish their heritability and genetic correlation with yield in target environments (Mir et al. 2012). The importance of roots for water and nutrient capture and to increase yield under drought stress has been rediscovered in recent years (Turner et al. 2014).

Physiologists and breeders developed a general model for drought adaptation of wheat at CIMMYT (Reynolds et al. 2005), that encompasses traits including: pre-anthesis growth, access to water as a result of rooting depth or intensity that would be expressed by a relatively cool canopy, water-use efficiency (WUE), transpiration efficiency (TE) indicated by carbon isotope discrimination (CID) of leaves, and WUE of spike photosynthesis associated with re-fixation of respiratory CO_2, and photoprotection including energy dissipation, antioxidant systems and anatomical traits such as leaf wax. This model is used to assist in taking breeding decisions by permitting a strategic approach of accumulating drought adaptive alleles by crossing parents with contrasting drought-adaptive mechanisms (Mir et al. 2012). Several studies have shown that breeding for an enhanced WUE, TE and various root traits allows reduced water loss via transpiration, more water acquisition from soil and comparatively higher yield under stress conditions. Though the identification of various physiological traits for drought-adaptation and yield stability is time consuming, but once successful, the benefits are likely to be significant.

14.3.1 Breeding for Improved Water Use Efficiency, Transpiration Efficiency and Carbon Isotope Discrimination

Various plant physiological traits have a major effect on plant water use and breeding for these important traits might be substantial for avoiding dehydration under water limited conditions. Under such situations, maximizing soil moisture use is a crucial component of drought avoidance, which is generally expressed by WUE. Where additional water is not available to the crop, higher water-use efficiency (WUE) appears to be an important strategy to improve crop performance (Araus et al. 2002). Condon et al. (2004) suggested three key processes which can be exploited in breeding for high water-use efficiency: (1) moving more of the available water through the crop rather than it being wasted as evaporation from the soil surface; (2) acquiring more carbon (biomass) in exchange for the water transpired by the crop, i.e. improving crop transpiration efficiency; (3) partitioning more of the achieved biomass into the harvested product. Agronomic parameters like photosynthetic rate, relative water content (RWC) and stomatal conductance show strong positive correlations with WUE, whereas transpiration rate expresses negative correlation with WUE under drought (Akram et al. 2013).

For WUE, CID seems to be the best estimate and is based on higher affinity of the carbon-fixing enzyme (Rubisco) for the more common ^{12}C isotope over the less common ^{13}C. $\Delta^{13}C$ (ratio of stable isotopes ^{13}C:^{12}C) has been used as a surrogate for WUE and has been successfully used for tomato (Martin and Thorstenson 1988), wheat (Rebetzke et al. 2002) and rice (Impa et al. 2005). Under drought conditions, CID is negatively correlated to transpiration efficiency (Cabuslay et al. 2002; Kondo et al. 2004) and WUE (Impa et al. 2005) at the leaf level. A lower discrimination value indicated higher WUE. For cereals such as wheat or barley grown in Mediterranean-type environments, higher yields have often been associated with high $\Delta^{13}C$, even in relatively dry locations (Araus et al. 2003). CID measured on non-stressed leaf tissue during early development was used to select for transpiration efficiency (TE) in environments where a conservative use of water early in the cycle is necessary to compensate for extremely limited water availability during grain filling (Condon et al. 2004). Thus, CID has been suggested as an indirect tool for selecting plants having higher WUE, TE and yield under drought (Akhter et al. 2010; Mohankumar et al. 2011).

TE (the ratio of mass accumulation to transpiration) is under genetic control (Masle et al. 2005) and is considered as a potential trait for drought stress

(Manavalan et al. 2009). Increased TE is often suggested as a critical opportunity for genetic improvement for increased crop yields in water-limited environments (Sinclair 2012). The use of CID is expanded as an indirect way to assess TE at the plant level. For instance, a negative relationship was found between TE and CID in wheat (Ehdaie et al. 1991), peanut (Rao et al. 1993; Wright et al. 1994), sunflower (Lambrides et al. 2004) and barley (Anyia et al. 2007). A similar conclusion was drawn from a study in wheat (Monneveux et al. 2006), where the relationship between the grain $\Delta^{13}C$ and yields showed a strong association only under post-anthesis water stress, whereas no or a weak relationship was found under conditions of residual moisture, pre-anthesis water stress, or full irrigation.

14.3.2 Breeding for Canopy Temperature and Canopy Temperature Depression

Canopy temperature (CT) measurements have been widely used in recent years to study genotypic response to drought. Blum et al. (1989) used canopy temperatures of drought stresses wheat genotypes to characterize yield stability under various moisture conditions. A positive correlation was found between a drought susceptibility index and canopy temperature in stressed environments. Drought susceptible genotypes which suffered relatively greater yield loss under stress tended to have warmer canopies at midday. CT had widespread application in stress breeding, as it readily integrate the effects of many plants within a crop canopy and hence reduce the errors associated with plant-to-plant and leaf-to-leaf variation. Genotypes with cooler canopy temperatures can be used to indicate a better hydration status. Cooler CT is positively associated with yield under drought stress and both physiological (Lopes and Reynolds 2010) and genetic (Pinto et al. 2010) evidence suggests this to be associated with a root capacity. It is used routinely, particularly for stress diagnostic and breeding selection of stress adapted genotypes. Under drought conditions, it is related to the capacity to extract water from deeper soil profiles and agronomic WUE, while under irrigated conditions it may indicate photosynthetic capacity, sink strength or vascular capacity depending on the genetic background, environment and developmental stage (Pietragalla 2012).

In plant breeding and selection for drought resistance the interest is in finding genotypes that maintain transpiration, gas exchange and therefore a lower canopy temperature as compared with other genotypes under the same field conditions. Relatively lower canopy temperature in drought stressed crop plants indicates a relatively better capacity for taking up soil moisture and for maintaining a relatively better plant water status by various plant constitutive or adaptive traits (Blum 2009). Recent data show CT to be associated with deeper roots under drought (Lopes and Reynolds 2010). Studies using various crops including wheat (Rebetzke et al. 2013), rice (Horie et al. 2006), sorghum (Mutava et al. 2011), potato (Prashar et al. 2013) and maize (Shaibu et al. 2015) have all reported that canopy temperatures can be associated with yield and could therefore be used as a selection technique (Grant et al. 2007; Zia et al. 2013).

Canopy temperature depression (CTD), the difference between air temperature (T_a) and canopy temperature (T_c) is another trait which is being used successfully as a selection criterion for tolerance to drought in breeding programs. CTD has played an important role to search physiological basis of grain yield where high CTD (CTD = T_a–T_c) value indicate cool canopy. It has been used in various practical applications including evaluation of plant response to environmental stress like drought (Blum et al. 1989; Rashid et al. 1999). CTD effected by biological and environmental factors like water status of soil, wind, evapotranspiration, cloudiness, conduction systems, plant metabolism, air temperature, relative humidity, and continuous radiation (Reynolds et al. 2001), has preferably been measured in high air temperature and low relative humidity because of high vapour pressure deficit conditions (Amani et al. 1996). CTD shows high genetic correlation with yield and high values of proportion of direct response to selection (Reynolds et al. 2001), indicating heritability and therefore amenability of this trait to early generation selection. Under dryland conditions, grain yield and mean CTD were correlated positively (Royo et al. 2002). Bilge et al. (2008) found that CTD was positively correlated with grain yield, spike yield, and grain numbers per spike. This positive correlation between CTD and grain yield showed that CTD can be used for selection criteria in breeding programs.

In an study conducted by Balota et al. (1993), CTD was measured in wheat under dryland and irrigation at preheading, anthesis and one to five weeks after anthesis and they observed that the best linear correlations between yield and CTD were obtained when CTD was sampled at anthesis and good estimates were also obtained from 1 to 3 weeks from anthesis. Balota et al. (2007) and Reynolds et al. (1997) suggested that during heading/anthesis stage would be the best time for measuring CTD regarding to high correlation with grain yield. According to Bilge et al. (2008), at late periods of heading in bread wheat, CTD value positively correlates with grain yield and grain numbers per spike. Abdipur et al. (2013) evaluated the efficiency of canopy temperature depression at different growth stages for screening drought tolerant wheat genotypes. They found that CTD had significant correlation with grain yield at anthesis half-way and medium milky stage and concluded that can be used as potential selection criterions for grain yield and wheat drought tolerance in breeding programs. The significant correlation of CT and CTD with mean productivity and stress tolerance index is well established (Guendouz et al. 2012). Mohammadi et al. (2012) showed that the lower canopy temperature under different water availability conditions caused higher grain yield. Therefore, CT and CTD can be used as a selected criterion in plant breeding for drought tolerance.

Guendouz et al. (2013) found significant correlation between flag leaf reflectance and canopy temperature and proved the efficiency of using leaf reflectance at RB (red and blue wavelength) in screening for drought tolerance in durum wheat cultivars. They observed that under non irrigated condition CTD correlated significantly and negatively with reflectance at Red and Blue; but under irrigated conditions canopy CTD correlated significantly and negatively with leaf reflectance at Red (654 nm). In addition, under non irrigated conditions there is a significant and positive correlation between canopy temperature (CT) and leaf reflectance at Red and Blue (450 nm).

14.3.3 Breeding for Light Interception and Radiation-Use Efficiency

Crop biomass production depends on the ability of the canopy to intercept the incoming photosynthetically active radiation (PAR), which is a function of leaf area index (LAI) and canopy architecture; and to convert this radiation into new biomass, i.e. radiation use efficiency (RUE) (Sinclair and Muchow 1999). RUE is the key factor determining the crop yield and is related to crop biomass and LAI. It is affected by abiotic factors such as drought (Jamieson et al. 1995; Ali et al. 2012). Irrigated crops allow RUE to remain relatively stable throughout the growth cycle; however, water deficits decrease RUE, particularly during early grain filling. If crops function in a continual adjustment phase to stress, there might be little benefit to reducing RUE in response to water stress. Water-stress related reductions in RUE are reported to occur in barley (Legg et al. 1979). Wajid et al. (2007) reported that when drought stress was imposed before or after anthesis, the primary cause of reduced RUE was a decrease in intercepted light, which ultimately reduced the photosynthetic products being sent to the economical organ of the plant. To obtain a high yield from a given cultivar under dryland conditions, it is necessary to achieve optimum RUE (Miranzadeh et al. 2011). Peter (2010) evaluated effects of different survival strategy (escape/tolerance) and canopy structure (tillering ability) on cereal RUE under drought during grain filling and concluded that higher tillering ability more efficiently utilized incident PAR for biomass production and yield under sufficient as well as insufficient water supply.

LAI or green area index (GAI) are precise ways of estimating the light-capturing capacity of a canopy and, although light interception tends to saturate at LAI >3, the distribution of leaves can effect RUE (Parry et al. 2011). Light interception (LI) reduces as the plant encounters water deficits, as leaf expansion is reduced or as leaves senesce (Bruce et al. 2002). Changes in the efficiency of light interception and in the costs for light harvesting along the light gradient from the top to the bottom of the plant canopy are the major means by which an efficient light harvesting is achieved (Gratani 2014). During prolonged water deficits, cassava reduces its canopy by shedding older leaves and forming smaller new leaves leading to less light interception, another adaptive trait to drought (El-Sharkawy 2007). The yield potential (YP), expressed as a function of the light intercepted (LI) and radiation-use efficiency (RUE) (whose product is biomass), the partitioning of biomass to yield (the HI) and the focus of improving all the three components should be undertaken through complex physiological trait -based breeding (Mir et al. 2012).

14.3.4 Breeding for Stomatal Characters

Stomata are specialized epidermal structures that control the exchange of water and carbon dioxide between the plant and atmosphere (Xu and Zhou 2008,). Stomatal resistance (SR) (s cm^{-1}) is a character leading to water regulation of plants.

This character has been used widely as a criterion to screen water-stress tolerant varieties by several researchers (Blum et al. 1981; Jones 1987; Gumuluru et al. 1989; Araghi and Assad 1998). Since most of the water escapes through the stomata (Wang and Clarke 1993), stomatal size and frequency are among factors which influence stomatal resistance. While selecting drought tolerant wheat genotypes, stomatal resistance is a better indicator than leaf water potential and plant resistance to water flow (Adjei and Kirkham 1980). Reduction of stomata frequency and size could be used in obtaining water stress resistance (Mehri et al. 2009). Stomatal conductance had a positive genotypic and phenotypic correlation with grain yield (Shaibu et al. 2015).

Theoretically, it is expected that plants with low stomatal resistance have more dry matter production due to more gas exchange. In water-stress conditions, the effect of stomatal resistance depends on intensity and type of water-stress. When the amount of water is limited in respect to the duration of water-stress, any factor that promotes transpiration can bring the plants to a lethal level of leaf water content at the end of the period of water-stress. In this situation, plants lose a considerable amount of their vegetative growth which could contribute to assimilation after stress. In this situation low SR does not contribute to plant production particularly when the duration of water-stress is long. Conversely, when there is a considerable amount of water in the soil so that water supply is adequate in respect to water-stress duration, low SR provides a situation for the plants that they can take up more water from the soil and also contribute to higher water content of tissues by the end of the stress period (Mohammady 2011).

With regard to stomatal resistance, another aspect to be considered is the duration of stomatal closure. Stomatal closure for a long period negatively affects potential crop yield (Venora and Calcagno 1991). Therefore, it seems that partial closure of stomata particularly at mid day when temperature is high and opening of stomata when temperature is not high are beneficial to plant yield. Under water-stress conditions, SR mainly plays its positive role through water conservation and consequently by reducing water loss. In this situation, plants endure water-stress without severe damage (Mohammady 2011).

The leaf stomata is a pivotal controlling the exchange of CO_2 and water vapor, although such processes may be affected by many environmental variables including light, water status, temperature and CO_2 concentration. Under water stress, photosynthesis limitation can result from both stomatal and non stomatal effects, depending on drought intensification and species (Boyer et al. 1997; Xu and Zhou, 2008). Stomatal behavior and density measurements have the advantage of being rapid, requiring little space and allowing precise control of environmental conditions. In general, tolerant cultivars efficiently decreased their water loss by means of the reduction in stomata density, dimensions and area and consequently avoid dehydration effects. According to Mehri et al. (2009), drought tolerant wheat genotypes have less stomata and sensitive genotypes have more stomata. Recently, Kusvuran et al. 2010 indicated that more tolerance to drought is related to less stomata density in leaf in control conditions. Thus, stomatal conductance has been proposed as a selection tool for drought tolerance. Interspecific differences occur in species in their response and relationship of stomatal conductance to leaf water potential as stomatal conductance is controlled by complex interaction of intrinsic and extrinsic factors and not soil water availability alone. Nevertheless, studies mainly show that stomata close with increasing drought. Therefore, measuring stomatal characters (size and frequency) and control of water loss can aid in identification of desirable genotypes. These screening methods can be used to phenotype large populations to identify chromosomal regions controlling stomatal opening and closing and toward breeding crops with optimal stomatal response with some plasticity in behavior, so that stomata remain open under ample water conditions but close as water deficit increases (Obidiegwu et al. 2015).

14.3.5 *Breeding for Improved Root Traits*

Plant root growth encompasses a remarkable genetic diversity in terms of growth patterns, architecture, and environmental adaptations. In order to harness this valuable diversity for improving rice response to drought, an understanding of key root traits and effective drought response mechanisms is necessary. The ability to grow deep roots is currently the most accepted target trait for improving drought resistance, but genetic variation has been reported for a number of traits that may affect drought response. Rice genotypes that have deep, coarse roots with a high ability of branching and penetration and higher root to shoot ratio are reported as component traits of drought avoidance (Gowda et al. 2011). Capacity for deep root growth and large xylem diameters in deep roots may improve root acquisition of water when ample water at depth is available. While small xylem diameters in targeted seminal roots save soil water deep in the soil profile for use during crop maturation (Comas et al. 2013).

Significant genetic variation exists among different plant genotypes for root morphological traits (O'Toole and Bland 1987) such as root diameter, root depth, root pulling force, deep root to shoot ratio, root number, root growth, and root penetration ability.

Breeding of new cultivars with excellent root quality ensures absorption of water from deeper soil layers under low soil moisture and help in more efficient utilization of water for potato production. Positive correlation between root mass, shoot mass and final tuber yield led to suggestion of using root mass in the plow layer as a selection criterion for potato (Iwama 2008). Another approach to the selection of deep rooting genotypes is to measure the pulling resistance (PR) of roots (Stalham and Allen 2004). Deeper/profuse roots were found to increase plant access to water from deeper soil layers and support greater crop growth under drought conditions (Price et al. 2002; Sinclair 2011). Therefore, in several crops such as chickpea (Silim and Saxena, 1993), wheat (Reynolds et al., 2007) and rice (Yadav et al. 1997; Price et al. 2002), deeper/profuse roots are targeted to improve grain yield under rainfed conditions. However, some recent studies (Zaman-Allah et al. 2011a, 2011b) reported that selection for yield under terminal drought conditions was not essentially dependent on deeper/profuse root systems, but rather on several other critical traits that contribute to soil moisture conservation during late season water deficits. These traits include low leaf conductance under non-limited water conditions during the vegetative stage, low leaf expansion rate when soil moisture is still non-limiting for plant growth and a restriction of plant growth under progressive exposure to stress and a higher fraction of transpirable soil water (FTSW) thresholds that reduce transpiration, thus avoiding rapid soil water depletion (Mir et al. 2012). Early vigorous root proliferation may be a useful selection trait for maintaining yield under restricted water level (Puértolas et al. 2014). In addition to these factors, the hydraulic characteristics of the plant, and its interaction with the soil environment is highly significant in drought adaptation (Vadez 2014).

The relationship between root growth and grain yield under drought is complex. Positive associations between root length and grain yield have been documented in rice (Mambani and Lal 1983; Lilley and Fukai 1994). In contrast, Ingram et al. (1994) found no significant association between the two traits. Recently, Kashiwagi et al. (2015) screened contrasting chickpea accessions which comprise rich diversity for root traits, such as root biomass and rooting depth under the terminal drought and concluded that increasing rooting depth/biomass will increase the uptake of water and yield in chickpea, although such an increase may be metabolically expensive.

14.4 Role of Major Drought Associated QTLs in Breeding

Tolerance to drought is a complex quantitative trait controlled by several small effect genes or QTLs and is often confounded by differences in plants phenology (Fleury et al. 2010). To address the complexity of plant responses to drought, it is vital to understand the molecular mechanisms of yield stability. A large number of studies are been conducted to characterize the genetic basis of drought resistance by analyzing the QTLs for yield. However, identification of most precise and consistent QTL across the environments and genetics backgrounds is essential for their successful use in Marker-assisted selection (MAS). Recently, MAStechnology in crop breeding has been deployed in breeding practices to improve crop DR. The common strategy is to introgress major QTLs for drought-resistant donor genotypes into high-yielding but less drought-resistant or drought-sensitive recipient parents. This means that the superior cultivars developed contain only the major QTLs from the donor, and the whole genome of the recurrent parent remains. For example, the drought tolerant variety PY84 was developed by introgressing root trait QTLs in the elite rice cultivars IR64 and Kalinga (Steele et al. 2006). Numerous QTLs for drought-tolerant traits have been identified in major crops, and many attempts have been made to use these major QTLs to develop drought-tolerant crops. However, very few have proven successful, owing mainly to the influence of genetic background and the environment (Hu and Xiong 2014).

A QTL mapping approach concerning both shoots and roots has been carried out in rice and a number of QTLs identified. Candidate genes and ESTs

(Expressed Sequence Tags) have been identified underlying QTLs for drought tolerance (Diab et al. 2004). Recent development of molecular linkage maps of rice and other advances in molecular biology offer new opportunities for drought resistance breeding. Several QTLs with large effects on grain yield and/or flowering unique to particular hydrological conditions is reported by several researchers (Bernier et al. 2007; Kumar et al. 2007; Venuprasad et al. 2009). Du et al. (2009) identified a total of 40 quantitative trait loci (QTLs) in soybean: 17 for leaf water status traits under drought stress and 23 for seed yield under well-watered and drought-stressed conditions in both field and greenhouse trials. Dixit et al. 2014a identified three QTL—$qDTY_{3.1}$ (RM168-RM468), $qDTY_{6.1}$ (RM586-RM217), and $qDTY_{6.2}$ (RM121-RM541)—for grain yield of rice under drought. $qDTY_{3.2}$ is another most consistent QTL identified for GY under drought (Ding et al. 2011; Dixit et al. 2012, 2014b). Effect of major GY QTL differ across varying drought intensities. For example, Bernier et al. (2009) reported an increasing effect of $qDTY_{12.1}$ on GY with increasing intensity of drought. Swamy et al. (2013) and Dixit et al. (2014a) reported the effect of specific combinations of QTL on GY underdrought.

Studies have also identified genomic regions associated with yield under stress. Chromosome regions (e.g. near umc11 on chromosome 1 and near csu133 on chromosome 2) with QTLs controlling a number of morpho-physiological traits and GY across populations and conditions of different water supply have been identified in maize (Tuberosa et al. 2002). Tuberosa et al. (2002) reviewed QTLs for abscisic acid (ABA) concentration and root traits, both of which are involved in the adaptive response to drought in maize. The rice grain yield QTL region on chromosome 2 was reported to contain QTLs for leaf rolling, leaf drying, canopy temperature, productive tiller number, and stress recovery in this mapping population (Gomez et al. 2010). Nine and 24 QTLs for physio-morphological and plant production traits were identified in managed and natural drought stress conditions in rice, respectively. Yield QTLs that were consistent in the target environment over seasons were identified on chromosomes 1, 4, and 6, which could stabilize the productivity in high-yielding rice lines in a water-limited rainfed ecosystem. These yield QTLs also govern highly heritable key secondary traits, such as leaf drying, canopy temperature, panicle harvest index and harvest index (Prince et al. 2015).Prince et al. (2015) observed Three QTL regions on chromosome 1 (RM8085), chromosome 4 (I12S), and chromosome 6 (RM6836) which harbor significant additive QTLs for various physiological and yield traits under drought stress in rice.

14.5 Conclusion

Drought is a major constraint for agriculture production worldwide, and continuously changing climatic conditions are making the situation even more worst. Thus there is urgent necessity to understand mechanism of crop production, so that effective strategy can be adopted to prevent expected food crises in future. Factors like WUE, TE, CID, CT, CTD, RUE, stomatal and root characters determine plant productivity under drought conditions and can be used as a selected criterion in plant breeding for improved drought tolerance. However, the progress in breeding for drought resistance is rather slow due to the complexity of the trait and poor understanding of the genetic basis and mechanism of drought resistance in real field conditions. Therefore, it's a challenge for the breeders to integrate effectively all those available strategies to get the desired level of tolerance to drought.

To address the complexity of plant responses to drought, it is also important to understand the molecular mechanisms of yield stability. Thus, a large number of studies conducted recently are focused on characterizing the genetic basis of drought resistance by analyzing the QTLs for yield. A large number of QTLs for drought-tolerant traits have been identified in major crops and many superior cultivars developed by introgressing them in plants through breeding. However, further studies are required to investigate new QTLs for yield and their role under stressed conditions.

References

Abdipur M, Ramezani HR, Bavei V, Talaee S. Effectiveness of canopy temperature and chlorophyll content measurements at different plant growth stages for screening of drought tolerant wheat genotypes. Am Euras J Agric Environ Sci. 2013;13(10):1325–38.

Adjei GB, Kirkham MB. Evaluation of winter wheat cultivars for drought resistance. Euphytica. 1980;29:155–60.

Akhter J, Monneveux P, Sabir SA, Ashraf MY, Lateef Z, Serraj R. Selection of drought tolerant and high water use efficient rice cultivars through ^{13}C isotope discrimination technique. Pak J Bot. 2010;42(6):3887–97.

Akram HM, Ali A, Sattar A, Rehman HSU, Bibi A. Impact of water deficit stress on various physiological and agronomic traits of three basmati rice (*Oryza sativa* L.) cultivars. J Anim Plant Sci. 2013;23(5):1415–23.

Ali H, Tariq N, Ahmad S, Chattha TH, Rasheed M, Hussain A. Growth and radiation use efficiency of wheat as affected by different irrigation levels and phosphorus application methods. J Anim Plant Sci. 2012;22(4):1118–25.

Amani I, Fischer RA, Reynolds MP. Canopy temperature depression association with yield of irrigated spring wheat cultivars in hot climate. J Agron Crop Sci. 1996;176:119–29.

Anjum SA, Xie X, Wang L, Saleem MF, Man C, Lei W. Morphological, physiological and biochemical responses of plants to drought stress. Afr J Agric Res. 2011;6(9):2026–32.

Anyia AO, Slaski JJ, Nyachiro JM, Archambault DJ, Juskiw P. Relationship of carbon isotope discrimination to water use efficiency and productivity of barley under field and greenhouse conditions. J Agron Crop Sci. 2007;193:313–23.

Araus JL, Slafer GA, Reynolds MP, Royo C. Plant breeding and water relations in C3 cereals: what should we breed for? Ann Bot London. 2002;89:925–40.

Araus JL, Villegas D, Aparicio N, Garcia del Moral LF, El Hani S, Rharrabti Y, Ferrio JP, Royo C. Environmental factors determining carbon isotope discrimination and yield in durum wheat under Mediterranean conditions. Crop Sci. 2003;43:170–80.

Balota M, Amani I, Reynolds MP, Acevedo E. Evaluation of membrane thermostability and canopy depression as screening traits for heat tolerance in wheat. Wheat special report No: 20. Mexico: CIMMYT, 1993.

Balota M, Payne WA, Evett SR, Lazar MD. Canopy temperature depression sampling to assess grain yield and genotypic differentiation in winter wheat. Crop Sci. 2007;47:1518–29.

Bänziger M, Edmeades GO, Beck D, Bellon M. Breeding for drought and nitrogen stress tolerance in maize: from theory to practice. Mexico: CIMMYT; 2000.

Bernier J, Kumar A, Ramaiah V, Spaner D, Atlin G. A large-effect QTL for grain yield under reproductive-stage drought stress in upland rice. Crop Sci. 2007;47:507–18.

Bernier J, Kumar A, Venuprasad R, Spaner D, Verulkar S, Mandal NP, Sinha PK, Peeraju P, Dongre PR, Mahto RN, Atlin G. Characterization of the effect of a QTL for drought resistance in rice, qtl12.1, over a range of environments in the Philippines and eastern India. Euphytica. 2009;166:207–17.

Bilge B, Yildirim M, Barutcular C, Genc I. Effect of canopy temperature depression on grain yield and yield components in bread and durum wheat. Notulae Botanicae Horti Agrobotanici. 2008;36:34–7.

Blum A. Effective use of water (EUW) and not water-use efficiency (WUE) is the target of crop yield improvement under drought stress. Field Crop Res. 2009;112:119–23.

Blum A, Gozlan G, Mater J. Manifestation of dehydration avoidance in wheat breeding germplasm. Crop Sci. 1981;21:495–9.

Blum A, Shipiler L, Golan G, Mayer J. Yield stability and canopy temperature of wheat genotypes under drought stress. Field Crop Res. 1989;22:289–96.

Boyer JS, Wong SC, Farquhar CD. CO_2, and water vapor exchange across leaf cuticle (epidermis) at various water potentials. Plant Physiol. 1997;114:185–91.

Bruce WB, Edmeades GO, Barker TC. Molecular and physiological approaches to maize improvement for drought tolerance. J Exp Bot. 2002;53(366):13–25.

Cabuslay GS, Ito O, Alejal AA. Physiological evaluation of responses of rice (*Oryza sativa* L.) to water deficit. Plant Sci. 2002;163(4):815–27.

Cattivelli L, Rizza F, Badeck FW, Mazzucotelli E, Mastrangelo AM, Francia E, Mare C, Tondelli A, Stanca AM. Drought tolerance improvement in crop plants: An integrated view from breeding to genomics. Field Crop Res. 2008;105:1–14.

Comas LH, Becker SR, Cruz VMV, Byrne PF, Dierig DA. Root traits contributing to plant productivity under drought. Front Plant Sci. 2013;4:442.

Condon AG, Richards RA, Rebetzke GJ, Farquhar GD. Breeding for high water-use efficiency. J Exp Bot. 2004;55:2447–60.

Diab AA, Teulat-Merah B, This D, Ozturk NZ, Benscher D, Sorrells ME. Identification of drought-inducible genes and differentially expressed sequence tags in barley. Theor Appl Genet. 2004;109:1417–25.

Ding X, Li X, Xiong L. Evaluation of near-isogenic lines for drought resistance QTL and fine mapping of a locus affecting flag leaf width, spikelet number, and root volume in rice. Theor Appl Genet. 2011;123:815–26.

Dixit S, Huang BE, Sta Cruz MT, Maturan PT, Ontoy JCE, Kumar A. QTLs for tolerance of drought and breeding for tolerance of abiotic and biotic stress: an integrated approach. PLoS One. 2014a;9(10), e109574.

Dixit S, MallikarjunaSwamy BP, Vikram P, Bernier J, Sta Cruz MT, et al. Increased drought tolerance and wider adaptability of *qDTY12.1* conferred by its interaction with *qDTY2.3* and *qDTY3.2*. Mol Breed. 2012;30:1767–79.

Dixit S, Singh A, Cruz MTS, Maturan PT, Amante M, Kumar A. Multiple major QTL lead to stable yield performance of rice cultivars across varying drought intensities. BMC Genet. 2014b;15:16.

Du W, Yu D, Fu S. Detection of quantitative trait loci for yield and drought tolerance traits in soybean using a recombinant inbred line population. J Integr Plant Biol. 2009;51(9):868–78.

Ehdaie B, Hall AE, Farquhar GD, Nguyen HT, Waines JG. Water use efficiency and carbon isotope discrimination in wheat. Crop Sci. 1991;31:1282–8.

El-Sharkawy MA. Physiological characteristics of cassava tolerance to prolonged drought in the tropics: implications for breeding cultivars adapted to seasonally dry and semiarid environments. Braz J Plant Physiol. 2007;19(4):257–86.

Fischer KS, Lafitte R, Fukai S, Atlin G, Hardy B, editors. Breeding rice for drought-prone environments. Los Baños (Philippines): International Rice Research Institute; 2003. 98 p.

Fischer RA, Wood JT. Drought resistance in spring wheat cultivars. III. Yield association with morpho-physiological traits. Aust J Agric Res. 1979;30:1001–20.

Fleury D, Jefferies S, Kuchel H, Langridge P. Genetic and genomic tools to improve drought tolerance in wheat. J Exp Bot. 2010;61:3211–22.

Araghi SG, Assad MT. Evaluation of four screening techniques for drought resistance and their relationship to yield reduction ratio in wheat. Euphytica. 1998;103:293–9.

Gomez MS, Boopathi NM, Kuma SS, Ramasubramanian T, Chengsong Z, Jeyaprakash P, Senthil A, Babu RC. Molecular mapping and location of QTLs for drought-resistance traits in *indica* rice (*Oryza sativa* L.) lines adapted to target environments. Acta Physiol Plant. 2010;32:355–64.

Gowda VRP, Henry A, Yamauchi A, Shashidhar HE, Serraj R. Root biology and genetic improvement for drought avoidance in rice. Field Crop Res. 2011;122:1–13.

Graham J, Ratnaparkhe MB, Powell W. Molecular mapping and breeding of physiological traits. In: Kole C, Abbott AG, editors. Principles and practices of plant genomics, Molecular breeding, vol. 2. Enfield: Science; 2008. p. 217–41.

Grant OM, Tronina L, Jones HG, Chaves MM. Exploring thermal imaging variables for the detection of stress responses in grapevine under different irrigation regimes. J Exp Bot. 2007;58:815–25.

Gratani L. Plant phenotypic plasticity in response to environmental factors. Adv Bot. 2014;208747:17 p.

Guendouz A, Guessoum S, Maamari BM, Hafsi M. Canopy temperature efficiency as indicators for drought tolerance in durum wheat (*Triticum durum* Desf.) in semi arid conditions. J Agric Sustain. 2012;1(1):23–38.

Guendouz A, Guessoum S, Benidirand M, Hafsi M. Relationship between flag leaf reflectance and canopy temperature in durum wheat (*Triticum durum* Desf.) cultivars under stressed and irrigated conditions. J Agric Sustain. 2013;3(1):78–86.

Gumuluru S, Hobbs SLA, Jana S. Genotypic variability in physiological characters and its relationship to drought tolerance in durum wheat. Can J Plant Sci. 1989;69:703–11.

Horie T, Matsuura S, Takai T, Kuwasaki K, Ohsumi A, Shiraiwa T. Genotypic difference in canopy diffusive conductance measured by a new remote-sensing method and its association with the difference in rice yield potential. Plant Cell Environ. 2006;29:653–60.

Hu H, Xiong L. Genetic engineering and breeding of drought-resistant crops. Annu Rev Plant Biol. 2014;65:715–41.

Impa SM, Nadaradjan S, Boominathan P, Shashidhar G, Bindumadhava H, Sheshshayee MS. Carbon isotope discrimination accurately reflects variability in WUE measured at a whole plant level in rice. Crop Sci. 2005;45(6):2517–22.

Ingram KT, Bueno FD, Namuco OS, Yambao EB, Beyrouty CA. Rice root traits for drought resistance and their genetic variation. In: Kirk GJD, editor. Rice Roots: Nutrient and Water Use. Manila, Philippines: International Rice Research Institute; 1994.

Iwama K. Physiology of the potato: new insights into root system and repercussions for crop management. Potato Res. 2008;51:333–53.

Jamieson PD, Martin RJ, Francis GS, Wilson DR. Drought effects on biomass production and radiation-use efficiency in barley. Field Crop Res. 1995;43:77–86.

Jones H. Breeding for stomatal characters. In: Zeiger E, Farquhar GD, Cowan IR, editors. Stomatal Function. Stanford, USA: Stanford University Press; 1987. p. 431–43.

Karamanos AJ, Papatheohari AY. Assessment of drought resistance of crop genotypes by means of the Water Potential Index. Crop Sci. 1999;39:1792–7.

Kashiwagi J, Krishnamurthy L, Purushothaman R, Upadhyaya HD, Gaur PM, Gowda CLL, Itoc O, Varshney RK. Scope for improvement of yield under drought through the root traits in chickpea (*Cicer arietinum* L.). Field Crop Res. 2015;170:47–54.

Kondo M, Pablico PP, Aragones DV, Agbisit R. Genotypic variations in carbon isotope discrimination, transpiration efficiency, and biomass production in rice as affected by soil water conditions and N. Plant Soil. 2004;267(1):165–77.

Kosová K, Vítámvás P, Urban MO, Kholová J, Prášil IT. Breeding for enhanced drought resistance in barley and wheat – drought-associated traits, genetic resources and their potential utilization in breeding programmes. Czech J Genet Plant Breed. 2014;50:247–61.
Kumar A, Bernier J, Verulkar S, Lafitte HR, Atlin GN. Breeding for drought tolerance: direct selection for yield, response to selection and use of drought-tolerant donors in upland and lowland-adapted populations. Field Crop Res. 2008;107:221–31.
Kumar R, Venuprasad R, Atlin GN. Genetic analysis of rainfed lowland rice drought tolerance under naturally-occurring stress in eastern India: heritability and QTL effects. Field Crop Res. 2007;103:42–52.
Kusvuran S, Dasgan HY, Kuçukkomurcu S, Abak K. Relationship between drought tolerance and stomata density in melon. Acta Hortic. 2010;871:291–300.
Lafitte HR, Li ZK, Vijayakumar CHM, Gao YM, Shi Y, et al. Improvement of rice drought tolerance through backcross breeding: evaluation of donors and selection in drought nurseries. Field Crop Res. 2006;97:77–86.
Lambrides CJ, Chapman SC, Shorter R. Genetic variation for carbon isotope discrimination in sunflower: association with transpiration efficiency and evidence for cytoplasmic inheritance. Crop Sci. 2004;44:1642–53.
Legg BJ, Day W, Lawlor DW, Parkinson KJ. The effects of drought on barley growth: models and measurements showing the relative importance of leaf area and photosynthetic rate. J Agric Sci Cambr. 1979;92:703–16.
Lilley JM, Fukai S. Effect of timing and severity of water deficit on four diverse rice cultivars. Rooting pattern and soil water extraction. Field Crop Res. 1994;37:205–13.
Lopes MS, Reynolds MP. Partitioning of assimilates to deeper roots is associated with cooler canopies and increased yield under drought in wheat. Funct Plant Biol. 2010;37:147–56.
Mambani B, Lal R. Response of upland rice varieties to drought stress. Relation between the root system development and leaf water potential. Plant Soil. 1983;73:59–72.
Manavalan LP, Guttikonda SK, Tran LSP, Nguyen HT. Physiological and molecular approaches to improve drought resistance in soybean. Plant Cell Physiol. 2009;50:1260–76.
Martin B, Thorstenson YR. Stable carbon isotope composition (delta-c-13), water-use efficiency, and biomass productivity of *Lycopersicon esculentum*, *Lycopersicon pennellii*, and the F1-hybrid. Plant Physiol. 1988;88:213–7.
Masle J, Gilmore SR, Farquhar GD. The ERECTA gene regulates plant transpiration efficiency in *Arabidopsis*. Nature. 2005;436:866–70.
Mehri N, Fotovat R, Sba J, Jabbari F. Variation of stomata dimensions and densities in tolerant and susceptible wheat cultivars under drought stress. J Food Agric Environ. 2009;7(1):167–70.
Mir RR, Zaman-Allah M, Sreenivasulu N, Trethowan R, Varshney RK. Integrated genomics, physiology and breeding approaches for improving drought tolerance in crops. Theor Appl Genet. 2012;125(4):625–45.
Miranzadeh H, Emam Y, Seyyed H, Zare S. Productivity and radiation use efficiency of four dryland wheat cultivars under different levels of nitrogen and chlormequat chloride. J Agric Sci Tech. 2011;13:339–51.
Mohammadi MR, Sabaghnia KN, Shefazadeh MK. Effective application of canopy temperature for wheat genotypes screening under different water availability in warm environments. Bulg J Agric Sci. 2012;18:934–41.
Mohammady DS. Physiological characters associated with water-stress tolerance under pre anthesis water-stress conditions in wheat. 2011. http://www.shigen.nig.ac.jp/ewis/article/html/17/article.html.Retreived. Accessed Sept 10, 2015.
Mohankumar MV, Sheshshayee MS, Rajanna MP, Udayakumar M. Correlation and path analysis of drought tolerance traits on grain yield in rice germplasm accessions. J Agric Biol Sci. 2011;6(7):70–7.
Monneveux P, Rekika D, Acevedo E, Merah O. Effect of drought on leaf gas exchange, carbon isotope discrimination, transpiration efficiency and productivity in field grown durum wheat genotypes. Plant Sci. 2006;170:867–72.

Mutava RN, Prasad PVV, Tuinstra MR, Kofoid MD, Yu J. Characterization of sorghum genotypes for traits related to drought tolerance. Field Crop Res. 2011;123:10–8.

Turner NC, Blum A, Cakir M, Steduto P, Tuberosaand R, Young N. Strategies to increase the yield and yield stability of crops under drought – are we making progress?Functional. Plant Biol. 2014;41:1199–206.

O'Toole JC, Bland WL. Genotypic variation in crop plant root systems. Adv Agron. 1987;41: 91–145.

Obidiegwu JE, Bryan GJ, Jones HG, Prashar A. Coping with drought: stress and adaptive responses in potato and perspectives for improvement. Front Plant Sci. 2015;6:542.

Ouk M, Basnayake J, Tsubo M, Fukai S, Fischer KS, et al. Use of drought response index for identification of drought tolerant genotypes in rainfed lowland rice. Field Crop Res. 2006;99:48–58.

Pandey V, Shukla A. Acclimation and tolerance strategies of rice (*Oryza sativa* L.) under drought stress. Rice Sci. 2015;22(4):147–61.

Parry MAJ, Reynolds M, Salvucci ME, Raines C, Andralojc PJ, Zhu XG, Price GD, Condon AG, Furbank RT. Raising yield potential of wheat. II. Increasing photosynthetic capacity and efficiency. J Exp Bot. 2011;62:453–67.

Peter F. Radiation use efficiency in spring barley under drought: a crosstalk between survival strategy and canopy structure. J Cent Eur Agric. 2010;11(1):83–92.

Pietragalla J. Canopy temperature. In: Pask AJD, Pietragalla J, Mullan DM, Reynolds MP, editors. Physiological Breeding II: A field guide to Wheat phenotyping. Mexico: CIMMYT; 2012. p. 10–6.

Pinto RS, Reynolds MP, Mathews KL, McIntyre CL, Olivares-Villegas JJand Chapman SC. Heat and drought adaptive QTL in a wheat population designed to minimize confounding agronomic effects. Theor Appl Genet. 2010;121:1001–21.

Prashar A, Yildiz J, McNicol JW, Bryan GJ, Jones HG. Infra-red thermography for high throughput field phenotyping in *Solanum tuberosum*. PLoS One. 2013;8, e65816.

Price AH, Cairns JE, Horton P, Jones HG, Griffiths H. Linking drought-resistant mechanisms to drought avoidance in upland rice using a QTL approach: progress and new opportunities to integrate stomatal and mesophyll responses. J Exp Bot. 2002;53:989–1004.

Prince SJ, Beena R, Gomez SM, Senthivel S, Babu RC. Mapping consistent rice (*Oryza sativa* L.) yield QTLs under drought stress in target rainfed environments. Rice. 2015;8:25.

Puértolas J, Ballester C, Elphinstone ED, Dodd IC. Two potato (*Solanum tuberosum*) varieties differ in drought tolerance due to differences in root growth at depth. Funct Plant Biol. 2014;41:1107–18.

Rao RCN, Williams JH, Wadia KDR, Hubick KT, Farquhar GD. Crop growth, water use efficiency and carbon isotope discrimination in groundnut (*Arachis hypogeae* L.) genotypes under end of season drought conditions. Ann Appl Biol. 1993;122:357–67.

Rashid A, Stark JC, Tanveer A, Mustafa T. Use of canopy temperature measurements as a screening tool for drought tolerance in spring wheat. J Agron Crop Sci. 1999;182:231–7.

Rebetzke GJ, Condon AG, Richards RA, Farquhar GD. Selection for reduced carbon isotope discrimination increases aerial biomass and grain yield of rainfed bread wheat. Crop Sci. 2002;42:739–45.

Rebetzke GJ, Rattey AR, Farquhar GD, Richards RA, Condon AG. Genomic regions for canopy temperature and their genetic association with stomatal conductance and grain yield in wheat. Funct Plant Biol. 2013;40:14–33.Reynolds MP, Dreccer F, Trethowan R. Drought-adaptive traits derived from wheat wild relatives and landraces. J Exp Bot. 2007;58:177–186.

Reynolds MP, Mujeeb-Kazi A, Sawkins M. Prospects for utilising plant-adaptive mechanisms to improve wheat and other crops in drought- and salinity-prone environments. Ann Appl Biol. 2005;146:239–59.

Reynolds MP, Dreccer F, Trethowan R. Drought-adaptive traits derived from wheat wild relatives and landraces. J Exp Bot. 2007;58:177–86.

Reynolds MP, Nagarajan S, Razzaque MA and Ageeb OAA (1997) Using canopy temperature depression to select for yield potential of wheat in heat-stressed environments. Wheat special report No: 42. Mexico: CIMMYT.

Reynolds MP, Nagarajan S, Razzaque MA, Ageeb OAA. Heat tolerance. In: Reynolds MP, Ortiz-Monasterio JI, McNab A, editors. Application of Physiology in Wheat Breeding. Mexico: CIMMYT; 2001. p. 124–35.

Richards RA. Crop improvement for temperate Australia: Future opportunities. Field Crop Res. 1991;26:141–69.

Royo C, Villegas D, Garcia Del Moral LF, Elhani S, Aparicio N, Rharrabti Y, Araus JL. Comparative performance of carbon isotope discrimination and canopy temperature depression as predictors of genotypes differences in durum wheat yield in Spain. Aust J Agric Res. 2002;53:561–9.

Shaibu AS, Adnan AA, Rabiu IU. Genetic correlation and contribution of some physiological traits to yield in some selected maize genotypes. Int J Res Sci Technol. 2015;5(4):1–10. Silim SN, Saxena MC. Adaptation of spring-sown chickpea to the Mediterranean basin: I. Response to moisture supply. Field Crops Res. 1993;34:121–136.

Silim SN, Saxena MC. Adaptation of spring-sown chickpea to the Mediterranean basin: I. Response to moisture supply. Field Crops Res. 1993;34:121–36.

Sinclair TR. Challenges in breeding for yield increase for drought. Trends Plant Sci. 2011;16:289–93.

Sinclair TR. Is transpiration efficiency a viable plant trait in breeding for crop improvement? Funct Plant Biol. 2012;39(5):359–65.

Sinclair TR, Muchow RC. Radiation use efficiency. Adv Agron. 1999;65:215–65.

Stalham MA, Allen EJ. Water uptake in the potato (*Solanum tuberosum*) crop. J Agric Sci. 2004;142:373–93.

Steele KA, Price AH, Shashidhar HE, Witcombe JR. Marker-assisted selection to introgress rice QTLs controlling root traits into an Indian upland rice variety. Theor Appl Genet. 2006;112:208–21.

Swamy BPM, Ahmed HU, Henry A, Mauleon R, Dixit S, Vikram P, Tilatto R, Verulkar SB, Perraju P, Mandal NP, Variar M, Robin S, Chandrababu R, Singh ON, Dwivedi JL, Das SP, Mishra KK, Yadaw RB, Aditya TL, Karmakar B, Satoh K, Moumeni A, Kikuchi S, Leung H, Kumar A. Genetic, physiological, and gene expression analyses reveal that multiple QTL enhance yield of rice mega-variety IR64 under drought. PLoS One. 2013;8(5), e62795.

Tuberosa R, Salvi S, Sanguineti MC, Land PI, Maccaferr M, Cont SI. Mapping QTLs regulating morpho-physiological traits and yield: case studies, shortcomings and perspectives in drought-stressed maize. Ann Bot. 2002;89:941–63.

Vadez V. Root hydraulics: the forgotten side of roots in drought adaptation. Field Crop Res. 2014;165:15–24.

Venora G, Calcagno F. Study of stomatal parameters for selection of drought resistant varieties in *Triticum durum* DESF. Euphytica. 1991;57:275–83.

Venuprasad R, Bool ME, Dalid CO, Bernier J, Kumar A, Atlin GN. Genetic loci responding to two cycles of divergent selection for grain yield under drought stress in a rice breeding population. Euphytica. 2009;167:261–9.

Venuprasad R, Sta Cruz MT, Amante M, Magbanua R, Kumar A, Atlin GN. Response to two cycles of divergent selection for grain yield under drought stress in four rice breeding populations. Field Crop Res. 2008;107:232–44.

Wajid A, Hussain K, Maqsood M, Ahmad A, Hussain A. Influence of drought on water use efficiency in wheat in semi-arid regions of Punjab. Soil Environ. 2007;26:64–8.Wang H, Clarke JM. Relationship of excised-leaf water-loss and stomatal frequency in wheat. Can J Plant Sci. 1993;73:93–99.

Wang H, Clarke JM. Relationship of excised-leaf water-loss and stomatal frequency in wheat. Can J Plant Sci. 1993;73:93–9.

Wright GC, Rao RCN, Farquhar GD. Water use efficiency and carbon isotope discrimination in peanut under water deficit conditions. Crop Sci. 1994;34:92–7.

Xu Z, Zhou G. Responses of leaf stomatal density to water status and its relationship with photosynthesis in a grass. J Exp Bot. 2008;59(12):3317–25.

Yadav R, Courtois B, Huang N, McLaren G. Mapping genes controlling root morphology and root distribution in a doubled-haploid population of rice. Theor Appl Genet. 1997;94:619–32.

Yu SB, Xu WJ, Vijayakumar CHM, Ali J, Fu BY, et al. Molecular diversity and multilocus organization of the parental lines used in the International Rice Molecular Breeding Program. Theor Appl Genet. 2003;108:131–40.

Zaman-Allah M, Jenkinson DM, Vadez V. Chickpea genotypes contrasting for seed yield under terminal drought stress in the field differ for traits related to the control of water use. Funct Plant Biol. 2011a;38:270–81.

Zaman-Allah M, Jenkinson DM, Vadez V. A conservative pattern of water use, rather than deep or profuse rooting, is critical for the terminal drought tolerance of chickpea. J Exp Bot. 2011b;62:4239–52.

Zia S, Romano G, Spreer W, Sanchez C, Cairns J, Araus JL, et al. Infrared thermal imaging as a rapid tool for identifying water-stress tolerant maize genotypes of different phenology. J Agron Crop Sci. 2013;199:75–84.

Chapter 15
Photosynthesis, Antioxidant Protection, and Drought Tolerance in Plants

Irada M. Huseynova, Samira M. Rustamova, Durna R. Aliyeva, Hasan G. Babayev, and Jalal A. Aliyev

15.1 Introduction

Drought tolerance is considered to be a quantitative trait manifesting complex phenotypic and genetic control (McWilliam 1989; Saleh et al. 2014). Because of global environmental changes and increase in the world population, maintaining plant productivity under drought conditions is of great importance (Takeda and Matsuoka 2008). Interest in research of physiological and biochemical processes improving plant tolerance against adverse environmental factors has been increasing recently (Bray et al. 2000; Wang et al. 2003). From this point of view the study of stress effects on physiological and biochemical processes occurring in higher plants is considered to be actual.

The main food product for more than 35 % of the world population is wheat. Among crops wheat more attracts the attention of researchers because of its genetic properties and tolerance against water deficiency. From this point of view one of the major duties of selectionists is finding ways for improving drought tolerance and productivity of wheat. Drought, which is one of the stress factors, adversely affects plant growth and development and sharply reduces its productivity. Drought

I.M. Huseynova (✉) • S.M. Rustamova • D.R. Aliyeva • H.G. Babayev
Department of Fundamental Problems of Biological Productivity, Institute of Molecular Biology and Biotechnology, Azerbaijan National Academy of Sciences, Baku 1073, Azerbaijan
e-mail: huseynova-i@botany-az.org

J.A. Aliyev
Department of Fundamental Problems of Biological Productivity, Institute of Molecular Biology and Biotechnology, Azerbaijan National Academy of Sciences, Baku 1073, Azerbaijan

Department of Plant Physiology and Biotechnology, Research Institute of Crop Husbandry, Ministry of Agriculture of Azerbaijan Republic, Pirshagi Vil., Sovkhoz 2, Baku 1098, Azerbaijan

M.A. Hossain et al. (eds.), *Drought Stress Tolerance in Plants, Vol 1*,
DOI 10.1007/978-3-319-28899-4_15

changes colloidal-chemical composition of cytoplasm leading to water deficiency, protein decomposition, and the decrease in organic compounds accumulated in plants (Taiz and Zeiger 2006). The decline in water potential caused by high temperature and drought leads to the induction of osmotic stress (Molinari et al. 2007). ROS produced under drought damage vital macromolecules of cells and weaken their functions (Foyer et al. 1994; Noctor and Foyer 1998; Mittler 2002), cause peroxidation of membrane lipids (Mead 1976), inactivate enzymes (Fucc et al. 1983), and inhibit cell cycle (Rehman et al. 2005). The plant response to stress is dependent on the degree and duration of stress. Drought causes an excessive accumulation of ROS leading to the induction of antioxidant enzymes such as superoxide dismutase (SOD, EC 1.15.1.1), catalase (CAT, EC 1.11.1.6), and peroxidase (POX, EC 1.11.1.7). The enzymes of ascorbate-glutathione cycle ascorbate peroxidase (APX, EC 1.11.1.11), glutathione reductase (GR, EC 1.8.1.7), dehydroascorbate reductase (DHAR, EC 1.8.5.1), and monodehydroascorbate reductase (MDAR, EC 1.6.5.4) are also activated. There have been recent reports about connection between drought stress and the antioxidant ability of wheat, maize, and rice plants (Lascano et al. 2001; Jiang and Zhang 2002; Guo et al. 2006).

Understanding the genetic and physiological bases of drought tolerance in crop plants is necessary for developing drought tolerant genotypes through conventional breeding. Plant response to drought has very complex nature and must be viewed as a whole system and large scale identification of probable dehydration stress-related genes or QTLs is necessary. The appropriate technology and resources are required for QTL cloning, which is a very time-consuming procedure. Therefore, marker-assisted selection (MAS) and highly productive cultivar development provide great advantages. Molecular maps, developed with DNA markers, are used in the identification of QTLs. The establishment of the molecular maps became possible due to the recent achievements in functional genomics including bacterial artificial chromosomes (BACs), gene sequence data, molecular marker technology, and bioinformatic tools for comparative genomics (Budak et al. 2013).

15.2 Antioxidant Defence System of Wheat Under Drought Stress

Reactive oxygen species (ROS) are formed in various compartments of plant cells under normal as well as stress conditions. The main reason of cell damage caused by various stressors is intensification of ROS generation (Sivamani et al. 2000). Plant tolerance against stressor effects is dependent on the ratio of ROS and the activity of antioxidant system (AOS). ROS include hydrogen peroxide (H_2O_2), anion-radical ($O_2^{\cdot -}$), hydroxyl radical ($HO^{\cdot}$), hyperoxide radical ($HO_2^{\cdot}$), singlet oxygen (O_2), and ozone (O_3). Chloroplasts, mitochondria, and peroxisomes are considered to be major intracellular generators of ROS. The photosyntheticelectron transport chain is thought to be the main source of ROS in plant tissues (Asada 1994). $O_2^{\cdot -}$ and H_2O_2 are mainly formed in chloroplasts due to the electron acceptor of PS I. Wherein singlet oxygen is formed as a result of the electron transfer from the excited chlorophyll molecule to molecular oxygen. ROS are formed also in the electron transport chain of mitochondria. Mitochondria are very sensitive to oxidative stress. Lenaz (1998) studied lipid and protein peroxidation and mutations of mitochondrial DNA caused by oxidative stress. If plants have no defence mechanism, ROS can cause serious damage in various cell structures and disturb their functions. For example, superoxide radical forms hydroxyl radical as a result of the Fenton reaction. Hydroxyl radical is very toxic and capable of lipid peroxidation and destroying DNA and proteins (Arora et al. 2002). Moreover, recently an opinion formed about a signal role of ROS in closing stomata under stress. Abscisic acid (ABA) is formed leading to the formation of the signaling cascade in cells and stomatal closure in response to drought stress. This process is very important under stress conditions, because stomata localized in epidermis control CO_2 uptake for photosynthesis and water loss through transpiration. AOS is one of the major chains of the signaling cascade localized in the plasmatic membrane, which activates Ca channels. Thus, the sequence of processes occurring in cells under drought stress is as follows: physiological water deficiency-stomatal closure regulated by ABA-limiting CO_2 uptake-weakening electron transport chain and ROS formation. Thus, ROS perform dual function in cells: they play a role of signaling molecule at low concentrations, while they are too harmful and can cause plant death at higher concentrations (Slesak et al. 2007).

H_2O_2 and MDA are considered to be major indicators of drought stress in plants. Their amount during stress is dependent on plant species, stress duration, and plant age. Comparative study of tolerance of C3 and C4 plants exposed to 5 and 10 day drought showed that H_2O_2 and MDA were more accumulated in C3 plants than in C4 ones (Uzildaya et al. 2012). This confirms that C3 plants are more sensitive to drought. After 35-day exposure to water stress H_2O_2 amount increased in 30-day-old *T. durum* seedlings (Miller et al. 2010), whereas in 3-month-old *T. aestivum* H_2O_2 amount decreased and MDA increased (Simova-Stoilova et al. 2009). The amount of peroxide groups is considered to be one of the major markers of oxidative stress. Numerous experimental data obtained recently show that one of the universal response reactions of plant cells to extreme environmental conditions is the activation of the lipid peroxidation (LPO) process (Da Costa and Huang 2007; Pandey et al. 2010). LPO reaction occurs in all cells of living organisms especially in lipid structures of the cell membrane. This process consists of three phases: initiation, propagation, and termination. In the first phase organic radicals (R·) are produced under the influence of various stressors. At the next stage these radicals immediately react with O_2 molecules forming peroxide radicals (RO_2·). Then they affect unsaturated lipids and produce organic peroxides and new radicals. In other words, lipid peroxidation occurs and its main sign is an increase in MDA amount. Peroxides are chemically active compounds and their structures are similar to H_2O_2 structure. Termination occurs due to the interaction between radicals and antioxidants. Peroxidation products—ROS formed as a result of the stressor effect are considered to be initial mediators of the antioxidant system activation. A high concentration of ROS in plant cells is considered to be one of the major signs of stress and plants try to maintain their amount in a level necessary for normal cell homeostasis. In cells the amount of ROS is regulated by AOS. The antioxidant protection system includes mainly antioxidant enzymes—superoxide dismutase, catalase, peroxidase, and also fat-soluble antioxidants—tocopherol, ubiquinone, retinol, carotenoids; water-soluble antioxidants-glutathione, ascorbic acid, etc. (Alscher et al. 1997). Thus, AOS is the main line of cell defense against toxic influences of oxygen. Induction of antioxidants is dependent on the degree of stress. High or low light intensities, high salt concentrations, long-term drought accompanied by high temperature, etc. lead to the induction of antioxidants and activities of the oxidative stress enzymes. Antioxidants play an important role in the plant protection against oxidative damage under stress. Major ROS scavenging antioxidants are tocopherols, carotenoids, phenolic compounds and also proline and glycine betaine (Allakhverdiev et al. 2007). Osmolytes neutralize ROS produced under stress as well as control the regulation of protein structures.

Carotenoids are mainly localized in chloroplasts and participate in the regulation of the plant pigment system, maintaining functional activity of chlorophyll and protection of photosynthetic apparatus against singlet oxygen under stress conditions. According to Krieger-Liszkay (2004), β-carotene with α-tocopherol is the primary defense line for the protection against AOS. α- and γ-tocopherols play an important role in $^1O_2$ neutralizing. They participate in the protection of PS II and its D1 protein especially under stress conditions (Kruk et al. 2005).

Ascorbic acid controls physiological processes such as growth, differentiation of tissue and organs, and metabolism in plants. Its main function under stress is decreasing concentrations of superoxide and hydroxyl radicals.

Glutathione protects thiol groups against oxidation, inactivates radicals, reacts with ROS, and destroys peroxides. The reduced form of glutathione (GSH) is converted into the oxidized form (GSSG) by the enzyme glutathione reductase (GR).

Glycine betaine (GB) is mainly accumulated in higher plants under drought and salt stress (Rhodes and Hanson 1993). GB synthesis in higher plants is implemented as follows:

$$\text{Choline} \rightarrow \text{betaine aldehyde} \rightarrow \text{glycine betaine}$$

The first step is catalyzed by choline monooxygenase (Brouquisse et al. 1989) and the second one by betaine aldehyde dehydrogenase (Weigel et al. 1986).

The changes in activities of the oxidative enzymes occurring under stress are dependent on plant species. For example, in sunflower seedlings and a herbaceous plant *Aegilops squarrosa* the activity of superoxide dismutase decreased under water stress (Badiani et al. 1990). In contrast, significant increases of the enzyme activity were observed in wheat (Sairam et al. 1998) and rice (Sharma and Dubey 2005) under drought conditions. Simova-Stoilova et al. (2010) detected increases in catalase activity under drought and the enzyme activity appeared to be higher in sensitive genotypes compared with tolerant ones. Another research revealed decreases of CAT activity in rice seedlings under drought (Sharma and Dubey 2005).

Activities of ascorbate peroxidase and glutathione reductase increased in wheat seedlings (Keles and Oncel 2002) and alfalfa plants (Rubio et al. 2002) exposed to water deficiency.

One of the major enzymes of oxidative stress glutathione reductase converts oxidized form of glutathione (GSSG) into reduced one (GSH) (Alscher et al. 1997).

Experiments performed under field conditions with wheat plants exposed to moderate drought at the stage of germination showed that tolerant genotypes were better adapted to stress conditions than sensitive ones (Khanna-Chopra and Selote 2007). Selote and Khanna-Chopra (2004) detected that rice plants exposed to drought stress had high values of RWC and turgor potential and low concentrations of H_2O_2 due to very high activities of SOD and APO. A tolerant genotype (N22) was observed to be more productive compared with a sensitive genotype (N118).

Some researchers showed that CAT played the main role in the detoxification of H_2O_2 (Lei et al. 2006). However, Sofo and co-authors (2005) confirmed that CAT was less important compared with APO for the detoxification of H_2O_2 in the olive tree roots exposed to a long-term stress. Some biochemical parameters were studied for drought tolerant (*M. prunifolia*) and sensitive (*M. hupehensis*) varieties of apple under 12-day water deficiency (Wang et al. 2012). There were no significant changes in antioxidant parameters and lipid peroxidation in control variants. Under drought conditions H_2O_2 and MDA increased more in *M. hupehensis* compared with the tolerant variety. However, the activities of SOD, POD, APO, GR, and DHAR were higher in the *M. prunifolia* variety. But no marked differences were observed in the activities of CAT and MDHAR. Moreover, ultrastructures of organoids were better maintained and amounts of H_2O_2 and MDA were less in the tolerant variety. This shows that antioxidant enzymes can decrease harmful effects induced by drought in the tolerant *M. prunifolia* variety.

Thus, the main purpose of the presented work was the study of effects of long-term soil drought on physiological, biochemical, and molecular processes occurring in contrasting wheat genotypes during the generative development periods. The obtained data can contribute to the more comprehensive understanding of mechanisms affecting physiological, biochemical, and molecular processes and finding ways for improving productivity under long-term soil drought conditions.

15.2.1 Plant Material and Experimental Conditions

Two contrasting durum wheat (*Triticum durum* Desf.) genotypes from the gene pool of the Research Institute of Crop Husbandry were used in this study. Barakatli-95—short stature, with vertically oriented small leaves and grain yield of 6.0–7.0 th^{-1}, drought tolerant and Garagylchyg-2—short stature, with vertically oriented small leaves and grain yield of 7.0–8.0 th^{-1}, drought sensitive. Plants were grown in the field over a wide area under normal water supply and drought conditions. Dehydration was achieved by cessation of watering. Control variants of wheat varieties were watered till the end of the vegetation, while for experimental variants drought was imposed at the stage of intensive growing from April to June. Experiments were carried out in phases of flowering, milk ripeness, and wax ripeness, as this period is the most sensitive to water deficiency and plants are more exposed to water stress.

15.2.2 Determination of Glycine Betaine Content

Glycine betaine content was determined using the method of Grieve and Grattan (1983). Optical density of the stained solution was measured with spectrophotometer at 365 nm. Betaine amount was defined from calibration curve, using commercial preparation (Serva) as a standard.

15.2.3 Determination of Hydrogen Peroxide Content

Hydrogen peroxide content was assayed with the redox active indicator xylenol orange according to Bellincampi et al. (2000). Supernatant absorbance was measured at 560 nm. Values were calculated using standard curve with known amount of H_2O_2.

15.2.4 Determination of Malondialdehyde Content

Lipid peroxidation was estimated as TBARS (Heath and Packer 1968). MDA concentration was estimated by subtracting the nonspecific absorption at 600 nm from the absorption at 532 nm, using an absorbance coefficient of extinction (155 mM^{-1} cm^{-1}).

15.2.5 Isolation of the Enzyme Extract

To obtain total cell extract, wheat leaves and roots were homogenized in a medium containing 1 mM EDTA, 2 mM phenylmethylsulfonyl fluoride (PMSF), 1 % PVP, 100 mM Na-phosphate buffer (pH 7.8), 0.1 % Triton×-100, then filtered and centrifuged for 20 min at 15,000×*g*. The resulting supernatant was used for analysis of antioxidant enzymes.

15.2.6 Determination of the Isoenzyme Spectrum of Antioxidant Enzymes

Qualitative changes in the enzyme activities were determined using a native PAGE electrophoresis according to the method of Davis (Davis 1964). A separating gel of 8 % acrylamide was used for visualization of CAT, BPO and 10 % for GR isoenzymes. Enzyme extract in 50 % glycerol with 1 % bromophenol blue was applied to the gel. Electrophoresis was carried out for 3 h at 4 °C with a steady current of 30 mA, using the device SE 250 (Amersham Biosciences, USA). Following electrophoretic separation, the gels were stained for different isoenzymes.

15.2.7 Determination of Isoenzyme Spectrum of CAT

Staining of catalase lines was performed by the method of Anderson (Anderson et al. 1995). Gels were incubated in 0.003 % H_2O_2 for 10 min and developed in a 1 % (w/v) $FeCl_3$ and 1 % $K_3Fe(CN_6)$ (w/v) solution for 10 min.

15.2.8 Determination of Isoenzyme Spectrum of BPO

Staining of benzidine peroxidase lines was performed by the method of Cuypers et al. (2002). For analysis of benzidine peroxidase isoenzymes, the gel was incubated for 1 h at 35 °C in the solution containing 0.1 g benzidine/100 mL 0.2 M sodium acetate buffer (pH 5.0) and 2.5 mL 3 % H_2O_2/100 mL benzidine solution.

15.2.9 Determination of Isoenzyme Spectrum of GR

The spectra of GR isozymes were visualized according to the modified method of Rao et al. (1996). Identification of the isoforms occurred as a result of the reaction of glutathione, reduced by 3-(4,5-dimethyl-2-thiazolyl)-2,5-diphenyl-2-tetrazolium bromide and 2,6-dichlorophenol indophenol when shaking in the dark for 1 h.

15.2.10 Statistical Analysis

The paper presents data of three experiments carried out in three biological replicates. Calculations, graphing, and their descriptions were performed using applications: Microsoft Office Word 7 and Excel 7 for Windows XP.

Leaves and roots of the plant were taken for the experiment. One of the major plant organs—root is known to be directly exposed to stressors such as soil drought and salt. Accumulation of H_2O_2 and MDA is considered to be the main indicator of oxidative stress. The long-term soil drought led to the H_2O_2 increase in leaves and roots (Table 15.1). There were no marked changes in the amount of H_2O_2 in the control variants during the active development period. The lowest H_2O_2 content was observed in Barakatli-95, and the highest in Garagylchyg-2 at all stages both under well-watered and drought stress conditions. These results are consistent with previous reports about tomato (Behnamnia et al. 2009) and wheat (Alexieva et al. 2001; Luna et al. 2005) plants.

Malondialdehyde is one of the most important indicators of plant tolerance against stress. There have been a number of reports confirming that different stress effects inhibit biochemical processes in cells, which is accompanied by the intensification of lipid peroxidation and MDA accumulation (Shao et al. 2005a, b; Tatar

Table 15.1 Dynamics of H_2O_2 accumulation (μmol/g dry weight) in leaves and roots of wheat genotypes under soil drought conditions

Genotypes	Variants	Flowering		Milk ripeness		Wax ripeness	
		Leaf	Root	Leaf	Root	Leaf	Root
Barakatli-95	Watering	3.61±0.18	1.11±0.05	4.18±0.20	1.13±0.05	4.96±0.25	1.13±0.05
	Drought	7.53±0.37	1.31±0.06	7.93±0.39	1.69±0.08	9.68±0.48	1.93±0.09
Garagylchyg-2	Watering	4.91±0.24	1.12±0.04	5.14±0.25	1.14±0.06	5.94±0.29	1.13±0.04
	Drought	9.27±0.46	1.44±0.07	10.3±0.51	2.29±0.12	12.45±0.62	2.52±0.12

Table 15.2 Dynamics of MDA accumulation (μmol/g fresh weight) in leaves and roots of wheat genotypes under soil drought conditions

Genotypes	Variants	Flowering		Milk ripeness		Wax ripeness	
		Leaf	Root	Leaf	Root	Leaf	Root
Barakatli-95	Watering	6.32±0.31	1.12±0.056	5.95±0.27	0.29±0.014	5.03±0.25	0.29±0.014
	Drought	7.24±0.36	1.23±0.061	6.81±0.34	0.42±0.021	6.02±0.31	0.66±0.033
Garagylchyg-2	Watering	2.45±0.12	0.58±0.029	7.73±0.38	0.25±0.012	5.44±0.26	0.24±0.012
	Drought	3.46±0.17	0.91±0.045	8.39±0.42	0.36±0.018	7.95±0.39	0.96±0.048

Table 5.3 Dynamics of GB accumulation (μg/g fresh weight) in leaves and roots of wheat genotypes under soil drought conditions

Genotypes	Variants	Flowering		Milk ripeness		Wax ripeness	
		Leaf	Root	Leaf	Root	Leaf	Root
Biakatli-95	Watering	1.79±0.01	5.81±0.29	7.16±0.35	8.65±0.43	9.05±0.45	5.09±0.25
	Drought	2.60±0.13	17.5±0.86	6.69±0.33	17.5±0.87	27.1±1.35	8.61±0.43
Garagylchyg -2	Watering	3.18±0.16	8.30±0.41	3.91±0.19	10.3±0.51	8.21±0.41	6.20±0.31
	Drought	6.10±0.31	11.2±0.56	7.21±0.36	11.7±0.57	16.8±0.84	9.83±0.49

and Gevrek 2008). Under stress conditions the level of MDA accumulation is different in wheat genotypes with contrasting tolerance. The level of MDA content under water deficiency was found to increase in leaves and roots of both tolerant and sensitive lines compared with control variants (Table 15.2).

In leaves of Garagylchyg-2 under progressive drought conditions, significant accumulation of MDA occurred compared to the beginning of the vegetation, while MDA content decreased in the tolerant Barakatli-95 variety, indicating a less damaging impact of drought on the leaves of this variety. In susceptible wheat genotypes an increase of lipid peroxidation level was more significant, probably due to the characteristic properties of the membrane structures of plant cells. The highest MDA content was observed in the sensitive Garagylchyg-2 genotype in the phase of milk ripeness in both control and stressed variants (7.7 and 8.3 μmol/g fresh weight).

The experiments showed that MDA amount was higher in roots at the beginning of drought. It declined a little at the milk ripeness stage, remains stable till the end of drought in control variants, and sharply increased in stress variants. Its amount 2.25 times increased in the Barakatli-95 variety and 3.88 times in the Garagylchyg-2 genotype compared with control variants. Thus, the tolerant genotypes accumulate less H_2O_2 and MDA than sensitive ones and are less subjected to oxidative stress effects. Similar results were obtained in experiments performed with olive (Sofo et al. 2004), sunflower (Bailly et al. 1996), and coffee (Queiroz et al. 1998) plants.

We have studied GB content in wheat leaves and roots under normal and water deficit conditions. Differences between tolerant and sensitive forms have been observed in the content of GB (Table 15.3).

The glycine betaine content of the shoots and roots was significantly different between genotypes, as was the response of the genotypes to drought. The results from this experiment suggest that Barakatli-95 may be more tolerant to drought than Garagylchyg-2.

The table shows dynamics of the level fluctuations of glycine betaine in wheat cells during ontogenesis. Glycine betaine content in cells increased exceeding the parameters measured under normal conditions and GB level appeared to be higher in the tolerant wheat genotype compared with the sensitive one. At the end of ontogenesis 10- and 2.75-fold increases in GB content occurred in the Barakatli-95 and Garagylchyg-2 varieties, respectively. The long-term drought caused threefold increase in GB accumulation in leaves of Barakatli-95, compared with the control variant and nearly tenfold increase compared with its content at the beginning of the vegetation. These data show that accumulation of GB in wheat leaves is plant response to water deficiency and it is one of the most effective protective reactions. Glycine betaine can simultaneously function as a compound: (a) regulating intracellular osmotic potential; (b) controlling pH of cytoplasm; (c) stabilizing the structure of cell membranes (Rajasekaran et al. 1997; Allakhverdiev et al. 2007). At the same time accumulation of GB could be the result of multidirectional processes: increase of biosynthesis; decrease of degradation; changes in transport; decomposition of

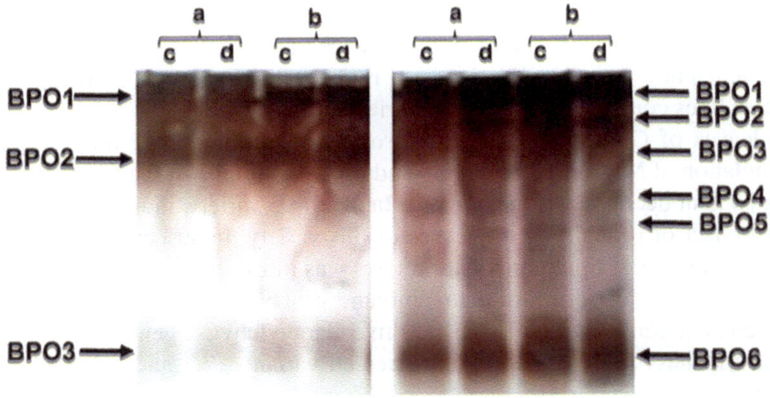

Fig. 15.1 Electrophoretic spectra of benzidine peroxidase in leaves (*left*) and roots (*right*) of wheat grown under drought. (*c*) watering, (*d*) drought, (*a*) Barakatli-95, (*b*) Garagylchyg-2. Electrophoresis was performed in 7 % PAAG, in Tris–glycine buffer, pH 8.3, at 4 °C, for 3 h at a steady current of 30 mA. 45 μg of protein was applied per lane

GB enriched proteins. The first two processes are possible in the case of active cell metabolism and adaptation to environmental stress conditions. Based on the obtained experimental data we suggest that over accumulation of GB increases the tolerance of wheat plants to water deficit because of the regulation of ion homeostasis and neutralizing ROS.

Assessment of electropherograms for CAT, BPO, and GR in two wheat genotypes displayed one, three, seven isoforms in leaves and three, six, seven isoforms in roots, respectively.

There is a wide range of peroxidase isoforms in plants. Spectrum of peroxidase forms is characterized by a very high lability, which gives reason to classify them as markers of physiological state of the plant. Multiple forms of peroxidases perform different functions in plants: some are involved in the processes of growth, while others have a protective role, providing the opportunity to obtain energy required for plants under stress for sustaining their vital functions. Three isoforms of benzidine peroxidase in leaves and six isoforms in roots were detected in the spectrum of the enzyme with intensifying soil drought. Isoenzyme spectrum analysis revealed more intensive staining of some bands of peroxidase isoenzymes (BPO2, BPO5, and BPO6) under water deficiency (Fig. 15.1), which apparently indicates the possibility of de novo synthesis of the enzyme.

Jang et al. (2004) showed a link between peroxidase isoenzyme composition and participation of PO genes in the formation of the protective mechanism in potato against pathogen infection. Based on the data, the authors concluded that the stress caused by the penetration of the pathogen had a significant effect on the gene expression of peroxidase. Seven GR isoforms were found in leaves of drought stressed wheat, four in roots of Barakatli and seven in roots of Garagylchyg (Fig. 15.2).

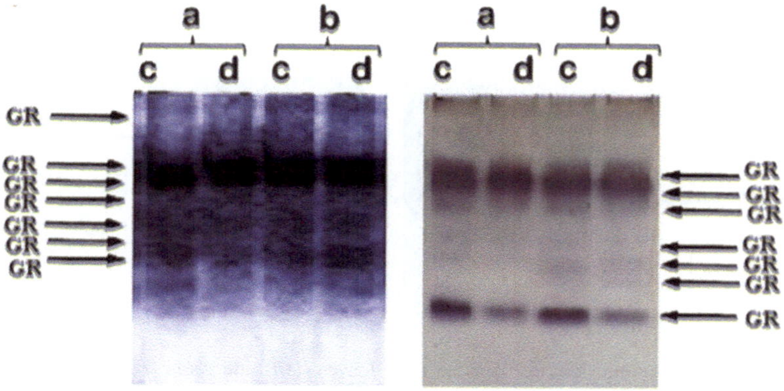

Fig. 15.2 Electrophoretic spectra of glutathione reductase in leaves (*left*) and roots (*right*) of wheat grown under soil drought. (*c*) Watering, (*d*) drought, (*a*) Barakatli-95, (*b*) Garagylchyg-2. Electrophoresis was performed in 10 % PAAG, in Tris–glycine buffer, pH 8.3, at 4 °C, for 3 h at a steady current of 30 mA. 40 μg of protein was applied per lane

There are conflicting data on changes in the activity of glutathione reductase under stress. When exposed to various concentrations of NaCl (50, 100, 150, 200 mM) GR activity in two wheat varieties decreased (Esfandiari et al. 2007). Long-term salinity caused an increase in the activity of GR in a perennial grass *Pennisetum clandestinum* (Muscolo et al. 2003). Yannarelli et al. (2007) found that treatment with Cd resulted in an increase of GR activity in leaves and roots of wheat (*Triticum aestivum* L.), grown at a moderately toxic cadmium concentration (100 μM). There were also reports about an increase in GR activity in rice seedlings (Sharma and Dubey 2005) and alfalfa plants (Rubio et al. 2002) under water deficiency. Lascano et al. (2001) observed an increase in GR activity and in amounts of the reduced form of glutathione and ascorbic acid under drought stress. The tolerant genotypes were less subjected to oxidative damage than sensitive ones.

Analysis of the catalase isoenzyme content revealed only one isoform of the enzyme with a low electrophoretic mobility in wheat leaves, both in stressed and control variants, which corresponds to the literature data (Racchi et al. 2001), whereas in the roots one isoform was found under normal and three isoforms under drought conditions. Stress associated with long-term soil drought in the roots of wheat has led to an increase in the heterogeneity due to the formation of two new sedentary forms of catalase: CAT2 and CAT3 (Fig. 15.3). It is assumed that the dynamics of the CAT activity increase is associated with a gradual increase in the concentration of hydrogen peroxide due to the dismutation reaction. Furthermore, catalase has a low substrate affinity and starts to operate at relatively high peroxide contents.

The results showed that intensification of drought in the active development period leads to the increase in H_2O_2 causing an adequate increase in catalase activity. Thus, the obtained data characterize the quantitative and qualitative changes in enzymes in different organs of wheat genotypes under long-term soil drought. The observed heterogeneity of individual forms of enzymes may have adaptive value and is a measure of resistance to water stressor.

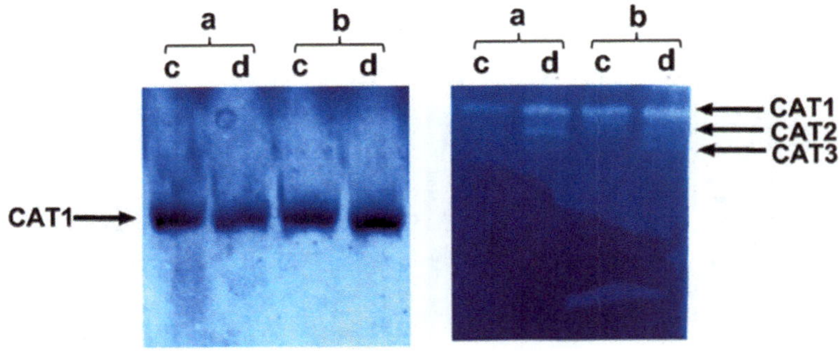

Fig. 15.3 Electrophoretic spectra of catalase in leaves (*left*) and roots (*right*) of wheat grown under soil drought. (*c*) Watering, (*d*) drought, (*a*) Barakatli-95, (*b*) Garagylchyg-2. Electrophoresis was performed in 7 % PAAG, in Tris–glycine buffer, pH 8.3, at 4 °C, for 3 h at a steady current of 30 mA. 40 μg of protein was applied per lane

15.3 Dynamics of Changes of Some Physicochemical and Kinetic Parameters of Carbonic Anhydrase in Leaf, Awn, and Root Cells of Wheat Under Drought

The first prediction of the existence of CA was based on theoretical considerations. Under usual conditions the reaction of the interconversion of CO_2 and bicarbonate proceeds very slowly, which led to the idea of the necessity of the catalytic factor in living organisms. Thus, carbonic anhydrase (CA; EC 4.2.1.1) was discovered by two research groups independently of each other (Meldrum and Roughton 1932; Stadie and O'Brien 1933) in erythrocytes of mammalians. Then it was detected in animals (Shpak 1980), plants, microalgae (Pronina 2000; Moroney et al. 2001), archae- and eubacteria (Supuran 2011; Smith and Ferry 2000).

According to the modern classification, all CAs are divided into three major families (α, β, γ). CAs involved in these families have no any homology in amino acid sequences and they evolved independently of each other (Ludwig 2011; Hewett-Emmett and Tashian 1996; Dudoladova et al. 2004; Smith and Ferry 2000). The family of α-CA is *evolutionarily younger* than β- and γ-CAs (Rowlett 2010; Smith and Ferry 2000). Spatial structures of CAs from various families are different. So α-Cas are mainly monomers (Smith and Ferry 2000; Supuran 2011), γ-CAs have three-dimensional structures (Ferry 2010), while β-CAs consist of proteins having structures from dimer to octamer (Smith and Ferry 2000; Zimmermann and Ferry 2008; Ferry 2010).

CAs of different origins differ not only structurally but also in their localization. So, CA is localized in cell membranes (Guliev et al. 2003), cytoplasm (Hiltonen et al. 1998; Guliev et al. 2003), chloroplast, mitochondria, and carboxysomes (Moroney et al. 2001; Yu et al. 1992).

The substrate of CA-inorganic carbon (C_i) is one of the major metabolites in living cells, useful for different forms of life functions and this may be the reason of such a wide distribution of CA. This enzyme functions as a C_i transporter for CO_2 and HCO_3^-. There are other enzymes in living organisms, which substrates and productions are CO_2 and HCO_3^-. CA provides a basis for the reaction by recovering CO_2/HCO_3^- concentration or by destroying CO_2/HCO_3^- production. Due to this function CA participates in various biological processes (Kupriyanova and Pronina 2011; Zabaleta et al. 2012; Fan et al. 2015). It was proved that CA participates in fundamental processes such as transport of C_i compounds and ions, carbon concentrating mechanisms in C3 and C4 plants, calcification (Xiao et al. 2015) and regulation of acid-alkaline balance in cells. The main functional role of CA in cells is the regulation of the C_i flow between cells and outer space and also between tissues and intracellular compartments (Pronina et al. 2002).

One of the major reasons of plant death under stress is the reduced photosynthetic CO_2assimilation. Currently researchers have controversial opinions about physiological and biochemical bases of this reduction. Some authors believe that the reduction of photosynthetic CO_2 assimilation occurs due to stomatal closure, which prevents water loss under drought and fulfills gas exchange, leading to the decrease of CO_2 concentration at the reaction center (Cornic and Massacci 1996; Chaves et al. 2002). Other researchers think that the reason of this process is phosphorylation in chloroplasts, reduced ATP synthesis under drought leading to reduction of the regeneration of ribulose-1,5-bisphosphate, which is the primary CO_2 acceptor in photosynthesis. As stomatal closure occurs in both cases, photosynthetic activity declines due to the decrease of the intracellular CO_2 concentration. Moreover, the irreversible decline in photosynthesis under severe drought occurs not only because of the decrease in CO_2 concentration, but also due to the reduced synthesis of photosynthetic enzymes.

Recent ecological problems, lack of sown areas due to the growth population, shortage of drinking water caused by drought, salinization of soils make necessary the developing highly productive plant varieties, tolerant to drought. Therefore, the study of physiological and biochemical mechanisms of drought tolerance is of great importance.

Durum wheat (*Triticum durum*) varieties Barakatli-95 and Garagylchyg-2 were chosen as the study objects. The Barakatli-95 variety is highly productive (70–80 cwt/ha) and drought tolerant; the Garagylchyg-2 variety is also highly productive (70–80 cwt/ha), but drought sensitive.

Wheat plants were grown at the experimental field of the Azerbaijan Research Institute of Crop Husbandry situated on Absheron peninsula. A group of plants was exposed to artificial soil drought created by ceasing watering in the phase of active growth of flag leaves and generative organs on May and at the beginning of June. Another group was watered till the end of the vegetation. The experiments were carried out during active growth phases—the end of tube formation, flowering, earing and ear filling with control and stressed plants. Samples were taken at the same day after 2–3 h of illumination.

To obtain the enzymatic extract, samples of roots, awns, and flag leaves were washed with distilled water and dried on the filter paper. 7 mL of homogenization medium –0.05 M Tris–HCl buffer (pH 8.4) containing 1 mM EDTA, 5 mM DTT, 0.01 M NaCl, 0.5 % Triton× 100, and 1 % PVP was added to every gram of the plant material. Homogenization was performed for 2 min with 30 s intervals at 4 °C. The homogenate was filtered through two layers of capron cloth and the obtained filtrate was centrifugated for 10 min at 500×*g*. The supernatant was centrifugated again for 30 min at 5000×*g* and the obtained supernatant was used for the research.

15.3.1 CA Activity

CA activity was measured using an electrometric method based on the releasing activity of H+ ions in the $CO_2 + H_2O \rightarrow H^+ + HCO_3^-$ reaction (Wilbur and Anderson 1948). The reaction was carried out in 0.05 M Tris–HCl buffer (pH 8.1) at 2–4 °C by adding 10–200 μL of the enzyme preparation. The reaction started by adding 3 mL of the saturated solution of CO_2. The final volume of the reaction mixture was 20 mL. The changes in pH were measured with pH meter (universal ionomer EB-74) and XY-RECORDER ENDIM 620.02.

The enzyme activity was estimated in conventional units according to the formula (Rickli et al. 1964).

$$U = 10(T_0 / T - 1)$$

where T_0 is non-enzymatic reaction (control) time, T—enzymatic reaction time, U—the enzyme activity in conventional units.

The saturated solution of CO_2 was prepared in bi-distilled water at 25°C and 1 atm pressure. The concentration of CO_2 was determined using reverse titration with HCl in the presence of $Ba(OH)_2$.

15.3.2 Relative Water Content

Relative water content (RWC) was determined in every phase of ontogenesis using the method described in Tambussi (2005). Calculations were performed according to the formula:

$$\mathrm{RWC} = 100\ \%\ (\mathrm{M}_F - \mathrm{M}_d) / (\mathrm{M}_T - \mathrm{M}_d)$$

where M_F is leaf fresh mass, M_d leaf dry mass, and M_T mass after saturation.

15.3.3 Determination of Kinetic Parameters

The study of kinetic parameters was carried out with purified enzyme preparations. The standard assay system consisted of 0.05 M Tris–HCl buffer (pH 8.1) and substrate CO_2. The reaction rate constant K_m (Michaelis–Menten constant) and maximum rate of (V_{max}) the reaction were calculated using the Lineweaver Burk procedure.

15.3.4 Determination of Chlorophyll Concentration

Chlorophyll amount was determined spectrophotometrically using 80 % acetonic extract (Sims and Gamon 2002).

15.3.5 Determination of Protein Concentration

Total protein content was determined according to Sedmak and Grossberg (1977). BSA was used as a standard protein marker for constructing calibration curve.

CA activity in wheat genotypes was very dependent on plant age. It was highest in the first leaf on the 7–8th days of the growth. On the following days CA activity decreased gradually in the first leaf and increased in the second leaf reaching the maximum value on the 14–16th days. This tendency was also observed in flag and preceding leaves.

The enzyme activity was also strongly dependent on pH of the reaction medium and the optimum pH was found to be between 8.3 and 8.5. The strong dependence of CA activity on pH was first shown in 1964 by Rougton and Bos. Contrary to the studied enzyme the optimum pH of CAs from lettuce (Walk and Metzner 1975), pea (Kisiel and Graf 1972), etc. leaves was neutral. CA of wheat leaves, unlike the enzyme of chick-pea leaves was thermolabile. The enzyme activity increased until temperature reached 40 °C, a further rise in temperature caused a decrease in CA activity and at 60 °C the enzyme was completely inactivated. However, the enzyme was found to be thermostable in bacteria and some dicot C3 and C4 plants (Di Fiore et al. 2015).

For the first time in our laboratory, wheat leaf CA was established to be homodimer of 55–60 kDa. It was detected that a symmetric pick of this isoform with Stokes radius of 32.25A° was localized in leaf chloroplasts (Aliyev et al. 1989).

It is very important to determine the subcellular localization of CA different molecular forms for the study of physiological functions and kinetic properties of the enzyme. The results of the research carried out in our laboratory showed that CA of wheat leaves is localized in chloroplasts of mesophyll cells and there is a strong correlation between the intensity of CO_2 assimilation and CA activity (Guliyev et al. 1985). The enzyme has only one isoform which does not change in leaves, roots, and awns under drought.

CA activity was found to be higher in the highly productive and drought tolerant Barakatli-95 variety compared with the highly productive and drought sensitive Garagylchyg-2 variety. It was reported earlier that genotypes with contrasting productivity and drought tolerance are different not only at the level of the primary photochemical reactions. They differ also in the primary photosynthetic C metabolism and transportation, and in the distribution of photoassimilates among different organs of wheat (Aliyev et al. 1996).

Thus, according to the obtained results the high CA activity along with the other carboxylating enzymes and factors contributed to high productivity of intensive wheat genotypes. Table 15.4 showed that CA activity in flag leaves of the Barakatli-95 variety increased until the end of the active vegetation phases, after a certain period of time it stabilized and decreased beginning from the grain filling stage. However, CA activity in Garagylchyg-2 increased faster than in Barakatli-95, then it decreased for a short period of time and the vegetation period of this variety was shorter. The rate of photosynthetic CO_2 assimilation, which was more pronounced in the flowering and earing phases, decreased sharply beginning from the grain filling stage. The vegetation period of Garagylchyg-2 decreased by 10–15 days compared with Barakatli-95 under drought conditions. It is thought to be an adaptive response for protection of this drought sensitive variety against long-term drought. We paid more attention to awns of both varieties as CA activity in awns is very different from that of glumes and grains (Table 15.4).

The dependence of the maximum reaction rate on the substrate concentration was studied for the kinetic characterization of the enzyme and the values of K_m and V_{max} were estimated for the reaction catalyzed by CA in roots and leaves. The obtained data are presented in Table 15.4 and Fig. 15.4.

For this purpose the leaf enzyme extract was applied to the DEAE-cellulose column. Then the obtained active fraction was ultrafiltrated through Ripor-4-25 membrane, dialyzed and applied to the Sephadex G-200 column. After the elution the obtained active fraction was ultrafiltrated again, precipitated and used for the study of the kinetic parameters.

The study of the kinetics of the CO_2 hydratation reaction shows high catalytic ability of wheat CA similar to CAs in other plants. According to the Lineweaver–Burk plots, the enzyme-substrate interaction follows the Michaelis-Menten kinetics in leaves of both varieties and in control as well as stressed variants (Fig. 15.4). As seen in Fig. 15.4a, bK_m values in leaves and roots were higher in drought variants compared with control ones and in flag leaves of the both variants these values were lower than in roots of the Barakatli-95 genotype. The values of V_{max} decreased in both organs and this decrease was more marked in flag leaves under drought. Figure 15.4c, d presented kinetic parameters of CA in Garagylchyg-2. These results show that CA used more CO2 for the possible catalytic reaction when approaching to the grain filling phase. This caused the decrease in Vmax of the CA reaction in leaves leading to the decline in photosynthetic intensity, which was more pronounced under drought. The values of Km and Vmax were close for control and stressed plant roots and they were 1.5–2 times less compared with the same parameters in leaves.

The similar tendency was observed in the active growth phase of the plant, but in Garagylchyg-2 the maximum rate of the reaction was lower compared with that in

Table 15.4 Time-dependent changes in CA activity of root, leaf, and ear elements of the wheat varieties under drought

			Tube formation		Beginning of the flowering		End of the flowering		Earing		End of the earing		Grain filling	
			CA activity											
Genotype	Variant	Object	1	2	1	2	1	2	1	2	1	2	1	2
Barakatli-95	C	Leaf	0.53	44.1	2.57	205.6	1.83	220.5	2.78	33.3	2.01	189.1	1.2	74.4
		Awn							4.07	269	4.04	253	3.8	242.8
		Root	0.44	38.0	0.62	104.2	1.02	156.0	1.01	149.3	0.91	162.8	2.1	201.3
	D	Leaf	2.1	166.1	7.0	525.0	1.7	131.8	3.08	259.0	1.9	142.6	0.44	33.7
		Awn							1.92	88.2	2.0	90.1	1.63	71.2
		Root	1.2	168.9	1.6	173.1	1.65	175.0	1.9	186.7	2.56	211.4	2.47	209.1
Garagylchyg-2	C	Leaf	0.97	62.1	6.7	452.3	2.5	152.6	3.06	269.0	1.99	128.1	0.72	54.5
		Awn							3.13	144.0	2.1	100.3	1.3	80.0
		Root	0.32	32.0	0.105	44.5	1.54	105.4	2.62	190.6	2.91	218.7	1.71	183.8
	D	Leaf	1.55	83.7	5.68	454.4	4.53	383.4	3.56	269.0	2.4	111.7	0.97	28.5
		Awn							2.84	125.0	1.87	92.4	0.9	59.8
		Root	0.55	42.8	1.19	63.2	1.66	121.3	2.73	202.1	2.02	265.8	1.69	191.4

Note: *C* control, *D* drought; (1) specific CA activity (conventional unit/mg protein·min), (2) CA activity based on fresh weight (mM $CO_2 \cdot g^{-1} \cdot FW$)

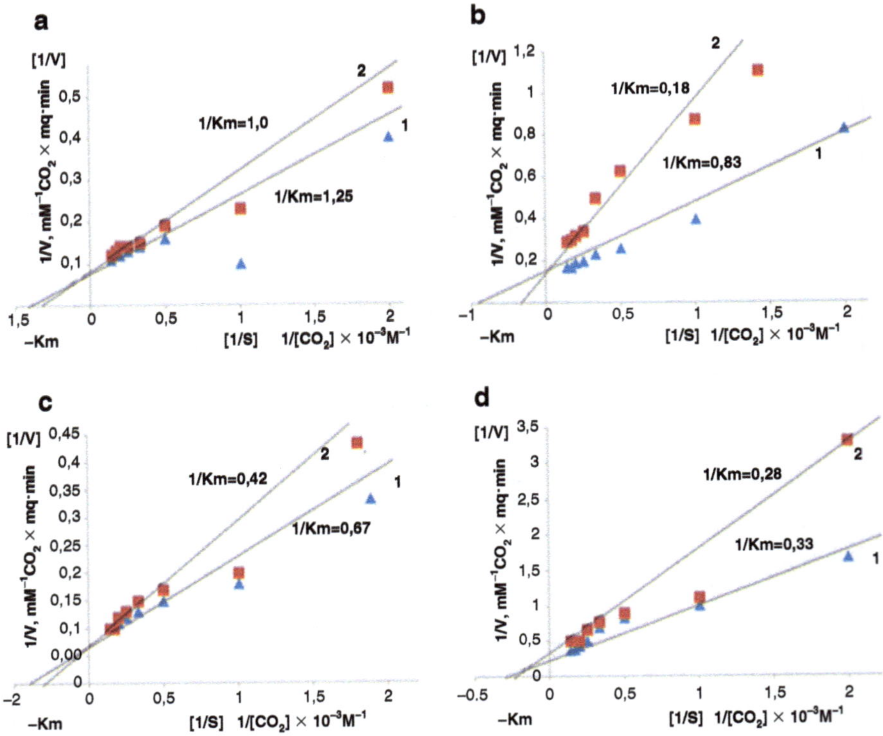

Fig. 15.4 The dependence between the rate of the reaction catalyzed by CA and the substrate concentration in the phase of flowering in leaf and root cells of the wheat varieties under drought conditions. (**a**) Barakatli-95 leaf, (**b**) Barakatli-95 root, (**c**) Garagylchyg-2 leaf, (**d**) Garagylchyg-2 root, 1-control (watered), 2-experiment (drought)

Table 15.5 Comparative study of the reaction catalyzed by CA in leaf and root cells of the wheat varieties in the flowering phase under drought

Genotype	Variant	Object	K_m (mM)	V_{max}, (mM/mg·min)	pK_m
Barakatli-95	Watered	Leaf	0.81	2.47	7.5
		Root	1.23	1.63	7.43
	Drought	Leaf	1.0	2.0	7.5
		Root	5.55	0.36	7.43
Garagylchyg-2	Watered	Leaf	1.49	1.34	7.5
		Root	3.33	0.6	7.45
	Drought	Leaf	2.38	0.84	7.5
		Root	3.57	0.56	7.45

Barakatli-95 in both variants and both organs. Ionization coefficient (Table 15.5) was also determined using graphical method of Dixon (1953). pK_m was found to be between 7.43 and 7.5 which corresponds only to the histidine imidazole ring. Similar data (Pocker and Ng 1973) were presented earlier including results obtained with amaranth (Guliev et al. 2003) and chick-pea (Aliyev et al. 1996) leaves. It suggests that the molecular organization of the active centers in leaves of C4 dicot amaranth, C3 dicot chick-pea, and C3 monocot wheat are similar and no significant changes occur in their active centers under drought.

15.4 Assessment of Wheat Genotypes Using TRAP Marker Linked to Cell Membrane Stability

The identification of several yield QTLs have been performed recently in wheat plants using linkage analysis and association mapping. Measurements related to plant productivity, which is the most crucial trait to breeders, have been used for the determination of the most QTLs for drought tolerance in wheat. However, these studies are impeded because of the complexity involving multiple loci and interactions between genotype and environment. As QTLs established in one environment may not be confirmed in other, it is difficult to describe productivity with respect to water use and its accurate phenotyping is also a challenge. Therefore it is necessary to take into consideration the environmental varieties and carry out large scale phenotyping in multiple fields. QTLs associated with specific components of drought response have been identified in studies performed with T. durum, *T. aestivum*, and *T. durum* × *T. dicoccoides* mapping populations. But the genomic regions associated with individual QTLs are still very large and unsuitable for screening in breeding programs (Budak et al. 2013).

For the identification of agronomically desirable alleles that exist at QTLs, genomics based methods contributing to the more effective improvement of the drought tolerance and high productivity under drought conditions are used. Identifying QTLs controlling important traits in wheat under drought stress is important for developing cultivars that are improved for those traits (Tuberosa and Salvi 2006). The recent study revealed some QTLs for physiological traits under drought stress in wheat (Barakat et al. 2013; Elshafei et al. 2013; Saleh et al. 2014).

Marker assisted selection reducing problems associated with genotype–environment interactions can improve the selection efficiency and combine different tolerance traits into a single efficient genotype. Molecular markers are considered to be an effective tool for obtaining genetic information and during the last few years their use in the assessment of genetic diversity in wheat (*Triticum aestivum* L.) has increased (Manifesto et al. 2001; Roy et al. 2004; Barakat et al. 2010). "Omics" studies or QTL mapping of productivity related traits can be used for the identification of potential markers for stress tolerance. Molecular markers providing methods for the determination of quantitative traits such as drought tolerance are very important for increasing selection efficiency. Molecular markers are plentiful, independent of tissue or environmental effects, and allow cultivar identification in early stages of plant development. Therefore, they are a useful complement to morphological and physiological characterization of cultivars. Molecular characterization of cultivars is also useful for evaluating potential genetic erosion (Manifesto et al. 2001). It is necessary to understand the genetic basis of phenotypic variability for improving wheat tolerance against stress factors. Target region amplification polymorphism (TRAP) was developed and used in genetic mapping recently (Li and Quiros 2001; Hu and Vick 2003; Liu et al. 2005; Wang et al. 2005; Al-Doss et al. 2011). TRAP is a relatively new PCR-based technique allowing the detection of the large numbers of loci in a single reaction without extensive pre-PCR processing of samples. For generating polymorphic markers around targeted putative candidate gene sequences, TRAP uses bioinformatics tools and the EST database information. So, this technique is very useful in the study of plant genomics involved in genetic mapping and marker-trait association (Liu et al. 2005). TRAP was also successfully used for the assessment of the genetic diversity in genetic stocks of wheat (Xu et al. 2003; Al-Doss et al. 2011; Barakat et al. 2013). There have been also reports on the identification of new SRAP and TRAP markers linked to leaf chlorophyll content, flag leaf senescence, and cell membrane stability traits in water-stressed wheat plants (Elshafei et al. 2013; Saleh et al. 2014; Moustafa et al. 2014). It was reported that these TRAP markers can be used in breeding for drought tolerance in wheat (Saleh et al. 2014).

Tolerance to drought stress is determined by the ability of plants to maintain membrane integrity under drought (Vieira Da Silva et al. 1974). Water deficiency can seriously impair membrane structure and disturb its function (Buchanan and Gruissem 2000). Drought causes a significant decrease in lipid content (Martins Júnior et al. 2008) and more pronounced changes are observed in polar lipids (Yordanov et al. 2000). Water deficiency leads to disruption of the interactions between lipids and proteins of membranes, as well as changes in the permeability of the membranes. Saneoka et al. (2004) and Azizi-e-Chakherchaman et al. (2009) studied the dependence of plasma membrane stability (obtained from EC measurement) from grain yield in Lentil under stress and non-stress conditions. According to these authors plasmatic membrane is stable in genotypes under non-stress conditions. The stability of cell membrane has been exclusively used as selection criteria for different abiotic stresses such as drought and high temperature in wheat, rice, cotton, and sorghum (Habibpor et al. 2011). Associations between cell membrane stability and different agronomic traits were established using polyethylene glycol (PEG-6000) in vitro (Dhanda et al. 2004). So, the ability of plants to maintain membrane integrity under drought is considered to be a determinant for plant drought tolerance (Abdullah et al. 2011).

15.4.1 Extraction of Plant DNA

DNA extraction was carried out using the CTAB method with some modifications (Murray and Thompson 1980). Fresh plant tissue as a fragment of leaf was minced in liquid nitrogen, suspended in 1000 μL of CTAB extraction buffer (100 mM Tris–HCl, pH 8.0; 20 mM EDTA, pH 8.0; 1.4 mM NaCl; 40 mMβ-mercaptoethanol), and pre-warmed in a water bath at 60 °C. Homogenization was completed by intense Vortex shaking. Then 400 mL of chloroform (99.8 %) was added into each tube and the tubes were gently mixed. Next the tubes were placed in a water bath and incubated for 10 min at 60 °C. After incubation, the tubes were centrifuged in an Eppendorf type benchtop centrifuge (15,000×*g*) for 10 min at room temperature. After centrifugation the supernatant was carefully selected (taking care not to capture sediment particles) and transferred to clean 1.5 mL Eppendorf type tubes and 600 mL of cold isopropanol was added, mixed well, and left at room temperature for 3–5 min. At this stage we can observe the dispersed DNA precipitate. The tube contents were centrifuged at room temperature in the Eppendorf type benchtop centrifuge (15,000×*g*) for 10 min. The precipitate was washed several times with 70 % ethanol, dried in a thermostat at 56 °C for 5 min, and dissolved in TE buffer (10 mM Tris–HCl, pH 8; 1 mM EDTA). Samples were left in a refrigerator at 4 °C for the complete dissolution of the DNA in a buffer.

15.4.2 DNA Quantification

After dissolution of the DNA the quantity was determined by optical density (OD) at $\lambda = 260$ using the ULTROSPEC 3300 PRO spectrophotometer ("AMERSHAM", USA). Purity of the genomic DNA was determined by the ratio of absorptions at A260/A280. Quality of the DNA was checked on the basis of performance of the extracted DNA samples in 0.8 % agarose gel stained with 10 mg/mL of ethidium bromide in 1× TBE (Tris base, Boric acid, EDTA) buffer. The gel was developed and photographed under ultraviolet light using "Gel Documentation System UVITEK" (UK).

15.4.3 DNA Amplification

Polymerase chain reaction was performed by Williams et al. (1990). DNA amplification was performed in a 25 μL reaction mixture volume, containing 10× buffer, 20 ng of the genomic DNA, 0.2 μM primer, 200 μM of each of the following: dATP, dCTP, dGTP, and dTTP, 2.5 mM $MgCl_2$, and 0.2 units of Taq-polymerase in the incubation buffer. TRAP9 primer was used for the test (Table 15.6).

Table 15.6 Nucleotide sequence of the TRAP9 primer used for the DNA amplification

Primers	Nucleotide sequences (5′→3′)	Localization in chromosomes	Expected fragment, bp
TRAP9F	TGAGTCCAAACCGGAGC	5A	290
TRAP9R	TCACCCGCACCTTCTTCC		

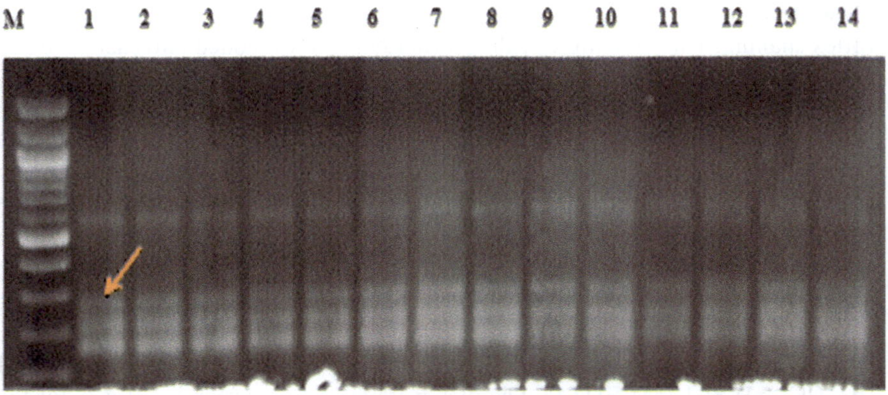

Fig. 15.5 PCR-profiles of *Triticum* plants induced by TRAP9 primer. *Arrow* indicates the 290 bp. M-molecular weight marker (1 kb plus DNA ladder). (*1*) Vugar, (*2*) Mugan, (*3*) Shiraslan-23, (*4*) Ag bugda, (*5*) Alinca-84, (*6*) Kakhraba, (*7*) Tartar, (*8*) Mirvari, (*9*) Sharq, (*10*) Gyzyl bugda, (*11*) Barakatli-95, (*12*) Garagylchyg-2, (*13*) Shirvan bugda, (*14*) Garabag

PCR was performed in the "Applied Biosystems 2720 Thermal Cycler" (Singapore) thermocycler under the following conditions: After incubation at 94 °C for 5 min, 5 cycles were performed with 94 °C for 1 min, 35 °C for 1 min, and 72 °C for 1 min 40 s. Further, the similar 35 cycles were performed with exception for the annealing temperature at 50 °C and a final extension at 72 °C for 7 min. The reaction products were separated by electrophoresis in a 1.2–2 % agarose gel in the HR-2025-High Resolution («IBI SCIENTIFIC», USA) horizontal electrophoresis machine with addition of ethidium bromide and documented using «Gel Documentation System UVITEK». Dimensions of amplified fragments were determined with respect to 1 kb DNA marker. Statistical analysis included binary matrix compilation, in which "presence" (1) or "absence" (0) of fragments with equal molecular weight on the electropherogram were noted.

Sixty-nine wheat genotypes collected in the Gene Pool of the Research Institute of Agriculture (Baku) acted as a research object (Table 15.7). Plants were cultivated in field conditions. TRAP9 markers linked to the cell membrane stability were used for the screening (Saleh et al. 2014).

According to the electrophoretic profiles obtained from PCR data (Fig. 15.5), 290 bp fragments characteristic for TRAP9 marker were successfully amplified in the studied 17 durum (*Triticum durum* Desf.) and 48 bread (*Triticum aestivum* L.) wheat genotypes. It suggests that QTL for cell membrane stability is present on chromosomes 5A of these genotypes. Only four bread wheat genotypes are exceptions, in other words, the presence of QTL of interest in the mentioned genotypes was not proved using TRAP9 marker (Table 15.7).

Membrane stability which is one of the physiological traits, correlating with plant performance under drought is considered to be a useful measure of wheat tolerance against stress (Blum and Ebercon, 1981). The associated molecular markers at a major locus contributing to water-stress tolerance can be identified for the indirect selection of wheat plants for water-stress tolerance (Visser 1994). However,

Table 15.7 PCR analysis using TRAP9 marker

Triticum durum Desf.			
Vugar	+	Mugan	+
Shiraslan-23	+	Ag bugda	+
Alinca-84	+	Kakhraba	+
Tartar	+	Mirvari	+
Sharq	+	Shirvan-5	+
Gyzyl bugda	+	Barakatli-95	+
Tartar-2	+	Garagylchyg-2	+
Garabag	+	Shirvan bugda	+
Shirvan-3	+		+
Triticum aestivum L.			
Akinchi-84	+	Giymatly-2/17	+
Pirshahin	+	Pirshahin 1	+
Gunashli	+	Ugur	+
Dagdaş	+	Parzivan-1	+
Shafag	+	Shaki-1	+
Mirbashir-128	+	Nurlu-99	+
Gobustan	+	Gyrmyzy gul	+
Yegana	+	Gyrmyzy gul-1	+
Zirva-80	+	Azamatli-95	+
Zirva-85	+	Tale-38	+
Aran	+	Ruzi-84	+
Azeri	+	12nd FAWWON №97 (130/21)	+
Murov	+	4th FEFWSN №50 (130/32)	+
Murov-2	+	Bezostaya	+
Səba	+	N-6	–
Taraggi	+	Layagatli-80	+
Bayaz	+	N-20	+
Shafag 2	+	Agali	+
Fatima	+	Farahim	+
Az 026	+	Gonam	+
Ni 447	+	Parvin	+
Sonmaz	+	Marxal	+
Baba75	+	N-17	–
Zager	+	N-50	+
N-8	–	Saratovskaya 29	+
N-9	–	Mahmud	+++

in most cases screening a relatively large number of individuals in the population is required for identifying molecular markers associated with important genes or traits (Lawson et al. 1994). Various factors, such as the genetic, physiological, and molecular bases of the traits including interactions among the different component traits with the environments must be taken into consideration in the breeding for complex traits (Tester and Langridge 2010). Due to their insensitivity to environment, DNA markers associated with the genomic regions are suitable for selecting genotypes with increased drought tolerance. Molecular markers linked to the drought tolerance trait are considered to be a more reliable tool for selecting drought tolerant genotypes at early stages (Saleh et al. 2014).

15.5 Conclusion and Future Perspectives

The obtained data characterize the quantitative and qualitative changes in antioxidant enzymes in different organs of wheat genotypes under long-term soil drought. The observed heterogeneity of individual forms of enzymes may have adaptive value and is a measure of resistance to water stressor. These results can be used as practical biochemical parameters for selection of drought tolerant wheat genotypes when selecting drought tolerant cultivars for breeding in arid regions.

The study of long-term drought effects on wheat productivity showed that productivity could be related to the tolerance of the variety. The differences in changes of CA activity in both durum wheat genotypes with similar productivity and contrasting tolerance, the increased activity in awns compared with flag leaves and roots during earing and grain filling phases, the increased activity in roots compared with flag leaves in the grain filling phase and slight differences between organs in changes of K_m and V_{max} values in the earing, flowering, and grain filling phases are thought to be adaptive traits against stress factors.

One of the major cellular targets characteristics for various stresses is the stability of the cell membrane. Therefore, application of quantitative trait loci (QTLs) analysis to study the physiological traits will improve our understanding genetic factors that influence these complex traits.

Acknowledgement This work was supported by a grant (EIF-2010-1(1)-40/24-M-20) of the Science Development Foundation under the President of the Republic of Azerbaijan.

References

Abdullah F, Hareri F, Naaesan M, Ammar MA, Kanbar OZ. Effect of drought on different physiological characters and yield component in different varieties of Syrian durum wheat. J Agric Sci. 2011;3(3):127–33.

Al-Doss AA, Saleh M, Moustafa KA, Elshafei AA, Barakat MN. Comparative analysis of diversity based on morpho-agronomic traits and molecular markers in durum wheat under heat stress. Afr J Biotechnol. 2011;10:3671–81.

Alexieva V, Sergiev I, Mapelli S, Karanov E. The effect of drought and ultraviolet radiation on growth and stress markers in pea and wheat. Plant Cell Environ. 2001;24:1337–44.

Aliyev J, Guliyev N, Hidayatov RB. Isolation and some properties of wheat leaf carbonic anhydrase. News Azerbaijan Acad Sci Ser Biol Sci. 1989;5:15–29 (in Russian).

Aliyev J, Krimov SC, Guliyev N, Ahmedova A. Features of C metabolism in wheat genotypes with contrasting photosynthetic signs. Plant Physiol. 1996;43(1):49–56 (in Russian).

Allakhverdiev SI, Los DA, Mohanty P, Nishiyama Y, Murata N. Glycinebetaine alleviates the inhibitory effect of moderate heat stress on the repair of photosystem II during photoinhibition. Biochim Biophys Acta. 2007;1767:1363–71.

Alscher RG, Donahue JL, Cramer CL. Reactive oxygen species and antioxidants: relationships in green cells. Physiol Plant. 1997;100:224–33.

Anderson M, Prasad T, Stewart C. Changes in isozyme profiles of catalase, peroxidase, and glutathione reductase during acclimation to chilling in mesocotyls of maize seedlings. Plant Physiol. 1995;109:1247–57.

Arora A, Sairam RK, Srivastava GC. Oxidative stress and antioxidative systems in plants. Curr Sci. 2002;82:1227–38.

Asada K. Mechanisms for scavenging reactive molecules generated in chloroplasts under light stress. In: Baker NR, Bowyer RJ, editors. Photoinhibition of photosynthesis From molecular mechanisms to the field. Oxford: Bios Scientific; 1994. p. 129–42.

Azizi-e-Chakherchaman S, Mostafaei H, Yari A, Hassanzadeh M, Jamaati-e-Somarin S, Easazadeh R. Study of relationships of leaf relative water content, cell membrane stability and duration of growth period with grain yield of lentil under rainfed and irrigated conditions. Res J Biol Sci. 2009;4:842–7.

Badiani M, De Biasi MG, Cologna M, Artemi F. Catalase, peroxidase and superoxide dismutase activities in seedlings submitted to increasing water deficit. Agrochimica. 1990;34:90–102.

Bailly C, Benamar A, Corbineau F, Côme D. Changes in malondialdehyde content and in superoxide dismutase, catalase and glutathione reductase activities in sun flower seed as related to deterioration during accelerated aging. Physiol Plant. 1996;97:104–10.

Barakat MN, Al-Doss AA, Moustafa KA, Ahmed EI, Elshafei AA. Morphological and molecular characterization of Saudi wheat genotypes under drought stress. J Food Agric Environ. 2010;8:220–8.

Barakat MN, Wahba LE, Milad SI. Molecular mapping of QTLs for wheat flag leaf senescence under water-stress. Biol Plant. 2013;57:79–84.

Behnamnia M, Kalantari KM, Rezanejad F. Exogenous application of brassino steroid alleviates drought-induced oxidative stress in *Lycopersicon esculentum* L. Gen Appl Plant Physiol. 2009;35:22–34.

Bellincampi D, Dipierro N, Salvi G, Cervone F, De Lorenzo G. Extracellular H_2O_2 induced by oligogalacturonides is not involved in the inhibition of the auxin-regulated *rolB* gene expression in tobacco leaf explants. Plant Physiol. 2000;122:1379–85.

Blum A, Ebercon A. Cell membrane stability as a measure of drought and heat tolerance in wheat. Crop Sci. 1981;21:43–7.

Bray EA, Bailey-Serres J, Weretilnyk E. Responses to abiotic stresses. In: Gruissem W, Buchannan B, Jones R, editors. Biochemistry and molecular biology of plants. Rockville: American Society of Plant Physiologists; 2000. p. 1158–249.

Brouquisse R, Weigel P, Rhodes D, Yocum CF, Hanson AD. Evidence for a ferredoxin-dependent choline monooxygenase from spinach chloroplast stroma. Plant Physiol. 1989;90:322–9.

Buchanan BB, Gruissem W, Jones RL. Biochemistry and molecular biology of plants. Rockville: American Society Plant Physiologists; 2000.

Budak H, Kantar M, Kurtoglu KY. Drought tolerance in modern and wild wheat. Sci World J. 2013;548246:16 p, doi:10.1155/2013/548246

Chaves MM, Pereira JS, Maroco J, Rodrigues ML, Ricardo CPP, Osorio ML, Carvalho I, Faria T, Pinheiro C. How plants cope with water stress in the field. Photosynthesis and growth. Ann Bot. 2002;89:907–16.

Cornic G, Massacci A. Leaf photosynthesis under drought stress. In: Baker NR, editor. Photosynthesis and the environment. Dordrecht: Kluwer; 1996. p. 347–66.

Cuypers A, Vangronsveld J, Ciijsters H. Peroxidases in roots and primary leaves of Pharsalus vulgaris copper and zinc phytotoxicity: a comparison. J Plant Physiol. 2002;159:869–76.

Da Costa M, Huang B. Changes in antioxidant enzyme activities and lipid peroxidation for bentgrass species in response to drought stress. J Am Soc Hortic Sci. 2007;132(3):319–26.

Davis B. Disc electrophoresis. I. Method and application to human serum proteins. Ann N Y Acad Sci. 1964;121:404–27.

Dhanda SS, Sethi GS, Behl RK. Indices of drought tolerance in wheat genotypes at early stages of plant growth. J Agron Crop Sci. 2004;190:6–12.

Di Fiore A, Alterio V, Monti SM, De Simone G, D'Ambrosio K. Thermostable carbonic anhydrases in biotechnological applications. Int J Mol Sci. 2015;16(7):15456–80.

Dixon BM. The effect of pH on the affinities of enzymes for substraties and inhibitors. Biochem. 1953;55(2):161–71.

Dudoladova MV, Markelova AG, Lebedeva NV, Pronina N. Compartmentalization of α-and β-carbonic anhydrases in cells of halo and alkalophilic cyanobateria *Rhabdoderma lineare*. Plant Physiol. 2004;51:896–904 (in Russian).

Elshafei A, Saleh M, Al-Doss AA, Moustafa KA, Al-Qurainy FH, Barakat MN. Identification of new SRAP markers linked to leaf chlorophyll content, flag leaf senescence and cell membrane stability traits in wheat under water- stressed condition. Aust J Crop Sci. 2013;7:887–93.

Esfandiari E, Shekari F, Esfandiari M. The effect of salt stress on antioxidant enzymes activity and lipid peroxidation on the wheat seedlings. Not Bot Agrobot Cluj. 2007;35:48–56.

Fan J, Xu H, Luo Y, Wan M, Huang J, Wang W, Li Y. Impacts of CO_2 concentration on growth, lipid accumulation, and carbon-concentrating-mechanism-related gene expression in oleaginous *Chlorella*. Appl Microbiol Biotechnol. 2015;99(5):2451–62.

Ferreira FJ, Guo C, Coleman JR. Reduction of plastid-localized carbonic anhydrase activity results in reduced Arabidopsis seeding survivorship. Plant Physiol. 2008;147(2):585–94.

Ferry JG. The gamma class of carbonic anhydrases. Biochim Biophys Acta. 2010;1804(2): 374–81.

Foyer CH, Lelandais M, Kunert KJ. Photo oxidative stress in plants. Plant Physiol. 1994;92: 696–717.

Fucc L, Oliver CN, Coon MJ. Inactivation of metabolic enzymes by mixed-function oxidation reaction: possible implication in protein turnover and aging. Proc Natl Acad Sci U S A. 1983;80:1521–5.

Greive CM, Grattan SR. Rapid assay for determination of water-soluble quaternary amino compounds. Plant Soil. 1983;70:303–7.

Guliev NM, Babaev GG, Bairamov SM, Aliev DA. Purification, properties, and localization of two carbonic anhydrase from *Amaranthus cruentus* leaves. Russ J Plant Physiol. 2003;50(2): 213–9.

Guliyev N, Mamedov TH, Aliyev J. Subunit structure of carbonic anhydrase of chick-pea leaves. Rep USSR Acad Sci. 1985;280(6):1466–8 (in Russian).

Guo Z, Ou W, Lu S, Zhong Q. Differential responses of antioxidative system to chilling and drought in four rice cultivars differing in sensitivity. Plant Physiol Biochem. 2006;44:828–36.

Habibpor M, Valizadeh M, Shahbazi H, Ahmadizadeh M. Study of drought tolerance with cell membrane stability testing and relation with the drought tolerance indices in genotypes of wheat (*Triticum aestivum* L.). World Appl Sci J. 2011;13(7):1654–60.

Heath RL, Packer L. Photoperoxidation in isolated chloroplast. I. Kinetics and stoichiometry of fatty acid peroxidation. Arch Biochem Biophys. 1968;125:189–98.

Hewett-Emmett D, Tashian RE. Functional diversity, conservation and convergence in the evolution of α-, β- and γ-carbonic anhydrase gene families. Mol Phylogenet Evol. 1996;5:50–70.

Hiltonen T, Bjorkbacka H, Forsman C, Clarke AK, Samuelsson G. İntracellular β-carbonic anhydrase of the unicellular green alga *Coccomyxa*: cloning of the cDNA and characterization of the functional enzyme overexpressed in *Escherichia coli*. Plant Physiol. 1998;117:1341–9.

Hu J, Vick BA. Target region amplification polymorphism: a novel marker technique for plant genotyping. Plant Mol Biol Rep. 2003;21:289–94.

Jang IC, Park SY, Kim KY, Kwon SY, Kim JG, Kwak SS. Differential expression of 10 sweet potato peroxidase genes in response to bacterial pathogen, *Pectobacterium chrysanthemi*. Plant Physiol Biochem. 2004;42:451–5.

Jiang M, Zhang J. Water stress induced abscisic acid accumulation triggers the increased generation of reactive oxygen species and upregulates the activities of antioxidant enzymes in maize leaves. J Exp Bot. 2002;53:2401–10.

Keles Y, Oncel I. Response of the antioxidative defence system to temperature and water stress combinations in wheat seedlings. Plant Sci. 2002;163:783–90.

Khanna-Chopra R, Selote DS. Acclimation to drought stress generates oxidative stress tolerance in drought-resistant than susceptible wheat cultivar under field conditions. Environ Exp Bot. 2007;60:276–83.

Kisiel W, Graf G. Purification and characterization of carbonic anhydrase from *Pisum sativum*. Phytochemistry. 1972;11(1):113–7.

Krieger-Liszkay A. Singlet oxygen production in photosynthesis. J Exp Bot. 2004;56:337–46.

Kruk J, Hollander-Czytko H, Oettmeier W, Trebst A. Tocopherol as singlet oxygen scavenger in photosystem II. J Plant Physiol. 2005;162:749–57.

Kupriyanova EV, Pronina N. Carbonic anhydrase—an enzyme that converted the biosphere. Plant Physiol. 2011;58(2):163–76 (in Russian).

Lascano HR, Antonicelli GE, Luna CM, Melchiorre MN, Gomez LD, Raca RW, Trippi VS, Casano LM. Antioxidant system response of different wheat cultivars under drought: Field and in vitro studies. Aust J Plant Physiol. 2001;28:1095–102.

Lawson WR, Henry RJ, Kochman JK, Kong GA. Genetic diversity in sunflower (*Helianthus annus* L.) as revealed by random amplified polymorphic DNA analysis. Aust J Agric Res. 1994;45: 1319–27.

Lei Y, Yin C, Li C. Differences in some morphological, physiological, and biochemical responses to drought stress in two contrasting populations of *Populus prezwalskii*. Physiol Plant. 2006;127:187–91.

Lenaz G. Role of mitochondria in oxidative stress and aging. Biochim Biophys Acta. 1998;1366: 53–67.

Li G, Quiros CF. Sequence-related amplified polymorphism (SRAP), a new marker system based on a simple PCR reaction: its application to mapping and gene tagging in Brassica. Theor Appl Genet. 2001;103:455–61.

Liu ZH, Anderson JA, Hu J, Friesen TL, Rasmussen JB, Faris JD. A wheat intervarietal genetic linkage map based on microsatellite and target region amplified polymorphism markers and its utility for detecting quantitative trait loci. Theor Appl Genet. 2005;111:782–94.

Ludwig M. The molecular evolution of beta-carbonic anhydrase in Flaveria. J Exp Bot. 2011;62(9):3071–81.

Luna CM, Pastori GM, Driscoll S, Groten K, Bernard S, Foyer CH. Drought controls on H_2O_2 accumulation, catalase (CAT) activity and CAT gene expression in wheat. J Exp Bot. 2005;56(411):417–23.

Manifesto MM, Schlatter A, Hopp HE, Suarez EY, Dubcovsky J. Quantitative evaluation of genetic diversity in wheat germplasm using molecular markers. Crop Sci. 2001;41:682–90.

Martins Júnior RR, Costa Oliveira MS, Baccache MA, Monteiro de Paula F. Effects of water deficit and rehydration on the polar lipid and membranes resistance leaves of Phaseolus vulgaris L. cv Pérola. Braz Arch Biol Technol. 2008;51:361–7.

McWilliam J. The dimensions of drought. In: Baker F, editor. Drought resistance in cereals. Wallingford: CAB International; 1989. p. 1–11.

Mead JF. Free radical mechanism of lipid damage and consequences for cellular membranes. In: Pryor WA, editor. Freed radicals in biology. New York: Academic; 1976. p. 51–68.

Meldrum NV, Roughton FJW. Some properties of carbonic anhydrase the CO_2 enzyme present in blood. J Physiol. 1932;75:15–6.

Miller G, Suzuki N, Ciftci-Yilmaz S, Mittler S. Reactive oxygen species homeostasis and signaling during drought and salinity stresses. Plant Cell Environ. 2010;33(4):453–67.

Mittler R. Oxidative stress, antioxidants and stress tolerance. Trends Plant Sci. 2002;7:405–10.

Molinari HBC, Marur CJ, Daros E, de Campos MKF, de Carvalho JFR, Filho LFP, Pereira JCB, Vieira LGE. Evaluation of the stress inducible production of proline in transgenic sugarcane (*Saccharum* spp.): osmotic adjustment, chlorophyll fluorescence and oxidative stress. Physiol Plant. 2007;130:218–29.

Moroney JV, Bartlett SG, Samuelsson G. Carbonic anhydrase in plants age. Plant Cell Environ. 2001;24:141–53.

Moustafa KA, Saleh M, Al-Doss AA, Elshafei AA, Salem AK, Al-Qurainy FH, Barakat MN. Identification of TRAP and SRAP markers linked with yield components under drought stress in wheat (*Triticum aestivum* L.). Plant Omics J. 2014;7(4):253–9.

Murray MG, Thompson WF. Rapid isolation of high molecular weight plant DNA. Nucleic Acids Res. 1980;8:4321–5.

Muscolo A, Sidari M, Panuccio M. Tolerance of kikuyu grass to long term salt stress is associated with induction antioxidant defenses. Plant Growth Regul. 2003;41:57–62.

Noctor G, Foyer CH. Ascorbate and glutathione: keeping active oxygen under control. Annu Rev Plant Physiol Plant Mol Biol. 1998;49:249–79.

Pandey HC, Baig MJ, Chandra A, Bhatt RK. Drought stress induced changes in lipid peroxidation and antioxidant system in genus *Avena*. J Environ Biol. 2010;31(4):435–40.

Pocker Y, Ng JSY. Plant carbonic anhydrase: Properties and carbon dioxide hydration kinetics. Biochem. 1973;12(25):5127–34.

Pronina NA. Organization and physiological role of CO_2-concentrating mechanism in photosynthesis of microalgae. Plant Physiol. 2000;47:801–10 (in Russian).

Pronina N, Allahverdiyev SI, Kupriyanova EV, Klyachko-Girvich GL, Klimov VV. Localization of carbonic anhydrase in pea subchloroplast particles. Plant Physiol. 2002;49(3):341–9 (in Russian).

Queiroz CGS, Alonso A, Mares-Guia M, Magalhães AC. Chilling-induced changes in membrane fluidity and antioxidant enzyme activities in *Coffea arabica* L. roots. Biol Planta. 1998;41:403–13.

Racchi ML, Bagnoli F, Balla I, Daut S. Differential activity of catalase and superoxide dismutase in seedlings and *in vitro* micro propagated Oak (*Quercus robur* L.). Plant Cell Rep. 2001;20:169–74.

Rajasekaran IR, Kriedemann PF, Aspinall D, Paleg IG. Physiological significance of proline and glycinebetaine: Maintaining photosynthesis during NaCl stress in wheat. Photosynthetica. 1997;34:357–66.

Rao MV, Paliyath G, Ormrod DP. Ultraviolet-B- and ozone-induced biochemical changes in antioxidant enzymes of *Arabidopsis thaliana*. Plant Physiol. 1996;110:125–36.

Rehman S, Harris PJC, Ashraf M. Stress environments and their impact on crop production. In: Ashraf M, Harris PJC, editors. Abiotic stresses: plant resistance through breeding and molecular approaches. New York: Haworth Press; 2005. p. 3–18.

Rhodes D, Hanson AD. Quaternary ammonium and tertiary sulfonium compounds in higher plants. Annu Rev Plant Physiol Plant Mol Biol. 1993;44:357–84.

Rickli EE, Chazanfar SAS, Gibbons BH, Edsall JT. Carbonic anhydrase from human erythrocytes. Preparation and properties of two enzymes. J Biol Chem. 1964;239:1065–78.

Roughton FJW, Both VH. The effect of substrate concentration, pH and other factors upon the activity of carbonic anhydrase. Biochem J. 1964;40(40):319–20.

Rowlett RS. Structure and catalytic mechanism of the beta-carbonic anhydrases. Biochim Biophys Acta. 2010;1804(2):362–73.

Roy JK, Lakshmikumaran MS, Balyan HS, Gupta PK. AFLP-based genetic diversity and its comparison with diversity based on SSR, SAMPL, and phenotypic traits in bread wheat. Bio Genet. 2004;42:43–59.

Rubio MC, González EM, Minchin FR, Webb KJ, Arrese-Igor C, Ramos J, Becana M. Effects of water stress on antioxidant enzymes of leaves and nodules of transgenic alfalfa overexpressing superoxide dismutases. Physiol Planta. 2002;115:531–40.

Sairam R, Deshmukh P, Saxena D. Role of antioxidant systems in wheat genotype tolerance to water stress. Biol Planta. 1998;41:387–94.

Saleh MS, Al-Doss AA, Elshafei AA, Moustafa KA, Al-Qurainy FH, Barakat MN. Identification of new TRAP markers linked to chlorophyll content, leaf senescence, and cell membrane stability in water-stressed wheat. Biol Planta. 2014;58:64–70.

Saneoka H, Moghaieb REA, Premachandra GS, Fujita K. Nitrogen nutrition and water stress effects on cell membrane stability and leaf water relations in *Agrostis palustris* Huds. Environ Exp Bot. 2004;52:131–8.

Sedmak JJ, Grossberg SE. A rapid, sensitive, and versatile assay for protein using coomassie Brilliant Blue G-250. Anal Biochem. 1977;79:544–52.
Selote DS, Khanna-Chopra R. Drought-induced spikelet sterility is associated with an inefficient antioxidant defence in rice panicles. Physiol Plant. 2004;121(3):462–71.
Shao HB, Liang ZS, Shao MA. Changes of some anti-oxidative enzymes under soil water deficits among 10 wheat genotypes at maturation stage. Colloids Surf B Biointerfaces. 2005a;45:7–13.
Shao HB, Liang ZS, Shao MA. Adaptation of higher plants to stresses and stress signal transduction. Acta Ecol Sin. 2005b;25:1871–82.
Sharma P, Dubey RS. Modulation of nitrate reductase activity in rice seedlings under aluminium toxicity and water stress: role of osmolytes as enzyme protectant. J Plant Physiol. 2005;162:854–64.
Shpak G. On the biological role of carbonic anhydrase in animal organisms. Successes Modern Biol. 1980;89:18–27 (in Russian).
Simova-Stoilova L, Dermirevska K, Petrova T, Tsenov N, Feller U. Antioxidative protection and proteolytic activity in tolerant and sensitive wheat (*Triticum aestivum L.*) varieties subjected to long-term field drought. Plant Growth Regul. 2009;58:107–17.
Simova-Stoilova L, Vaseva I, Grigorova B, Demirevska K, Feller U. Proteolytic activity and cysteine protease expression in wheat leaves under severe soil drought and recovery. Plant Physiol Biochem. 2010;48:200–6.
Sims DA, Gamon JA. Relationships between leaf pigment content and spectral reflectance across a wide range of species, leaf structures and developmental stages. Remote Sensing of Environ. 2002;81:337–54.
Sivamani E, Bahieldin A, Wraith IM, Al-Neimi T, Dyer WE, Ho TD, Qu R. Improved biomass productivity and water use efficiency under water deficit conditions in transgenic wheat constitutively expressing the barley HVA1 gene. Plant Sci. 2000;155:1–9.
Slesak I, Libik M, Karpinska B, Miszalski Z. The role of hydrogen peroxide in regulation of plant metabolism and cellular signaling in response to environmental stresses. Acta Biochim Pol. 2007;54(1):39–50.
Smith KS, Ferry JG. Prokaryotic carbonic anhydrase. FEMS Microbiol Rev. 2000;24:335–66.
Sofo AB, Dichio C, Xiloyannis C, Masia C. Lipoxygenase activity and proline accumulation in leaves and roots of olive trees in response to drought stress. Physiol Plant. 2004;121:58–65.
Sofo A, Dichio B, Xiloyannis C, Masia A. Antioxidant defenses in olive trees during drought stress: changes in activity of some antioxidant enzymes. Plant Biol. 2005;32:45–53.
Stadie WC, O'Brien H. The catalysis of the hydration of carbonic dioxide and dehydration of carbonic acid by the enzyme from red blood cells. J Biochem. 1933;103:521–9.
Supuran CT. Bacterial carbonic anhydrases as drug targets: toward novel antibiotics? Front Pharmacol. 2011;2:34.
Taiz L, Zeiger E. Plant physiology. 4th ed. Sunderland: Sinecure; 2006. 764 p.
Takeda S, Matsuoka M. Genetic approaches to crop improvement: responding to environmental and population changes. Nat Rev Genet. 2008;9:444–57.
Tambussi EA, Nogues S, Araus JL. Ear of durum wheat under water stress: water relations and photosynthetic metabolism. Planta. 2005;19:1–25.
Tatar O, Gevrek M. Influence of water stress on proline accumulation, lipid peroxidation and water content of wheat. Asian J Plant Sci. 2008;7:409–12.
Tester M. Langridge P. Breeding technologies to increase crop production in a changing world. Science. 2010;327(5967):818–22.
Tuberosa R, Salvi S. Genomics-based approaches to improve drought tolerance of crops. Trends Plant Sci. 2006;11:405–12.
Uzildaya B, Turkana I, Sekmena AH, Ozgura R, Karakayab HC. Comparison of ROS formation and antioxidant enzymes in *Cleome gynandra* (C4) and *Cleome spinosa* (C3) under drought stress. Plant Sci. 2012;182:59–70.
Vieira da Silva J, Naylor AW, Kramer PJ. Some ultrastructural and enzymatic effects of drought stress in cotton (*Gossypium hirsutum* L.) leaves. Proc Natl Acad Sci U S A. 1974;71: 3243–324.

Visser B. Technical aspects of drought tolerance. Biotechnology Dev Monitor. 1994;18:5.

Walk AR, Metzner H. Purification and characterization of chloroplast carbonate dehydratase from leaves of *Lactuca sativa*. Hoppe Seylers Z Physiol Chem. 1975;356(11):1733–41.

Wang W, Vinocur B, Altman A. Plant responses to drought, salinity and extreme temperatures: towards genetic engineering for stress tolerance. Planta. 2003;218:1–14.

Wang G, Pan J, Li X, He H, Wu A, Cai R. Construction of a cucumber genetic linkage map with SRAP markers and location of the genes for lateral branch traits. Sci China Ser. 2005;48:213–20.

Wang S, Liang D, Li C, Hao Y, Ma F, Shu H. Influence of drought stress on the cellular ultrastructure and antioxidant system in leaves of drought-tolerant and drought-sensitive apple rootstocks. Plant Physiol Biochem. 2012;51:81–9.

Weigel P, Weretilnyk EA, Hanson AD. Betaine aldehyde oxidation by spinach chloroplasts. Plant Physiol. 1986;82:753–9.

Wei-Hong S, Wu Y-Y, Zhen-Zhen S, Wu Q-X, Xin-Yu W. Enzymatic characteristics of higher plant carbonic anhydrase and its role in photosynthesis. J Plant Stud. 2014;3(2):39–44.

Wilbur KM, Anderson NG. Electrometric and colometric determination of carbonic anhydrase. J Biol Chem. 1948;176:147–51.

Williams JG, Kubelik KJ, Livak JA, Tingey SV. DNA polymorphisms amplified by arbitrary primers are useful genetic markers. Nucleic Acids Res. 1990;18:6531–5.

Xiao L, Lian B, Hao J, Liu C, Wang S. Effect of carbonic anhydrase on silicate weathering and carbonate formation at present day CO_2 concentrations compared to primordial values. Sci Rep. 2015;5:7733.

Xu SS, Hu J, Faris JD. Molecular characterization of Langdon durum wheat Triticum dicoccoides chromosome substitution lines using TRAP (target region amplification polymorphism) markers. In: Pogna NE, Romano EA, Pogna EA, Galterio G, editors. Proceedings of the 10th international wheat genetic symposium, vol. 1. Rome: Institute Sperimentale per la Cerealicoltura; 2003. p. 91–4.

Yannarelli GG, Fernandez-Alvarez AJ, Santa-Cruz DM, Tomaro ML. Glutathione reductase activity and isoforms in leaves and roots of wheat plants subjected to cadmium stress. Phytochemistry. 2007;68:505–12.

Yordanov I, Velikova V, Tsonev T. Plant responses to drought, acclimation, and stress tolerance. Photosynthetica. 2000;38:171–86.

Yu JW, Price GD, Song L, Badger MR. Isolation of a putative carboxysomal carbonic anhydrase gene from the cyanobacterium *Synechococcus PCC 7942*. Plant Physiol. 1992;100:794–800.

Zabaleta E, Martin MV, Braun HP. A basal carbon concentrating mechanism in plants? Plant Sci. 2012;187:97–104.

Zimmerman SA, Ferry JG. The beta and gamma classes of carbonic anhydrase. Curr Pharm Des. 2008;14:716–21.

Chapter 16
Glyoxalase Pathway and Drought Stress Tolerance in Plants

Mohammad Rokebul Hasan, Ajit Ghosh, Charanpreet Kaur, Ashwani Pareek, and Sneh Lata Singla-Pareek

16.1 Introduction

Crop plants are constantly exposed to a broad range of environmental stresses. Of which, drought is the most devastating one that barriers agroecosystem productivity (Lambers et al. 2008; Farooq et al. 2011). It adversely affects plant metabolism, growth, development, and survival, and thus, is a constraint for plant productivity worldwide (Ahuja et al. 2010; Hasanuzzaman and Fujita 2011; Hasanuzzaman et al. 2012). In addition, climate prediction models indicate more severe and frequent droughts in future, thereby drastically impacting global crop production (IPCC 2008; Manavalan et al. 2009). Being sessile and sensitive organisms, plants have evolved a wide range of molecular programs to readily sense, respond, and cope with changing environments in order to protect themselves from these unforeseen

M.R. Hasan • S.L. Singla-Pareek (✉)
Plant Molecular Biology Group, International Centre for Genetic Engineering and Biotechnology (ICGEB), Aruna Asaf Ali Marg, New Delhi 110 067, India
e-mail: sneh@icgeb.res.in; snehpareek@gmail.com

A. Ghosh
Plant Molecular Biology Group, International Centre for Genetic Engineering and Biotechnology (ICGEB), Aruna Asaf Ali Marg, New Delhi 110 067, India

Department of Biochemistry and Molecular Biology, Shahjalal University of Science and Technology, Sylhet 3114, Bangladesh

C. Kaur
Plant Molecular Biology Group, International Centre for Genetic Engineering and Biotechnology (ICGEB), Aruna Asaf Ali Marg, New Delhi 110 067, India

Stress Physiology and Molecular Biology Laboratory, School of Life Sciences, Jawaharlal Nehru University, New Delhi 110 067, India

A. Pareek
Stress Physiology and Molecular Biology Laboratory, School of Life Sciences, Jawaharlal Nehru University, New Delhi 110 067, India

M.A. Hossain et al. (eds.), *Drought Stress Tolerance in Plants, Vol 1*,
DOI 10.1007/978-3-319-28899-4_16

variations (Ahuja et al. 2010). Response mechanisms to drought stress involve changes at morphological, physiological, and biochemical levels (Zhu 2001). Re-programming in gene expression occurs under stress conditions causing alterations in plant biochemical, transcriptomic, and proteomic machinery (Cohen et al. 2010; Ahuja et al. 2010). In such situations, tolerance to stress can be achieved through modulation of several genes or by organizing the action of different genes from various cellular biochemical pathways (Sasaki-Sekimoto et al. 2005).

As a common phenomenon, stress leads to excessive production of certain deleterious chemical entities such as reactive oxygen species (ROS) and methylglyoxal (MG) in plants (Yadav et al. 2007; Hossain and Fujita 2010; Hossain et al. 2011a). MG is a ubiquitous metabolite generated as a concomitant of intracellular metabolism and, therefore, exists in all cells during normal physiological growth and development conditions and accumulates to higher concentrations under many environmental stresses (Yadav et al. 2008). It is responsible for oxidative stress either through increased production of ROS or by forming advanced glycation end products (AGEs) with macromolecules (Kalapos 2008; Sousa Silva et al. 2013). As it accumulates at higher concentrations under stress conditions, plants have evolved several detoxification mechanisms to combat the so-called dicarbonyl and oxidative stress caused by MG. The primary route for MG detoxification is the thiol-dependent glyoxalase system which catalyzes the conversion of cytotoxic MG (2-oxopropanal) to D-lactic acid via S-D-lactoylglutathione (SLG) (Fig. 16.1). The presence of the

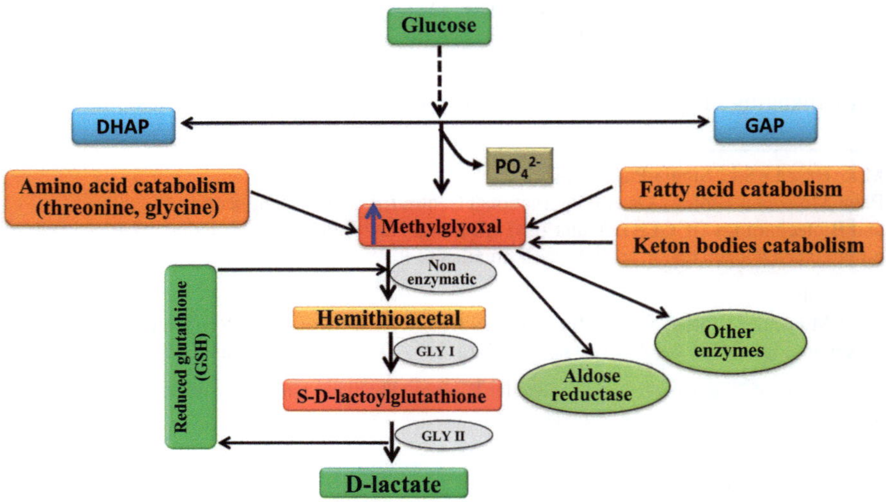

Fig. 16.1 Different routes of methylglyoxal formation and detoxification system in plants. Nonenzymatic generation of MG through β-elimination of phosphate group from enediolate phosphate intermediate is the central route of MG synthesis in plants. Besides, metabolism of amino acids, fatty acids, and ketone bodies contribute to MG formation; first enzyme GLY I converts hemithioacetal formed from spontaneous combination of MG and GSH into S-D-lactoylglutathione which is then converted to D-lactate by GLY II, regenerating GSH in the system. *DHAP* dihydroxyacetone phosphate, *GAP* D-glyceraldehyde-3-phosphate

glyoxalase pathway has been reported in several plant species and involves two enzymes, GLY I and GLY II, which have been purified as well as physiologically and biochemically characterized and functionally validated from various plant species (Yadav et al. 2007; Hoque et al. 2007; Hasanuzzaman and Fujita 2011; Hossain et al. 2014). The efficient role of this pathway in stress management has been extensively studied in various living organisms, including prokaryotes to eukaryotes, and has been shown to be associated with abiotic stress adaptation (Kaur et al. 2014a). Here, we discuss basic molecular programs suggested to confer tolerance to drought stress alongside their envisaged approaches. Special emphasis will be given on molecular mechanisms of glyoxalase pathway mediated drought stress tolerance in plants.

16.2 Effects of Drought on Plant Health

Drought is harmful for the plant growth and development with varying effects based on the severity of the stress. The plants also display a variety of responses on exposure to drought conditions causing alterations at both morphological and molecular levels (Farooq et al. 2009). Drought condition in plants results in alterations in relative water content, water and nutrient relations, photosynthesis, assimilate partitioning and respiration thereby, limiting economic yield (Farooq et al. 2009). Siddique et al. (2001) reported that the relative water content, transpiration rate of wheat and rice under drought stress was lower than control ones. Nutrient contents such as P and PO_4^{3-} in the plant tissue decreased significantly under drought conditions, because of lowered PO_4^{3-} mobility as a result of lower water availability (Peuke and Rennenberg 2004). Drought negatively affects plant photosynthetic efficiency caused by a reduction in leaf expansion, hampered photosynthetic machinery, and early leaf senescence (Wahid and Rasul 2005). The metabolism of carbohydrate, concentration of sucrose in leaves and their export rate decreased due to an increase in the acid invertase activity caused by drought stress (Kim et al. 2000). Liu and Li (2005) observed that the biomass of shoot and root, photosynthesis, and respiration rate of root reduced sharply in wheat exposed to severe drought conditions. Drought-induced yield reduction has been reported in pigeon pea also where a 40–55 % decrease in seed yield was observed at the flowering stage (Nam et al. 2001).

Environmental factors activate a variety of plant responses to drought stress, from altered gene expression and cellular metabolism to adjustment in proper growth and development, thus enabling them to survive under such conditions (Yamaguchi-Shinozaki and Shinozaki 2006; Rampino et al. 2006; Perera et al. 2008; Oh et al. 2009; Wilson et al. 2009). Under drought conditions, gene expression related to various processes such as signaling which includes transcription factors (like NAC family genes, basic leucine zippers, MYB-type transcription factors, zinc fingers, and ethylene-responsive factors) and protein kinases (like calcium-dependent protein kinase and CBL-interacting protein kinase); osmolyte biosynthesis (e.g., trehalose biosynthesis); accumulation of antioxidants (like Mn-superoxide

Table 16.1 Loss in plant yield due to drought stress in some important field crops

Sl No.	Crop	Growth stage	Yield reduction (%)	References
1	Rice	Reproductive (severe stress)	48–94	Lafitte et al. (2007)
2	Rice	Grain filling (severe stress)	60	Basnayake et al. (2006)
3	Wheat	Stem elongation + anthesis	22	Akram et al. (2011)
4	Barley	Seed filling	49–57	Samarah (2005)
5	Maize	Grain filling	79–81	Monneveux et al. (2006)
6	Sunflower	Reproductive	60	Mazahery-Laghab et al. (2003)
7	Soybean	Reproductive	46–71	Samarah et al. (2006)
8	Chickpea	Reproductive	45–69	Nayyar et al. (2006)

dismutase); and several other processes are known to be affected (Sahoo et al. 2013). It has been reported that the severity as well as duration of the drought stress is determinate for economic yield reduction in many commercial field crop species (Table 16.1). In order to survive under stressful conditions plants must upregulate MG and ROS detoxification processes to avoid cellular damage and also to maintain steady state in different plant physiological processes. In this article we shall discuss the effect of drought stress at biochemical and molecular levels only.

16.3 Methylglyoxal Synthesis, Toxicity, and Accumulation Under Drought Conditions

MG is unavoidably produced during metabolism even under normal physiological conditions (Yadav et al. 2005; Hossain et al. 2009). The generation rate of MG varies depending upon the organism, tissue, cell, and physiological conditions (Yadav et al. 2005) and is formed via different nonenzymatic and enzymatic pathways (Richard 1993). In plants, spontaneous synthesis of MG by nonenzymatic mechanisms is considered to be the central route for its generation under normal and stress circumstances (Fig. 16.2). The nonenzymatic formation of MG occurs via removal of phosphoryl group through β-elimination from 1,2-enediolate of triose sugars, dihydroxyacetone phosphate (DHAP) and D-glyceraldehyde-3-phosphate (GAP), during glycolysis (Phillips and Thornalley 1993; Richard 1993). Under stress, in order to maintain metabolic homeostasis, the glycolysis increases resulting in disproportion in the pathway. As a result, excessive MG is inevitably produced as a byproduct of glycolysis during such conditions. Apart from glycolysis, several other sources for MG generation have also been reported and include oxidation of aminoacetone (Lyles and Chalmers 1992), ketone bodies (Aleksandrovskii 1992), and acetone (Casazza et al. 1984; Koop and Casazza 1985) (Fig. 16.1).

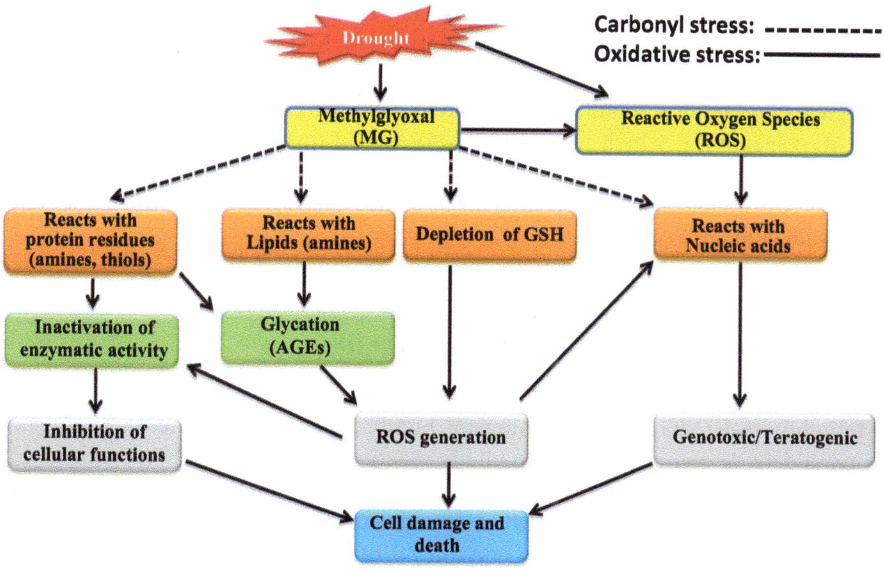

Fig. 16.2 Correlation between MG and ROS generation and their effects on cellular functions during drought stress in plants. MG exhibits direct inhibitory effects on proteins, lipids, and nucleic acids, resulting in carbonyl stress. Generation of ROS and depletion of glutathione is an indirect effect, causing cell damage or death and has been referred to as oxidative stress

Besides these, Maillard (Thornalley et al. 1999) and lipoperoxidation (Esterbauer et al. 1982) reactions also contribute to nonenzymatic sources of MG.

Excessive MG is toxic to the cell inhibiting cell proliferation (Ray et al. 1994). It can easily react with amine groups of proteins, nucleic acids, and lipids in an irreversible manner and form methylglyoxal-derived Advance Glycation End Products (MAGE). MG forms hydroimidazolone derivate (three related structural isomers; MG-H1, MG-H2 and MG-H3), argpyrimidine and tetrahydropyrimidine (THP) with arginine residues (Gomes et al. 2006) and also with lysine residues forming CEL [N*ε*-(carboxyethyl)lysine] and MOLDs (methylglyoxal–lysine dimers) (Gomes et al. 2006), and upon reaction with nucleic acids it generates MGdG {3-(2-deoxyribosyl)-6,7-dihydro-6,7-dihydroxy-6-methylimidazo-[2,3-b] purine-9(8)-one} and CEdG [*N*2-(1-carboxyethyl)-deoxyguanosine] adducts (Thornalley 2003a). In addition, amine-containing basic phospholipids (phosphatidylethanolamine and phosphatidylserine) react with MG and form lipid linked AGEs (carboxymethylethanolamine) (Brown et al. 2005). Furthermore, MG has also been shown to induce ROS formation and apoptosis by activation of signal-regulating kinase (ASK1) (Du et al. 2001). The toxicity of MG is also evident from its ability to cause increased sister chromatin exchange, endoreduplication, DNA strand breaks as well as inducing point mutations (Chaplen 1998). Moreover, it is associated with inhibition of normal growth and development (Hoque et al. 2012c) and results in a number of diverse detrimental effects including the formation of advanced glycation end products (AGEs) and influencing the antioxidant defense system (Wu and Juurlink 2002; Hoque et al. 2010). MG levels rise to toxic concentrations in plants on exposure to drought stress. In rice, MG concentration at physiological conditions is about 27.5 ± 1.2 and 62.3 ± 3.2 μmol/g fresh weight in root and shoot, respectively, which increase two- to sixfold in response to drought (Yadav et al. 2005). In another study, MG concentration is reported to increase 1.63-fold as compared to control condition after 24 h of drought stress in pumpkin seedlings (Hossain et al. 2009).

16.4 Methylglyoxal Detoxification Pathways

Methylglyoxal (MG) is a physiological highly reactive genotoxic and cytogenic α-oxoaldehyde compound. Due to highly reactive properties of MG, its concentrations must be kept below the threshold levels to sustain cellular homeostasis. Whatever route through which MG is produced, it is primarily detoxified by the ubiquitous glyoxalase pathway (Thornalley 1993). Recent investigations in plants have demonstrated the involvement of the glyoxalase system in drought stress tolerance (Hossain et al. 2009; Hasanuzzaman and Fujita 2011). Apart from glyoxalase pathway, there are other enzymes involved in the detoxification process as well (Kalapos 1999).

16.4.1 Glyoxalase Pathway

The glyoxalase pathway is a ubiquitous mechanism for cellular metabolism of MG in the living systems and operates in the cytoplasm of cells in both prokaryotes and eukaryotes. At the time of its discovery in 1913 (Neuberg 1913; Dakin and Dudley 1913), it was believed to be a single enzyme. Later in 1951, involvement of two enzymes for MG detoxification was reported (Racker 1951). The thiol-dependent glyoxalase system comprises two enzymes, glyoxalase I (GLY I; S-D lactoylglutathione lyase; EC 4.4.1.5) and glyoxalase II (GLY II; hydroxyacylglutathione hydrolase; EC 3.1.2.6). The first enzyme of the pathway, GLY I, catalyzes the conversion of MG to S-D-lactoylglutathione with the help of reduced glutathione (GSH), while the second enzyme, GLY II, converts S-D-lactoylglutathione to D-lactic acid and regenerates GSH back to the system (Racker 1951) (Fig. 16.1). MG detoxification is highly dependent on the availability and concentration of endogenous GSH and thus, insufficiency of cellular GSH leads to the accumulation of MG. The overexpression studies of glyoxalase enzymes have demonstrated that glyoxalases can prevent excessive accumulation of MG in plants under stress conditions, acting primarily by maintaining intracellular antioxidant pools (Singla-Pareek et al. 2003; Hoque et al. 2007; Hasanuzzaman and Fujita 2011; Hasanuzzaman et al. 2011; El-Shabrawi et al. 2010; Ghosh et al. 2014). Additional information on the biological function of glyoxalase system comes from the molecular engineering studies of the corresponding genes. Several investigations provide a potential framework for understanding the physiological roles of the glyoxalase system in higher plants in response to various stresses. However, underexpression of glyoxalase I in tobacco showed increased levels of MG leading to cytotoxicity resulting in failure of seed germination (Yadav et al. 2005). In addition, it was reported that glyoxalase enzymes increased the tolerance of plants to drought-induced oxidative damage by maintaining the GSH/GSSG ratio (Hasanuzzaman and Fujita 2011). Further, upregulation of GLY I and GLY II can confer stress tolerance to plants. It was reported that drought stress enhanced GLY II transcript expression in *Brassica* and rice (Yadav et al. 2007; Saxena et al. 2005).

16.4.2 Non-glyoxalase Pathways

In addition to glyoxalases, there are other ways in which MG can be detoxified in the plant system. Since MG contains both ketone and aldehyde groups, it can readily undergo oxidation or reduction reactions (Kalapos 1999; Yadav et al. 2008). Consequently, the enzymes which are involved in oxido-reduction can catalyze the conversion of MG to either acetol or lactaldehyde. Enzymes such as aldo-reductases and dehydrogenases catalyze such reactions (Fig. 16.1). ALR1 (Alcohol; NADP-oxido-reductase, EC. 1.1.1.2), ALR2 (alditol: NAD poxido-reductase, EC. 1.1.1.21), and ALR3 (carbonyl reductase; EC. 1.1.1.184) are representatives of reductase family involved in MG detoxification. These ALRs have been shown to possess broad substrate specificity and are potentially involved in MG detoxification in the plants. Overexpression of aldose/aldehyde reductase (*ALR*) in tobacco plants has been shown to confer tolerance against drought stress. The transgenic plants exhibited reduced loss of photosynthetic efficiency and decreased lipid peroxidation, thiobarbituric acid reactive species (TBRS) and H_2O_2 accumulation as compared to non-transgenic plants (Hideg et al. 2003). Further, pyruvate dehydrogenases are found in abundance in plants and have also been shown to catalyze MG detoxification (Baggetto and Lehninger 1987). Therefore, efficient detoxification of MG might be a sustainable strategy for tolerance against various stresses (Hasanuzzaman and Fujita 2011).

16.5 Correlation Between MG and ROS Production

In plants, stress is generally associated with increased levels of MG and ROS such as superoxide radical (O_2^-), singlet oxygen ($^1O_2$), hydrogen peroxide (H_2O_2), and hydroxyl radical ($OH^·$) (Van Breusegem et al. 2001; Chaves et al. 2003; Reddy et al. 2004). Being a potent and highly reactive glycating agent, it accelerates inactivation of antioxidant defense mechanism (Martins et al. 2001; Thornalley 2003b). MG is

Table 16.2 Correlation between MG and ROS generation

Sl No.	Reaction	Catalyst	Reference
1	Aminoacetone + O_2 → MG + NH_4 + H_2O_2	Semicarbazide sensitive amine oxidase (SSAO)	Yu et al. (2003)
2	Aminoacetone + O_2 → MG + NH4 + O_2^-	Fe^{2+}	Dutra et al. (2001)
3	Acetol + O_2 → MG + H_2O_2	Galactose oxidase	Johnson et al. (1985)
4	MG + H_2O + O_2 → Pyruvate + H_2O_2	Glyoxal oxidase	Kersten and Kirk (1987)
5	MG + e^- → MG^- + O_2 → MG + O_2^-	PSI	Saito et al. (2011)

interlinked with ROS, as evident from the generation of ROS during both formation and decomposition of MG in different cellular reactions (Table 16.2). In plants, MG hampers normal physiological metabolic functions either directly or indirectly through generation of ROS in cells (Hoque et al. 2012a, c) (Fig. 16.2). ROS can readily react with various biologically important macromolecules such as proteins, lipids, and DNA, resulting in oxidative damage and impedes the normal cellular metabolic functions (Apel and Hirt 2004; Foyer and Noctor 2005). Prolonged drought stress accelerates overproduction of MG as well as ROS resulting in oxidative damage (Yadav et al. 2005; Smirnoff 1993). Thus, excessive accumulation of ROS can overcome the antioxidant defense system and results in alteration in metabolic processes, reduction in photosynthesis, interruptions in cellular coordination leading to growth retardation, reduced fertility, causes premature senescence and death of plants (Hossain et al. 2011b; Saito et al. 2011; Krasensky and Jonak 2012). Therefore, ROS should be regulated in plants through the synchronization of ROS production and ROS scavenging systems to withstand oxidative damage by homeostatic regulation of signaling events (Foyer and Noctor 2005).

16.6 Impairment of Cellular Functions by MG and ROS

MG and ROS are generated during the course of metabolism in vivo and are highly reactive glycation agents. The probable involvement of ROS in reactions between MG and macromolecules was first reported by Szent-Györgyi in the 1960s (Szent-Györgyi 1968). Later in the 1970s, the interaction between protein amines and MG was investigated (Kon and Szent-Györgyi 1973). Moreover, the ROS generating ability of MG was also reported (Kalapos et al. 1993). Currently, little information is available regarding the role of MG and ROS under drought stress conditions in plants. However, relationship of MG and ROS with cellular macromolecules had been studied over the course of time and can be utilized to study the emerging role of glyoxalase in plant drought tolerance. This section concentrates on understanding the mechanism of MG and ROS toxicity in the plants.

16.6.1 MG-Mediated Disruption in Cellular Functioning

Being a strong electrophile, MG can readily modify functional groups of various macromolecules and thus, influences their biological activity (Kalapos 1994). It disturbs cellular metabolism upon excessive accumulation and is directly involved in imposing carbonyl stress (Fig. 16.2), when MG levels supersede detoxification capability of glyoxalase I and other related enzymes, then carbonyls bind to protein, lipids, and other macromolecules, thereby leading to ROS generation and that advanced to apoptosis or to malfunction (Kalapos 2008). It is also inhibiting the activity of various important cellular enzymes, including glycolytic enzymes, intramitochondrial enzymes, Na^+-K^+-ATPase, transport proteins and enzymes participating in cell defense (Leoncini et al. 1980; Kun 1950; Mira et al. 1991; Ferguson et al. 1998; Vander Jagt et al. 1997; Amicarelli et al. 2003). Further, it is also capable of reacting with nucleic acids, and is suggested to be a carcinogenic, mutagenic, and teratogenic agent (Hasegawa et al. 1995; Sugimura and Sato 1983; Chaplen 1998; Brambilla et al. 1985). GSH is a well-known intracellular antioxidant agent involved in the protection of cells from oxidative stress (Sen 1997). It may be trapped as S-2-hydroxyacylglutathione at excessive accumulation of MG and subsequently causing GSH depletion (Kalapos et al. 1992). However, MG can act as directly as cytotoxic agent affecting various cellular machineries or it can reduce GSH concentration under stress condition. It is reported that a significant decrease in GSH levels occur in the presence of various concentrations of MG (Kalapos et al. 1992). Additionally, MG also decreased the thiol containing proteins level in isolated mitochondria (Kun 1950). Finally, MG inhibits the activity of several enzymes (Kalapos 1994) and also depletes GSH levels both in vivo and in vitro (Amicarelli et al. 2003).

16.6.2 ROS-Mediated Disruption in Cellular Functioning

Despite their toxic nature, ROS actually have a double role in vivo depending on their concentration, duration and site of action, preceding encounter to stress, etc. (Miller et al. 2010). In general, low doses are treated as signals that mediate at least some part of the responses towards stress while at certain levels of phytotoxicity, they cause a great threat that may in due course lead to programmed cell death (Gechev and Hille 2005). When the cellular ROS concentration exceeds beyond the threshold levels, then living systems can be said to be in a state of "oxidative stress" (Fig. 16.2). Abiotic stress such as drought leads to excessive accumulation of ROS due to imbalance in cellular homeostasis (Sharma and Dubey 2005). They can pose cellular damage by triggering oxidation of proteins, peroxidation of lipids, damage to nucleic acids, inhibition of enzyme activities, activation of programmed cell death (PCD) eventually leading to death of the cells (Reddy et al. 2004; Sharma and Dubey 2005; de Carvalho 2008; Ahuja et al. 2010; Karuppanapandian et al. 2011).

16.7 Drought Induced Alteration in Expression of Glyoxalase Genes

The role of glyoxalase genes has been demonstrated under abiotic stress conditions through various transcriptomic studies. Stress-induced alterations in glyoxalase gene expression clearly suggest a direct role of glyoxalase genes in stress adaptation and acclimation pathway. Upon mannitol treatment, a two- to threefold upregulation in GLY I expression has been observed in different tissues such roots, stems, and leaves (Espartero et al. 1995). GLY I preferentially accumulates in the phloem sieve elements as revealed through immunohistochemical localization analysis. Further, a dose-dependent GLY I transcript analysis has also been performed in *Brassica juncea* in response to salt, drought, and heavy metal stresses (Veena and Sopory 1999). A significant two- to threefold enhancement in the level of GLY I transcript was observed in response to 400 mM mannitol. In order to identify novel genes involved in desiccation tolerance in the foliage of the grass *Sporobolus stapfianus*, Blomstedt et al. 1998 prepared a cDNA library from the desiccated leaf tissue. After differential screening, six clones including GLY I have been identified that show increased transcript abundance and thus might be associated with desiccation tolerance. Northern blot analysis showed a threefold increase in GLY I transcript in response to dehydration as compared to the fully hydrated tissue and a twofold increase in response to subsequent drying. In *S. stapfianus*, GLY I transcripts are also induced by 1.6-fold after treatment with ABA. Moreover, microarray analysis of transgenic plants overexpressing NAC transcription factor genes shows upregulation of several stress-inducible genes including GLY I and resulting transgenic plants show significant tolerance towards drought stress (Tran et al. 2004). Further, a sharp fourfold upregulation in GLY I expression has been observed after transcriptome profiling of wild type and co-suppressed MSI1 (chromatin assembly factor 1) *Arabidopsis* lines (Alexandre et al. 2009). Apart from activation of GLY I transcripts, co-suppressed MSI1 plants have increased levels of free proline and showed enhanced tolerance towards drought. A noticeable increase in the GLY I transcript was also observed in pumpkin seedlings in response to different stresses including drought, salinity, heavy metal, and heat (Hossain et al. 2009). Moreover, genome wide expression analysis of *Arabidopsis* and rice using microarray data identified several glyoxalase members with altered expression in response to drought stress (Mustafiz et al. 2011). An upregulation in expression of *AtGLYI3*, *AtGLYI6*, and *AtGLYI7* genes occurs in a time-dependent manner under drought conditions in Arabidopsis seedlings, whereas *AtGLYI2*, *AtGLYI4*, and *AtGLYI9* are downregulated under such conditions. Similarly, rice glyoxalase genes, *OsGLYI2*, *OsGLYI6*, and *OsGLYI11*, are induced, but *OsGLYI5* and *OsGLYI10* are downregulated in response to drought stress in the rice seedlings. Expression of rice GLY I transcripts were further analyzed in the 2 weeks rice seedlings in response to different abiotic stresses such as heat, cold, dehydration, wounding, MG, salt, and oxidative stress by qRT-PCR (Kaur et al. 2013). A 4.5-fold upregulation in *OsGLYI-11.2* expression was observed, followed by *OsGLYI-7.1* under drought conditions; while other members *OsGLYI-2*, *OsGLYI-8*, and *OsGLYI-11.3* showed sharp decline in gene expression. Furthermore, differential gene expression studies in soybean leaf tissues revealed upregulation of GLY I family members along with other regulatory and functional genes under drought stress (Le et al. 2012).

Like GLY I, expression of GLY II transcript was also found to vary under different stresses. Expression of rice GLY II gene was analyzed in response to various abiotic stresses such as desiccation, salinity, heat, cold, and ABA and SA treatment (Yadav et al. 2005). Significant accumulation of GLY II transcript was found in response to all stress agents. Desiccation stress resulted in the accumulation of GLY II transcript in a short duration of 15 min followed by gradual increase in accumulation with time till 2 h (Yadav et al. 2007). Genome wide transcript analysis of rice GLY II transcripts showed strong induction of all GLY II members in response to drought stress (Mustafiz et al. 2011). Amongst the Arabidopsis GLY II genes, the expression of *AtGLYII1* and *AtGLYII2* was found to be highly upregulated in response to drought stress in both shoot and root tissues (Mustafiz et al. 2011). However *AtGLYII3*, *AtGLYII4*, and *AtGLYII5* were downregulated in response to drought stress in both shoot and root tissues in Arabidopsis.

16.8 Drought Induced Alteration in Levels of Glyoxalase Proteins

Proteins are vital components of living organisms that are directly involved in various physiological and metabolic pathways of cells. Hence, studying variations in levels of glyoxalase proteins or their enzyme activities will give more precision in understanding the role of these enzymes in stress adaptation and in efficient monitoring of the stress response. Activity of glyoxalase has been monitored by various research groups under different environmental stimuli. Initial reports have revealed an increase in GLY I activity during cell division (Deswal et al. 1993) and proliferative callus cultures of groundnut (*Arachis hypogaea* L.cv. JL24) (Jain et al. 2002). To identify the altered proteins during drought stress, functional proteome studies have been performed and have secured an important place in the era of comparative and functional genomics. To investigate the mechanism of plants' osmotic stress response, rice protein profiles were monitored from mannitol-treated plants using proteomics approach (Zang and Komatsu 2007). Proteins from the basal part of leaf sheaths showed strong induction in levels of GLY I protein in response to stress. To study the changes in wheat grain proteome in response to drought, two-dimensional gel electrophoresis among three wheat genotypes with different genetic background was performed under well-watered and drought conditions (Hajheidari et al. 2007). The overall effect of drought was highly significant and about 650 spots were reproducibly detected and analyzed. Mass spectrometry analysis using MALDITOF/TOF led to the identification of 57 proteins with significant alteration. A significant downregulation (twofold) in GLY I protein levels was observed in the susceptible genotypes, with no or insignificant changes in the tolerant counterpart.

Further, GLY I protein was also identified in a two-dimensional gel electrophoresis experiment carried out in two distinct sunflower genotypes in response to drought (Castillejo et al. 2008). The susceptible genotype showed a decrease in the intensity of the 17 spots out of 28 altered proteins. The proteins that showed a decline in their levels included a GLY I protein, along with some other important proteins such as photosystem II oxygen-evolving complex protein 1, carbonic anhydrase, RubisCO large and small subunits, ferredoxin-$NADP^+$ reductase, phosphoglycerate kinase, glyceraldehyde-3-phosphate dehydrogenase, aldolase and superoxide dismutase. Furthermore, comparative proteomic analysis of differentially expressed chickpea and rice extracellular matrix proteins also led to the identification of a GLY I protein during dehydration stress (Bhushan et al. 2007; Pandey et al. 2010). GLY I protein was also found to be significantly upregulated in the nuclear fraction of chickpea in response to dehydration stress (Pandey et al. 2008). In addition, analysis of drought responsive leaf proteome of a C3 xerophyte, *Citrullus lanatus* also revealed alteration in levels of GLY I protein (Akashi et al. 2011).

Significant increase in levels of GLY I protein and GLY I activity was observed in onion bulb in response to various stress treatments (Hossain et al. 2007). An induction of 1.3- to 1.4-fold was observed in both the levels of GLY I protein and activity in response to drought stress. A sharp increase in GLY I activity (1.27-fold) was observed after 24 h of drought stress in pumpkin seedlings (Hossain et al. 2009). A similar pattern of induction was observed in GLY II enzyme activity in response to drought. The potential role of various chemical compounds in increasing drought tolerance by enhancing glyoxalase enzyme activity has been determined by different studies (Hasanuzzaman and Fujita 2011; Alam et al. 2013). For instance, drought stress induced oxidative damage of rapeseed seedlings could be reversed by the pretreatment of selenium that enhances the activities of antioxidant and MG detoxifying enzymes (Hasanuzzaman and Fujita 2011). Selenium pretreated rapeseed seedlings exposed to various degrees of drought stress showed a sharp rise in their ascorbic acid level, reduced glutathione content, and maintained a high GSH/GSSG ratio as compared with the drought-stressed plants without selenium treatment. It has been reported that pretreatment with 25 mM of selenium resulted in a 23 % increase in GLY I activity and also a significant increase in GLY II activity in rapeseed seedlings as compared to control. A similar study showed that exogenous addition of salicylic acid in mustard seedlings mediates short-term tolerance against drought stress by upregulating the antioxidant defense and glyoxalase pathway (Alam et al. 2013). Drought stress resulted in a sharp decline in the level of ascorbate, relative water content, and chlorophyll content in the mustard seedlings, but increased their proline, malondialdehyde, and H_2O_2 levels. However, salicylic acid supplementation in the drought stressed seedlings enhanced ascorbate, reduced glutathione, chlorophyll, and relative water content, as well as decreased the GSSG level to maintain the ratio of GSH/GSSG. Salicylic acid supplemented drought stressed seedlings also enhanced the enzyme activities of GLY I, GLY II, and different antioxidant enzymes as compared to drought-stressed plants without salicylic acid supplementation. Moreover, temperature (either heat or cold)-shock positively modulates the oxidative protection in salinity and drought stressed mustard

(*Brassica campestris* L.) seedlings in a very similar mechanism by increasing glyoxalase activity (Hossain et al. 2013a, b). Seedlings pre-exposed to either heat-shock or cold-shock conditions positively modulate the activities of GLY I and GLY II, and maintain lower levels of GSSG, H_2O_2, and malondialdehyde as compared to the control as well as non-treated drought stressed seedlings.

16.9 Signaling Roles of MG in Regulation of Stomatal Closure and Stress Responsive Gene Expression

Despite having inhibitory effects on cell growth, MG has been shown to possess signaling roles in bacteria (Campbell et al. 2007), humans (Kang et al. 1996; Akhland et al. 2001), and yeast (Maeta et al. 2005; Takatsume et al. 2006). However in plants, role of MG in signal transduction is less studied. Nonetheless, it has been reported that MG induces ROS formation (Hoque et al. 2012a) and that ROS mediates abscisic acid (ABA) and methyl jasmonate (MeJA) signaling pathways in guard cells related to stomatal regulation (Munemasa et al. 2007). Hoque and coworkers have shown that MG induces stomatal closure in a reversible manner and also induces generation of ROS in *Arabidopsis* (Hoque et al. 2012a). It was found that MG induced significant accumulation of ROS and also increased cytosolic Ca^{2+} oscillations in the guard cells which were suppressed by pretreatment with SHAM (salicylhydroxamic acid). SHAM-sensitive peroxidases diffuse extracellular oxidative burst into the intracellular space contributing to intracellular ROS accumulation in the guard cells and trigger stomatal closure via a Ca^{2+}-dependent pathway (Hoque et al. 2012a). Additionally, it was also observed that MG was also engaged in inhibiting light-induced stomatal opening via the modification of C-terminal region of KAT1, an inward-rectifying potassium channel thereby inhibiting K^+ influx into the guard cells (Hoque et al. 2012b). The involvement of MG in regulation of stomatal movements indicates towards its role in signal transduction pathways in drought stress adaptation. Because of closure of stomata is the primary response of almost all plants to drought to prevent transpirational water loss (Mansfield and Atkinson 1990). Regulation of stomata may result in response to decrease in leaf turgor or low humidity atmosphere (Ludlow and Muchow 1990; Maroco et al. 1997). In response to drought, MG levels have been reported to increase up to sixfold depending upon the crop species (Yadav et al. 2005).

Further, MG is capable of altering expression of genes known to be involved in drought stress adaptation. For instance, MG was found to affect the transcript levels of ABA-dependent genes, RD29B and RAB18, which are generally induced in response to dehydration. MG could significantly induce RD29B (fivefold) and RAB18 (threefold) gene expression that too in a concentration-dependent manner (Hoque et al. 2012c). In addition, MG has also been shown to enhance expression of triose phosphate isomerase (*OscTPI*) and *OsETHE1* in rice (Sharma et al. 2012; Kaur et al. 2014b). Moreover, global gene expression profiles in rice in response to exogenous MG showed its involvement in signal transduction. MG affected the expression of various genes involved in stress-induced signal transduction cascades such as protein kinases (mitogen-activated protein kinase, calcium/calmodulin-dependent protein kinases, Ser/Thr protein kinase, histidine kinase, and receptor-like kinase) and transcription factors (bZIP, AP2 domain-containing protein, NAM, WRKY, and zinc finger proteins), which were significantly represented in the perturbed transcriptomes, indicating an interlink between MG and stress-responsive signal transduction pathways (Kaur et al. 2015). Collectively, MG plays a significant role in signal transduction possibly acting as a stress signal molecule in plants, where it conveys signals to the cellular machinery to maintain the cellular homeostasis towards adaptation in drought stress.

16.10 Conclusion and Future Perspective

The pathways involved in drought stress adaptation in plants are regulated at both physiological and molecular levels. Molecular information of response and tolerance mechanisms is likely to pave way for engineering plants that could make them withstand drought stress. Many achievements have been made over the last few years in understanding the protective role of glyoxalases in MG detoxification under drought conditions (Fig. 16.3). Drought stress leads to increased accumulation of MG and MG-derived ROS. It is now well known that MG has deleterious effects on

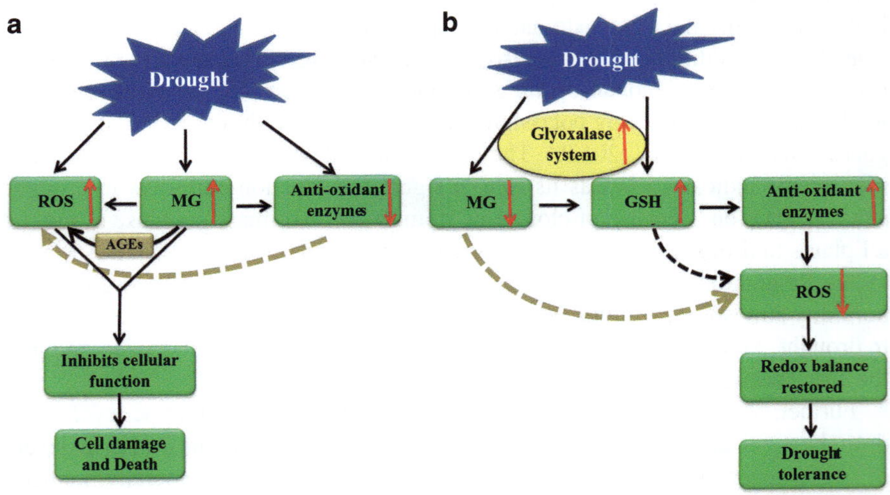

Fig. 16.3 Role of glyoxalase pathway in drought stress adaptation. During drought, MG and ROS levels increase which then impair the redox balance of cell. MG levels also induce ROS generation through the formation of AGEs, resulting in ROS-mediated cellular injury and death (**a**). Increase in glyoxalase activity through overexpression helps in maintaining cellular redox homoeostasis under drought stress by reducing MG levels and regenerating GSH back into the system, thereby decreasing ROS generation which leads to improved drought tolerance (**b**)

plant growth and development and that glyoxalase pathway serves an important detoxification role in the living systems. Several transcriptome and proteome studies carried out to identify genes involved in drought stress response have revealed a link between glyoxalases and drought stress adaptation indicating glyoxalase pathway to be a crucial intracellular component of plant stress response. Further, MG transmits signals to the cellular machinery for inducing changes in plant transcriptome, transcription factors, protein kinase as well as regulation of stomatal movements for adaptation to drought stress conditions. However, the specific role of MG as a signal molecule itself or as a component in signaling cascade in plants needs further investigation for deeper understanding of its role in stress response and tolerance.

References

Ahuja I, de Vos RCH, Bones AM, Hall RD. Plant molecular stress responses face climate change. Trends Plant Sci. 2010;12:664–74.

Akashi K, Yoshida K, Kuwano M, Kajikawa M, Yoshimura K, Hoshiyasu S, Inagaki N, Yokota A. Dynamic changes in the leaf proteome of a C3 xerophyte, *Citrullus lanatus* (wild watermelon), in response to water deficit. Planta. 2011;233:947–60.

Akhland AA, Hossain K, Mitsui H, Kato M, Miyata T, Inagi R, Du J, Takeda K, Kawamoto Y, Suzuki H, Kurokawa K, Nakashima I. Glyoxal and methylglyoxal trigger distinct signals for MAP family kinases and caspase activation in human endothelial cells. Free Radic Biol Med. 2001;31:20–30.

Akram M. Growth and yield components of wheat under water stress of different growth stages. Bangladesh J Agric Res. 2011;36:455–68.

Alam MM, Hasanuzzaman M, Nahar K, Fujita M. Exogenous salicylic acid ameliorates short-term drought stress in mustard (*Brassica juncea* L.) seedlings by up-regulating the antioxidant defense and glyoxalase system. Aust J Crop Sci. 2013;7:1053–63.

Aleksandrovskii YA. Antithrombin III, C1 inhibitor, methylglyoxal, and polymorphonuclear leukocytes in the development of vascular complications in diabetes mellitus. Thromb Res. 1992;67:179–89.

Alexandre C, Möller-Steinbach Y, Schönrock N, Gruissem W, Hennig L. Arabidopsis MSI1 is required for negative regulation of the response to drought stress. Mol Plant. 2009;2(4): 675–87.

Amicarelli F, Colafarina S, Cattani F, Cimini A, Di Ilio C, Ceru MP, Miranda M. Scavenging system efficiency is crucial for cell resistance to ROS mediated methylglyoxal injury. Free Radic Biol Med. 2003;35:856–71.

Apel K, Hirt H. Reactive oxygen species: metabolism, oxidative stress, and signal transduction. Annu Rev Plant Biol. 2004;55:373–99.

Baggetto LG, Lehninger AL. Isolated tumoral pyruvate dehydrogenase can synthesize acetone which inhibits pyruvate oxidation as well as other aldehydes. Biochem Biophys Res Commun. 1987;145:153–9.

Basnayake J, Fukai S, Ouk M. Contribution of potential yield, drought tolerance and escape to adaptation of 15 rice varieties in rainfed lowlands in Cambodia. Proceedings of the Australian Agronomy Conference. Brisbane: Australian Society of Agronomy; 2006.

Bhushan D, Pandey A, Choudhary MK, Datta A, Chakraborty S, Chakraborty N. Comparative proteomics analysis of differentially expressed proteins in chickpea extracellular matrix during dehydration stress. Mol. Cell Proteom. 2007;6:1868–84.

Blomstedt CK, Gianello RD, Hamill JD, Neale AD, Gaff DF. Drought-stimulated genes correlated with desiccation tolerance of the resurrection grass *Sporobolus stapfianus*. Plant Growth Regul. 1998;24:153–61.

Brambilla G, Sciaa L, Faggin P, Finollo R, Bassi AM, Ferro M, Marinari UM. Methylglyoxal-induced DNA-protein cross-links and cytotoxicity in Chinese hamster ovary cells. Carcinogenesis. 1985;6:683–6.

Brown BE, Dean RT, Davies MJ. Glycation of low-density lipoproteins by methylglyoxal and glycolaldehyde gives rise to the in vitro formation of lipid-laden cells. Diabetologia. 2005; 48:361–9.

Campbell AK, Naseem R, Holland IB, Matthews SB, Wann KT. Methylglyoxal and other carbohydrate metabolites induce lanthanum-sensitive Ca^{2+} transients and inhibit growth in *E. coli*. Arch Biochem Biophys. 2007;468:107–13.

Casazza JP, Felver ME, Veech RL. The metabolism of acetone in rat. J Biol Chem. 1984; 259:231–6.

Castillejo MA, Maldonado AM, Ogueta S, Jorrın JV. Proteomic analysis of responses to drought stress in sunflower (*Helianthus annuus*) leaves by 2DE gel electrophoresis and mass spectrometry. Open Proteom J. 2008;1:59–71.

Chaplen FWR. Incidence and potential implications of the toxic metabolite methylglyoxal in cell culture: A review. Cytotechnology. 1998;26:173–83.

Chaves MM, Maroco JP, Pereira JS. Understanding plant responses to drought—from genes to the whole plant. Funct Plant Biol. 2003;30:239–64.

Cohen D, Bogeat-Triboulot MB, Tisserant E, Balzergue S, Martin-Magniette ML, Lelandais G, Ningre N, Renou JP, Tamby JP, Le Thiec D, Hummel I. Comparative transcriptomics of drought responses in Populus: a meta-analysis of genome-wide expression profiling in mature leaves and root apices across two genotypes. BMC Genomics. 2010;11:630.

Dakin HD, Dudley HW. An enzyme concerned with the formation of hydroxyl acids from ketonic aldehydes. J Biol Chem. 1913;14:155–7.

de Carvalho MHC. Drought stress and reactive oxygen species: Production, scavenging and signaling. Plant Signal Behav. 2008;3:156–65.

Deswal R, Chakravarty TN, Sopory SK. The glyoxalase system in higher plants: regulation in growth and differentiation. Biochem Soc Trans. 1993;21:527–30.

Du J, Suzuki H, Nagase F, Akhand AA, Ma XY, Yokoyama T, Miyata T, Nakashima I. Superoxide-mediated early oxidation and activation of ASK1 are important for initiating methylglyoxal-induced apoptosis process. Free Radic Biol Med. 2001;31:469–78.

Dutra F, Knudsen FS, Curi D, Bechara EJ. Aerobic oxidation of aminoacetone, a threonine catabolite: Iron catalysis and coupled iron release from ferritin. Chem Res Toxicol. 2001;14: 1323–9.

El-Shabrawi H, Kumar B, Kaul T, Reddy MK, Singla-Pareek SL, Sopory SK. Redox homeostasis, antioxidant defense, and methylglyoxal detoxification as markers for salt tolerance in Pokkali rice. Protoplasma. 2010;245:85–96.

Espartero J, Sanchez-Aguayo I, Pardo JM. Molecular characterization of glyoxalase-I from a higher plant: upregulation by stress. Plant Mol Biol. 1995;29:1223–33.

Esterbauer H, Cheeseman KH, Dianzani MU, Poli G, Slater TF. Separation and characterization of the aldehydic products of lipid peroxidation stimulated by ADP-Fe^{2+} in rat liver microsomes. Biochem J. 1982;208:129–40.

Farooq M, Wahid A, Kobayashi N, Fujita D, Basra SMA. Plant drought stress: effects, mechanisms and management. Agron Sustain Dev. 2009;29(1):185–212.

Farooq M, Bramley H, Palta JA, Siddique KHM. Heat stress in wheat during reproductive and grain filling phases. Crit Rev Plant Sci. 2011;30:491–507.

Ferguson GP, Tötemeyer V, MacLean MJ, Booth IR. Methylglyoxal production in bacteria: suicide or survival? Arch Microbiol. 1998;170:209–19.

Foyer CH, Noctor G. Redox homeostasis and antioxidant signaling: a metabolic interface between stress perception and physiological responses. Plant Cell. 2005;17:1866–75.

Gechev TS, Hille J. Hydrogen peroxide as a signal controlling plant programmed cell death. J Cell Biol. 2005;168(1):17–20.

Ghosh A, Pareek A, Sopory SK, Singla-Pareek SL. A glutathione responsive rice glyoxalase II, OsGLYII-2, functions in salinity adaptation by maintaining better photosynthesis efficiency and anti-oxidant pool. Plant J. 2014;80(1):93–105.

Gomes RA, Vicente Miranda H, Silva MS, Graca G, Coelho AV, Ferreira AE, Cordeiro C, Freire AP. Yeast protein glycation in vivo by methylglyoxal. Molecular modification of glycolytic enzymes and heat shock proteins. FEBS J. 2006;273:5273–87.

Hajheidari M, Eivazi A, Buchanan BB, Wong JH, Majidi I, Salekdeh GH. Proteomics uncovers a role for redox in drought tolerance in wheat. J Proteome Res. 2007;6:1451–60.

Hasanuzzaman M, Fujita M. Selenium pretreatment upregulates the antioxidant defense and methylglyoxal detoxification system and confers enhanced tolerance to drought stress in rapeseed seedlings. Biol Trace Elem Res. 2011;143:1758–76.

Hasanuzzaman M, Hossain MA, Fujita M. Selenium induced upregulation of the antioxidant defense and methylglyoxal detoxification system reduces salinity-induced damage in rapeseed seedlings. Biol Trace Elem Res. 2011;143:1704–21.

Hasanuzzaman M, Hossain MA, da Silva JAT, Fujita M. Plant Responses and tolerance to abiotic oxidative stress: Antioxidant defenses is a key factors. In: Bandi V, Shanker AK, Shanker C, Mandapaka M, editors. Crop stress and its management: Perspectives and strategies. Berlin: Springer; 2012. p. 261–316.

Hasegawa R, Ogiso T, Imaida K, Shirai T, Ito N. Analysis of the potential carcinogenicity of coffee and its related compounds in a medium-term liver bioassay of rats. Food Chem Toxicol. 1995;33:15–20.

Hideg E, Nagy T, Oberschall A, Dudits D, Vass I. Detoxification function of aldose/aldehyde reductase during drought and ultraviolet-B (280-320) stresses. Plant Cell Environ. 2003;26: 513–22.

Hoque MA, Banu MN, Nakamura Y, Shimoishi Y, Murata Y. Proline and glycinebetaine enhance antioxidant defense and methylglyoxal detoxification systems and reduce NaCl induced damage in cultured tobacco cells. J Plant Physiol. 2007;165:813–24.

Hoque MA, Uraji M, Banu MNA, Mori IC, Nakamura Y, Murata Y. The effect of methylglyoxal on glutathione *S*-transferase from *Nicotiana tabacum*. Biosci Biotechnol Biochem. 2010;74:2124–6.

Hoque TS, Uraji M, Ye W, Hossain MA, Nakamura Y, Murata Y. Methylglyoxal-induced stomatal closure accompanied by peroxidase-mediated ROS production in Arabidopsis. J Plant Physiol. 2012a;169:979–86.

Hoque TS, Okuma E, Uraji M, Furuichi T, Sasaki T, Hoque MA, Nakamura Y, Murata Y. Inhibitory effects of methylglyoxal on light-induced stomatal opening and inward K^+ channel activity in Arabidopsis. Biosci Biotechnol Biochem. 2012b;76:617–9.

Hoque TS, Uraji M, Tuya A, Nakamura Y, Murata Y. Methylglyoxal inhibits seed germination and root elongation and up-regulates transcription of stress-responsive genes in ABA-dependent pathway in Arabidopsis. Plant Biol. 2012c;14:854–8.

Hossain MA, Fujita M. Evidence for a role of exogenous glycinebetaine and proline in antioxidant defense and methylglyoxal detoxification systems in mung bean seedlings under salt stress. Physiol Mol Biol Plants. 2010;16:19–29.

Hossain MD, Rohman MM, Fujita M. Comparative investigation of glutathione S-transferases, glyoxalase I and allinase activities in different vegetable crops. J Crop Sci Biotechnol. 2007;10:21–8.

Hossain MA, Hossain MZ, Fujita M. Stress-induced changes of methylglyoxal level and glyoxalase I activity in pumpkin seedlings and cDNA cloning of glyoxalase I gene. Aust J Crop Sci. 2009;3:53–64.

Hossain MA, Hasanuzzaman M, Fujita M. Coordinate induction of antioxidant defense and glyoxalase system by exogenous proline and glycinebetaine is correlated with salt tolerance in mung bean. Front Agric China. 2011a;1:1–14.

Hossain MA, Teixeira da Silva JA, Fujita M. Glyoxalase system and reactive oxygen species detoxification system in plant abiotic stress response and tolerance: An intimate relationship. In: Shanker AK, Venkateswarlu B, editors. Abiotic Stress/Book 1. Croatia: INTECH-Open Access; 2011b. p. 235–66.

Hossain MA, Mostofa MG, Fujita M. Cross protection by cold-shock to salinity and drought stress-induced oxidative stress in mustard (*Brassica campestris* L.) seedlings. Mol Plant Breed. 2013a;4:50–70.

Hossain MA, Mostofa MG, Fujita M. Heat-shock positively modulates oxidative protection of salt and drought-stressed mustard (*Brassica campestris* L.) seedlings. J Plant Sci Mol Breed. 2013b;2:2.

Hossain MA, Mostofa MG, Burritt DJ, Fujita M. Modulation of reactive oxygen species and methylglyoxal detoxification systems by exogenous glycinebetaine and proline improves drought tolerance in mustard (*Brassica juncea* L.). Int J Plant Biol Res. 2014;2(2):1014.

Jain M, Choudhary D, Kale RK, Bhalla-Sarin N. Salt and glyphosphate-induced increase in glyoxalase I activity in cell lines of groundnut (*Arachis hypogaea*). Physiol Plant. 2002;114: 499–505.

Johnson JM, Halsall HB, Heinemen WR. Redox activation of galactose oxidase: thin-layer electrochemical study. Biochemistry. 1985;24:1579–85.

Kalapos MP. Methylglyoxal toxicity in mammals. Toxicol Lett. 1994;73:3–24.

Kalapos MP. Methylglyoxal in living organisms: chemistry, biochemistry, toxicology and biological implications. Toxicol Lett. 1999;110:145–75.

Kalapos MP. The tandem of free radicals and methylglyoxal. Chem Biol Interact. 2008;171(3): 251–71.

Kalapos MP, Garzó T, Antoni F, Mandl J. Accumulation of S-D-lactoylglutathione and transient decrease of glutathione level caused by methylglyoxal load in isolated hepatocytes. Biochim Biophys Acta. 1992;1135:159–64.

Kalapos MP, Littauer A, de Groot H. Has reactive oxygen a role in methylglyoxal toxicity? A study on cultured rat hepatocytes. Arch Toxicol. 1993;67:369–72.

Kang Y, Edwards LG, Thornalley PJ. Effect of methylglyoxal on human leukaemia 60 cell growth: modification of DNA, G1 growth arrest and induction of apoptosis. Leuk Res. 1996;5: 397–405.

Karuppanapandian T, Wang HW, Prabakaran N, Jeyalakshmi K, Kwon M, Manoharan K, Kim W. 2,4-dichlorophenoxyacetic acid-induced leaf senescence in mung bean (*Vigna radiata* L. Wilczek) and senescence inhibition by co-treatment with silver nanoparticles. Plant Physiol Biochem. 2011;49:168–77.

Kaur C, Vishnoi A, Ariyadasa TU, Bhattacharya A, Singla-Pareek SL, Sopory SK. Episodes of horizontal gene-transfer and gene-fusion led to co-existence of different metal-ion specific glyoxalase I. Sci Rep. 2013;3:3076.

Kaur C, Ghosh A, Pareek A, Sopory SK, Singla-Pareek SL. Glyoxalases and stress tolerance in plants. Biochem Soc Trans. 2014a;42(2):485–90.

Kaur C, Mustafiz A, Sarkar A, Ariyadasa TU, Singla-Pareek SL, Sopory SK. Expression of abiotic stress inducible ETHE1-like protein from rice is higher in roots and is regulated by calcium. Physiol Plant. 2014b;152(1):1–16. doi:10.1111/ppl.12147.

Kaur C, Kushwaha HR, Mustafiz A, Pareek A, Sopory SK, Singla-Pareek SL. Analysis of global gene expression profile of rice in response to methylglyoxal indicates its possible role as a stress signal molecule. Front Plant Sci. 2015;6:682.

Kersten PJ, Kirk TK. Involvement of a new enzyme, glyoxal oxidase, in extracellular H_2O_2 production by *Phanerochaete chrysosporium*. J Bacteriol. 1987;169:2195–201.

Kim JY, Mahé A, Brangeon J, Prioul JL. A maize vacuolur invertase, IVR2, is induced by water stress. Organ/tissue specificity and diurnal modulation of expression. Plant Physiol. 2000; 124:71–84.

Kon H, Szent-Györgyi A. Charge transfer between amine and carbonyl. Proc Natl Acad Sci U S A. 1973;70:3139–40.

Koop DR, Casazza JP. Identification of ethanol-inducible P-450 isozyme 3a as the acetone and acetol monooxygenase of rabbit microsomes. J Biol Chem. 1985;260:13607–12.

Krasensky J, Jonak C. Drought, salt, and temperature stress-induced metabolic rearrangements and regulatory networks. J Exp Bot. 2012;63:1593–608.

Kun K. Inhibition of succinic dehydrogenase by methylglyoxal. J Biol Chem. 1950;187:289–97.

IPCC. Kundzewicz ZW, Palutikof J, Wu S, editors. Climate change and water. Technical paper of the Intergovernmental Panel on Climate Change. Cambridge: Cambridge University Press; 2008.

Lafitte HR, Yongsheng G, Yan S, Lil ZK. Whole plant responses, key processes, and adaptation to drought stress: the case of rice. J Exp Bot. 2007;58:169–75.

Lambers H, Chapin FS, Pons TL. Plant physiological ecology. 2nd ed. New York: Springer; 2008.

Le DT, Nishiyama R, Watanabe Y, Tanaka M, Seki M, Ham LH, Yamaguchi-Shinozaki K, Shinozaki K, Tran LSP. Differential gene expression in soybean leaf tissues at late developmental stages under drought stress revealed by genome-wide transcriptome analysis. PLoS One. 2012;7, e49522.

Leoncini G, Maresca M, Bonsignore A. The effect of methylglyoxal on the glycolytic enzymes. FEBS Lett. 1980;117:17–8.

Liu HS, Li FM. Root respiration, photosynthesis and grain yield of two spring wheat in response to soil drying. Plant Growth Regul. 2005;46:233–40.

Ludlow MM, Muchow RC. A critical evaluation of traits for improving crop yields in water-limited environments. Adv Agron. 1990;43:107–53.

Lyles GA, Chalmers J. The metabolism of aminoacetone to methylglyoxal by semicarbazide-sensitive amine oxidase in human umbilical artery. Biochem Pharmacol. 1992;43:1409–14.

Maeta K, Izawa S, Inoue Y. Methylglyoxal, a metabolite derived from glycolysis, functions as signal initiator of the high osomolarity glycerol-miotgen-activated protein kinase cascade and calcineurin/Crz1-mediated pathway in *Sacchromyces cerevisiae*. J Biol Chem. 2005;280: 253–60.

Manavalan LP, Guttikonda SK, Tran LS. Nguyen HT (2009) Physiological and molecular approaches to improve drought resistance in soybean. Plant Cell Physiol. 2009;50:1260–76.

Mansfield TJ, Atkinson CJ. Stomatal behaviour in water stressed plants. In: Alscher RG, Cumming JR, editors. Stress responses in plants: adaptation and acclimation mechanisms. New York: Wiley-Liss; 1990. p. 241–64.

Maroco JP, Pereira JS, Chaves MM. Stomatal responses to leaf-to-air vapour pressure deficit in Sahelian species. Aust J Plant Physiol. 1997;24:381–7.

Martins AMTBS, Cordeiro CAA, Freire AMJP. In situ analysis of methylglyoxal metabolism in Saccharomyces cerevisiae. FEBS Lett. 2001;499:41–4.

Mazahery-Laghab H, Nouri F, Abianeh HZ. Effects of the reduction of drought stress using supplementary irrigation for sunflower (*Helianthus annuus*) in dry farming conditions. Pajouheshva-Sazandegi Agron Hort. 2003;59:81–6.

Miller G, Suzuki N, Ciftci-Yilmaz S, Mittler R. Reactive oxygen species homeostasis and signalling during drought and salinity stresses. Plant Cell Environ. 2010;33(4):453–67.

Mira ML, Martinho F, Azevedo MS, Manso CF. Oxidative inhibition of red blood cell ATPases by glyceraldehyde. Biochim Biophys Acta. 1991;1060:257–61.

Monneveux P, Sánchez C, Beck D, Edmeades GO. Drought tolerance improvement in tropical maize source populations: evidence of progress. Crop Sci. 2006;46:180–91.

Munemasa S, Oda K, Watanabe-Sugimoto M, Nakamura Y, Shimoishi Y, Murata Y. The coronatine-insensitive 1 mutation reveals the hormonal signaling interaction between abscisic acid and methyl jasmonate in Arabidopsis guard cells. Specific impairment of ion channel activation and second messenger production. Plant Physiol. 2007;143:1398–407.

Mustafiz A, Singh AK, Pareek A, Sopory SK, Singla-Pareek SL. Genome-wide analysis of rice and Arabidopsis identifies two glyoxalase genes that are highly expressed in abiotic stresses. Funct Integr Genomics. 2011;11(2):293–305.

Nam NH, Chauhan YS, Johansen C. Effect of timing of drought stress on growth and grain yield of extra-short-duration pigeonpea lines. J Agric Sci. 2001;136:179–89.

Nayyar H, Kaur S, Singh S, Upadhyaya HD. Differential sensitivity of Desi (small-seeded) and Kabuli (large-seeded) chickpea genotypes to water stress during seed filling: effects on accumulation of seed reserves and yield. J Sci Food Agr. 2006;86:2076–82.

Neuberg C. The destruction of lactic aldehyde and methylglyoxal by animal organs. Biochem J. 1913;49:502–6.

Oh SJ, Kim YS, Kwon CW, Park HK, Jeong JS, Kim JK. Overexpression of the transcription factor AP37 in rice improves grain yield under drought conditions. Plant Physiol. 2009;150(3): 1368–79.
Pandey A, Chakraborty S, Datta A, Chakraborty N. Proteomics approach to identify dehydration responsive nuclear proteins from chickpea (*Cicer arietinum* L.). Mol. Cell Proteom. 2008; 7:88–107.
Pandey A, Rajamani U, Verma J, Subba P, Chakraborty N, Datta A, Chakraborty S, Chakraborty N. Identification of extracellular matrix proteins of rice (*Oryza sativa* L.) involved in dehydration-responsive network: a proteomic approach. J Proteome Res. 2010;9:3443–64.
Perera IY, Hung CY, Moore CD, Stevenson-Paulik J, Boss WF. Transgenic Arabidopsis plants expressing the type 1 inositol 5-phosphatase exhibit increased drought tolerance and altered abscisic acid signaling. Plant Cell. 2008;20(10):2876–93.
Peuke AD, Rennenberg H. Carbon, nitrogen, phosphorus, and sulphur concentration and partitioning in beech ecotypes (*Fagus sylvatica* L.): phosphorus most affected by drought. Trees. 2004;18:639–48.
Phillips SA, Thornalley PJ. The formation of methylglyoxal from triose phosphates. Investigation using a specific assay for methylglyoxal. Eur J Biochem. 1993;212:101–5.
Racker E. The mechanism of action of glyoxalase. J Biol Chem. 1951;190:685–96.
Rampino P, Pataleo S, Gerardi C, Mita G, Perrotta C. Drought stress response in wheat: physiological and molecular analysis of resistant and sensitive genotypes. Plant Cell Environ. 2006; 29(12):2143–52.
Ray S, Dutta S, Halder J, Ray M. Inhibition of electron flow through complex I of the mitochondrial respiratory chain of Ehrlich ascites carcinoma cells by methylglyoxal. Biochem J. 1994;303:69–72.
Reddy AR, Chaitanya KV, Vivekanandan M. Drought-induced responses of photosynthesis and antioxidant metabolism in higher plants. J Plant Physiol. 2004;161:1189–202.
Richard JP. Mechanism for the formation of methylglyoxal from triosephosphates. Biochem Soc Trans. 1993;21:549–53.
Sahoo KK, Tripathi AK, Pareek A, Singla-Pareek SL. Taming drought stress in rice through genetic engineering of transcription factors and protein kinases. Plant Stress. 2013;7(1): 60–72.
Saito R, Yamamoto H, Makino A, Sugimoto T, Miyake C. Methylglyoxal functions as Hill oxidant and stimulates the photoreduction of O(2) at photosystem I: a symptom of plant diabetes. Plant Cell Environ. 2011;34:1454–64.
Samarah NH. Effects of drought stress on growth and yield of barley. Agron Sustain Dev. 2005;25:145–9.
Samarah NH, Mullen RE, Cianzio SR, Scott P. Dehydrin-like proteins in soybean seeds in response to drought stress during seed filling. Crop Sci. 2006;46:2141–50.
Sasaki-Sekimoto Y, Taki N, Obayashi T, Aono M, Matsumoto F, Sakurai N, Suzuki H, Hiraj MY, Noji M, Saito K, Masuda T, Takamiya KI, Shibata D, Ohta H. Coordinated activation of metabolic pathways for antioxidants and defence compounds by jasmonates and their roles in stress tolerance in *Arabidopsis*. Plant J. 2005;44:563–668.
Saxena M, Bisht R, Roy DS, Sopory SK, Bhalla-Sarinn M. Cloning and characterization of a mitochondrial glyoxalase II from *Brassica juncea* that is upregulated by NaCl, Zn and ABA. Biochem Biophys Res Commun. 2005;336:813–9.
Sen CK. Nutritional biochemistry of cellular glutathione. Nutr Biochem. 1997;8:660–72.
Sharma P, Dubey RS. Drought induces oxidative stress and enhances the activities of antioxidant enzymes in growing rice seedlings. Plant Growth Regul. 2005;46(3):209–21.
Sharma S, Mustafiz A, Singla-Pareek SL, Shankar Srivastava P, Sopory SK. Characterization of stress and methylglyoxal inducible triose phosphate isomerase (OscTPI) from rice. Plant Signal Behav. 2012;7:1337–45.
Siddique MRB, Hamid A, Islam MS. Drought stress effects on water relations of wheat. Bot Bull Acad Sin. 2001;41:35–9.

Singla-Pareek SL, Reddy MK, Sopory SK. Genetic engineering of the glyoxalase pathway in tobacco leads to enhanced salinity tolerance. Proc Natl Acad Sci U S A. 2003;100:14672–7.

Smirnoff N. The role of active oxygen in the response of plants to water deficit and desiccation. New Phytol. 1993;125:27–58.

Sousa Silva M, Gomes RA, Ferreira AE, Ponces Freire A, Cordeiro C. The glyoxalase pathway: the first hundred years… beyond. Biochem J. 2013;453(1):1–15.

Sugimura T, Sato S. Mutagens–carcinogens in foods. Cancer Res. 1983;43:2415–21.

Szent-Györgyi A. Bioelectronics. New York: Academic; 1968.

Takatsume Y, Izawa S, Inoue Y. Methylglyoxal as a signal initiator for activation of the stress-activated protein kinase cascade in the fission yeast *Schizosaccharomyces pombe*. J Biol Chem. 2006;281:9086–92.

Thornalley PJ. The glyoxalase system in health and disease. Mol Asp Med. 1993;14:287–371.

Thornalley PJ. Protecting the genome: defence against nucleotide glycation and emerging role of glyoxalase I overexpression in multidrug resistance in cancer chemotherapy. Biochem Soc Trans. 2003a;31:1372–7.

Thornalley PJ. Glyoxalase I-structure, function and a critical role in the enzymatic defence against glycation. Biochem Soc Trans. 2003b;31:1343–8.

Thornalley PJ, Langborg A, Minhas HS. Formation of glyoxal, methylglyoxal and 3-deoxyglucosone in the glycation of proteins by glucose. Biochem J. 1999;344:109–16.

Tran L-SP, Nakashima K, Sakuma Y, Simpson SD, Fujita Y, Maruyama K, Fujita M, Seki M, Shinozaki K, Yamaguchi-Shinozaki K. Isolation and functional analysis of Arabidopsis stress-inducible NAC transcription factors that bind to a drought-responsive cis-element in the early responsive to dehydration stress 1 promoter. Plant Cell. 2004;16:2481–98.

Van Breusegem F, Vranova E, Dat JF, Inze D. The role of active oxygen species in plant signal transduction. Plant Sci. 2001;161:405–14.

Vander Jagt DL, Hunsaker LA, Vander Jagt TJ, Gomez MS, Gonzales DM, Deck LM, Royer RE. Inactivation of glutathione reductase by 4-hydroxynonenal and other endogenous aldehydes. Biochem Pharmacol. 1997;53:1133–40.

Veena RVS, Sopory SK. Glyoxalase I from Brassica juncea: molecular cloning, regulation and its over-expression confer tolerance in transgenic tobacco under stress. Plant J. 1999;17:385–95.

Wahid A, Rasul E. Photosynthesis in leaf, stem, flower and fruit. In: Pessarakli M, editor. Handbook of photosynthesis. 2nd ed. Boca Raton: CRC Press; 2005. p. 479–97.

Wilson PB, Estavillo GM, Field KJ, Pornsiriwong W, Carroll AJ, Howell KA, Woo NS, Lake JA, Smith SM, Harvey Millar A, von Caemmerer S, Pogson BJ. The nucleotidase/phosphatase SAL1 is a negative regulator of drought tolerance in Arabidopsis. Plant J. 2009;58(2):299–317.

Wu L, Juurlink BH. Increased methylglyoxal and oxidative stress in hypertensive rat vascular smooth muscle cells. Hypertension. 2002;39(3):809–14.

Yadav SK, Singla-Pareek SL, Ray M, Reddy MK, Sopory SK. Methylglyoxal levels in plants under salinity stress are dependent on glyoxalase I and glutathione. Biochem Biophys Res Commun. 2005;337:61–7.

Yadav SK, Singla-Pareek SL, Kumar M, Pareek A, Saxena M, Sarin NB, Sopory SK. Characterization and functional validation of glyoxalase II from rice. Protein Expr Purif. 2007;51(1):126–32.

Yadav SK, Singla-Pareek SL, Sopory SK. An overview on the role of methylglyoxal and glyoxalases in plants. Drug Metabol Drug Interact. 2008;23(1-2):51–68.

Yamaguchi-Shinozaki K, Shinozaki K. Transcription regulatory networks in cellular responses and tolerance to dehydration and cold stresses. Annu Rev Plant Biol. 2006;57:781–803.

Yu PH, Wright S, Fan EH, Lun DZR, Gubisne-Harberle D. Physiological and pathological implications of semicarbazide sensitive amine oxidase. Biochim Biophys Acta. 2003;1647:193–9.

Zang X, Komatsu S. A proteomics approach for identifying osmotic-stress-related proteins in rice. Phytochemistry. 2007;68:426–37.

Zhu JK. Plant salt tolerance. Trends Plant Sci. 2001;6(2):66e71.

Chapter 17
Drought Tolerant Wild Species Are the Important Sources of Genes and Molecular Mechanisms Studies: Implication for Developing Drought Tolerant Crops

Imrul Mosaddek Ahmed, Umme Aktari Nadira, Guoping P. Zhang, and Feibo B. Wu

17.1 Introduction

Drought is the single largest abiotic stress factor leading to reduced crop yields and dramatically threatens the food supply worldwide. Furthermore, global climate change is projected to have a significant impact on temperature and precipitation profiles, with increasing incidence and severity of drought stress. Increasing demand by municipal and industrial users further reduces the amount of water available for irrigated crops. One of the sustainable and cost-effective solutions for increasing crop stability and productivity is genetic improvement for higher tolerance to drought stress (Ashraf et al. 2009; Blum 1988). While natural selection has favoured mechanisms for adaptation and survival, breeding activity has directed selection towards increasing the economic yield of crop species. However, tolerance to drought stress is a quantitative trait controlled by many different genes. Thus, improvement of tolerance of crop plants to drought is proved to be somewhat elusive to plant breeders (Lopes et al. 2011).

A long process of domestication, especially modern breeding and cultivation programs, primitive landraces has been replaced by modern cultivars. This process of "genetic erosion" in many domesticated plants has been under way for decades,

I.M. Ahmed
Department of Agronomy, College of Agriculture and Biotechnology, Zijingang Campus, Zhejiang University, Hangzhou 310058, China

Plant Physiology Division, Bangladesh Agricultural Research Institute, Joydebpur, Gazipur 1701, Bangladesh

U.A. Nadira • G.P. Zhang • F.B. Wu (✉)
Department of Agronomy, College of Agriculture and Biotechnology, Zijingang Campus, Zhejiang University, Hangzhou 310058, China
e-mail: wufeibo@zju.edu.cn

M.A. Hossain et al. (eds.), *Drought Stress Tolerance in Plants, Vol 1*,
DOI 10.1007/978-3-319-28899-4_17

many other crop plants, many of the ancient landraces have vanished, and the genetic diversity of the cultivated forms has become significantly reduced (Ellis et al. 2000). Due to the rapid loss of genetic variation through cultivar replacement, modern crop varieties have become more sensitive to abiotic and biotic stresses, and their monotonous genetic background has been a bottleneck to the breeding of improved cultivars.

Development of crop varieties for improved drought resistance requires the knowledge of physiological mechanisms and genetic control of the contributing traits at different plant developmental stages. Water deficit accelerates abscisic acid (ABA) biosynthesis, which decreases stomatal conductance to minimize transpirational losses. To cope with such challenges, understanding the effects of drought on plants and morphological and physiological adaptations is crucial. Recently, the utilization of drought tolerant wild species and the rapid advances in molecular biological, functional genomics, and transgenic technologies have facilitated drought-related studies, resulting in significant progress in the identification of related genes and gene regions and dissection of some of its molecular aspects. The existing literature on drought stress was reviewed to assess the present position and to identify research gaps to address future research needs. This chapter provides an overview of present understanding of drought response and tolerance of wild species and summarizes current research on the enhancement of the growth and yield ability currently unrecognized in water-limited environments. Strategies used previously to achieve progress in drought environments are analyzed, improvements were proposed, and attempts have been made to assess the potential impacts of current research endeavor.

17.2 Mechanism of Drought Stress Tolerance

Three main mechanisms reduce crop yield by soil water deficit: (1) reduced canopy absorption of photosynthetically active radiation, (2) decreased radiation-use efficiency, and (3) reduced harvest index (Earl and Davis 2003). Desiccation is a much more extensive loss of water that can potentially lead to gross disruption of metabolism and cell structure and eventually to the cessation of enzyme catalyzing reactions. Drought is characterized by reduction of water content, turgor, total water potential, wilting, closure of stomata, and decrease in cell enlargement and growth. The isolation and characterization of genes conferring tolerance to stress by expression in genetic modified (GM) crops requires the previous, in-depth understanding of the mechanisms plants use as a response to stress, which together with the academic interest of the topic has stimulated the study of these mechanisms over the last two decades. These studies have revealed a series of basic, conserved stress response pathways, apparently used by all plants tolerant as well as sensitive which are activated at the cellular level in response to different types to abiotic stress, and include: (1) the control of water transport, ion transport and ion homeostasis, to prevent cellular dehydration and to maintain osmotic balance, including the compartmentalization of toxic ions in the vacuole and the synthesis and accumulation of compatible solutes or osmolytes in the cytosol; these osmolytes have additional functions as osmoprotectants, directly stabilizing proteins and cellular structures under dehydration conditions and protecting the cell against oxidative stress as scavengers of reactive oxygen species (ROS); (2) synthesis of specific protective proteins, such as osmotine, heat-shock proteins, and late embryogenesis abundant (LEA) proteins; and (3) synthesis of antioxidant compounds (reduced glutathione, flavonoids and other phenolics, carotenoids, vitamins C and E, etc.) and activation of enzymatic antioxidant systems (superoxide dismutase, catalase, ascorbate peroxidase, glutathione peroxidase, glutathione reductase, etc.), induced in response to oxidative stress generated either directly (e.g., by ozone or high UV irradiation) or secondarily by other stressful environmental conditions (Ashraf et al. 2009; Hussain et al. 2008; Munns 2002; Vinocur and Altman 2005; Wang et al. 2003; Zhu 2001).

17.3 Drought Tolerance in Wild Species for Cultivated Variety Improvement

Over the next few decades we must boost crop productivity if we are to feed a growing world population, which will reach more than 9×10^9 people by 2050; and we should do it in the frame of a sustainable agriculture, with an increasing scarcity of new arable land and of water for irrigation. To meet the needs of the growing world population, it is essential to effectively utilize dehydrated and salted soil by developing crop varieties that are well adapted to drought and salt stress. However, the progress toward developing drought- and salt-tolerant crops is significantly hampered by the physiological and genetic complexity of these traits. Considering the limitations of traditional plant breeding, the most promising strategy to achieve this goal will rely on the generation of transgenic plants expressing genes conferring tolerance. We propose wild stress tolerant species as more suitable models to investigate these mechanisms, as well as a possible source of biotechnological tools ("stress tolerance" genes, stress-inducible promoters) for the genetic engineering of stress tolerance in crop plants.

17.3.1 Rice

Common wild rice (*O. rufipogon* Griff.), as the progenitor of cultivated rice (*O. sativa* L.), constitutes the primary gene pool for rice genetic improvement (Second 1982; Oka 1988; Wang et al. 1992). During the course of domestication from wild rice to cultivated rice, profound changes of agronomic traits and genetic diversity occurred, and the number of alleles of cultivated rice was only 60 % that of wild rice, suggesting that many alleles of wild rice were lost during the course of domestication, which led to lower genetic diversity of the cultivated rice (Sun et al. 2001). Because of the narrow genetic base of cultivated rice, it is necessary to explore primitive and broad genetic resources. The wild rice may offer abundant genetic resources in drought tolerance research as it has more novel alleles than cultivated rice.

17.3.2 Wheat

Wild emmer wheat is important for its high drought tolerance, and some of *T. dicoccoides* genotypes are fully fertile in arid desert environments. Wild emmer wheat accessions were shown to thrive better under water-limited conditions in terms of their productivity and stability, compared to durum wheat. The wild emmer gene pool was shown to offer a rich allelic repertoire of agronomically important traits including drought tolerance (Saranga et al. 2008; Peleg et al. 2005; Peng et al. 2011; Peng et al. 2013). Hence, *T. dicoccoides* is an important source of drought-related genes and highly suitable as a donor for improving drought tolerance in cultivated wheat species.

17.3.3 Barley

Wild barley *H.spontaneum,* thereafter named *Hsp*, is the direct ancestor of cultivated barley and contains valuable novel genes for barley improvement (Baum et al. 2003; Ceccarelli et al. 2004; Forster et al. 1997; Grando et al. 2001; Ivandic et al. 2000). Because there is no biological isolation barriers with the cultivated barley, crossing between *H. spontaneum* and *H. vulgare* was possible in nature (Harlan 1976) and it makes *H. spontaneum* accessible for immediate use in barley breeding, and allows for improved agronomic characteristics of cultivated varieties such as drought and salt tolerance. Many studies have described the variation in populations of wild barley and its potential contribution of useful alleles for crop improvement (Grando and Ceccarelli 1995; Ivandic et al. 2003; Volis et al. 2002; Baum et al. 2003), and Nevo and Chen (2010) summarized drought tolerances in wild relatives for barley improvement.

17.3.4 Sunflower

Helianthus annuus ssp. *Annuus* L. constitutes a potential genetic resource because it has naturalized in the semiarid zone of central Argentina. The assessment of these genetic materials for tolerance to water deficit matters because they represent a source of genes for drought tolerance, useful to sunflower breeding. Drought-resistant genotypes should be achieved using phenotypic traits easy to identify. Parameters such as leaf area are widely used to characterize the performance under stress. Leaf temperature is an easily measured physiological parameter that allows an indirect way to estimate plant transpiration and is well correlated with water availability. Relative water content indicates the ability to retain water from the soil and expresses plant osmotic adjustment capacity (Blum 1989).

17.3.5 Soybean

The wild ancestor of soybean, *Glycinesoja* (Sieb. and Zucc.) also known as wild soybean, is perhaps the most genetically diverse resource available to soybean breeders that will cross freely with the domesticate (Harlan and de Wet 1971). An annual plant, it is commonly found throughout China and most of Asia, including its arid regions (Hymowitz and Singh 1987). Wild soybean has been the object of two previous water deficit studies; drought resistance was found to be superior to soybean in one case (Chen et al. 2006) but similar to soybean in the other (James et al. 2008). Physiological processes of wild and domesticated soybean were not compared, so the basis for the putative drought resistance is unresolved. There are no reports of using wild soybean as a source of drought resistance in breeding, in large part because no wild soybean accessions have been verified as resistant.

17.3.6 Chickpea

The genus *Cicer* consists of nine annual species including the cultivated chickpea, *Cicerarietinum* L., 33 wild perennials, and one unspecified wild (van der Maesen 1987). The world annual *Cicer* species is limited to 572 accessions held in nine gene banks, many of which are possibly duplicates or misidentified (Shan et al. 2005). So far, the existing germplasm has been evaluated for drought resistance, mainly using cultivated chick peas (Serraj et al. 2004; Kashiwagi et al. 2006). Although wild *Cicer* species have been evaluated for different biotic and abiotic stresses (Singh et al. 1998; Collard et al. 2001; Toker 2005; Sharma et al. 2006), there are only a few reports regarding drought resistance in annual wild *Cicer* species (Kashiwagi et al. 2005), and none in perennial wild *Cicer* species. The purpose of this study was to evaluate perennial wild *Cicer* species for drought resistance and compare them with annual wild *Cicer* species and cultivated chickpeas (Table 17.1).

17.4 Conventional Breeding for Drought Tolerance

Conventional improvement to obtain new individuals is based on their genetic variation and uses the selection to incorporate better characteristics into the progeny. For this purpose, two plants processing desirable traits are selected and then crossed to exchange their genes, so that the offspring has new genetics arrangement. Conventional breeding has been going on for hundreds of years, and is still commonly used today. During the last century, conventional breeders at different renowned international research centers have made considerable strides in developing drought tolerant lines/cultivars of some important food crops. For example, breeding approach started at the International Maize and Wheat Improvement

Table 17.1 Drought tolerant cultivars/lines of different crops developed through conventional breeding at different centers/institutions

Crop	Cultivars/ Line	How developed	Centers/institutions involved	Reference
Wheat (*Triticum aestivum* L.)	Willow Creek	Through breeding in single replication observation (SROB) nurseries	Montana Agricultural Experiment Station, Sydney	Cash et al. (2009)
	NE01643	A bulk breeding procedure was used and approximately 50 % of F3 population was visually selected on the basis of agronomic appearance	Nebraska Agricultural Experiment Station and the USDA-ARS	Baenziger et al. (2008)
Barley (*Hordeum vulgare* L.)	Giza 126	Selected for drought resistance in an F3 population received from ICARDA, initially originating from a single cross Baladi Bahteem/SD729-Por 12762BC	International Center for Agricultural Research in the Dry Areas (ICARDA)	Noaman et al. (1995)
	Giza 2000	The pedigree breeding method was used for development and it was originated from the cross between the Egyptian local cultivar Giza 121 and the line 366/13/1 (Giza 117/ Bahteem 52//Giza 118/FAO 86)	Barley Research Department, Agricultural Research Center at Giza, Egypt	Noaman et al. (2007)
Maize (*Zea mays* L.)	Oh605	22 selected full-sib progenies from the AAE population were intermated with 30 full-sib progenies obtained from OhS3267LAN plants	Ohio State University (OSU), Ohio Agricultural Research and development Center, USA	Pratt and Casey (2006)
Chickpea (*Cicer arietinum* L.)	FLIP 87-59C	Developed by crossing ILC3843 with FLIP87	International Center for Agricultural Research in the Dry Areas (ICARDA)	Singh et al. (1998)

Center (CIMMYT), Mexico in the 1970s for developing drought tolerant maize is worth mentioning. A number of maize hybrids developed by the CIMMYT scientists were found superior to all those developed by private enterprises, in terms of growth and grain yield under drought-prone environments (Bänziger et al. 2004). Moreover, at CIMMYT, a new synthetic hexaploid has been developed by crossing the diploid wild ancestor, *Aegilops tauschii* (goat grass), with tetraploid durum wheat (*Triticum turgidum* var. *durum*). These hexaploid synthetics containing a complete D-genome from *A. tauschii* have provided a significant new variation for tolerance to both biotic and abiotic stresses (Villareal et al. 1994; Valkoun 2001). At CIMMYT, more than 1000 accessions of *A. Tauschii* have been evaluated and new hexaploid lines developed. A significant new genetic variation in these newly developed hexaploid wheat has been observed for abiotic stresses including drought stress (Valkoun 2001). Useful variation for drought tolerance has also been identified in *Triticum urartu, T. boeticum, T. dicoccoides* (Valkoun 2001), and *Aegilops geniculata* (Zaharieva et al. 2001). However, in view of Skovmand et al. (2001) *A. tauschii* is the predominant source of variation for drought tolerance. Furthermore, breeding approach started at ICARDA, Aleppo, Syria in the 1980s for developing drought tolerant barley is worth mentioning. In barley, many elite barley cultivars are produced by conventional breeding and some of them are still used as good materials for studies and barley breeding. For instance, Giza 126, selected for drought resistance in an F3 population received from ICARDA, initially originating from a single cross Baladi Bahteem/SD729-Por 12762BC (Noaman et al. 1995). Giza 132 derived from an F3 population and the pedigree method of breeding was used and Giza 132 originated from the cross Rihane-05//As46/Aths/3/Aths/Lignee 686 (Noaman et al. 2007). Plant breeders, at the Field Crops Development Centre (FCDC) Lacombe, AB, Canada, have developed a drought tolerant barley cultivar "Bentley" in 2009, which is derived from crossing I92125 with TR229 (Juskiw et al. 2009). These drought tolerant cultivar/lines of barley provide a sound testament that conventional plant breeding played a considerable role during the last century not only for improving the quality and yield of crops, but also for improving abiotic stress tolerance including drought tolerance. However, now there is a general consensus of the plant breeders that empirical plant breeding is a highly time-consuming as well as a cost- and labor-intensive approach. While transferring desired genes from one plant to other through the conventional plant breeding, a number of undesired genes are also transferred simultaneously; and reproductive barriers limit transfer of favorable alleles from interspecific and inter-generic sources. As an alternative strategy to conventional breeding, marker-assisted selection (MAS) breeding and genetic engineering breeding are being employed emphatically worldwide for improving stress tolerance of most crops.

17.5 Marker-Assisted Selection Breeding for Drought Tolerance

Through marker-assisted selection (MAB) breeding it is now possible to examine the usefulness of thousands of genomic regions of a crop germplasm under drought stress, which was, in fact, previously not possible. By examining the breeding value of each of the genomic regions, the breeder can coalesce genes of multifarious origins in novel ways, which was not possible previously with conventional breeding tools and protocols (Table 17.2).

Drought tolerance is a complex quantitative trait controlled by several combinations of genes and gene families which are not easy to select simultaneously. Quantitative trait loci (QTL) analysis has proven to be a valuable method in discovering the genetic basis of quantitative traits. Dissection of drought tolerance, a complex quantitative phenotype, affected by multiple loci requires the identification of related QTLs (Ashraf, 2010).

Table 17.2 Enhancing drought tolerance in different crop lines/varieties using marker-assisted selection

Crop	QTL used	QTL donor line/cultivar	Recipient line/cultivar	line/cultivar developed	Trait improved	Reference
Rice (*Oryza sativa* L.)	QTL9 (on chromosome 9)	Azucena	Kalinga III	Near-isogenic lines (NILs)	Improved root length and thickness under both irrigated and drought-stressed field conditions	Steele et al. (2006)
	QTL9 (on chromosome 9)	Azucena	Kalinga III	Near-isogenic lines (NILs)	Under field conditions, NILs out-performed Kalinga III for grain and straw yield. However, the lines had higher straw yield than grain yield as introgressed genes involved in partitioning of biomass to the roots and stems, rather than to the grain	Steele et al. (2007)
Barley (*Hordeum vulgare* L.)	81 QTLs were used, out of which 6 (1H-3, 2H-1, 3H-2, 4H-3, 1H-5, 3H-1, and 3H-4) were for grain yield	*Hordeum spontaneum*	*Hordeum vulgare*	Back cross population	Improved grain yield, and reduced negative impact of drought on grain filling	Baum et al. (2003), Talame et al. (2004), Tuberosa and Salvi (2006)
Maize (*Zea mays* L.)	Five QTLs located on chromosome 1, 2, 3, 8, and 10	Ac7643	CML 247	Marker-assisted back cross (MABC)-derived BC2F3hybrid	Under severe drought stress conditions, MAS-derived varieties were about 50 % better in yield and showed reduced asynchrony between male and female flowering	Ribaut and Ragot (2006)

In rice, for example, a number of drought-related QTL have been identified for different growth and physiological traits involved in drought tolerance (Lafitte et al. 2004). For example, Courtois et al. (2003) found 28 QTL responsible for various root characteristics involved in drought resistance. Similarly, 14 QTL related to osmotic adjustment have been identified in an independent study on rice (Robin et al. 2003). In another study pertaining to identification of QTL related to root traits and osmotic adjustment in rice, 36 QTL related to some key root traits and 5 related to osmotic adjustment were identified (Zhang et al. 2001). While assessing the role and genetic mechanism of leaf water potential (LWP) in japonica rice (*Oryza sativa* L. sub sp. *japonica*) under various water-limited regimes of upland and lowland environments, Yan-Ying et al. (2008) detected 6 QTL for LWP. Of the 6 QTL, the two for LWP at predawn in upland (wpiu 1and wpiu 4) and one for LWP at midday in upland (wpu 6) explained 5.4 %, 7.9 %, and 10.0 % of the phenotypic variation, respectively.

In wheat, the position of genes exhibiting a significant effect on ABA accumulation due to drought stress was identified using a series of single chromosome substitution lines and populations obtained from a cross between a high-ABA-producing cv. Ciano 67 and a low-ABA-producing cv. Chinese Spring (Quarrie et al. 1994). They observed that chromosome 5A carries gene(s) for ABA accumulation. MAPMAKER-QTL showed that the ABA quantitative trait locus is located between the two loci Xpsr575 and Xpsr426, approximately 8 cM from Xpsr 426. Genetic loci known to be involved in the control of specific traits in cultivated barley can now be targeted and investigated in the wild barley gene pool in the search for novel and rare alleles. Recent advances in molecular genetics are creating exciting new approaches for testing and development of more drought resistance crops. The efficient use of genetic variation from *H. vulgare* ssp. *spontaneum* in breeding programs depends on the analysis of important traits and the establishment of genetic relationships. The genetic diversity and genetic marker associations with plant traits and site of origin eco-geography in wild barley have been studied using many markers, including storage proteins, isozyme polymorphisms, restriction fragment length polymorphism (RFLP), random amplified polymorphic DNAs (RAPDs), simple sequence repeats (SSR), amplified fragment length polymorphism (AFLP) markers, and single nucleotide polymorphisms (SNPs). *H. vulgare* ssp. *spontaneum* possesses more variation than cultivated barley and many alleles are associated with specific environments (Forster et al. 2000). For instance, with some efforts a set of drought tolerant accessions may be identified that carries tolerant alleles at different QTLs. Such accessions can be used as donor parents in marker-assisted back crossing (MABC) program to transfer the tolerant alleles in elite barley lines. The near-isogenic lines generated that way can be characterized further to confirm the QTLs, provide more precise estimates of allele effects and assess whether individual QTL are equally effective to drought stress (Roy et al. 2010). In another study, Rostoks et al. (2006) used high-throughput SNP genotyping assays and demonstrated that the linkage disequilibrium (LD) present in elite barley germplasm, after accounting for population structure, can be effectively exploited to map traits by using whole-genome association scans with several hundreds to thousands of markers. By using DArT

(Diversity Array Technology) markers on a collection of 192 barley genotypes that represented landraces, old- and contemporary cultivars sampling key regions around the Mediterranean basin and Europe, Comadran et al. (2009) investigated patterns of genetic diversity and LD and found that the collection was a suitable resource for association mapping. By combining the DArT marker data with the yield data in mixed model analyses, Comadran et al. (2011) identified several QTLs for yield under drought conditions in Mediterranean basin. Many efforts are presently aimed at building drought tolerance into barley, but many challenges still remain. Therefore, barley improvement under drought stress environments can be further enhanced through the use of wild barley as a rich source of genetic variation that could be transferred into high yielding barley varieties. Thus, there is a need to seek more efficient approaches for genetically tailoring crops for enhanced drought tolerance

17.6 Proteins and Genes Associated with Drought Tolerance

To study the dynamics of plant metabolism under stress and unravel regulatory mechanisms in place, it is important to combine the traditional more descriptive physiological approaches with the techniques of functional genomics, namely the high-throughput methods for transcriptomic, proteomic, metabolomic, and ionomic analysis. In wheat and barley, transcript, protein, metabolite profiling studies conducted in the last couple of years are shown in Table 17.3. For example, changes in protein profile of barley in response to drought stress were analyzed using a proteomics technique (Kausar et al. 2013; Ashoub et al. 2013). Alterations in proteins related to the energy balance and chaperons were the most characteristic features to explain the differences between the drought-tolerant and the drought-sensitive accessions. Further alterations in the levels of proteins involved in metabolism, transcription, and protein synthesis were identified under drought stress in barley (Ashoub et al. 2013). Kausar et al. (2013) observed that metabolism-related proteins decreased in sensitive, but increased in tolerant genotype under drought stress. Photosynthetic-related proteins were decreased and increased among the three sensitive and three tolerant genotypes, respectively. These results suggest that chloroplastic metabolism and energy-related proteins might play a significant role in the adaptation process of barley seedlings under drought stress.

Probable drought-related genes and QTLs, identified in "omics" and "QTL mapping" studies, should be further characterized, prior to their use in the development of better yielding cultivars (Fig. 17.1). Elucidation of these components includes analyzing their gene and protein structure, and determining their roles and interactions in the complex network of stress response signaling (Farooq et al. 2009). Their functional relevance to drought tolerance should be shown and eventually confirmed with transgenic studies. This section summarizes the recent research regarding the characterization of drought-related genes and its proteins, and functional genomics studies.

Table 17.3 Transcript and protein profiling studies conducted on wheat and barley in the last few years under drought stress conditions

Cultivars	Tissue	Drought stress application	Method	Key findings	References
Plainsman and Cappelle Desprez	Root	Moderate drought stress applied on tillering stage	cDNA microarray	Comparison of these two wheat genotypes demonstrates the importance of complex transcriptional regulation in the adaptation to abiotic stresses like limited water supply	Se˘cenji et al. (2010)
Excalibur, RAC875, Kukri	Leaf	Cyclic drought applied after first flag leaf formation mimicking field conditions	SCX column HPLC and MS	Two genes, *Td1ITF1* and *Td6TM1*, code for receptor proteins homologous to putative receptor protein kinase, that have been found to be involved in the plant defense and adaptation response	Rampino et al. (2012)
Cultivar Vinjett	Grain	Drought applied at terminal spikelet or at anthesis	2D gel and MS	Identified drought-responsive proteins included proteins involved in primary metabolism, storage and stress response such as late embryogenesis abundant proteins, peroxiredoxins and α-amylase/trypsin inhibitors	Yang et al. (2011)
Om Rabia3, Mahmoudi	Embryo	Drought applied at final development stage of seed maturity	2D gel and HPRP column and MS	Several proteins belonging to the seed storage family, LEA-type/heat-shock proteins, enzyme metabolism, and radical scavengers were identified by analysis of trypsin digested peptides via mass spectrometry	Irar et al. (2010)
Wild emmer wheat (Y12-3, A24-39)	Leaf	Terminal drought applied at inflorescence emergence stage	Transcript profiling	The mechanisms of drought tolerance were identified in roots of wild emmer wheat, the potential of this gene pool as a valuable source for novel candidate genes to improve drought tolerance in cultivated wheat	Krugman et al. (2011)
Morex	Lemma, palea, awn, seed	Barley plants were exposed to drought treatment for 4 days at the grain-filling stage by withholding water	Barley1 Genome Array	The lemma and the palea are more resistant to drought stress compared with the awn	Abebe et al. (2010)

(continued)

Table 17.3 (continued)

Cultivars	Tissue	Drought stress application	Method	Key findings	References
Golden Promise, Basrah	Leaf and root	Drought applied at seedling stage for 7 days by withholding water	2D DIGE and MS	This study shows that the enhanced drought tolerance of variety Basrah is driven by an enhanced regulation of ROS under drought	Wendelboe-Nelson and Morris (2012)
Yeongyang Bori	Leaf	Germinated seedlings were kept on the growth chamber for 24 h without filter papers	Gene Fishing technique	The increases of the mRNA expression of genes encoding dehydrins, receptor kinase, and jasmonic acid induced protein suggest that multiple adaptive-mechanisms are involved in plants to cope in drought stress	Lee et al. (2011)
Martin, HS41-1, and Moroc9-75	Leaf	Drought treatment was started by withholding water at the flowering stage; available water content in the soil reached 10 % to allow drought measurements	22K Affymetrix Barley 1 microarray	*CSMO* and an *AAP* in the biosynthesis and translocation of glycine-betaine for osmoprotection; *ADOR, ALDH, GST* and *SPDS* in scavenging ROS for detoxification; and *HSP17.8* and *DHN3* in the stability of proteins and membranes for protecting the cell from injury under drought stress	Guo et al. (2009)
Arta and Keel	Leaf	The soil water content (SWC) of potted plants was adjusted to 50 % of the FC. For the drought treatment, the SWC was reduced to 15 % FC by controlled withholding of water; pots were weighed daily and watered to match the weight of the heaviest pot	2D DIGE and MS	In this study suggested that barley has adapted to nonlethal drought by avoidance mechanisms, such as the reduction of growth which allowed the plants to maintain a cellular homeostasis as seen in the stability of photosynthesis and of the proteome under drought	Rollins et al. (2013)

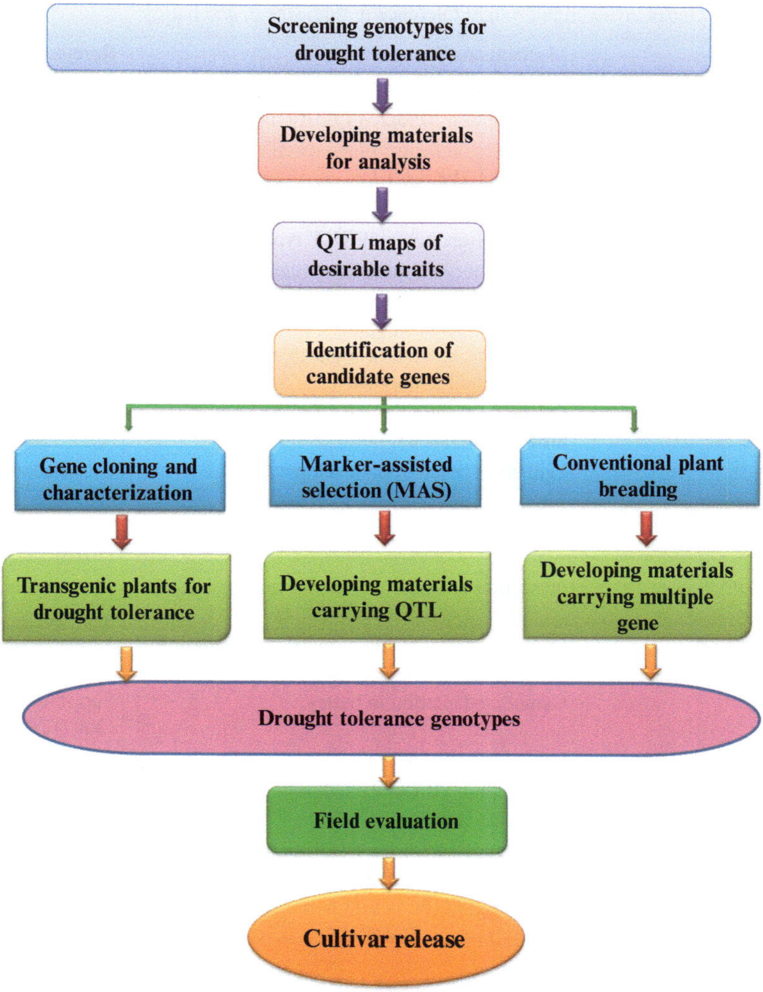

Fig. 17.1 Schematic framework for evolving and developing drought tolerant crop cultivars. Using developed materials, QTL analysis and gene mapping are conducted. For gene cloning, identified genes or major QTL are analyzed in detail using a large size population. A cloned gene for drought tolerance is transferred into widely adapted varieties. To develop the materials carrying the gene or QTL for drought tolerance, DNA markers which linked to the gene or QTL are used for marker-assisted selection. Similarly, marker-assisted selection and conventional breeding approaches are used to incorporate desirable traits in crops. Promising genotypes are evaluated in laboratory and field and used for release of crop cultivars. Cited from Ahmed et al. (2015).

Several families of transcription factors, including DREB/CBF, ERF, MYK, MYB, AREB/ABF, NAC and HDZip, have been shown to be involved in the regulation of drought response in plants (Yamaguchi-Shinozaki and Shinozaki 2006). The advent of genomics has offered a comprehensive profiling for the changes in gene expression resulting from exposure to drought. A number of genes have shown their

Table 17.4 Genes involved in drought tolerance in rice, wheat, and barley

Gene	Mechanism of action	Reference
TaPIMP1	Transcription factor: R2R3 type MYB TF	Liu et al. (2011)
TaMYB3R1	Transcription factor: MYB3R type MYB TF	Cai et al. (2011)
TaNAC (NAM/ATAF/ CUC)	Transcription factor: plant-specific NAC (NAM/ATAF/CUC) TF	Tang et al. (2012)
TaMYB33	Transcription factor: R2R3 type MYB TF	Qin et al. (2012)
TaWRKY2, TaWRKY19	Transcription factor: WRKY type TF	Niu et al. (2012)
DHNs	Stability of plant membrane (dehydration tolerance)	Karami et al. (2013)
Dhn3, Dhn9	Improved Chl a, b contents, osmotic adjustment, stomatal conductance, plant biomass, and grain yield	Karami et al. (2013)
LEA (*HVA1*)	Overaccumulation of LEA proteins increases drought tolerance	Liang et al. (2012)
HvNACs	Leaf senescence, root development	Christiansen et al. (2011)
MYB	Growth and development	Tombuloglu et al. (2013)
CBF/DREB	protection of cell from damage and desiccation	Morran et al. (2011)
HvWRKY38	Improved survival and biomass accumulation following dehydration stress	Xiong et al. (2010)
Hsdr4	Osmotic adaptation in barley	Suprunova et al. (2007)
eibi1	Leaf water conservation	Chen et al. (2004)
OsABA8ox3	Controlling ABA level and drought stress resistance in rice	Cai et al. (2015)
OsDHODH1	Overexpression of the OsDHODH1 gene in rice increased the DHODH activity and enhanced plant tolerance to drought stresses	Liu et al. (2009)
OsMAPK5	Overexpression *CaMV35SP*, survivability	Xiong and Yang (2003)
OsCDPK7	Calcium-dependent protein kinase, overexpression *CaMV35SP*, plant growth, F_v/F_m	Saijo et al. (2000)
OsCIPK12	CBL-interacting protein kinase, Overexpression *CaMV35SP*, survivability	Xiang et al. (2007)
OsSIK1	Improved drought and salt tolerance, with increased activities of peroxidase and superoxide and lower accumulation of H_2O_2	Ouyang et al. (2010)

involvement in drought response mechanism (Table 17.4). A brief description of these functionally characterized barley genes is given as follows.

Dehydrins (*Dhns*), peripheral membrane proteins which functionally protect the cell from drought stress or temperature change, are among the most frequently observed proteins in plants under water stress (Suprunova et al. 2004). A total of 13

Dhn genes were found on four barley chromosomes (Choi et al. 1999 ; Choi and Close 2000; Rodriguez et al. 2005). *Dhn1, Dhn2,* and *Dhn9* (previously reported for *Dhn4a*) were mapped to the long arm of chromosome 5H; *Dhn3, Dhn4, Dhn5, Dhn7, Dhn8,* and *Dhn12* were allocated to the long arm of chromosome 6H; *Dhn6* and *Dhn13* were allocated to chromosome 4SH; *Dhn10* and *Dhn11* were identified on chromosome 3HL. *Dhn1* and *Dhn2* are completely linked, together with *Dhn9*, and are located in the same QTL region of salt tolerance, freezing tolerance, and ABA accumulation. The drought-tolerant or drought-resistant mechanism of these genes is different, and the expression of some stress-related genes was shown to be linked to stress-tolerant QTLs (Cattivelli et al. 2002). This may be due to differential expression patterns and furthermore indicates that each member in this family has a specific function in the process of plant response to drought. *Dhn* genes (*Dhn* 1, 3, 5, 6, and 9) were also found in wild barley (*H. spontaneum*), and these genes were not expressed in well-watered plants. High polymorphism with no geographic structure was found in *Dhn5* in a collection of wild barley from the Mediterranean across the Zagros Mountains and into Southwest Asia, and moderate polymorphism associated with geographic structure was found in *Dhn9* locus (Morrell et al. 2003). In wild barley, the role of *Dhn1* in drought tolerance is supported by several reports on co-localization of such QTLs with *Dhn* genes, e.g., QTLs for RWC (Relative Water Content) (Teulat et al. 2003) and winter-hardiness (Pan et al. 1994; van Zee et al. 1995) overlapping with a cluster of *Dhn* genes on chromosome 5H. Wide allelic variation was found at the *Dhn4* locus in *H. spontaneum* germplasm from Israel (Close et al. 2000). High polymorphism with no geographic structure was found in *Dhn5* in a collection of wild barley from the Mediterranean across the Zagros Mountains and into south west Asia, and moderate polymorphism associated with geographic structure was found in *Dhn9* locus (Morrell et al. 2003). Karami et al. (2013) reported that, under drought stress condition, *Dhn1, Dhn3, Dhn5, Dhn7* and *Dhn9* were exclusively induced in drought-tolerant barley variety Yousef, and the relative gene expression of *Dhn3, Dhn9* had the direct correlations with Chl a, b contents, osmotic adjustment, stomatal conductance, plant biomass and grain yield, and the negative correlations with MDA and electrolyte leakage levels. The evaluation of the relative expression of the *Dhn4* gene showed more sensitive protective reactions in more resistant genotypes. In Okal (cold tolerant) and Tadmor (drought tolerant), a higher relative expression after ABA application was observed by Melišová et al. (2011).

NAC (NAM, ATAF1/2, and CUC2) domain proteins are transcriptional factors conserved in plant species and reported to play diverse roles in various processes including plant developmental, abiotic, and biotic stress responses (Zheng et al. 2009). In a recent study, *T. aestivum* NAC (NAM/ATAF/CUC) transcription factors (TFs) were identified in silico, phylogenetically classified and characterized, and their expression profiles were monitored in response to ABA and drought stress. In response to these treatments, *TaNAC4a* and *TaNAC6* exhibited similar expression trends, suggesting an ABA-dependent regulation of drought, while in the case of *TaNTL5* and *TaNAC2a*, the changes in the expression were not parallel (Tang et al. 2012). Molecular characterization of novel *HvNACs* genes in barley suggests conserved functions in the areas of secondary cell wall biosynthesis, leaf senescence, root development, seed development, and hormone-regulated stress responses (Christiansen et al. 2011). Stress-responsive *NAC1* (*SNAC1*) is predominantly induced in the guard cells by water scarcity condition (Hu et al. 2006).

Expression of the late embryogenesis abundant (*LEA*) gene is usually associated with plant response to dehydration. In barley, products of *LEA* genes (*HVA1*) might play a role in vegetative growth of Tibetan hulless barley (Liang et al. 2012). Barley *HVA1* confers drought and salt tolerance in transgenic maize (Nguyen and Sticklen 2013), dehydration tolerance in transgenic rice via cell membrane protection (Chandra Babu et al. 2004) and in transgenic wheat by improving biomass productivity and water use efficiency under water deficit conditions (Sivamani et al. 2000). Overexpression of *HVA1* generates tolerance to salinity and water stress in transgenic mulberry (*Morus indica*) (Lal et al. 2008).

Myeloblastosis oncogenes (MYB) are involved in several processes of growth and development, and response to stress in plants. One of the major classes of TFs involved in ABA-dependent stress responses is MYB TFs, and in the recent years, there has been a focus on the elucidation of bread wheat R2R3 and MYB3R type MYB TFs, known to be involved in ABA signaling of drought. The R2R3 type MYB *TaPIMP1* was originally described as the first defense related MYB in wheat; however, detailed analyses indicated that *TaPIMP1* is also induced by abiotic stresses, particularly drought. In addition, the induction of its expression by ABA and its inability to bind to the DRE-box element as indicated by EMSA suggest that TaPIMP1 acts in the ABA-dependent pathways of drought response (Zhang et al. 2012). Similarly, *TaMYB33*, another drought responsive R2R3 type MYB, was shown to be induced by ABA treatment, and the overexpression in Arabidopsis plants could not detect a significant increase in DREB2, suggesting that *TaMYB33* is also involved in ABA-dependent mechanisms (Qin et al. 2012). In barley, *MYB* gene has been identified which encodes for R2R3-type MYB protein and possibly involved in both boron stress and divergent regulation mechanisms in plants (Tombuloglu et al. 2013). The dehydration-responsive element-binding proteins (DREBs) or C-repeat-binding proteins (CBFs) were responsible for gene regulation under water deficit condition. A number of DREB homologs have deen identified in wheat and, although DREB2-mediated drought response is not fully elucidated yet, enhanced drought tolerance through DREB-mediated pathways is considered to involve LEA proteins (Egawa et al. 2006). However, two different wheat DREB (*TaDREB2* and *TaDREB3*) factors strongly regulate many different *CBF/DREB* genes from barley, which leads to the substantial improvement of barley capacity to survive during severe drought and frost stresses (Morran et al. 2011). WRKY transcription factors are key regulators of many plant processes, which are involved in several stages of growth of plant, response to stress and developmental stages. These WRKY genes have been reported to take part in dehydration tolerance. A constitutive expression of the barley *HvWRKY38* transcription factor enhances drought tolerance in turf and forage grass (Xiong et al. 2010). In another study, WRKY type transcription factors (*TaWRKY2* and *TaWRKY19*) which are known to be involved in plant abiotic stress response and ABA signaling were identified computationally, localized to the nucleus and shown to bind specifically to *cis*-element, W box. This report revealed that *WRKY19* as a component of both ABA and DREB pathways, showing *WRKY19* expression level, was responsive to ABA application, and in transgenic *WRKY19* deficient plants, the expression levels of DREB pathway components were altered (Niu et al. 2012).

In search of drought-resistant genes in wild barley, a novel gene *Hsdr4* was identified by Suprunova et al. (2007). Analysis of the *Hsdr4* promoter region revealed a new putative miniature inverted-repeat transposable element (MITE) and several potentially stress-related binding sites for transcription factors (MYC, MYB, LTRE, and GT-1), suggesting a role of *Hsdr4* in plant tolerance to dehydration stress (Nevo 2013). The *Hsdr4* was mapped to the 3HL within a region that was previously shown to affect osmotic adaptation in barley.

Seiler et al. (2011) reported that under terminal drought stress, ABA and its degradation products (phaseic acid and diphaseic acid) increased in barley flag leaves and 19 of the 41 ABA metabolism genes exhibited differential regulation in flag leaves. The incidental discovery of a spontaneous wilty mutant (*eibi1*), hypersensitive to drought in a desert wild barley in Israel, led to the identification of a major gene contributing to the generation of cutin and enabling land life (Chen et al. 2004). *eibi1* expressed the highest relative water-loss rate among the known wilty mutants, showing to be one of the most drought-sensitive mutants. *eibi1* had the same abscisic acid (ABA) level, the same ability to accumulate stress-induced ABA, and the same stomatal movement in response to light, dark, drought, and exogenous ABA as the wild type. Thus, *eibi1* was neither an ABA-deficient nor an ABA-insensitive mutant. The transpiration rate of *eibi1* was closer to the chlorophyll efflux rate than to stomatal density. A fine-scale genetic mapping of the *eibi1* locus on chromosome 3H is perfectly collinear with the equivalent region on rice chromosome 1 (Chen et al. 2011).

Some TFs and drought-responsive proteins need to be phosphorylated/dephosphorylated or modified by posttranslational regulation to become active, and genes encoding mitogen-activated protein kinase (MAPK) cascades, calcineurin B-like protein interacting protein kinase (CIPK), calcium-dependent protein kinase (CDPK or CPK), and receptor-like kinases have roles in drought stress signaling and regulation pathways (Table 17.4). *OsMAPK5* was the first characterized protein kinase in rice for regulating drought and other abiotic stresses, but it negatively regulates biotic stress (Xiong and Yang 2003). Two types of Ca^{2+}-sensing protein kinases, CIPK and CDPK, are also involved in stress signaling. *OsCDPK7* and *OsCIPK12* overexpressing rice showed increased tolerance to drought (Saijo et al. 2000; Xiang et al. 2007). The observed DT in rice conferred by *OsCIPK12* correlated with a significant increase in the proline and soluble sugar content after exposure to drought stress conditions (Xiang et al. 2007). Among the other cases, overexpression of the receptor-like protein kinase gene *OsSIK1* resulted in improved drought and salt tolerance, with increased activities of peroxidase and superoxide and lower accumulation of H_2O_2 (Ouyang et al. 2010).

17.7 Genetic Transformation for Enhanced Drought Tolerance—Transgenic Approach

The capacity of stably inserting a wide collection of drought-related genes to plant genomes has opened amazing opportunities for crop improvement. Transgenic approach is being pursued actively throughout the world to improve traits including tolerance to biotic and abiotic stresses in a number of crops (Ashraf et al. 2008). As with salt stress, plant responses to drought stress are complex, because it involves many genes with additive effects, so the prospects of improving drought tolerance in crops seem not to be very bright. Despite this, efforts have been made during the last few decades to generate transgenic lines of different crops, which have shown improved tolerance to drought stress. Currently, barley transformation is in a developmental phase, and various barley genotypes, alternative target tissues and methodologies are tested for developing an efficient technique for barley genetic transformation (Forster et al. 2000). The first fertile transgenic barley plants were produced by particle bombardment of immature embryos, and a cultivar Golden Promise features prominently in this study (Wan and Lemaux 1994). As following, Golden Promise has become a standard genotype for barley transformation with various tissues including immature embryos, callus, and microspores for plasmid bombardment. On the other hand, Golden Promise is an extremely important cultivar, which has been widely used for genetic studies including genetic map construction and development of doubled haploidy (DH) populations.

So far, some important genes isolated from barley were proved with function involving in salt tolerance. For instance, *HVA1* was isolated and characterized from *Hordeum vulgare* L. aleurone layers and was found to be stress induced (Hong et al. 2005). Similarly, a LEA gene *HVA1* (which encodes a group 3 LEA protein) from barley was engineered in rice (Xu et al. 1996), and wheat (Sivamani et al. 2000). Both rice and wheat transformed lines so produced showed enhanced tolerance to drought stress. In another experiment, overexpression of barley *HvCBF4* enhances tolerance to drought stress in transgenic rice (Oh et al. 2007). Constitutive expression of the barley *HvWRKY38* transcription factor enhances drought tolerance in turf and forage grass (Xiong et al. 2010).

Like in the case of plant salt tolerance (Ashraf and Akram 2009), most of the drought tolerant transgenic lines of different crops developed are based on only a single gene transformation, whereas the claims of the scientists regarding the performance of the lines with respect to drought tolerance seem to be overstated as earlier reported in the case of salt tolerance (Flowers 2004; Ashraf and Akram 2009). Thus, manipulation of a number of genes predominantly involved in stress tolerance to transgenic plants seems to be a plausible approach. This will certainly allow pyramiding of desirable traits to achieve considerable advance in barley drought tolerance.

17.8 Conclusions and Future Perspectives

Drought is one of the most severe stresses limiting plant growth and yield. A broad range of biochemical and physiological traits have been implicated in drought stress tolerance. Considerable advances have been made in understanding the plant's adaptation in stress environments and complex genetics involving multitude of gene and stress tolerance mechanisms. There is a great potential of genetic breeding for drought tolerance through the contribution of wild species to the identification of drought QTLs and functional markers. Importantly, several QTLs for key morphophysiological characteristics and yield were identified under water-limited conditions through creation of linkage maps using parents with different drought coping abilities. In recent decades, application of high-throughput screening, "omics" strategies on wild species different crops with differential drought tolerance coping abilities, has revealed several stress-related candidate gene (s) or gene block (s). Furthermore, using a variety of bioinformatics, molecular biology, and functional genomic tools, drought-related candidates were characterized, and their roles in drought tolerance were studied. Major drought-related molecules were revealed to be signal transduction pathway components and transcription factors.

To explain the quantitative differences in the responses to environmental stress of tolerant and sensitive species, it could be assumed that certain proteins encoded by "stress tolerance" genes playing essential roles in the mechanisms of tolerance have a higher intrinsic activity in the tolerant species than the homologous proteins from the sensitive one. These proteins could include, among others, ion transporters of the plasma membrane or the tonoplast, enzymes involved in osmolyte biosynthesis, enzymatic antioxidant systems, or proteins regulating the expression or activity of any of them. On the other hand, the differences in the response could also be due to differences in the level of expression of the corresponding genes, either because of the relative strength of their promoters or because of their regulatory mechanisms. For example, the expression of a particular gene could be stress-inducible in a tolerant species, but not in a related non-tolerant taxon.

Therefore, stress-tolerant wild plants also represent a possible source of new, more efficient biotechnological tools for the genetic improvement of stress tolerance in crop plants: genes conferring higher levels of tolerance by overexpression in transgenic plants (as compared to homologous genes isolated from stress-sensitive species), or stress-regulated promoters which could be used for the controlled expression of any putative stress tolerance gene. Research exploiting recent advances in genomics technologies has made it possible to dissect and resynthesize molecular regulation of drought and manipulate crop genomes for drought tolerance. The future efforts will be to integrate and translate these resources into practical higher yielding field products.

References

Abebe T, Melmaiee K, Berg V, Wise RP. Drought response in the spikes of barley: gene expression in the lemma, palea, awn, and seed. Funct Integr Genomics. 2010;10(2):191–205.

Ahmed IM, Nadira UA, Zhang GP, Wu FB. Exploration and Utilization of Drought-Tolerant Barley Germplasm. In: Zhang GP , Li CD, editors. Exploration, Identification and Utilization of Barley Germplasm. Zhejiang University Press: AP Elsevier; 2015. p. 115–152.

Ashoub A, Beckhaus T, Berberich T, Karas M, Brüggemann W. Comparative analysis of barley leaf proteome as affected by drought stress. Planta. 2013;237(3):771–81.

Ashraf M, Athar HR, Harris PJC, Kwon TR. Some prospective strategies for improving crop salt tolerance. Adv Agron. 2008;97:45–110.

Ashraf M, Öztürk MA, Athar H. Salinity and water stress: improving crop efficiency, vol. 44. Berlin: Springer; 2009.

Ashraf M. Inducing drought tolerance in plants: Recent advances. Biotechnol Adv. 2010; 28, 169–183.

Ashraf M, Akram NA. Improving salinity tolerance of plants through conventional breeding and genetic engineering: an analytical comparison. Biotechnol Adv. 2009;27:744–52.

Baenziger PS, Beecher B, Graybosch RA, Ibrahim AMH, Baltensperger DD, Nelson LA, et al. Registration of 'NEO1643' wheat. J Plant Registr. 2008;2(1):36–42.

Bänziger M, Setimela PS, Hodson D, Vivek B. Breeding for improved drought tolerance in maize adapted to southern Africa. Proceedings of the 4th International Crop Science Congress. Brisbane, Australia. Published on CDROM. 2004.

Baum M, Grando S, Backes G, Jahoor A, Sabbagh A, Ceccarelli S. QTLs for agronomic traits in the Mediterranean environment identified in recombinant inbred lines of the cross 'Arta' × *H. spontaneum* 41-1. Theor Appl Genet. 2003;107(7):1215–25.

Blum A. Plant breeding for stress environments. Boca Raton: CRC Press; 1988.

Blum A. Breeding methods for drought resistance. In: Jones HG, Flowrs TJ, Jones MB, editors. Plant under stress. Cambridge: Cambridge University Press; 1989. p. 197–215.

Cai H, Tian S, Liu C, Dong H. Identification of a MYB3R gene involved in drought, salt and cold stress in wheat (*Triticum aestivum* L.). Gene. 2011;485(2):146–52.

Cai S, Jiang G, Ye N, Chu Z, Xu XZ, Zhang JH, Zhu GH. A key ABA catabolic gene, OsABA8ox3, is involved in drought stress resistance in rice. PLoS One. 2015;10(2), e0116646.

Cash SD, Bruckner PL, Wichman DM, Kephart KD, Berg JE, Boyner R, et al. Registration of Willow Creek' forage wheat. J Plant Registr. 2009;3(2):185–90.

Cattivelli L, Baldi P, Crosatti C, Di Fonzo N, Faccioli P, Grossi M, Mastrangelo AM, Pecchioni N, Stanca AM. Chromosome regions and stress-related sequences involved in resistance to abiotic stress in Triticeae. Plant Mol Biol. 2002;48(5-6):649–65.

Ceccarelli S, Grando S, Baum M, Udupa SM. Breeding for drought resistance in a changing climate. In: Rao SC, Ryan J, editors. Challenges and strategies of dryland agriculture. Madison: Crop Science Society of America and American Society of Agronomy; 2004. p. 167–90.

Chandra Babu R, Zhang J, Blum A, David Ho T-H, Wu R, Nguyen H. *HVA1*, a *LEA* gene from barley confers dehydration tolerance in transgenic rice (*Oryza sativa* L.) via cell membrane protection. Plant Sci. 2004;166(4):855–62.

Chen, YY, Chen PY, de los Reyes BG. Differential responses of the cultivated and wild species of soybean to dehydration stress. Crop Sci. 2006;46:2041–6.

Chen G, Komatsuda T, Ma JF, Nawrath C, Pourkheirandish M, Tagiri A, Hu Y-G, Sameri M, Li X, Zhao X. An ATP-binding cassette subfamily G full transporter is essential for the retention of leaf water in both wild barley and rice. Proc Natl Acad Sci U S A. 2011;108(30):12354–9.

Chen G, Sagi M, Weining S, Krugman T, Fahima T, Korol AB, Nevo E. Wild barley eibi1 mutation identifies a gene essential for leaf water conservation. Planta. 2004;219(4):684–93.

Choi D-W, Close T. A newly identified barley gene, Dhn12, encoding a YSK2 DHN, is located on chromosome 6H and has embryo-specific expression. Theor Appl Genet. 2000;100(8): 1274–8.

Choi D-W, Zhu B, Close T. The barley (*Hordeum vulgare* L.) dehydrin multigene family: sequences, allele types, chromosome assignments, and expression characteristics of 11 Dhn genes of cv Dicktoo. Theor Appl Genet. 1999;98(8):1234–47.

Christiansen MW, Holm PB, Gregersen PL. Characterization of barley (*Hordeum vulgare* L.). BMC Res Notes. 2011;4(1):302.

Close T, Choi D, Venegas M, Salvi S, Tuberosa R, Ryabushkina N, Turuspekov Y, Nevo E (2000) Allelic variation in wild and cultivated barley at the Dhn4 locus, which encodes a major drought-induced and seed protein, DHN4. In: 8th International Barley Genetics Symposium, Adelaide, SA, South Australia

Collard BCY, Ades PK, Pang ECK, Brouwer JB, Taylar PWJ. Prospecting for sources of resistance to Ascochyta blight in wild Cicer species. Austral Plant Pathol. 2001;30:271–6.

Comadran J, Thomas WT, van Eeuwijk FA, Ceccarelli S, Grando S, Stanca AM, Pecchioni N, Akar T, Al-Yassin A, Benbelkacem A, Ouabbou H, Bort J, Romagosa I, Hackett CA, Russell JR. Patterns of genetic diversity and linkage disequilibrium in a highly structured *Hordeum vulgare* association-mapping population for the Mediterranean basin. Theor Appl Genet. 2009;119:175–87.

Comadran J, Russell JR, Booth A, Pswarayi A, Ceccarelli S, Grando S, Stanca AM, Pecchioni N, Akar T, Al-Yassin A, Benbelkacem A, Ouabbou H, Bort J, van Eeuwijk FA, Thomas WTB, Romagosa I. Mixed model association scans of multi-environmental trial data reveal major loci controlling yield and yield related traits in *Hordeum vulgare* in Mediterranean environments. Theor Appl Genet. 2011;122:1363–73.

Courtois B, Shen L, Petalcorin W, Carandang S, Mauleon R, Li Z. Locating QTLs controlling constitutive root traits in the rice population IAC 165-Co39. Euphytica. 2003;134:335–45.

Earl H, Davis RF. Effect of drought stress on leaf and whole canopy radiation use efficiency and yield of maize. Agron J. 2003;95:688–96.

Egawa C, Kobayashi F, Ishibashi M, Nakamura T, Nakamura C, Takumi S. Differential regulation of transcript accumulation and alternative splicing of a *DREB2* homolog under abiotic stress conditions in common wheat. Genes Genet Sys. 2006;81(2), 77–91.

Ellis RP, Forster BP, Robinson D, Handley LL, Gordon DC, Russell JR, Powell W. Wild barley: a source of genes for crop improvement in the 21st century? J Exp Bot. 2000;51:9–17.

Farooq M, Wahid A, Kobayashi N, Fujita D, Basra SMA. Plant drought stress: effects, mechanisms and management. Agron Sustain Dev. 2009; 29, 185–212.

Flowers TJ. Improving crop salt tolerance. J Exp Bot. 2004;55:307–19.

Forster B, Rzussell J, Ellis R, Handley L, Robinson D, Hackett C, Nevo E, Waugh R, Gordon D, Keith R. Locating genotypes and genes for abiotic stress tolerance in barley: a strategy using maps, markers and the wild species. New Phytol. 1997;137(1):141–7.

Forster BP, Ellis RP, Thomas WTB, Newton AC, Tuberosa R, This D, El-Enein RA, Bahri MH, Salem MB. The development and application of molecular markers for abiotic stress tolerance in barley. J Exp Bot. 2000;51:19–27.

Grando S, Ceccarelli S. Seminal root morphology and coleoptile length in wild (*Hordeum vulgare* ssp. *spontaneum*) and cultivated (*Hordeum vulgare* ssp. *vulgare*) barley. Euphytica. 1995;86(1):73–80.

Grando S, Von Bothmer R, Ceccarelli S. Genetic diversity of barley: use of locally adapted germplasm to enhance yield and yield stability of barley in dry areas. Broadening the genetic base of crop production CABI/FAO/IPGRI:351-372. 2001.

Guo P, Baum M, Grando, S, Ceccarelli S, Bai G, Li R, von Korff M, Varshney RK, Graner A, Valkoun J. Differentially expressed genes between drought-tolerant and drought-sensitive barley genotypes in response to drought stress during the reproductive stage. J Exp Bot. 2009; 60:3531–3544.

Harlan J. Genetic resources in wild relatives of crops. Crop Sci. 1976;16(3):329–33.

Harlan JR, de Wet JMJ. Toward a rational classification of cultivated plants. Taxon. 1971;20:509–17.

Hussain TM, Chandrasekhar T, Hazara M, Sultan Z, Saleh BK, Gopal GR. Recent advances in salt stress biology-a review. Biotechnol Mol Biol Rev. 2008;3:8–13.

Hong BS, Zong-Suo L, Ming-An S. LEA proteins in higher plants: structure, function, gene expression and regulation. Colloids Surf B Biointerfaces. 2005;45:131–5.

Hu H, Dai M, Yao J, Xiao B, Li X, Zhang Q, Xiong L. Overexpressing a NAM, ATAF, and CUC (NAC) transcription factor enhances drought resistance and salt tolerance in rice. Proc Natl Acad Sci U S A. 2006;103(35):12987–92.

Hymowitz T, Singh RJ. Taxonomy and speciation. In: Wilcox JR, editor. Soybeans: Improvement, production, and uses. 2nd ed. Madison: American Society of Agronomy; 1987. p. 23–48.

Irar S, Brini F, Goday A, Masmoudi K, Pagès M. Proteomic analysis of wheat embryos with 2-DE and liquid-phase chromatography (Proteome Lab PF-2D)—a wider perspective of the proteome. J Proteomics. 2010;73(9):1707–21.

Ivandic V, Hackett C, Zhang Z, Staub J, Nevo E, Thomas W, Forster B. Phenotypic responses of wild barley to experimentally imposed water stress. J Exp Bot. 2000;51(353):2021–9.

Ivandic V, Thomas W, Nevo E, Zhang Z, Forster B. Associations of simple sequence repeats with quantitative trait variation including biotic and abiotic stress tolerance in *Hordeum spontaneum*. Plant Breed. 2003;122(4):300–4.

James AT, Lawn RJ, Cooper M. Genotypic variation for drought stress response traits in soybean. I. variation in soybean and wild *Glycine* spp. for epidermal conductance, osmotic potential, and relative water content. Aust J Agric Res. 2008;59:656–69.

Juskiw PE, Helm JH, Oro M, Nyachiro JM, Salmon DF. Registration of 'Bentley' barley. J Plant Registr. 2009;3(2):119–23.

Kashiwagi J, Krishnamurthy L, Upadhyaya HD, Krishna HS, Vandez CV, Serraj R. Genetic variability of drought-avoidance root traits in the mini-core germplasm collection of chickpea (*Cicer arietinum* L.). Euphytica. 2005;146:213–22.

Kashiwagi J, Krishnamurthy L, Crouch JH, Serraj R. Variability of root length density and its contributions to seed yield in chickpea (*Cicer arietinum* L.) under terminal drought stress. Field Crops Res. 2006;95:171–81.

Karami A, Shahbazi M, Niknam V, Shobbar ZS, Tafreshi RS, Abedini R, Mabood HE. Expression analysis of dehydrin multigene family across tolerant and susceptible barley (*Hordeum vulgare* L.) genotypes in response to terminal drought stress. Acta Physiol Planta. 2013;35(7):2289–97.

Kausar R, Arshad M, Shahzad A, Komatsu S. Proteomics analysis of sensitive and tolerant barley genotypes under drought stress. Amino Acids. 2013;44(2):345–59.

Krugman T, Peleg Z, Quansah L. Alteration in expression of hormone-related genes in wild emmer wheat roots associated with drought adaptation mechanisms. Funct Integr Genomics. 2011;11(4):565–83.

Lafitte HR, Price AH, Courtois B. Yield response to water deficit in an upland rice mapping population: associations among traits and genetic markers. Theor Appl Genet. 2004;109:1237–46.

Lal S, Gulyani V, Khurana P. Overexpression of *HVA1* gene from barley generates tolerance to salinity and water stress in transgenic mulberry (*Morus indica*). Transgenic Res. 2008;17(4):651–63.

Lee KW, Choi GJ, Kim KY, Ji HC, Zaman R, Lee SH. Identification of drought induced differentially expressed genes in barley leaves using the annealing control-primer-based GeneFishing technique. Aust J Crop Sci. 2011;11(5):1364–9.

Liang J, Deng G, Long H, Pan Z, Wang C, Cai P, Xu D, Nima Z-X, Yu M. Virus-induced silencing of genes encoding LEA protein in Tibetan hulless barley (*Hordeum vulgare* ssp. *vulgare*) and their relationship to drought tolerance. Mol Breed. 2012;30(1):441–51.

Liu WY, Wang MM, Huang J, Tang HJ, Lan HX, Zhang HS. The OsDHODH1 gene is involved in salt and drought tolerance in rice. J Integr Plant Biol. 2009;51(9):825–33.

Liu H, Zhou X, Dong N, Liu X, Zhang H, Zhang Z. Expression of a wheat MYB gene in transgenic tobacco enhances resistance to *Ralstonia solanacearum*, and to drought and salt stresses. Funct Integr Genomics. 2011;11(3):431–43.

Lopes MS, Araus JL, Van Heerden PD, Foyer CH. Enhancing drought tolerance in C4 crops. J Exp Bot. 2011;62(9):3135–53.

Melišová L, Holková L, Ullmannováet K, Hrstková P, Hronková M. Indirect evaluation of drought tolerance of barley. Tagungsband der 61. Jahrestagung der Vereinigung der Pflanzenzüchter und Saatgutkaufleute Österreichs, 23–25 November 2010, Raumberg-Gumpenstein, Österreich. Ertrag vs. Qualität bei Getreide, Öl und Eiweisspflanzen. Wheat stress, Höhere Bundeslehr- und Forschungsanstalt für Landwirtschaft Raumberg-Gumpenstein. 2011.

Munns R. Comparative physiology of salt and water stress. Plant Cell Environ. 2002;25:239–50.

Morran S, Eini O, Pyvovarenko T, Parent B, Singh R, Ismagul A, Eliby S, Shirley N, Langridge P, Lopato S. Improvement of stress tolerance of wheat and barley by modulation of expression of DREB/CBF factors. Plant Biotechnol J. 2011;9(2):230–49.

Morrell PL, Lundy KE, Clegg MT. Distinct geographic patterns of genetic diversity are maintained in wild barley (*Hordeum vulgare* ssp. *spontaneum*) despite migration. Proc Natl Acad Sci U S A. 2003;100(19):10812–7.

Nevo E. Evolution of wild barley and barley improvement. In: Zhang G, Li C, Xu L, editors. Advances in barley sciences. Dordrecht: Springer; 2013. p. 1–23.

Nevo E, Chen G. Drought and salt tolerances in wild relatives for wheat and barley improvement. Plant Cell Environ. 2010;33(4):670–85.

Nguyen T, Sticklen M. Barley *HVA1* gene confers drought and salt tolerance in transgenic maize (*Zea mays* L.). Adv Crop Sci Technol. 2013;1(105):2.

Niu CF, Wei W, Zhou QY. Wheat WRKY genes *TaWRKY2* and *TaWRKY19* regulate abiotic stress tolerance in transgenic Arabidopsis plants. Plant Cell Environ. 2012;35(6):1156–70.

Noaman MM, El Sayad AA, Asaad FA, El Sherbini AM, El Bawab AO, El Moselhi MA. Registration of 'Giza 126' barley. Crop Sci. 1995;35(6):1710.

Noaman MM, Ahmed IA, El-Sayed AA, Abo-El-Enin RA, El-Gamal AS, El-Sherbiny AM. Registration of 'Giza 2000' drought-tolerant six-rowed barley for rainfed and new reclaimed areas in Egypt. Crop Sci. 2007;47:440.

Oh SJ, Kwon CW, Choi DW, Song SI, Kim JK. Expression of barley *HvCBF4* enhances tolerance to abiotic stress in transgenic rice. Plant Biotechnol J. 2007;5:646–56.

Oka HI. Origin of cultivated rice, Developments in crop science, vol. 14. Amsterdam: Elsevier Science; 1988.

Ouyang S-Q, Liu Y-F, Liu P, Lei G, He S-J. Receptor-like kinase *OsSIK1* improves drought and salt stress tolerance in rice (*Oryza sativa*) plants. Plant J. 2010;62:316–29.

Pan A, Hayes P, Chen F, Chen T, Blake T, Wright S, Karsai I, Bedö Z. Genetic analysis of the components of winterhardiness in barley (*Hordeum vulgare* L.). Theor Appl Genet. 1994;89(7-8):900–10.

Peleg Z, Fahima T, Abbo S. Genetic diversity for drought resistance in wild emmer wheat and its ecogeographical associations. Plant Cell Environ. 2005;28(2):176–91.

Peng JH, Sun DF, Nevo E. Domestication evolution, genetics and genomics in wheat. Mol Breed. 2011;28(3):281–301.

Peng JH, Sun DF, Peng YL, Nevo E. Gene discovery in *Triticum dicoccoides*, the direct progenitor of cultivated wheats. Cer Res Commun. 2013;41(1):1–22.

Pratt RC, Casey MA. Registration of maize germplasm line Oh605. Crop Sci 2006;46:1004–5.

Qin Y, Wang M, Tian Y, He W, Han L, Xia G. Over-expression of *TaMYB33* encoding a novel wheat MYB transcription factor increases salt and drought tolerance in Arabidopsis. Mol Biol Rep. 2012;39(6):7183–92.

Quarrie SA, Gulli M, Calestani C, Steed A, Marmiroli N. Location of a gene regulating drought-induced abscisic acid production on the long arm of chromosome 5A of wheat. Theor Appl Genet. 1994;89:794–800.

Rampino P, Mita G, Fasano P. Novel durumwheat genes up-regulated in response to a combination of heat and drought stress. Plant Physiol Biochem. 2012;56:72–8.

Ribaut JM, Ragot M. Marker-assisted selection to improve drought adaptation in maize: theback-cross approach, perspectives, limitations, and alternatives. J Exp Bot. 2006;58:351–60.

Robin S, Pathan MS, Courtois B, Lafitte R, Carandang S, Lanceras S. Mapping osmotic adjustment in an advanced back-cross inbred population of rice. Theor Appl Genet. 2003;107(7):1288–96.

Rodriguez E, Svensson J, Malatrasi M, Choi D-W, Close T. Barley *Dhn13* encodes a KS-type dehydrin with constitutive and stress responsive expression. Theor Appl Genet. 2005;110(5): 852–8.

Rollins J, Habte E, Templer S, Colby T, Schmidt J, von Korff M. Leaf proteome alterations in the context of physiological and morphological responses to drought and heat stress in barley (*Hordeum vulgare* L.). J Exp Bot. 2013;64:3201–3212.

Rostoks N, Ramsay L, MacKenzie K, Cardle L, Bhat PR, Roose ML, Svensson JT, Stein N, Varshney RK, Marshall DF, Graner A, Close TJ, Waugh R. Recent history of artificial outcrossing facilitates whole-genome association mapping in elite inbred crop varieties. Proc Natl Acad Sci U S A. 2006;103:18656–61.

Roy JK, Smith KP, Muehlbauer GJ, Chao S, Close TJ, Steffenson BJ. Association mapping of spot blotch resistance in wild barley. Mol Breed. 2010;26:243–56.

Saijo Y, Hata S, Kyozuka J, Shimamoto K, Izui K. Over-expression of a single Ca^{2+}-dependent protein kinase confers both cold and salt/drought tolerance on rice plants. Plant J. 2000;23:319–27.

Saranga ZY, Krugman T, Abbo S, Nevo E, Fahima T. Allelic diversity associated with aridity gradient in wild emmer wheat populations. Plant Cell Environ. 2008;31(1):39–49.

Sečenji M, Hideg E, Bebes A, Gyorgyey J. Transcriptional differences in gene families of the ascorbate-glutathione cycle in wheat during mild water deficit. Plant Cell Rep. 2010;29(1):37–50.

Second G. Origin of the genetic diversity of cultivated rice (*Oryza* spp.), study of the polymorphism scored at 40 isozyme loci. Jpn J Genet. 1982;57:25–57.

Seiler C, Harshavardhan VT, Rajesh K, Reddy PS, Strickert M, Rolletschek H, Scholz U, Wobus U, Sreenivasulu N. ABA biosynthesis and degradation contributing to ABA homeostasis during barley seed development under control and terminal drought-stress conditions. J Exp Bot. 2011;62(8):2615–32.

Serraj R, Krishnamurthy L, Kashiwagi J, Kumar J, Chandra S, Crouch JH. Variation in root traits of chickpea (*Cicer arietinum* L.) grown under terminal drought. Field Crops Res. 2004;88:115–27.

Shan F, Clarke HC, Plummer JA, Yan G, Siddique KHM. Geographical patterns of genetic variation in the world collections of wild annual Cicer characterized by amplified fragment length polymorphisms. Theor Appl Genet. 2005;110:381–91.

Sharma HC, Bhagwat MP, Pampapathy G, Sharma JP, Ridsdill-Smith TJ. Perennial wild relatives of chickpea as potential sources of resistance to Helicoverpa armigera. Genet Resour Crop Evol. 2006;53:131–8.

Singh KB, Ocampo B, Robertson LD. Diversity for abiotic and biotic stress resistance in the wild annual Cicer species. Genet Resour Crop Evol. 1998;45:9–17.

Sivamani E, Bahieldin A, Wraith JM, Al-Niemi T, Dyer WE, Ho T-HD QR. Improved biomass productivity and water use efficiency under water deficit conditions in transgenic wheat constitutively expressing the barley *HVA 1* gene. Plant Sci. 2000;155(1):1–9.

Skovmand B, Reynolds MP, DeLacy IH. Searching genetic resources for physiological traits with potential for increasing yield. In: Reynolds MP, Ortiz-Monasterio I, McNab A, editors. Application of physiology in wheat breeding. Mexico: CIMMYT; 2001. p. 17–28.

Steele KA, Price AH, Shashidar HE, Witcombe JR. Marker-assisted selection to introgress rice QTLs controlling root traits into an Indian upland rice variety. Theor Appl Genet 2006; 112:208–21.

Steele KA, Virk DS, Kumar R, Prasad SC, Witcombe JR. Field evaluation of upland rice lines selected for QTLs controlling root traits. Field Crops Res 2007;101:180–6.

Sun CQ, Wang XK, Yoshimura A, Iwata N. Comparison of the genetic diversity of common wild rice (*Oryza rufipogon* Griff.) and cultivated rice (*O. sativa* L.) using RFLP markers. Theor Appl Genet. 2001;102:157–62.

Suprunova T, Krugman T, Fahima T, Chen G, Shams I, Korol A, Nevo E. Differential expression of dehydrin genes in wild barley, *Hordeum spontaneum*, associated with resistance to water deficit. Plant Cell Environ. 2004;27,1297–1308.

Suprunova T, Krugman T, Distelfeld A, Fahima T, Nevo E, Korol A. Identification of a novel gene (*Hsdr4*) involved in water-stress tolerance in wild barley. Plant Mol Biol. 2004;64(1–2):17–34.

Talame V, Sanguineti MC, Chiapparino E, Bahri H, Ben Salem M, Forster BP, et al.Identification of *Hordeum spontaneum* QTL alleles improving field performance ofbarley grown under rainfed conditions. Ann Appl Bot. 2004;144:309–20.

Tang Y, Liu M, Gao S. Molecular characterization of novel TaNAC genes in wheat and overexpression of *TaNAC2* a confers drought tolerance in tobacco. Plant Physiol. 2012;144(3):210–24.

Teulat B, Zoumarou-Wallis N, Rotter B, Salem MB, Bahri H, This D. QTL for relative water content in field-grown barley and their stability across Mediterranean environments. Theor Appl Genet. 2003;108(1):181–8.

Toker C. Preliminary screening and selection for cold tolerance in annual wild Cicer species. Genet Resour Crop Evol. 2005;52:1–5.

Tombuloglu H, Kekec G, Sakcali MS, Unver T. Transcriptome-wide identification of *R2R3-MYB* transcription factors in barley with their boron responsive expression analysis. Mol Genet Genom. 2013;1–15.

Tuberosa R, Salvi S. Genomics approaches to improve drought tolerance in crops.Trends Plant Sci. 2006;11:405–12.

van der Maesen LJG. Origin, history and taxonomy of chickpea. In: Saxena MC, Singh KB, editors. The chickpea. Wallingford: CAB Int; 1987. p. 11–34.

Vinocur B, Altman A. Recent advances in engineering plant tolerance to abiotic stress: achievements and limitations. Curr Opin Biotechnol. 2005;16:123–32.

Valkoun JJ. Wheat pre-breeding using wild progenitors. Euphytica. 2001;119:17–23.

van Zee K, Chen FQ, Hayes PM, Close TJ, Chen TH. Cold-specific induction of a dehydrin gene family member in barley. Plant Physiol. 1995;108(3):1233–9.

Villareal RL, Mujeeb-Kazi A, Rajaram S, Toro ED. Morphological variability in some synthetic hexaploid wheats derived from *Triticum turgidum* × *T. tauschii*. J Genet Breed. 1994;48:7–16.

Volis S, Mendlinger S, Ward D. Differentiation in populations of *Hordeum spontaneum* along a gradient of environmental productivity and predictability: life history and local adaptation. Biol J Linn Soc. 2002;77(4):479–90.

Wang ZY, Second G, Tanksley SD. Polymorphism and phylogenetic relationships among species in the genus Oryza as determined by analysis of nuclear RFLPs. Theor Appl Genet. 1992;83:565–81.

Wan Y, Lemaux PG. Generation of large numbers of independently transformed fertile barley plants. Plant Physiol. 1994;104:37–48.

Wang W, Vinocur B, Altman A. Plant responses to drought, salinity and extreme temperatures: towards genetic engineering for stress tolerance. Planta. 2003;218:1–14.

Wendelboe-Nelson C, Morris PC. Proteins linked to drought tolerance revealed by DIGE analysis of drought resistant and susceptible barley varieties. Proteomics. 2012;12:3374–85.

Xiang Y, Huang Y, Xiong L. Characterization of stress-responsive CIPK genes in rice for stress tolerance improvement. Plant Physiol. 2007;144:1416–28.

Xiong L, Yang Y. Disease resistance and abiotic stress tolerance in rice are inversely modulated by an abscisic acid–inducible mitogen-activated protein kinase. Plant Cell. 2003;15:745–59.

Xiong X, James VA, Zhang H, Altpeter F. Constitutive expression of the barley *HvWRKY38* transcription factor enhances drought tolerance in turf and forage grass (*Paspalum notatum* Flugge). Mol Breed. 2010;25(3):419–32.

Xu D, Duan X, Wang B, Hong B, Ho THD, Wu R. Expression of a late embryogenesis abundant protein gene, *HVA1*, from barley confers tolerance to water deficit and salt stress in transgenic rice. Plant Physiol. 1996;110:249–57.

Yang F, Jørgensen AD, Li H. Implications of high-temperature events and water deficits on protein profiles in wheat (*Triticum aestivum* L. cv. *Vinjett*) grain. Proteomics. 2011;11(9):1684–95.

Yan-Ying QU, Ping MU, Xue-Qin L, Yu-Xiu T, Feng W, Hong-Liang Z. QTL mapping and correlations between leaf water potential and drought resistance in rice under upland and lowland environments. Acta Agron Sin. 2008;34(2):198–206.

Yamaguchi-Shinozaki K, Shinozaki K. Transcriptional regulatory networks in cellular responses and tolerance to dehydration and cold stresses. Annu Rev Plant Biol. 2006;57:781–803.

Zaharieva M, Gaulin E, Havaux M, Acevedo E, Monneveux P. Drought and heat responses in the wild wheat relative *Aegilops geniculata* Roth: potential interest for wheat improvement. Crop Sci. 2001;41:1321–9.

Zhang J, Zheng HG, Aarti A, Pantuwan G, Nguyen TT, Tripathy JN. Locating genomic regions associated with components of drought resistance in rice: comparative mapping within and across species. Theor Appl Genet. 2001;103:19–29.

Zhang Z, Liu X, Wang X. An R2R3 MYB transcription factor in wheat, *TaPIMP1*, mediates host resistance to Bipolaris sorokiniana and drought stresses through regulation of defense-and stress-related genes. New Phytol. 2012;196(4):1155–70.

Zheng X, Chen B, Lu G, Han B. Overexpression of a NAC transcription factor enhances rice drought and salt tolerance. Biochem Biophys Res Commun. 2009;379(4):985–9.

Zhu J-K. Plant salt tolerance. Trends Plant Sci. 2001;6:66–71.

Chapter 18
Tailored Responses to Simultaneous Drought Stress and Pathogen Infection in Plants

Aanchal Choudhary, Prachi Pandey, and Muthappa Senthil-Kumar

18.1 Introduction

With the changing global climate, a series of environmental factors are modified concurrently, along with changes in their intensity and timing. Thus plants are exposed to combinations of abiotic and biotic stressors whose combined impact can adversely affect crop performance and survival (Mittler 2006; Atkinson and Urwin 2012). Of the possible biotic and abiotic stress combinations, simultaneous drought stress and pathogen infection is one of the best studied combinations (Mayek-Perez et al. 2002; McElrone et al. 2003; Sharma et al. 2007; Király et al. 2008; Xu et al. 2008; Carter et al. 2009; Wang et al. 2009; Ramegowda et al. 2013). Drought is one of the most damaging and frequently occurring abiotic factors that can potentially alter the outcome of plant–pathogen interactions (Sharma and Pande 2013). Phenotypic responses of plants exposed to drought stress and pathogen infection vary depending on the severity and duration of each stress and also differs with pathogen type, e.g., fungi, oomycetes, bacteria, or viruses (Olson et al. 1990; McElrone and Forseth 2004; Achuo et al. 2006; Xu et al. 2008). Based on these factors, the combination of drought and pathogen infection can have two outcomes. In the first scenario, the two stressors can act additively, and result in enhanced damage to the plant. For example, drought has been shown to aggravate many fungal (Mayek-Perez et al. 2002), bacterial (McElrone et al. 2001; Mohr and Cahill 2003), and viral (Olson et al. 1990; Prasch and Sonnewald 2013) infections in plants. The susceptibility is attributed to drought-induced increase in abscisic acid (ABA) in plants which then suppresses their defense against pathogens mediated by salicylic acid, jasmonic acid, and ethylene signaling. Few other drought-induced physiological changes like accumulation of osmolytes and nutrient leakage have been reported

A. Choudhary • P. Pandey • M. Senthil-Kumar (✉)
National Institute of Plant Genome Research,
JNU Campus, Aruna Asaf Ali Marg, P.O. Box 10531, New Delhi 110 067, India
e-mail: skmuthappa@nipgr.ac.in

M.A. Hossain et al. (eds.), *Drought Stress Tolerance in Plants, Vol 1*,
DOI 10.1007/978-3-319-28899-4_18

to enhance disease in plants by enriching nutrient supply for the pathogens (Mayek-Perez et al. 2002). Additionally, some pathogens can influence plant water relations leading to low water potential in plant cells, thereby, increasing the effects of water deficit (English-Loeb et al. 1997; Smit and Vamerali 1998; Audebert et al. 2000; Amtmann et al. 2008; Goel et al. 2008; Mittler and Blumwald 2010; Choi et al. 2013). In the second scenario, the simultaneous exposure to drought and pathogen infection can alleviate the effect of either or both the stresses thereby enhancing plants tolerance to the stresses. For example, drought stress has been shown to increase plant tolerance towards some pathogens like *Botrytis cinerea* and *Pseudomonas syringae* (Achuo et al. 2006; Ramegowda et al. 2013). Moreover, pathogen-mediated alleviation of drought stress has also been reported in some cases. For example, infection with *Cucumber mosaic virus* (CMV) led to improved drought tolerance of plants like *Capsicum annum, Solanum lycopersicum* and *Nicotiana tabacum* (Xu et al. 2008). This has been attributed to increased levels of osmoprotectants (trehalose) and antioxidants (anthocyanins and ascorbic acid) (Xu et al. 2008). Infection with *Tobacco mosaic virus* (TMV) enhanced ABA level in *N. tabacum* (Whenham et al. 1986), which points towards the probable role of ABA in virus infection-mediated drought resistance in plants. Thus, ABA might act as a global regulator of stress responses and facilitate fine-tuning of plant stress responses to focus on the more severe threat (Anderson et al. 2004; Yasuda et al. 2008; Ton et al. 2009).

18.2 Plant Responses Under Combined Stress: Tailored Responses

Plants have developed specific mechanisms that allow them to detect environmental changes and respond to complex stress conditions. Findings from the recent studies suggest that some of the responses triggered under combined stress are different from the responses seen in plants exposed to the same stressors individually (Rizhsky et al. 2002, 2004; Anderson et al. 2004; Mittler 2006; Asselbergh et al. 2008; Atkinson and Urwin 2012). Combinatorial stress results in novel interactions between signaling components, which makes the response of the plant distinct from its response to single stresses. Thus, under combined stress, plants exhibit "tailored adaptation strategies," which are customized specifically to the stress combinations (Atkinson and Urwin 2012). Rather than producing an additive response pertaining to each stress, plants instigate some entirely new responses specific for each stress combination (Atkinson and Urwin 2012; Atkinson et al. 2013; Prasch and Sonnewald 2013, 2015; Rasmussen et al. 2013; Rivero et al. 2013; Bostock et al. 2014; Kissoudis et al. 2014; Suzuki et al. 2014). This differential response is necessary to efficiently balance resource allocation between growth and defense and to help the plant respond to stress in a way that does not hamper its fitness (Herms and Mattson 1992; Smith 2007; Bechtold et al. 2010).

Plant's response to concurrently occurring biotic and abiotic stresses varies with the severity, timing, and duration of each of the stressors involved (Atkinson and Urwin 2012). The detailed study of plant responses to combined drought and pathogen infection has revealed that some of the responses shown were similar to that evoked under the individual stresses (Fig. 19.1a). Such responses are thus "shared" between a plant subjected to the two stressors separately and in combination. Apart from the shared responses, several unique responses (indicated in the Fig. 19.1a as "C") are also seen under combined stress, implying that the response is not merely the additive effect of single stress responses (Atkinson and Urwin 2012; Atkinson et al. 2013). In certain situations, plant prioritizes its response towards the more severe threat, i.e., the stress which is more damaging and requires immediate attention.

Thus, the adaptation strategies of plants under combined stress constitute different types of responses depending upon the nature and severity of the stresses (Fig. 19.1b). As mentioned above, the response can be new and not observed under either of the individual stress conditions (unique response) or be similar to the responses evoked by each of the single stresses (shared responses). However, these shared responses can be selectively activated or repressed under combined stress and thus be tailored according to the varying severity of the two stresses encountered (prioritized responses). In some cases, the stress combination can also lead to nullification of the effects of the two stresses on plants (canceled response). Therefore, in order to truly characterize the response of plants to simultaneously occurring stresses, each stress combination should be studied as an entirely new stress (Mittler and Blumwald 2010). A brief discussion on the different categories of tailored response is provided in the section below.

18.2.1 Unique Responses

Recent studies have indicated that the combination of drought and pathogen evokes unique responses in plants, which are not seen when each stress is imposed individually (Choi et al. 2013; Prasch and Sonnewald 2013). These unique responses have been studied only at the molecular level. For example, the exposure of *Vitis vinifera* plants to the combined drought and *Xylella fastidiosa* infection for 4 weeks led to the modulation of 90 transcripts out of which 39 were unique to the combined stress treatment (Choi et al. 2013). Similar results were reported in yet another study wherein the combined virus, drought, and heat treatment to *A. thaliana* plants led to differential expression of 776 unique transcripts (Prasch and Sonnewald 2013; Ramegowda and Senthil-Kumar 2015). These "unique" genes were not seen in transcriptional profile of the individually stressed plants. Re-analysis of the microarray results of Prasch and Sonnewald (2013) by Ramegowda and Senthil-Kumar (2015) revealed that these unique genes constitute several WRKY transcription factors, signaling proteins like receptor like kinases and protein phosphatases. These results suggest that combined stress treatment leads to a reprogramming of gene expression

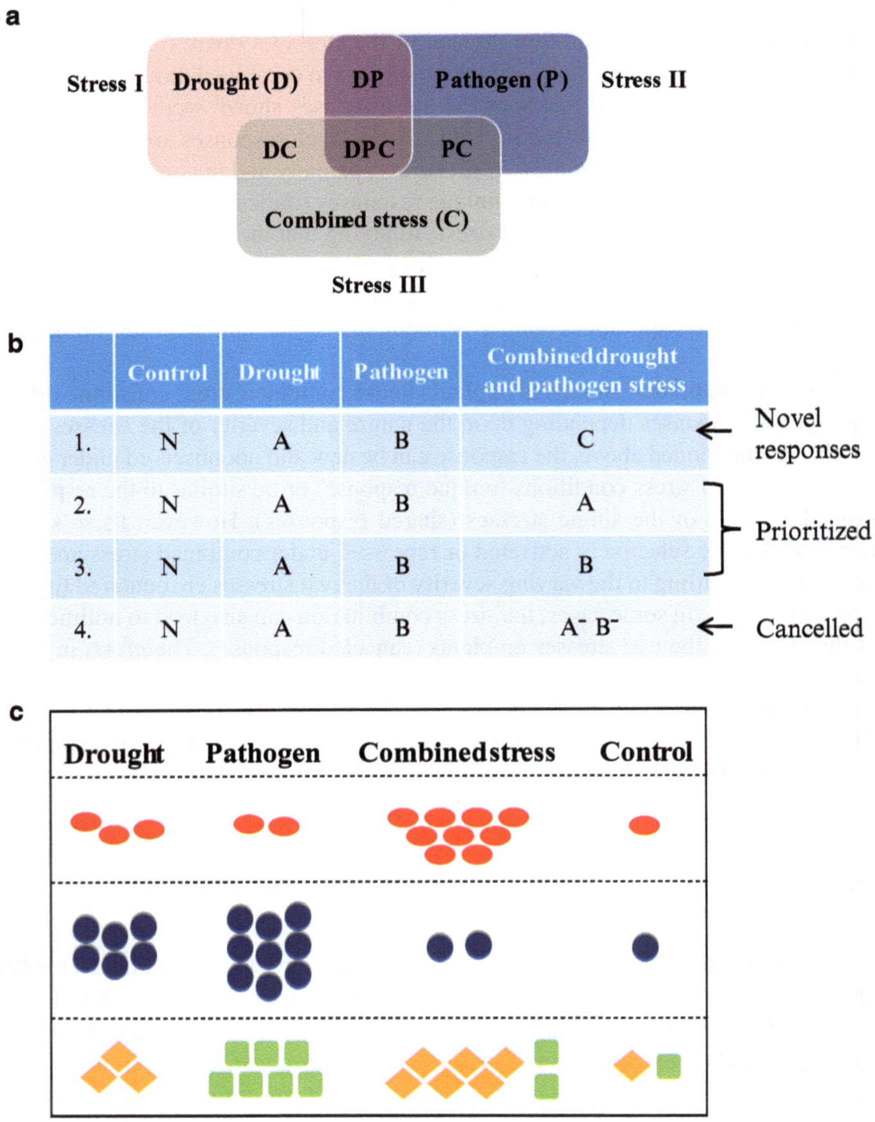

Fig. 19.1 Hypothetical model depicting tailored responses in plants exposed to combined drought stress and pathogen infection. (**a**) Venn diagram shows plant's response to drought (stress I), pathogen (stress II), and their combination (stress III, an altogether new stress). DC—responses shared between drought stress and combination stress, PC—responses shared between pathogen and combined stress, and DPC—responses shared among drought stress, pathogen and combined drought and pathogen. (**b**) Schematic representation of modulation of plant adaptation strategies under combined stress. N—response under optimal growth conditions, A—response to drought, B—response to pathogen infection, last column illustrates the three types of tailored responses under combined stress. Row 1—novel responses (**c**) induced only under combined stress. These responses are not seen under single stress situations. Row 2 and 3—under combined stress, the adaptation

of plants. The presence of combined stress specific unique molecular responses have also been seen in case of drought and heat stress combinations (Rizhsky et al. 2002, 2004; Rampino et al. 2012; Johnson et al. 2014) which further authenticates the tailoring of molecular responses to stress combinations. Although not much information is available in this regards, this preliminary information supported by further studies is useful to unravel the mechanism behind the unique responses seen under the combined stress conditions.

18.2.2 Prioritized Responses

Apart from the unique responses, certain responses, characteristic of the individual stresses, are also observed when plants are exposed to combined stresses. Being common to the two individual stress conditions, these responses are termed as shared responses. However, these shared responses are further attuned to the combined stress. Plants when challenged with two stresses simultaneously prioritize their response towards the more damaging stress, overriding the defense pathway for the less severe stress (Prasch and Sonnewald 2013; Rasmussen et al. 2013). This results in suppression of responses to the stress, which is less severe. For example, plants exposed to water deficit and pathogen infection simultaneously often show weakened defenses and enhanced susceptibility to the pathogen (Audebert et al. 2000; Amtmann et al. 2008; Goel et al. 2008; Mittler and Blumwald 2010). In the study conducted by Atkinson and Urwin (2012), the combined effect of drought and infection with root-knot nematodes *Meloidogyne incognita* on nutritional quality of tomato was investigated. The physiological responses of the plants were compared for different stress treatments and the levels of antioxidants in fruits were analyzed. Significantly higher levels of flavonoids were found in infected plants compared to controls, while a little or no change in flavonoid concentration was reported as a result of water stress only. Interestingly, when the two stresses were applied simultaneously, the heightened accumulation of flavonoids seen under nematode stress was reduced to a level which was not significantly different from the control and water-stressed plants. This can be explained by the drought-induced accumulation of ABA which in turn inhibits the transcription of defense and pathogen-responsive genes, thus preventing nematode-induced flavonoid accumulation (Anderson et al. 2004). The carotenoids (lycopene and β-carotene) concentration was significantly

Fig. 19.1 (continued) strategies are prioritized for the more severe stress among the two. In 2 and 3, response under combined stress resembles the response to drought and pathogen alone, respectively. Row 4—Responses evoked independently under single stresses are absent under combined stress. Fig. 19.1 (continued) (**c**) Illustration depicting tailored molecular responses under combined stress. Each symbol represents a gene product and the number represents their level relative to control. In row 1 and 2, level of the gene product shared between two stresses changes in magnitude under combined stress. Row 3 depicts response at the gene level prioritized for a particular stress, in this case drought. The proposed models are general and can be extended to few other stress combinations as well

reduced in water-stressed tomatoes but remained unaffected by nematode stress. However, under combined stress the expected reduction in the carotenoids level was not seen. The antagonism between drought-induced ABA and ethylene may be the reason for observed inhibition in carotenoid accumulation (Anderson et al. 2004). Additionally, when water deficit and nematode infection occurred in combination, the plant's physiological response was more similar to that of water stress alone in the early harvested tomatoes but to nematode stress alone in the late-harvested tomatoes. These results support the hypothesis that plant stress responses are specifically tailored to the exact combination of environmental stresses encountered, to the extent that the plant responds to whichever stress is most severe, overriding the pathway for the less severe stress (Anderson et al. 2004). The prioritization of responses as a mechanism to focus plants metabolism in deploying their adaptation strategies towards the high impacting stress can be seen as an effective strategy supporting the concept of growth-defense trade-offs in plants (Huot et al. 2014).

18.2.3 Canceled Responses

Interaction of two stresses can also lead to amplification of the tolerance responses, i.e., when two stresses are imposed simultaneously, their effect on plants get "canceled" resulting in enhanced plant tolerance to combined stress as compared to individual stress conditions (Rasmussen et al. 2013). Adaptation strategies that are not sufficient to protect the plants under individual stresses act in unison under the combined stress and the negative impact of the two stresses is canceled. Canceled responses were reported under salt and heat stress combination. For example, some proteins, induced during salt stress (e.g., choline monooxygenase, chloroplastic ATP synthase, V-type proton ATPase catalytic subunit A) and heat stress (e.g., heat shock 70 kDa protein) in *Suaeda salsa*, were unchanged during combined salt and heat treatment (Li et al. 2011). Canceled response in case of drought and pathogen stresses have not yet been reported.

In addition to the above mentioned types of responses that are exhibited by plants as a part of tailored adaptation strategy to counter the combined stress, the "tailoring" can also be observed at the molecular level. Combined stress may lead to the expression of a new set of genes, which are not expressed under individual stress conditions. The molecular response of plants to the two stress conditions and their combination also consists of several commonly regulated genes. However, a change in their expression level can be seen under combined stress (Prasch and Sonnewald 2013). Broadly there can be three different scenarios as indicated in Fig. 19.1c. In case I, the gene product reached beyond the additive level under combined stress, while in case II, the relative level declined and reached closer to that seen under control. Case III depicts prioritization of responses towards a particular stress, wherein the gene product related to plant response to one stress (in this case, drought) is upregulated at the cost of the gene product involved in defense against the other (pathogen stress) (Fig. 19.1c).

18.3 Tailored Responses of Plants to Combined Drought and Pathogen Stress

18.3.1 Morphophysiological Responses

A study comparing the responses of ten ecotypes of *Arabidopsis thaliana* under two individual and combined abiotic stresses revealed that there were no unique morphophysiological responses evoked under combined stress. The responses observed under combined stress were shared and majorly prioritized for one of the stresses (Vile et al. 2012). Some recent reports have indicated the prioritization of stomatal defense responses under simultaneously imposed biotic and abiotic stimuli. When *Vicia faba* and *A. thaliana* were subjected to a combination of biotic stress (*Escherichia coli* or *Pseudomonas syringae*) and several abiotic stresses including water deficit, stomatal responses to abiotic stresses were found to override the responses to biotic stresses (Ou et al. 2014). Similar inferences were obtained from another study on the effect of combined drought and virus infection on *A. thaliana* plants. The microscopic analysis of length-width ratio of stomata of *A. thaliana* plants subjected to concurrent *Turnip mosaic virus* (TuMV) infection, heat, and drought stress in single, double, and triple combinations revealed that stomata were closed under combined treatments of virus and drought, and virus and heat, as well as during the triple stress, while heat stress alone or virus infection resulted in stomatal opening (Prasch and Sonnewald 2013). Also, *X. fastidiosa,* a wilt causing pathogen, influenced the water status (indicated by measurement of leaf water potential, stomatal conductance and transpiration rate) of *V. vinifera* and thus aggravated the effect of drought on the plants (Choi et al. 2013).

18.3.2 Transcriptomic and Metabolic Responses

Till date only four studies have documented the global transcriptome and metabolome changes in plants simultaneously exposed to combinations of various biotic and abiotic stresses (Atkinson et al. 2013; Choi et al. 2013; Prasch and Sonnewald 2013; Rasmussen et al. 2013). The recurrent observation from all these studies is that the adaptation strategies of a plant are specifically tailored in accordance with the combination of stresses it encounters and their severity. As mentioned earlier, the molecular responses can be either unique or shared. A study undertaken by Rasmussen et al. (2013) revealed that 61 % of the transcriptome changes in *A. thaliana* in response to combined stress were not predictable from the responses to single stress treatments (cold, heat, high light, salt, and flagellin). The uniqueness in molecular response seen under combined stress stems from the induction of certain unique transcripts and from selective activation or repression of transcripts responsive to a particular stress. A total of 23 genes were specifically expressed when *A. thaliana* plants were subjected to a combination of drought, heat, and TuMV

(Prasch and Sonnewald 2013). Among these most of the genes encoded stress responsive proteins. Ramegowda and Senthil-Kumar (2015) reanalyzed the transcriptomic data from the above experiment using Bio Conductor package in R statistical program and reported a total of 1370 genes differentially expressed under combined drought and virus infection. Interestingly, out of 1370 genes, 98 genes were unique to virus stress and 157 were unique to drought stress, while 776 were unique to combined drought stress and virus infection. The stress-specific genes upregulated under individual drought and virus infection were 16 and 29, respectively, and the number increased to 72 under combined stress (Prasch and Sonnewald 2013; Pandey et al. 2015). Most of the stress combination specific genes belonged to the category of transcription factors and other regulatory genes including dehydration responsive element binding 2A (DREB2A) and genes encoding zinc finger proteins. Other combined stress associated genes reported were those encoding pentatricopeptide repeat containing protein, abi5 binding protein (AFP1), cold-regulated 47, and universal stress protein family protein. A time-dependent modulation was shown in the transcriptome of *V. vinifera* plants upon exposure to combined drought and *X. fastidiosa* infection (Choi et al. 2013). No significant change in the transcriptome was seen in the early phase (4 weeks posttreatment); however, the number of differentially expressed genes increased with increasing stress exposure (8 weeks posttreatment) and a total of 90 unique transcripts were seen in combined stressed plants. An early upregulation of 9-cis epoxycarotenoid dioxygenase (NCED), an ABA biosynthesis gene, was also reported only under combined stress. These genes are characteristic examples of unique responses under combined stress.

Apart from unique responses, prioritized molecular responses have also been observed under combined stresses. Rasmussen et al. (2013) reported that among the transcripts resulting in antagonistic responses under combined stress, 5–10 % are prioritized under combined stress. In another report, the transcript profile of *A. thaliana*, under simultaneously imposed drought and *Heterodera schachtii*, was shown to be more similar to the expression profile of the plants exposed to water deficit alone than that of the nematode infected plant (Atkinson et al. 2013). Prasch and Sonnewald (2013) also provided evidence for the prioritization of plant's responses towards abiotic stress at the cost of defense responses against biotic stress. The enhanced expression of defense genes that mediates basal as well as *R*-gene-mediated resistance in virus infected *A. thaliana* was abolished under combined virus, heat, and drought stress. In the combined stress situation, only six *R* genes were differentially regulated and none of them were commonly regulated between virus and combined stress, indicating changes in the defense program. One of the genes exclusively downregulated under combined stress was ribosomal protein S6 (*RPS6*). *RPS6* has been shown to mediate resistance via enhanced disease susceptibility (*EDS1*) against *P. syringae* pv. *syringae* effector HopA1 (Kim et al. 2009) as well as against fungal pathogens. These observations indicate the differential response of plants towards abiotic and biotic stresses, which resulted in preferential deactivation of defense responses against various pathogens.

The cytoplasmic protein response (CPR) marker genes constitute another class of shared molecular response under combined drought and virus infection in *A. thaliana* (Prasch and Sonnewald 2013). It is speculated that enhanced CPR supports viral-replication and systemic cell-to-cell spread of the virus in the plant, resulting in increased susceptibility of the host plants (Mayer and Bukau 2005; Prasch and Sonnewald 2013).

Metabolic profiling of plants subjected to combined drought and TuMV treatment revealed the preferential accumulation of tricarboxylic acid (TCA) cycle intermediates and the amino acids derived from them (Prasch and Sonnewald 2013). Under drought stress, increased levels of proline help to protect against osmotic stress (Hanson and Hitz 1982). Interestingly, a combination of drought and virus infection resulted in increased proline accumulation (Prasch and Sonnewald 2013). Altogether, the results obtained from transcriptomic and metabolic studies reflect upon the complexity in plant's responses under the combined stress scenario and highlight the fact that the mechanism of plants' tolerance to combined stresses cannot be completely understood from single stress studies.

18.4 Conclusions and Future Perspectives

The changing climatic conditions impact plants both in terms of stress severity and number of stresses. Hence, understanding the effect of the combined abiotic and biotic stresses on growth and development of plants is important. It has been shown in recent studies that plants tailor some of their responses to the stress combination. This either involves complete reprogramming of plant molecular responses leading to the prioritization of responses towards the more severe stress, or modulation in the magnitude of the shared responses. The tailored responses depends on the nature and intensity of the stresses involved, the age of the plant at which the stress in encountered, and the inherent stress tolerance nature of the plant species.

Recent studies have shed preliminary, but useful information on the combined stress response of plants. Further identification of the genes involved in tailored response and their complete mechanistic understanding can help in formulating the signaling networks and pathways involved in combined stress response. This can not only help in strengthening our knowledge about the unconventional and unique plant adaptation strategies but can also provide important leads for the development of crops that can efficiently tolerate simultaneously occurring drought and pathogen stresses.

Acknowledgements Projects at MS-K Lab are supported by National Institute of Plant Genome Research core funding and DBT-Ramalingaswami re-entry fellowship grant (BT/RLF/re-entry/23/2012). A.C. and P.P. acknowledge DBT-Junior Research Fellowship (JRF, DBT/2014/NIPGR/261) and DST-SERB young scientist grant (SB/YS/LS-71/2014), respectively, for the financial support. Authors thank Mrs. Urooj Fatima for critical reading of the manuscript.

References

Achuo EA, Prinsen E, Höfte M. Influence of drought, salt stress and abscisic acid on the resistance of tomato to *Botrytis cinerea* and *Oidium neolycopersici*. Plant Pathol. 2006;55:178–86.

Amtmann A, Troufflard S, Armengaud P. The effect of potassium nutrition on pest and disease resistance in plants. Physiol Plant. 2008;133:682–91.

Anderson JP, Badruzsaufari E, Schenk PM, Manners JM, Desmond OJ, Ehlert C. Antagonistic interaction between abscisic acid and jasmonate-ethylene signaling pathways modulates defense gene expression and disease resistance in *Arabidopsis*. Plant Cell. 2004;16:3460–79.

Asselbergh B, De Vleesschauwer D, Höfte M. Global switches and fine-tuning-ABA modulates plant-pathogen defense. Mol Plant Microbe Interact. 2008;21:709–19.

Atkinson NJ, Urwin PE. The interaction of plant biotic and abiotic stresses: from genes to the field. J Exp Bot. 2012;63:3523–43.

Atkinson NJ, Lilley CJ, Urwin PE. Identification of genes involved in the response of *Arabidopsis* to simultaneous biotic and abiotic stresses. Plant Physiol. 2013;162:2028–41.

Audebert A, Coyne D, Dingkuhn M, Plowright R. The influence of cyst nematodes (*Heterodera sacchari*) and drought on water relations and growth of upland rice in Côte d'Ivoire. Plant Soil. 2000;220(1–2):235–42.

Bechtold U, Lawson T, Mejia-Carranza J, Meyer RC, Brown IR, Altmann T, Ton J, Mullineaux PM. Constitutive salicylic acid defences do not compromise seed yield, drought tolerance and water productivity in the *Arabidopsis* accession C24. Plant Cell Environ. 2010;33:1959–73.

Bostock RM, Pye MF, Roubtsova TV. Predisposition in plant disease: exploiting the nexus in abiotic and biotic stress perception and response. Annu Rev Phytopathol. 2014;52:517–49.

Carter AH, Chen XM, Garland-Campbell K, Kidwell KK. Identifying QTL for high-temperature adult-plant resistance to stripe rust (*Puccinia striiformis f. sp.tritici*) in the spring wheat (*Triticum aestivum* L.) cultivar 'Louise'. Theor Appl Genet. 2009;119:1119–28.

Choi H-K, Iandolino A, da Silva FG, Cook DR. Water deficit modulates the response of *Vitis vinifera* to the Pierce's disease pathogen *Xylella fastidiosa*. Mol Plant Microbe Interact. 2013;26(6):643–57.

English-Loeb G, Stout MJ, Duffey SS. Drought stress in tomatoes: changes in plant chemistry and potential nonlinear consequences for insect herbivores. Oikos. 1997;79:456–68.

Goel AK, Lundberg D, Torres MA, Matthews R, Akimoto-Tomiyama C, Farmer L, Dangl JL, Grant SR. The *Pseudomonas syringae* type III effector HopAM1 enhances virulence on water-stressed plants. Mol Plant Microbe Interact. 2008;21:361–70.

Hanson AD, Hitz WD. Metabolic responses of mesophytes to plant water deficits. Annu Rev Plant Physiol Plant Mol Biol. 1982;33:163–203.

Herms DA, Mattson WJ. The dilemma of plants: to grow or defend. Q Rev Biol. 1992;67:283–335.

Huot B, Yao J, Montgomery BL, He SY. Growth-defense tradeoffs in plants: a balancing act to optimize fitness. Mol Plant. 2014;7(8):1267–87.

Johnson SM, Lim FL, Finkler A, Fromm H, Slabas AR, Knight MR. Transcriptomic analysis of *Sorghum bicolor* responding to combined heat and drought stress. BMC Genomics. 2014;15:456.

Kim SH, Kwon SI, Saha D, Anyanwu NC, Gassmann W. Resistance to the *Pseudomonas syringae effector* HopA1 is governed by the TIR-NBSLRR protein RPS6 and is enhanced by mutations in SRFR1. Plant Physiol. 2009;150:1723–32.

Király L, Hafez YM, Fodor J, Király Z. Suppression of *Tobacco mosaic virus*-induced hypersensitive-type necrotization in tobacco at high temperature is associated with downregulation of NADPH oxidase and superoxide and stimulation of dehydroascorbate reductase. J Gen Virol. 2008;89:799–808.

Kissoudis C, van de Wiel C, Visser RGF, van der Linden G. Enhancing crop resilience to combined abiotic and biotic stress through the dissection of physiological and molecular crosstalk. Front Plant Sci. 2014;5:207.

Li W, Zhang C, Lu Q, Wen X, Lu C. The combined effect of salt stress and heat shock on proteome profiling in *Suaeda salsa*. J Plant Physiol. 2011;168(15):1743–52.

Mayek-Perez N, Garcia-Espinosa R, Lopez-Castaneda C, Acosta-Gallegos JA, Simpson J. Water relations, histopathology and growth of common bean (*Phaseolus vulgaris* L.) during pathogenesis of Macrophomina phaseolina under drought stress. Physiol Mol Plant Pathol. 2002;60:185–95.

Mayer MP, Bukau B. Hsp70 chaperones: cellular functions and molecular mechanism. Cell Mol Life Sci. 2005;62:670–84.

McElrone AJ, Forseth IN. Photosynthetic responses of a temperate liana to *Xylella fastidiosa* infection and water stress. J Phytopathol. 2004;152:9–20.

McElrone AJ, Sherald JL, Forseth IN. Effects of water stress on symptomatology and growth of *Parthenocissus quinquefolia* infected by *Xylella fastidiosa*. Plant Dis. 2001;85:1160–4.

McElrone AJ, Sherald JL, Forseth IN. Interactive effects of water stress and xylem-limited bacterial infection on the water relations of a host vine. J Exp Bot. 2003;54:419–30.

Mittler R. Abiotic stress, the field environment and stress combination. Trends Plant Sci. 2006;11:15–9.

Mittler R, Blumwald E. Genetic engineering for modern agriculture: challenges and perspectives. Annu Rev Plant Biol. 2010;61:443–62.

Mohr PG, Cahill DM. Abscisic acid influences the susceptibility of *Arabidopsis thaliana* to *Pseudomonas syringae* pv. tomato and *Peronospora parasitica*. Funct Plant Biol. 2003;30:461–9.

Olson AJ, Pataky JK, D'Arcy CJ, Ford RE. Effects of drought stress and infection by maize dwarf mosaic virus on sweet corn. Plant Dis. 1990;74:147–51.

Ou X, Gan Y, Chen P, Qiu M, Jiang K, et al. Stomata prioritize their responses to multiple biotic and abiotic signal inputs. PLoS One. 2014;9(7), e101587.

Pandey P, Ramegowda V, Senthil-Kumar M. Shared and unique responses of plants to multiple individual stresses and stress combinations: physiological and molecular mechanisms. Front Plant Sci. 2015;6:723.

Prasch CM, Sonnewald U. Simultaneous application of heat, drought and virus to *Arabidopsis thaliana* plants reveals significant shifts in signaling networks. Plant Physiol. 2013;162(4):1849–66.

Prasch CM, Sonnewald U. Signaling events in plants: stress factors in combination change the picture. Environ Exp Bot. 2015;114:4–14. doi:10.1016/j.envexpbot.2014.06.020.

Ramegowda V, Senthil-Kumar M. The interactive effects of simultaneous biotic and abiotic stresses on plants: mechanistic understanding from drought and pathogen combination. J Plant Physiol. 2015;176:47–54.

Ramegowda V, Senthil-Kumar M, Ishiga Y, Kaundal A, Udayakumar M, Mysore KS. Drought stress acclimation imparts tolerance to *Sclerotinia sclerotiorum* and *Pseudomonas syringae* in *Nicotiana benthamiana*. Int J Mol Sci. 2013;14(5):9497–513.

Rampino P, Mita G, Fasano P, Maria G, Aprile A, Dalessandro G, De Bellis L, Perrotta C. Novel durum wheat genes up-regulated in response to a combination of heat and drought stress. Plant Physiol Biochem. 2012;56:72–8.

Rasmussen S, Barah P, Suarez-Rodriguez MC, Bressendorff S, Friis P, Costantino P, Bones AM, Nielsen HB, Mundy J. Transcriptome responses to combinations of stresses in Arabidopsis. Plant Physiol. 2013;161:1783–94.

Rivero RM, Mestre TC, Mittler R, Rubio F, Garcia-Sanchez F, Martinez V. The combined effect of salinity and heat reveals a specific physiological, biochemical and molecular response in tomato plants. Plant Cell Environ. 2013;37:1059–73.

Rizhsky L, Liang HJ, Mittler R. The combined effect of drought stress and heat shock on gene expression in tobacco. Plant Physiol. 2002;130:1143–51.

Rizhsky L, Liang H, Shuman J, Shulaev V, Davletova S, Mittler R. When defense pathways collide. The response of *Arabidopsis* to a combination of drought and heat stress. Plant Physiol. 2004;134:1683–96.

Sharma M, Pande S. Unravelling effects of temperature and soil moisture stress response on development of dry root rot [*Rhizoctonia bataticola* (Taub.)] butler in chickpea. Am J Plant Sci. 2013;4:584–9.

Sharma RC, Duveiller E, Ortiz-Ferrara G. Progress and challenge towards reducing wheat spot blotch threat in the Eastern Gangetic Plains of South Asia: is climate change already taking its toll? Field Crop Res. 2007;103:109–18.

Smit AL, Vamerali T. The influence of potato cyst nematodes (*Globodera pallida*) and drought on rooting dynamics of potato (*Solanum tuberosum* L.). Eur J Agron. 1998;9(2–3):137–46.

Smith AM. Stitt M (2007) Coordination of carbon supply and plant growth. Plant Cell Environ. 2007;30:1126–49.

Suzuki N, Rivero RM, Shulaev V, Blumwald E, Mittler R. Abiotic and biotic stress combinations. New Phytol. 2014;203(1):32–43.

Ton J, Flors V, Mauch-Mani B. The multifaceted role of ABA in disease resistance. Trends Plant Sci. 2009;14:10–317.

Vile D, Pervent M, Belluau M, Vasseur F, Bresson J, Muller B, et al. *Arabidopsis* growth under prolonged high temperature and water deficit: independent or interactive effects? Plant Cell Environ. 2012;35:702–18.

Wang Y, Bao Z, Zhu Y, Hua J. Analysis of temperature modulation of plant defense against biotrophic microbes. Mol Plant Microbe Interact. 2009;22:498–506.

Whenham RJ, Fraser RSS, Brown LP, Payne JA. *Tobacco-mosaic virus*-induced increase in absicisic acid concentration in tobacco leaves: intracellular location in light and dark-green areas, and relationship to symptom development. Planta. 1986;168:592–8.

Xu P, Chen F, Mannas JP, Feldman T, Sumner LW, Roossinck MJ. Virus infection improves drought tolerance. New Phytol. 2008;180:911–21.

Yasuda M, Ishikawa A, Jikumaru Y, Seki M, Umezawa T, Asami T. Antagonistic interaction between systemic acquired resistance and the abscisic acid-mediated abiotic stress response in *Arabidopsis*. Plant Cell. 2008;20(6):1678–92.

Chapter 19
Manipulation of Programmed Cell Death Pathways Enhances Osmotic Stress Tolerance in Plants: Physiological and Molecular Insights

Thi My Linh Hoang, Brett Williams, and Sagadevan G. Mundree

19.1 Introduction

Programmed cell death (PCD) is a physiological and genetically controlled process that is evolutionarily conserved across kingdoms. PCD allows multicellular organisms to eliminate excessive or damaged cells which arise during development and in response to abiotic and biotic stress (Williams and Dickman 2008; Fomicheva et al. 2012). Programmed cell death has been studied extensively in animals and the underlying mechanisms in plants are gradually being discovered.

The roles of PCD during the development of animals were thoroughly reviewed in Fuchs and Steller (2011), especially in regulation of structure sculpting and driving morphogenesis, deletion of unwanted or redundant transient functional structures, control of cell numbers and elimination of unwanted and potentially dangerous cells. In plants, PCD is involved in many stages of development from the embryo to reproduction and ageing such as embryogenesis, somatic embryogenesis (Giuliani et al. 2002; Suarez et al. 2004; Hill et al. 2013), sex determination in unisexual species (Dellaporta and Calderon-Urrea 1994; Beers 1997), seed development (Young and Gallie 2000) and senescence (Greenberg 1996; Yen and Yang 1998; Simeonova et al. 2000; Yoshida 2003). PCD also plays an important role in the elicitation of defence mechanisms. For example, the hypersensitive response, which occurs at the site of pathogen attack and involves programmed cell death of infected as well as uninfected adjacent bystander cells, is one of the strategies that plants employ to prevent pathogen invasion (Lam et al. 2001; Lam 2004).

Although programmed cell death plays important roles during development and in response to environmental stimuli, it may be beneficial or detrimental to the plant depending on the context (Williams and Dickman 2008). Being sessile, plants are

T.M.L. Hoang • B. Williams • S.G. Mundree (✉)
Centre for Tropical Crops and Biocommodities, Queensland University of Technology (QUT), PO Box 2434, Brisbane, QLD 4001, Australia
e-mail: sagadevan.mundree@qut.edu.au

M.A. Hossain et al. (eds.), *Drought Stress Tolerance in Plants, Vol 1*,
DOI 10.1007/978-3-319-28899-4_19

particularly vulnerable to aberrant environmental conditions including saline soils and water deficit. To mitigate osmotic stress, plants implement a range of strategies; however, if these mechanisms are unable to cope with the prolonged stress imposed the plant will implement selective PCD as a last ditch effort to survive; sacrifice a few cells for the greater good of the organism as a whole (Hara-Nishimura et al. 1991; Greenberg 1996). Paradoxically, studies have shown that inhibition of PCD during stress promotes survival (Dickman et al. 2001b; Awada et al. 2003; Shabala et al. 2007; Wang et al. 2009b; Li et al. 2010; Hoang et al. 2014, 2015). In the following sections we will provide an overview of programmed cell death and the physiological and molecular basis of the enhancement of tolerance to osmotic stress associated with drought and salinity through the prevention of programmed cell death in plants.

19.2 Programmed Cell Death: An Overview

19.2.1 PCD: A Physiological Mechanism for Normal Development

Regulation of homeostatic balance between cell division and cell death is fundamental for proper development and well-being of all multicellular organisms (Rudin and Thompson 1997). Genetically regulated mechanisms in multicellular organisms not only determine which cells live but also which cells die (Raff 1992; Chinnaiyan and Dixit 1996). To keep balance with the number of new cells arising from the body's stem cell populations, about ten billion cells die every day in adulthood. This normal homeostasis is regulated through apoptosis—one form of programmed cell death (Renehan et al. 2001). Apoptosis is extremely important during various developmental processes and normal physiology (Elmore 2007). The sculpturing of shape during developing limb to form foetal fingers and toes together with the resorption of the tadpole tail during metamorphosis into a frog are two well-known examples of the programmed cell death involvement in normal development (Zuzarte-Luís and Hurlé 2002). Evidence indicates that abnormal regulation of programmed cell death especially apoptosis is associated with a wide range of diseases. Insufficient apoptosis results in excessive cell accumulation causing autoimmunity or cancer; inappropriate cell death can lead to chronic degenerative diseases, heart failure, cerebral ischemia, Alzheimer disease, infertility and immunodeficiency (Kondo 1988; Leijon et al. 1994; Edwards 1998; Danial and Korsmeyer 2004; Rami et al. 2008; Lukiw and Bazan 2010; Whelan et al. 2010; King and Cidlowski 1998).

19.2.2 PCD: As a Host Defence Mechanism Against Biotic and Abiotic Stresses

In addition to development, PCD pathways are also used for adaptation to environmental stresses (Vaux et al. 1994; Mittler and Lam 1996; Vaux and Strasser 1996). The Hypersensitive response (HR) of plants to pathogen infection is one example.

Plants lack of an active immune system which can produce specialized cells, such as T cells in animal systems that can attack, disable and eliminate pathogen, they instead, induce programmed cell death as one of general defence strategies (Lam et al. 2001; Lam 2004). During interaction between biotrophic pathogens and host plants, programmed cell death in the form of HR helps plants to prevent infection as biotrophy by definition require living cells for growth and colonization. However, in some instances plants infected by necrotrophic pathogens, e.g. *Sclerotinia sclerotiorum*, cell death is disadvantageous for the plant as necrotrophic pathogens require dead or dying cells for nutrients. The role of PCD in plant–pathogen interaction, therefore, depends upon the context and in some circumstances the host is involved in the process as a passive participant (Williams and Dickman 2008).

Cell death in response to abiotic stress provides an advantage to plants in some circumstances but not in others. For example, programmed cell death during hypoxia-induced aerenchyma formation in root of maize enables the plants to survive and develop in wetlands where there is limited or no oxygen present (Drew et al. 2000). However, in response to most of other abiotic stresses such as drought, salinity, heat, cold, wounding, UV radiation, aluminium, acifluorfen, sulfentrazone, menadione and hydrogen peroxide, prevention of cell death brings more benefit to the plants than execution of cell death as evidenced in many studies (Dickman et al. 2001b; Lincoln et al. 2002; Qiao et al. 2002; Li and Dickman 2004b; Xu et al. 2004; Doukhanina et al. 2006; Shabala et al. 2007; Wang et al. 2009a, b; Kabbage et al. 2010; Li et al. 2010; Hoang et al. 2014, 2015).

19.2.3 PCD: A Conserved Mechanism

Programmed cell death particularly apoptosis, the physiological form of PCD, has been studied for more than 40 years and is known to occur in many species across all kingdoms. For example, human Bcl-2 can partially complement *Caenorhabditis elegans* Ced-9 mutants even though the two genes have limited sequence homology. The animal-derived anti-apoptotic genes *Ced-9* and *Bcl-2* confer tolerance to a wide range of biotic and abiotic stresses, upon overexpression in plants (Qiao et al. 2002; Chen and Dickman 2004; Shabala et al. 2007; Wang et al. 2009a; Paul et al. 2011).

Since the first evidence that a genetic programme existed for physiological (programmed) cell death came from studying development in the nematode *Caenorhabditis elegans* (Kerr et al. 1972; Horvitz et al. 1982; Ellis and Horvitz 1986; Vaux et al. 1988) the study of pathways and regulation of programmed cell death has been carried out on several model systems including *C. elegans*, the vinegar fly *Drosophila melanogaster* and the mouse. The conservation of the core apoptotic machinery has been found across vast evolutionary distances from worm to human; however, it is somewhat obscure in plants (Williams and Dickman 2008; Fuchs and Steller 2011).

As the core apoptotic machinery is conserved across kingdoms, details of a well-studied programmed cell death model would be helpful to establish an understanding of programmed cell death in plants. In the next section we will review the literature of mammalian PCD pathways.

19.3 Apoptosis: A Genetically Controlled Cell Death

Three types of programmed cell death have been categorized in mammals based on morphological criteria: apoptosis (type I), autophagy (type II) and necrosis (type III) (Kourtis and Tavernarakis 2009; Kroemer et al. 2009). Other forms of cell death in mammals related to inflammation response during pathogen invasion have also been observed. This includes pyroptosis (or caspase-1-dependent cell death) and necroptosis (or programmed necrosis) (see review by Bergsbaken et al. (2009) and Vandenabeele et al. (2010)). Amongst the aforementioned types, apoptosis has been the most studied and the best understood form of PCD in mammals. Apoptosis is a genetically controlled and highly orchestrated cell death. Cells undergoing apoptosis have distinct morphological changes including cell shrinkage, membrane blebbing, chromatin condensation, apoptotic body formation and fragmentation, minor modification of cytoplasmic organelles, and the apoptotic bodies were engulfed by resident phagocytes in vivo (Gilchrist 1998; Bredesen 2000; Collazo et al. 2006; Kroemer et al. 2009).

19.3.1 *Execution of Apoptosis*

The execution of apoptosis in mammals relies on the activation of caspases (cysteine aspartic acid specific proteases), a family of highly specific cysteine proteases that are ubiquitously expressed, as inactive precursors (zymogens) with little or no protease activities (Fuchs and Steller 2011). Caspases can be thought of as the central executioners of apoptotic pathways because they bring about most of the visible changes that characterize apoptotic cell death. For example, hallmarks of apoptosis such as DNA fragmentation and membrane blebbing are associated with caspase-3 activities (Hengartner 2000; Zimmermann et al. 2001). Genetic evidence also showed that caspases and their activators play central roles in apoptosis (Cecconi et al. 1998; Los et al. 1999; Zheng et al. 1999; Luthi and Martin 2007).

The mammalian caspase family can be divided into two subfamilies. The first one is involved in inflammation, where caspases act as pro-cytokine activators and include members of caspases-1, -4, -5, -11, -12, -13 and -14. The other subfamily is involved in apoptosis and includes caspase-2, -3, -6, -7, -8, -9 and -10. The apoptotic subfamily can be further categorized into two subgroups: initiator caspases caspase-2, -8, -9 and -10; and executioners or effector caspases caspase-3, -6 and -7 (Zimmermann et al. 2001; Shi 2002; Boatright and Salvesen 2003; Fomicheva et al. 2012).

Since unregulated caspase activity would be lethal for a cell, caspases are synthesized as single-chain zymogens and stored in the cytoplasm as relatively inactive precursors (pro-caspases). Pro-caspases must undergo an activation process during apoptosis to become active caspases (Srinivasula et al. 1998; Yang et al. 1998; Chen and Wang 2002; Boatright and Salvesen 2003; Shi 2004).

The activation of caspases during apoptosis has been reported to occur through three signalling pathways defined as the extrinsic, intrinsic and perforin/granzyme pathways (Elmore 2007). The extrinsic pathway is associated with a group of transmembrane proteins, "death receptors", which act as surface sensors for the presence of specific extracellular death signals from ligands of tumour necrosis factor (TNF) family (Fomicheva et al. 2012). Death receptors transmit apoptotic signals initiated by specific death ligands and can activate the caspase cascade within seconds of ligand binding (Vaux and Korsmeyer 1999). The extrinsic pathway of caspase activation is initiated by the ligation of the respective ligand (FasL) to the death receptor (Fas) to form microaggregates at the cell surface. This complex allows the adaptor molecule FADD (Fas-associated protein with death domain) to be recruited to it cytosolic tail by a multi-step mechanism. FADD recruits pro-caspase-8 or pro-caspase-10 by protein–protein interaction via homologous death effector domain (DED) to assemble a death-inducing signalling complex (DISC). During DISC assembly pro-caspase-8 or pro-caspase-10 is activated and released to cytoplasm where it cleaves and hence activates downstream caspase, typically caspase-3. The active caspase-3 cleaves several death substrates leading to the well-known apoptotic hallmarks including nuclear fragmentation, DNA fragmentation, membrane blebbing and other morphological and biochemical changes (Chinnaiyan et al. 1995; Algeciras-Schimnich et al. 2002; Boatright and Salvesen 2003; Yin et al. 2006; Portt et al. 2011; Fomicheva et al. 2012). The extrinsic pathway is responsible for elimination of unwanted cells during development, immune system education and immune system-mediated tumour removal (immune-surveillance) (Boatright and Salvesen 2003).

The intrinsic pathway involves the participation of mitochondrion as a central organelle; therefore, it is also termed as mitochondrial pathway. The mitochondrial pathway is induced by several stimuli such as UV radiation, DNA damage, voltage changes, oxidative stress [hydrogen peroxide (H_2O_2) or nitrogen oxide (NO)] or growth factor withdrawal (starvation), resulting in the dissipation of mitochondrial membrane potential and increased permeability. The permeabilization of the mitochondrial outer membrane leads to the release of apoptogenic molecules and proteins including cytochrome c, certain caspases, endonuclease G, Smac/Diablo and apoptosis inducing factor (AIF) from the inter-membrane space of mitochondrion to cytoplasm, resulting in both caspase-dependent and caspase-independent PCD (Brenner and Mak 2009; Paul 2009). The release of cytochrome c into cytosol and the presence of dATP are essential requirements for apoptosis mediated by mitochondria (Liu et al. 1996; Goldstein et al. 2000; Purring-Koch and McLendon 2000). Upon releasing, cytochrome c binds to Apaf-1 (apoptotic protease activating factor-1, the mammalian homolog of *C. elegans* Ced-4) in the presence of dATP to form an Apaf-1 complex (Zou et al. 1997; Hu et al. 1999) which then binds to pro-caspase-9 to assemble an oligo-protein complex termed "apoptosome" (Cain et al. 2000; Gupta 2001; Acehan et al. 2002; Gewies 2003). The apoptosome activates caspase-9 by dimerization (Purring-Koch and McLendon 2000; Pop et al. 2006). Active caspase-9 activates downstream caspase, typically caspase-3, resulting in apoptosis. The intrinsic pathway is used to eliminate cells in response to chemotherapeutic drugs, ionizing radiation, mitochondrial damage and certain developmental cues (Boatright and Salvesen 2003).

The perforin/granzyme pathway involves the cytotoxic T cells and secretion of the transmembrane pore-forming molecule perforin with a subsequent release of cytoplasmic granules which contains two most important components: serine protease granzyme A and B (Elmore 2007). Granzyme B can activate pro-caspase-10 through the cleavage of this protein at aspartate residues; it can also cleave factors like inhibitor of caspase activated DNAse (ICAD) (Sakahira et al. 1998). Granzyme B can also cleave and activate Bid causing a release of cytochrome c, thereby activating the intrinsic pathway of cell death (Russell and Ley 2002).

Although each pathway is capable of functioning independently, cross-talk between pathways is common. For example, three pathways cooperate to enhance apoptosis through a BH3-only protein member of Bcl-2 pro-apoptotic protein, Bid (Li et al. 1998; Barry and Bleackley 2002); and more importantly, these pathways converge, leading to the activation of the effector caspase-3 (Schimmer 2004; Williams and Dickman 2008).

19.3.2 Regulation of Apoptosis

Apoptosis can be regulated in a number of ways including regulators of the death receptors (extrinsic pathway), regulators of mitochondrial-driven PCD (intrinsic pathway) and direct regulator of caspases through Inhibitor of Apoptosis (IAP) proteins. The regulation of apoptosis mediated by death receptors occurs at multiple levels including regulation of expression of ligands and death receptors and regulation of intracellular regulatory molecules (Chen and Wang 2002). Meanwhile members of B cell lymphoma 2 (Bcl-2) protein family provides a critical role in regulation of mitochondrial-driven PCD pathway. They can either disrupt or maintain the integrity of mitochondrial membranes, thereby promote or prevent the release of apoptogenic proteins such as cytochrome c from inter-mitochondrion membrane space which can activate pro-caspase-9 through assembling of apoptosome leading to apoptosis (Zheng et al. 1998; Heiden et al. 1999; Chen and Wang 2002; Youle and Strasser 2008; Fuchs and Steller 2011; Martinou and Youle 2011). Bcl-2 family members are characterized by the presence of one or more conserved sequence motifs within α helical segments known as Bcl-2 homology (BH) domains designated BH1, BH2, BH3 and BH4. These BH domains are the only areas of sequence conservation between family members and strongly influence whether the family member is pro- or anti-apoptotic (Danial 2007; Williams and Dickman 2008). Many members of the Bcl-2 family have a conserved C-terminal transmembrane region (TM) that is responsible for their localization on the outer mitochondrial membrane, endoplasmic reticulum and nuclear envelope to the cytosolic aspect (Strasser et al. 2000; Soriano and Scorrano 2010). Bcl-2 family members can be divided into two groups: pro-apoptotic and anti-apoptotic depending upon their functions. At least four models of how Bcl-2 family members regulate apoptosis have been proposed [see review by Strasser et al. (2000)]. However exact mechanistic details of how Bcl-2 proteins regulate cell death remain unknown (García-Sáez 2012).

Although pro-caspases have a low protease activity, this activity is significant; and since pro-caspases are widely expressed in living cells, unregulated caspase activation would be lethal. Therefore cells must have an efficient mechanism to prevent unnecessary caspase activation. Inhibitor of apoptosis (IAP) protein is one of an important family of caspase inhibitors (Fuchs and Steller 2011). The first member of the IAP family was identified by Crook et al. (1993) from the baculovirus *Cyndia pomonella*. Since then several IAPs have been characterized (Birnbaum et al. 1994; Clem and Miller 1994; Hay et al. 1995; Rothe et al. 1995; Roy et al. 1995; Deveraux et al. 1997; Huang et al. 2000).

IAP family members are characterized by the presence of one to three baculoviral IAP repeat (BIR) domains, a region of approximately 70 amino acids. In some IAP members, BIR domains allow them to bind to and inhibit initiator and effector caspases as well as downstream proteases, thereby preventing apoptosis (Deveraux and Reed 1999; Vaux and Silke 2005). Unlike FLIP [FLICE (other name of caspase-8)-inhibitory protein] or Bcl-2 anti-apoptotic proteins which can only regulate death receptor or mitochondrial-driven PCD pathways, respectively, IAPs are unique in that they are capable of inhibiting both extrinsic and intrinsic pathways due to their inhibition of caspase cleavage at the initial phase of the cascade (Straszewski-Chavez et al. 2004).

The activity of IAP family members is regulated by IAP antagonists, a protein family whose members can bind to the BIR domain of IAP and inactivate the anti-apoptotic function. In Drosophila the anti-apoptotic activity of DIAP1 has been reported to be blocked by *reaper*, *hid* and *grim* encoded proteins (Goyal et al. 2000). In mammalian systems the three well-known IAP antagonists are Smac (second mitochondria-derived activator of caspases), Diablo (Direct IAP binding protein with low pI) and HtrA2/Omi identified by Du et al. (2000), Verhagen et al. (2000) and Suzuki et al.(2001), respectively.

19.4 Programmed Cell Death in Plants

19.4.1 Plant PCD During Development and with Abiotic and Biotic Stress

Most of the functions of PCD (apoptosis and autophagy) that were witnessed in other multicellular organisms such as in animals are also observed in plants. For example, the involvement of PCD in tissue remodelling has been reported in leaf shape remodelling of the lance plant (Gunawardena et al. 2004). PCD functions in deletion of temporary functional structures that are no longer required for the plant development such as suspensor and aleurone layer cells (Pennell and Lamb 1997; Bozhkov et al. 2005). The aleurone is the outer surrounding layer of endosperm, a store of nutrients materials, in mature seeds. The death of aleurone layer cells during seed germination in cereals is an example of the function of PCD in removing a no longer required structure during plant development. Nutrients required for the growth of the embryo during seed germination are initially obtained from the store in the embryo and subsequently from mobilization of the materials stored in the endosperm. The hydrolytic process of materials stored in endosperm required hydrolytic enzymes which are synthesized in aleurone cells (Kuo et al. 1996; Wang et al. 1996b; Fath et al. 2000). However, aleurone layer cells are not required for young plants and are therefore programmed to die after contributing their hydrolytic enzymes usually a few days after seed germination (Wang et al. 1996b; Bethke et al. 1999; Fath et al. 2000, 2002).

PCD also plays a key role in the specialization of cells including the development of xylem tracheary elements (Fukuda et al. 1998; Groover and Jones 1999) or cell death in root cap cells which protect the root meristem (Wang et al. 1996a). Additionally, PCD plays a role in the redistribution of nutrients, for example, cell death during senescence recycles nutrients from older to younger organs (Greenberg 1996; Yen and Yang 1998; Simeonova et al. 2000; Yoshida 2003). PCD occurs throughout the plant life cycle in many sites of the plants (Pennell and Lamb 1997).

In terms of defence, as mentioned in previous section, PCD is induced as general defence strategies in plants to compensate for the absence of immune system as well as the inability to move to escape environmental challenges during pathogens invasion. The decision to kill adjacent uninfected cells to create a "barrier of death" separating the pathogen from healthy tissues help plants minimize the detrimental effects of pathogens invasion (Dangl and Jones 2001; Lam et al. 2001; Lam 2004). Hypoxic conditions in maize triggered cell death in the cortex of the roots and stem to form aerenchyma which facilitates an efficient transportation of oxygen from aerial organs to waterlogged stem bases and roots is another example of PCD function to enable plants to cope with unfavourable environmental conditions (Pennell and Lamb 1997; Drew et al. 2000).

The functions of PCD in plants and animals appear to be conserved with some typical morphological features of PCD in animals such as cell shrinkage, DNA cleavage and DNA fragmentation were also observed during plant PCD, the question about the similarity of molecular mechanisms involved in PCD between the two kingdoms remain unanswered (Fomicheva et al. 2012). Plant cells display several unique features compared to their animal counterparts including the presence of chloroplast, vacuoles and totipotency. Additionally, unlike animal cells, plant cells are held together by rigid cell walls which prevent active phagocytosis; plants also lack "true" caspases. Despite intense searches, caspases, which are the most characteristic proteases and known to have essential functions in the initiation and execution of apoptosis in animal cells, have yet to be found in plants (Vartapetian et al. 2011; Domínguez and Cejudo 2012). However a number of caspase-like proteases in plants have been identified including metacaspases (Uren et al. 2000), vacuolar processing enzymes (VPE) (Hatsugai et al. 2004) and subtilisin-like proteases (saspases and phytapases) (Chichkova et al. 2004, 2010; Coffeen and Wolpert 2004). Although plant caspase-like proteases have been identified, their target proteins and the way in which they are activated, regulated and participate in plant PCD pathways awaits further investigation.

It has been suggested in the literature that plants do not exhibit "classical" apoptosis (van Doorn 2011). Van Doorn (2011) therefore proposed a classification of plants PCD in which two categories of PCD were described: vacuolar plant cell death and necrotic plant cell death. There are many cases of plant PCD however, not falling within either of the proposed categories. Classification of plant PCD therefore should base on other criteria such as molecular mechanisms and basic components of PCD apparatus rather than morphology alone (Fomicheva et al. 2012). Other authors, Reape et al.(2008), described three different modes of programmed cell death in plants including apoptotic-like PCD (AL-PCD), autophagy and necrosis. Reape et al. (2008) also proposed an apoptotic-like regulation of PCD in plants in which mitochondrial membrane permeabilization plays a central role via the forming of permeability transition pore (PTP), which is induced by the changes in phosphate and/or ATP level, build-up of Ca^{2+} and ROS production following cellular stress (Reape and McCabe 2010).

19.4.2 Plant PCD Regulators

Similar to the case of true caspases, attempts to identify plant homologues of mammalian core regulators of apoptosis using informatics tools at the primary sequence level such as BLAST or FASTA have failed. A search for functional similarity based on prediction from structural similarity has been conducted with the assumption that distantly related proteins may have limited overall (undetectable) sequence homology but key features such as helical structure, hydrophobicity, water accessible surfaces, electrostatic potential, fold and catalytic sites may be conserved; and functional predictions can be made independently of the primary sequence (Doukhanina et al. 2006; Kabbage and Dickman 2008). Using this approach, a family of *Bcl-2* associated gene product (BAG) proteins of *Arabidopsis* was identified by profile-sequence (PFAM) and profile-profile (FFAS) algorithms (Doukhanina et al. 2006). The BAG family has been identified in yeast and animals, and is believed to function through a complex interaction with signalling molecules and molecular chaperones; under stress conditions, the BAG proteins recruit molecular chaperones to target proteins and modulate their functions by altering protein conformation (Sondermann et al. 2001; Takayama and Reed 2001). The search of the *Arabidopsis thaliana* genome sequence resulted in recognition of seven homologues of the BAG proteins family with limited sequence but high structural similarity to their human counterparts and contained putative Hsp70 biding sites (Doukhanina et al. 2006). Of the seven homologues of BAG family in *Arabidopsis thaliana,* four are with domain organization similar to animal BAGs including AtBAG1-4 which are predicted to localize in cytosol, and three (AtBAG5-7) contain a calmodulin-binding motif near the BAG domain. This is a novel feature associated with plant BAG family and possibly reflecting differences between animal and plant PCD

(Kabbage and Dickman 2008). AtBAGs have been speculated to bind Hsp70 in a manner similar to their animal counterparts; this is at least the case of AtBAG4. AtBAG4 conferred tolerance to a wide range of abiotic stress in transgenic tobacco. AtBAG6 may have a role in basal resistance by limiting disease development in *Botrytis cinerea.* The functional differences between AtBAG4 and AtBAG6 lead to a hypothesis that the BAG family has developed specialized roles for cell regulation (Kabbage and Dickman 2008). Similarly to their mammalian counterparts, the proposed function of plant BAG proteins is to coordinate signals for cell growth and to induce cell survival or cell death pathways in response to stress (Doukhanina et al. 2006). *Arabidopsis* BAG family members are localized to a variety of subcellular organelles for a range of cellular functions including the important function in PCD pathways and cytoprotection (Williams et al. 2010).

Despite limited understanding of the molecular mechanisms driving programmed cell death in plants, there is no doubt that PCD occurs in plants during development and during the interaction between plants, the environment and pathogen challenge.

19.4.3 Plant PCD-Induced Factors

PCD has been reported to be triggered in many plant species during abiotic and biotic stress. For example, salinity stress-induced PCD has been reported in barley (Hatsugai et al. 2006), *Arabidopsis* (Huh et al. 2002), rice (Li et al. 2007; Liu et al. 2007; Jiang et al. 2008), tobacco (Doukhanina et al. 2006; Shabala et al. 2007) and tomato (Li et al. 2010); Drought-induced PCD in tobacco (Awada et al. 2003); fungi-induced PCD in tobacco (Dickman et al. 2001a). Reactive oxygen species (ROS) signals that originate from different organelles such as chloroplast and mitochondria can also trigger PCD (Foyer and Noctor 2005; Rhoads et al. 2006). In plants, ROS can play a dual role acting as both toxic compounds and secondary messengers in signal transduction pathways in a variety of scenarios (Miller et al. 2008, 2010). ROS levels were reported to increase in plants resulting in significant cellular damage during drought and salinity stress (Borsani et al. 2005; Zhu et al. 2007; Xu et al. 2010). Other factors such as UV radiation, DNA damage, voltage changes, oxidative stress [hydrogen peroxide (H_2O_2) or nitrogen oxide (NO)] or growth factor withdrawal (starvation) can also trigger cell death in plants (Lam 2004; Roos and Kaina 2006; Nawkar et al. 2013).

19.5 Physiological Basis of Anti-apoptotic Genes Enhance Tolerance to Osmotic Stress in Plants

Anti-apoptotic genes have been reported to enhance tolerance to a range of abiotic and biotic stresses including drought and salinity for more than a decade. However, the physiological basis of stress tolerance especially cell membrane integrity, ion homeostasis, photosynthesis efficiency and relative water content in plant expressing anti-apoptotic genes exposed to osmotic stress associated with salinity was reported recently (Hoang et al. 2014, 2015). The expression of anti-apoptotic gene in plants suppresses programmed cell death induced by stress, thereby promoting survival. The decision of whether a given cell should live or die is essential for the well-being of all multi-cellular organisms (Metazoan). Under several stimuli, this decision depends on the result of a battle between anti-apoptotic (pro-survival) and pro-apoptotic proteins or signals (Li and Dickman 2004a; Williams and Dickman 2008). The ratio of anti-apoptotic (pro-survival) versus pro-apoptotic (pro-death) proteins also regulates PCD sensitivity (Fulda et al. 2010). The master switch of the cell life/death decision during osmotic stress associated with salinity stress is the "balance of the pro-death and pro-survival signals" of the system. By exogenous expression of an anti-apoptotic (pro-survival) gene, researchers have pushed the plant to make the "life decision" at the onset of a given stress. Expression of pro-survival genes coincided with reduced pro-death signals such as ROS levels which in turn supported the maintenance of cell membrane integrity and ion homeostasis. This maintenance promoted sustained photosynthetic efficiency which in turn provided energy for growth. Well-maintained growth further dilutes the ion concentration in cells which helps maintain ion homeostasis leading to the increased membrane integrity, relative water content, net photosynthesis and finally growth and yield (Fig. 19.1).

19.5.1 Suppression of Stress-Induced Cell Death in Plants

Hallmark features of apoptotic-like cell death in plants have been observed during drought and salinity stress. Exogenous expression of a range of PCD related genes from different sources have shown evidence of cell death suppression, thereby enhancing tolerance to those stresses in plants (Li et al. 2010; Hoang et al. 2014, 2015). One of the established methods for detecting apoptotic hallmarks is the Terminal deoxynucleotidyl transferase dUTP Nick End Labelling (TUNEL) assay (Fig. 19.2). TUNEL assay is a broad use assay for detecting the nick end of DNA resulted from the DNA fragmentation during apoptosis. Nucleic acid in TUNEL positive cells are selectively stained and fluoresces green, indicating the presence of apoptotic-like bodies, whereas all nucleic acid is counter-stained with propidium iodide and fluoresces red.

19.5.2 Reactive Oxygen Species, Water Retention and Cell Membrane Integrity

Homeostasis of cellular ROS levels promotes maintenance of cellular membrane integrity. Studies have shown that ROS-induced cell death can result from oxidative processes such as membrane lipid peroxidation, protein oxidation, enzyme inhibition

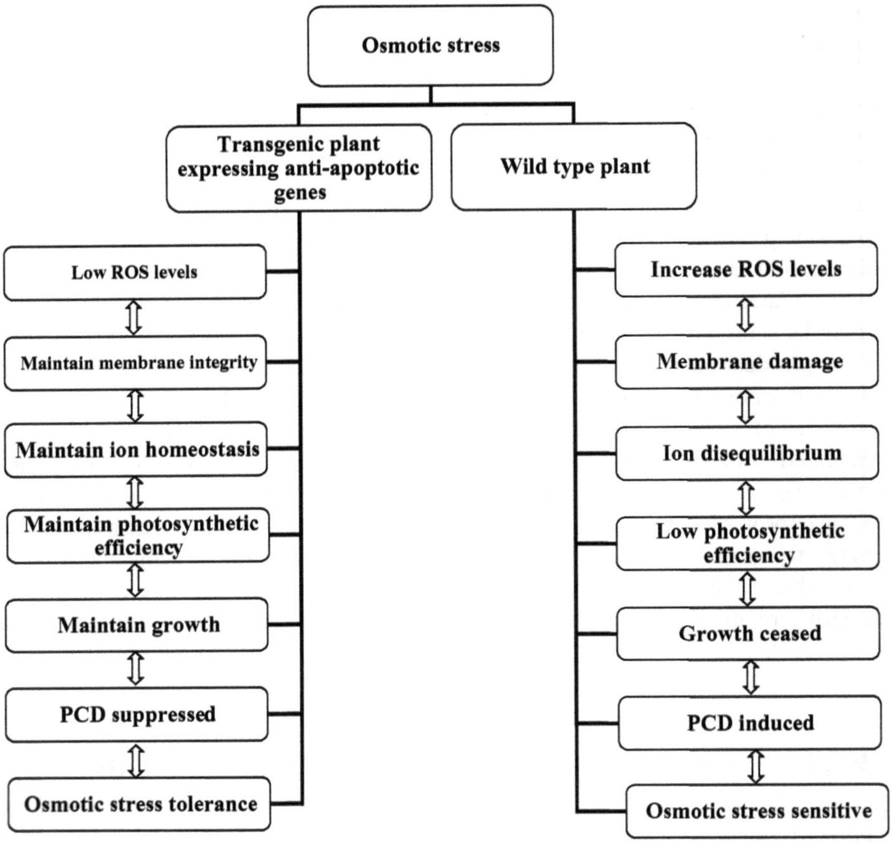

Fig. 19.1 Schematic proposing salinity-induced cell death switch for salinity stress tolerance in plant

and DNA, RNA damage (Mittler 2002). The cell membrane is the first site of signal perception as well as the primary defence against abiotic stresses including salinity and it is one of the most vulnerable targets for ROS due to the predominance of lipids (Ghosh et al. 2011). The maintenance of cell membrane integrity and stability under water stress is an important component of tolerance against water deficit (caused by drought and salinity stress) in plants.

ROS levels in plants expressing anti-apoptosis genes from different sources were maintained at significantly lower levels compared to those in wild-type plants (Li et al. 2010; Hoang et al. 2015). The low level of ROS causes less damage to membrane of plant under osmotic stress associated with salinity (Hoang et al. 2015). Among the four types of ROS ($O^{\cdot}_2$, $OH^{\cdot}$, $NO^{\cdot}$ and H_2O_2), H_2O_2 is a relative long-life molecule (1 ms) and it can diffuse some distance cross-linking cell wall structural proteins and more importantly H_2O_2 itself can stimulate further ROS accumulation and function as a local trigger of PCD (Levine et al. 1994; Dat et al. 2000). H_2O_2 can

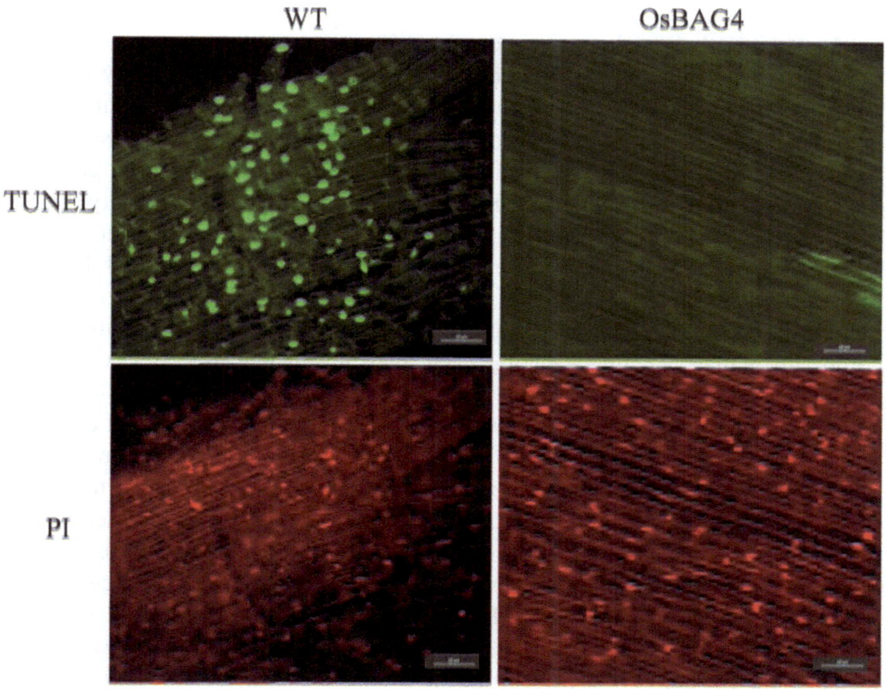

Fig. 19.2 Overexpression of anti-apoptotic gene OsBAG4 showed evidence of cell death suppression during osmotic stress associated with salinity stress (100 mM NaCl) in rice (*Oryza sativa* L.). *WT* Wild type, *PI* propidium iodide, *TUNEL* terminal deoxynucleotidyl transferase dUTP nick end labelling. Images were taken under a confocal microscope. Magnifications as indicated

originate from photosynthesis, photorespiration, respiration and many other cellular processes. It is a potent inhibitor of photosynthesis as it can inhibit CO_2 fixation up to 50 % (Foyer and Shigeoka 2011).

19.5.3 Ion Homeostasis

Transgenic plants expressing anti-apoptotic gene from different sources accumulate low Na^+, high K^+ and maintain low Na^+/K^+ ratios during salinity stress (Hoang et al. 2014, 2015). This is probably a result of the maintenance of cell membrane integrity in transgenic plants expressing the anti-apoptotic gene during salinity stress. High Na^+ levels are toxic to cells because Na^+ has similar physicochemical properties to K^+, it can compete with K^+ for major binding sites in key metabolic processes such as enzymatic reactions, ribosome functions and proteins biosynthesis in the cytoplasm leading to disturbance in metabolism (Shabala and Cuin 2008; Marschner 2011; Wang et al. 2013). In addition Na^+ can displace Ca^+ from plasma membranes inducing K^+ leaks out of the cytoplasm across the plasma membrane (Cramer et al. 1985). This results in a decrease in cytosolic K^+ concentration and effects the Na^+/K^+ ratio, hence leads to a disturbance of metabolism. Under typical physiological condition, the influx of Na^+ into plant cells is through the H^+-ATPase channel which is responsible for general transport of ions and nutrients through the plasma membrane; plants maintain a low cytosolic Na^+/K^+ ratio as it is necessary for providing favourable conditions for continued physiological and metabolic activity. During salinity stress increased extracellular Na^+ concentrations create a large electrochemical gradient that favours the passive transport of Na^+ into the cell through K^+ transporters result in high cytosolic Na^+ concentration (Blumwald 2000). To maintain low cytosolic Na^+ concentrations, plant cells need to extrude Na^+ of the cell or compartmentalize Na^+ into vacuoles. The main mechanism for Na^+ extrusion in plant cells is mediated by the plasma membrane H^+-ATPase (Sussman 1994). As the cell membrane in wild-type plants was damaged during salinity stress it could not use this strategy to pump Na^+ out of the cell; hence the Na^+ concentration was recorded at high levels in leaf cells of those plants. On the contrary, transgenic plants expressing anti-apoptotic genes can maintain cell membrane integrity and therefore could use the H^+-ATPase to extrude Na^+ thus maintaining a low concentration of Na^+ in cytoplasm. The high maintenance of low cytosolic Na^+ concentrations facilitates a high concentration of K^+ therefore ensuring a low Na^+/K^+ ratio that could offer an optimal cellular environment for enzymes function thus supporting metabolism. The high cytosolic K^+ concentration in plants expressing anti-apoptotic genes enables the plants to inhibit PCD. Cytosolic K^+ have been suggested to be related to the PCD process as it can affect caspases and caspases-like activities in animal and plants, respectively. Low cytosolic K^+ content in animal tissue correlates with high caspase activity; and the activation of K^+ efflux, the main cause of cytosolic K^+ content decrease, in plant cells leads to PCD hydrolase activation (Hughes and Cidlowski 1999; Shabala 2009; Demidchik et al. 2010).

19.5.4 Chlorophyll Content, Maximal Photochemical Efficiency, Photosynthetic Rate and Growth Under Osmotic and Ionic Stress Associated with Salinity

Photosynthesis is a fundamental physiological process that provides a source of energy for plants to grow and cope with environmental stresses. Under drought and salinity stress, sensitive cultivars usually display chlorophyll damage, less efficiency of PSII and low photosynthesis efficiency meanwhile the tolerant cultivars can maintain these parameters quite well (Dionisio-Sese and Tobita 2000; Ismail et al. 2007; Cha-Um et al. 2009). The expression of anti-apoptotic genes in transgenic tobacco led to the maintenance of chlorophyll content as well as the maximal photochemical efficiency of PSII (Shabala et al. 2007). Photosynthetic rate, growth and yield components were also maintained higher in rice expressing anti-apoptotic genes during salinity stress (Hoang et al. 2015). Transgenic rice expressing anti-apoptotic genes *AtBAG4, Hsp70, OsBAG4, p35* and *SfIAP* maintain growth rate (shoot growth, dry weight, number of tillers) and yield components (number of panicles per plant and number of spikelets per panicle) during salinity stress. This is probably also a result of the maintenance of high cytosolic K^+ in transgenic rice plants expressing *AtBAG4, Hsp70, OsBAG4, p35* and *SfIAP* (Hoang 2014). It is well known that salinity causes two types of stress on plants: (1) osmotic stress which affects plant growth immediately and is caused by excess salt outside the roots and (2) ionic stress which develops over time and is due to a combination of ion accumulation in the shoot and an inability to tolerate the ions that have accumulated (Munns 2002; Munns et al. 2006; Munns and Tester 2008). In low salt environments plant cells can take up water and nutrients from the soil solution to support higher osmotic pressures compared to that of soil solution. However, in high salt environments, the osmotic pressure of the soil exceeds that of plant cells (osmotic stress) and reduces the ability of plants cells to take up soil water and minerals (Kader and Lindberg 2010). In response to osmotic stress, shoot growth rate decreases immediately (Munns and Tester 2008). High cytosolic K^+ in transgenic plants expressing pro-survival genes helped the plants to adjust osmotic stress and maintain high growth rates because one of the important cellular roles of K^+ is to contribute to adjustment of osmotic pressure, hence maintain cell turgor (Maathuis and Amtmann 1999). The maintenance of growth rate leads to higher yield components in transgenic rice expressing anti-apoptotic genes in comparison to wild-type plants which had very low cytosolic K^+ under salinity stress condition (Hoang 2014). Another factor that causes reduced growth rates in high salt environments is inadequate photosynthesis due to limited carbon dioxide uptake as a consequence of stomatal closure (Zhu 2001). Transgenic rice plants expressing anti-apoptotic genes such as *AtBAG4, Hsp70, OsBAG4, p35* and *SfIAP* maintained high net photosynthesis which provided ample energy for their growth and development.

19.6 Molecular Basis of Anti-apoptotic Genes Enhance Tolerance to Abiotic Stress in Plant

19.6.1 Prevention of Protein Misfolding

Abiotic stresses usually cause protein dysfunction; therefore, one of the most important strategies for cells survival under stress is maintaining proteins in their conformations and preventing the aggregation of non-native proteins (Timperio et al. 2008). Members of the highly conserved heat shock protein family are chaperones that play a key role within the promotion of correct protein folding and proteostasis control (Hartl et al. 2011). A definitive feature of the BAG (Bcl2 anthanogen gene) family of proteins is their ability to bind and facilitate the function of HSPs (Doukhanina et al. 2006; Williams et al. 2010). The expression of anti-apoptotic genes such as *Hsp70*, *AtBAG4* and *OsBAG4* may assist in the folding of proteins and prevention of protein denaturation in high ROS environments, thus maintaining efficiency of cellular processes and mitigating the production of ROS and plant damage under stress condition (Hoang 2014). Portt et al. (2011) proposed a schematic representation of the processes involved in inducing stress-mediated cell death and its inhibition by key anti-apoptotic proteins. In that scheme stress induced an unknown substrate that mediated activation of BH3-only BcL-2 proteins, mitochondria (or other ROS producing system such as NADPH oxidase) and sphingomyelinase. This activation led to the action of at least three pro-apoptotic messengers including active Bax, increased ROS and sphingolipid ceramide, thereby causing cell death. Heat shock proteins (HSPs) were proposed to function as anti-apoptotic proteins by blocking that unknown substrate, thereby preventing the generation of active Bax, increasing ROS and sphingolipid ceramide.

19.6.2 Direct Sequestering ROS

Abiotic stresses cause enhanced generation of ROS in plants due to disruption of cellular homeostasis (Sharma et al. 2012). In plants, ROS are versatile molecules playing dual roles as both toxic compounds and signal transduction molecules that mediate responses to environmental stresses, pathogen infection, developmental stimuli and even PCD (Miller *et al.* 2008, 2010). The onset of PCD pathways is triggered by increased ROS levels, among other signals, that originate from a variety of organelles including the chloroplast and mitochondria (Foyer and Noctor 2005; Rhoads *et al.* 2006). During salinity stress, ROS levels have been reported to increase causing significant injury and eventual death (Borsani et al. 2005; Zhu et al. 2007; Chawla et al. 2013). If left unchecked, copious ROS production can denature proteins and damage cell membrane through the lipid peroxidation. Evidence showed that expression of the anti-apoptotic gene *p35* inhibited H_2O_2-induced PCD in insect cells by directly sequestering ROS. The antioxidant function of p35 has been attributed to the presence of metal-binding sites in the proteins that could enhance its antioxidant property and/or its three-dimensional structure contains some amino acids that confer electro-dynamically stable configuration conducive to ROS-trapping. The antioxidant role of p35 was also supported by the chemical radio-protectors formed by six cysteine residues in its sequence which can react with certain ROS in a constant rate (Sah et al. 1999).

19.6.3 Selective Degradation of Cellular Proteins

The ability to confer tolerance to salinity stress of the anti-apoptotic gene *SfIAP* was attributed to its E3 ubiquitin ligase activity (Kabbage et al. 2010). *SfIAP* has been transformed into tobacco and tomato and reported to confer tolerance to salinity, heat, fumonisin B1 and resistance to necrotrophic fungus *Alternaria alternate* (Kabbage et al. 2010; Li et al. 2010). All aspects of a plant's life are controlled by the regulated synthesis of new polypeptides and the precise degradation of pre-existing proteins (proteolysis). Ubiquitin/26S proteasome is arguably the dominant proteolytic system in plants (Smalle and Vierstra 2004). Proteolysis via Ubiquitin/26S proteasome pathway requires sequential enzyme activities including a ubiquitin activating enzyme (E1) which forms a thioester bond with the C terminus of ubiquitin in the presence of ATP and then transfers the activated ubiquitin to a ubiquitin conjugating enzyme (E2), E2 then transfers ubiquitin directly to a ubiquitin-ligating enzyme (E3) which transfer ubiquitin to the targeted substrate (Smalle and Vierstra 2004; Kabbage et al. 2010). The expression of *SfIAP* in tobacco resulted in accumulation of ubiquitinated proteins that assist the selective degradation of cellular damaged proteins generation during salt stress. In a presence of a proteasome inhibitor, no significant accumulation of ubiquitylated proteins in plant expressing the anti-apoptotic gene *SfIAP* was observed and *SfIAP* showed no protection during salinity stress in those plants (Kabbage et al. 2010)

In summary, expression of anti-apoptotic genes enhances tolerance to environmental stresses in plants through a number of approaches ranging from cellular to the whole plant. These approaches facilitate the plants to maintain normal physiological and cellular process, thereby successfully coping with stresses.

19.7 Implication and Future Directions

In the next 50 years there will be a massive challenge to sustain an ever-increasing global population. Between the years 1980 and 2000 global population boomed from 4.4 billion through to 6.1 billion, however, food production increased by 50 %. By 2050 this problem will be exacerbated with world population predicted to reach 9.6 billion. In order to sustain this increased population, global food within the next 50 years will have to match that which occurred in the last 10,000 years combined. This is a challenge because there is very little potential for future expansion of arable lands whilst climate predictions suggest that a larger portion of the globe will be subjected to erratic environmental conditions and abiotic stress (Eckardt 2009; FAO 2009, 2012; Cominelli et al. 2013). Two abiotic stress factors that significantly hinder world crop production are soil water deficit and salinization (Munns 2011). Researchers have shown that manipulation of PCD pathways can be applied to monocots and dicots for enhancing stress tolerance to a range of abiotic and biotic stresses. Currently we are able to produce crops with enhanced drought and salinity tolerance that survive in the glasshouse, however, once applied in the field the tolerance fails due to combined stresses. One approach with prospective application for the generation of the "next frontier of crop plants" with broad-spectrum tolerance is the exogenous expression of genes that suppress innate Programmed Cell Death (PCD) pathways.

Acknowledgement The authors thank Australian Government for financial support through Endeavour Postgraduate Awards.

References

Acehan D, Jiang X, Morgan DG, Heuser JE, Wang X, Akey CW. Three-dimensional structure of the apoptosome: implications for assembly, Procaspase-9 binding, and activation. Mol Cell. 2002;9:423–32. doi:http://dx.doi.org/10.1016/S1097-2765(02)00442-2.

Algeciras-Schimnich A, Shen L, Barnhart BC, Murmann AE, Burkhardt JK, Peter ME. Molecular ordering of the initial signaling events of CD95. Mol Cell Biol. 2002;22:207–20.

Awada T, Dunigan DD, Dickman MB. Animal anti-apoptotic genes ameliorate the loss of turgor in water-stressed transgenic tobacco. Can J Plant Sci. 2003;83:499–506.

Barry M, Bleackley RC. Cytotoxic T lymphocytes: all roads lead to death. Nat Rev Immunol. 2002;2:401–9.

Beers EP. Programmed cell death during plant growth and development. Cell Death Differ. 1997;4:649–61.

Bergsbaken T, Fink SL, Cookson BT. Pyroptosis: host cell death and inflammation. Nat Rev Micro. 2009;7:99–109.

Bethke P, Lonsdale J, Fath A, Jones R. Hormonally regulated programmed cell death in barley aleurone cells. Plant Cell. 1999;11:1033–46.

Birnbaum MJ, Clem RJ, Miller LK. An apoptosis-inhibiting gene from a nuclear polyhedrosis virus encoding a polypeptide with Cys/His sequence motifs. J Virol. 1994;68:2521–8.

Blumwald E. Sodium transport and salt tolerance in plants. Curr Opin Cell Biol. 2000;12:431–4. doi:http://dx.doi.org/10.1016/S0955-0674(00)00112-5.

Boatright KM, Salvesen GS. Mechanisms of caspase activation. Curr Opin Cell Biol. 2003;15:725–31. doi:http://dx.doi.org/10.1016/j.ceb.2003.10.009.

Borsani O, Zhu J, Verslues PE, Sunkar R, Zhu J-K. Endogenous siRNAs derived from a pair of natural cis-antisense transcripts regulate salt tolerance in Arabidopsis. Cell. 2005;123:1279–91. doi:http://dx.doi.org/10.1016/j.cell.2005.11.035.

Bozhkov P, Filonova L, Suarez M. Programmed cell death in plant embryogenesis. Curr Top Dev Biol. 2005;67:135–79.

Bredesen DE. Apoptosis: overview and signal transduction pathways. J Neurotrauma. 2000; 17:801–10.

Brenner D, Mak TW. Mitochondrial cell death effectors. Curr Opin Cell Biol. 2009;21:871–7. doi:http://dx.doi.org/10.1016/j.ceb.2009.09.004.

Cain K, Bratton SB, Langlais C, Walker G, Brown DG, Sun X-M, Cohen GM. Apaf-1 oligomerizes into biologically active ~700-kda and inactive ~1.4-mda apoptosome complexes. J Biol Chem. 2000;275:6067–70. doi:10.1074/jbc.275.9.6067.

Cecconi F, Alvarez-Bolado G, Meyer BI, Roth KA, Gruss P. Apaf1 (CED-4 Homolog) regulates programmed cell death in mammalian development. Cell. 1998;94:727–37. doi:http://dx.doi.org/10.1016/S0092-8674(00)81732-8.

Cha-Um S, Supaibulwattana K, Kirdmanee C. Comparative effects of salt stress and extreme pH stress combined on glycinebetaine accumulation, photosynthetic abilities and growth characters of two rice genotypes. Rice Sci. 2009;16:274–82. doi:10.1016/s1672-6308(08)60091-8.

Chawla S, Jain S, Jain V. Salinity induced oxidative stress and antioxidant system in salt-tolerant and salt-sensitive cultivars of rice (*Oryza sativa* L.). J Plant Biochem Biotechnol. 2013;22:27–34. doi:10.1007/s13562-012-0107-4.

Chen S, Dickman MB. Bcl-family members localize to tobacco chloroplasts and inhibit programmed cell death induced by chloroplast-targeted herbicides. J Exp Bot. 2004;55:2617–23.

Chen M, Wang J. Initiator caspases in apoptosis signaling pathways. Apoptosis. 2002;7:3139. doi:http://dx.doi.org/10.1023/A:1016167228059.

Chichkova NV, Kim SH, Titova ES, Kalkum M, Morozov VS, Rubtsov YP, Kalinina NO, Taliansky ME, Vartapetian AB. A plant caspase-like protease activated during the hypersensitive response. Plant Cell. 2004;16:157–71. doi:10.1105/tpc.017889.
Chichkova NV, Shaw J, Galiullina RA, Drury GE, Tuzhikov AI, Kim SH, Kalkum M, Hong TB, Gorshkova EN, Torrance L, Vartapetian AB, Taliansky M. Phytaspase, a relocalisable cell death promoting plant protease with caspase specificity. EMBO J. 2010;29:1149–61.
Chinnaiyan AM, Dixit VM. The cell-death machine. Curr Biol. 1996;6:555–62. doi:http://dx.doi.org/10.1016/S0960-9822(02)00541-9.
Chinnaiyan AM, O'Rourke K, Tewari M, Dixit VM. FADD, a novel death domain-containing protein, interacts with the death domain of Fas and initiates apoptosis. Cell. 1995;81:505–12. doi:http://dx.doi.org/10.1016/0092-8674(95)90071-3.
Clem RJ, Miller LK. Control of programmed cell death by the baculovirus genes p35 and IAP. Mol Cell Biol. 1994;14:5212–22.
Coffeen WC, Wolpert TJ. Purification and characterization of serine proteases that exhibit caspase-like activity and are associated with programmed cell death in *Avena sativa*. Plant Cell. 2004;16:857–73.
Collazo C, Chacón O, Borrás O. Programmed cell death in plants resembles apoptosis of animals. Biotecnol Apl. 2006;23:1–10.
Cominelli E, Conti L, Tonelli C, Galbiati M. Challenges and perspectives to improve crop drought and salinity tolerance. New Biotechnol. 2013;30:355–61. doi:http://dx.doi.org/10.1016/j.nbt.2012.11.001.
Cramer GR, Läuchli A, Polito VS. Displacement of Ca^{2+} by Na^{+} from the plasmalemma of root cells: a primary response to salt stress? Plant Physiol. 1985;79:207–11.
Crook NE, Clem RJ, Miller LK. An apoptosis-inhibiting baculovirus gene with a zinc finger-like motif. J Virol. 1993;67:2168–74.
Dangl JL, Jones JDG. Plant pathogens and integrated defence responses to infection. Nature. 2001;411:826–33. doi:http://dx.doi.org/10.1038/35081161.
Danial NN. BCL-2 family proteins: Critical checkpoints of apoptotic cell death. Clin Cancer Res. 2007;13:7254–63. doi:10.1158/1078-0432.ccr-07-1598.
Danial NN, Korsmeyer SJ. Cell death: critical control points. Cell. 2004;116:205–19. doi:http://dx.doi.org/10.1016/S0092-8674(04)00046-7.
Dat J, Vandenabeele S, Vranová E, Van Montagu M, Inzé D, Van Breusegem F. Dual action of the active oxygen species during plant stress responses. Cell Mol Life Sci. 2000;57:779–95. doi:10.1007/s000180050041.
Dellaporta SL, Calderon-Urrea A. The sex determination process in maize. Science. 1994; 266:1501–5. doi:10.2307/2885174.
Demidchik V, Cuin TA, Svistunenko D, Smith SJ, Miller AJ, Shabala S, Sokolik A, Yurin V. Arabidopsis root K+-efflux conductance activated by hydroxyl radicals: single-channel properties, genetic basis and involvement in stress-induced cell death. J Cell Sci. 2010;123:1468–79. doi:10.1242/jcs.064352.
Deveraux QL, Reed JC. IAP family proteins—suppressors of apoptosis. Genes Dev. 1999;13: 239–52.
Deveraux QL, Takahashi R, Salvesen GS, Reed JC. X-linked IAP is a direct inhibitor of cell-death proteases. Nature. 1997;388:300–4. doi:10.1038/40901.
Dickman MB, Park YK, Oltersdorf T, Li W, Clemente T, French R. Abrogation of disease development in plants expressing animal antiapoptotic genes. Proc Natl Acad Sci U S A. 2001a;98:6957–62. doi:10.1073/pnas.091108998.
Dickman MB, Park YK, Oltersdorf T, Li W, Clemente T, French R. Abrogation of disease development in plants expressing animal antiapoptotic genes. Proc Natl Acad Sci U S A. 2001b;98: 6957–62.
Dionisio-Sese ML, Tobita S. Effects of salinity on sodium content and photosynthetic responses of rice seedlings differing in salt tolerance. J Plant Physiol. 2000;157:54–8. doi:http://dx.doi.org/10.1016/S0176-1617(00)80135-2.

Domínguez F, Cejudo FJ. A comparison between nuclear dismantling during plant and animal programmed cell death. Plant Sci. 2012;197:114–21. doi:10.1016/j.plantsci.2012.09.009.
Doukhanina EV, Chen S, Van Der Zalm E, Godzik A, Reed J, Dickman MB. Identification and functional characterization of the BAG protein family in *Arabidopsis thaliana*. J Biol Chem. 2006;281:18793–801.
Drew MC, He C-J, Morgan PW. Programmed cell death and aerenchyma formation in roots. Trends Plant Sci. 2000;5:123–7. doi:http://dx.doi.org/10.1016/S1360-1385(00)01570-3.
Du C, Fang M, Li Y, Li L, Wang X. Smac, a mitochondrial protein that promotes cytochrome c-dependent caspase activation by eliminating IAP inhibition. Cell. 2000;102:33–42. doi:http://dx.doi.org/10.1016/S0092-8674(00)00008-8.
Eckardt NA. The future of science: food and water for life. Plant Cell. 2009;21:368–72.
Edwards MJ. Apoptosis, the heat shock response, hyperthermia, birth defects, disease and cancer. Where are the common links? Cell Stress Chaperones. 1998;3:213–20.
Ellis HM, Horvitz HR. Genetic control of programmed cell death in the nematode C. elegans. Cell. 1986;44:817–29. doi:http://dx.doi.org/10.1016/0092-8674(86)90004-8.
Elmore S. Apoptosis: a review of programmed cell death. Toxicol Pathol. 2007;35:495–516. doi:10.1080/01926230701320337.
FAO. Declaration of the world summit on food security. 16–18 November 2009, Rome. http://www.fao.org/wsfs/world-summit/wsfs-challenges/en/
FAO. World water day 2012 celebration. UN Conference Centre, Bangkok, 22 March 2012. http://www.fao.org/asiapacific/rap/home/meetings/list/detail/en/?meetings_id=637&year=2012
Fath A, Bethke P, Lonsdale J, Meza-Romero R, Jones R. Programmed cell death in cereal aleurone. Plant Mol Biol. 2000;44:255–66. doi:10.1023/A:1026584207243.
Fath A, Bethke P, Beligni V, Jones R. Active oxygen and cell death in cereal aleurone cells. J Exp Bot. 2002;53:1273–82. doi:10.1093/jexbot/53.372.1273.
Fomicheva AS, Tuzhikov AI, Beloshistov RE, Trusova SV, Galiullina RA, Mochalova LV, Chichkova NV, Vartapetian AB. Programmed cell death in plants. Biochemistry (Mosc). 2012;77:1452–64. doi:10.1134/S0006297912130044.
Foyer CH, Noctor G. Redox homeostasis and antioxidant signaling: a metabolic interface between stress perception and physiological responses. Plant Cell Online. 2005;17:1866–75. doi:10.1105/tpc.105.033589.
Foyer CH, Shigeoka S. Understanding oxidative stress and antioxidant functions to enhance photosynthesis. Plant Physiol. 2011;155:93–100. doi:10.1104/pp.110.166181.
Fuchs Y, Steller H. Programmed cell death in animal development and disease. Cell. 2011;147:742–58. doi:http://dx.doi.org/10.1016/j.cell.2011.10.033.
Fukuda H, Watanabe Y, Kuriyama H, Aoyagi S, Sugiyama M, Yamamoto R, Demura T, Minami A. Programming of cell death during xylogenesis. J Plant Res. 1998;111:253–6. doi:10.1007/BF02512179.
Fulda S, Gorman AM, Hori O, Samali A. Cellular stress responses: cell survival and cell death. Int J Cell Biol. 2010;2010:23.
García-Sáez AJ. The secrets of the Bcl-2 family. Cell Death Differ. 2012;19:1733–40.
Gewies A. Introduction to apoptosis. ApoReview. 2003;1–26.
Ghosh N, Adak MK, Ghosh PD, Gupta S, Sen Gupta DN, Mandal C. Differential responses of two rice varieties to salt stress. Plant Biotechnology Reports 2011;5:89–103.
Gilchrist DG. Programmed cell death in plant disease: the purpose and promise of cellular suicide. Annu Rev Phytopathol. 1998;36:393–414.
Giuliani C, Consonni G, Gavazzi G, Colombo M, Dolfini S. Programmed cell death during embryogenesis in maize. Ann Bot. 2002;90:287–92. doi:10.1093/aob/mcf173.
Goldstein JC, Waterhouse NJ, Juin P, Evan GI, Green DR. The coordinate release of cytochrome c during apoptosis is rapid, complete and kinetically invariant. Nat Cell Biol. 2000;2:156–62.
Goyal L, Mccall K, Agapite J, Hartwieg E, Steller H. Induction of apoptosis by Drosophila reaper, hid and grim through inhibition of IAP function. EMBO J. 2000;19:589–97.

Greenberg JT. Programmed cell death: a way of life for plants. Proc Natl Acad Sci U S A. 1996;93:12094–7.

Groover A, Jones AM. Tracheary element differentiation uses a novel mechanism coordinating programmed cell death and secondary cell wall synthesis. Plant Physiol. 1999;119:375–84. doi:10.2307/4278635.

Gunawardena AH, Greenwood JS, Dengler NG. Programmed cell death remodels lace plant leaf shape during development. Plant Cell. 2004;16:60–73.

Gupta S. Molecular steps of death receptor and mitochondrial pathways of apoptosis. Life Sci. 2001;69:2957–64. doi:http://dx.doi.org/10.1016/S0024-3205(01)01404-7.

Hara-Nishimura I, Inoue K, Nishimura M. A unique vacuolar processing enzyme responsible for conversion of several proprotein precursors into the mature forms. FEBS Lett. 1991;294:89–93. doi:http://dx.doi.org/10.1016/0014-5793(91)81349-D.

Hartl FU, Bracher A, Hayer-Hartl M. Molecular chaperones in protein folding and proteostasis. Nature. 2011;475:324–32.

Hatsugai N, Kuroyanagi M, Yamada K, Meshi T, et al. A plant vacuolar protease, VPE, mediates virus-induced hypersensitive cell death. Science. 2004;305:855–8.

Hatsugai N, Kuroyanagi M, Nishimura M, Hara-Nishimura I. A cellular suicide strategy of plants: vacuole-mediated cell death. Apoptosis. 2006;11:905–11. doi:http://dx.doi.org/10.1007/s10495-006-6601-1.

Hay BA, Wassarman DA, Rubin GM. Drosophila homologs of baculovirus inhibitor of apoptosis proteins function to block cell death. Cell. 1995;83:1253–62. doi:http://dx.doi.org/10.1016/0092-8674(95)90150-7.

Heiden MGV, Chandel NS, Schumacker PT, Thompson CB. Bcl-xL prevents cell death following growth factor withdrawal by facilitating mitochondrial ATP/ADP exchange. Mol Cell. 1999;3:159–67. doi:http://dx.doi.org/10.1016/S1097-2765(00)80307-X.

Hengartner MO. The biochemistry of apoptosis. Nature. 2000;407:770–6. doi:http://dx.doi.org/10.1038/35037710.

Hill RD, Huang S, Stasolla C. Hemoglobins, programmed cell death and somatic embryogenesis. Plant Sci. 2013;211:35–41. doi:http://dx.doi.org/10.1016/j.plantsci.2013.06.010.

Hoang TML. Engineering salinity tolerance in rice by manipulation of exogenous expression of cell death regulators. Ph.D. Thesis. Brisbane: Queensland University of Technology; 2014.

Hoang TML, Williams B, Khanna H, Dale J, Mundree SG. Physiological basis of salt stress tolerance in rice expressing the antiapoptotic gene SfIAP. Funct Plant Biol. 2014;41:1168–77. doi:http://dx.doi.org/10.1071/FP13308.

Hoang TML, Moghaddam L, Williams B, Khanna H, Dale J, Mundree SG. Development of salinity tolerance in rice by constitutive-overexpression of genes involved in the regulation of programmed cell death. Front Plant Sci. 2015;6:175. doi:10.3389/fpls.2015.00175.

Horvitz HR, Ellis HM, Sternberg PW. Programmed cell death in nematode development. Neurosci Comment. 1982;1:56–65.

Hu Y, Benedict MA, Ding L, Núñez G. Role of cytochrome c and dATP/ATP hydrolysis in Apaf-1 mediated caspase-9 activation and apoptosis. EMBO J. 1999;18:3586–95.

Huang Q, Deveraux QL, Maeda S, Salvesen GS, Stennicke HR, Hammock BD, Reed JC. Evolutionary conservation of apoptosis mechanisms: lepidopteran and baculoviral inhibitor of apoptosis proteins are inhibitors of mammalian caspase-9. Proc Natl Acad Sci U S A. 2000;97:1427–32.

Hughes FM Jr., Cidlowski JA. Potassium is a critical regulator of apoptotic enzymes in vitro and in vivo. Adv Enzyme Regul. 1999;39:157–71. doi:http://dx.doi.org/10.1016/S0065-2571(98)00010-7.

Huh G-H, Damsz B, Matsumoto TK, Reddy MP, Rus AM, Ibeas JI, Narasimhan ML, Bressan RA, Hasegawa PM. Salt causes ion disequilibrium-induced programmed cell death in yeast and plants. Plant J. 2002;29:649–59. doi:10.1046/j.0960-7412.2001.01247.x.

Ismail A, Heuer S, Thomson M, Wissuwa M. Genetic and genomic approaches to develop rice germplasm for problem soils. Plant Mol Biol. 2007;65:547–70. doi:10.1007/s11103-007-9215-2.

Jiang A-L, Cheng Y, Li J, Zhang W. A zinc-dependent nuclear endonuclease is responsible for DNA laddering during salt-induced programmed cell death in root tip cells of rice. J Plant Physiol. 2008;165:1134–41. doi:10.1016/j.jplph.2007.12.008.

Kabbage M, Dickman MB. The BAG proteins: a ubiquitous family of chaperone regulators. Cell Mol Life Sci. 2008;65:1390–402.

Kabbage M, Li W, Chen S, Dickman MB. The E3 ubiquitin ligase activity of an insect anti-apoptotic gene (SfIAP) is required for plant stress tolerance. Physiol Mol Plant Pathol. 2010;74:351–62. doi:http://dx.doi.org/10.1016/j.pmpp.2010.06.002.

Kader MA, Lindberg S. Cytosolic calcium and pH signaling in plants under salinity stress. Plant Signal Behav. 2010;5:233–8.

Kerr JF, Wyllie AH, Currie AR. Apoptosis: a basic biological phenomenon with wide-ranging implications in tissue kinetics. Br J Cancer. 1972;26:239–57.

King KL, Cidlowski JA. Cell cycle regulation and apoptosis. Annu Rev Physiol. 1998;60: 601–17.

Kondo S. Altruistic cell suicide in relation to radiation hormesis. Int J Radiat Biol. 1988;53:95–102. doi:10.1080/09553008814550461.

Kourtis N, Tavernarakis N. Autophagy and cell death in model organisms. Cell Death Differ. 2009;16:21.

Kroemer G, Galluzzi L, Vandenabeele P, Abrams J, Alnemri ES, Baehrecke EH, Blagosklonny MV, El-Deiry WS, Golstein P, Green DR, Hengartner M, Knight RA, Kumar S, Lipton SA, Malorni W, Nuñez G, Peter ME, Tschopp J, Yuan J, Piacentini M, Zhivotovsky B, Melino G. Classification of cell death: recommendations of the Nomenclature Committee on Cell Death 2009. Cell Death Differ. 2009;16:3–11. doi:http://dx.doi.org/10.1038/cdd.2008.150.

Kuo A, Cappelluti S, Cervantes-Cervantes M, Rodriguez M, Bush DS. Okadaic acid, a protein phosphatase inhibitor, blocks calcium changes, gene expression, and cell death induced by gibberellin in wheat aleurone cells. Plant Cell. 1996;8:259–69.

Lam E. Controlled cell death, plant survival and development. Nat Rev Mol Cell Biol. 2004; 5:305–15.

Lam E, Kato N, Lawton M. Programmed cell death, mitochondria and the plant hypersensitive response. Nature. 2001;411:848–53.

Leijon K, Hammarström B, Holmberg D. Non-obese diabetic (NOD) mice display enhanced immune responses and prolonged survival of lymphoid cells. Int Immunol. 1994;6:339–45. doi:10.1093/intimm/6.2.339.

Levine A, Tenhaken R, Dixon R, Lamb C. H_2O_2 from the oxidative burst orchestrates the plant hypersensitive disease resistance response. Cell. 1994;79:583–93. doi:http://dx.doi.org/10.1016/0092-8674(94)90544-4.

Li W, Dickman M. Abiotic stress induces apoptotic-like features in tobacco that is inhibited by expression of human Bcl-2. Biotechnol Lett. 2004a;26:87–95. doi:10.1023/B:BILE.0000012896.76432.ba.

Li W, Dickman MB. Abiotic stress induces apoptotic-like features in tobacco that is inhibited by expression of human Bcl-2. Biotechnol Lett. 2004b;26:87–95.

Li H, Zhu H, Xu C-J, Yuan J. Cleavage of BID by Caspase 8 mediates the mitochondrial damage in the Fas pathway of apoptosis. Cell. 1998;94:491–501. doi:http://dx.doi.org/10.1016/S0092-8674(00)81590-1.

Li J-Y, Jiang A-L, Zhang W. Salt stress-induced programmed cell death in rice root tip cells. J Integr Plant Biol. 2007;49:481–6. doi:10.1111/j.1744-7909.2007.00445.x.

Li W, Kabbage M, Dickman MB. Transgenic expression of an insect inhibitor of apoptosis gene, SfIAP, confers abiotic and biotic stress tolerance and delays tomato fruit ripening. Physiol Mol Plant Pathol. 2010;74:363–75. doi:10.1016/j.pmpp.2010.06.001.

Lincoln JE, Craig R, Overduin B, Smith K, Bostock R, Gilchrist DG. Expression of the antiapoptotic baculovirus p35 gene in tomato blocks programmed cell death and provides broad-spectrum resistance to disease. Proc Natl Acad Sci U S A. 2002;99:15217–21.

Liu X, Kim CN, Yang J, Jemmerson R, Wang X. Induction of apoptotic program in cell-free extracts: requirement for dATP and cytochrome c. Cell. 1996;86:147–57. doi:http://dx.doi.org/10.1016/S0092-8674(00)80085-9.

Liu S-H, Fu B-Y, Xu H-X, Zhu L-H, Zhai H-Q, Li Z-K. Cell death in response to osmotic and salt stresses in two rice (*Oryza sativa* L.) ecotypes. Plant Sci. 2007;172:897–902. doi:10.1016/j.plantsci.2006.12.017.

Los M, Wesselborg S, Schulze-Osthoff K. The role of caspases in development, immunity, and apoptotic signal transduction: lessons from knockout mice. Immunity. 1999;10:629–39. doi:http://dx.doi.org/10.1016/S1074-7613(00)80062-X.

Lukiw WJ, Bazan NG. Inflammatory, apoptotic, and survival gene signaling in Alzheimer's disease. A review on the bioactivity of Neuroprotectin D1 and apoptosis. Mol Neurobiol. 2010;42:10–6.

Luthi A, Martin S. The CASBAH: a searchable database of caspase substrates. Cell Death Differ. 2007;14:641–50.

Maathuis FJM, Amtmann A. K^+ nutrition and Na^+ toxicity: the basis of cellular K^+ Na^+ ratios. Ann Bot. 1999;84:123–33. doi:http://dx.doi.org/10.1006/anbo.1999.0912.

Marschner P. Marschner's mineral nutrition of higher plants. Amsterdam: Elsevier Science; 2011.

Martinou J-C, Youle RJ. Mitochondria in apoptosis: Bcl-2 family members and mitochondrial dynamics. Dev Cell. 2011;21:92–101.

Miller G, Shulaev V, Mittler R. Reactive oxygen signaling and abiotic stress. Physiol Plant. 2008;133:481–9. doi:10.1111/j.1399-3054.2008.01090.x.

Miller G, Suzuki N, Ciftci-Yilmaz S, Mittler R. Reactive oxygen species homeostasis and signalling during drought and salinity stresses. Plant Cell Environ. 2010;33:453–67. doi:10.1111/j.1365-3040.2009.02041.x.

Mittler R. Oxidative stress, antioxidants and stress tolerance. Trends Plant Sci. 2002;7:405–10.

Mittler R, Lam E. Sacrifice in the face of foes: pathogen-induced programmed cell death in plants. Trends Microbiol. 1996;4:10–5. doi:http://dx.doi.org/10.1016/0966-842X(96)81499-5.

Munns R. Comparative physiology of salt and water stress. Plant Cell Environ. 2002;25:239–50.

Munns R. Plant adaptations to salt and water stress: differences and commonalities. In: Ismail T, editor. Advances in botanical research. Waltham: Academic Press; 2011. p. 1–32.

Munns R, Tester M. Mechanisms of salinity tolerance. Annu Rev Plant Biol. 2008;59:651–81.

Munns R, James RA, Läuchli A. Approaches to increasing the salt tolerance of wheat and other cereals. J Exp Bot. 2006;57:1025–43.

Nawkar GM, Maibam P, Park JH, Sahi VP, Lee SY, Kang CH. UV-induced cell death in plants. Int J Mol Sci. 2013;14:1608–28. doi:http://dx.doi.org/10.3390/ijms14011608

Paul J-Y. The manipulation of apoptosis-related genes to generate resistance to fusarium wilt and water stress in banana. Ph.D. Thesis. Brisbane: Queensland University of Technology; 2009.

Paul J-Y, Becker DK, Dickman MB, Harding RM, Khanna HK, Dale JL. Apoptosis-related genes confer resistance to Fusarium wilt in transgenic 'Lady Finger' bananas. Plant Biotechnol J. 2011;9:1141–8. doi:10.1111/j.1467-7652.2011.00639.x.

Pennell RI, Lamb C. Programmed cell death in plants. Plant Cell. 1997;9:1157–68.

Pop C, Timmer J, Sperandio S, Salvesen GS. The apoptosome activates Caspase-9 by dimerization. Mol Cell. 2006;22:269–75. doi:http://dx.doi.org/10.1016/j.molcel.2006.03.009.

Portt L, Norman G, Clapp C, Greenwood M, Greenwood MT. Anti-apoptosis and cell survival: a review. Biochim Biophys Acta. 2011;1813, 238-259. doi:http://dx.doi.org/10.1016/j.bbamcr.2010.10.010.

Purring-Koch C, Mclendon G. Cytochrome c binding to Apaf-1: the effects of dATP and ionic strength. Proc Natl Acad Sci. 2000;97:11928–31. doi:10.1073/pnas.220416197.

Qiao J, Mitsuhara I, Yazaki Y, Sakano K, Gotoh Y, Miura M, Ohashi Y. Enhanced resistance to salt, cold and wound stresses by overproduction of animal cell death suppressors Bcl-xl and Ced-9 in tobacco cells—their possible contribution through improved function of organella. Plant Cell Physiol. 2002;43:992–1005.

Raff MC. Social controls on cell survival and cell death. Nature. 1992;356:397–400.

Rami A, Bechmann I, Stehle JH. Exploiting endogenous anti-apoptotic proteins for novel therapeutic strategies in cerebral ischemia. Prog Neurobiol. 2008;85:273–96. doi:http://dx.doi.org/10.1016/j.pneurobio.2008.04.003.

Reape T, Mccabe P. Apoptotic-like regulation of programmed cell death in plants. Apoptosis. 2010;15:249–56. doi:10.1007/s10495-009-0447-2.

Reape TJ, Molony EM, Mccabe PF. Programmed cell death in plants: distinguishing between different modes. J Exp Bot. 2008;59:435–44. doi:10.1093/jxb/erm258.

Renehan AG, Booth C, Potten CS. What is apoptosis, and why is it important? Br Med J. 2001;322:1536–8.

Rhoads DM, Umbach AL, Subbaiah CC, Siedow JN. Mitochondrial reactive oxygen species. Contribution to oxidative stress and interorganellar signaling. Plant Physiol. 2006;141:357–66. doi:10.1104/pp.106.079129.

Roos WP, Kaina B. DNA damage-induced cell death by apoptosis. Trends Mol Med. 2006;12:440–50. doi:http://dx.doi.org/10.1016/j.molmed.2006.07.007.

Rothe M, Pan M-G, Henzel WJ, Ayres TM, Goeddel DV. The TNFR2-TRAF signaling complex contains two novel proteins related to baculoviral inhibitor of apoptosis proteins. Cell. 1995;83:1243–52.

Roy N, Mahadevan MS, Mclean M, Shutter G, Yaraghi Z, Farahani R, Baird S, Besner-Johnston A, Lefebvre C, Kang X, Salih M, Aubry H, Tamai K, Guan X, Ioannou P, Crawford TO, De Jong PJ, Surh L, Ikeda J-E, Korneluk RG, Mackenzie A. The gene for neuronal apoptosis inhibitory protein is partially deleted in individuals with spinal muscular atrophy. Cell. 1995;80:167–78. doi:http://dx.doi.org/10.1016/0092-8674(95)90461-1.

Rudin CM, Thompson CB. Apoptosis and disease: regulation and clinical relevance of programmed cell death. Annu Rev Med. 1997;48:267–81. doi:10.1146/annurev.med.48.1.267.

Russell JH, Ley TJ. Lymphocyte-mediated cytotoxicity. Annu Rev Immunol. 2002;20:323–70. doi:http://search.proquest.com/docview/201638982?accountid=13380.

Sah NK, Taneja TK, Pathak N, Begum R, Athar M, Hasnain SE. The baculovirus antiapoptotic p35 gene also functions via an oxidant-dependent pathway. Proc Natl Acad Sci U S A. 1999; 96:4838–43.

Sakahira H, Enari M, Nagata S. Cleavage of CAD inhibitor in CAD activation and DNA degradation during apoptosis. Nature. 1998;391:96–9. doi:http://dx.doi.org/10.1038/34214.

Schimmer AD. Inhibitor of apoptosis proteins: translating basic knowledge into clinical practice. Cancer Res. 2004;64:7183–90. doi:10.1158/0008-5472.can-04-1918.

Shabala S. Salinity and programmed cell death: unravelling mechanisms for ion specific signalling. J Exp Bot. 2009;60:709–12. doi:10.1093/jxb/erp013.

Shabala S, Cuin TA. Potassium transport and plant salt tolerance. Physiol Plant. 2008;133:651–69. doi:10.1111/j.1399-3054.2007.01008.x.

Shabala S, Cuin TA, Prismall L, Nemchinov LG. Expression of animal CED-9 anti-apoptotic gene in tobacco modifies plasma membrane ion fluxes in response to salinity and oxidative stress. Planta. 2007;227:189–97. doi:10.1007/s00425-007-0606-z.

Sharma P, Jha AB, Dubey RS, Pessarakli M. Reactive oxygen species, oxidative damage, and antioxidative defense mechanism in plants under stressful conditions. J Bot. 2012;2012:26. doi:10.1155/2012/217037.

Shi Y. Caspase activation: revisiting the induced proximity model. Cell. 2004;117:855–8. doi:http://dx.doi.org/10.1016/j.cell.2004.06.007.

Shi Y. Mechanisms of caspase activation and inhibition during apoptosis. Mol Cell. 2002;9:459–70. doi:http://dx.doi.org/10.1016/S1097-2765(02)00482-3.

Simeonova E, Sikora A, Charzyńska M, Mostowska A. Aspects of programmed cell death during leaf senescence of mono- and dicotyledonous plants. Protoplasma. 2000;214:93–101. doi:10.1007/BF02524266.

Smalle J, Vierstra RD. The ubiquitin 26s proteasome proteolytic pathway. Annu Rev Plant Biol. 2004;55:555–90.

Sondermann H, Scheufler C, Schneider C, Hohfeld J, Hartl FU, Moarefi I. Structure of a Bag/Hsc70 complex: convergent functional evolution of Hsp70 nucleotide exchange factors. Science. 2001;291:1553–7.

Soriano ME, Scorrano L. The interplay between BCL-2 family proteins and mitochondrial morphology in the regulation of apoptosis. In: Hetz C, editor. BCL-2 protein family. New York: Springer; 2010. p. 97–114.

Srinivasula SM, Ahmad M, Fernandes-Alnemri T, Alnemri ES. Autoactivation of Procaspase-9 by Apaf-1-mediated oligomerization. Mol Cell. 1998;1:949–57. doi:http://dx.doi.org/10.1016/S1097-2765(00)80095-7.

Strasser A, O'Connor L, Dixit VM. Apoptosis signaling. Annu Rev Biochem. 2000;69:217–45.

Straszewski-Chavez SL, Abrahams VM, Funai EF, Mor G. X-linked inhibitor of apoptosis (XIAP) confers human trophoblast cell resistance to Fas-mediated apoptosis. Mol Hum Reprod. 2004;10:33–41. doi:10.1093/molehr/gah001.

Suarez MF, Filonova LH, Smertenko A, Savenkov EI, Clapham DH, Von Arnold S, Zhivotovsky B, Bozhkov PV. Metacaspase-dependent programmed cell death is essential for plant embryogenesis. Curr Biol. 2004;14:R339–40. doi:http://dx.doi.org/10.1016/j.cub.2004.04.019.

Sussman M. Molecular analysis of protein in the plant plasma membrane. Annu Rev Plant Physiol Plant Mol Biol. 1994;45:211–34.

Suzuki Y, Imai Y, Nakayama H, Takahashi K, Takio K, Takahashi R. A serine protease, HtrA2, is released from the mitochondria and interacts with XIAP, inducing cell death. Mol Cell. 2001;8:613–21. doi:http://dx.doi.org/10.1016/S1097-2765(01)00341-0.

Takayama S, Reed JC. Molecular chaperone targeting and regulation by BAG family proteins. Nat Cell Biol. 2001;3:E237–41.

Timperio AM, Egidi MG, Zolla L. Proteomics applied on plant abiotic stresses: Role of heat shock proteins (HSP). J Proteom. 2008;71:391–411. doi:http://dx.doi.org/10.1016/j.jprot.2008.07.005.

Uren AG, O'Rourke K, Aravind L, Pisabarro MT, Seshagiri S, Koonin EV, Dixit VM. Identification of paracaspases and metacaspases: two ancient families of caspase-like proteins, one of which plays a key role in MALT lymphoma. Mol Cell. 2000;6:961–7.

Van Doorn WG. Classes of programmed cell death in plants, compared to those in animals. J Exp Bot. 2011;62:4749–61. doi:10.1093/jxb/err196.

Vandenabeele P, Galluzzi L, Vanden Berghe T, Kroemer G. Molecular mechanisms of necroptosis: an ordered cellular explosion. Nat Rev Mol Cell Biol. 2010;11:700–14. doi:http://dx.doi.org/10.1038/nrm2970.

Vartapetian AB, Tuzhikov AI, Chichkova NV, Taliansky M, Wolpert TJ. A plant alternative to animal caspases: subtilisin-like proteases. Cell Death Differ. 2011;18:1289–97.

Vaux DL, Korsmeyer SJ. Cell death in development. Cell. 1999;96:245–54. doi:http://dx.doi.org/10.1016/S0092-8674(00)80564-4.

Vaux DL, Silke J. IAPs, RINGs and ubiquitylation. Nat Rev Mol Cell Biol. 2005;6:287–97. doi:http://dx.doi.org/10.1038/nrm1621.

Vaux DL, Strasser A. The molecular biology of apoptosis. Proc Natl Acad Sci U S A. 1996;93:2239–44.

Vaux DL, Cory S, Adams JM. Bcl-2 gene promotes haemopoietic cell survival and cooperates with c-myc to immortalize pre-B cells. Nature. 1988;335:440–2.

Vaux DL, Haecker G, Strasser A. An evolutionary perspective on apoptosis. Cell. 1994;76:777–9. doi:http://dx.doi.org/10.1016/0092-8674(94)90350-6.

Verhagen AM, Ekert PG, Pakusch M, Silke J, Connolly LM, Reid GE, Moritz RL, Simpson RJ, Vaux DL. Identification of DIABLO, a mammalian protein that promotes apoptosis by binding to and antagonizing IAP proteins. Cell. 2000;102:43–53. doi:http://dx.doi.org/10.1016/S0092-8674(00)00009-X.

Wang H, Li J, Bostock R, Gilchrist D. Apoptosis: a functional paradigm for programmed plant cell death induced by a host-selective phytotoxin and invoked during development. Plant Cell. 1996a;8:375–91.

Wang M, Oppedijk B, Lu X, Duijn B, Schilperoort R. Apoptosis in barley aleurone during germination and its inhibition by abscisic acid. Plant Mol Biol. 1996b;32:1125–34. doi:10.1007/BF00041396.

Wang W, Pan J, Zheng K, Chen H, Shao H, Guo Y, Hongwu B, Han N, Wang J, Zhu M. Ced-9 inhibits Al-induced programmed cell death and promotes Al tolerance in tobacco. Biochem Biophys Res Commun. 2009a;383:141–5.

Wang Z, Song J, Zhang Y, Yang B, Chen S. Expression of baculovirus anti-apoptotic p35 gene in tobacco enhances tolerance to abiotic stress. Biotechnol Lett. 2009b;31:585–9.

Wang M, Zheng Q, Shen Q, Guo S. The critical role of potassium in plant stress response. Int J Mol Sci. 2013;14:7370–90.

Whelan RS, Kaplinskiy V, Kitsis RN. Cell death in the pathogenesis of heart disease: mechanisms and significance. Annu Rev Physiol. 2010;72:19–44. doi:10.1146/annurev.physiol.010908.163111.

Williams B, Dickman M. Plant programmed cell death: can't live with it; can't live without it. Mol Plant Pathol. 2008;9:531–44.

Williams B, Kabbage M, Britt R, Dickman MB. AtBAG7, an Arabidopsis Bcl-2–associated athanogene, resides in the endoplasmic reticulum and is involved in the unfolded protein response. Proc Natl Acad Sci U S A. 2010;107:6088–93.

Xu P, Rogers SJ, Roossinck MJ. Expression of antiapoptotic genes *bcl-xL* and *ced-9* in tomato enhances tolerance to viral-induced necrosis and abiotic stress. Proc Natl Acad Sci U S A. 2004;101:15805–10.

Xu Z, Zhou G, Shimizu H. Plant responses to drought and rewatering. Plant Signal Behav. 2010;5:649–54.

Yang X, Chang HY, Baltimore D. Autoproteolytic activation of pro-caspases by oligomerization. Mol Cell. 1998;1:319–25. doi:http://dx.doi.org/10.1016/S1097-2765(00)80032-5.

Yen C-H, Yang C-H. Evidence for programmed cell death during leaf senescence in plants. Plant Cell Physiol. 1998;39:922–7.

Yin Q, Park HH, Chung JY, Lin S-C, Lo Y-C, Da Graca LS, Jiang X, Wu H. Caspase-9 Holoenzyme is a specific and optimal Procaspase-3 processing machine. Mol Cell. 2006;22:259–68. doi:http://dx.doi.org/10.1016/j.molcel.2006.03.030.

Yoshida S. Molecular regulation of leaf senescence. Curr Opin Plant Biol. 2003;6:79–84. doi:http://dx.doi.org/10.1016/S1369526602000092.

Youle RJ, Strasser A. The BCL-2 protein family: opposing activities that mediate cell death. Nat Rev Mol Cell Biol. 2008;9:47–59. doi:http://dx.doi.org/10.1038/nrm2308.

Young TE, Gallie DR. Programmed cell death during endosperm development. Plant Mol Biol. 2000;44:283–301. doi:http://dx.doi.org/10.1023/A:1026588408152.

Zheng L, Trageser CL, Willerford DM, Lenardo MJ. T cell growth cytokines cause the superinduction of molecules mediating antigen-induced T lymphocyte death. J Immunol. 1998;160:763–9.

Zheng TS, Hunot S, Kuida K, Flavell RA. Caspase knockouts: matters of life and death. Cell Death Differ. 1999;6:1043–53.

Zhu JK. Plant salt tolerance. Trends Plant Sci. 2001;6:66–71. doi:http://dx.doi.org/10.1016/S1360-1385(00)01838-0.

Zhu J, Fu X, Koo YD, Zhu JK, Jenney FEJ, Adams MW, Zhu Y, Shi H, Yun DJ, Hasegawa PM, Bressan RA. An enhancer mutant of Arabidopsis salt overly sensitive 3 mediates both ion homeostasis and the oxidative stress response. Mol Cell Biol. 2007;27:5214–24.

Zimmermann KC, Bonzon C, Green DR. The machinery of programmed cell death. Pharmacol Ther. 2001;92:57–70. doi:http://dx.doi.org/10.1016/S0163-7258(01)00159-0.

Zou H, Henzel WJ, Liu X, Lutschg A, Wang X. Apaf-1, a human protein homologous to C. elegans CED-4, participates in cytochrome c-dependent activation of caspase-3. Cell. 1997;90:405–13. doi:http://dx.doi.org/10.1016/S0092-8674(00)80501-2.

Zuzarte-Luís V, Hurlé JM. Programmed cell death in the developing limb. Int J Dev Biol. 2002;46:871–6.

Chapter 20
Antioxidant Signaling and Redox Regulation in Drought- and Salinity-Stressed Plants

Ananya Chakrabarty, Manashi Aditya, Nivedita Dey, Nabanita Banik, and Soumen Bhattacharjee

20.1 Introduction

Drought and salinity are two of the most important environmental cues limiting crop production worldwide. Globally, the annual losses in agricultural production from salt-affected land are approximately US$12 billion (Qadir et al. 2008; Flowers et al. 2010). In addition, the worldwide cost of drought events to agriculture is at least an order of magnitude higher. As a consequence of global land salinization, the frequency of drought events is also expected to increase. Thus, breeding crops for salinity and drought stress tolerance is absolutely essential for future food security.

Plants need to regulate an intricate metabolic balance of multiple pathways for the maintenance of cellular homeostasis, especially under environmental stress. Unfavorable environmental cues, particularly drought and salinity, were shown to disrupt the redox balance by aggravating the production of reactive oxygen species (ROS), causing oxidative injury (Miller et al. 2010; Ben Hamed et al. 2013). In fact, as a consequence of the uncoupling of different pathways, high-energy electrons are often transferred to molecular oxygen to form ROS (Mittler 2002; Miller et al. 2010; Bhattacharjee 2005, 2014). The uncontrolled generation of ROS ($^1O_2$, $O_2^{\bullet-}$, H_2O_2,$^{\bullet}OH$, $RO^{\bullet}$, RCO, etc.) causes nonspecific oxidative damage to almost every important class of cellular macromolecules (Apel and Hirt 2004; Bhattacharjee 2012; Mittler et al. 2004). However, under optimal environmental conditions, the

A. Chakrabarty • M. Aditya • N. Dey • N. Banik
Plant Physiology and Biochemistry Research Laboratory, UGC Centre for Advanced Study, Department of Botany, The University of Burdwan, Burdwan 712101, West Bengal, India
e-mail: soumen1995@yahoo.com

S. Bhattacharjee, Ph D (✉)
Professor & Head, Cetntre of Advanced Studies, Department of Botany, University of Burdwan, West Bengal, India
e-mail: soumen1995@yahoo.com

M.A. Hossain et al. (eds.), *Drought Stress Tolerance in Plants, Vol 1*,
DOI 10.1007/978-3-319-28899-4_20

production of ROS is low and under tight control of the antioxidative defense system.

The fundamental relationship between ROS production and stress tolerance is not as straightforward as one may expect. The past decade witnessed several studies suggesting that ROS integrate signaling pathways involved in plant growth, development, gravitropism, hormonal action, and many other physiological phenomena (Mittler 2002, 2004; Apel and Hirt 2004; Foyer and Noctor 2005a, b; Miller et al. 2008; Bhattacharjee 2012; Hossain et al. 2015). In a number of these cases, the generation of ROS is genetically programmed, and ROS are used as second messengers or signaling intermediates (Foyer and Noctor 2005a, b; Bhattacharjee 2012). Thus, it appears that ROS have pleiotropic effects in plants (Storey 1996; Bhattacharjee 2012). When ROS are produced in a well-ordered manner within specific compartments, they have key roles in plant stress response, growth, and development. However, when ROS are produced in excess, the resulting uncontrolled oxidation leads to cellular damage and eventual cell death. Under this situation, to prevent oxidative damage yet allow the beneficial functions of ROS to continue, the antioxidant defenses must be sure to control the titer of ROS (Noctor and Foyer 1998a, b).

Depending on the severity and duration of drought and salinity stress as well as on the developmental stage during which the plant is exposed to those unfavorable environmental cues, the redox homeostasis of the plant will change. Overenergization and a reduction of the redox components in the Z-scheme of photosynthesis and mitochondrial electron transport chain (ETC) are the main causes of redox imbalance in plant cells (Fig. 20.1). The reactions of the photosynthetic carbon oxidation cycle in the peroxisome also significantly contribute to the formation of ROS (Fig. 20.1). Another important source of ROS in plant cells, particularly upregulated under stress, is membrane-bound *Rboh*. Plants use an efficient ROS-detoxification mechanism involving antioxidative enzymes, molecules, and quenchers. The components of the antioxidative defense system is found in almost all cellular compartments, demonstrating the significance of ROS detoxification in cellular homeostasis and survival (Fig. 20.1).

Drought and salinity are the most important environmental constraints that have been found to enhance the generation and accumulation of ROS, causing oxidative deterioration and thus limiting plant productivity (Abbasi et al. 2007; Koussevitzky et al. 2008; Miller et al. 2010). Accordingly, under this situation the role of the antioxidative defense system is found to be extremely crucial for maintaining redox homeostasis and resuming normal growth and development in plants. While oxidative stress is triggered by the accumulation of ROS under drought and salinity, potentiating oxidative damage, several other studies have implicated the central role of ROS in the signaling network associated with drought and salinity stress acclimation and adaptation (Mittler et al. 2004; Miller et al. 2010; Hossain et al. 2015).). Therefore, the two somewhat opposing functions of ROS, that is, as toxic metabolite and as beneficial signaling component under drought and salinity stresses, underscore the need to elucidate the mechanism that influences the ROS–antioxidant interaction at the metabolic interface to determine the subsequent fate of the

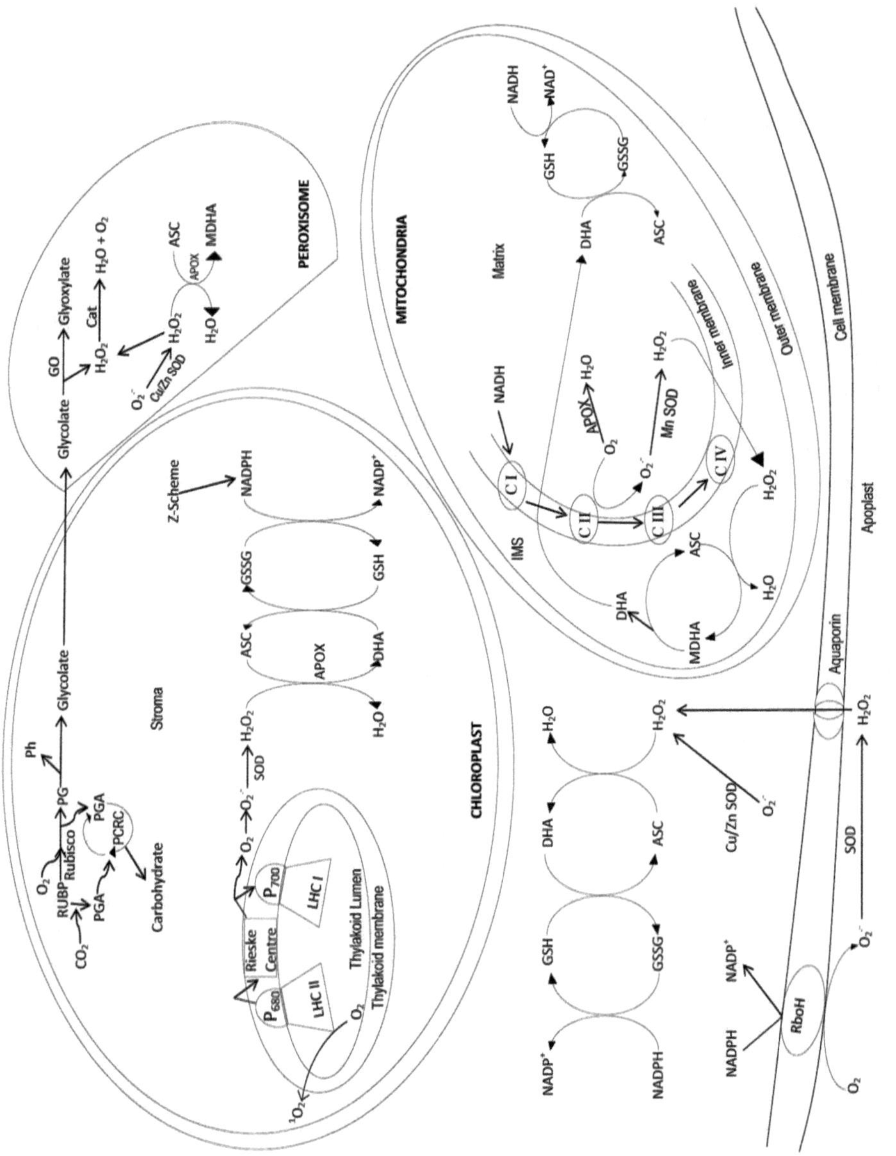

Fig. 20.1 Impact of oxidant–antioxidant interaction associated with maintenance of redox homeostasis in plant cell (details in text)

cell. Elucidation of the redox-regulatory metabolism and ROS–antioxidant interaction and their associated signaling under drought and salinity stresses will not only provide an insight into plant response to drought and salinity stresses but will also help us to map out strategies to enhance the tolerance of crops under those environmental constraints.

20.2 Redox Homeostasis and Regulation in Plants

Redox reaction, or the transfer of electrons, is a natural process for cell metabolism and occurs during biological energy transduction in the inner mitochondrial and thylakoid membranes. Any condition that leads to loss of redox homeostasis because of the overaccumulation of prooxidants may be referred to as oxidative stress. As a consequence, cellular compounds undergo redox modification resulting from the imbalance between prooxidant–antioxidant ratios caused by different unfavorable environmental cues. This disequilibrium has been correlated with altered physiological conditions and many diseases in the plant kingdom. Despite the fluctuations in external environment, plants always thrive to maintain a constant internal environment. Redox regulation is the key in adjusting plant metabolism and development to the prevailing environmental conditions. In mitochondria and chloroplasts, the ETC is associated with carbon metabolism through NAD(P)H and ATP. Alterations in carbon metabolism and energy balance during stress have been reported in both these organelles. To avoid excessive accumulation of ROS and oxidative damage, a highly regulated system is needed to maintain coordination between these organelles, including reversible redox regulation of proteins by thiol–disulfide exchange, regulation of phosphoproteins, activation of signaling pathways by ROS-responsive regulatory genes, and ROS–antioxidant interactions (Foyer and Noctor 2009; Suzuki et al. 2011).

The redox state of a chloroplast is primarily regulated through the plastoquinone (PQ) pool and involved in activation of the state transition that balances and maintains the photosynthetic energy distribution between PSI and PSII in the short term. In the long term, imbalances in energy distribution between the two photosystems are counteracted by adjusting the photosystem stoichiometry, changing the abundance of reaction center and light-harvesting proteins (Zer and Ohad 2003; Dietzel and Pfannschmidt 2008).

Mitochondrial functions during increased respiratory activities and photorespiratory metabolism are sensitive to oxidative damage. Oxidized lipids such as polyunsaturated fatty acids generated under these conditions were known to inhibit tricarboxylic acid cycle activity, disturbing carbon and nitrogen metabolism (Taylor et al. 2002, 2004; Mueller 2004).In the inner mitochondrial membrane, the activities of alternative oxidase (AOX), type II NAD(P)H dehydrogenase, and uncoupling proteins optimize the flow of electrons, preventing overreduction of the mitochondrial ETC (mtETC) and generation of excess ROS. These are regarded as regulators of the mitochondrial redox state and ROS generation (Noctor et al. 2007; Rasmusson and Wallstrom 2010).

Redox regulation and ROS metabolism are interlinked and involve the coordinated function of mitochondria, chloroplasts, and other organelles (Dinakar et al. 2010). Photorespiration, carbon and nitrogen metabolism, redox state of NAD(P)H, and other factors can influence the energy flow and redox fluctuations among chloroplasts, mitochondria, and the cytosol (Noctor et al. 2007; Foyer and Noctor 2009). Therefore, these organelles are interconnected to a wider redox network and require a high regulation and coordination under abiotic stress conditions. To initiate redox regulation and relative contribution of each organelle to the redox network would require monitoring the redox states in chloroplasts and mitochondria simultaneously. ROS network genes and redox regulatory enzymes such as the ascorbate–glutathione (ASC–GSH) cycle enzymes monodehydroascorbate reductase (MDHAR), glutathione reductase (GR), and glutathione peroxidase (GPX) are known to express both in chloroplasts and in mitochondria (Mittler et al. 2004). These two organelles are also sources of retrograde signaling. A deviation from regular redox homeostasis can be sensed in the chloroplast and mitochondria and transmitted to the nucleus by retrograde signaling cascades, and the nucleus will subsequently modulate anterograde control (Woodson and Chory 2008).

In plants, ubiquitous small-molecule redox couples are used as signaling molecules that limit damage from potentially harmful ROS (Sierla et al. 2013). Cellular redox pairs include nicotinamide adenine dinucleotide phosphate ($NADP^+$/NADPH), and reduced/oxidized GSH and ASC. These factors are crucial signaling molecules for plant responses to stress.

ROS-induced cellular redox modifications lead to oxidation of free thiol side chains on cysteines of regulatory proteins. Reversible modifications of the cysteine thiol include the covalent attachment of nitric oxide (S-nitrosylation), thiol hydroxylation (S-sulfonation), disulfide bridge formation (S-thiolation), covalent attachment of glutathione (S-glutathionylation), and further oxidation of sulfonic groups to the sulfenic and sulfonic states. ROS also directly damage proteins by altering amino acids, oxidizing tyrosine, tryptophan, histamine, and methionine (Moller et al. 2007). Hence, the modified proteins appear to be ideal candidates for a signaling event that yields different modified products resulting in the activation of different sets of genes (Apel and Hirt 2004). In the ASC–GSH cycle, the newly formed hydrogen peroxide (H_2O_2) can be decomposed by the mitochondrial peroxidase activities dependent on the antioxidants ASC for the enzyme ascorbate peroxidase (APX), thiol reductant GSH for the GPXs, and thioredoxin/peroxiredoxin system (Trx/Prx). The oxidized forms of ASC generated are then reduced by the flavin adenine dinucleotide (FAD)–containing MDHAR in an NAD(P)H-dependent pathway and by dehydroascorbate reductase (DHAR) using GSH as electron donor. Oxidized GSH (GSSG) is reduced by GR and oxidized by thioredoxin reductase, both processes NADPH dependent (Martí et al. 2009) (Fig. 20.1).

20.3 ROS Metabolism and Their Regulation Under Drought and Salinity Stresses

Plant metabolism is sensitive to environmental changes; an imbalance in its path can induce oxidative stress in cells by the generation and accumulation of ROS, which cause the oxidation of cellular components, interfere in metabolic activities, and affect membrane integrity. ROS production in mitochondria has been reported to increase under salinity and drought conditions. Furthermore, oxidative damage induced by NaCl stress can affect different cellular targets selectively: Complex I of the ETC was found to be damaged via oxidative stress while complex II was damaged directly by salt (Hamilton and Heckathorn 2001). Hence, changes in ROS levels caused by the perturbation of the respiratory complex I have been proposed to trigger a mitochondrial retrograde signal (Rhoads and Subbaiah 2007). The adaptive response of plants induced by salt stress is well documented in *Arabidopsis*. Out of 300 salt stress–induced genes, more than half had a predicted mitochondrial localization (Heazlewood et al. 2007). In general, an induced expression of antioxidant defense genes is usually correlated with enhanced salt stress tolerance (Attia et al. 2008).

The ROS are produced in plant cells via a number of routes, and most of the cellular compartments have the potential to produce ROS (Bhattacharjee 2005 (Fig. 20.1). Drought and salinity are the most important constraints in crops, resulting in large yield losses and limiting the average yield increase (Chen et al. 2013; Agarwal et al. 2013). During drought and salinity stresses, ROS can produce in different cell organelles that are very important for plant survival. In chloroplasts, photosystems I and II (PSI and PSII) are the major sites for the production of singlet oxygen ($^1O_2$) and superoxide ions ($O_2^{\bullet-}$). In mitochondria, complex I, ubiquinone, and complex III of the ETC are the major sites for the generation of $O_2^{\bullet-}$ (Gill and Tuteja 2010). The cell organelles follow different patterns for ROS metabolism and regulation (Miller et al. 2010).

Chloroplast: The reaction centers of photosystems I (PSI) and II (PSII) in a chloroplast's thylakoids are a major ROS source in plant cells. ROS production in the chloroplast is largely based on overreduction of the photosynthetic redox carriers under excess excitation energy conditions, or when the energy exceeds the amount required for photosynthetic CO_2 assimilation (Mühlenbock et al. 2008; Asada 2006). Under water stress conditions (drought and salinity stresses), reduced CO_2 availability due to stomatal closure and exposure to continuous excessive light direct higher electron leakage to molecular oxygen, thus generating $O_2^{\bullet-}$ at PSI by the Mehler reaction either directly or via ferredoxin, a stromal protein (Asada 2006) (Fig. 20.1). Actually, the limitation on CO_2 fixation will reduce $NADP^+$ regeneration through the Calvin cycle (PCRC). As a result, an overreduction of the photosynthetic ETC takes place (Cruz and de Carvalho 2008). A membrane-attached copper/zinc superoxide dismutase (Cu/ZnSOD) in the vicinity of PSI detoxifies the $O_2^{\bullet-}$ into H_2O_2. A membrane-bound thylakoid ascorbate peroxidase (tylAPX) reduces the H_2O_2 to water (H_2O); this is also referred to as the water–water cycle. In this process, ASC is oxidized to monodehydroascorbate (MDHAs) radicals. MDHAs are reduced back to ASC via GSH (Pfannschmidt 2003) (Fig. 20.1). Prx and Trx were shown to play an important role providing antioxidative protection via the detoxification of photochemically produced H_2O_2 in chloroplasts during drought and oxidative stresses (Dietz et al. 2006; Vieira Dos Santos and Rey 2006). The leakage of electrons to O_2 in the Mehler reaction was increased approximately 50 % in drought-stressed wheat than in unstressed wheat, as estimated in previous studies; with sunflowers, electron leakage of thylakoid membrane to O_2 was also increased under drought stress (Cruz and de Carvalho 2008). $^1O_2$ is generated at PSII by excited triplet-state chlorophyll at the P680 reaction center and in the light-harvesting complex (LHC) when the PQ pool becomes overreduced (Krieger-Liszkay 2005; Asada 2006).Under drought stress a real threat for the chloroplast is the production of the hydroxyl radical ($OH^•$) through "iron-catalzsed" reduction of H_2O_2 by both SOD and ASC, which can damage the thylakoid membrane and photosynthetic apparatus (Vranova et al. 2002; Cruz and de Carvalho 2008). H_2O_2 has a positive role in reducing the $^1O_2$, and treatment with exogenous H_2O_2 promotes the oxidation of quinone A (QA), the primary PQ electron acceptor, which increases the photosynthetic electron transport flow and decreases the generation of $^1O_2$ during stress (Karpinska et al. 2000; Asada 2006). A transmembrane protein with an ankyrin-repeat motif has been identified as a component of signal transduction pathways that influences the abscisic acid (ABA)–induced accumulation of ROS under salinity stress (Sakamoto et al. 2008; Witzel et al. 2009).Water-stressed conditions also cause retrograde signaling. $^1O_2$ accumulated in the chloroplast is sensed or mediated to the nucleus via a concerted action of two chloroplast proteins, EXECUTER1 and EXECUTER2. In the nucleus or cytosol, the blue light photoreceptor cry1 is involved in the $^1O_2$-mediated stress response.

Peroxisome: Peroxisomes produce H_2O_2 at high rates in drought and salinity stresses. In these stressed conditions, reduced water availability and stomatal closure decrease the CO_2/O_2 ratio in mesophyll cells and increase photorespiration and the production of glycolate. Under drought stress the photorespiratory pathway is enhanced, especially when maximum RuBP oxygenation has taken place because of a decrease in CO_2 fixation (Cruz and de Carvalho 2008). According to Noctor and collaborators (2002a, b), the oxidation of glycolate by glycolate oxidase in peroxisomes produces the majority of H_2O_2 during photorespiration. It was had estimated that over 70 % of the total H_2O_2 was produced by photorespiration under drought stress (Cruz and de Carvalho 2008). Catalases are the major antioxidative enzymes that detoxify H_2O_2 in peroxisomes, under increased photorespiration conditions (Mittler et al. 2004; Vandenabeele et al. 2004). During photorespiration under drought stress, H_2O_2 is produced in peroxisomes, where the thiol enzyme of PCRC inhibits H_2O_2 (Cruz and de Carvalho 2008). APX and the ASC–GSH cycle can also help to scavenge H_2O_2 in peroxisomes (Jiménez et al. 1997). Sometimes salinity can decrease AsA and GSH content and induce lipid peroxidation in peroxisomes (Mittova et al. 2003). Peroxisomal polyamine oxidase (POX) may help in the regulation of drought-responsive genes by balancing ROS generation and scavenging (Kamada-Nobusada et al. 2008; Miller et al. 2010). Photorespiration itself is also beneficial for plants during drought stress. Due to a reduced rate of RuBP carboxylation, it can protect the photosynthetic apparatus from photoinhibition by dissipating energy through an alternative path (Cruz and de Carvalho 2008).

Mitochondria and apoplast: ROS production in mitochondria has been shown to increase significantly under drought and salinity stresses (Bartoli et al. 2004; Pastore et al. 2007). Increased mitochondrial respiration during water stress causes generation of ROS including $O_2^{\bullet-}$, which, in turn, reduces to H_2O_2 during the transfer of lectrons from the cytochrome electron transport system to O_2 (Norman et al. 2004; Rhoads et al. 2006). Overreduction of the ubiquinone (UQ) pool caused by perturbation in mitochondrial ETC function increases ROS production (Rhoads et al. 2006). Mitochondrial AOX maintains the reduction state of the UQ pool, lowers ROS production in mitochondria, and uncouples ATP production, preventing programmed cell death (PCD) induced by downregulation of the cytochrome pathway (Norman et al. 2004). Manganese-SOD (Mn-SOD) converts $O_2^{\bullet-}$ to H_2O_2 and O_2 in the initial step of the ROS detoxification (Arnholdt-Schmitt et al. 2006; Rhoads et al. 2006).

The apoplast is also an important site for H_2O_2 production in drought and salinity stresses (Hu et al. 2006; Jubany-Marí et al. 2009). *AtRbohD* and *AtRbohF* of *Arabidopsis* encode two major NADPH oxidases expressed in guard and mesophyll cells generating ROS that are required for ABA-induced stomatal closure (Kwak et al. 2003; Torres and Dangl 2005). The accumulation of H_2O_2 in the apoplast is thought to be involved in acclimation responses of plants (e.g., growth and cell wall strengthening) to drought and salinity stresses (Ros Barceló 2005; Jubany-Marí et al. 2009).

20.3.1 Hormonal Regulation

ROS-induced responses and signaling under dehydration stress induced by salinity and drought are intertwined with plant hormonal responses. Ethylene (ET) biosynthesis is an early response, and later, salicylic acid (SA), jasmonic acid (JA), and ABA are produced. ET and SA signaling promote enhanced ROS production and PCD, whereas JA attenuates this by reducing ROS production. The connections between oxidative stress and auxin were recently discovered. Defects in the antioxidative capacity of a Trx and GSH mutant resulted in altered auxin homeostasis and development (Bashandy et al. 2010). Iglesias et al. (2010) have shown that auxin receptor mutants were more tolerant to H_2O_2 under salinity stress. Gibberellin signaling is linked with ROS by stimulating the destruction of the nuclear growth–repressing DELLA proteins that regulate transcript levels of antioxidant enzymes (Achard et al. 2008).

ABA, on the other hand, plays an important role in integrating dehydration stress signals and controlling downstream stress responses. ABA levels are continuously adjusted by the plant in response to the changing water status of the soil. The mechanism includes both ABA-dependent and ABA-independent processes. It has been suggested that various stress signals and ABA share common elements in the signaling pathway that crosstalk to maintain cellular homeostasis (Thomashow 1999; Shinozaki and Yamaguchi-Shinozaki 2000).

In *Arabidopsis* several ABA-deficient mutants such as *aba1, aba2,* and *aba3* have been reported (Koornneef et al. 1998). Environmental conditions such as dehydration and salt stress activate ABA-dependent and -independent gene expression systems involving ABA responsive element (ABRE) binding factors/ABA-responsive element binding proteins, MYC/MYB, drought-responsive element binding factors (DREB2), and NAC (NAM, ATAF2, and CUC) transcription factors (TFs) (Agarwal and Jha 2010)

During drought stress, cytokinin (CK) levels decrease, increasing shoot responses to ABA, leading to stomatal closure (Goicoechea et al. 1997). These stress-induced changes in CKs and ABA level promote early leaf senescence, which leads to leaf abscission and results in reduction of water loss (Pospíšilová et al. 2000).

Ethylene response factors (ERFs) were suggested to enhance plant tolerance to dehydration stress. Transgenic tobacco expressing JERF3, an osmotic- and oxidative-stress responsive ERF, showed enhanced tolerance to drought and salinity stresses and decreased accumulation of ROS (Wu et al. 2008).

20.3.2 Osmolyte-Mediated Regulation

To facilitate water uptake during drought and saline conditions, plants accumulate solutes such as proline and glycine betaine (GB) (Ashraf and Foolad 2007). These osmolytes were suggested to be important for protecting cells against increased levels of ROS accumulation under stress conditions. Dehydration stress leads to proline production in the cytosol and the vacuole (Aubert et al. 1999; Miller et al. 2010) and was shown to facilitate the defense against harmful ROS. By quenching $^1O_2$ and directly scavenging $HO^•$, proline might be able to protect proteins, DNA, and membranes (Matysik et al. 2002). According to one study, transgenic wheat plants that accumulated higher proline than wild type exhibited less membrane lipid peroxidation during drought, indicating a role for proline in reducing ROS damage during drought (Vendruscolo et al. 2007). Impaired proline accumulation has been found to enhance the accumulation of ROS, which subsequently not only enhances plant sensitivity to salinity and dehydration stress but also potentiates oxidative damage (Miller et al. 2010).

On the other hand, GB is known to accumulate mainly in the chloroplast and maintain PSII efficiency under dehydration stress conditions (Ben Hassine et al. 2008) Exogenous GB treatment prevents salinity-induced structural damage to ROS-producing organelles, (Ashraf and Foolad 2007). These results suggest role of proline and GB in the regulation of ROS metabolism under salinity and drought stress.

Table 20.1 Stimulation of antioxidant enzymes in response to drought and salinity-induced oxidative stress

Stresses	Antioxidant enzymes	Plant species	References
Drought	SOD, GPX, APX, MDHAR, DHAR, and GR	*Oryza sativa*	Bhattacharjee (2012), Sharma and Dubey (2005), Sharma et al. (2012)
	SOD, CAT, and GPX	*Beta vulgaris*	Sayfzadeh and Rashidi (2011), Sharma et al. (2012)
	SOD, APX and GR	*Triticum sativum*	Sgherri et al. (1994), Sharma et al. (2012)
Salinity	SOD, CAT, GPX, APX, GR	*O. sativa*	Sharma et al. (2012)
	GPX	*O. sativa*	Mittal and Dubey (1991)
	APX, GST, GLX I, SAM synthase	*Hordeum vulgare*	Witzel et al. (2009)

20.4 Enzymatic and Nonenzymatic ROS Scavenging Under Drought and Salinity Stress

It is now well accepted that ROS accumulation is crucial to plant development as well as in defense mechanisms (Foyer and Noctor 2005a, b). The excessive formation of ROS, which disrupts redox homeostasis of the cell, is called *oxidative stress*. Since these oxy free radicals are strong oxidants and very toxic, organisms have developed different systems to detoxify these radicals that involve various antioxidative enzymes, small-molecular-weight antioxidants, quenchers, and more. Strong regulation of these enzymes is essential to keep the content of ROS under tight control (del Río et al. 2002a, b) (Table 20.1). In addition to the antioxidative enzyme, there is also a nonenzymatic component of the scavenging system consisting of antioxidative molecules such as ASC, GSH, α-tocopherol, and carotenoids (Diaz-Vivancos et al. 2008). ROS scavenging enzymes such as SOD, catalase (CAT), APX, and GR, in combination with ASC and GSH, play a pivotal role in ROS detoxification in plant cells. Many redox-active phenolics, carotenoids, and tocopherols are also essential for ROS detoxification.

Superoxide dismutase: SOD is ubiquitous and a primary scavenger of ROS in plant. In transgenic plants, SOD is upregulated under higher salt or drought stress, which leads to overexpression of SOD (Badawi et al. 2004). Thus, SOD plays a critical role in the survival of plants under environmental stress. SOD made the first line of defense against ROS. Four different isoenzymes of SOD, namely, Cu/ZnSOD (dimmers, found in the cytosolic fraction and also in chloroplasts in higher plants), MnSOD enzymes (tetramers, found in the mitochondria and peroxysome), and FeSOD (present in chloroplasts), are found in plant cell. The activity of SOD isoenzymes can be detected by their sensitivity to potassium cyanide and H_2O_2. The upregulation of SOD occurs to cope with oxidative stress caused by abiotic stress and has a critical role in the survival of plants under salt and drought stress (Gambarova and Gins 2008; Kukreja et al. 2005; Gapinska et al. 2008). The distribution of SOD isoenzymes is also distinctive (Torres 2010). A comparison of the effects of drought and water stress on wheat genotypes suggests that different mechanisms participate in ROS detoxification. ROS-detoxifying enzymes have been shown to be inefficient in plants subjected to drought-induced oxidative stress (Chen et al. 2010). Eyidogan and Oz (2005) noted three SOD activity bands (MnSOD, FeSOD, and Cu/ZnSOD) in *Cicer arietinum* under salt stress. Furthermore, a significant increase in the activities of Cu/ZnSOD and MnSOD isozymes under salt stress was observed by Pan et al. (2006). An increase in SOD activity following drought stress was noted in three cultivars of *Phaseolus vulgaris* (Zlatev et al. 2006), *Alternanthera philoxeroides* (Wang et al. 2008), and *Oryza sativa* (Sharma and Dubey 2005). Under high light condition, drought significantly increased the SOD activity in comparison to low light. In an interesting study, Rossa et al. (2002) observed that transgenic rice plants overexpressing OsMT1a demonstrated enhanced drought tolerance (Yang et al. 2009). The overexpression of MnSOD in transgenic *Arabidopsis* plants also showed increased salt tolerance (Wang et al. 2004). The overexpression of MnSOD in transformed *Lycopersicon esculentum* plants also showed enhanced tolerance against salt stress (Wang et al. 2007).

Catalase: CAT, the first antioxidant enzyme to be discovered and characterized (Mhamdi et al. 2010), is a heme-containing enzyme that catalyzes the dismutation of H_2O_2 into H_2O and O_2. The enzyme occurs in all aerobic eukaryotes, and its function is to remove the H_2O_2 generated in peroxisomes by oxidases involved in the β-oxidation of fatty acids, photorespiration, and purine catabolism and during oxidative stress (Mittler 2002; Vellosillo et al. 2010). This is also a result of the proliferation of peroxisomes during stresses, which might help in the scavenging of H_2O_2 that diffuses from the cytosol (Lopez-Huertas et al. 2000). CAT has one of the highest turnover rates for all enzymes: One molecule of CAT can convert about six million molecules of H_2O_2 to H_2O and O_2 per minute. Various isoforms of CAT have been described in several plant species (Dat et al. 2003; Vandenabeele et al. 2004). The three isoforms present in *Arabidopsis*, namely, CAT-1, CAT-2, and CAT-3, have been noticed to function under stress (Frugoli et al. 1996). The *CAT1* gene is mainly expressed in pollen and seeds, *CAT2* in photosynthetic tissues but also in roots and seeds, and *CAT3* in vascular tissues and in leaves. The isozymes CAT1 and CAT2 are localized in peroxisomes and the cytosol, whereas CAT3 is a mitochondrial isozyme. Stress conditions that reduce the rate of protein turnover, such as salinity and drought, reduce CAT activity (Karuppanapandian et al. 2006a, b; Karuppanapandian and Manoharan 2008; Chen et al. 2010; Hojati et al. 2010). CAT is a light-sensitive protein that has a high rate of turnover; environmental stresses that reduce the rate of protein turnover, such as salinity and drought, cause the depletion of CAT activity (Boguszewska et al. 2010; Mhamdi et al. 2010). Kukreja et al. (2005) reported an increase in CAT activity in *C. arietinum* roots under salinity stress, whereas, in another study, Sharma and Dubey (2005) reported a decrease in CAT activity in rice seedlings under drought stress. The *Escherichia coli* CAT encoded by the *katE* gene overexpressed in *O. sativa* conferred tolerance to transgenic rice plants under salt stress (Nagamiya et al. 2007). Eyidogan and Oz (2005) reported a significant increase in CAT activity in *C. arietinum* leaves under salt treatment. Similarly, an increase in the CAT activity in *C. arietinum* roots following salinity stress was noted by Kukreja et al. (2005). Srivastava et al. (2005)reported a decrease in CAT activity in *Anabaena doliolum* under NaCl salinity stress. Simova-Stoilova et al. (2010) reported increased CAT activity in wheat under drought stress, especially in sensitive varieties. In another study, Sharma and Dubey (2005) reported a decrease in CAT activity in rice seedlings following drought stress. Pan et al. (2006) studied the combined effect of salt and drought stresses and found that it decreases the CAT activity in *Glycyrrhiza uralensis* seedlings. Transgenic rice plants overexpressing *OsMT1a* showed increased CAT activity and thus enhanced tolerance to drought (Yang et al. 2009). Thus, all the experimental evidence strongly supports the role of CAT associated with redox buffering, which has a putative role in stress tolerance under heat and dehydration stress in plant cell.

Ascorbate peroxidase: APX is thought to play the most essential role in scavenging ROS and protecting cells in higher plants, particularly under environmental stress. APX is involved in scavenging of H_2O_2 in water–water and ASC–GSH cycles and utilizes ASC as the electron donor. APX uses ASC as a hydrogen donor to break down H_2O_2 to form H_2O and MDHA Asada 2000) (Table 20.2). $O_2^{\cdot-}$ generated at the membrane surface can thus be trapped and converted immediately to H_2O_2, which is scavenged by membrane-bound APX (Asada 2000, 2006). The APX family consists of at least five different isoforms, including the thylakoid (tAPX) and glyoxisome membrane (gmAPX) forms as well as the chloroplast stromal soluble form (sAPX) and the cytosolic form (cAPX) (Noctor and Foyer 1998a, b). Enhanced activity of APX was also found in salt-stressed *A. doliolum* (Srivastava et al. 2005). Sharma and Dubey (2005) found that mild drought–stressed plants had higher chloroplastic APX activity than control grown plants, but the activity declined at the higher level of drought stress. According to Koussevitzky et al. (2008), cytosolic APX1 plays a key role in the protection of plants from a combination of drought and heat stress. It has also been noted that the overexpression of APX in *Nicotiana tabacum* chloroplasts enhanced plant tolerance to salt and dehydration stress (Badawi et al. 2004). Yang et al. (2009) correlated the enhanced tolerance of *OsMT1a* overexpressing transgenic rice plants to drought stress with the increase in APX activity. In one study, the expression patterns of APX were analyzed in roots of etiolated *O. sativa* seedlings under NaCl stress and the mRNA levels for two cytosolic APXs (*OsAPX1* and *OsAPX2*), two peroxisomal APXs (*OsAPX3* and *OsAPX4*), and four chloroplastic APXs (*OsAPX5–OsAPX8*) were quantified in the rice genome. It was noted that 150 and 200 mM NaCl increased the expression of *OsAPX8* and the activities of APX, but there was no effect on the expression of *OsAPX1–OsAPX7* in rice roots (Hong et al. 2007). Transgenic *Arabidopsis* plants overexpressing *OsAPXa* or *OsAPXb* exhibited increased salt tolerance. It was found that the overproduction of *OsAPXb* enhanced and maintained APX activity to a much higher extent than *OsAPXa* in transgenic plants under different NaCl concentration (Lu et al. 2007). $O_2^{\cdot-}$ generated at the membrane surface can thus be trapped and converted immediately to H_2O_2, which is scavenged by membrane-bound APX (Asada 2000, 2006). The mRNA of cytosolic APX showed upregulation during drought stress in the alfalfa nodule (Naya et al. 2007). Transgenic *Arabidopsis* plants overexpressing cytosolic APXs

Table 20.2 Differential antioxidant competence between salt-resistant and -susceptible varieties

Salt-resistant cultivar	Salt-susceptible cultivar	Differential antioxidant competence	References
Oryza sativa L, Cultivar SR26 B	*O.ryza sativa* L, Cultivar Ratna	Better antioxidative competence in Cv. SR 26 B compared to Ratna in terms of activities of antioxidative defense enzymes (CAT, SOD, APOX, GR), ROS accumulation, oxidative damages to membrane components	Bhattacharjee (2012), Chakraborty and Bhattacharjee (2014)
O. sativa L, Cultivar SR26 B	*O. sativa* L, Cultivar Ratna	Increased transcript abundance of antioxidative defense enzymes (CAT, SOD, APOX, GR)	Chakraborty and Bhattacharjee (2014)
Hordeum marinum	*Hordeum vulgare*	Increased SOD, CAT activity in the halophyte	Seckin et al. (2010)
Lycopersicon pennellii	*Lycopersicon esculentum*	Increased SOD activities in chloroplast, mitochondria, and peroxisomes of the halophyte	Shalata et al. (2001)
L. pennellii	*L.n esculentum*	Constitutively higher peroxisome CAT activity and further increased during salt stress in the halophyte	Mittova et al. (2000, 2003), Shalata et al. (2001)
L. pennellii	*L. esculentum*	Constitutively higher APX activity and further increased during salt stress in chloroplast, peroxisome, and mitochondrial fraction of the halophyte	Mittova et al. (2000, 2003), Shalata et al. (2001)
H. marinum	*H. vulgare*	Increased POX activity in the halophyte	Seckin et al. (2010)
L. pennellii	*L. esculentum*	MDAR activity increased in the halophyte but decreased in the glycophyte	Mittova et al. (2000, 2003), Shalata et al. (2001)
H. marinum	*H. vulgare*	Activities of isoenzymes GR1, 3, 6, 7 increased in halophyte	Seckin et al. (2010)
L. pennellii	*L. esculentum*	Constitutively higher GR activity in mitochondrial and peroxisome fractions of the halophyte but decreased during stress in both the glycophyte and halophyte	Mittova et al. (2003)

exhibited an increased tolerance to salt stress compared to wild-type plants (Lu et al. 2007). All the experimental data conclusively support the significant role of APX in redox status and associated oxidative stress tolerance under dehydration stress.

Guaiacol peroxidase (GuPX): Guaiacol is a heme-containing protein, and an important peroxidase group, found in the cytoplasm and apoplast, oxidizes a large number of organic compounds such as phenols, aromatic amines, hydroquinones, and others. In most plants, about 90 % of the peroxidase activity is referred to as GuPX (Foyer et al. 1994). Among the various antioxidants, GuPX can be considered one of the key ones since both its extra- and intracellular forms participate in the breakdown of H_2O_2. Induction in GuPX activity has been reported in common bean (*P. vulgaris*) nodules under salinity stress conditions (Jebara et al. 2005). An initial increase in GuPX activity in both the leaf and root tissues of green gram (*Vigna radiata*) (Panda 2001), cowpea (Cavalcanti et al. 2007), and rice (*O. sativa*) (Koji et al. 2009) has been reported under salinity stress. Some studies reported the increased GuPX activity under drought stress conditions in various plants, including sunflower (Gunes et al. 2008) and poplar (Xiao et al. 2008). Under sublethal salinity conditions, the level of GuPX activity has been used as a potential biomarker to evaluate the intensity of stress. An enhancement in GuPX activity suggests that this enzyme serves as an intrinsic defense tool to resist stress-induced oxidative damage in plants (Cavalcanti et al. 2007; Koji et al. 2009). A concomitant increase in GuPX activity in both the leaf and root tissues of *V. radiata* (Panda 2001) and *O. sativa* (Koji et al. 2009) has also been reported under salinity stress.

Glutathione reductase: GR, a flavoprotein oxidoreductase found in both prokaryotes and eukaryotes, is a potential enzyme of the ASC–GSH cycle and plays an essential role in the defense system against ROS by sustaining the reduced status of GSH (Edwards et al. 1990; Creissen et al. 1994). GR catalyzes the reduction of GSH, a molecule involved in many metabolic regulatory and antioxidative processes in plants, because GR catalyzes the NADPH-dependent reaction of the disulfide bond of GSSG and is thus important for maintaining the GSH pool (Rao and Reddy 2008). Eyidogan and Oz (2005) reported increased GR activity in the leaf tissue of *C. arietinum* L. cv. Gokce under salt stress, whereas Kukreja et al. (2005) noted increased GR activity in *C. arietinum* roots following salt stress. Sharma and Dubey (2005) noted a significant increase in GR activity in drought-stressed *O. sativa* seedlings.

Glutathione peroxidase: GPXs are a large family of diverse isozymes that use GSH to reduce H_2O_2 and organic and lipid hydroperoxides and therefore help plant cells from oxidative stress (Noctor et al. 2002a, b). Millar et al. (2003) identified a family of seven related proteins in the cytosol, chloroplasts, mitochondria, and endoplasmic reticulum, named AtGPX1–AtGPX7, in *Arabidopsis*. It has also been reported that *PHGPx* mRNA levels show increase in plant tissues under salt stress (Sreenivasulu et al. 2004), heavy metal stress (Li et al. 2011), oxidative stress (Li et al. 2000, 2004), and mechanical stimulation (Depège et al. 1998). It was noted that GPX activity in transgenic *Gossypium hirsutum* seedlings was 30–60 % higher under normal conditions, but was not different than GPX activity in WT seedlings under salt stress conditions (Light et al. 2005).

Monodehydroascorbate reductase: In plants, MDHAR is an enzyme of the GSH–ASC cycle that is a major antioxidant protecting cell damage against ROS. This enzyme is found in chloroplasts, the cytosol, mitochondria, glyoxysomes, and leaf peroxisomes (del Río et al. 2002a, b; Mittler 2002). MDHAR is a FAD enzyme that is mainly present as chloroplastic and cytosolic isozymes. Two enzymes are involved in the regeneration of reduced ascorbate, namely MDHAR, which uses NADPH directly to recycle ASC, and DHAR. However, MDHA is itself an efficient electron acceptor (Noctor and Foyer 1998a, b; Asada 2000) Sharma and Dubey (2005) reported that the activities of enzymes involved in the regeneration of ASC, that is, MDHAR, DHAR, and GR, were higher as compared to untreated control in drought-stressed rice seedlings. The overexpression of MDHAR in transgenic tobacco increased the tolerance against salt and osmotic stresses (Eltayeb et al. 2007). MDHAR exhibits a high specificity for MDHA as the electron acceptor, preferring NADH rather than NADPH as the electron donor. Asada (1999) studied the multi-step reduction of FAD in detail. The first step is the reduction of the enzyme FAD to form a charge transfer complex. The reduced enzyme donates electrons successively to MDHA, producing two molecules of ASC via a semiquinone form. It is well established that the disprotonation by photoreduced ferrodoxin (redFd) in the thylakoids is of great importance. Since redFd can reduce MDHA more effectively than NADPH, MDHAR cannot participate in the reduction of MDHA in the thylakoidal scavenging system. Therefore, MDHAR only functions in the presence of NAD(P)H (Asada 1999).

Dehydroascorbate reductase: DHAR is a major antioxidant in plants that detoxifies ROS and maintains photosynthesis. DHAR helps in regeneration of ascorbic acid (ASA) from its oxidized state. MDHA formation is followed by the univalent oxidation of ASA. Then MDHA is converted to dehydroascorbate (DHA), which is a divalent oxidation product. DHA is then reduced to ASA by DHAR in a reaction that requires GSH (Eltayeb et al. 2007). DHAR has been found to strongly influence the control stomatal opening and closing and hence plays a vital role in water use economy of plants under dehydration stress (Chen and Gallie 2005). An increase in DHAR activity was found to be associated with salt tolerance in *Arabidopsis* (Ushimaru et al. 2006; Eltayeb et al. 2006) and with drought and ozone stress tolerance in tobacco (Eltayeb et al. 2006)

Glutathione S-transferases: Glutathione transferases are now known as glutathione S-transferases (GSTs). GSTs are used in the detoxification of herbicides, hormone homeostasis, vacuolar sequestration of anthocyanin, tyrosine metabolism, hydroxyperoxide detoxification, regulation of apoptosis, and plant responses to biotic and abiotic stresses (Dixon et al. 2010).GSTs have the potential to remove cytotoxic or genotoxic compounds that can react with DNA, RNA, and protein. Gapinska et al. (2008) observed that under salinity stress in *L. esculentum* roots, the activity of GST was increased. In an experiment drought-tolerant and drought-sensitive sorghum were treated with 150 mM NaCl for 72 h; the result was significantly higher activity of GST and CAT for scavenging H_2O_2 (Jogeswar et al. 2006).

$$\text{Superoxide dismutase}(\text{SOD}, \text{EC}1.15.1.1): \ O^{2\bullet-} + O^{2\bullet-} + 2H+ \rightarrow 2H_2O_2 + O_2$$
$$\text{Catalase}(\text{CAT}, \text{EC}1.11.1.6): \ H_2O_2 \rightarrow H_2O + \%O_2$$
$$\text{Ascorbate peroxidase}(\text{APX}, \text{EC}1.11.1.11): \ H_2O_2 + \text{ASA} \rightarrow 2H_2O + \text{DHA}$$
$$\text{Glutathione peroxidase}(\text{GPX}, \text{EC}1.11.1.9): \ H_2O_2 + \text{GSH} \rightarrow H_2O + \text{GSSG}$$
$$\text{Monodehydroascorbate reductase}(\text{MDHAR}, \text{EC}1.6.5.4):$$
$$\text{MDHA} + \text{NAD}(\text{P})\text{H} \rightarrow \text{ASA} + \text{NAD}(\text{P})^{+}$$
$$\text{Dehydroascorbate reductase}(\text{DHAR}, \text{EC}1.8.5.1):$$
$$\text{DHA} + 2\text{GSH} \rightarrow \text{ASA} + \text{GSSG}$$

20.4.1 Nonenzymatic Antioxidants

α-Tocopherol: α-Tocopherol is present in the cell membrane and functions as a chain-breaking antioxidant (Blake et al. 1987). Once the tocopherol radical is formed, it can migrate to the membrane surface and is reconverted to α-tocopherol by reaction with ASC or GSH. Tocopherol is a lipid-soluble antioxidant known as a potential scavenger of ROS and lipid radical. In biomembranes, it has both antioxidant and nonantioxidant functions. Tocopherol acts as a $^1O_2$ scavenger and gives thylakoid membrane stability. It is located in the thylakoid membrane of the chloroplast. Tocopherol has four isomers, among which α-tocopherol has the highest antioxidative activity.(Kamal-Eldin and Appelqvist 1996). In the leaves of many plant species, α-tocopherol is found in high levels, but the leaves are low in γ-tocopherol. It is estimated that one molecule of α-tocopherol can lead to energy transfer (Munné-Bosch 2005). Under severe dehydration stress, which affects membrane lipids, the role of tocopherol seems very promising for stabilizing membranes.

Ascorbic acid: ASA is one of the most powerful antioxidants in most plant cell types, organelles, and the apoplast (Horemans et al. 2000; Smirnoff 2000). It occurs in all plant tissues, usually at higher levels in photosynthetic cells and meristems. Its concentration is reported to be highest in mature leaves with a fully developed chloroplast and highest chlorophyll. About 30–40 % of the total ASC is in the chloroplast, and stomatal concentrations as high as 50 mM have been reported (Foyer and Noctor 2005a, b). In the aqueous phase, AA donates electron in a wide range of enzymatic and nonenzymatic Enzymatic ROS-scavenging reactions under drought and salinity stress reactions and in detoxification of ROS compound. ASA detoxifies $O_2^{\bullet-}$, $OH^{\bullet}$, and $^1O_2$ and H_2O_2 to H_2O via APX reaction (Noctor and Foyer 1998a, b). ASA regenerates TOC from the tocopheroxyl radical, which provides membrane protection (Horemans et al. 2000; Smirnoff 2000). Oxidation of ASA occurs in two steps: First, MDHA is produced and if this compound is not rapidly reduced to form ASCs, it disproportionates into ASA and DHA. ASA is also implicated in the pH-mediated modulation of PSII activity, and its downregulation is associated with zeaxanthin formation. The role of ASA under salinity and drought stress in operating several antioxidative defense mechanisms, including the ASC–GSH pathway, is extremely significant.

Carotenoids: Carotenoids are pigments found in plants and microorganisms. They are lipid-soluble antioxidants that have a multitude of functions in plant metabolism, including oxidative stress tolerance. They are also referred to as antenna molecules; they absorb 450–570 nm of the visible spectrum and pass the captured energy to the chlorophyll. Another important role is to detoxify various forms of ROS. Carotenoids can protect the photosystem by reacting with liquid peroxidation products to terminate chain reactions, by scavenging $^1O_2$ and dissipating the energy as heat, by reacting excited chlorophyll molecules to prevent the formation of $^1O_2$, or by dissipating excess excitation energy through the xanthophyll cycle. β-carotene in photosynthetic tissue may be accomplished via the direct quenching of chlorophyll, which prevents $^1O_2$ generation and thereby inhibits oxidative damage caused by abiotic stresses, including salinity and drought (Collins 2001). After the quenching of chlorophyll, energy is transformed into carotenoids from chlorophyll, which leads to the formation of ROS in low concentrations (Mortensen et al. 2001) .

20.4.2 Mannitol

In response to salinity, mannitol is accumulated in many plant species (Stoop et al. 1996). Mannitol accumulation does not affect photosynthesis. The mannitol-synthesizing ability was introduced into transgenic tobacco by the *E. coli mt1D* gene encoding mannitol dehydrogenase. Transgenic tobacco plants accumulate more mannitol, but these plants were said to be more salt-tolerant. (Tarczynski et al. 1993). Mannitol mainly scavenges hydroxyl radicals from cells (Smirnoff 2000). This might be significant for plants exposed to drought and high salinity as there is strong evidence that the oxidative generation of active oxygen species increases under such conditions (Biehler and Fock 1996).

20.5 ROS–Antioxidant Interaction in Redox Signaling Under Drought and Salinity Stress

It is obvious that various abiotic stresses lead to the overproduction of ROS in plants, which is toxic; damages carbohydrates, proteins, lipids, and DNA; and ultimately results in oxidative stress (Gill and Tuteja 2010). Drought and salinity stresses are two of the most important abiotic stresses that crops can experience. Drought and salinity response in different plants depend on the species' inherent "strategy" and the duration and severity of the period of drought, which essentially is the outcome of the plant's oxidant–antioxidant interaction (Fig. 20.1). A prolonged period of drought stress will result in oxidative damage from the overproduction of ROS (Cruz and de Carvalho 2008). In stressed conditions, the production and elimination of ROS generally maintain a balance (Foyer and Noctor 2000; Porcel et al. 2003). In optimized conditions, ROS are not toxic; rather, they are helpful in signaling processes and in expressing a number of genes useful in numerous physiological plant processes like root hair growth, stomatal closure, and PCD. If the amount of ROS exceeds the amount that can be removed by enzymatic (SOD, CAT, APX, GR, MDHAR, DHAR, GPX, GuPX, and GST) or nonenzymatic (ASA, GSH, phenolic compounds, alkaloids, nonprotein amino acids, and α-tocopherols) antioxidant systems (Table 20.1) or if the normal amount of antioxidants is depleted, then damage appears in the photosynthetic apparatus and leads to cell destruction because of oxidative stress (Gill and Tuteja 2010). Differential antioxidant capacities were found to be absolutely associated with changes to the competence of redox-regulatory metabolisms of closely related germplasms (Table 20.2).

Drought: Drought stress affects vital metabolic functions and maintenance of turgor pressure (cell expansion and formation). In this situation, to minimize water loss, plants naturally close their stomata. But stomatal closure also decreases the CO_2 supply within the plant leaves and CO_2 fixation in photosynthesis, which disturb the well-tuned balance between ROS production and scavenging processes (Mittler 2002). ROS including highly reactive $^1O_2$, $OH^{\bullet}$, O_2^-, and H_2O_2 can damage cells under drought stress by different oxidative processes such as membrane lipid peroxidation, enzyme inhibition, protein oxidation, and damage of nucleic acids (Grene 2002). In different plants, diverse antioxidase activities are found against drought stress. SOD, peroxidase (POD), CAT, and APX form the antioxidase system (Chen et al. 2013). SOD is the center of the antioxidase system, which is widespread in the plant body (Scandaliol 1993). Again, as members of the antioxidase system, POD and CAT play a very important role in drought stress (Chen et al. 2013). Previous research showed that the protective enzymes of rice seedlings were significantly increased by pretreating seeds with 10–50 % polyethylene glycol than by pretreating seeds with water (Sun et al. 2010). In wheat, rice, pea, tepary bean, and olive trees, water stress increases SOD activity, causing increased oxidative stress tolerance of plants (Cruz and de Carvalho 2008). In sunflower seedlings, however, SOD activity decreased under water stress conditions, which might help to keep stomata slightly open and avoid complete inhibition of CO_2 fixation (Cruz and de Carvalho 2008). It is an adaptive feature of plants in drought stress (Cruz and de Carvalho 2008). Multiple previous studies of drought stress effects in plants showed that APX, CAT, GR, and GPX/GST were increased during drought stress in different plants. APX activity increased in rice, wheat, cotton, maize, pea, and so on; CAT increased in alfalfa, drought-tolerant rice, maize, pea, and so on; increased activity of GR was found in maize, wheat, cowpea, tepary bean, common bean, cotton, spring wheat, and more; and GPX/GST activity was found to increase in drought-tolerant varieties of rice. Increased APXs scavenge elevated intracellular H_2O_2 under drought stress, whereas increased GPXs protect biomembranes reacting with

Trx (Cruz and de Carvalho 2008). According to Malan et al. 1990, in maize, enhanced activity of antioxidative enzymes shows increased drought tolerance of the plants. Antioxidants like ASA, GSH, carotenoid, and others play an important role in balancing the concentration of ROS in plant cells by directly or indirectly quenching the ROS (Shao et al. 2008). Several plasma membrane–anchored protein kinases (receptor-like kinases, or RLKs) act as signal transducers under drought stress (Xiong et al. 2002; Marshall et al. 2012). RLKs are plant-specific, containing a transmembrane domain and intracellular kinase domain with or without an extracellular domain or only containing an intracellular kinase domain (Shiu and Bleecker 2001, 2003; Jurca et al. 2008; Vij et al. 2008). RLKs sense environmental stimuli leading to homodimerization or heterodimerization, which is followed by autophosphorylation of the cytoplasmic kinase domain, and downstream signaling components are activated by transphosphorylation (Morris and Walker 2003). A cytoplasmic RLK gene of rice named *GROWTH UNDER DROUGHT KINASE* (*GUDK*) is drought inducible and required for grain yield. Loss-of-function *gudk* mutant lines show a reduction in photosynthesis under controlled drought stress at the vegetative stage and show sensitivity to salinity and osmotic stress at the seedling stage. Transactivation assays confirmed that *GUDK* is required for activation of stress genes by TF *OsAP37* and regulation of the grain yield in rice (Ramegowda et al. 2014). For many drought-related genes, miRNAs function as critical posttranscriptional modulators controlling their expression. These miRNAs, combined with their target genes, constitute large regulatory networks that control metabolic pathways in the response to drought stress. They are involved in drought stress responses, including ABA response, osmoprotection, and antioxidant defense, by downregulating the respective target genes encoding regulatory and functional proteins (Ding et al. 2013).

Salinity: Plant response under salt stress showed decreasing stomatal conductance, which can avoid excessive water loss and decreases the intracellular CO_2 concentration. It depletes the oxidized $NADP^+$ (final acceptor of electrons in PSI) and increases the leakage of electrons to O_2 (Abogadallah 2010). In salinity stress, the ratio of Ca^{2+} or K^+ to Na^+ becomes unfavorable, and toxic intracellular Na^+ concentration and membrane lipid peroxidation are enhanced (Abogadallah 2010). At a concentration above 100 mM, Na^+ severely inhibits many enzymes, including photosynthetic ones (Munns et al. 2006). In salt-acclimated plants, increased primary metabolites linked to amino acid and nitrogen or carbohydrate and polyol metabolism have been found that play a great role in scavenging ROS and excess accumulated ammonium ions, osmotic adjustment, membrane and protein protection, and so forth (Ashraf and Akram 2009). Depletion of organic acids under salt stress following decreased photosynthesis may be involved in compensating for ionic imbalance .A salt-sensitive *O. sativa* cultivar Taipei 309 can tolerate the oxidative stress of 0.35 M NaCl by the mechanism of tolerance existing in its shoot tissue. It showed elevated SOD and H_2O_2 activity. Activities of APX and CAT were normal as in control condition; an early increase in GR activity and no extensive increases in lipid peroxidation were found. It also balanced the decline of the GSH concentration, increased the GSSG concentration, and returned to the concentration seen in normal control culture. In barley, the salt-tolerant cultivar induces high expression of S-adenosylmethionine (SAM) synthase, which catalyzes SAM, the universal donor of methyl group for carbohydrates, membrane lipids, flavonoids, DNA, and protein (Roje 2006; Witzel et al. 2009). Another one is a carboxymethylenebutenolidase-like protein with a dienelactone hydrolase (DLH) domain. In plants, DLH substrates are unknown, but the presence of it in plants under salt stress suggests that it has a regulatory role in redox metabolism. The higher abundance of peroxidase in the tolerant variety suggests that cell wall modifications may occur there to reduce ion influx under salinity stress. The ROS scavenging activity was performed by GST and lactoylglutathione lyase (Glyoxalase I) in this plant (Witzel et al. 2009).

A high concentration of salt in the soil solution can prevent water uptake, which induces water-deficit effects as found in drought stress. So it is simple to find similarities in the signaling and response of salinity and drought-stressed plants including ABA biosynthesis, closing of stomata, and increased production of compatible osmoprotectants and antioxidants (Tippmann et al. 2006). There is also significant signaling crosstalk between drought and salinity stresses (Shinozaki et al. 2003). The stress is recognized by ROS and modulation of intracellular calcium (Ca^{2+}) directly or mediated by receptor. Then the signals through the specific phosphorylation cascades like mitogen-activated protein kinases (MAPKs) and calcium-dependent protein kinases (CDPKs) interact with TFs and response genes and respond to stress (Fig. 20.1). Due to limitation of photosynthesis in drought and salinity stresses, the redox status of mitochondria is changed and promotes increased ROS production along with downstream signaling pathway (Tippmann et al. 2006). For the dual nature of ROS, antioxidant systems play an additional role in redox signaling to optimize the amount of but not entirely remove ROS. Enhanced production of ROS during stress acts as a signal to initiate the stress response. Those signaling pathways include MAPK cascade, the Ca^{2+} binding calmodulin, histidine kinase sensor, TFs, and so on (Apel and Hirt 2004). Another important messenger in drought stress is ABA. All these signaling pathways are interconnected. ABA biosynthesis is essential for the activation of many protective measures in drought and salinity stresses and starts with the decrease in water potential (Tippmann et al. 2006; Cutler et al. 2010; Raghavendra et al. 2010; Weiner et al. 2010). ABA initiates several genes mediated through ABA-responsive *cis-* and *trans*-acting factors (Chaves et al. 2003; Golldack et al. 2014). In the promoters of some of the genes, a specific ABRE is found that is upregulated during drought stress (Shinozaki et al. 2003; Yamaguchi-Shinozaki and Shinozaki 2005, 2006). There is also an ABA-independent signaling pathway for drought-induced gene expression containing a DRE (Tippmann et al. 2006). Drought-induced genes encode proteins with repair and damage control functions along with the proteins related to metabolic, osmotic, and structural adjustment (Ingram and Bartels 1996). Proline biosynthesis increases the concentration of compatible osmoprotectants in the cells, and ROS scavenging proteins limit damage by secondary oxidative stress (Chaves et al. 2003). In the case of salinity stress, the salt overly sensitive (SOS) pathway was studied in *Arabidopsis* (Zhu et al. 2007). Excess Na^+ triggers a cytosolic Ca^{2+} signal. Ca^{2+} change in salt-stressed condition is sensed by a calcium binding protein encoded by SOS3. The Ca^{2+}-bound SOS3 then interacts with SOS2 and forms a complex to regulate the expression of SOS1 (the plasma membrane Na^+/H^+ antiporter) to reestablish ionic homeostasis within cells (Zhu et al. 2007). AtNHX (vacuolar Na^+/H^+ antiporter) helps to compartmentalize Na^+ in the vacuole to maintain the intracellular K^+ status, and overexpression enhances the salt tolerance of *Arabidopsis* plants (Apse et al. 1999). In drought and salinity stresses, five signal transduction pathways that are both ABA dependent and ABA independent are followed. MYB2 and MYC2 express the ABA-inducible gene *RD22*. The RD26 NAC TF also expresses the ABA-responsive stress gene (Fig. 20.1). The bZIP-type AREB/ABF TFs AREB1, AREB2, and AREB3 cooperatively target ABRE-dependent gene expression (Yoshida et al. 2010). DRE mainly regulates genes of drought and salinity stresses. DREB2s are important TFs in dehydration and high salinity stress-responsive gene expression. The *ERD1* gene is expressed by NAC and HD-ZIP TFs (Shinozaki and Yamaguchi-Shinozaki 2007).

20.6 Antioxidant and Redox Sensing Under Drought and Salinity Stresses

ROS-mediated signaling under abiotic stress including drought and salinity stresses largely involves heterotrimeric G-proteins (Joo et al. 2005) and protein phosphorylation regulated by specific MAP kinases and protein Tyr phosphatases (Kovtun et al. 2000; Gupta and Luan 2003; Rentel et al. 2004). The biochemical and structural bases of kinase pathway activation by ROS remain to be established in plants, but thiol oxidation likely plays a key role. The best-characterized redox signal transduction system in plants is the stromal ferredoxin–Trx system, which functions in the regulation of photosynthetic carbon assimilation (Fig. 20.2). Signal transmission involves disulfide–thiol conversion in target enzymes and is probably achieved by a light-induced decrease in the Trx redox potential (Setterdahl et al. 2003). Thiol groups are likely important in other types of redox signal transduction, including ROS sensing by receptor kinases, such as ETR1 (Desikan et al. 2005). Thiol-based regulation may be important in plant acclimation to environmental change, particularly where redox interactions play a key role in the orchestration of the dehydration and salinity stress response. In plants, as in other organisms, thiol-containing domains are oxidized by ROS, either directly or indirectly, to give relatively stable oxidation products with modified physical conformations or biochemical activities (Bauer et al. 1999; Delauney et al. 2002). In addition to disulfides, other oxidized species of Cys sulfur that may be important in redox sensing include sulfenic acid and glutathionylated Cys, sulfenyl amide groups, and sulfur–metal bonds. All these signaling mechanisms are outcomes of the direct effect of ROS on TFs or proteins of signal transduction pathways operating under salinity and dehydration stress.

Overexpression of TFs like Zat10, Zat12, and JERF3 enhance the expression of ROS-scavenging genes and tolerance to salinity and dehydration stress (Wu et al.

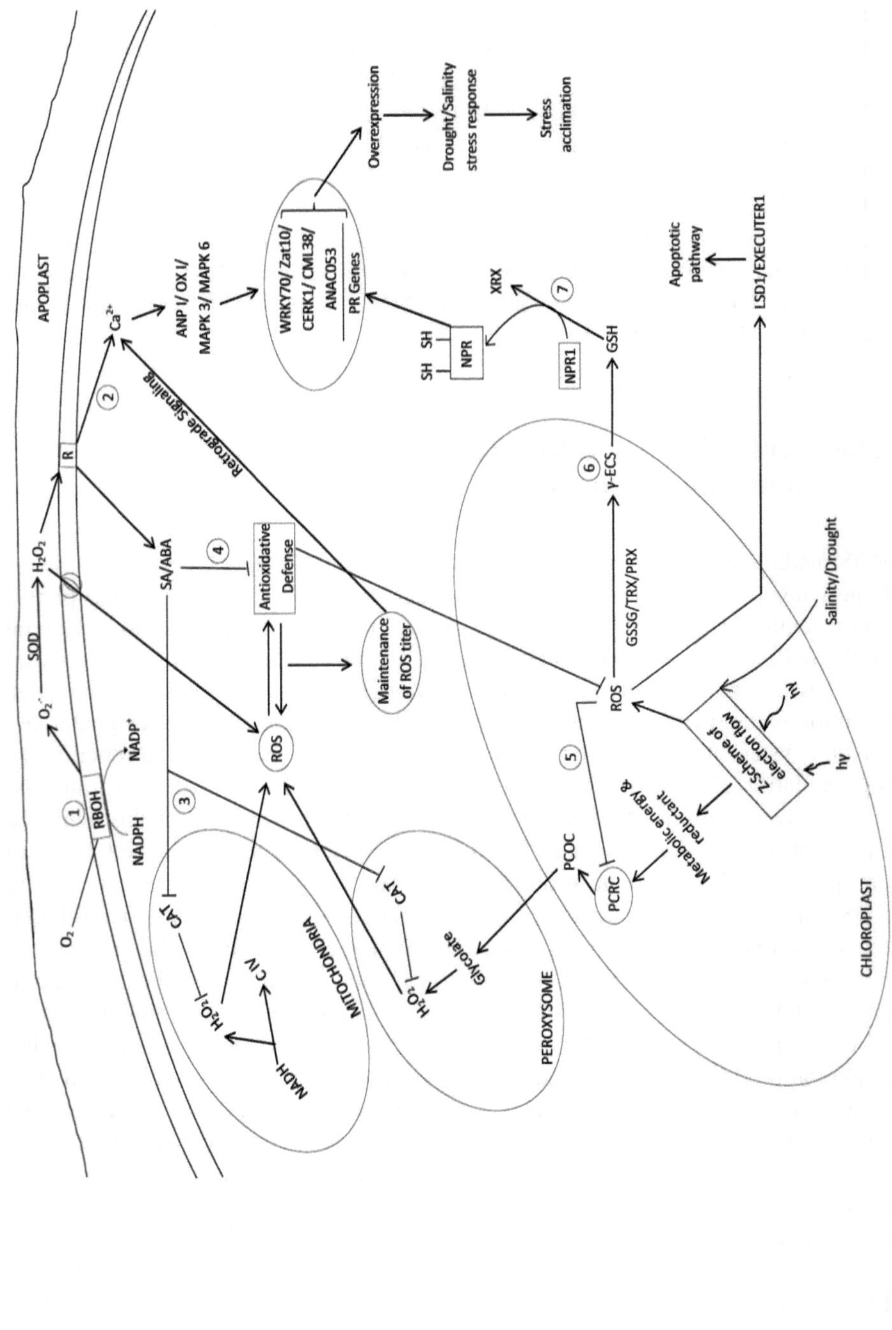
APOPLAST
O_2
RBOH
NADPH
$NADP^+$
$O_2^{\cdot-}$
SOD
H_2O_2
R
Ca^{2+}
Retrograde Signaling
ANP I/ OX I/ MAPK 3/ MAPK 6
WRKY70/ Zat10/ CERK1/ CML38/ ANAC053
PR Genes
Overexpression
Drought/Salinity stress response
Stress acclimation
SA/ABA
Antioxidative Defense
Maintenance of ROS titer
ROS
CAT
MITOCHONDRIA
NADH
C IV
PEROXYSOME
Glycolate
PCOC
PCRC
Metabolic energy & reductant
Z-Scheme of electron flow
hγ
Salinity/Drought
GSSG/TRX/PRX
γ-ECS
GSH
NPR1
NPR
SH SH
XRX
Apoptotic pathway
LSD1/EXECUTER1
CHLOROPLAST
1
2
3
4
5
6
7

Fig. 20.2 Hypothetical scheme showing redox redox-regulatory elements with oxidant–antioxidant signaling associated with drought and NaCl stress responses. Plasma membrane–bound NADPH oxidase (RBOH) are activated by drought and salinity as membrane-mediated signaling and produce apoplastic H_2O_2. Apoplastic H_2O_2 can cause oxidative damage to membrane-bound receptors and is buffered/opposed by ascorbateperoxidase (*1*). ROS production, particularly H_2O_2, activates signaling through MAP kinases (2), downstream of which exhibit upregulation of defense genes. Downregulation of antioxidative defense could also occur because of posttranslational modulation of hemefunctions by ABA and SA (*3*) and programmed withdrawal of antioxidative enzyme expression (*4*). Uncontrolled generation of ROS and oxidative modification of chloroplast metabolism by glutathionylation and interaction with thioredoxin, peroxiredoxin- mediated enzymes (*5*). Activation of GSH synthesis under oxidative stress (*6*) triggered by unfavorable environmental cues, which is further linked to activation of thioredoxin or glutaredoxin, which subsequently reduces MPR 1 (*7*), received for activation of PR genes. Enhanced level of ROS (chloroplastic) may activate apoplastic pathway as well

2008; Miller et al. 2010). Furthermore, overexpression of mitogen-activated kinase kinase (MAPKK1) in *Arabidopsis* enhances the activities of the MAPK cascade, which is also activated by the ROS–antioxidant signaling network under salinity and dehydration stress. The net outcome of ROS–antioxidant interaction is that deficiency in MKK1 resulted in increased ROS production and enhanced stress sensitivity (Xing et al. 2008).

Recently, the role of antioxidative enzymes apart from their traditional scavenging function has become evident in signaling (Fig. 20.2). Knockout of the *APX1* gene was shown to perform and grow better than wild-type plants under salinity stress (Miller et al. 2007). Similarly, the *APX1*-deficient mutant of *Arabidopsis* that eventually caused elevated accumulation of H_2O_2 caused induction of several defense genes under stress (Davletova et al. 2005a, b). Likewise the antisense *CAT1* and *APX1* in tobacco are found to be suffering from oxidative damage, but their double antisense lines became more tolerant (Rizhsky et al. 2002). In fact, the overexpression of *Zat7* was found to be associated with the knocked-out *Apx* plant, which resulted in enhanced expression of defense transcripts such as *WRKY 70, NHX1, AOX1*, and others, offering dramatic tolerance of plants to dehydration and salinity stress (Ciftci-Yilmaz et al. 2007).

The role of antioxidants and redox-sensing mechanisms in conferring stress tolerance of plants under salinity and dehydration stress can be further corroborated by examining the role of several ROS-responsive transcripts encoding TFs and signaling proteins (Fig. 20.2). A good number of ROS-responsive transcripts encoding TFs are found to be upregulated ($\geq$2) in *apx1* plants under oxidative stress, salinity (*WRKY 70, CERK1*), and dehydration stress (*CML 38, ANACO53*). Although the role of these ROS-responsive regulators are not precisely known, experimental evidence supports their role in specific acclimation to each salinity and dehydration stress in accordance with the typical role of redox regulation.

20.7 Conclusions and Perspective

The redox signal transduction with antioxidants and ROS interacting at the metabolic interface is a universal feature of plants. We have presented an interpretation of how plant cells, through a series of interacting components with different antioxidant buffering capacities, determine the fate of the cell under salinity and dehydration stresses. The unfavorable environmental cues trigger on specific signaling that is controlled or buffered by antioxidants through modulation of redox status and sensing mechanisms. As a consequence, several redox-sensitive signal transductions in locations such as thylakoids, apoplast, cytosol, and ER occur. In fact, both prooxidants and antioxidants contribute their signaling role using several cell signaling pathways (kinase-dependent and -independent pathways) and redox-sensing mechanisms. In these signaling episodes, antioxidants not only are the passive onlookers, but rather function as key signaling compounds that establish a dynamic metabolic boundary between stress perception and physiological response. The current data suggest the role of both nonenzymatic and enzymatic antioxidants as key arbitrators of intracellular redox status and potential with their differential status between cellular compartments permitting antioxidant-driven vectorial signaling. How antioxidants and ROS quenchers initiate and control redox signal transduction and subsequently trigger differential expression of genes associated with the performance of the plant under salinity and dehydration stress will be the most fascinating subject of future research in plant redox biology.

Acknowledgments A.C., N.D. acknowledges UGC, New Delhi, and Govt. of West Bengal, respectively, for Research fellowship. Research facility of CAS (UGC) Govt. of India to the Department of Botany, University of Burdwan is also gratefully acknowledged

References

Abbasi AR, Hajirezaei M, Hofius D, Sonnewald U, Voll LM. Specific roles of alpha- and gamma-tocopherol in abiotic stress responses of transgenic tobacco. Plant Physiol. 2007;143:1720–38.

Abogadallah GM. Antioxidative defense under salt stress. Plant Signal Behav. 2010;5(4):369–74.

Achard P, et al. Plant DELLAs restrain growth and promote survival of adversity by reducing the levels of reactive oxygen species. Curr Biol. 2008;18:656–60.

Agarwal PK, Jha B. Transcription factors in plants and ABA dependent and independent abiotic stress signaling. Biol Planta. 2010;54:201–12.

Agarwal PK, Shukla PS, Gupta K, Jha B. Bioengineering for salinity tolerance in plants: state of the art. Mol Biotechnol. 2013;54:102–23.

Apel K, Hirt H. Reactive oxygen species: metabolism, oxidative stress, and signal transduction. Annu Rev Plant Biol. 2004;55:373–99.

Apse MP, Aharon GS, Snedden WA, Blumwald E. Salt tolerance conferred by overexpression of a vacuolar Na^+/H^+ antiport in *Arabidopsis*. Science. 1999;285:1256–8.

Arnholdt-Schmitt B, Costa JH, De Melo DF. AOX—a functional marker for efficient cell reprogramming under stress? Trends Plant Sci. 2006;11:281–7.

Asada K. The water-water cycle in chloroplasts: scavenging of active oxygens and dissipation of excess photons. Annu Rev Plant Physiol Plant Mol Biol. 1999;50:601–39.

Asada K. The water-water cycle as alternative photon and electron sinks. Philos Trans R Soc Lond B Biol Sci. 2000;355:1419–31.

Asada K. Production and scavenging of reactive oxygen species in chloroplasts and their functions. Plant Physiol. 2006;141:391–6.

Ashraf M, Akram NA. Improving salinity tolerance of plants through conventional breeding and genetic engineering: analytical comparison. Biotechnology. 2009;27:744–52.

Ashraf M, Foolad MR. Roles of glycine betaine and proline in improving plant abiotic stress resistance. Environ Exp Bot. 2007;59:206–16.

Attia H, Arnaud N, Karray N, Lachaâl M. Long-term effects of mild salt stress on growth, ion accumulation and superoxide dismutase expression on *Arabidopsis* rosette leaves. Physiol Plant. 2008;132:293–305.

Aubert S, Hennion F, Bouchereau A, Gout E, Bringly R, Dorne AJ. Subcellular compartmentation of proline in the leaves of the subantarctic Kerguelen cabbage *Pringlea antiscorbutica* R. Br. in vivo 13C-NMR study. Plant Cell Environ. 1999;22:255–9.

Badawi GH, Yamauchia Y, Shimada E, Sasaki R, Kawano N, et al. Enhanced tolerance to salt stress and water deficit by overexpressing superoxide dismutase in tobacco (*Nicotiana tabacum*) chloroplasts. Plant Sci. 2004;66:919–28.

Bartoli CG, Gomez F, Martinez DE, Guiamet JJ. Mitochondria are the main target for oxidative damage in leaves of wheat (*Triticum aestivum* L.). J Exp Bot. 2004;403:1663–9.

Bashandy T, Guilleminot J, Vernoux T, Caparros-Ruiz D, Ljung K, Meyer Y, Reichheld JP. Interplay between the NADP-linked thioredoxin and glutathione systems in Arabidopsis auxin signaling. Plant Cell. 2010;22:376–91.

Bauer CE, Elsen S, Bird TH. Mechanisms for redox control of gene expression. Annu Rev Microbiol. 1999;53:495–523.

Ben Hamed K, Ellouzi H, Talbi OZ, Hessini K, Slama I, Ghnaya T, Bosch SM, Savoure A, Abdelly C. Physiological response of halophytes to multiple stresses. Funct Plant Biol. 2013;40:883–96.

Ben Hassine A, Ghanem ME, Bouzid S, Lutts S. An inland and a coastal population of the Mediterranean xerohalophyte species *Atriplex halimus* L. differ in their ability to accumulate proline and glycinebetaine in response to salinity and water stress. J Exp Bot. 2008;59:1315–26.

Bhattacharjee S. Reactive oxygen species and oxidative burst: roles in stress, senescence and signal transduction in plants. Curr Sci. 2005;89(7):1113–21.

Bhattacharjee S. An inductive pulse of hydrogen peroxide pretreatment restores redox- homeostasis and mitigates oxidative membrane damage under extremes of temperature in two rice cultivars (*Oryza sativa* L., Cultivars Ratna and SR 26B). Plant Growth Regul. 2012;68:395–410.

Bhattacharjee S. Membrane lipid peroxidation and its conflict of interest: two faces of oxidative stress. Curr Sci. 2014;107(11):1811–23.

Biehler K, Fock H. Evidence for the contribution of the Mehler-peroxidase reaction in dissipating excess electrons in drought stressed wheat. Plant Physiol. 1996;112:265–72.

Blake DR, Allen RE, Lunec J. Free radicals in biological systems—a review oriented to inflammatory process. Br Med Bull. 1987;43:371–85.

Boguszewska D, Grudkowska M, Zagdańska B. Drought responsive antioxidant enzymes in potato (*Solanum tuberosum* L.). Potato Res. 2010;53:373–82.

Cavalcanti FR, Lima JPMS, Ferreira-Silva SL, Viegas RS, Silveira JAG. Roots and leaves display contrasting oxidative response during salt stress and recovery in cowpea. J Plant Physiol. 2007;164:591–600.

Chakraborty A, Bhattacharjee S. Differential competence of redox-regulatory mechanism under extremes of temperature determines growth performances and cross tolerance in two indica rice cultivars. J Plant Physiol. 2014;176:65–77.

Chaves MM, Maroco JO, Pereira JS. Understanding plant responses to drought—from genes to the whole plant. Funct Plant Biol. 2003;30:239–64.

Chen Z, Gallie DR. Increasing tolerance to ozone by elevating folia ascorbic acid confers greater protection against ozone than increasing avoidance. Plant Physiol. 2005;138:1673–89.

Chen Q, Zhang M, Shen S. Effect of salt on malondialdehyde and antioxidant enzymes in seedling roots of Jerusalem artichoke (*Helianthus tuberosus* L.). Acta Physiol Plant. 2010;33:273–8.

Chen Q, Tao S, Bi X, Xu X, Wang L, Li X. Research progress in physiological and molecular biology mechanism of drought resistance in rice. Am J Mol Biol. 2013;3(2):102–7.

Ciftci-Yilmaz S, Morsy MR, Song L, Coutu A, Krizek BA, Lewis MW, Warren D, Cushman J, Connolly EL, Mittler R. The ear-motif of the C2H2 zinc-finger protein ZAT7 plays a key role in the defense response of *Arabidopsis* to salinity stress. J Biol Chem. 2007;282:9260–8.

Collins A. Carotenoids and genomic stability. Mutat Res. 2001;475:1–28.

Creissen GP, Broadbent P, Kular B, Reynolds H, Wellburn AR, Mullineaux PM. Manipulation of glutathione reductase in transgenic plants: implications for plant responses to environmental stress. Proc R Soc Edinb. 1994;102B:167–75.

Cutler SR, Rodriguez PL, Finkelstein RR, Abrams SR. Abscisic acid: emergence of a core signalling network. Annu Rev Plant Biol. 2010;61:651–79.

Dat JF, Pellinen R, Beeckman T, Van De Cotte B, Langebartels C, Kangasjarvi J, Inze D, Van Breusegem F. Changes in hydrogen peroxide homeostasis trigger an active cell death process in tobacco. Plant J. 2003;33:621–32.

Davletova S, Rizhsky L, Liang H, Shengqiang Z, Oliver DJ, Coutu J, Shulaev V, Schlauch K, Mittler R. Cytosolic ascorbate peroxidase 1 is a central component of the reactive oxygen gene network of *Arabidopsis*. Plant Cell. 2005a;17:268–81.

Davletova S, Schlauch K, Coutu J, Mittler R. The zinc-finger protein Zat12 plays a central role in reactive oxygen and abiotic stress signaling in *Arabidopsis*. Plant Physiol. 2005b;139:847–56.

Cruz M, de Carvalho H. Drought stress and reactive oxygen species production, scavenging and signalling. Plant Signal Behav. 2008;3(3):156–65.

del Río LA, Corpas FJ, Sandalio LM, Palma JM, Gómez M, Barroso JB. Reactive oxygen species, antioxidant systems and nitric oxide in peroxisomes. J Exp Bot. 2002a;53:1255–72.

del Río LA, Sandalio LM, Palma JM, Corpas FJ, LópezHuertas E, Romero-Puertas MC, McCarthy I. Peroxisomes, reactive oxygen metabolism, and stress-related enzyme activities. In: Baker A, Graham I, editors. Plant peroxisomes biochemistry, cell biology and biotechnological applications. Dordrecht: Kluwer Academic; 2002b. p. 221–58.

Delauney A, Pflieger D, Barrault MB, Vinh J, Toledano MB. A thiol peroxidase is an H_2O_2 receptor and redoxtransducer in gene activation. Cell. 2002;111:1–11.

Depège N, Drevet J, Boyer N. Molecular cloning and characterization of tomato cDNAs encoding glutathione peroxidase-like proteins. Eur J Biochem. 1998;253:445–51.

Desikan R, Hancock JT, Bright J, Harrison J, Weir I, Hooley R, Neill SJ. A role for ETR1 in hydrogen peroxide signaling in stomatal guard cells. Plant Physiol. 2005;137:831–4.

Diaz-Vivancos P, Clemente-Moreno MJ, Rubio M, Olmos E, García JA, Martínez-Gómez P, Hernández JA. Alteration in the chloroplastic metabolism leads to ROS accumulation in pea plants in response to plum pox virus. J Exp Bot. 2008;59:2147–60.

Dietz KJ, Jacob S, Oelze ML, Laxa M, Tognetti V, De Miranda SM, Baier M, Finkemeier I. The function of peroxiredoxins in plant organelle redox metabolism. J Exp Bot. 2006;57:1697–709.

Dietzel L, Pfannschmidt T. Photosynthetic acclimation to light gradients in plant stands comes out of shade. Plant Signal Behav. 2008;3:1116–8.

Dinakar C, Abhaypratap V, Yearla SR, Raghavendra AS, Padmasree K. Importance of ROS and antioxidant system during the beneficial interactions of mitochondrial metabolism with photosynthetic carbon assimilation. Planta. 2010;231:461–74.

Ding Y, Tao Y, Zhu C. Emerging roles of microRNAs in the mediation of drought stress response in plants. J Exp Bot. 2013;64(11):3077–86.

Dixon DP, Skipsey M, Edwards R. Roles for glutathione transferases in plant secondary metabolism. Phytochemistry. 2010;71(4):338–50. doi:10.1016/j.phytochem.2009.12.012.

Edwards EA, Rawsthorne S, Mullineaux PM. Subcellular distribution of multiple forms of glutathione reductase in leaves of pea (*Pisum sativum* L.). Planta. 1990;180:278–84.

Eltayeb AE, Kawano N, Badawi GH, Kaminaka H, Sanekata T, Morishima I, Shibahara T, Inanaga S, Tanaka K. Enhanced tolerance to ozone and drought stresses in transgenic tobacco overexpressing dehydroascorbate reductase in cytosol. Physiol Plant. 2006;127:57–65.

Eltayeb AE, Kawano N, Badawi GH, Kaminaka H, Sanekata T, Shibahara T, Inanaga S, Tanaka K. Overexpression of monodehydroascorbate reductase in transgenic tobacco confers enhanced tolerance to ozone, salt and polyethylene glycol stresses. Planta. 2007;225:1255–64.

Eyidogan F, Oz MT. Effect of salinity on antioxidant responses of chickpea seedlings. Acta Physiol Plant. 2005;29:485–93.

Flowers TJ, Gaur PM, Gowda CLL, Krishnamurthy L, Srinivasan S, Siddique KHM, Turner NC, Vadez V, Varshney RK, Colmer TD. Salt sensitivity in chickpea. Plant Cell Environ. 2010;33:490–509.

Foyer CH, Noctor G. Oxygen processing in photosynthesis: regulation and signaling. New Phytol. 2000;146(3):359–88.

Foyer CH, Noctor G. Oxidant and antioxidant signaling in plants: a re-evaluation of the concept of oxidative stress in a physiological context. Plant Cell Environ. 2005a;28:1056–71.

Foyer CH, Noctor G. Redox homeostasis and antioxidant signaling: a metabolic interface between stress perception and physiological responses. Plant Cell. 2005b;17:1866–75.

Foyer CH, Noctor G. Redox regulation in photosynthetic organisms: signaling, acclimation and practical implications. Antioxid Redox Signal. 2009;11:861–905.

Foyer CH, Lelandais M, Kunert KJ. Oxidative stress in plants. Physiol Plant. 1994;92:696–717.

Frugoli JA, Zhong HH, Nuccio ML, McCourt P, McPeek MA, Thomas TL, McClung CR. Catalase is encoded by a multigene family in *Arabidopsis thaliana* (L.) Heynh. Plant Physiol. 1996;112:327–36.

Gambarova NG, Gins MS. Characteristics of oxidative stress of plants with C3 and C4 photosynthesis during salinization. Russ Agric Sci. 2008;34:77–80.

Gapinska M, Skłodowska M, Gabara B. Effect of short- and long-term salinity on the activities of antioxidative enzymes and lipid peroxidation in tomato roots. Acta Physiol Plant. 2008;30:11–8.

Gill SS, Tuteja N. Reactive oxygen species and antioxidant machinery in abiotic stress tolerance in crop plants. Plant Physiol Biochem. 2010;48(12):909–30.

Goicoechea N, Antolin MC, Sánchez-Díaz M. Gas exchange is related to the hormone balance in mycorrhizal or nitrogen-fixing alfalfa subjected to drought. Physiol Plant. 1997;100:989–97.

Golldack D, Li C, Mohan H, Probst N. Tolerance to drought and salt stress in plant: unravelling the signalling networks. Front Plant Sci. 2014;5:151.

Grene R. Oxidative stress and acclimation mechanisms in plants. In: Somerville CR, Meyerowitz EM, editors. The Arabidopsis book. Rockville: The American Society of Plant Biologists; 2002. p. 1–20.

Gunes A, Pilbeam D, Inal A, Coban S. Influence of silicon on sunflower cultivars under drought stress, I: growth, antioxidant mechanisms and lipid peroxidation. Commun Soil Sci Plant Anal. 2008;39:1885–903.

Gupta R, Luan S. Redox control of protein tyrosine phosphatases and mitogen-activated protein kinases in plants. Plant Physiol. 2003;132:1149–52.

Hamilton EW, Heckathorn SA. Mitochondrial adaptations to NaCl. Complex I is protected by antioxidants and small heat shock proteins, whereas Complex II is protected by proline and betaine. Plant Physiol. 2001;126:1266–74.

Heazlewood JL, Verboom RE, Tonti-Filippini J, Small I, Millar AH. SUBA: the Arabidopsis subcellular database. Nucleic Acids Res. 2007;35:D213–8.

Hojati M, Modarres-Sanavy SAM, Karimi M, Ghanati F. Responses of growth and antioxidant systems in *Carthamus tinctorius* L. under water deficit stress. Acta Physiol Plant. 2010;33:105–12.

Hong CY, Hsu YT, Tsai YC, Kao CH. Expression of *ASCORBATE PEROXIDASE 8* in roots of rice (*Oryza sativa* L.) seedlings in response to NaCl. J Exp Bot. 2007;58:3273–83.

Horemans N, Foyer CH, Asard H. Transport and action of ascorbate at the plant plasma membrane. Trends Plant Sci. 2000;5:263–7.

Hossain MA, Bhattacharjee S, Armin S-M, Quin P, Xin W, Li H-Y, Buritt D J, Fujita M, Tran Lam Son. Hydrogen peroxide priming modulates abiotic oxidative stress: insight from ROS detoxification and scavenging. Frontiers in Plants Sci. doi:10.3389/fpls..2015.00420.

Hu X, Zhang A, Zhang J, Jiang M. Abscisic acid is a key inducer of hydrogen peroxide production in leaves of maize plants exposed to water stress. Plant Cell Physiol. 2006;47:1484–95.

Iglesias MJ, Terrile MC, Bartoli CG, D'Ippólito S, Casalongué CA. Auxin signaling participates in the adaptative response against oxidative stress and salinity by interacting with redox metabolism in *Arabidopsis*. Plant Mol Biol. 2010;74:215–22.

Ingram J, Bartels D. The molecular basis of dehydration tolerance in plants. Annu Rev Plant Physiol Plant Mol Biol. 1996;47:377–403.

Jebara S, Jebara M, Limam F, Aouani ME. Changes in ascorbate peroxidase, catalase, guaiacol peroxidase and superoxide dismutase activities in common bean (*Phaseolus vulgaris*) nodules under salt stress. J Plant Physiol. 2005;162:929–36.

Jiménez A, Hernandez JA, Del Río LA, Sevilla F. Evidence for the presence of the ascorbate-glutathione cycle in mitochondria and peroxisomes of pea leaves. Plant Physiol. 1997;114:275–84.

Jogeswar G, Pallela R, Jakka NM, Reddy PS, Rao JV, Sreenivasulu N, Kavi Kishor PB. Antioxidative response in different sorghum species under short-term salinity stress. Acta Physiol Plant. 2006;28:465–75.

Joo JH, Wang S, Chen JG, Jones AM, Fedoroff NV. Different signaling and cell death roles of heterotrimeric G protein a and b subunits in the *Arabidopsis* oxidative stress response to ozone. Plant Cell. 2005;17:957–70.

Jubany-Marí T, Munné-Bosch S, Lopez-Carbonell M, Alegre L. Hydrogen peroxide is involved in the acclimation of the Mediterranean shrub, *Cistus albidus* L., to summer drought. J Exp Bot. 2009;60:107–20.

Jurca ME, Bottka S, Fehér A. Characterization of a family of Arabidopsis receptor-like cytoplasmic kinases (RLCK class VI). Plant Cell Rep. 2008;27(4):739–48.

Kamada-Nobusada T, Hayashi M, Fukazawa M, Sakakibara H, Nishimura M. A putative peroxisomal polyamine oxidase, AtPAO4, is involved in polyamine catabolism in *Arabidopsis thaliana*. Plant Cell Physiol. 2008;49:1272–82.

Kamal-Eldin A, Appelqvist LA. The chemistry and antioxidant properties of tocopherols and tocotrienols. Lipids. 1996;31:671–701.

Karpinska B, Wingsle G, Karpinski S. Antagonistic effects of hydrogen peroxide and glutathione on acclimation to excess excitation energy in *Arabidopsis*. IUBMB Life. 2000;50:21–6.

Karuppanapandian T, Manoharan K. Uptake and translocation of tri- and hexa-valent chromium and their effects on black gram (*Vigna mungo* L. Hepper cv. Co4) roots. J Plant Biol. 2008;51:192–201.

Karuppanapandian T, Sinha PB, Kamarul Haniya A, Manoharan K. Differential antioxidative responses of ascorbate-glutathione cycle enzymes and metabolites to chromium stress in green gram (*Vigna radiata* L. Wilczek) leaves. J Plant Biol. 2006a;49:440–7.

Karuppanapandian T, Sinha PB, Premkumar G, Manoharan K. Chromium toxicity: correlated with increased in degradation of photosynthetic pigments and total soluble protein and increased peroxidase activity in green gram (*Vigna radiata* L.) seedlings. J Swamy Bot Cl. 2006b;23:117–22.

Koji Y, Shiro M, Michio K, Mitsutaka T, Hiroshi M. Antioxidant capacity and damages caused by salinity stress in apical and basal regions of rice leaf. Plant Prod Sci. 2009;12:319–26.

Koornneef M, Leon-Kloosterziel KM, Schwartz S, Hand Zeevaart JAD. The genetic and molecular dissection of abscisic acid biosynthesis and signal transduction in *Arabidopsis*. Plant Physiol Biochem. 1998;36:83–9.

Koussevitzky S, Suzuki N, Huntington S, Armijo L, Sha W, Cortes D, Shulaev V, Mittler R. Ascorbate peroxidase 1 plays a key role in the response of *Arabidopsis thaliana* to stress combination. J Biol Chem. 2008;283:34197–203.

Kovtun Y, Chiu W-L, Tena G, Sheen J. Functional analysis of oxidative stress-activated mitogen-activated protein kinase cascade in plants. Proc Natl Acad Sci U S A. 2000;97:2940–5.

Krieger-Liszkay A. Singlet oxygen production in photosynthesis. J Exp Bot. 2005;56(411): 337–46.

Kukreja S, Nandval AS, Kumar N, Sharma SK, Unvi V, Sharma PK. Plant water status, H_2O_2 scavenging enzymes, ethylene evolution and membrane integrity of *Cicer arietinum* roots as affected by salinity. Biol Plant. 2005;49:305–8.

Kwak JM, Mori IC, Pei ZM, Leonhardt N, Torres MA, Dangl JL, Bloom RE, Bodde S, Jones JDG, Schroeder JI. NADPH oxidase AtrbohD and AtrbohF genes function in ROS dependent ABA signaling in Arabidopsis. EMBO J. 2003;22:2623–33.

Li WJ, Feng H, Fan JH, Zhang RQ, Zhao NM, Liu JY. Molecular cloning and expression of a phospholipid hydroperoxide glutathione peroxidase homolog in *Oryza sativa*. Biochim Biophys Acta. 2000;1493:225–30.

Li WJ, Gueta-Dahan Y, Lev-Yadun S, Gollop R, Ben-Hayyim G. The salt-stress signal transduction pathway that activates the gpx1 promoter is mediated by intracellular H_2O_2, different from the pathway induced by extracellular H_2O_2. Plant Physiol. 2004;135:1685–96.

Li JT, Qiu ZB, Zhang XW, Wang LS. Exogenous hydrogen peroxide can enhance tolerance of wheat seedlings to salt stress. Acta Physiol Plant. 2011;33:835–42. doi:10.1007/s11738-010-0608-5.

Light GG, Mahan JR, Roxas VP, Allen RD. Transgenic cotton (*Gossypium hirsutum* L.) seedlings expressing a tobacco glutathione S-transferase fail to provide improved stress tolerance. Planta. 2005;222:346–54.

Lopez-Huertas E, Charlton WL, Johnson B, Graham I, Baker A. Stress induces peroxisome biogenesis genes. EMBO J. 2000;19(24):6770–7.

Lu ZQ, Liu D, Liu SK. Two rice cytosolic ascorbate peroxidases differentially improve salt tolerance in transgenic *Arabidopsis*. Plant Cell Rep. 2007;26:1909–17.

Malan C, Greyling M, Gressel J. Correlation between Cu Zn superoxide dismutase and glutathione reductase and environmental and xenobiotic stress tolerance in maize inbreds. Plant Sci. 1990;69:157–66.

Marshall A, Aalen RB, Audenaert D, Beeckman T, Broadley MR, Butenko MA, Caño-Delgado AI, de Vries S, Dresselhaus T, Felix G, et al. Tackling drought stress: receptor-like kinases present new approaches. Plant Cell. 2012;24(6):2262–78.

Martí MC, Olmos E, Calvete JJ, Díaz I, Barranco-Medina S, Whelan J, et al. Mitochondrial and nuclear localization of a novel pea thioredoxin: identification of its mitochondrial target proteins. Plant Physiol. 2009;150:646–57.

Matysik J, Alia A, Bhalu B, Mohanty P. Molecular mechanisms of quenching of reactive oxygen species by proline under stress in plants. Curr Sci. 2002;82:525–32.

Mhamdi A, et al. Catalase function in plants: a focus on Arabidopsis mutants as stress-mimic models. J Exp Bot. 2010;61:4107–320.

Millar AH, Mittova V, Kiddle G, Heazlewood JL, Bartoli CG, Theodoulou FL, Foyer CH. Control of ascorbate synthesis by respiration and its implication for stress responses. Plant Physiol. 2003;133:443–7.

Miller G, Suzuki N, Rizhsky L, Hegie A, Koussevitzky S, Mittler R. Double mutants deficient in cytosolic and thylakoid ascorbate peroxidase reveal a complex mode of interaction between reactive oxygen species, plant development, and response to abiotic stresses. Plant Physiol. 2007;144:1777–85.

Miller G, Shulaev V, Mittler R. Reactive oxygen signaling and abiotic stress. Physiol Plant. 2008;133:481–9.

Miller G, Suzuki N, Ciftci-Yilmaz S, Mittler R. Reactive oxygen species homeostasis and signalling during drought and salinity stresses. Plant Cell Environ. 2010;33:453–67.

Mittal R, Dubey RS. Behaviour of peroxidases in rice: changes in enzymatic activity and isoforms in relation to salt tolerance. Plant Physiol Biochem. 1991;29:31–40.

Mittler R. Oxidative stress, antioxidants, and stress tolerance. Trends Plant Sci. 2002;7:405–10.

Mittler R, Vanderauwera S, Gollery M, Van Breusegem F. Reactive oxygen gene network of plants. Trends Plant Sci. 2004;10:490–8.

Mittova V, Tal M, Volokita M, Guy M. Up-regulation of the leaf mitochondrial and peroxisomal antioxidative systems in response to salt-induced oxidative stress in the wild salt-tolerant tomato species *Lycopersicon pennellii*. Plant Cell Environ. 2003;26:845–56.

Mittova V, Volokita M, Guy M, Tal M. Activities of SOD and the ascorbate–glutathione cycle enzymes in subcellular compartments in leaves and roots of the cultivated tomato and its wild salt-tolerant relative Lycopersicon pennellii. Physiol Plant. 2000;110:42–51.

Moller IM, Jensen PE, Hansson A. Oxidative modifications to cellular components in plants. Annu Rev Plant Biol. 2007;58:459–81.

Morris ER, Walker JC. Receptor-like protein kinases: the keys to response. Curr Opin Plant Biol. 2003;6(4):339–42.

Mortensen A, Skibsted LH, Truscott TG. The interaction of dietary carotenoids with radical species. Arch Biochem Biophys. 2001;385:13–9.

Mueller MJ. Archetype signals in plants: the phytoprostanes. Curr Opin Plant Biol. 2004;7:441–8.

Munné-Bosch S. The role of a-tocopherol in plant stress tolerance. J Plant Physiol. 2005;162:743–8.

Munns R, James RA, Läuchli A. Approaches to increasing the salt tolerance of wheat and other cereals. J Exp Bot. 2006;57(5):1025–43.

Mühlenbock P, et al. Chloroplast signaling and *LESION SIMULATING DISEASE1* regulate crosstalk between light acclimation and immunity in *Arabidopsis*. Plant Cell. 2008;20:2339–56.

Nagamiya K, Motohashi T, Nakao K, Prodhan SH, Hattori E, Hirose S, Ozawa K, Ohkawa Y, Takabe T, Takabe T, Komamine A. Enhancement of salt tolerance in transgenic rice expressing an *Escherichia coli* catalase gene, *katE*. Plant Biotechnol Rep. 2007;1:49–55.

Naya L, Ladrera R, Ramos J, González EM, Arrese-Igor C, Minchin FR, Becana M. The response of carbon metabolism and antioxidant defenses of alfalfa nodules to drought stress and to the subsequent recovery of plants. Plant Physiol. 2007;144:1104–14.

Noctor G, Foyer CH. A re-evaluation of the ATP: NADPH budget during C3 photosynthesis. A contribution from nitrate assimilation and its associated respiratory activity? J Exp Bot. 1998a;49:1895–908.

Noctor G, Foyer CH. Ascorbate and glutathione: keeping active oxygen under control. Annu Rev Plant Physiol Plant Mol Biol. 1998b;49:249–79.

Noctor G, Gomez L, Vanacker H, Foyer CH. Interactions between biosynthesis, compartmentation, and transport in the control of glutathione homeostasis and signalling. J Exp Bot. 2002a;53:1283–304.

Noctor G, Veljovic-Jovanovic S, Driscoll S, Novitskaya L, Foyer CH. Drought and oxidative load in wheat leaves: a predominant role for photorespiration? Ann Bot. 2002b;89:841–50.

Noctor G, De Paepe R, Foyer CH. Mitochondrial redox biology and homeostasis in plants. Trends Plant Sci. 2007;12:125–34.

Norman C, Howell KA, Millar AH, Whelan JM, Day DA. Salicylic acid is an uncoupler and inhibitor of mitochondrial electron transport. Plant Physiol. 2004;134:492–501.

Pan Y, Wu LJ, Yu ZL. Effect of salt and drought stress on antioxidant enzymes activities and SOD isoenzymes of liquorice (*Glycorhiza uralensis* Fisch). J Plant Growth Regul. 2006;49:157–65.

Panda SK. Response of green gram seeds under salinity stress. Indian J Plant Physiol. 2001;6:438–40.

Pastore D, Trono D, Laus MN, Di Fonzo N, Flagella Z. Possible plant mitochondria involvement in cell adaptation to drought stress. A case study: durum wheat mitochondria. J Exp Bot. 2007;58:195–210.

Pfannschmidt T. Chloroplast redox signals: how photosynthesis controls its own genes. Trends Plant Sci. 2003;8(1):33–41.

Porcel R, Barea JM, Ruiz-Lozano JM. Antioxidant activities in mycorrhizal soybean plants under drought stress and their possible relationship to the process of nodule senescence. New Phytol. 2003;157(1):135–43.

Pospíšilová J, Synková H, Rulcová J. Cytokinins and water stress. Biol Planta. 2000;43:321–8.

Qadir M, Tubeileh A, Akhtar J, Larbi A, Minhas PS, Khan MA. Productivity enhancement of salt-affected environments through crop diversification. Land Degrad Dev. 2008;19(4):429–53.

Raghavendra AS, Gonugunta VK, Christmann A, Grill E. ABA perception and signalling. Trends Plant Sci. 2010;15:395–401.

Ramegowda V, Basu S, Krishnan A, Pereira A. Rice growth under drought kinase is required for drought tolerance and grain yield under normal and drought stress conditions. Plant Physiol. 2014;166(3):1634–45.

Rao ASVC, Reddy AR. Glutathione reductase: a putative redox regulatory system in plant cells. In: Khan NA, Singh S, Umar S, editors. Sulfur assimilation and abiotic stresses in plants. Dordrecht: Springer; 2008. p. 111–47.

Rasmusson AG, Wallstrom SV. Involvement of mitochondria in the control of plant cell NAD(P)H reduction levels. Biochem Soc Trans. 2010;38:661–6.

Rentel MC, Lecourieux D, Ouaked F, Usher SL, Peterson L, Okamoto H, Knight H, Peck SC, Grierson CS, Hirt H, Knight MR. OXI1 kinase is necessary for oxidative burst-mediated signaling in *Arabidopsis*. Nature. 2004;427:858–61.

Rhoads DM, Subbaiah CC. Mitochondrial retrograde regulation in plants. Mitochondrion. 2007;7:177–94.

Rhoads DM, Umbach AL, Subbaiah CC, Siedow JN. Mitochondrial reactive oxygen species. Contribution to oxidative stress and interorganellar signaling. Plant Physiol. 2006;141:357–66.

Rizhsky L, Hallak-Herr E, Van Breusegem F, Rachmilevitch S, Barr JE, Rodermel S, Inze D, Mittler R. Double antisense plants lacking ascorbate peroxidase and catalase are less sensitive to oxidative stress than single antisense plants lacking ascorbate peroxidase or catalase. Plant J. 2002;32:329–42.

Roje S. S-Adenosyl-L-methionine: beyond the universal methyl group donor. Phytochemistry. 2006;67:1686–98.

Ros Barceló A. Xylem parenchyma cells deliver the H_2O_2 necessary for lignification in differentiating xylem vessels. Planta. 2005;220:747–56.

Rossa MM, De Oliveira MC, Okamoto OK, Lopes PF, Colepicolo P. Effect of visible light on superoxide dismutase (SOD) activity in the red alga *Gracilariopsis tenuifrons* (Gracilariales, Rhodophyta). J Appl Phycol. 2002;14:151–7.

Sakamoto M, Munemura I, Tomita R, Kobayashi K. Involvement of hydrogen peroxide in leaf abscission signaling, revealed by analysis with an in vitro abscission system in Capsicum plants. Plant J. 2008;56:13–27.

Sayfzadeh S, Rashidi M. Response of antioxidant enzymes activities of sugar beet to drought stress. J Agric Biol Sci. 2011;6:27–33.

Scandaliol JG. Oxygen stress and super oxide dimutases. Plant Physiol. 1993;101(1):7–12.

Seckin B, Turkan I, Sekmen AH, Ozfidan C. The role of antioxidant defense systems at differential salt tolerance of *Hordeum marinum* Huds. (sea barley grass) and *Hordeum vulgare* L. (cultivated barley). Environ Exp Bot. 2010;69(1):76–85.

Setterdahl AT, Chivers PT, Hirasawa M, Lemaire SD, Keryer E, Miginiac-Maslow M, Kim SK, Mason J, Jacquot JP, Longbine CC, de Lamotte-Guery F, Knaff DB. Effect of pH on the oxidation-reduction properties of thioredoxins. Biochemistry. 2003;42:14877–84.

Sgherri CLM, Loggini B, Puliga S, Navari-Izzo F. Antioxidant system in *Sporobolus stapfianus*: changes in response to desiccation and rehydration. Phytochemistry. 1994;35:561–5.

Shalata A, Mittova V, Volokita M, Guy M, Tal M. Response of cultivated tomato and its wild salt-tolerant relative *Lycopersicon pennelli* to salt-dependent oxidative stress: the root antioxidative system. Physiol Planta. 2001;122:487–94.

Shao HB, Chu LY, Shao MA, Jaleel CA, Mi HM. Higher plant antioxidants and redox signaling under environmental stresses. C R Biol. 2008;331(6):433–41.

Sharma P, Dubey RS. Modulation of nitrate reductase activity in rice seedlings under aluminium toxicity and water stress: role of osmolytes as enzyme protectant. J Plant Physiol. 2005;162:854–64.

Sharma P, Jha AB, Dubey RS, Pessarakli M. Reactive oxygen species, oxidative damage, and antioxidative defense mechanism in plants under stressful conditions. J Bot. 2012:217037. doi:http://dx.doi.org/10.1155/2012/217037.

Shinozaki K, Yamaguchi-Shinozaki K. Molecular responses to dehydration and low temperature: differences and cross-talk between two stress signaling pathways. Curr Opin Plant Biol. 2000;3:217–23.

Shinozaki K, Yamaguchi-Shinozaki K. Gene networks involved in drought stress response and tolerance. J Exp Bot. 2007;58(2):221–7.

Shinozaki K, Yamaguchi-Shinozaki K, Seki M. Regulatory network of gene expression in the drought and cold stress responses. Curr Opin Plant Biol. 2003;6:410–7.

Shiu SH, Bleecker AB. Receptor-like kinases from Arabidopsis form a monophyletic gene family related to animal receptor kinases. Proc Natl Acad Sci U S A. 2001;98(19):10763–8.

Shiu SH, Bleecker AB. Expansion of the receptor-like kinase/Pelle gene family and receptor-like proteins in Arabidopsis. Plant Physiol. 2003;132:530–43.

Sierla M, Rahikainen M, Salojarvi J, Kangasjarvi J, Kangasjarvi S. Apoplastic and chloroplastic redox signaling networks in plant stress responses. Antioxid Redox Signal. 2013;18:2220–39.

Simova-Stoilova L, Vaseva I, Grigorova B, Demirevska K, Feller U. Proteolytic activity and cysteine protease expression in wheat leaves under severe soil drought and recovery. Plant Physiol Biochem. 2010;48:200–6.

Smirnoff N. Ascorbic acid: metabolism and functions of a multi-facetted molecule. Curr Opin Plant Biol. 2000;3:229–35.

Sreenivasulu N, Miranda M, Prakash HS, Wobus U, Weschke W. Transcriptome changes in foxtail millet genotypes at high salinity: identification and characterization of a PHGPx gene specifically upregulated by NaCl in a salt-tolerant line. J Plant Physiol. 2004;161:467–77.

Srivastava AK, Bhargava P, Rai LC. Salinity and copper-induced oxidative damage and changes in antioxidative defense system of *Anabaena doliolum*. W J Microb Biotechnol. 2005;22: 1291–8.

Stoop JMH, Williamson JD, Pharr DM. Mannitol metabolism in plants: a method for coping with stress. Trends Plant Sci. 1996;1:139–55.

Storey KB. Oxidative stress: animal adaptations in nature. Braz J Med Biol Res. 1996;29:1715–33.

Sun YY, Sun YJ, Wang MT, Li XY, Guo X, Hu R, Ma J. Effects of seed priming on germination and seedling growth of rice under water stress. Acta Agron Sin. 2010;36(11):1931–40.

Suzuki N, Koussevitzky S, Mittler R, Miller G. ROS and redox signaling in the response of plants to abiotic stresses. Plant Cell Environ. 2011;10:1365–3040.

Tarczynski MC, Jensen RG, Bohnert HJ. Stress protection of transgenic tobacco by production of the osmolyte mannitol. Science. 1993;259:508–10.

Taylor NL, Day DA, Millar AH. Environmental stress causes oxidative damage to plant mitochondria leading to inhibition of glycine decarboxylase. J Biol Chem. 2002;277:42663–8.

Taylor NL, Day DA, Millar AH. Targets of stress-induced oxidative damage in plant mitochondria and their impact on cell carbon/nitrogen metabolism. J Exp Bot. 2004;55:1–10.

Thomashow MF. Plant cold acclimation: freezing tolerance genes and regulatory mechanisms. Annu Rev Plant Physiol Plant Mol Biol. 1999;50:571–99.

Tippmann HF, Schluter U, Collinge DB. Common themes in biotic and abiotic stress signalling in plants. Floric Ornam Plant Biotechnol. 2006;3:52–67.

Torres MA. ROS in biotic interactions. Physiol Plant. 2010;138:414–29.

Torres MA, Dangl JL. Functions of the respiratory burst oxidase in biotic interactions, abiotic stress and development. Curr Opin Plant Biol. 2005;8:397–403.

Ushimaru T, Nakagawa T, Fujioka Y, Daicho K, Naito M, Yamauchi Y, Nonaka H, Amako K, Yamawaki K, Murata N. Transgenic Arabidopsis plants expressing the rice dehydroascorbate reductase gene are resistant to salt stress. J Plant Physiol. 2006;163:1179–84.

Vandenabeele S, Vanderauwera S, Vuylsteke M, Rombauts S, Langebartels C, Seidlitz HK, Zabeau M, Van Montagu M, Inze D, Van Breusegem F. Catalase deficiency drastically affects gene expression induced by high light in Arabidopsis thaliana. Plant J. 2004;39:45–58.

Vellosillo T, Vicente J, Kulasekaran S, Hamberg M, Castresana C. Emerging complexity in reactive oxygen species production and signaling during the response of plants to pathogens. Plant Physiol. 2010;154:444–8.

Vendruscolo EC, Schuster I, Pileggi M, Scapim CA, Molinari HB, Marur CJ, Vieira LG. Stress-induced synthesis of proline confers tolerance to water deficit in transgenic wheat. J Plant Physiol. 2007;164:1367–76.

Vieira Dos Santos C, Rey P. Plant thioredoxins are key actors in the oxidative stress response. Trends Plant Sci. 2006;11:329–34.

Vij S, Giri J, Dansana PK, Kapoor S, Tyagi AK. The receptor-like cytoplasmic kinase (OsRLCK) gene family in rice: organization, phylogenetic relationship, and expression during development and stress. Mol Plant. 2008;1(5):732–50.

Vranova E, Inzé D, Van Breusegem F. Signal transduction during oxidative stress. J Exp Bot. 2002;53:1227–36.

Wang Y, Ying Y, Chen J, Wang XC. Transgenic Arabidopsis overexpressing Mn-SOD enhanced salt-tolerance. Plant Sci. 2004;167:671–7.

Wang Y, Wisniewski M, Meilan R, Uratsu SL, Cui MG, Dandekar A, Fuchigami L. Ectopic expression of Mn-SOD in *Lycopersicon esculentum* leads to enhanced tolerance to salt and oxidative stress. J Appl Horticult. 2007;9:3–8.

Wang L, Zhou Q, Ding L, Sun Y. Effect of cadmium toxicity on nitrogen metabolism in leaves of *Solanum nigrum* L. as a newly found cadmium hyperaccumulator. J Hazard Mater. 2008;154:818–25.

Weiner JJ, Peterson FC, Volkman BF, Cutler SR. Structural and functional insights into core ABA signaling. Curr Opin Plant Biol. 2010;13:495–502.

Witzel K, Weidner A, Surabhi GK, Börner A, Mock HP. Salt stress-induced alterations in the root proteome of barley genotypes with contrasting response towards salinity. J Exp Bot. 2009;60(12):3545–57.

Woodson JD, Chory J. Coordination of gene expression between organellar and nuclear genomes. Nat Rev Genet. 2008;9:383–95.

Wu L, Zhang Z, Zhang H, Wang XC, Huang R. Transcriptional modulation of ethylene response factor protein JERF3 in the oxidative stress response enhances tolerance of tobacco seedlings to salt, drought, and freezing. Plant Physiol. 2008;148:1953–63.

Xiao X, Xu X, Yang F. Adaptive responses to progressive drought stress in two *Populus cathayana* populations. Silva Fenn. 2008;42:705–19.

Xing Y, Jia W, Zhang J. AtMKK1 mediates ABA-induced CAT1 expression and H_2O_2 production via AtMPK6-coupled signaling in Arabidopsis. Plant J. 2008;54:440–51.

Xiong L, Schumaker KS, Zhu JK. Cell signaling during cold, drought, and salt stress. Plant Cell. 2002;14 Suppl 1:S165–83.

Yamaguchi-Shinozaki K, Shinozaki K. Organization of cis-acting regulatory elements in osmotic- and cold-stress-responsive promoters. Trends Plant Sci. 2005;10:88–94.

Yamaguchi-Shinozaki K, Shinozaki K. Transcriptional regulatory networks in cellular responses and tolerance to dehydration and cold stresses. Annu Rev Plant Biol. 2006;57:781–803.

Yang Z, Wu Y, Li Y, Ling HQ, Chu C. OsMT1a, a type 1 metallothionein, plays the pivotal role in zinc homeostasis and drought tolerance in rice. Plant Mol Biol. 2009;70:219–29.

Yoshida T, FujitaY SH, Kidokoro S, Maruyama K, Mizoi J. AREB1, AREB2, and ABF3 are master transcription factors that cooperatively regulate ABRE-dependent ABA signalling involved in drought stress tolerance and require ABA for full activation. Plant J. 2010;61:672–85.

Zer H, Ohad I. Light, redox state, thylakoid-protein phosphorylation and signaling gene expression. Trends Biochem Sci. 2003;28:467–70.

Zhu J, Fu X, Koo YD, et al. An enhancer mutant of Arabidopsis salt overly sensitive 3 mediates both ion homeostasis and the oxidative stress response. Mol Cell Biol. 2007;27:5214–24.

Zlatev ZS, Lidon FC, Ramalho JC, Yordanov IT. Comparison of resistance to drought of three bean cultivars. Biol Plant. 2006;50:389–94.

Chapter 21
Determination of Compositional Principles for Herbaceous Plantings in Dry Conditions

Dagmar Hillová

21.1 Introduction

The current processes of urban growth and urbanization in synergy with climate change have multiple negative impacts on all environmental parameters (Fernández-Canero et al. 2011). In urban areas, in contrast to peri-urban and rural areas, this pressure is more extreme, related to higher atmospheric pollution levels caused by traffic and other anthropogenic emissions, but also to limiting water availability and higher temperatures typical of the city microclimate (Ferrini and Fini 2013). Practices such as building green infrastructure and water conservation are increasingly being seen as the best practices in climate adaptation. Green technologies and infrastructure solutions are often implemented such as managing storm-water, filter water, slow runoff, or reducing local ambient heat, and clean air, helping to reduce greenhouse gas emissions, noise control, glare and reflection reduction, and can provide shade that reduces man-made cooling needs and hence electricity demand, wind control, privacy, screening of objectionable views and objects (Foster et al. 2011; Robinette 1984). However, drought condition in urban environment leads to poor vitality and plant decline. There is an urgent need to develop and apply comprehensive concepts for sustainable plantings in dry conditions (Pauleit 2003).

This chapter deals with approaches that have been useful in herbaceous planting design specifically designed to reduce water use relative to uniform turfgrass landscapes (St. Hilaire et al. 2008), water-intensive herbaceous plantings or ornamental annual flower bed plantings. The main approach in too many cities, especially when it is needed to reduce or completely eliminate the use of irrigation water, has been to substitute water-intensive herbaceous plantings by new drought resistant and low

D. Hillová (✉)
Department of Planting Design and Maintenance, Faculty of Horticulture and Landscape Engineering, Slovak University of Agriculture, Nitra, Slovak Republic
e-mail: dagmar.hillova@uniag.sk

M.A. Hossain et al. (eds.), *Drought Stress Tolerance in Plants, Vol 1*,
DOI 10.1007/978-3-319-28899-4_21

maintenance concept of herbaceous planting. We must realize that turf is not necessary for aesthetic enhancement and can be replaced with low ground cover and other plant materials to achieve the same affect. The turf must be used (really functional), for example, for lawn-dependent activities (sport, active recreation), and maintenance practices could be to break away from the fine, precise clipped lawn in large parks in favour of meadow-like turf (Robinette 1984).

21.2 New Approaches in Planning Herbaceous Perennials in Urban Environment

The last years have seen a tide of interest in the development of nature in cities, and an increasing amount of landscape development in urban areas has involved the use of naturalistic plantings and ecologically inspired planting design, in contrast to more traditional, formal planting styles (Özgüner et al. 2007). The new aesthetically pleasing and ecologically sound approach, not to mention economically appealing develops as a consequence of decline of nature, lack of funding to adequately maintain and develop urban green spaces and extreme climate mainly of arid regions and microregions.

Low water use landscaping, water efficient landscaping and Xeriscaping are key water conservation approaches promoted in not only periodically water-deficit regions (Smith and St. Hilaire 1999). Also, in humid continental climate, in intensively developed urban cities are formed small green unit-structures, often associated with paved surfaces (traffic roundabouts and stripes along paths, roads and other corridors) with different ecosystems (Sjöman et al. 2015), similar to arid regions, requiring similar approaches. Research of low water use landscaping or water-efficient landscaping is strongly needed all around the world to develop herbaceous planting design, which can reduce water consumption without compromising landscape functionality or aesthetics (St. Hilaire et al. 2008). Nevertheless, these new drought tolerant, sustainable planting types or xeriscaping require to develop a new landscape ethic and aesthetic consciousness in planting the landscape (Robinette 1984). This requirement resonates in the professional community for several years. Dunnett and Hitchmough (2004) and Dunnett (2004) emphasize that in urban context, designed nature-like vegetation must be strongly informed by aesthetic principles if it is to be understood and valued by the public at large. All planting if it is to be successful involves compromise between what is aesthetically desirable and what is possible in the ecological reality.

In the United States, there is an important awareness of garden design and its impact on water consumption embodied in concept such as water-wise landscaping and xeriscaping (Fernández-Canero et al. 2011). The awareness of drought and critical economic incentive thresholds can significantly influence participation in a xeriscape conversion programme (Sovocool et al. 2006). Xeriscape incorporates seven key principles: (a) sound landscape planning and design, (b) limitation of turf to appropriate, functional area, (c) use of water-efficient plants, (d) efficient irrigation (usually by a means other than spray irrigation), (e) soil amendments, (f) use of mulches and (g) appropriate landscape maintenance (Sovocool et al. 2006, Fernández-Canero et al. 2011).

Xeriscape approach in American arid areas are replaced in European countries by (a) German school of low maintenance herbaceous plantings with traditionally used perennials plants known as 'SilberSommer' style, (b) North American prairie planting design for urban green space representative by Cassian Schmidt and (c) 'The Sheffield School' of natural plant communities planting design as aesthetic and ecological models for low maintenance design.

In all cases and in all regions the specification, selection and installation of xeric adapted or low water use plants should always be one of the first strategies suggested when considered about planning a new landscape or renovating or redesigning an older one for use in water-efficient landscaping (Robinette 1984; Sun et al. 2012). Choice of plant species to urban environment should not be entirely dictated by what survives in dry conditions, but consideration should be given for supporting those species that provide the greatest ecosystem service potential (Blanusa et al. 2013). The xeric adapted or drought plants should in this respect also be: (a) plant with a cooling effect (Wolf and Lundholm 2008), (b) fire-retardant vegetation which can reduce the frequency and severity of brush fires, (c) erosion control plants—protect the soil facilitate infiltration of rainwater for recharging the groundwater supply, and reduce run-off, (d) shade effect, shade reduces the need for irrigation by decreasing the loss of water from both the soil and plants (Robinette 1984). A number of studies worldwide have investigated stress tolerance, particularly survival and growth rates, ecological function and cooling potential of herbaceous plant selection on extensive and semi-extensive green roof (Blanusa et al. 2013), but other herbaceous landforms (meadows, undergrowth, ornamental beds and borders) are unclear.

21.3 Defining the Selection Criteria and Screening Techniques

Evidently, ornamental plants are very much undervalued in targeting research on drought resistance (compared with crop plants). Recently, landscape architects, in the conception of herbaceous plantings, selected suitable plants according to habitat classification defined by Hansen and Stahl (1993), or used identification and selection potential species from natural vegetation systems and habitats, where plants are exposed to environmental conditions similar to those in inner-city environments (Sjöman et al. 2015). In spite of the responsible selection of perennials, herbaceous plant composition in urban plantings look like unaesthetic, often because of negative morphological changes of plants in the dry conditions. Extreme conditions of the urban environment in interaction with climate change puts pressure on reviewing previously processed systems of herbaceous perennial habitats. The fact that global warming much has changed does not mean that previous knowledge is worthless, only that the old lessons must be updated if they are to modern templates for drought tolerant landscaping (Werick and Palmer 2008).

Thus, there is a strong need to increase systematic work on selection of herbaceous perennials adapted to dry urban conditions. Selection criteria are directed toward growth, survival and sufficiently aesthetic appearance for at least 6–10 years. Screening techniques should (a) mass field selection to discover sources of resistant (Sjöman et al. 2015), (b) trial planting over a wide area and under highly variable conditions, (c) assess plant performance at the critical developmental stage, (d) be completed in a relatively short periodical time, (e) use relatively small quantities of plant material and (f) be capable of screening large group of taxa. Although many techniques are available for examining plant water relationships, most of these are too laborious and time consuming (Johnson 1980; Ferrini and Fini 2013) for use in plant selection for drought tolerant landscaping.

When assessing the drought response of ornamental herbaceous perennials were used experimental methods of analysing the *growth traits* (leaf area, dry weights), physiological traits evaluated *water regime* (relative water content, water use efficiency, osmotic potential, water potential, turgor potential), physiological traits evaluated *gas exchange* (stomatal conductance, net photosynthesis, photosynthetic capacity, CO_2 assimilation, transpiration), *anatomical traits* (content of leaf pigments, accumulation and mobilization of proline, ammonium, soluble carbohydrates, soluble sugars—sucrose, glucose, fructose, and their degree of polymerization), *appearance traits* (visual quality rating, growth index) (Zollinger et al. 2006; Hillová et al. 2011; Prevete et al. 2000; Chapman and Auge 1994; Garland et al. 2012; Anjum et al. 2011; Paganová et al. 2015). Laboratory methods and techniques are often time consuming and usually destructive for the determination of physiological state of plants. Sample may by analysed only one time, so monitoring of changes during vegetation period is practically impossible. For the long-term monitoring of state of plants directly in field conditions, apparatuses for non-destructive measurements are needed (Sochor et al. 2014).

In the last years, chlorophyll fluorescence analysis has become one of the most powerful and widely used techniques available to plant physiologists and ecophysiologists (Maxwell and Johnson 2000), which permits early detection of stress before physical signs of deterioration become evident (Percival and Sheriffs 2002). Application of chlorophyll *a* fluorescence has been applied in drought stress measurement, but the conclusions are controversial, especially on the usefulness of maximum quantum efficiency of PSII (F_v/F_m) (Guo and Tan 2015). Steady-state chlorophyll fluorescence (F_s) measurements are an easy means to monitor changes in plant photosynthesis, and therefore, provide a rapid assessment of plant stress (Cendrero et al. 2012). Chlorophyll fluorescence values obtained from excised leaves of plants subjected to dehydration in vitro (Percival 2005; Faraloni, et al 2011) provided a measurable indicator of whole-plant performance following drought in situ and to gain a greater understanding of alteration in leaf photosynthetic properties between species (Percival and Sheriffs 2002). The fast measurement of steady-state chlorophyll fluorescence (F_s) on excised leaves of plants gives the landscape architects the tools for selecting and sorting a wide range of traditional use of perennials, as well as the ever-expanding assortment new varieties.

Table 21.1 Sorting herbaceous perennials into different habitats according to Hansen and Stahl (1993), and their Φ_{PSII} values obtained from excised leaves of plants subjected to dehydration stress for 24 h

Species	HS identification	HS description	ΦPSII (no units)
Salvia officinalis 'Purpurascens'	3.3.1.	Drought-resistant dwarf and sub-shrubs	0.78 ± 0.021^{g}
Geranium macrorrhizum	2.1.2.	Widely spreading plants for sun and semi-shade on the woodland edge	0.77 ± 0.014^{g}
Geranium × magnificum 'Rosemoor'	2.1.2.	Widely spreading plants for sun and semi-shade on the woodland edge	0.77 ± 0.028^{g}
Salvia officinalis	3.3.1.	Drought-resistant dwarf and sub-shrubs	0.75 ± 0.060^{g}
Perovskia atriplicifolia	3.3.1.	Drought-resistant dwarf and sub-shrubs	0.68 ± 0.088^{f}
Stachys officinalis	2.1.8.	Plants for an open woodland edge on dry to moist, sandy, silica-rich soils	$0.67 \pm 0.095^{e,f}$
Geranium 'Philippe Vapelle'	2.1.2.	Widely spreading plants for sun and semi-shade on the woodland edge	$0.65 \pm 0.060^{e,f}$
Geranium maculatum 'Chatto'	2.1.2.	Widely spreading plants for sun and semi-shade on the woodland edge	0.60 ± 0.218^{e}
Stachys officinalis 'Hummelo'	2.1.8.	Plants for an open woodland edge on dry to moist, sandy, silica-rich soils	0.53 ± 0.210^{d}
Waldsteinia geoides	1.1.1.	Low, shade-tolerant plants	0.40 ± 0.223^{c}
Stachys grandiflora	2.2.2.	Plants loosely bound to the woodland edge	0.37 ± 0.101^{c}
Rudbeckia fulgida var. *speciosa*	3.4.6.	North American wild perennials with border character	0.36 ± 0.071^{c}
Brunnera macrophylla	2.2.2.	Perennials loosely bound to the woodland edge	0.22 ± 0.101^{b}
Epimedium × versicolor 'Sulphureum'	1.1.1.	Low, shade-tolerant perennials	$0.20 \pm 0.078^{a,b}$
Alchemilla mollis	2.2.2.	Plants loosely bound to the woodland edge	0.14 ± 0.043^{a}

Fast and affordable method of measurement Φ_{PSII} (quantum efficiency of PSII) was used for comparing 15 taxa sorted into different habitats reflecting their ecological requirements and proper garden habitat according to Hansen and Stahl (1993), identified by the numbers of subdivision within the main garden habitats (HS identification) and short habitat description (HS description) referred to in Table 21.1. Depending on the Φ_{PSII} values obtained from excised leaves of plants subjected to dehydration stress for 24 h, herbaceous perennials could be ranked in five main groups, as shown Table 21.1 and Fig. 21.1: (a) a group of four taxa (*Salvia officinalis* 'Purpuranscens', *Salvia officinalis*, *Geranium macrorrhizum*, *Geranium × manificum* 'Rosemoor'), which was found to be the most 'resistant' and which retained Φ_{PSII} values more than 0.75, and difference in values compared to the initial values are less than 0.05, (b) a group of four taxa (*Stachys officinalis*, *Geranium* 'Philippe Vapelle',

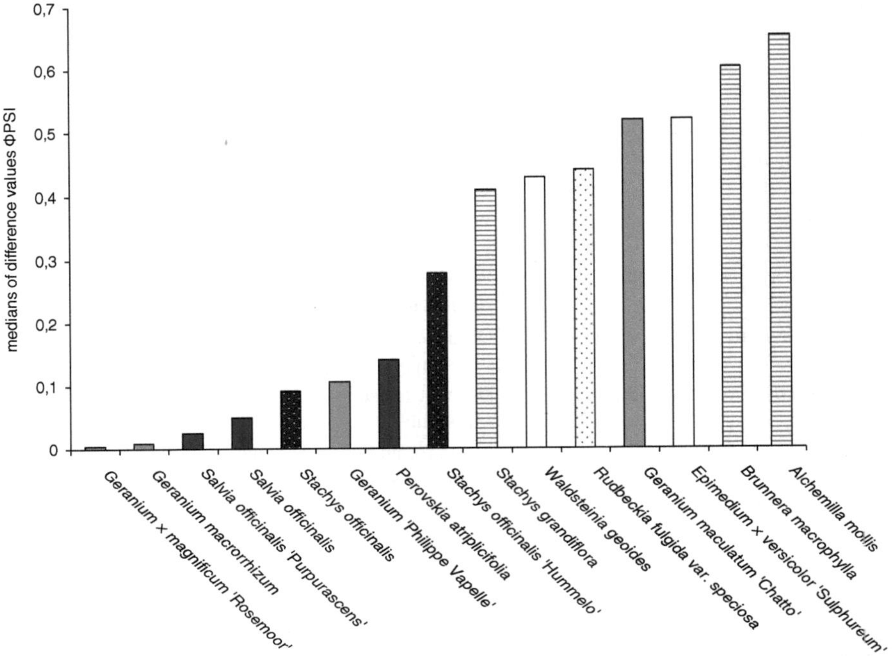

Fig. 21.1 The difference in values Φ_{PSII} obtained from excised leaves of plants subjected to dehydration stress for 24 h compared to the initial values

Geranium maculatum 'Chatto'*, *Perovskia atriplicifolia*) which maintained Φ_{PSII} values between 0.74 and 0.60, and difference in values compared to the initial values are 0.05–0.15 (0.52*), (c) a group of one taxa (*Stachys officinalis* 'Hummelo'), which maintained Φ_{PSII} values between 0.59 and 0.50, and difference in values compared to the initial values are 0.16 and 0.30, (d) a group of three taxa (*Stachys grandiflora*, *Waldsteinia geoides*, *Rudbeckia fulgida* var. *speciosa*), which maintained Φ_{PSII} values between 0.49 and 0.35, and difference in values compared to the initial values are 0.31–0.45 and (e) a group of three taxa (*Epimedium × versicolor* 'Sulphureum', *Brunnera macrophylla*, *Alchemilla mollis*), the most 'susceptible' to dehydration, which maintained Φ_{PSII} values between 0.15 and 0.34, and difference in values compared to the initial values are more than 0.5.

Mentioned groups do not fully correspond with the traditional use of perennials sorting into groups according to Hansen and Stahl (1993). The group A, which was found to be the most 'resistant', also contains typical woodland edge plants (*Geranium macrorrhizum*, *Geranium × manificum* 'Rosemoor'), often enough to create a weed-proof groundcover, thrive on nutrient-rich, predominantly moist but also intermittently dry soils (Hansen and Stahl 1993). The group B also represents a mixture of typical xerophytic plants and typical woodland edge plants (*Geranium* 'Philippe Vapelle', *Geranium maculatum* 'Chatto').

In horticultural practice, the assortment of the genus *Geranium* L. is considered particularly suited to the urban environment (Brtáňová 2015). It is worth noting also inclusion of varieties and native species (*Salvia officinalis* 'Purpuranscens', *Salvia officinalis*) uniformly in Group A, and conversely (*Stachys officinalis*, *Stachys officinalis* 'Hummelo') in different groups A and B, which is in accordance that 'border perennials' (those plants, that are the result of many years breeding and selection), demand the intensive maintenance (Hansen and Stahl 1993; Hillová 2012; Scarfone 2007), including irrigation. Also based on these results, we can conclude that the chlorophyll fluorescence measurements carried out in vitro on dehydrated detached leaves could be used as a valid tool for the rapid screening of different herbaceous and woody perennials (Faraloni et al. 2011; Percival and Sheriffs 2002; Percival 2005) resistant to drought stress.

21.4 The Draft Standard for Planning and Management Dry Tolerant Landscaping

Clear guidance is needed for landscape architects, city planners, local authorities and other practitioners on how best to manage public urban green spaces in order to respond to climatic change (Ferrini and Fini 2013) or for planning and management dry tolerant landscaping. The landscape designers must be aware of responsible landscape practices, the landscape installation contractors must be aware of what can and should be done to save the maximum amount of water, the landscape maintenance personnel need to be aware of what can be done, who can do what, and when it can be done, whether before, during or after installation (Robinette 1984).

Recently, the practice used only a recommendation for xeriscape households but recommendations for the public sector are missing. The standard for planning and management dry tolerant landscaping should be defined guidance for (a) site assessment and their modification, (b) plant selection, (c) dry tolerant planting design and (d) management dry tolerant landscaping, in order to ensure extend life, maintain or improve the aesthetic appearance, vitality and health, and therefore the functionality of herbaceous plantings in urban areas.

21.4.1 The Guidance for Site Assessment and Modification

It's important to improve all aspects about plant placement (Morrison 2004), what in case drought resistant plantings mainly means providing the erection of wind barriers to reduce the dehydration of the soil around the plants (Robinette 1984).

Urban soils often have low organic matter content, low and unbalanced nutrient contents and or low nutrient availability due to a high soil pH (Ferrini and Fini 2013) (Fig. 21.2). Soils which have been stripped of topsoil or compacted during construction do not retain water well (Robinette 1984). Adding organic matter (Fig. 21.3)

Fig. 21.2 Naturalistic design of drought resistant herbaceous perennials (1 year old) with gravel mulching on original anthropogenic substrate after construction (Nitra city center, Slovak Republic)

Fig. 21.3 Naturalistic design of drought resistant herbaceous perennials (1 year old) with gravel mulching on complete replacement of the original substrate by growing medium (Nitra city center, Slovak Republic)

or moisture retaining materials (Dunnett and Kingsbury 2004), like a lignite and zeolite that are able to retain the water in soil, improves the water efficiency of the soil and provides better values of physiological and growth parameters (Sochor et al. 2014; Robinette 1984).

Very discussed access to drought resistant herbaceous planting is the application of mulches, which are able to conserve soil moisture (Robinette 1984) through little heat radiation of the soil surface and kept evaporation to a minimum (Ferrini and Fini 2013). Except this, the application of mulches reduces soil compaction and suppresses weed growth and enriching the soil (Robinette 1984; Fenzl and Kircher 2009; Baroš and Martinek 2011; Baroš et al. 2014). Low-maintenance plantings should be mulched by gravels fraction (4/6–8/22 mm) of minimum thickness 5–7 cm (Trevisan-Smykalova 2004, Fenzl and Kircher 2009; Baroš and Martinek 2011; Hillová 2012).

21.4.2 *The Guidance for Plant Selection*

Obviously the first step in conserving water and thus energy in park areas is to select those plants which will require less water through their effective lifetime (Robinette 1984). But which are these plants? We are able to sort their responsibly? The research was mainly concerned to arid regions, but what we need to do in the urban environment? Landscape architects and their research delay behind the threats of climate change that brings extreme drought conditions in cities. The plant species must be used fulfilling site needs as well as water conservation goals.

Currently landscape architects applied the certified methodologies which recommended herbaceous perennials plant list suitable for low-maintenance plantings (Fenzl and Kircher 2009; Baroš and Martinek 2011; Trevisan-Smýkalová 2004) or rural plantings (Baroš et al. 2014). Herbaceous perennial plant lists contained in the methodologies have the ability to make the work of landscape architects more effective and ensure long performance plantings based on professional proposal, establishment and maintenance. Drawing up (development) of similar herbaceous perennial sets focused on drought conditions in urban environment became the goal of research on Slovak Agricultural University.

Composition of those drought resistant sets of plants must always be based on knowledge of the ecological and growing requirements of plants (Hansen and Stahl 1993) correlated with drought-stress tolerance measurement (Paganová et al. 2015), knowledge about sociability of plants (Hansen and Stahl 1993; Baroš and Martinek 2011; Hillová 2012) and life strategies of plants (Grime 2002, Hitchmough and Dunnet 2004) and last but not least understanding and acceptance of self-regulatory principles in the composition of herbaceous perennials sets.

Fig. 21.4 Uniform conventional planting of drought resistant herbaceous perennials (2 years old), with bare gaps left from dead plants (Nitra city center, Slovak Republic)

21.4.3 The Guidance for Dry Tolerant Planting Design

The conception of using drought resistant herbaceous planting is based on changing the appearance of an area in the landscape so that it will be more naturally manicured and thus will require the use of less water (Robinette 1984). Uniform plantings of a limited number of species must be avoided (Fig. 21.4). Through careful selection and grouping plants communities of aesthetically compatible plants can be created which can satisfy our desire for visual uniformity. Increasing biodiversity and keeping good species diversity in plantings is always a wise management decision (Ferrini and Fini 2013). The efficiency of water use can be improved by compositional principles of herbaceous planting in dry conditions: (a) increasing in the density of plant cover (Ferrini and Fini 2013), (b) proportional planting of succulent plants with high resistance to drought and other perennial plants and grasses with rapid regenerative capacity (Wolf and Lundholm 2008), (c) proportional planting of different growth form (Hitchmough and Dunnet 2004), (d) proportional planting of plant growth strategy (Grime 2002) and (e) proportional planting of performance criteria (longevity, vegetative spread, competitiveness, speed of establishment, self-seeding) (Kingsbury 2008).

The herbaceous perennial plantings designed under the guidance for dry tolerant planting design should be able to long-term resilience (6–10 years) without supplementary irrigation or exploring the options of sprinkler or drip irrigation systems in establishing watering priorities and altering irrigation practices to use irrigation water more efficiently (Robinette 1984).

21.4.4 The Guidance for Management Dry Tolerant Landscaping

The long-term maintenance plan is essential to the water conservation process. Herbaceous planting cultivation practices to conserve water would include the (a) minimize nitrogen levels of fertilize, (b) maximize weeding, to remove competition from ornamental landscape planting areas (Robinette 1984) and (c) regularly renewal of mulch layer (Robinette 1984; Hillová 2012).

The specificity of maintenance drought resistant planting is based mainly on the concept of naturalistic or ecologically based ornamental design with higher degree of self-regulation system in maintenance, which creates an attractive and dynamic planting in time and space (Baroš and Martinek 2011; Hillová 2012). The plantings maintaining dynamic balance and the landscape maintenance personnel only partially regulate plantings. The man who provides the maintenance of planting must be equipped with knowledge and experience with the herbaceous perennials assortment and skill and experience to selectively manage those dynamic community requiring selective maintenance management of individual taxa.

21.5 Conclusion and Future Perspectives

Climate changes and extreme conditions of the urban environment have been forced landscape architects, landscape managers and practitioners of green areas to look for new approaches of the herbaceous perennials planting design, especially when the demand for greening of urban areas were related to non-compliant growth conditions, and traditional assortment is unable to provide long-lasting and functional plantings. The most important steps in the functional application of herbaceous perennials are (a) to select compositional sets of herbaceous perennials (from the traditional species and varieties, new varieties, and native species) suitable to the various types of the urban environment conditions (pedestrian zone, traffic roundabouts and stripes along paths, roads and other corridors, green roof), (b) to verify from those sets different variants of compositional grouping of plants (different participation of plant-growth forms, plant strategies and plant performances), (c) to verify from those sets different variants of biotechnic principles for establishment (applying additives and mulches) and maintenance (type and timing of the cut) of drought resistant planting.

The research team of Department of Planting Design and Maintenance, Faculty of Horticulture and Landscape Engineering, Slovak University of Agriculture, aims to process certified methods of using herbaceous plantings in dry conditions of urban environment, based on an extensive study.

Acknowledgment This work was supported by the Ministry of Education, Science, Research and Sport of the Slovak Republic *VEGA 1/0246/13*.

References

Anjum SA, Xie XY, Wang LCH, Saleem NF, Man CH, Lei W. Morphological, physiological and biochemical responses of plants to drought stress. Afr J Agric Res. 2011;6(9):2026–32.

Baroš A, Barošová I, Boček S, Businský R, Demková K, Dokoupil L, Kašková M, Kučera Z, Medková L, Šantrůčková M. Metodika pro výběr vhodných druhů dřevin a bylin pro venkovská sídla. VÚKOZ, v.v.i. a Univerzita Karlova v Praze, Přírodovědecká fakulta, Průhonice a Praha. 2014.

Baroš A, Martinek J. Trvalkové výsadby s vyšším stupněm autoregulace a extenzivní údržbou: plánování, zakládání, údržba, doporučené směsi: certifikovaná metodika. Průhonice: Výzkumný ústav Silva Taroucy pro krajinu a okrasné zahradnictví, 84 s. 2011. ISBN:978-80-85116-88-5.

Blanusa T, Monteiro MMV, Fantozzi F, Vysini E, Li Y, Cameron RW. Alternatives to Sedum on green roofs: can broad leaf perennial plants offer better 'cooling service'? Build Environ. 2013;59:99–106.

Brtáňová M. Stanovenie fyziologických parametrov *Geranium macrorrhizum* L. v podmienkach sucha. Thesis. Nitra: Slovak Agriculture University; 2015.

Cendrero MM, et al. Steady-state chlorophyll fluorescence (Fs) as a tool to monitor plant heat and drought stress. In: AGU Fall Meeting Abstracts, 2012. p. 0421.

Chapman DS, Auge RM. Physiological mechanisms of drought resistance in four native ornamental perennials. J Am Soc Hortic Sci. 1994;119(2):299–306.

Dunnett N, Hitchmough J. More than nature. Landscape Design. 2004;28–30.

Dunnett N. The dynamic nature of plant communities—pattern and process in designed plant communities. In: Dunnett N, Kircher W, Kingsbury N, editors. The dynamic landscape: design, ecology and management of naturalistic urban planting. London: Taylor & Francis; 2004. p. 127.

Dunnett N, Kingsbury N. Planting green roofs and living walls, vol. 254. Portland, OR: Timber Press; 2004.

Faraloni C, et al. Chlorophyll fluorescence technique as a rapid tool for in vitro screening of olive cultivars (*Olea europaea* L.) tolerant to drought stress. Environ Exp Bot. 2011;73:49–56.

Fenzl J, Kircher W. Bernburger Staudenmix–Ein Forschungsprojekt der Hochschule Anhalt. 2009.

Ferrini F, Fini A. Planning the green city of 2050: species selection in a global change scenario. In: Proceedings of the 89th International Congress of the International Society of Arboriculture, August, 2013, Toronto, Canada, p. 3–7.

Fernández-Canero R, Ordovás J, Machuca MÁH. Domestic gardens as water-wise landscapes: a case study in southwestern Europe. HortTechnology. 2011;21(5):616–23.

Foster J, Lowe A, Wilkenman S. The value of green infrastructure for urban climate adaptation. Washington, DC: Center for Clean Air Policy; 2011.

Garland KF, Burnett SE, Day ME, van Iersel MW. Influence of substrate water content and daily light integral on photosynthesis, water use efficiency, and morphology of Heuchera americana. J Am Soc Hortic Sci. 2012;137(1):57–67.

Grime JP. Plant strategies, vegetation processes, and ecosystem properties. 2nd ed. Chichester: Wiley; 2002. 456 p. ISBN 978-0-470-85040-4.

Guo Y, Tan J. Recent advances in the application of chlorophyll a fluorescence from photosystem II. Photochem Photobiol. 2015;91(1):1–14.

Hansen R, Stahl F. Perennials and their garden habitats. Cambridge: Cambridge University Press; 1993, 449 p. ISBN 0-521-35194-4.

Hillová D. Tvorba v sadovníckom kvetinárstve. 1. vyd. Nitra: Slovenská Poľnohospodárska Univerzita; 2012. 86 s. ISBN 978-80-552-0885-5.

Hillová D, Lichtnerová H, Dragúňová M. Drought tolerance of open woodland edge herbaceous perennials = Tolerancija na sušu zeľjastich višegodišňich biľaka sa rubnich delova otvorenich šumskich područja. In: Environmental protection of urban and suburban settlements. Ecological movement of Novi Sad: Novi Sad; 2011. p. 425–32. ISBN 978-86-83177-44-8.

Hitchmough J, Dunnet N, editors. Naturalistic herbaceous vegetation for urban landscapes. In: The dynamic landscape: design, ecology and management of naturalistic urban planting. London: Taylor & Francis; 2004. p. 172.

Johnson DA. Improvement of perennial herbaceous plants for drought-stressed western rangelands. In: Turner NC, Kramer PJ, editors. Adaptation of plants to water and high temperature. New York: Wiley; 1980. p. 419–33.

Kingsbury N. An investigation into the performance of species in ecologically based ornamental herbaceous vegetation, with particular reference to competition in productive environments. Doctoral dissertation. Sheffield: University of Sheffield; 2008.

Maxwell K, Johnson GN. Chlorophyll fluorescence—a practical guide. J Exp Bot. 2000; 51(345):659–68.

Morrison D. A methodology for ecological landscape and planting design—site planning and spatial design. In: Dunnett N, Kircher W, Kingsbury N, editors. The dynamic landscape: design, ecology and management of naturalistic urban planting. London: Taylor & Francis; 2004.

Özgüner H, Kendle AD, Bisgrove RJ. Attitudes of landscape professionals towards naturalistic versus formal urban landscapes in the UK. Landscape and Urban Planning. 2007;81(1):34–45.

Paganová V, Hillová D, Lichtnerová H, Moravčík L, Raček M, Šajbidorová V. Rastlinné druhy pre urbanizované prostredie v podmienkach nedostatku vody. Nitra: Slovenská Poľnohospodárska Univerzita; 2015. ISBN 978-80-552-1398-9.

Pauleit S. Urban street tree plantings: identifying the key requirements. Proc Inst Civ Eng Municip Eng. 2003;156(1):43–50.

Percival GC, Sheriffs C. Identification of drought-tolerant woody periennials using chlorophyll fluorescence. J Arboricult. 2002;28(5):215–23.

Percival GC. Identification of foliar salt tolerance of woody perennials using chlorophyll fluorescence. HortScience. 2005;40(6):1892–7.

Prevete KJ, Fernandez RT, Miller WB. Drought response of three ornamental herbaceous perennials. J Am Soc Hortic Sci. 2000;125(3):310–7.

Robinette GO. Water conservation in landscape design & management. New York: Van Nostrand Reinhold; 1984.

Scarfone SC. Professional planting design an architectural and horticultural approach for creating mixed bed plantings. Hoboken: Wiley; 2007. 272p.

Sjöman H, Bellan P, Hitchmough J, Oprea A. Herbaceous plants for climate adaptation and intensely developed urban sites in northern Europe: a case study from the eastern Romanian Steppe. Ekologia. 2015;34(1):39–53.

Smith C, Hilaire R. Xeriscaping in the urban environment. New Mexico J Sci. 1999;39:241–50.

Sochor J, et al. Environmental monitoring of soil conditioner effects on photosynthetic parameters of Acer campestre L. Afr J Biotechnol. 2014;11(17):4007–21.

Sovocool KA, Morgan M, Bennett D. An in-depth investigation of xeriscape as a water conservation measure. AWWA. 2006;98(2):82–93.

St. Hilaire R, Arnold MA, Devitt DA, Hurd BH, Lesiker BJ, Lohr VI, Martin CA, McDonald GV, Morris RL, Pittenger DR, Shaw DA, Wilkerson DC, Zoldoske DF. Efficient water use in residential urban landscapes. HortScience. 2008;43:2081–92.

Sun H, Kopp K, Kjelgren R. Water-efficient urban landscapes: integrating different water use categorizations and plant types. HortScience, 2012;47(2):254–63.

Trevisan-Smýkalová J. Optimalizace trvalkových výsadeb. Zahrada-Park-Krajina, Roč. 2004;15, č. 6, příl. s. VI–X.

Werick WJ, Palmer RN. It's time for standards of practice in water resources planning. J Water Res Plan Manage. 2008;134(1):1–2.

Wolf D, Lundholm JT. Water uptake in green roof microcosms: Effects of plant species and water availability. Ecol Eng. 2008;33(2):179–86.

Zollinger N, Kjelgren R, Cerny-Koening T, Kopp K, Kening R. Drought responses of six ornamental herbaceous perennials. Sci Hortic. 2006;109:267–74.

Index

M.A. Hossain et al. (eds.), *Drought Stress Tolerance in Plants, Vol 1*,
DOI 10.1007/978-3-319-28899-4

E

Q

R

MIX
Papier aus verantwortungsvollen Quellen
Paper from responsible sources
FSC® C105338

If you have any concerns about our products,
you can contact us on
ProductSafety@springernature.com

In case Publisher is established outside the EU,
the EU authorized representative is:
Springer Nature Customer Service Center GmbH
Europaplatz 3, 69115 Heidelberg, Germany

Printed by Libri Plureos GmbH
in Hamburg, Germany